ANNUAL REVIEW OF NEUROSCIENCE

EDITORIAL COMMITTEE (1983)

ANNUAL REVIEW OF NEUROSCIENCE

VOLUME 6, 1983

W. MAXWELL COWAN, *Editor*
Salk Institute for Biological Studies

ERIC M. SHOOTER, *Associate Editor*
Stanford University School of Medicine

CHARLES F. STEVENS, *Associate Editor*
Yale University School of Medicine

RICHARD F. THOMPSON, *Associate Editor*
Stanford University

ANNUAL REVIEWS INC.   4139 EL CAMINO WAY   PALO ALTO, CALIFORNIA 94306 USA

ANNUAL REVIEWS INC.
Palo Alto, California, USA

International Standard Serial Number: 0147-006X
International Standard Book Number: 0-8243-2406-4

PRINTED AND BOUND IN THE UNITED STATES OF AMERICA

Annual Review of Neuroscience
Volume 6, 1983

CONTENTS

(continued)

CONTENTS (*continued*)

SOME RELATED ARTICLES IN OTHER *ANNUAL REVIEWS*

Special Announcement: New From Annual Reviews

Volume 1 of the *Annual Review of Immunology* (Editors: William E. Paul, C. Garrison Fathman, and Henry Metzger) will be published in April, 1983.

Some Historical and Modern Aspects of Amino Acids, Fermentations and Nucleic Acids, Proceedings of a Symposium held in St. Louis, Missouri, June 3, 1981, edited by Esmond E. Snell. Published October, 1982. 141 pp.; softcover; $10.00 USA/$12 elsewhere, postpaid per copy

---

From the *Annual Review of Biochemistry,* Volume 51 (1982)

The Conformation, Flexibility, and Dynamics of Polypeptide Hormones, *T. Blundell and S. Wood*

Intermediate Filaments: A Chemically Heterogeneous, Developmentally Regulated Class of Proteins, *E. Lazarides*

The Nicotinic Cholinergic Receptor: Correlation of Molecular Structure with Functional Properties, *B. M. Conti-Tronconi and M. A. Raftery*

The Biology and Mechanism of Action of Nerve Growth Factor, *B. A. Yankner and Eric M. Shooter*

From the *Annual Review of Biophysics and Bioengineering,* Volume 11 (1982)

Physiological and Pharmacological Manipulations with Light Flashes, *H. A. Lester and J. M. Nerbonne*

X-Ray Computed Tomography: From Basic Principles to Applications, *R. A. Robb*

Diffusion-Enhanced Fluorescence Energy Transfer, *L. Stryer, D. D. Thomas, and C. F. Meares*

(*continued*)

(*continued*)

Ann. Rev. Neurosci. 1983. 6:1–42

NOBEL LAUREATES IN NEUROSCIENCE: 1904–1981

Herbert H. Jasper

Centre de recherche en sciences neurologiques de l'Université de Montréal
and The Montreal Neurological Institute of McGill University, Montreal,
Canada H3C 3J7

Theodore L. Sourkes

Departments of Psychiatry and Biochemistry, Faculty of Medicine, McGill
University, Montreal, Canada H3A 1A1

Introduction

Prompted by the award of the Nobel Prize in Physiology or Medicine to
three of our most distinguished neuroscientists in 1981, David Hubel, Torsten Wiesel, and Roger Sperry, it would seem an appropriate time to present
a brief review of Nobel Laureates in Neuroscience over the years. This
provides an interesting thumbnail sketch of some of the highlights in the
historical development of neuroscience since the turn of the century.[1]

The field of neuroscience has become so broad during recent years as to
involve aspects of most biological and many medical sciences, as well as
many discoveries in physics and chemistry. We confine our selection to the
Prizes in Physiology or Medicine for discoveries directly related to the
nervous system. This account is further condensed by the format of this
Prefatory Chapter, but we try to overcome this limitation to some extent
by giving references to some of the principal publications of each of the
Prize winners in the bibliography. We present our own evaluations and

[1]Our principal source of information for this brief review has been the four-volume publication of the Nobel Foundation in 1972 of *Nobel Lectures, Physiology or Medicine 1901–1970.*
This publication contains the citations and evaluations of the Nobel Committee, in addition
to biographies and lectures of the Prize winners themselves. Information for the period of
1970–1981 was obtained from accounts published in *Science* and elsewhere, in addition to our
personal acquaintance with the neuroscientists themselves and with their work.

1

0147-006X/83/0301-0001$02.00

comments on the significance of the work of each Laureate for neuroscience, in addition to those of the Nobel Committee.

Out of the total of 74 Nobel Awards in Physiology or Medicine, we select 22 that we consider direct contributions to neuroscience. There were 129 individual laureates in Physiology or Medicine, some sharing an award with one or two others. Of these we have chosen 41 for this review of Nobel Laureates in neuroscience.

1904   Ivan Petrovich Pavlov

In recognition of his work on the physiology of digestion, through which knowledge of vital aspects of the subject has been transformed and enlarged.

Ivan Pavlov was born in Ryazan, a small city just south of Moscow, in 1849. He studied at St. Petersburg University, where he graduated from its Medico-Chirurgical Academy in 1879 and where he began his research in physiology. After two years of training in Germany with Ludwig and Heidenhain, he returned to become professor of pharmacology at the Military Medical Academy. He later held the chair of physiology. In 1891 he was appointed Director of the Department of Physiology of the Institute of Experimental Medicine in St. Petersburg, where for over 30 years he devoted himself to studies of conditioned reflexes. He continued his work and lectures there and abroad until his death in 1936 at the age of 87.

Pavlov's Nobel Lecture was entitled "Physiology of Digestion," although he is better known to neuroscience for his work on conditioned reflexes. This research developed from his observations on the "psychic" control of salivary and gastric secretion, which he was able to demonstrate by the use of indwelling salivary and gastric fistulae in the chronic, alert dog preparation. The sight or smell of food caused profuse secretion, which was arrested by section of the vagus nerves, and could be reproduced by electrical stimulation of these nerves in chronic preparations. A bell, the conditioned stimulus (CS) was then sounded. Eventually, the sound of the bell alone could provoke secretion.

Pavlov was greatly influenced by Sechenov's theory of the reflex mechanisms of psychic activity. His initial work, however, appeared to be concerned mainly with the psychic influences upon reflex activity, the central or "psychic" control of functions of the autonomic nervous system. With his colleagues he elaborated a series of complex forms of conditioned reflex (CR) (delayed, trace, cyclic, etc), together with second- and third-order CR, which he considered to be the mechanism for the acquisition of language and symbolic thought, or "higher nervous activity." For example, he delivered an address to the International Medical Congress in Madrid (1903) on "The Experimental Psychology and Psychopathology of Animals."

The work of Pavlov[2] and his school has had a long-lasting influence upon concepts of learning and of the brain mechanisms underlying behavior and mental function, both normal and abnormal. The precise measurements made possible by the various forms of CR paradigm still form the methods of choice for most studies on physiological mechanisms of learning.

1906   Camillo Golgi and Santiago Ramón y Cajal

In recognition of their work on the structure of the nervous system.

Camillo Golgi (1843–1926) was born in Córteno (Brescia), Italy. He obtained his medical training at the University of Padua, the second oldest university in Italy, from which he graduated in 1865. He was first attracted to psychiatry by Lombroso, and then to cellular pathology by Virchow. He began working in the histology laboratories at the University of Pavia in northern Italy, where he studied neuroglia. In 1872, unfortunate circumstances forced him to take a position in a small hospital for incurables at Abbiategrasso, where, in spite of the lack of laboratory facilities, in 1873 he developed his famous silver impregnation method for the selective staining of nerve cells. He later used gold as well. He was appointed professor of anatomy at the University of Siena in 1879, but returned to Pavia the following year to become Extraordinary Professor of Histology and to take the chair of general pathology. For his work in neurohistology and neuropathology and his studies of malaria he received many awards and honorary degrees, becoming one of the best known Italian medical scientists of his time. He became Rector of the University of Pavia before his death.

Golgi's Nobel Lecture was entitled "The Neuron Doctrine, Theory and Facts." The neuron doctrine was then a controversial subject. In spite of the elegance of his "black staining" of individual nerve cells, with all their axonal and dendritic appendages, Golgi was never convinced of the structural discreteness of nerve cells. In his Nobel Lecture he stated, "In my opinion, we cannot draw any conclusion, one way or the other, from all that has been said on the importance which different structures identified in ganglion cells have, in being for or against the neurone doctrine. . . . Anatomical arguments . . . do not offer any basis firm enough to uphold this doctrine." He believed that nerve cells were interconnected by a neurofibrillary continuity, a viewpoint strongly opposed by Cajal.

Golgi is best known for his silver staining techniques and for identifying several types of nerve cells that bear his name: The Golgi cells of the

---

[2]An English translation of Pavlov's *The Work of the Digestive Glands* was provided by W. H. Thompson in 1902. His work on *Conditioned Reflexes* was translated by Anrep and published in 1927. His colleague, Boris P. Babkin, wrote his biography in 1949, *Pavlov: A Biography.*

cerebellum; Golgi II interneurons with axons that remain within the local gray matter; Golgi tendon organs; and the "Golgi apparatus." His work was summarized in four large volumes, *Opera Omnia,* published in 1903.

Cajal (1852–1934) was born in Petilla, a small village in the Spanish Pyrenees. He studied medicine in Zaragoza, where he graduated in 1873, and obtained his MD at Madrid, before returning to Zaragoza where he became professor of anatomy in 1877. He later held the chairs of anatomy at Valencia and Barcelona. It was in those universities that he began his work with the Golgi silver stain, which culminated in his incomparable two volume work on the histology of the nervous sytem. In 1892 he was appointed Professor of Normal Histology and Pathological Anatomy at the University of Madrid. He remained there until his death at the age of 82.

Cajal's Nobel Lecture was entitled "The Structure and Connections of Neurons." With his own reduced silver method and his refinements of the Golgi technique, Cajal studied all kinds of cells and their interconnections throughout the nervous system, including the sense organs. His microscopic preparations, and especially his drawings, are works of precision and art that have never been surpassed. Contrary to the opinion of Golgi, Cajal gave strong support to the "neuron doctrine" that synaptic relationships between neurons are by contiguity rather than by continuity, and that each nerve cell is a separate independent biological unit. He described the dynamic polarization of the neuron and speculated on the significance of the integrative capacity of the large receptor surface of dendrites. Equally outstanding were his pioneering studies of embryonic development and degeneration and regeneration in the nervous system (Cajal 1928).

Cajal is truly one of the giants of neuroscience. His studies continue to provide a firm foundation for our knowledge of the cellular morphology of the nervous system, in spite of the modern revolution in neuroanatomy made possible with the new techniques of electron microscopy, radioautography, and cytochemistry.

Much of Cajal's work was published in a two-volume edition in French in 1909 entitled *Histologie du Système Nerveux de l'Homme et des Vertébrés* (republished in 1952–1955 as *Histologie du Système Nerveux* by L. Azoulay, Instituto Ramón y Cajal). Most of his original slides and elegant pen-and-ink drawings can still be seen at the Cajal Museum in Madrid.

Cajal has been honored and commemorated in many ways: among them were an international symposium held in his honor by the Karolinska Institute in Sweden in 1952 (*Acta Physiol. Scand.* Vol. 29, Suppl. 106) and the formation of the Cajal Club of the American Anatomical Society. A fascinating autobiography in two volumes, translated by Craigie & Cano, *Recollections of My Life,* was published by the American Philosophical Society in 1937.

1911   Allvar Gullstrand

For his work on the dioptrics of the eye.

Allvar Gullstrand was a Swedish ophthalmologist, born in Landskrona in 1862. He studied at the Universities of Uppsala, Vienna, and Stockholm. He defended his doctorate thesis at Stockholm University in 1890, was named Docent in Ophthalmology in 1891, and was called later to the new chair of ophthalmology at Uppsala. He was self-taught in geometrics and physiological optics. His thesis in 1890, "A Contribution to the Theory of Astigmatism," contained the foundation of his most notable later work on the "intracapsular mechanism of accommodation."

Gullstrand was able to determine from his mathematical analysis and the dioptric investigation of the lens during accommodation that only about two-thirds of accommodation is extracapsular, and one third is due to rearrangement of the internal structure of the lens, i.e. "intracapsular." His sophisticated studies of physiological optics enabled him also to explain astigmatism and monochromatic aberrations, and contributed to our knowledge of the structure and function of the cornea. His Nobel Lecture was entitled, "How I Found the Mechanisms of Intracapsular Accommodation."

Gullstrand designed and improved many ophthalmological instruments, including glasses for use after removal of cataracts and a reflex-free ophthalmoscope. His most important invention was the slit-lamp, which, when combined with a binocular microscope, became an indispensable part of every ophthalmological examination. An account of Gullstrand's work was published as an appendix to Helmholtz's *Physiological Optics* (1924).

1914   Robert Bárány

For his work on the physiology and pathology of the vestibular apparatus.

Bárány (1876–1936), born in Vienna, studied medicine at the University there, graduated in 1900, and took up otology as his specialty. During the First World War, he was a surgeon in the Austrian army, and was a prisoner in a Russian prisoner-of-war camp when the announcement of his Nobel Prize came in 1914. Through the intervention of the King of Sweden and the Red Cross he was released and received his Prize in 1915. While in prison he wrote a treatise on consciousness and the mind-body problem. After the war he left Vienna because of a controversy with his Austrian colleagues. He emigrated to Uppsala, Sweden, and established the Otological Institute at the Uppsala University.

Bárány's interest in the vestibular system began when he observed the effect of temperature on the nystagmus produced by irrigation of the ears, the "caloric reaction." He then developed the Bárány rotating chair test and other tests of vestibular and cerebellar function, in order to establish relationships between the vestibular system and cerebellum in the control of muscle tone and posture. He was among the first to use local cooling as a means of producing reversible paralysis of a specific small area of the brain.

1922   Archibald Vivian Hill

For his discovery relating to the production of heat in muscles (Hill).

Archibald Vivian Hill (b. 1886), was one of the most distinguished British biophysicists. He studied mathematics at Trinity College, Cambridge, but then joined F. G. Hopkins in the Department of Biochemistry, where he investigated the formation of lactic acid in muscle. He then studied with Bürker in Germany (1910–1911). After the First World War he was appointed to the chair of physiology at Manchester, and later moved to London as the Jodrell Professor of Physiology at University College. In 1926 he became Foulerton Professor of the Royal Society. He was Secretary (and Foreign Secretary) of the Society (1935–1946) and served as Secretary-General of the International Council of Scientific Unions (1952–1956).

Hill's Nobel Lecture was entitled "The Mechanism of Muscular Contraction." With the help of his very able technician, Peter Downing, he was able to construct an ultrasensitive miniature thermopile and galvanometers capable of rapid measurement of the heat produced by the contraction of muscle, of the order of $0.003°C$, in a few hundredths of a second. He found that the initial heat was anaerobic, only the later, or "recovery," heat requiring oxygen. The actual contraction did not need oxygen, so the heat produced during that phase could not have come from the oxidation of lactic acid. This raised doubts about the role of lactate in the mechanism of muscular contraction. Hill found that the oxygen debt was paid off and the lactic acid removed only during the phase of recovery from contraction. His discovery of glycolysis during muscular contraction had far-reaching consequences for our understanding of metabolic events in muscle as well as other tissues of the body, including the nervous system. His studies of muscle were published by the Royal Society, London (1938), and summarized in an article in the *British Medical Bulletin* (1956).

After this work Hill turned his attention to studies of the excitability, conduction, and metabolism of nerve fibers. He developed a mathematical theory of electrical excitability of nerve that accounted for the rising and falling phase of the spike, the strength-duration curve of excitability, the mechanisms of "accommodation" to slowly rising currents, and the shorter time constant of large diameter fibers.

He then proceeded, in collaboration with Peter Downing, Ralph Gerard, and Y. Zotterman, to measure the heat produced by various nerves. With Downing's ultrasensitive thermopile and galvanometer, they were able to measure heat production even in the resting nerve (about 6×10^{-5} cal/g/-sec). It was approximately doubled during maximum excitation. As in muscle, there were an initial evolution of heat and two phases of delayed or recovery heat that followed excitation. This first phase lasted 20–30 sec. This was followed by the much longer and larger phase of recovery heat, which reached 15 to 50 times the initial heat, a far greater increase than that observed in muscle. Both initial and delayed heat in nerve occurred undiminished in the absence of oxygen, though nerve became inexcitable more quickly than did muscle. Oxygen consumption was increased during excitation and recovery when present, but increased much more if the nerve had been excited in the absence of oxygen. It showed an oxygen debt as did muscle. These important studies of nerve excitability and metabolism were described by Hill in his Liversidge Lecture at Cambridge in 1932.

1927   Julius Wagner-Jauregg

For his discovery of the therapeutic value of malaria inoculation in the treatment of dementia paralytica.

Wagner-Jauregg (1857–1940) was born in Wels, Austria. He completed his medical education at the University of Vienna in 1880. Following studies in experimental pathology and internal medicine, he took up psychiatry. Except for a short period at the University of Graz (1889–1892) where he was Professor of Psychiatry, he spent his career at the University of Vienna. He became Professor of Neurology and Psychiatry there in 1902 (following Krafft-Ebing), and held that position until his retirement in 1928.

Wagner-Jauregg had noticed as early as 1887 that some patients with general paralysis of the insane (GPI) underwent remission when suffering from an infection accompanied by fever. At that time it occurred to him that the induction of an infectious disease to duplicate these "natural experiments" might be useful in the treatment of GPI, and he thought malaria, which could be interrupted with quinine, and erysipelas might be suitable diseases. In 1917 he carried out that study when a wounded soldier with malaria was admitted to his ward. He inoculated nine patients suffering from GPI with malaria: six patients had extensive remission, three of them lasting at least ten years. Fever therapy in a variety of forms came to be used in GPI, but was supplanted when it was found in the 1940s that the disease responds to large doses of penicillin.

Wagner-Jauregg's discovery of treatment of GPI by means of a heroic measure provided a new therapy for an important mental disease. In the area of medical theory it provided a convincing measure of support for the organic nature of mental disease. In this connection it should also be noted that he was a very early proponent of fortification of table salt with iodine, for use in areas of endemic goiter, as a means of prevention against cretinism.

1929   Christiaan Eijkman

For his discovery of the antineuritic vitamin.

Christiaan Eijkman (1858–1930) was born in Nijkerk, The Netherlands. He studied physiology at the University of Amsterdam starting in 1875 and, having received his medical degree there, he joined the Dutch army for service in the East Indies. He spent two periods in Batavia (1883–1885 and 1886–1896), and received training in bacteriology in Holland during the interval. In 1886 he became Director of the Laboratory of Bacteriology and Pathology, where he carried out his investigations on beri-beri, an endemic disorder with neurological, cardiac, and other manifestations. From 1896 to 1928 he was Professor of Hygiene at the University of Utrecht. He shared the Nobel Prize in 1929 with (Sir) Frederick Gowland Hopkins.

In 1889 Eijkman observed an outbreak of avian disease in the chicken-run of his Batavian laboratory, and was struck by its resemblance to human beri-beri: unsteady gait, muscular weakness causing falls, progressive paralysis, cyanosis, hypothermia, stupor, and finally death. Pathological examination showed it to be a polyneuritis. When the disease suddenly cleared up he learned that the attendant had changed the birds' diet from polished rice to rice containing the pericarp (hull) and much of the germ. A deliberate feeding experiment with the two types of rice confirmed the nutritional etiology of the disorder (Eijkman 1930). Initially Eijkman thought that a nutrient in the hull neutralized the damage caused by a starch-rich diet, but he later gave up that idea. Colleagues related to him their experience with the various rice diets in the prisons and among the native population generally, and its relation to the incidence of beri-beri. Conclusive experiments were finally carried out by Eijkman's successor, C. Grijns, with prisoners. Grijns defined beri-beri as a deficiency disease and showed that protection was endowed by a factor in an extract of beans. He was the first to provide the correct interpretation of the connection between consumption of the polished rice diet and the etiology of beri-beri. Subsequently, the prophylactic effect of extracts of rice hulls and of beans was demonstrated, and ultimately shown to be due to the content of thiamine, then known as vitamin B_1. In the 1930s, R. A. Peters showed that in thiamine deficiency the tissues metabolize pyruvate defectively; K. Lohmann and P. Schuster proved that the vitamin is the organic portion of cocarboxylase. In 1939,

I. Banga, S. Ochoa, and R. A. Peters showed that the thiamine coenzyme is active in correcting pyruvate metabolism in vitro.

Beri-beri is now recognized as resulting from the chronic deficiency of thiamine in the diet. The acute deficiency gives rise to Wernicke's encephalopathy. Eijkman's investigations, which led to the description of the nutritional origin of certain neurological disorders, eliminated the mass of conflicting theories about the etiology of beri-beri, such as the involvement of toxic, microbiological, climatic, and other factors. The possible nervous system functions of a large number of vitamins (and mineral cofactors) continue to interest many biochemists as a fruitful area of research.

1932 Sir Charles Scott Sherrington and Edgar Douglas (Lord) Adrian

For their discoveries regarding the function of neurons.

Sherrington was born in Islington, London, in 1857. He obtained his scientific and medical education at Cambridge and at St. Thomas's Hospital Medical School in London. His interest in the physiology and pathology of the nervous system developed during his medical studies, as well as through contacts with the work of Goltz in Strasbourg and David Ferrier on the localization of function in the brain, and by his association with Virchow in neuropathology. His own researches began while he was a lecturer at St. Thomas's and a fellow at Cambridge. He was responsible for inviting Cajal to deliver the Croonian Lecture (1894) of the Royal Society in London. He was appointed in 1895 to the chair of physiology at University College, Liverpool, where he did much of the work for which the Nobel Prize was awarded. He remained in Liverpool until 1913 when he became Waynflete Professor of Physiology at Oxford, a post that he held until his retirement in 1936. He died in 1952, at the age of 95 years.

Sherrington's Nobel Lecture of 1932 was entitled "Inhibition as a Coordinative Factor." He described the form and plasticity of spinal reflexes, including reflex stepping movements, as based upon a balance of central inhibitory and excitatory states and emphasized the importance of inhibition in the sculpturing of posture and in the control of the reciprocal action of antagonistic muscles in movement and locomotion.

Sherrington traveled to Canada and the United States in 1903, and the next year delivered the Silliman Memorial Lectures at Yale University. His book, *The Integrative Action of the Nervous System,* which was based upon those lectures, has remained a classic of neuroscience ever since. After many publications and lectures on the reflex organization of the spinal cord, he was invited to present the Rede Lecture at Cambridge in 1933. He delivered the Gifford Lectures in Edinburgh in 1937–1938; this resulted in the publication of still another classic, *Man on His Nature* (1940). In this book he elaborated on his dualistic view of man's nature: body and mind.

Adrian (1889–1977), who shared the prize with Sherrington, though at a younger age, was born in London. He studied medicine in London and Cambridge, receiving his MD degree from Trinity College in 1915. He began his work in neurophysiology with Keith Lucas at Cambridge, where he was Research Professor of the Royal Society from 1929–1937. He became professor, succeeding Barcroft as director of the Department of Physiology, in 1937. He retired in 1951 to become Master of Trinity. He also served as President of the Royal Society, the Royal Society of Medicine, and the British Association for the Advancement of Science. He retired as Master of Trinity in 1965 to become Vice-Chancellor, and then Chancellor, of Cambridge, 1968–1976.

Adrian's Nobel Lecture was entitled simply, "The Activity of Nerve Fibres." He was the first to record action potentials from single sensory and motor nerve fibers by a delicate dissection technique, and by the use of vacuum tube amplification and a more rapidly moving oscillograph, the capillary electrometer and later the Matthews mirror oscillograph.

In 1925 Adrian and Y. Zotterman were the first to record the impulse traffic in a single sensory fiber stimulated by the adequate—physiological —stimulation of its nerve ending. From their records they could state that the transmission in sensory nerve fibers occurs according to what today would be called "impulse frequency modulation," which 20 years later was introduced in radio technique (FM) as the safest way of communication. Strength of contraction in muscle was also related to frequency of firing in motor nerve fibers (with D. Bronk). He proposed that there were long duration potentials in nerve centers and in sense organs causing repetitive firing of nerve fibers (now known as "generator potentials").

Adrian explored all forms of sensation, including auditory, olfactory, and visual, as well as somatic sensations from skin and muscle. He then became interested in sensory representation in the cerebral cortex and cerebellum, which he recorded with the technique of evoked potentials. He discovered an additional somatic sensory area, II, and topographically organized connections between the sensory cortex and cerebellum. He also studied the spontaneous electrical activity in the water beetle and gold fish. He was one of the first to confirm, with B. H. C. Matthews, the work of Hans Berger on the human electroencephalogram.

Some of Adrian's principal publications were *The Basis of Sensation, The Action of Sense Organs* (1927), *The Mechanism of Nervous Action* (1932), and *The Physical Background of Perception* (1947).

1935   Hans Spemann

For his discovery of the organizer effect in embryonic development.

Hans Spemann (1869–1941) was born in Stuttgart, Germany and studied medicine at the Universities of Heidelberg, Munich, and Wurzburg. He worked at the Zoological Institute at Wurzburg until 1908, when he accepted the chair of zoology and comparative anatomy at Rostock. In 1914 he became Director of the Kaiser Wilhelm Institute for Biology in Berlin-Dahlem. He then went to the University of Freiburg-im-Breisgau as professor until his retirement in 1935.

Spemann's experiments were carried out on young amphibian embryos. By the transposition of tissue from one part of the developing embryo to another, he was able to show that groups of cells, originally destined for one part of the body, changed their direction of development when transplanted to a different region. For example, he was able to show that the optic cup is able to bring about the formation of a lens from epidermis transplanted from a distant part of the body that is not normally involved in lens production. Similarly, the covering epidermis clears, as in the formation of a transparent cornea, despite its "foreign" noncorneal origin. Division of the embryo or transplantation of pieces of the medullary plate showed that rearranged cells develop according to their new local environment. In 1918 Spemann presented his concept of the "organizers" to explain his observation that the morphogenesis of the embryo is the result of interactions between different regions of tissue. He became the leader of a school of experimental embryologists who provided many examples of embryonic induction in development.

Spemann's findings were confirmed in other species and led to studies of the chemistry of embryonic development by Joseph Needham and Jean Brachet, with particular regard to the chemical nature of the organizer.

1936   Sir Henry Hallett Dale and Otto Loewi

For their discoveries relating to chemical transmission of the nerve impulse.

Born in London in 1875, Sir Henry Dale trained as a pharmacologist at Trinity College, Cambridge. In 1906 he became Director of the Wellcome Physiological Research Laboratories, where he spent many years. He became Director of the National Institute for Medical Research in London in 1928. During his career he also served as President of the Royal Society and of the Royal Society of Medicine.

Dale's Nobel Lecture was entitled "Some Recent Extensions of the Chemical Transmission of the Effects of Nerve Impulses." The extensions he referred to were the demonstration, with his colleagues Feldberg, Gaddum, Brown, and Vogt, of evidence that neuromuscular transmission in striate muscle, as well as synaptic transmission in the superior cervical ganglion, is mediated by the liberation of acetylcholine at the prejunctional axonal terminals. Dale summarized this work further in his *Harvey Lectures* (1937) and in his autobiography (1953).

Loewi had shown, with the help of Dale's work on acetylcholine, that this substance is liberated from the vagus nerve to cause slowing of the isolated frog heart. It was assumed that adrenaline mediated the action of sympathetic nerves to the heart, although this was questioned by Walter B. Cannon. Cannon proposed instead that two substances, sympathin E and sympathin I, are involved as transmitters of the peripheral sympathetic system, and that these are not identical with adrenaline.

Otto Loewi (1873–1961) was born in Frankfurt-am-Main. He met Dale while working in Starling's laboratory in London. He worked also with Elliott in Cambridge. He then was appointed Professor of Pharmacology at the University of Graz, Austria, but was forced to leave his post when the country was invaded by the Germans in 1938. He went first to Brussels, then to Oxford, and finally settled in New York City, where he was appointed Professor in the College of Medicine of New York University in 1940, and adopted American citizenship. His Nobel Lecture was entitled, "The Chemical Transmission of Nerve Action."

Chemical transmission had long been suspected to occur in the peripheral autonomic neuromuscular system. In 1904, T. R. Elliott proposed that adrenaline might be involved in the sympathetic system. A few years later, W. E. Dixon suggested that an unknown chemical mediator might be involved in the action of the vagus nerve on the heart. There was no definitive proof for these hypotheses, until Loewi performed his elegantly simple experiments. Utilizing the frog heart preparation he was able to show that stimulation of the vagus nerve liberated a substance into the Ringer's solution bathing the heart, and that this substance was able to slow a second isolated "test" heart beating in the same solution. This action was blocked by atropine, prolonged by eserine, and duplicated by acetylcholine, in accordance with the muscarinic action of acetylcholine previously described by Dale.

The extension of the principle of chemical transmission to the neuromuscular junction in striate muscle, and to synapses in the cervical ganglion, was a much greater surprise, since it was generally believed in the 1930s that transmission in these, as in central synapses, was electrical in nature. One of the strongest proponents of the electrical theory was J. C. Eccles, then working in Sherrington's laboratory at Oxford. For several years the meetings of the Physiological Society were enlivened by the heated debates between Dale and Eccles, but Eccles subsequently became convinced of the inadequacy of the electrical theory, and proceeded to provide elegant evidence for chemical transmission at synapses on the anterior horn cells of the spinal cord. This is discussed below in relation to Eccles' own Nobel Prize award. (See also Eccles' chapter in Volume 5 of the *Annual Review of Neuroscience.*)

The establishment of chemical transmission throughout the central nervous system, as well as for ganglionic and junctional synapses in the peripheral autonomic system and at striate neuromuscular junctions, represented a revolutionary advance in neuroscience, the importance of which has become more fully appreciated in recent years with the discovery of many new transmitter substances in the central and peripheral nervous systems.

1938 Corneille Jean François Heymans

For the discovery of the role played by sinus and aortic mechanisms in the regulation of respiration.

Heymans (1892–1968) was born in Ghent, Belgium. He was the son of Dr. J. F. Heymans, Professor of Pharmacology and Rector of the University of Ghent, as well as founder of the Heymans Institute of Pharmacology, Pharmacodynamics and Toxicology. Heyman's father was his first and principal teacher and it was with him that the original experiments, that led to the award of the Nobel Prize, were begun. Corneille Heymans also studied with E. Gley in Paris, N. M. Arthus in Lausanne, H. H. Meyer in Vienna, and E. H. Starling in London. He spent one year (1927–1928) in the United States. He succeeded his father as Director of the Heymans Institute of Pharmacology and also became Rector of the University.

Heyman's Nobel Lecture was entitled "The Part Played by Presso- and Chemoreceptors in Respiratory Control." By use of a crossed perfusion from one dog to the isolated head of another he was able to show that there were chemo- and pressoreceptors in the carotid sinus and aortic arch capable of controlling respiration and blood pressure, as well as cerebral circulation. In 1937 the sensory nature of this control was demonstrated independently by Rijlant, Stella, and Zotterman, who recorded action potentials from carotid sinus nerves in response to controlled pressor and chemical stimuli administered to the carotid body. These studies represented a major advance in our knowledge of the reflex mechanisms that control respiration and blood pressure and are of critical importance for the regulation of the cerebral circulation (Heymans & Neil 1958).

1944   Joseph Erlanger and Herbert Spencer Gasser

For their discoveries regarding the highly differentiated functions of single nerve fibers.

Joseph Erlanger (1874–1965) was born in San Francisco. He received his BS degree in chemistry from the University of California and in 1899 his MD degree from Johns Hopkins University. He was appointed assistant in the Department of Physiology and soon became Associate Professor at Johns Hopkins University Medical School. He then joined the faculty of the newly founded University of Wisconsin as their first professor of physiology. Herbert Gasser was one of his pupils there. In 1910, Erlanger was appointed Professor and Head of the Department of Physiology at Washington University in St. Louis, where Gasser joined him in the joint research that earned them the Nobel Prize. Erlanger became Professor Emeritus in 1944. His Nobel Lecture was entitled, "Some Observations on the Response of Single Nerve Fibers."

Herbert Gasser (1888–1963) was born in Platteville, a small town in Wisconsin. He studied at the State Normal School and at the State University, where he was introduced to physiology by Erlanger. He graduated in medicine from Johns Hopkins in 1915. He returned to Wisconsin for a year in pharmacology, and then joined Erlanger in St. Louis, where he became Professor of Pharmacology in 1921. With the help of the Rockefeller Foundation, he spent two years studying in Europe and then returned to St. Louis. In 1931 he was appointed Professor of Physiology at Cornell Medical School in New York City. In 1935 he became Director of the Rockefeller Institute for Medical Research until his retirement from that post in 1953. He continued to carry out electrophysiological research into the properties of ultrafine nerve fibers until his death. His Nobel Lecture was entitled, "Mammalian Nerve Fibers."

Before the work of Erlanger & Gasser in the early 1920s, the true form of the nerve action potential was unknown, owing to the lack of sufficiently rapid and sensitive recording instruments. Erlanger & Gasser, with the help of George Bishop, solved this problem by the combination of high gain vacuum tube amplification with the relatively inertialess cathode-ray oscilloscope, which is an adaptation of the Braun tube long known to physics. With their new instrumentation they could discern various groups of fibers in a mixed nerve by their different thresholds and velocities of conduction. The fibers differed also in their refractory periods and after-potentials. A near-linear relationship was found between fiber diameter and velocity, the most rapid A fibers conducting at over 100 m/sec, whereas the velocity of the unmyelinated C fibers was only about 1 m/sec. The large fibers had a much greater spike height as well as the lowest threshold and shortest time constant (chronaxie) for electrical stimulation. The widely different delays in the transmission of sensory information to the CNS were very important to our understanding of central processing of sensory information. Muscle afferents were most rapid, while pain was found to be conducted by two groups of fibers, the A delta and C fibers, having very different velocities, which accounted for "rapid" and "slow" pain perception.

Erlanger concluded his Nobel Lecture with the statement that he had pursued his studies because, "In the investigation of this comparatively simple structure, the nerve fiber, lies the hope of finding clues to an understanding of the much more complicated mechanisms that determine the activities of peripheral and central nervous mechanisms." Erlanger & Gasser did not realize, at the time, how "much more complicated" these mechanisms of the central nervous system would become after the discovery of chemical transmission at central synapses and the highly specialized properties of cells and dendrites as compared to their axons. Their work was summarized in their Johnson Foundation Lectures (1937) and in Gasser's Harvey Lecture (1937).

1949   Walter Rudolf Hess and Antonio Egas Moniz

For his discovery of the functional organization of the interbrain as a coordinator of the activities of the internal organs (Hess).

For his discovery of the therapeutic value of prefrontal leucotomy in certain psychoses (Moniz).

Born in Frauenfeld in eastern Switzerland, Walter Hess (1881–1973) studied medicine in Lausanne, Berne, Berlin, Kiel, and Zurich, where he obtained his MD in 1906. He practiced ophthalmology for a few years, but then decided to devote himself to the study of physiology. He was appointed Director of the Physiological Institute in Zurich in 1917. After the First World War he studied with Langley in England, where he became particularly interested in the autonomic nervous system. He also worked in the laboratories of Sherrington, Starling, Hopkins, and Dale before returning to his position at the Physiological Institute in Zurich. After becoming Professor Emeritus in 1951, Hess continued to pursue his major interest in the central autonomic system in relation to behavior and mental functions, attempting to bridge the gap between physiology and psychiatry. His Nobel Lecture was entitled, "The Central Control of the Activity of the Internal Organs."

Hess developed a chronic implanted electrode technique in order to study cinematographically the responses of unanesthetized, freely moving cats to electrical stimulation of local points in the diencephalon. The exposed tips (0.2 mm) of the fine insulated steel needle electrodes were then located by histological studies to establish anatomical relationships to behavioral and physiological responses. A very large series of such experiments formed the basis for his functional atlas of the diencephalon. This made it possible for the first time to observe not only specific autonomic responses in unanesthetized animals, but also to observe their interrelationships with each other and their integration with the behavior of the animal in a relatively normal environmental situation.

Hess mapped out diencephalic regions controlling blood pressure, respiration, pupillary dilatation, defecation and micturition, vomiting, and bulimia, as well as coordinated "extrapyramidal" motor responses such as head and eye turning, licking and chewing movements, and coordinated emotional behavior such as fear or rage. Sleep was produced by stimulation of the medial thalamus. Hypothalamic responses were classified as either ergotropic (dynamogenic) or trophotropic, corresponding roughly to their mediation by the sympathetic or parasympathetic systems, respectively. The integration of autonomic responses with appropriate somatomotor and emotional behavior made these pioneering studies of far-reaching importance to neuroscience in general and particularly to clinical neuropsychiatry. A summary of Hess' work was published in English in 1954 entitled *Diencephalon: Autonomic and Extrapyramidal Functions.*

Egas Moniz (1874–1955), who shared the prize with Walter Hess, was a leading Portugese neurologist and politician, born in Avança. He studied in the medical faculty of Coimbra, where he was appointed professor in 1902. In 1911, he became the first occupant of the chair of neurology in Lisbon. In addition to his important contributions to neurology and to medical literature, Moniz was active in the political life of Portugal, serving as Minister of Foreign Affairs and as President of the Portugese delegation to the Paris Peace Conference in 1918.

Moniz made two important contributions to neurology in collaboration with Almeida Lima: (*a*) they developed the x-ray technique of cerebral angiography; (*b*) they introduced the practice of prefontal leucotomy (for which Moniz was cited in the Nobel Prize). Moniz published a large monograph on the diagnosis of cerebral tumors by angiography in 1931; his first memoir on prefrontal leucotomy appeared in 1936.

Moniz was seeking some way to treat patients with intractable mental diseases, which he believed "are deeply rooted in the synaptic complex which regulates matters of knowledge in the consciousness. . . . It is necessary to alter the synaptic arrangements . . . thereby the corresponding thoughts are altered and forced into other channels." His interest in the frontal lobes was derived from animal experiments of Bechterew & Luzaro in Russia, and especially the experiments of Fulton & Jacobsen in the U.S. on the effects of bifrontal lesions in the chimpanzee. He was also influenced by the effects of neurosurgical removal of the frontal lobes in man, as reported by Brickner & Dandy, and by Penfield.

Moniz & Lima performed their first prefrontal leucotomy in 1935, thus inaugurating psychosurgery. The operation consisted of lesions restricted to the white matter of both frontal lobes, usually performed by the use of a leucotome. During the following years thousands of severely disturbed psychiatric patients were "treated" by this method with some beneficial results in the most severe cases, but with many serious and sometimes disastrous side effects. Fortunately, the use of this method has been almost entirely discontinued with the advent of effective psychopharmacology.

1957   Daniel Bovet

For his discoveries relating to synthetic compounds that inhibit the action of certain body substances, and especially their action on the vascular system and the skeletal muscles.

Daniel Bovet was born in Neuchâtel, Switzerland in 1907. He attended the University of Geneva, from which he received the DSc degree in 1929. He at once joined the laboratory of Ernest Fourneau, an eminent chemist at the Pasteur Institute in Paris, and remained there until 1947, when he moved to the Istituto Superiore di Sanità in Rome to open a new laboratory of chemotherapeutics. He was aided in this venture by his wife, Philomena Nitti, also a chemist, with whom he wrote an extensive monograph on drug structure in relation to pharmacodynamic activity on the autonomic nervous system (1958). Bovet's Nobel Lecture was entitled, "The Relationships Between Isosterism and Competitive Phenomena in the Field of Drug Therapy of the Autonomic Nervous System and that of the Neuromuscular Transmission."

Bovet's experience in Fourneau's laboratory introduced him to the modern practice of medicinal chemistry: the synthesis of large numbers of new molecules expected to exert biological activity, along with extensive screening in biological tests for such activities. Fourneau was seeking agonists and antagonists of the biogenic amines, particularly epinephrine and histamine; Bovet's work with him led to the discovery of some of the early antihistaminic drugs.

By the early 1940s, experience with antivitamins, especially with the then dramatic new substance sulfanilamide as an antagonist of the bacterial vitamin p-aminobenzoic acid, lent great credence to the "antimetabolite hypothesis." The hypothesis quickly became the justification for research programs in medicinal chemistry in hundreds of centers.

Bovet was a pioneer in the study of synthetic compounds affecting the metabolism and the postsynaptic action of biogenic amines and neurotransmitters such as adrenaline, noradrenaline, 5-hydroxytryptamine, acetylcholine, and histamine. He was particularly interested in noradrenergic blocking agents, antihistamines, and curariform anticholinergic compounds such as decamethonium and succinylcholine, synthetic atropine-like substances, and derivatives of lysergic acid. With his colleague, Longo, he made important observations on the pharmacology of the reticular system in relation to the electroencephalogram. This type of research produced a great increase in the size and efficacy of the physician's armamentarium, and freed therapeutics from dependence upon the handful of natural drugs that are useful in human disease.

1961   George von Békésy

For his discoveries concerning the physical mechanisms of stimulation within the cochlea.

Von Békésy was a Hungarian physicist, born in Budapest in 1899. After obtaining his primary education in many countries (his father was a diplomat), he obtained his PhD degree in physics at the University of Budapest. He worked in the Hungarian Telephone Research Laboratory, where he developed an ingenious and rapid method for discovering defects in the network of telephone lines by their resonant responses to click stimuli, a principle he later used to analyze the properties of the cochlea of the inner ear.

Von Békésy became Professor of Experimental Physics at the University of Budapest in 1939, and held that post until 1946, when he emigrated to Stockholm to work at the Royal Caroline Medical Institute and at the Royal Institute of Technology. He moved again, in 1947, to the Psycho-Acoustic Laboratories of Harvard University in Cambridge, where he pursued much of the work for which he received the Nobel award. He retired to the University of Hawaii in Honolulu, where he died in 1972.

Von Békésy was the first physicist to receive the Nobel Prize in Physiology or Medicine, for his ingenious applications of physical principles to sensory physiology, especially to our knowledge of the middle and internal ear in audition. His Nobel Lecture was entitled, "Concerning the Pleasures of Observing, and the Mechanics of the Inner Ear."

Von Békésy measured the physical properties of the inner ear and cochlear membrane with a highly sensitive microelectrode technique. This enabled him to apply mechanical stimuli to local regions of the basilar membrane while recording the electrical potentials in the hair cells that give rise to action potentials in different fibers of the auditory nerve. He was able to measure how sound waves are transmitted by the ossicles of the middle ear to the ear drum and then by traveling waves in the endolymph to excite frequency-selective portions of the basilar membrane. With these data he was able to construct a physical model of the auditory apparatus that explained the function of the inner ear. He then found that some of the principles could be applied to skin sensation, which also involved the phenomenon of "surround inhibition," i.e. increased contrast between the site stimulated and the surrounding area of skin. This principle was found to apply to the retina as well. Von Békésy's ingenious application of the principles of physics to sensory physiology, his direct measurements by the use of high gain amplifiers and microscopic stroboscopy, and the testing of his theories of psychophysical techniques did much to solve many problems in acoustics, as well as to improve our understanding of other modalities of sensation (tactile, visual, and olfactory). He invented several diagnostic instruments, such as the audiometer that bears his name, which enables the otologist to distinguish between cochlear and nerve deafness. A translation of von Békésy's work on the inner ear was published in the *Journal of the Acoustical Society of America* in 1949 and summarized in his book, *Experiments in Hearing,* in 1960. An account of the more general applications to other sensory modalities was published in 1967.

1963 John Carew Eccles, Alan Lloyd Hodgkin, and Andrew Fielding Huxley

For their discoveries concerning the ionic mechanisms involved in excitation and inhibition in the peripheral and central portions of the nerve cell membrane.

John Eccles (Sir John), born in Melbourne, Australia in 1903, received his degree in medicine at Melbourne University in 1925 and continued his studies in Sherrington's Department of Physiology at Oxford, where he obtained his PhD in 1929. He studied excitation and inhibition at Oxford until 1937 when he was appointed director of a medical research laboratory in Sydney. In 1944 he went to the University of Otago in New Zealand, but then returned to Australia to become Professor of Physiology at the newly-founded National University in Canberra. Upon retiring from Canberra, he came to the Institute for Biomedical Research in Chicago (1966–1968). He was subsequently appointed Professor of Physiology at the State University of Buffalo in New York (1968–1975), where he continued his neurophysiological studies of the cerebellum. He has continued his writing and lecturing from his retirement retreat in Ticino, Switzerland.

John Eccles has been one of the most productive and prolific contributors to neuroscience in our century. His Nobel Lecture was entitled, "The Ionic Mechanisms of Postsynaptic Inhibition." Although while at Oxford, Eccles was a strong defender of the electrical theory of transmission in central synapses, he convinced himself of the inadequacy of this theory by his elegant intracellular microelectrode studies of spinal neurones. [See Eccles (1982) for his own account.] He then provided convincing evidence for the chemical nature of central synaptic transmission, by the use of ultrafine glass micropipettes, first devised by Ling & Gerard in Chicago. He used them both for intracellular recording of membrane and synaptic potentials as well as for the iontophoretic ejection of chemical substances. He was able to determine the ionic mechanisms involved in the response of nerve cell membranes to neurotransmitter substances as measured by excitatory (Na^+ and K^+) and inhibitory (Cl^- and K^+) postsynaptic potentials (EPSP and IPSP). He was able to demonstrate the importance of inhibitory interneurons, such as the Renshaw cells in the spinal cord, and to describe the mechanisms and properties of presynaptic inhibition.

Eccles then turned his attention to suprasegmental portions of the nervous system, where he applied his microphysiological techniques to a comprehensive study of synaptic mechanisms of the cerebellum (Eccles, Ito & Szentágothai 1967). The thalamus, cerebral cortex, and hippocampus were also explored in research directed toward an understanding of what Eccles termed "the neurophysiological basis of the mind," which was the title of his Waynflete Lectures at Oxford in 1952. Among many other important publications of Eccles are his Herter Lectures delivered at Johns Hopkins 1955, published under the title of *The Physiology of Nerve Cells,* The Ferrier Lecture of the Royal Society in 1960, and his most comprehensive review of central synaptic mechanisms published in 1964 under the title, *The Physiology of Synapses.* He delivered the Gifford Lectures in Edinburgh (1977–1979).

Alan Hodgkin was born in Banbury, Oxfordshire, in 1914. He began his scientific studies at Cambridge in 1932. His work on the physiology of nerve fibers led to an invitation by Gasser to work at the Rockefeller Institute in New York (1937–1938). While in the United States he worked at the Woods Hole Marine Biological Laboratory with K. S. Cole, who, with H. J. Curtis and J. Z. Young, introduced him to the use of the squid giant axon, a preparation that became of critical importance to his Nobel Prize research. During the Second World War he was a scientific officer in the Air Ministry. He returned to the Department of Physiology at Cambridge in 1945, where A. F. Huxley was among his students. Hodgkin became a Fellow and Foulerton Research Professor of the Royal Society, winning its Royal Medal in 1958.

Andrew Huxley was born in London in 1917. He entered Cambridge in 1935 and specialized at first in the physical sciences. Becoming interested in physiology in 1939, he joined Hodgkin at the Plymouth Marine Biological Laboratory, where he carried out his first research in 1939. After the war he returned to Cambridge, where he worked until 1960, when he was appointed Professor of Physiology and Head of the Department at University College in London. He was elected to the Royal Society in 1955.

The classical work of Hodgkin & Huxley on the ionic basis of excitation and conduction in nerve fibers was carried out in close collaboration. Hodgkin's Nobel Lecture was entitled, "The Ionic Basis of Nervous Conduction," while that of Huxley was entitled "The Quantitative Analysis of Excitation and Conduction in Nerve."

Hodgkin & Huxley set out to test Bernstein's theory of the ionic basis for the resting and action potentials in the "semipermeable" nerve membrane. By the use of a glass micropipette inserted into the axoplasm, and with the voltage clamp technique developed by Cole, Hodgkin & Huxley were able to measure separately the inward Na^+ current responsible for the overshoot of the action potential well above the level of the resting potential. The outward K^+ current was measured directly by the use of radioactive K^+ ions. Ionic concentrations could be manipulated both inside and outside the axon, and measured in the extruded axoplasm.

A mathematical model was then developed, with the aid of electronic computers, which would predict the form of the action potential at various ionic concentrations inside and outside the axon. The predictions corresponded closely with their experimental results. Huxley concluded his Nobel Lecture with the statement,

> I would not like to leave you with the impression that the particular equations we produced in 1952 are definitive. . . . Both Hodgkin and I feel that these equations should be regarded as a first approximation which needs to be refined and extended in many ways in the search for the actual mechanisms of permeability changes on the molecular scale.

Their theory has stood the test of time as applied to many nerves, and we are approaching an understanding of nerve membrane on the "molecular scale." [See Hodgkin (1964) for a summary of the work of Hodgkin & Huxley.]

1967   Ragnar Granit, Haldane Keffer Hartline, and George Wald

For their discoveries concerning the primary physiological and chemical visual processes in the eye.

In his evaluation of the work of Granit, Hartline & Wald, Professor Bernhard of the Nobel Committee stated that this work provided "a deepened insight into the subtle processes in the eye which form the basis for our ability to perceive light and to distinguish brightness, color, form and movement, . . . of paramount importance to understanding sensory processes in general."

Ragnar Granit was born in Finland in 1900, and graduated from Helsingfors University in 1919. While in school he took part in Finland's War of Liberation and was decorated with the Cross of Freedom IV Class "with sword" in 1918. Granit first became interested in psychology while studying at the Ålbo Academy. He graduated Mag. Phil. in 1923 from the University of Helsingfors. He proceeded to take his MD (1927) there in order to pursue his research on vision, which had interested him from the beginning. In 1929, he became Docent in Tigerstedt's Physiological Institute.

Granit was a Fellow in Medical Physics at the Johnson Foundation (1929–1931) with Detlev Bronk, where he became associated with Keffer Hartline. He studied in Sherrington's laboratories in Oxford as a Fellow of the Rockefeller Foundation, and then returned to Helsinki as Professor of Physiology. In 1940 he accepted an appointment at the Royal Caroline Institute in Stockholm. His laboratories became a part of the Nobel Medical Institute, where he received a personal research chair and was Director of the Institute of Neurophysiology (1945–1967). He was member of the Swedish Medical Research Council and president of the Royal Swedish Academy of Sciences.

Granit retired as Professor Emeritus in 1967, but has remained active in writing and lecturing and is still a roving ambassador of neuroscience. He was visiting professor at the Rockefeller University in New York and at Oxford, and Fogarty Scholar at the National Institutes of Health in Bethesda. He was visiting professor at Düsseldorf in 1974 and at the Max Planck Institute in 1976. He has been consultant to Moruzzi's Institute in Pisa for several years.

Granit's Nobel Lecture was entitled, "The Development of Retinal Neurophysiology." This was his first research interest, begun in the 1920s, but only one of many important contributions he has made over the years to our more general understanding of sensory perception and to mechanisms of sensorimotor control.

Granit's early studies were concerned with an analysis of the significance of the various components of the electroretinogram (ERG) in relation to excitatory and inhibitory processes of the retina. He found that flicker fusion was not only related to light intensity but also to the area of the visual field illuminated; this he attributed to surround inhibition.

Influenced by Adrian's work on sensory messages in individual nerve fibers, he began a long series of studies on the response of individual ganglion cells in the mammalian retina to controlled light and color stimuli. He discovered "on" and "off" as well as "on/off" cells. In his analysis of unit retinal responses to color he discovered "modulator bands" of interaction between absorption spectra, sharpening contrast, "crispening of information by interaction, largely inhibitory in nature." He also found that some cells in the vertebrate retina are hyperpolarized rather than depolarized by light, and that inhibitory interactions between retinal elements play an important role, as shown by Kuffler and Hartline, accounting for the "off" discharge as well as for visual contrast. He made extensive studies of the retinal mechanisms of light and dark adaptation and the spectral properties of retinal elements in a variety of animals (frog, cat, guinea pig, etc). These studies were summarized by Granit in 1947 and in his Silliman Memorial Lectures at Yale published in 1955.

By the time Grant received his Nobel award he had abandoned the visual system and launched his "second career" in studies of muscle afferents and mechanisms of alpha and gamma motor control in spinal cord, cerebellum, and motor cortex. This work was partially reviewed in his Silliman Lectures, which described also his thoughts about sensory discrimination and integration in general.

Granit received many honors and delivered many honorary lectures, including the Sherrington Memorial Lectures of the Royal Society of Medicine in London in 1967 and in Liverpool in 1970. He initiated an important series of Nobel Symposia on *Muscular Afferents and Motor Control* (1966). He also published two books on this subject: *Basis of Motor Control* (1970) and *Mechanisms Regulating the Discharge of Motoneurons* (1972). His more philosophical thoughts were expressed in his recent book (1977), entitled *The Purposive Brain.*

Keffer Hartline was born in Bloomsburg, Pennsylvania, in 1903. He received his MD at Johns Hopkins and then studied physics on an NRC fellowship there and abroad. Upon his return to the United States, he worked at the Johnson Foundation for Medical Physics at the University of Pennsylvania. For a time, he was Associate Professor of Physiology at Cornell Medical College in New York City, but returned to the Johnson Foundation until 1949 when he became Professor of Biophysics at Johns Hopkins University. Here he was joined by his colleague Floyd Ratliff in 1954. He then moved his laboratories in 1963 to the Rockefeller University in New York City where he remained as Professor until retirement.

Hartline's Nobel Lecture was entitled, "Visual Receptors and Retinal Interaction." He began his studies by single fiber analysis of the optic nerve of the frog, but most of his research was carried out on the eye and optic nerve of the horseshoe crab (*Limulus*). In this relatively simple eye, the electrical activity of individual nerve fibers and photoreceptors (ommatidia) could be recorded by dissection and by intracellular microelectrodes. The photoreceptors could also be stimulated individually so that the generator potentials and action potentials to local light stimulation could be recorded, as well as the interaction between adjacent receptors. Surround inhibition was thus subjected to quantitative study, which provided direct evidence for retinal mechanisms underlying the phenomena of visual contrast and Mach's bands in human vision. [See Hartline (1941–1942) on "The Neural Mechanisms of Vision".]

George Wald was born in New York City in 1906. He studied at New York University and pursued graduate studies in zoology at Columbia with Selig Hecht (PhD 1932). He then worked with Otto Warburg in Berlin-Dahlem on an NRC Fellowship, at which time he first identified vitamin A in the retina. After further studies in Heidelberg, Zurich, and Chicago, Wald returned to Harvard (1935) as a tutor in biochemistry, then became Professor in the Department of Biology (1944–1948) and later University Professor (1948–1977) until 1977, when he became Professor Emeritus. In 1953 he received the Lasker Award.

Wald's principal research was on the photochemistry and biochemistry of visual pigments in the retina. His Nobel Lecture was entitled, "The Molecular Basis of Visual Excitation."

Wald determined the chemical composition of photopigments of the retina: the retinals, 1 and 2, which, in vertebrates, combine with the rod opsin to form rhodopsin and porphyropsin, and with cone opsin to form iodopsin and cyanopsin. The retinals are aldehydes derived from vitamins A_1 and A_2 by dehydrogenation of the alcoholic group, and they act as visual pigments when the side chain is in the 11-*cis* configuration, one that facilitates union with the opsins. On exposure to light the pigments are converted to a steady-state mixture of carotenoids, with various proportions of the geometrical isomers of the retinals. The pigments must be reisomerized to the 11-*cis* configuration before they can combine again with the opsins to regenerate visual pigment. The manner in which these reactions are transduced into electrical potentials and action potentials in ganglion cells must still be clarified. Wald's original work was summarized in the *Annual Review of Biochemistry* in 1953 (Vol. 22).

1970   Julius Axelrod, Ulf Svante von Euler, and Bernard Katz

For their discoveries concerning the humoral transmitters in the nerve terminals and the mechanisms for their storage, release, and inactivation.

Axelrod was born in New York City in 1912. He obtained the BS degree from the College of the City of New York in 1933, and then worked at various jobs during the depression. He returned to graduate studies, receiving the MS degree from New York University in 1941. One of his jobs brought him under the leadership of B. B. Brodie, whom he followed to Bethesda, when Brodie established a research laboratory at the National Heart Institute in the early 1950s. While working there, Axelrod earned the PhD degree from George Washington University (1955). The work for which he was recognized by the Nobel Committee was performed at the Heart Institute and is described in his Nobel Prize Lecture, "Noradrenaline:

Fate and Control of its Biosynthesis." He continues his research at the National Institute of Mental Health, where he is Chief of the Section on Pharmacology at the Laboratory of Clinical Science.

Axelrod's work mentioned in the citation stemmed from his studies of the metabolism of catecholamines containing tritium in high specific activity. It was thought that noradrenaline labeled in this way would be eminently useful in detecting its pathways of intermediary metabolism. The experiments yielded an unexpected result: the tracer amine was rapidly taken up by organs with a sympathetic innervation, and specifically into the postganglionic nerve endings, where it was then stored just like endogenously synthesized amine and was released under the influence of nerve impulses. With his colleagues, Axelrod demonstrated the factors determining the "uptake," storage, and release of catecholamines and the role of many important drugs in affecting these processes. His investigations of the metabolism of catecholamines led to extensive study of transmethylation processes in their inactivation, and carried over to the methylation of a derivative of serotonin in the course of biosynthesis of melatonin in the pineal gland. Axelrod has also studied the neural and endocrine regulation of catecholamine biosynthesis in the adrenal medulla, and has demonstrated the transsynaptic induction of tyrosine hydroxylase and dopamine beta-hydroxylase. Most recently, he has investigated the role of methylation of phospholipids in membrane transformations.

Ulf Svante von Euler was born in Stockholm in 1905. Both his parents were scientists: his father had shared the Nobel Prize in Chemistry with A. Harden in 1929. Ulf von Euler received his medical degree from the Karolinska Institute in 1930, and later studied with Sir Henry Dale, Sir John Gaddum, and I. de Burgh Daly in England, and with Corneille Heymans in Belgium. In 1939 he became Professor of Physiology at the Karolinska, where he has remained for his entire professional life. At one time he was chairman of the Nobel Committee for the Prize in Physiology or Medicine, and from 1965–1975 was President of the Nobel Foundation.

Von Euler's outstanding contribution for which he received the Nobel citation was his discovery in 1946 that the transmitter at sympathetic postganglionic nerve endings is noradrenaline (not adrenaline, as had been suggested by others), and that it is released upon stimulation and partly excreted in the urine. The results of his studies of catecholamine release from the adrenal gland in a wide variety of experimental and clinical states, including stress, have long been incorporated into standard textbooks. Much of his earlier research was summarized in a monograph entitled simply, *Noradrenaline* (1956), and in his Nobel Prize Lecture, "Adrenergic Neurotransmitter Functions" (1971). In 1951 he demonstrated (with A. Engel) the increased output of urinary catecholamines in pheochromocytoma, and the value of that measurement in diagnosing the presence of chromaffin tumours. In 1956 he showed with A. Hillarp that noradrenaline is stored in the nerves in subcellular particles from which it is released on stimulation.

Von Euler has made other important contributions to physiology. While at the National Institute for Medical Research in London during the early 1930s, he and Gaddum discovered Substance P. A few years later he discovered prostaglandin. These contributions laid the basis for the current intensive investigations of polypeptides as hormones and pharmacological agents and of the prostaglandins and thromboxanes.

Von Euler is a member of the Royal Society of London and the National Academy of Science in Washington. He has received many honorary degrees from universities in Europe and North and South America. A symposium, edited by Wolstenholme & O'Connor, was held in his honor by the Ciba Foundation in 1968, entitled "Adrenergic Neurotransmission." Von Euler also organized and edited with B. Pernow a Nobel Symposium on Substance P in Stockholm (1976) and published a book on prostaglandins (with R. Eliasson) in 1967.

Bernard Katz was born in Leipzig, Germany, in 1911. He obtained his MD from the University of Leipzig in 1935 and then studied with A. V. Hill at University College, London, where he received the PhD in 1938. He worked with Eccles and Kuffler on neuromuscular physiology on a Carnegie Research Fellowship in Sydney, Australia and returned to University College for his DSc in 1942. He was appointed the first professor of biophysics at University College in 1952. He is a Fellow of the Royal Society and of the Royal Society of Medicine. He was knighted in 1967.

The title of Katz's Nobel Lecture was, "On the Quantal Mechanism of Neural Transmitter Release." His classical microphysiological studies of the release of acetylcholine at the neuromuscular junction were carried out in collaboration with Paul Fatt, J. del Castillo, and R. Miledi. They discovered miniature end-plate potentials, which they proved to be due to the spontaneous quantal release of packets of acetylcholine molecules presumably stored in synaptic vesicles in the axonal terminals at the end-plate. They showed that this release was Ca^{2+}-dependent. Depolarization of the axonal terminals by the action potential caused an influx of Ca^{2+} that caused the release by exocytosis of many vesicles. The acetylcholine then acted upon receptors in the post-junctional membrane to allow a sudden influx of Na^+ and K^+ to set up a large end-plate potential sufficient to excite adjacent muscle fibers. The award winning experiments were described in the *Proceedings of the Royal Society* by Katz & Miledi in 1965 (Katz & Miledi 1965a–c).

Katz showed that there was considerable plasticity in the ACh transmitter mechanism as shown by "desensitization and potentiation," making it an important model synapse, as described by Katz in his books *Nerve, Muscle, and Synapse* (1966) and *The Release of Neural Transmitter Substances* (1969).

1973   Karl von Frisch, Konrad Lorenz, and Nikolaas Tinbergen

For their discoveries concerning organization and elicitation of individual and social behavior patterns.

This was the first Nobel Prize to recognize the importance of the naturalistic approach to an understanding of animal behavior. The systematic observation and recording of the development of characteristic patterns of behavior in animals raised in their natural habitat, or with systematic alterations in their environment during early development, has given much insight into the interaction of social and environmental factors upon the development of inborn patterns of behavior.

Karl von Frisch was a German zoologist born in Vienna in 1886. His zoological studies were carried out in the universities of Vienna and Munich. He became Professor and Director of the Zoological Institute at Rostock University in 1921, but later moved to Breslau, Munich and to Graz before settling at the University of Munich (1950–1958). He received international renown for his studies of the language of bees, published in his remarkable book *The Dancing Bees* (1961) and in *The Dance Language and Orientation of Bees* (1967). He has received many honors, including membership in the Academies of Sciences in Washington, Uppsala, Munich, Vienna, and Stockholm. He was able to show that the pattern of the flying "dance" of the bees could inform the swarm of the direction, distance, location, and even the quality of food to be sought there. He has made detailed studies of the sensory capacity of fish and insects.

Konrad Lorenz, an Austrian anatomist and animal psychologist, was born in Vienna in 1903. He studied at the University of Vienna, where he was an Assistant in the Institute of Anatomy (1928–1935) and Lecturer in Comparative Anatomy and Animal Psychology (1937–1940). He then became head of the Department of Psychology at Königsberg (1940–1945) and Director of the Max Planck Institute for Physiology and Behavior (1961–1973) at Seewiesen in Bavaria. He became Director of the Department of Animal Sociology at the Institute of Comparative Ethology of the Austrian Academy of Sciences in 1973. Among his many honors were membership in the Royal Society, London, and the Kalinga Prize from UNESCO. Lorenz is the acknowledged founder of the science of ethology.

Lorenz studied animal behavior in natural habitats and the modification of instinctive behavior patterns by the environment and by social contacts, especially with man. His favorite animal of study was the greylag goose, in which he discovered the phenomenon of "imprinting." This rapid form of learning is possible only at a critical period of the gosling's life, when the bird may become attached to man as a parental object and remain so. He also discovered "innate release mechanisms," which conditioned an animal's reaction to environmental and social stimuli. His many other observations of animal behavior had important consequences for our understanding of developmental and social aspects of human behavior as well. His publications include *King Solomon's Ring* (1952), *Evolution and Modification of Behavior* (1965), and *Studies in Animal and Human Behavior* (1970).

Tinbergen began naturalistic studies of animal behavior in The Netherlands, where he was born in 1907. He studied Zoology at the Universities of Leiden, Vienna, and Yale, returning to join the Faculty of Zoology at Leiden in 1936. It was at Leiden that he did much to develop the new science of ethology, and where he became Professor Emeritus of Experimental Zoology in 1949. He then moved to Oxford, became a British citizen in 1955, and was appointed Professor of Animal Behavior (1966–1974).

Tinbergen studied the behavior of insects, fish, and birds in their natural habitat. He analyzed the innate or instinctive factors as well as the sensory and social influences controlling their behavior. His principal publications include *Study of Instinct* (1951), *Animal Behavior* (1965), and *Social Behavior in Animals* (1953).

1976 Baruch S. Blumberg[3] and D. Carleton Gajdusek

For their discoveries concerning new mechanisms for the origin and dissemination of infectious diseases.

Daniel Carleton Gajdusek was born in Yonkers, New York in 1923. He took his undergraduate degree at the University of Rochester in 1943 and his medical degree at Harvard in 1946. His specialty training was initially in pediatrics, but he ultimately turned to virology. During an extended visit to the Walter and Eliza Hall Institute, headed by Sir MacFarlane Burnet (Nobel Prize, 1960), in Melbourne, Australia, Gajdusek learned about kuru, which led to his extensive studies in New Guinea. He has had experience with infectious disease in the field also in the United States and Iran. Since 1958 he has been at the National Institute of Neurological and Communicative Disorders and Stroke where, as Chief of the Laboratory of Central

[3]Blumberg is omitted from this review of neuroscience.

Nervous System Studies, he has developed knowledge of latent viruses and the diseases they cause.

Gajdusek's visit to New Guinea followed by only two years the discovery of a new neurological disorder, kuru, among members of a tribe of the Fore people living in the interior of the island, north of Port Moresby. In his Nobel Lecture he describes this disease as being "characterized by cerebellar ataxia and a shivering-like tremor that progresses to complete motor incapacity and death in less than one year from onset" (1977). The circumstances of the disease challenged Gajdusek's anthropological propensities, as well as his medical interest, and he spent the better part of a year living with the tribe. He was able to obtain tissue specimens for later analysis. The clinical and laboratory results were published when Gajdusek returned to the United States (1957), but the cause of kuru was still unknown. Soon after this, W. Hadlow drew attention to the similarities between kuru and a disease of sheep, scrapie, that belongs to a group of diseases known as "slow viral infections," i.e. infections that require prolonged incubation periods before generating pathology. Gajdusek, by this time established at the National Institute of Neurological and Communicative Disorders and Stroke, then succeeded in transmitting kuru to a subhuman primate in 1963. This was the first chronic degenerative disease of man demonstrated to be due to a slow virus. Transmission of the virus among the tribal people was attributed to the consumption of the infected brain of the deceased by the mourners. This success prompted Gajdusek to postulate that other degenerative diseases of man, or diseases occurring in a familial pattern, might also be due to slow viruses. Acting on this hypothesis he was able to transmit Creutzfeldt-Jakob disease to chimpanzees (1968). His efforts to transmit other human degenerative dementias (Alzheimer's disease, Pick's disease, Huntington's chorea, and Parkinsonism-dementia) to laboratory animals have not yet been successful.

1977 Roger C. L. Guillemin, Andrew V. Schally, and Rosalyn S. Yalow

For discoveries concerning the peptide hormones of the brain (Guillemin and Schally).

For the development of radioimmunoassays of peptide hormones (Yalow).

Roger Guillemin was born in Dijon, France in 1924. He received his undergraduate education there. After completing medical studies in Lyons (1949), he went to the Université de Montréal where he carried out research under the direction of Hans Selye, and received the PhD in 1952. He moved to Baylor University College of Medicine in 1953 and remained there for the next 17 years, except for a few years' interlude at the Collège de France in Paris. In 1970 he became a Fellow and Director of the Neuroendocrinology Laboratories at the Salk Institute for Biological Studies, where he still carries on his research.

Andrew Victor Schally was born in Wilno, Poland in 1926. He worked at the National Institute of Medical Research in London for some time, but emigrated to Canada where he completed his undergraduate studies at McGill University in 1955. While still an undergraduate, he had begun research under the direction of Murray Saffran at the Allan Memorial Institute of Psychiatry—work that led to the description of the first releasing hormone, CRF—and received the PhD degree in 1957. He then took a position with Guillemin at Baylor University College of Medicine, but in 1962 was appointed Chief of the Endocrine and Polypeptide Laboratories of the Veterans Administration Hospital in New Orleans. He is currently a Professor of Medicine at the Tulane University School of Medicine.

When the late Geoffrey Harris, a British endocrinologist and anatomist, proposed (1955) that the activities of the anterior pituitary gland are controlled by the brain, he was laying the foundation for the present-day recognition of the brain as an endocrine organ. Nervous control of endocrine secretion had been recognized many years earlier in respect to secretion of adrenal catecholamines, but the concept of nervous regulation of the secretion of the trophic hormones of the pituitary gland was entirely novel. The description by the Scharrers of hypothalamic neurons with an apparently secretory function (1963) made Harris's suggestion much more readily acceptable, as did Harris's demonstration of the hypothalamo-hypophyseal portal system and the deleterious consequences for pituitary function of its interruption. In 1955 Saffran & Schally described the first hypothalamic activity that could be attributed to a hormone, corticotropin-releasing factor (CRF). Five years later, Harris in Great Britain and S. M. McCann in the United States reported on the luteinizing hormone-releasing factor (LRF). These findings were the actual starting-point for the investigations that eventually brought the Nobel Prize to Guillemin and Schally.

Following his research at McGill University, Schally worked with Guillemin in Houston. Their initial research dealt with CRF, but the task was a forbidding one because of the great difficulty in purification and characterization of the molecule. Indeed, it was not until the end of 1981 that this task was accomplished (Vale et al 1981). In 1962, Schally established his own laboratory in New Orleans. Both he and Guillemin, now working independently, turned the efforts of their respective teams to the thyrotropin-releasing factor (TRF). Schally chose to work with pig hypothalamus, Guillemin continued with ovine tissue. After seven years' work, they provided the structure of TRF, an unusual tripeptide. They then took up the investigation of LRF: the two teams succeeded in isolating sheep and pig

LRF in 1971 and demonstrated that they are identical. Schally's group defined the structure of LRF at that time.

The factor causing release of growth hormones (GH) was the next in line of attack, but before much headway could be made, Guillemin discovered a substance in the hypothalamus that prevents the release of GH from the anterior pituitary gland, now known as somatostatin, and successfully identified its structure. The efforts to unravel the structures of these hypothalamic hormones are described in detail in their respective Nobel Lectures (1978).

In the succeeding years Guillemin has investigated endorphins. In 1976, his laboratory reported on the primary structure of α-endorphin, a hexadecapeptide, and γ-endorphin, a peptide with one more amino acid residue. These two endorphins contain metenkephalin as the amino-terminal sequence.

Rosalyn Sussman Yalow was born in New York in 1921. She studied at Hunter College (AB 1941), and then moved to the University of Illinois for graduate studies in physics (MS 1942, PhD 1945). She returned to Hunter College as Assistant Professor of Physics, but in 1950 took a position at the Veterans Administration Hospital of the Bronx, New York where she has been ever since. She is now Chief of the Radioimmunoassay Reference Laboratory there and the Nuclear Medicine Service.

In the mid-1950s, Yalow and the late S. Berson were testing a hypothesis proposed earlier by I. A. Mirsky about the biochemical defect in diabetes. Their experiments required the determination of the rate of metabolism of insulin, and they did this by administering ^{131}I-labeled insulin to diabetic and nondiabetic subjects, as well as to others who had received large amounts of insulin for the treatment of schizophrenia. Subjects receiving insulin therapeutically metabolized the labeled material more slowly, as judged by the rate of disappearance of the ^{131}I from the plasma. They postulated that the labeled insulin was being bound to antibodies, raised by the introduction of the exogenous insulin. Because of the low titer of such antibodies and their failure to precipitate, Yalow & Berson devised isotopic methods that would detect a soluble antigen-antibody complex. By electrophoresis of labeled insulin from plasma of their subjects, they were able to detect such antibodies, and this was reported in 1956. In her Nobel Prize address (1978), "Radioimmunoassay: A Probe for the Fine Structure of Biologic Systems," Yalow defined the new technique (RIA) as follows:

> The concentration of the unknown labeled antigen is obtained by comparing its inhibitory effect on the binding of radioactively labeled antigen to specific antibody with the inhibitory effect of known standards. . . . The RIA principle . . . can be extended to other systems in which in place of the specific antibody there is a specific reaction or binding substance.

These extensions took place slowly over the next decade, in relation to the binding of ^{60}Co-labeled vitamin B_{12} to intrinsic factor, as developed by V. Herbert, S. P. Rothenberg, and R. P. Ekins; with ^{131}I-labeled thyroxine through its binding by a specific serum globulin, through the studies of Ekins and B. E. P. Murphy; and the binding of plasma corticoids to trans-cortin, the serum corticosteroid-binding globulin, also by Murphy. These hormone studies verified the great sensitivity of what now became known as competitive protein-binding radioassays. These radioassays are receiving increasing attention by neuroscientists for mapping and detection of brain hormones. In fact, Yalow pointed out in her Nobel Lecture that the joint recognition of work in brain peptides and RIA signalized these new applications of the principle she had elaborated to the study of brain and nerve.

1979   Allan MacLeod Cormack and Godfrey Newbold Hounsfield

For the development of a revolutionary X-ray technique, computer axial tomography (CAT) scan.

Allen MacLeod Cormack was born in South Africa in 1924. After studying nuclear physics at Cambridge University, he returned to Capetown, where he became a member of the University faculty. In 1956 he was asked to serve as a part-time medical physicist at the Groote Schuur Hospital there. This gave him the opportunity to observe how the dose of radiation that patients were to receive in therapy was selected. Because of the variable attentuation of the x-ray beam by the skull and the soft tissues lying in its path, the radiologist was compelled to make an estimate of the dose that would ultimately reach the target, rather than a precise calculation. This was the genesis of his thinking about the problem that later came to be known as computerized tomography. After a sabbatical leave at the Harvard Cyclotron Laboratory, he began to work on the theory of image reconstruction and proceeded to test his ideas on models. In 1957, Cormack returned to the United States, taking a post in the Physics Department at Tufts University, where he has served as departmental chairman.

Godfrey Newbold Hounsfield was born in England in 1919. He is an electronics engineer at the Central Research Laboratories of EMI, (Electronic Musical Industries), a firm that he joined in the early 1950s. He directed the group that designed the first large solid state computer in the United Kingdom. He became interested in problems of pattern recognition in 1967, and soon undertook to extend the work to use of radiation, starting with γ-rays. Hounsfield is now the Head of the Medical Systems Department at the EMI Laboratories.

When a penetrating beam (e.g. X-rays or γ-rays) passes through an object, the coefficient of attentuation is the logarithm of the ratio of incident energy to transmitted energy. For homogeneous objects, this coefficient g is readily measurable as an exponential function, but for a nonhomogeneous body it is a composite of differential attentuations along the line of the beam. Cormack reasoned that if one could determine the attentuation coefficients of tissues by measurements made external to the body, the dose adjustments in radiation therapy would be much simpler to make. Such information would have a further dramatic consequence. If one thinks of the tissue to be imaged as composed of numerous thin slices layered one on top of the other, with the x-ray beam passed through the sections edgewise, the resulting data could be synthesized into a two-dimensional reconstruction of the internal structure of the organ. But in order to do so one would have to solve the problem of how to sort out the fluctuations in attenuation mathematically. Cormack showed that if g were determined for all lines intersecting the body, this could be achieved in two dimensions. His mathematical solution was published in two papers in the *Journal of Applied Physics* in 1963 and 1964; they met with little interest at the time. Yet his solution lies at the basis of the imaging technique used in computed tomography (CT), or computerized axial tomography (CAT scanning).

Hounsfield was intent upon improving the conventional x-ray technique to make it serviceable for imaging of hidden (internal) structures. Like Cormack, he constructed models for testing, but used a somewhat different mathematical solution. The successes with phantoms showed the way to clinical application. Hounsfield successfully devised a clinical tomograph, completed in 1971. Its computer made use of Cormack's algorithm. In effect, this was a method for resolving the more than 28,000 measurements, each the dependent variable of an integral equation. Hounsfield's earlier experience with γ-rays convinced him to turn to the more powerful X-rays for scanning of the brain. He later adapted the instrument for examining other parts of the body.

Cormack and Hounsfield were not alone in conceiving of tomographic principles. In 1961, W. H. Oldendorf, a California neurologist, concerned with improving the imaging of the brain without employing invasive procedures such as angiography and pneumoencephalography, suggested that the variable absorbances of the brain could be used to obtain an image of its internal structure. He actually constructed a model, but without computer reconstruction, and obtained a patent for its use with γ-rays. Oldendorf's paper (1961) on the utilization of these "radiodensity discontinuities" was the first to deal with radiographic tomography. For this work he received the Lasker Award for 1975 with Hounsfield.

Sometime after 1970, Cormack learned that his mathematical solution had been determined in 1917 by J. H. Radon. As it turned out the imaging problem was not the only one that needed the answer Cormack had provided: it was also important for problems of statistics, radio astronomy, electron microscopy, and optics (Cormack 1980).

With the introduction of CAT scanning, the need for pneumoencephalography and cerebral angiography has been sharply reduced. The technique serves in the diagnosis of brain tumors, brain hemorrhages, lesions of various types, and hydrocephalus, and is being used in studies of aging, dementia, and degenerative disorders. The principles of computed tomography can be applied to the use of nuclear magnetic resonance instead of X-rays, and this should provide information about the chemical composition of tissues. Hounsfield (1980) sees the two methods as becoming complementary to one another.

The 1979 citation recalls the award of the first Nobel Prize (in Physics) to Röntgen, a physicist, for his discovery of X-rays.

1981   David Hunter Hubel, Torsten Nils Wiesel, and Roger Wolcott Sperry

For discovering how sight stimulation in infancy is tied to future vision and how the brain interprets signals from the eye (Hubel & Wiesel).

For his research into the specialized functions of each side of the brain (Sperry).

David Hubel was born in Windsor, Ontario, Canada in 1926. He obtained his BSc in Physics at McGill University in Montreal in 1947 and his MD in 1951. He was a fellow at the Montreal Neurological Institute 1952–1954, where he studied with H. Jasper and W. Penfield. He was a resident in neurology at Johns Hopkins in 1954 and, as a US citizen, he was drafted to the Walter Reed Army Medical Research Institute in Washington (1955–1958), where he developed his tungsten microelectrode technique for recording from single cells in the cerebral cortex. He returned to Johns Hopkins where he became Associate Professor of Neurophysiology and Neuropharmacology, and later Professor of Neurophysiology. It was here that he became associated with Vernon Mountcastle, Stephen Kuffler, and Torsten Wiesel. He moved to Harvard Medical School with Kuffler in 1959, became Professor of Physiology in 1965, and was appointed George Packer Berry Professor of Neurobiology at the Harvard Medical School in 1968.

Torsten Wiesel was born in Uppsala, Sweden in 1924 and obtained his MD at the Karolinska Institute in 1954. He became Instructor in Physiology at the Royal Caroline Medical-Surgical Institute and Assistant in the Department of Child Psychiatry. He was then appointed Assistant Professor of Ophthalmology and Physiology at Johns Hopkins University, 1958–

1959. He moved with Kuffler to Harvard Medical School in 1959, where he later became Robert Winthrop Professor of Neurobiology. He succeeded Kuffler as Head of the Department of Neurobiology.

Most of the work done by Hubel and Wiesel was in such close collaboration that it can be described jointly. Their work was based upon extracellular microelectrode records of the firing of single cells in the visual cortex and lateral geniculate body in response to different points, patterns, movement, and color of photic stimulation. Electrophysiological studies were supplemented by anatomical studies using modern techniques of autoradiography, intracellular injection of fluorescent dyes, and Sokoloff's deoxyglucose method for determining the uptake of glucose in local areas of the visual cortex. This combination of precise microphysiological and histological study of the visual system in cats and monkeys, together with their studies of factors affecting the normal and abnormal development of this system in young animals, has provided a remarkably clear picture of the modular and hierarchical organization of the visual system from the retina, through the lateral geniculate bodies, to the primary and secondary areas of the visual cortex. Hubel & Wiesel demonstrated how this synaptic organization can be disturbed or permanently deformed by lack of appropriate visual experience early in life. As described by the Nobel Prize Committee, Hubel & Wiesel's work "represents a breakthrough in research into the ability of the brain to interpret the message from the eyes." (See Hubel 1982 for a brief historical account of the development of cortical neurobiology.)

Radially oriented columns of cells were found in the visual cortex that responded selectively to the orientation and direction of movement of lines and patterns in the visual field. These cells were described as "simple," "complex," or "hypercomplex" depending upon the complexity of visual pattern to which they were attuned. The more complex cells were found in the upper layers of the primary visual cortex, and especially in the parastriate cortex (areas 18 and 19).

Alternating columns of cells were also found for ocular dominance and confirmed by anatomical studies making use of the transynaptic axonal transport of radioactive amino acids injected into one eye. These results were confirmed by the use of Sokoloff's ^{3}H-deoxyglucose technique to study the effect of illumination of one eye upon glucose uptake in the striate cortex.

Suturing of one eye during the first few weeks of life in the cat or monkey caused a marked decrease in the ocular dominance columns for that eye, from which the animals might not recover. The synaptic organization of the various "feature detectors" in the visual system was found to be dependent upon visual experience during an early critical period of development in infancy.

Vernon Mountcastle, who first demonstrated the columnar organization of the cortex in the somatosensory system, has shown the general importance of this discovery for the specialized "modular" organization of all cortical areas. At the same time, he has pointed out that cortical areas cannot function as isolated units, but require "distributed systems" of integration for the organization of perception and motor behavior in the intact organism (Mountcastle 1978).

David Hubel delivered the Ferrier Lecture of the Royal Society in 1971 and has received several additional prizes and honors, including membership in the National Academy of Sciences and the Karl Lashley Prize of the American Philosophical Society in 1977. Torsten Wiesel is also a member of the National Academy of Sciences, and (with Hubel) was the recipient of the Lewis Rosenstiel Prize of Brandeis University and the Louisa Horwitz Prize of Columbia University.

Roger Sperry was born in Hartford, Connecticut, in 1913. He began his secondary education at Oberlin College in the early 1930s, majoring in English literature. He took his Master's degree at Oberlin in psychology and then went on to earn his PhD in zoology at the University of Chicago in 1941. He did postdoctoral work in psychobiology at Harvard University and later at the Yerkes Laboratories of Primate Biology in Georgia. He was Assistant Professor of Anatomy and Associate Professor of Psychology at the University of Chicago from 1946 to 1952. He was Section Chief of Developmental Neurology at the NIH in 1952–1953, but was then appointed Hixon Professor of Psychobiology at the California Institute of Technology in Pasadena in 1954, where he carried out his Nobel Prize research. He might well have received another Nobel award for equally distinguished work in the developmental biology of the nervous system, which he carried out chiefly at the University of Chicago, beginning with his work as a graduate student in Paul Weiss' laboratories there in 1940.

Sperry's (1945) previous work in developmental neurobiology was summarized in an important paper in the *Proceedings of the National Academy of Sciences* in 1963 (Vol. 50) in a paper entitled, "Chemoaffinity in the Orderly Growth of Nerve Fiber Patterns of Connections." His Nobel Prize research started about 20 years later. The functional significance of the great mass of nerve fibers connecting the right and left hemispheres, known as the corpus callosum, in man was unknown except for its ability to conduct epileptic discharge from one hemisphere to the other. Roger Sperry and his student, Ronald Myers, had already shown in animals that section of the corpus callosum prevented the transfer of training from one hemisphere to the other. An opportunity to study the effects of callosal section in man was provided for Sperry by the epileptic patients operated upon by the neurosurgeon Joseph Bogen for the control of intractable seizures.

Following what is now known as the "split brain" operation, Sperry and his students were able to demonstrate also a splitting of conscious awareness: objects presented to the right hemisphere alone were not recognized verbally by the language-dominant left hemisphere, even though they produced appropriate responses of the left hand. Verbal expression of conscious awareness was possible only for objects presented to the left hemisphere, which also seemed to excel in certain thought processes. However, the right hemisphere was found superior to the left in nonverbal processes such as spatial orientation, and probably some other abilities.

In spite of this lack of communication between the two hemispheres, patients with section of corpus callosum appeared quite normal and only occasionally had difficulty in coordinating the behavior of the hands. Sperry described these observations in his Harvey Lectures (1968), "Mental Unity Following Surgical Disconnection of the Cerebral Hemispheres."

Sperry was always concerned with the problems of the mechanisms of conscious awareness, though opposed to the dualist views expressed by some neuroscientists. The Nobel Award Committee cited Sperry for succeeding "brilliantly in extracting the secrets from both hemispheres of the brain and demonstrating that they are highly specialized and also that many higher functions are centered in the right hemisphere." Sperry wrote the Prefatory Chapter to Volume 4 of the *Annual Review of Neuroscience.*

Commentary

This brief review of Nobel Laureates in Neuroscience has highlighted many of the most important discoveries and developments in this area of scientific inquiry. The range of topics is wide, but they are all concerned with knowledge and understanding of the structure and function of the nervous system that have been gained since the turn of the century. Neuroscience has become such a wide-ranging conglomerate of biological, physical, mathematical, and medical sciences during the past 15–20 years that we are no longer surprised to find significant new discoveries and concepts coming from unexpected quarters. For example, the Neuroscience Research Program, pioneered by Frank Schmitt some 20 years ago, has brought dozens of different disciplines to bear on particular problems of neuroscience. In fact, before that time there was no formally organized discipline of neuroscience, except perhaps through the activities of the interdisciplinary International Brain Research Organization (IBRO), established in 1960.

In view of the responsibilities of the Nobel Committee for Physiology or Medicine to select only one prize each year in the whole field of biological and medical sciences, the fact that 22 out of the 74 awards made have been for discoveries considered of importance to neuroscience is a remarkable record. Even more significant is the number of individual scientists cited, since frequently an annual prize has been shared by two or three investigators for different, though often related, contributions or discoveries. Of the 129 individual Laureates in Physiology or Medicine, the work of 47 has had direct relevance to basic or clinical neuroscience. It is apparent, with the passage of time and evolution of new experience, that the contributions of some of those selected have been minor, although the contributions of most have been outstanding and lasting. Revolutionary discoveries, such as have been made recently in molecular biology, are rare, and when they do occur in neuroscience they are often the result of a team effort by several investigators. This has made selection of the prize-winners even more difficult. Despite these difficulties, it is remarkable how many of the most important discoveries in neuroscience have been recognized and honored by the Nobel Committee over the years, and how many of the leading neuroscientists involved have become Nobel Laureates.

Appreciation of the work of these Laureates must be viewed in its historical perspective. The work of Pavlov and his school of conditioned reflex physiology, for example, represented at the turn of the century a true breakthrough in our understanding of the nervous control of the autonomic system. Furthermore, it provided objective measurements and a conceptual framework for the learning process, and for the analytic and integrative functions of the cerebral cortex, although Pavlov's oversimplified notions of brain mechanisms may not now be acceptable in the light of present knowledge. Even the original and brilliant work of Cajal, which provided the foundation for our knowledge of the cellular morphology of the nervous system, has been surpassed during recent years by revolutionary advances in ultramicroscopic, autoradiographic, and histochemical techniques of neuroanatomy and neuropathology.

The development of techniques by Erlanger & Gasser in the 1920s for the accurate visualization of the action potential from individual nerve fibers with wide-ranging conduction velocities, coupled with Adrian's recordings from individual sensory and motor nerve fibers, represented a true breakthrough in neuroscience at the time. Sherrington, who shared the prize with Adrian, did not make any revolutionary discoveries but had a profound influence upon the development of neuroscience by his thoughtful analysis of the balance of excitatory and inhibitory states in the organization of the reflex and integrative functions of the nervous system. This brings us to another major advance in the early 1930s, with the discovery by Otto Loewi and by Sir Henry Dale and co-workers of the chemical transmission of the nerve impulse at neuromuscular and synaptic junctions.

It was not until after the second world war, about 30 years after Loewi and Dale's discovery, that the classical work of Eccles and of Hodgkin & Huxley was honored by the Nobel Committee for their discoveries of the ionic mechanisms involved in the generation and conduction of the action potential in peripheral axons as well as in the chemical transmission of excitatory and inhibitory postsynaptic potentials in spinal motoneurons. A clear picture of ionic mechanisms involved in electrical conduction of nerve impulses and in chemically mediated synaptic transmission in the peripheral and central nervous systems and at neuromuscular junctions was becoming firmly established. The identification of many new chemical transmitter substances had just begun.

Major advances in our understanding of sensory mechanisms during the post war period were recognized by the awards to von Békésy and to Granit, Hartline, and Wald. These were followed shortly by acknowledgment of major discoveries in our understanding of the metabolism and mechanisms of storage and release of humoral transmitters achieved through the work of Axelrod, von Euler, and Katz. The broadening view of neuroscience, from the cellular and molecular to the behavioral, was marked by the award to von Frisch, Lorenz, and Tinbergen for their careful naturalistic studies of animal behavior, and in the recent recognition of the work of Roger Sperry on the biological basis of conscious awareness and voluntary movement in man as revealed by carefully controlled studies of functions of the left and right hemisphere in patients with section of the corpus callosum.

Discoveries of major clinical significance have been recognized in the work of Gullstrand, Bárány, Wagner-Jauregg, Eijkman, Heymans, Moniz, Bovet, Gajdusek, and the most recent award to Cormack and Hounsfield for their development of computerized axial tomography.

Two of the most important recent developments in neuroscience are (a) the proliferation of many new transmitter or modulator substances, especially the neuropeptides, and (b) the cellular physiology of the brain in relation to complex perceptual and motor behavior. The "peptide revolution" was started by von Euler by his discovery of Substance P; more recent developments have been partially recognized by awards to Guillemin, Schally, and Yalow. Advances in our understanding of the cellular organization of the brain have been made possible by the perfection of microelectrode techniques introduced over 20 years ago, techniques that make it possible to record from anatomically identified single cells throughout the brain in relation to simple and complex perceptual and motor behavior. These studies have been well exemplified in the work of Hubel & Wiesel, who not only demonstrated the columnar, or modular, organization of assemblies of cortical cells for the detection of specific features of visual environment, but have shown how this organization can be modified by early visual experience. Their microphysiological studies were confirmed and complemented by the use of recently developed anatomical techniques.

ACKNOWLEDGMENTS

In the course of this review we have been impressed once again with the great contribution the Nobel Foundation has made to Neuroscience over the years. Not only have their most prestigious and generous prizes served to stimulate and to gain world wide recognition for many of the most outstanding discoveries and developments in experimental techniques and concepts, but the Foundation has fostered and supported research laboratories and the exchange of information in the neurosciences through their Symposia and workshops. The wisdom of the Committee on Prizes in Physiology or Medicine has been vindicated, with few exceptions, by the far-reaching and lasting importance of the contributions made by Nobel Laureates in neuroscience to its increasingly rapid growth during recent years.

Finally we would like to express our appreciation to Professor Ulf von Euler and Professor Yngve Zotterman and to the Editor, Dr. Max Cowan, for their critical reviews of this manuscript.

Literature Cited

Adrian, E. D. 1927. *The Basis of Sensation, The Action of Sense Organs.* London: Christophers. 122 pp.
Adrian, E. D. 1932. *The Mechanism of Nervous Action.* London: Oxford Univ. Press. 103 pp.
Adrian, E. D. 1947. *The Physical Background of Perception.* Oxford: Clarendon. 95 pp.
Axelrod, J. 1971. Noradrenaline: Fate and control of its biosynthesis. *Science* 173:598–606
Bárány, R. 1907. *Physiologie und Pathologie des Bogengangsapparats beim Menschen.* Vienna: Deuticke. 76 pp.
Bovet, D. 1958. Rélations d'isostérie et phénomènes compétitifs dans le domaine des médicaments du système nerveux végétatif et dans celui de la transmission neuromusculaire. In *Les Prix Nobel en 1957.* Stockholm: Nobel Found.
Cormack, A. M. 1980. Early two-dimensional reconstruction and recent topics stemming from it. *Science* 209:1482–86
Dale, H. H. 1937. Transmission of nervous effects by acetylcholine. *Harvey Lect.* 32:229–45
Dale, H. H. 1953. *Adventures in Physiology.* London: Pergamon. 652 pp.
Eccles, J. C. 1952. *The Neurophysiological Basis of the Mind, the Principles of Neurophysiology* (Waynflete Lect.). Oxford: Clarendon. 314 pp.

Eccles, J. C. 1955. *The Physiology of Nerve Cells* (Herter Lect.). Baltimore: Johns Hopkins Univ. Press. 270 pp.
Eccles, J. C. 1964. *The Physiology of Synapses.* New York: Academic. 316 pp.
Eccles, J. C. 1982. The synapse: From electrical to chemical transmission. *Ann. Rev. Neurosci.* 5:325–39
Eccles, J. C., Ito, M., Szentágothai, J. 1967. *The Cerebellum as a Neuronal Machine.* Berlin: Springer-Verlag. 335 pp.
Eijkman, C. 1930. Antineuritisches Vitamin und Beri-beri. In *Les Prix Nobel en 1929.* Stockholm: Nobel Found.
Erlanger, J., Gasser, H. S. 1937. *Electrical Signs of Nervous Activity* (Johnson Found. Lect.). Philadelphia: Univ. Penn. Press. 221 pp.
Gajdusek, D. C. 1977. Unconventional viruses and the origin and disappearance of kuru. *Science* 197:943–60
Gajdusek, D. C., Zigas, V. 1957. Degenerative disease of the central nervous system in New Guinea. The endemic occurrence of "kuru" in the native population. *N. Engl. J. Med.* 257:974–78
Gasser, H. S. 1937. The control of excitation in the nervous system. *Harvey Lect.* 32:169–93
Golgi, C. 1903. *Opera Omnia.* Milan: Univ. Hoepli. 4 vol.
Granit, R. 1947. *Sensory Mechanisms of the Retina.* London: Oxford Univ. Press. 412 pp.

Granit, R. 1955. *Receptors and Sensory Perception* (Silliman Mem. Lect.). New Haven: Yale Univ. Press. 369 pp.

Granit, R., ed. 1966. *Muscular Afferents and Motor Control. First Nobel Symposium, 1965.* Stockholm: Almqvist & Wiksell. 466 pp.

Guillemin, R. 1978. Peptides in the brain: The new endocrinology of the neuron. *Science* 202:390–402

Gullstrand, A. 1924. *Mechanism of Accommodation, an appendix to Helmholtz's Treatise on Physiological Optics,* Trans. J. P. C. Southall, 1:388–90. Rochester, NY: Opt. Soc. Am.

Harris, G. W. 1955. *Neural Control of the Pituitary Gland. Monogr. Phys. Soc.* 3. London: Arnold. 298 pp.

Hartline, H. K. 1941–1942. Neural mechanisms of vision. *Harvey Lect.,* Ser. 37

Hess, W. R. 1954. *Diencephalon, Autonomic and Extrapyramidal Functions.* New York: Grune & Stratton. 79 pp.

Heymans, C., Neil, E. 1958. *Reflexogenic Areas of the Cardiovascular System.* Boston: Little, Brown. 271 pp.

Hill, A. V. 1932. *Chemical Wave Transmission in Nerve.* London: Cambridge Univ. Press. 74 pp.

Hill, A. V. 1938. The heat of shortening and the dynamic constants of muscle. *Proc. R. Soc. London Ser. B* 126:136

Hill, A. V. 1956. Thermodynamics of muscle. *Br. Med. Bull.* 12

Hodgkin, A. L. 1964. *The Conduction of the Nervous Impulse* (Sherrington Lect.). Liverpool: Liverpool Univ. Press. 108 pp.

Hounsfield, G. N. 1980. Computed medical imaging. *Science* 210:22–28

Hubel, D. H. 1982. *Ann. Rev. Neurosci.* 5:363–70

Hubel, D. H., Wiesel, T. N. 1968. Receptive fields and functional architecture of the monkey striate cortex. *J. Physiol.* 195:215–43

Hubel, D. H., Wiesel, T. N. 1974. Sequence regularity and geometry of orientation columns in the monkey striate cortex. *J. Comp. Neurol.* 146:421–50

Karolinska Institutet, Department of Physiology (Stockholm). 1953. In honor of S. Ramón y Cajal: On the centenary of his birth, 1952, by members of a research group in neurophysiology. *Acta Physiol. Scand. Suppl.* 29:106. 651 pp.

Katz, B., Miledi, R. 1965a. Propagation of electrical activity in motor nerve terminals. *Proc. R. Soc. London Ser. B* 161:453–82

Katz, B., Miledi, R. 1965b. The measurement of synaptic delay and time course of acetylcholine: Release at the neuromuscular junction. *Proc. R. Soc. London Ser. B* 161:483–95

Katz, B., Miledi, R. 1965c. The effect of calcium on acetylcholine release from motor nerve terminals. *Proc. R. Soc. London Ser. B* 161:496–503

Katz, B. 1966. *Nerve, Muscle, and Synapse.* New York: McGraw-Hill. 193 pp.

Katz, B. 1969. *The Release of Neural Transmitter Substances* (Sherrington Lect.). Liverpool: Liverpool Univ. Press. 60 pp.

Lorenz, K. 1970. *Studies in Animal and Human Behavior,* Vols. 1, 2. Cambridge: Harvard Univ. Press

Mountcastle, V. B. 1978. An organizing principle for cerebral function: The unit module and the distributed system. In *The Mindful Brain,* ed. G. M. Edelman, V. B. Mountcastle, pp. 7–50. Cambridge: MIT Press

Nobel Foundation. 1972. *Nobel Lectures, Physiology or Medicine, 1901–1970.* Amsterdam: Elsevier. 4 vols.

Oldendorf, W. H. 1961. Isolated flying spot detection of radiodensity discontinuities displaying the internal structural pattern of a complex object. *IRE Trans. BioMed. Electron.* 8:68–72

Pavlov, I. P. 1927. *Conditioned Reflexes: An Investigation of the Physiological Activity of the Cerebral Cortex.* London: Oxford Univ. Press. 430 pp.

Ramón y Cajal, S. 1909. *Histologie du Système Nerveux de l'Homme et des Vertébrés.* Paris: Maloine (Reprinted in 1952, Madrid: Consejo Superior de Investigaciones Científicas)

Ramón y Cajal, S. 1928. *Degeneration and regeneration of the nervous system.* Transl. R. M. May, 1959. New York: Hafner

Saffran, M., Schally, A. V. 1955. The release of corticotropin by anterior pituitary tissue in vitro. *Can. J. Biochem. Physiol.* 33:408–15

Santini, M., ed. 1975. *Perspectives in Neurobiology: Golgi Centennial Symposium.* New York:Raven. 678 pp.

Schally, A. V. 1978. Aspects of hypothalamic regulation of the pituitary gland. *Science* 202:18–28

Scharrer, E., Scharrer, B. 1963. *Neuroendocrinology.* New York: Columbia Univ. Press. 289 pp.

Sherrington, C. S. 1906. *The Integrative Action of Nervous System.* New York: Scribner. 411 pp. (Reprinted in 1947, London: Cambridge Univ. Press. 433 pp)

Sherrington, C. S. 1933. *The Brain and Its Mechanisms* (Rede Lect.). London: Cambridge Univ. Press. 36 pp.

Sherrington, C. S. 1940. *Man on His Nature.* London: Cambridge Univ. Press. 413 pp.

Spemann, H. 1938. *Embryonic Development and Induction.* New Haven: Yale Univ. Press. 401 pp.

Sperry, R. W. 1945. Restoration of vision after crossing of optic nerves and after contralateral transplantation of eye. *J. Neurophysiol.* 8:15–28

Sperry, R. W. 1963. Chemoaffinity in the orderly growth of nerve fiber patterns of connections. *Proc. Natl. Acad. Sci. USA* 50:703–10

Sperry, R. W. 1968. Mental unity following surgical disconnection of the cerebral hemispheres. *Harvey Lect.* 62:292–322

Sperry, R. W. 1981. Changing priorities. *Ann. Rev. Neurosci.* 4:1–15

Tinbergen, N. 1951. *Study of Instinct.* Oxford: Clarendon. 228 pp.

Tinbergen, N. 1953. *Social Behavior in Animals.* London: Methuen. 150 pp.

Tinbergen, N. 1965. *Animal Behavior.* New York: Time-Life. 200 pp.

Vale, W., Spiess, Rivier, C., Rivier, J. 1981. Characterization of a 41-residue ovine hypothalamic peptide that stimulates secretion of corticotropin and β-endorphin. *Science* 213:1394–97

von Békésy, G. 1949. The vibration of cochlear partitions in anatomical preparations and in models of the inner ear. On the resonance curve and decay period at various points on the cochlear partition. *J. Acoust. Soc. Am.* 21:233–54 (Trans. from German).

von Békésy, G. 1960. *Experiments in Hearing.* (Trans. and ed. E. G. Weaver). New York: McGraw. 745 pp.

von Békésy, G. 1967. *Sensory Inhibition* (Langfeld Lect). Princeton Univ. Press. 265 pp.

von Euler, U. S. 1956. *Noradrenaline.* Springfield, Ill.: Thomas. 382 pp.

von Euler, U. S., Eliasson, R. 1967. *Prostaglandins,* Med. Monogr. Ser., Vol. 18. New York: Academic. 164 pp.

von Euler, U. S. 1971. Adrenergic neurotransmitter functions. *Science* 173: 202–6

von Euler, U. S., Pernow, B. eds., 1977. *Substance P, Nobel Symposium, Stockholm, Sweden, 1976.* New York: Raven. 344 pp.

von Frisch, K. 1961. *The Dancing Bees: An Account of the Life and Senses of the Honey Bee,* trans. D. Ilse. New York: Harcourt, Brace/World. 182 pp.

von Frisch, K. 1967. *The Dance Language and Orientation of Bees.* Cambridge, Mass: Belknap Press, Harvard Univ. Press. 566 pp.

Waddington, C. H. 1961. *The Nature of Life.* London: Allen & Unwin. 131 pp.

Wagner-Jauregg, J. 1928. Nobel-Vortrag von Julius Wagner-Jauregg. In *Les Prix Nobel en 1927.* Stockholm: Nobel Found.

Wald, G. 1953. The biochemistry of vision. *Ann. Rev. Biochem.* 22:497–526

Wiesel, T. N., Hubel, D. H. 1974. Ordered arrangement of orientation columns in monkeys lacking visual experience. *J. Comp. Neurol.* 158:307–18

Wiesel, T. N., Hubel, D. H., Lam, D. M. K. 1974. Autoradiographic demonstration of ocular-dominance columns in the monkey striate cortex by means of transneuronal transport. *Brain Res.* 79:273–79

Wolstenholme, G. E. W., O'Connor, M., eds. 1968. *Adrenergic Transmission: In Honour of U. S. Von Euler.* Study Group on Adrenergic Neurotransmission, London, CIBA Found. Study Group No. 33. London: Churchill. 123 pp.

Yalow, R. S. 1978. Radioimmunoassay: A probe for the fine structure of biologic systems. *Science* 200:1236–45

Ann. Rev. Neurosci. 1983. 6:43–71

THE CLASSIFICATION
OF DOPAMINE RECEPTORS:
Relationship to Radioligand
Binding

*Ian Creese, David R. Sibley, Mark W. Hamblin,
and Stuart E. Leff*

Department of Neurosciences, University of California at San Diego,
School of Medicine, La Jolla, California 92093

INTRODUCTION

The past 25 years has seen our appreciation of the function of dopamine
(DA) in the brain elevated from that of a precursor for norepinephrine to
a neurotransmitter in its own right. The association of disturbances of
dopaminergic neurotransmission with neurological and psychiatric disorders has further emphasized the crucial role of this neurotransmitter in
normal brain function. Dopaminergic agonists have a firmly established role
in the treatment of Parkinson's disease and may be of value in the therapy
of tardive dyskinesia. Dopaminergic antagonists have a longer history in the
treatment of schizophrenia, Huntington's disease, and Gilles de la Tourette's syndrome.

Since 1975, the elegantly simple radioligand binding technique has allowed direct examination of the interactions of agonists and antagonists
with putative dopamine receptors (DARs), and these studies form the major
focus of this review. Such studies have complemented investigations of DA
regulation of adenylate cyclase activity and hormone release in various
tissues. Although problems remain, the correspondence between such radioligand binding sites and functional DARs is steadily being established.
These experiments have clearly divided DARs into distinct subtypes, much
as was done earlier for the alpha and beta adrenergic receptors. These
findings will have a profound effect on our understanding of dopaminergic

43

0147-006X/83/0301-0043$02.00

pharmacology and neurotransmission and the role of DA in psychiatric and neurological disease.

PHARMACOLOGICAL CHARACTERIZATION OF DOPAMINE RECEPTORS

The Dopamine-Stimulated Adenylate Cyclase or "D-1" Dopamine Receptor

Cyclic adenosine monophosphate (cAMP) is a second messenger for a number of neurotransmitters and hormones. Greengard initially demonstrated the presence of a dopamine-sensitive adenylate cyclase in the bovine superior cervical ganglion (reviewed in Greengard 1976) and then identified a similar enzyme in the CNS (Kebabian et al 1972) that was also found in the retina (Brown & Makman 1972). Dopamine elicited maximal stimulation of cAMP accumulation at 100 μM concentrations with half maximal effects at about 2 μM. The regional distribution of the enzyme in brain tissue suggested an association with dopaminergic transmission since it was highest in brain regions richest in DA innervation.

Antipsychotic drugs, or neuroleptics, used in the treatment of schizophrenia, had long been suspected as being DAR antagonists. Greengard's group, and later Iversen and colleagues (Iversen 1975), evaluated this hypothesis directly. The phenothiazines, such as chlorpromazine, were effective competitive inhibitors of dopamine's stimulation of this enzyme (Clement-Cormier et al 1974, 1975, Miller et al 1974, Iversen et al 1976). In studies of an extensive series of phenothiazines there was a general parallel between their pharmacological potencies as DA antagonists in animals or their antipsychotic activities and their influences on the adenylate cyclase. However, this correlation between in vivo and in vitro activity did not extend to the butyrophenones (BUTYs) such as haloperidol and spiroperidol (Iversen 1975, Snyder et al 1975), which were too weak in antagonizing DA stimulated cAMP production compared to their potent in vivo activities. Surprisingly, the fairly potent antipsychotic benzamides, such as sulpiride, are almost devoid of inhibitory potency.

These discrepancies, initially overlooked, raised the possibility that BUTYs might not block DARs at all, but reduce dopaminergic activity indirectly. This would account for the marked difference in chemical structure between phenothiazines and BUTYs despite their pharmacological similarities. An alternative hypothesis, not considered initially, was that more than one type of DAR existed. Thus, BUTYs would exhibit weak affinity for the DAR responsible for eliciting an increase in cAMP, whereas they would exhibit higher potencies at those DARs responsible for their behavioral and clinical effects. It is now apparent that this hypothesis is more tenable. Indeed, Kebabian & Calne (1979) have written a seminal review of the pharmacological classification of DARs. They divided DARs into two general categories. D-1 receptors are responsible for stimulating dopamine-sensitive adenylate cyclase activity upon agonist activation (Table 1). The location for the prototype D-1 receptor is the parathyroid gland, where DA agonists stimulate cAMP synthesis concomitantly with parathyroid hormone release (Brown et al 1977, Brown et al 1980, Attie et al 1980). For a more detailed discussion of the D-1 DAR, see reviews by Miller & McDermed (1979) and Schmidt (1979).

The Dopamine-Inhibited Adenylate Cyclase or "D-2" Dopamine Receptor

In contrast to D-1 receptors, D-2 receptors were functionally classified as not enhancing adenylate cyclase activity upon agonist occupation. Instead, the consequences of D-2 receptor stimulation are either to decrease or to have no effect on the formation of cAMP (Table 1). Prototype D-2 receptors exist in the anterior and intermediate pituitary glands, where they inhibit the release of prolactin and alpha-MSH, respectively (Weiner & Ganong 1978, Cote et al 1980, 1981). Kebabian has shown that in the intermediate pituitary DA inhibits the beta adrenergic agonist stimulated synthesis of

Table 1 Functional classification of dopamine receptor subtypes[a]

	D–1	D–2
Prototype receptor location	Parathyroid gland	Anterior and intermediate pituitary glands
Adenylate cyclase linkage	Stimulatory	Inhibitory or unlinked
Agonists		
Dopamine	Full agonist (micromolar potency)	Full agonist (nanomolar potency)
Apomorphine	Partial agonist (micromolar potency)	Full agonist (nanomolar potency)
Antagonists		
Phenothiazines	Nanomolar potency	Nanomolar potency
Thioxanthenes	Nanomolar potency	Nanomolar potency
Butyrophenones	Micromolar potency	Nanomolar potency
Substituted benzamides	Inactive	Micromolar potency
Dopaminergic ergots	Antagonists or partial agonists (micromolar potency)	Full agonists (nanomolar potency)

[a] Modified from Kebabian & Calne (1979).

cAMP, leading to a dimunition of hormone release (Cote et al 1981), while Onali et al (1981) have shown a similar antagonism of VIP-stimulated adenylate cyclase in the anterior lobe.

The pharmacological profile of D-2 receptors is clearly distinct from that of D-1 receptors (Table 1). Agonists consistently demonstrate higher affinities in eliciting a biochemical or physiological response at D-2 receptors than at D-1 receptors. Apomorphine is a potent agonist with full intrinsic activity at D-2 receptors in contrast to its partial agonist activity at D-1 receptors. Similarly, various dopaminergic ergots (e.g. bromocryptine, lisuride, lergotrile) are full, potent (nanomolar) agonists at D-2 receptors but only weak, partial agonists or antagonists at D-1 receptors. With respect to antagonists, phenothiazines and thioxanthenes are potent antagonists at both D-1 and D-2 receptors and thus they are not useful for discriminating between these subtypes. In contrast, BUTYs and related drugs are very potent antagonists of D-2 receptors but exhibit only weak affinity for D-1 receptors. Similarly, substituted benzamides such as sulpiride are inactive at D-1 receptors but exhibit moderate affinity at D-2 receptors.

The preceding pharmacological profiles should allow one to predict theoretically which drugs or classes of drugs would be suitable to use in radioligand binding studies of DAR subtypes. High affinity ($K_d < 10$ nM) is an important constraint in radioligand binding experiments that use filtration to separate bound from free ligand (for discussion see Bennett 1978). Therefore, phenothiazines and thioxanthenes are the only drugs that could be expected to label D-1 receptors. Indeed, ^{3}H-flupenthixol (^{3}H-FLU) and ^{3}H-piflutixol, two of the most potent thioxanthenes, appear to label D-1 as well as D-2 receptors in the striatum (Hyttel 1978a,b, Cross & Owen 1980, Hyttel 1981). Butyrophenone or BUTY-like antagonists [e.g. ^{3}H-spiroperidol (^{3}H-SP), ^{3}H-haloperidol (^{3}H-HAL), ^{3}H-domperidone (^{3}H-DOM)] have been found to label preferentially D-2 receptors. At first glance, one would predict that only the D-2 receptor would be labeled with ^{3}H-agonists (Table 1). However, the situation is more complex, as one must consider the possibility that some conformations of D-1 receptors may have higher affinity for agonists in membrane preparations.

CHARACTERISTICS OF RADIOLIGAND BINDING TO DOPAMINE RECEPTORS

Dopamine Receptors in the Pituitary

The analysis of radioligand binding data in the pituitary is more straightforward than that of data obtained in the brain. This results from the presence of only a single DAR subtype (D-2) in the pituitary, in contrast to multiple receptor types in the brain. Thus, as an introduction to CNS studies the pituitary provides a good and readily interpretable starting point.

Several groups (Creese et al 1977b, Caron et al 1978, Cronin et al 1978, Calabro & MacLeod 1978) have used radioactive DA agonists and antagonists to identify a high affinity, stereoselective and saturable DAR in anterior pituitary membrane preparations. The rank order of agonists and antagonists for competing with radioligand binding to this DAR agrees closely with their rank order in inhibiting (agonists) or disinhibiting (antagonists) prolactin release (for reviews on DA regulation of prolactin release, see Weiner & Ganong 1978, MacLeod et al 1980). In addition, one group has provided immunocytochemical evidence that these DARs are largely confined to mammotrophs (Weiner et al 1979, Goldsmith et al 1979).

Since the anterior pituitary contains a potentially homogeneous population of D-2 receptors, we have investigated the radioligand-receptor binding characteristics in this tissue in detail. ^{3}H-SP has previously been shown to bind exclusively to DARs in the anterior pituitary of cattle (Creese et al 1977b), sheep (Cronin & Weiner 1979), and rats (Stefanini et al 1980). In bovine anterior pituitary membranes, the specific binding of ^{3}H-SP is homogeneous, saturable (B_{max} = 4 pmol/g tissue) and of high affinity (K_d = 0.3 nM). Using ^{3}H-SP, it can be demonstrated that antagonist competition curves exhibit monophasic, mass-action characteristics with pseudo-Hill coefficients equal to one (Sibley et al 1982). For example, Figure 1 shows the experimental data and the resulting computer modeled competition curve for the antagonist (+)butaclamol. The computer analysis employed is a nonlinear least squares curve fitting program that can analyze the data in terms of one or more classes of binding sites (De Lean et al 1980, Munson & Rodbard 1980). The (+)butaclamol curve fits best the model of a single homogeneous receptor state with a K_d of 1.1 nM.

In contrast, agonist/^{3}H-SP competition curves exhibit heterogenous characteristics in membrane preparations with pseudo-Hill coefficients less than unity (Sibley et al 1982). As shown in Figure 1, in the absence of guanine nucleotides, the apomorphine (APO)/^{3}H-SP curve is shallow (pseudo-Hill coefficient = 0.58), and computer analysis indicates that the data are best explained by a two-site/state-binding model. The K_d for the high and the low affinity binding sites/states (R_H and R_L) have been designated K_H and K_L, respectively. Interestingly, the two sites/states are present in about equal proportions in the membranes. In the presence of a saturating concentration of Gpp(NH)p, a nonmetabolizable analogue of GTP, the APO curve is shifted to the right and is steepened (pseudo-Hill coefficient = 0.94). Moreover, computer analysis of the data now indicates a single homogeneous population of binding sites in the membranes whose affinity for APO is not significantly different from the K_L value of the control curve (Figure 1).

We have also characterized the binding of the radiolabeled agonist ^{3}H-NPA to DARs in bovine anterior pituitary membrane preparations

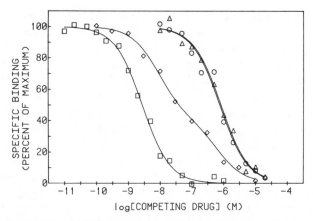

Figure 1   Computer modeled competition curves for ³H-spiroperidol binding to anterior pituitary preparations. The experimentally determined data points are shown by the *open symbols* while the *solid lines* are computer generated curves fitting the observed data points. The (+)butaclamol curve (□), performed using membranes, models best to a single binding site. The (–)apomorphine curve (◇) fits best to two binding states of high ($K_H = 5$ nM) and low ($K_L = 350$ nM) affinity in about equal proportions ($R_H = 55\%$, $R_L = 45\%$) in the membranes. With the addition of 0.1 mM GppNHp to the membranes (△), the (–)apomorphine curve fits best to a single receptor state ($K = 360$ nM). Using whole cells (○), the (–)apomorphine curve also models best to a single binding component ($K = 330$ nM).

(Sibley & Creese 1979, 1980a, Sibley et al 1982). One of the more striking findings with this radioligand is that its B_{max} is approximately 50% of that of ³H-SPs, suggesting that it labels only the high affinity agonist site/state (R_H) seen in agonist/³H-SP curves. This is also suggested by the finding that agonist/³H-NPA competition curves are homogeneous with single affinities that are not significantly different from the K_H values obtained from the corresponding agonist/³H-SP curve. Furthermore, saturating concentration of guanine nucleotides completely abolish the specific ³H-NPA binding to pituitary membrane preparations.

Recently, we have investigated the dopamine receptor-radioligand binding characteristics in intact, viable bovine anterior pituitary cells (Sibley et al 1981). In these experiments, bovine anterior pituitaries were first dispersed into single whole cells via collagenase treatment and then used directly in the binding experiments. Strikingly, agonist/³H-SP competition curves are homogeneous with pseudo-Hill coefficients of approximately one (Figure 1). Moreover, in Figure 1, one can see that the (–)APO/³H-SP curve in the cells is superimposable with the (–)APO/³H-SP + Gpp(NH)p curve in membrane preparations. Additionally, exogenously added guanine nucleotides no longer affect the (–)APO/³H-SP curve. This suggests that in intact cells, endogenous GTP regulates agonist binding in a fashion identical to that of exogenously added GTP in membrane preparations. It should also be noted that specific ^{3}H-NPA binding is not detectable in intact cells; this directly confirms the absence of a detectable R_H state in these cells. However, membranes prepared from these cells exhibit identical binding properties as membranes directly prepared from the whole gland, thus indicating that the lack of high affinity agonist binding is not the result of receptor degradation occurring during the collagenase-mediated dispersion. There is no difference in antagonist/^{3}H-SP curves or affinities between intact cell or membrane preparations.

Lefkowitz and co-workers have recently examined in detail the radioligand-receptor binding characteristics of the frog erythrocyte beta-adrenergic receptor system (reviewed in Lefkowitz 1980, Hoffman & Lefkowitz 1980). Their data with the frog erythrocyte beta receptor are qualitatively identical to our anterior pituitary DAR data. These investigators have proposed a ternary complex model to explain the binding data in the frog erythrocyte system (Kent et al 1980, De Lean et al 1980). This model is similar to the floating receptor or two-step models described previously by Jacobs & Cuatrecasas (1976) and Boeynaems and Dumont (1977).

$$L + R \rightleftharpoons LR \tag{1}$$

$$LR + X \rightleftharpoons LRX \tag{2}$$

Briefly, agonists or antagonists can bind to the receptor to form an initial drug-receptor complex (LR). The binding of agonists, however, induces a conformational change in the receptor so that it can now couple to a third membrane component (X). This ternary complex (LRX) is responsible for the high affinity agonist binding state (R_H). Limbird et al (1980a) have provided evidence that the third component is the guanine nucleotide-binding protein of the adenylate cyclase complex. The ternary complex of agonist, receptor, and nucleotide-binding protein is responsible for activating adenylate cyclase in the presence of GTP. This complex is formed as a transient intermediate, however, since the endogenous GTP rapidly induces its dispersal in intact cells.

The application of this model to the anterior pituitary DAR system is extremely attractive. However, DA does not appear to elicit an increase in anterior pituitary adenylate cyclase activity (Schmidt & Hill 1977, Clement-Cormier et al 1977, Mowles et al 1978, MacLeod et al 1980; however, see Ahn et al 1979). On the contrary, recent evidence suggests that DA may actually decrease cAMP formation in the anterior pituitary (DeCamilli et al 1979, Pawlikowski et al 1979, 1981, Labrie et al 1980, Giannattasio et al 1981, Ray & Wallis 1980, Onali et al 1981). Thus, the consequences of dopamine receptor-ternary complex formation may be to decrease mammo-troph cAMP content and thus decrease prolactin release. Lefkowitz et al (1981) recently suggested that the ternary complex model is applicable to alpha$_2$-adrenergic receptors that negatively modulate adenylate cyclase.

Some of the biochemical mechanisms involved in the dopaminergic regulation of hormone release have been better elucidated in the intermediate pituitary. It is known that the intermediate pituitary is predominantly composed of corticotrophic cells that synthesize and secrete a variety of peptides related to beta-lipotropin and ACTH, including beta-endorphin and alpha-MSH. Detailed studies of intermediate pituitary corticotroph regulation have been performed by Kebabian and colleagues (Cote et al 1980, 1981, Munemura et al 1980a,b). Using dispersed cells from rat intermediate pituitaries, they demonstrated that beta-adrenergic agonists, cAMP analogs, and phosphodiesterase inhibitors enhanced the secretion of alpha-MSH. Activation of the beta-receptor was accompanied by an increase in corticotrophic cAMP. DA inhibited the basal and isoproterenol ISO-enhanced release of alpha-MSH as well as the ISO-induced accumulation of cAMP. When homogenates of the intermediate pituitary were prepared and adenylate cyclase activity was measured directly, DA agonists noncompetitively inhibited the basal as well as the ISO-stimulated cyclase activity.

We have directly labeled the DAR in bovine intermediate pituitary membranes with both ^{3}H-SP and ^{3}H-NPA (Sibley & Creese 1980b) and found that the DAR binding characteristics in this tissue are remarkably similar to those seen in the anterior pituitary. These observations suggest the presence of identical D-2 DARs in the anterior and intermediate pituitary glands.

Dopamine Receptors in the Striatum

The very first DAR binding studies utilized ^{3}H-DA and ^{3}H-HAL as ligands in the examination of receptors in mammalian striatum (Creese et al 1975, Burt et al 1975, Seeman et al 1975). Considerable difficulty has been encountered in radioligand binding studies of striatal dopamine receptors, not only because, as in the anterior pituitary, such sites exist in multiple states, but also because of the coexistence of several distinct DAR subtypes. Most dopaminergic radioligands are not subtype selective and the specificity of agonist binding is highly ion and temperature dependent. Thus, while ^{3}H-thioxanthene ligands label both D-1 and D-2 receptors, ^{3}H-agonists may label these and a third binding site, termed D-3, as well.

STRIATAL D-1 RECEPTORS   A high affinity ($K_d = 4$ nM) striatal binding site for ^{3}H-flupentixol (Hyttel 1978a,b, Hyttel 1980, Cross & Owen 1980) has been identified that appears from competition studies to be the D-1 receptor. The potencies of a number of dopaminergic antagonists from a variety of structural classes in inhibiting dopamine-stimulated adenylate cyclase activity correlates well with their potencies in displacing ^{3}H-flupentixol binding. For example, thioxanthenes, which possess very high affinity for ^{3}H-flupentixol binding sites, also have nanomolar potency in the inhibition of the dopamine-stimulated adenylate cyclase. Butyrophenone affinity for both the cyclase and ^{3}H-flupentixol binding sites are one to two orders of magnitude lower. Agonists are active in both displacing ^{3}H-flupentixol and in stimulating cyclase in the micromolar range. These sites are present in approximately three times the numbers seen for D-2 or D-3 binding sites (see below). Detailed displacement studies have revealed that a minor portion, about 20%, of ^{3}H-flupentixol binding is to the D-2 receptors (Cross & Owen 1980). ^{3}H-Flupentixol binding can be directed to label exclusively the putative D-1 receptor by the inclusion of an appropriate "masking" drug, i.e. a low concentration of unlabeled BUTYs, in the assay to saturate the D-2 receptors. Recently, a new thioxanthene ligand, ^{3}H-piflutixol, has been characterized that possesses the same receptor specificity as ^{3}H-flupentixol, but has somewhat higher affinity for D-1 sites (approximtely 0.4 nM) (Hyttel 1981). One of the chief obstacles to the further characterization of the D-1 ^{3}H-thioxanthene binding sites has been the high level of nonspecific binding (generally about 60% of the total) seen with both ^{3}H-flupentixol and ^{3}H-piflutixol. Characteristics such as guanine nucleotide sensitivity require a better D-1 ligand before their detailed study becomes feasible. Dopaminergic ^{3}H-agonists label a set of divalent cation dependent, GTP sensitive sites in striatum that display high affinity for thioxathines and phenothiazines, but show low affinity for BUTYs (Hamblin & Creese 1982c). Although these sites have in the past been included (along with other GTP insensitive sites) under the "D-3" category, their characteristics suggest that they may in fact be a high agonist affinity state of the D-1 receptor (see "D-3" below). Whether or not these D-1 sites display up-regulation in response to denervation or chronic administration of antagonists is controversial (see Creese & Sibley 1980).

STRIATAL D-2 RECEPTORS   Several lines of evidence suggest that at least the majority of high affinity binding sites for ^{3}H-BUTYs in the striatum are quite similar, if not identical, to the D-2 pituitary receptor. The K_d for ^{3}H-SP binding in striatum, determined under a variety of conditions in rat, bovine, and human striatal membranes, has been reported as 0.1–0.3 nM (Fields et al 1977, Creese et al 1977b, Howlett & Nahorski 1978, Leysen et al 1978a, Quik & Iversen 1979), in excellent agreement with the value obtained in bovine anterior pituitary. Early equilibrium studies produced linear Scatchard plots for ^{3}H-HAL (Burt et al 1976) and ^{3}H-SP (Creese et al 1977b), and kinetic analysis yielded association and dissociation rates consistent with the existence of homogeneous binding sites. This evidence indicated that, as in the pituitary, there existed a single D-2 receptor. As in pituitary, ^{3}H-agonist ligands can, under appropriate conditions, label these same sites with high affinity in a GTP sensitive fashion ($IC_{50} = 100$ μM) (see below). Striatal ^{3}H-agonist D-2 binding (defined using a low concentration of a BUTY to displace specific binding) is enhanced by preincubation of the tissue homogenate with magnesium or other divalent metal cations and inclusion of these ions in the assay (Hamblin et al 1981, Hamblin & Creese 1982c), and D-2 specific ^{3}H-dopamine binding is decreased by monovalent cations with the rank order $Na^+ \geq Li^+ > K^+$. Such ion and nucleotide effects are consistent with an inhibitory cyclase linkage (Limbird 1981) demonstrated for the pituitary D-2 receptor, and more recently, for a striatal D-2 receptor by Stoof & Kebabian (1981). These D-2 sites are present in considerably higher numbers in striatum than pituitary with B_{max} values for ^{3}H-BUTYs typically reported from 25 to 50 pmol/g tissue (Creese et al 1977b, Leysen et al 1978a) or 250 to 600 fmol/mg protein (Fields et al 1977, Howlett & Nahorski 1978, Quik & Iversen 1979).

It should be noted that the most commonly used BUTY ligand, ^{3}H-SP, also labels serotonergic 5-HT$_2$ receptors in both striatum and other brain areas (Leysen et al 1978b, Creese & Snyder 1978, Peroutka & Snyder 1979). Care must therefore be taken in studies utilizing BUTY ligands to direct binding to the desired receptors so as to avoid spurious identification of multiple "dopaminergic" ^{3}H-BUTY binding sites, sites that could be either DA or serotonin receptors. This may be accomplished either by using an appropriate DAR selective "blank" to determine specific binding, or by including a competing drug as a "mask" of the undesired site. ADTN appears to offer a dopaminergic receptor selective blank when used in appropriate concentration (Quik et al 1978). Several serotonergic antagonists such as mianserin (Withy et al 1980), R43448 (List & Seeman 1981), or ketanserin (Leysen et al 1981) appear suitable for use as masks.

The existence of another separate BUTY binding site ("D-4" receptor) in striatum, characterized by high affinity for BUTYs and other DA antagonists and low affinity for DA agonists, has been postulated (Martres et al 1980, Sokoloff et al 1980b). As these authors note, however, D-2 receptors appear to convert to the D-4 type with the addition of GTP. This strongly suggests that, rather than being a separate receptor subtype, the "D-4" site is merely the R_L state of the D-2 receptor. Some evidence exists, however, that ^{3}H-BUTY binding sites in pituitary and striatum might differ. Kainic acid lesions of intrinsic striatal neurons selectively remove guanine nucleotide sensitive ^{3}H-BUTY binding sites, leaving intact a population of nucleotide insensitive sites on cortical afferent terminals (Creese et al 1979c). It is unclear presently whether these presynaptic receptors constitute a separate class of sites or are D-2 receptors confined to the R_L state as an artifact of the lesion technique.

Receptor binding studies have allowed more rigorous identification of the behavioral correlates of D-2 receptor function. The affinities of a number of structurally diverse DA antagonists for ^{3}H-BUTY binding sites correlate highly with their molar potency in antagonism of APO and amphetamine-induced stereotyped behavior in rat (Creese et al 1978a, Ogren et al 1978), blockade of APO-induced emesis in dog (Creese et al 1976), and antipsychotic activity in man (Creese et al 1976, Seeman et al 1976). The affinity of an antagonist for ^{3}H-BUTY binding is thus a powerful predictor of in vivo DAR antagonism and antipsychotic activity. The nanomolar affinity of the antipsychotic drugs for DAR binding sites is also commensurate with the plasma concentrations of these drugs at therapeutic dose levels as measured by the neuroleptic radioreceptor assay (Creese & Snyder 1977) and by other methods. A similar analysis has indicated that the anti-Parkinsonian effects of DA agonists are also mediated through the BUTY-labeled D-2 receptors (Titeler & Seeman 1978).

Since antipsychotic agents exert their ameliorative effects via antagonism at D-2 receptors and since there is little evidence for increased DA turnover in schizophrenia, it seems reasonable to suggest that the primary lesion in this disease might be an inappropriate "supersensitivity" or "up-regulation" of these sites. Several studies have supported this hypothesis, reporting 50–200% increase in postmortem ^{3}H-BUTY binding in brains of schizophrenics (Crow et al 1978, Lee & Seeman 1980, Lee et al 1978, Mackay et al 1980a). Interpretation of these studies, however, is difficult, because animal experiments show that chronic in vivo treatment with antagonists (Creese et al 1978a, Muller & Seeman 1978, Owen et al 1980, Lautin et al 1980) or lesioning of dopaminergic innervation (Creese et al 1977a, Nagy et al 1978, Sokoloff et al 1980a) results in substantial increases in both ^{3}H-antagonist and ^{3}H-agonist binding to D-2 sites. The increased D-2 B_{max} then could be due to the effects of the neuroleptics used to treat almost all of the patients in these studies. Indeed, this drug-induced up-regulation may be involved in the pathogenesis of tardive dyskinesia (Klawans 1973). In the postmortem series of Crow et al (1978) and Lee & Seeman (1980), an increased B_{max} for ^{3}H-SP was observed in the small number of drug free cases examined, whereas contrary results were obtained by MacKay et al (1980b). The question of D-2 receptor changes in schizophrenia must, for now, still be considered open.

STRIATAL ^{3}H-AGONIST BINDING SITES: D-3 VS D-1 RECEPTORS   Under some conditions the ^{3}H-agonist ligands ^{3}H-DA, ^{3}H-APO, ^{3}H-NPA, and ^{3}H-ADTN can bind to D-2 receptors with high affinity, as they do in anterior pituitary. However, a subset of the ^{3}H-agonist binding sites differ from the BUTY labeled D-2 binding sites in that although phenothiazines and thioxanthenes display high (nanomolar) affinity, BUTYs have micromolar affinities. It has been proposed that these agonist binding sites represent yet another distinct DAR, the "D-3" receptor (Titeler et al 1979). It also remains a possibility that some of this D-3 agonist binding is actually to a state of the D-1 receptor showing high affinity for agonists.

The D-3 receptors are operationally defined as binding sites with high affinity for ^{3}H-dopamine, ^{3}H-APO, ^{3}H-NPA, or ^{3}H-ADTN and low affinity for BUTYs. In practice, D-2 specific ^{3}H-DA binding can be defined as that displaced by 30 nM SP, while D-3 binding is that displaced by 10 μM alpha-flupentixol, but not by the above concentration of SP (Hamblin & Creese 1982c). Although D-3 receptors are apparently absent in pituitary, they are present in mammalian striatum at 10–40 pmol/g wet weight tissue (Burt et al 1976, Thal et al 1978, Komisky et al 1978, Creese & Snyder 1978, Hamblin & Creese 1982c) or about 50–700 fmol/mg membrane protein (Seeman et al 1975, Cronin et al 1978, Titeler & Seeman 1979, List et al 1980), depending on the conditions employed. In *bovine* striatal membranes at 37° in the presence of "physiological" (extracellular) concentrations of ions, ^{3}H-DA specifically labels only D-3 sites, with a K_d of about 10–20 nM (Creese et al 1975, Burt et al 1976). Such sites have also been labeled in both calf and rat striatum under various conditions with ^{3}H-APO (Thal et al 1978, Seeman et al 1979), ^{3}H-NPA (Creese et al 1979a, Titeler & Seeman 1979), and ^{3}H-ADTN (Creese & Snyder 1978, Seeman et al 1979) with affinities (K_d) in the nanomolar range. Oddly, under these same conditions, high affinity ^{3}H-DA binding to *rat* striatal membranes is not reproduceably found (Creese et al 1979b), although Seeman and co-workers have been able to obtain such binding under other conditions (Titeler et al 1979, List et al 1980). We have recently explained these divergent results by characterizing the effects of temperature and ionic conditions on ^{3}H-DA binding in rat caudate membranes. As for the agonist binding to D-2 sites, binding to D-3 sites is enhanced by preincubation in the presence of divalent metal cations such as Mg^{2+} (Bacopoulos 1981, Hamblin & Creese 1982c), even if the free cation is subsequently washed out. In the absence of divalent metal cations, D-3 specific ^{3}H-DA binding is not sensitive to GTP inhibition. Addition of Ca^{2+}, Mg^{2+}, Mn^{2+}, or Co^{2+} (1–10 mM) to the assay promotes an additional increase in D-3 specific ^{3}H-DA binding, and this additional binding is GTP sensitive ($IC_{50} = 10$ μM). Such an ion dependency as well as the greater potency of GTP at these sites relative to D-2 receptors is consistent with the idea that the sites are a state of D-1 receptor that shows high affinity for agonists. Na^+ is less potent in the inhibition of D-3 than D-2 specific ^{3}H-DA binding, although some inhibition can be observed. ^{3}H-Dopamine binding to both D-2 and D-3 sites is also reduced by increasing incubation temperature, although ^{3}H-BUTY binding to D-2 sites is not. The combined effects of sodium and temperature are sufficient to place ^{3}H-DA affinity for both D-2 and D-3 sites in rat membranes when assayed at 37°C in the presence of Na^+ outside the range detectable in filtration assays. Under these conditions, the K_i for unlabeled DA in displacement of more potent ^{3}H-agonists is about 200–300 nM (Creese et al 1979b). At 22–25°C in the absence of sodium, however, ^{3}H-DA displays K_d of 1.2 nM for D-3 sites (Titeler et al 1978). The dual labeling of D-2 and D-3 receptors by these agonists leads to their biphasic displacement by BUTY antagonists with both high (nanomolar) and low (micromolar) affinity components (Creese et al 1978b, Sokoloff et al 1980b). Thus, depending on tissue preparation and incubation conditions, DA ^{3}H-agonists can label D-2 or D-3 receptors either selectively or together.

SUBSTITUTED BENZAMIDE BINDING SITES   Rat striatum also possesses a saturable binding site for the substituted benzamide, ^{3}H-sulpiride, that may be different from those labeled by other dopaminergic ligands. Although the high affinity ^{3}H-sulpiride binding site is found predominantly in dopaminergically innervated tissues (Woodruff & Freeman 1981, Memo et al 1980) and has highest affinity of DA among the neurotransmitters screened, the affinity of sulpiride for these sites, $K_d = 7$nM (Woodruff & Freedman 1981), $K_d = 17$ nM (Memo et al 1980), or $K_d = 27$ nM (Theodorou et al 1979), is much greater than its reported affinity for D-1 sites labeled by ^{3}H-flupentixol ($K_i > 10,000$ nM) (Hyttel 1978b, Cross & Owen 1980), D-2 sites labeled by ^{3}H-BUTYs ($IC_{50} = 100$–1000 nM) (Leysen et al 1978a, Creese et al 1979b, Seeman et al 1978), or D-3 sites labeled by ^{3}H-APO ($K_i = 100,000$ nM) (Sokoloff et al 1980b). Recent reports, however, have demonstrated that ^{3}H-sulpiride affinity for the benzamide binding site (Theodorou et al 1980) as well as that of unlabeled sulpiride for the pituitary D-2 receptor (Stefanini et al 1981) are dependent on sodium concentration. Other variables that might influence binding affinity such as species differences have not yet been investigated in detail. This cannot then be taken as conclusive evidence that the benzamide binding site is distinct from the other dopamine receptors described above. Indeed, several lines of evidence suggest that this site represents the D-2 receptor also labeled by ^{3}H-BUTY, or perhaps a subset thereof. HAL affinity for ^{3}H-sulpiride sites ($IC_{50} = 4.5$ nM) (Woodruff & Freedman 1981) is quite close to its affinity at the D-2 site, whereas the binding sites for ^{3}H-sulpiride and ^{3}H-SP are similarly regulated by GTP (Freedman et al 1981). Binding of these two ligands may be similarly effected by kainic acid, 6-OHDA, and cortical lesions (Theodorou et al 1981), although other investigators have obtained contrary results (Memo et al 1980). Finally, the behavioral actions in both man and animals of the substituted benzamides are similar to those for the BUTYs (Jenner & Marsden 1979), which almost certainly exert their effects through the classic D-2 site (see above).

IRREVERSIBLE MODIFICATION OF STRIATAL DOPAMINE RECEPTORS   Preincubation of bovine caudate homogenates with phenoxybenzamine (POB) rapidly results in an irreversible decrease of the B_{max} for D-2 ^{3}H-antagonist binding, with maximum effect observed by 10 min and with a pseudo-IC_{50} of 1 μM (Hamblin & Creese 1980, 1982a). Binding of ^{3}H-DA when assayed under conditions selective for D-3 sites, however, is inhibited with an IC_{50} 100-fold higher than that for ^{3}H-SP binding—thus, a homogenate treatment with 10 μM POB almost completely eliminates ^{3}H-SP binding while leaving ^{3}H-DA binding nearly unaffected. As suggested earlier by displacement studies, these results confirm the hypothesis that the binding sites for ^{3}H-dopamine and ^{3}H-BUTYs appear to be physically distinct and do not interconvert under the conditions of the assay. ^{3}H-SP binding sites are protected from POB attack by occupancy, both by agonists and antagonists; this indicates that the POB effect is mediated through site-directed attack and not merely through a nonspecific membrane effect.

Binding of the agonist ligand ^{3}H-APO assayed under identical conditions is affected to a degree intermediate to that of ^{3}H-SP and ^{3}H-dopamine. The increased sensitivity of ^{3}H-APO high affinity binding sites in comparison with those for ^{3}H-DA suggested that even under conditions where ^{3}H-DA binding is D-3 selective, ^{3}H-APO labels both the relatively POB-resistant D-3 site and the relatively POB-sensitive D-2 site. As anterior pituitary is believed to contain D-2 but not D-3 binding sites, one might anticipate that ^{3}H-APO binding will show identical POB sensitivity to ^{3}H-SP binding in this tissue. This is indeed the case, thus reinforcing the hypothesis that both agonist and antagonist ^{3}H-ligands label a single receptor in the anterior pituitary.

Phenoxybenzamine at higher concentrations also inactivates the dopamine-stimulated adenylate cyclase (Walton et al 1978). Marchais & Bockaert (1980) demonstrated that homogenate pretreatment with 10 μM POB, which completely eliminates ^{3}H-SP binding, leaves 35% of the dopamine-stimulated adenylate cyclase activity intact; this supports the hypothesis that ^{3}H-SP sites are not linked to the cyclase in a stimulatory fashion. ^{3}H-Flupentixol binding, like adenylate cyclase activity, is more resistant to POB attack than other ^{3}H-antagonists.

Neumeyer and co-workers (1980) have synthesized a nitrogen mustard derivative of APO, (–)N-chloroethyl-norapomorphine (NCA), which also irreversibly inhibits ^{3}H-ligand binding to DAR subtypes, but which may display a selectivity quite different from that of POB. Although active in decreasing binding to ^{3}H-NPA labeled sites (presumably including both D-2 and D-3 sites) with an "IC$_{50}$" of 1.8 μM, up to 25 μM NCA has no effect on ^{3}H-SP binding. A similar elimination of ^{3}H-NPA sites is observed after in vivo administration of NCA intrastriatally (Costall et al 1980b). The dopamine-sensitive adenylate cyclase is also irreversibly inhibited in vitro by NCA treatment at much higher doses (Neumeyer et al 1980, Baldessarini et al 1980), perhaps by attack on the D-1 receptor. NCA causes many of the behavioral effects expected of dopaminergic antagonists when administered in vivo (Costall et al 1980a,b). Several problems remain in the interpretation of these data. It is unclear why NCA should not prevent ^{3}H-SP binding at doses that nearly eliminate the binding of ^{3}H-NPA, a ligand that under at least some conditions labels D-2 and D-3 sites with nearly equal affinity (Titeler & Seeman 1979). It is also unclear why NCA should be such a potent antagonist in behavioral measurements, when these behavioral effects have been so well documented to be mediated by D-2 receptor function (for review see Creese & Snyder 1978).

Heat treatment of tissue membranes in vitro also produces irreversible changes in dopaminergic ligand binding that probably represent an alteration in a GTP binding regulatory "N" protein, rather than in the receptors themselves (Hamblin & Creese 1982b). Exposure of caudate homogenates to 53°C causes a rapid decrease in specific binding of the agonist ligands ^{3}H-APO and ^{3}H-DA, whether to D-2 or D-3 sites, with more than one-half eliminated within 30 sec. Since the binding of ^{3}H-SP to D-2 sites is nearly unaffected, this effect cannot be explained merely as a loss of the binding sites per se and is highly reminiscent of the effects of guanine nucleotides on agonist binding (Hamblin & Creese 1982b).

Heat treatment also causes a reduction in potency of unlabeled agonists in displacement of ^{3}H-SP with a simultaneous increase in pseudo-Hill slope —the IC$_{50}$s for DA and APO are shifted 10 to 15-fold higher after exposure of homogenates to 53°C for 4 min. Micromolar concentrations of GDP, GTP, or Gpp(NH)p also cause a similar decrease in agonist potency and an increase in pseudo-Hill coefficient (Zahniser & Molinoff 1978, Creese & Sibley 1979), and maximal GTP and heat treatments are not additive in their effect on agonist affinity, consistent with a common site of action.

A single explanation of these common, nonadditive effects of heat treatment and guanine nucleotides may be related to the involvement of GTP-binding regulatory proteins ("N" proteins) that modulate receptor function, both positively and negatively, linked to adenylate cyclase (Ross & Gilman

1980, Limbird 1981, Rodbell 1980). Such proteins, when coupled with beta receptors—and possibly many other neurotransmitter and hormone receptors—enable high affinity binding of ^{3}H-agonist ligands and potent displacement of ^{3}H-antagonists by agonists. In several beta receptor systems, when N/receptor association is prevented, either by the addition of GTP (Limbird et al 1980a) or manipulations eliminating N directly (Pike & Lefkowitz 1980, Ross et al 1977, Howlett et al 1978, Limbird et al 1980b), high affinity ^{3}H-agonist binding is lost, and agonist/^{3}H-antagonist displacements are right-shifted and steepened. Antagonist binding remains unaffected. Thus, heat denaturation of such a regulatory moiety, rather than the receptor itself, would explain the DAR binding changes observed in the striatum. These studies suggest that a similar, heat-labile factor or factors regulate binding at both the ^{3}H-BUTY labeled D-2 site and the ^{3}H-DA labeled D-3 site (Hamblin & Creese 1982b). Similar results for the striatal D-2 site employing NEM treatment have recently been presented (Huff & Molinoff 1981).

Dopamine Receptors in the Retina

Retinal neurons have been identified that store, synthesize, release, and metabolize DA (Kramer 1976, Ehinger 1976, Sarthy & Lam 1979), and, as might be expected, classes of DARs similar to those found in the rest of the CNS have also been identified. D-1 receptors were first identified by the demonstration of a dopamine-stimulated adenylate cyclase in homogenate and intact preparations (Brown & Makman 1972). Succeeding studies characterized the retinal D-1 receptor in a number of species and showed that it shared a similar pharmacological specificity to the D-1 receptor in rat striatum (Watling et al 1979, Redburn et al 1980, Watling & Dowling 1981, Dowling & Watling 1981).

Attempts to show high affinity binding of [^{3}H]DOM to retinal membranes were initially unsuccessful (Watling et al 1979, Redburn et al 1980). Since DOM has been shown to be weak ($IC_{50} = 0.2$ mM) in inhibiting dopamine-sensitive adenylate cyclase, it was postulated that all or nearly all DARs in retina were of the D-1 type (Watling et al 1979, Redburn et al 1980). However, several investigators have demonstrated that ^{3}H-SP specific binding sites in bovine (Magistretti & Schorderet 1979, Makman et al 1980a,b) and goldfish (Redburn et al 1980) retina are similar to D-2 sites found in mammalian striatum. Creese & Sibley (1979) suggested that the failure to label retinal membrane D-2 receptors with [^{3}H]DOM is probably the result of high levels of nonspecific binding and the slightly lowered affinity that this compound shows for retinal D-2 receptors ($K_i = 15$ nM in inhibiting ^{3}H-SP binding in bovine retinal membranes when compared to its striatal K_i of 1 nM), a conclusion subsequently reached by Watling & Iversen (1981).

High affinity [³H]ADTN binding to rat and bovine retinal membranes has also been reported (Makman et al 1980b). The identity of these binding sites is unclear. The affinity of [³H]ADTN binding (K_d = 9 nM) did not correlate well with its affinity to inhibit [³H]spiperone binding (K_i = 580 nM) in unwashed bovine retinal membrane homogenates. However, antagonist specificities for high affinity ³H-ADTN binding and DA stimulated adenylate cyclase were similar and some, but not all, of these sites were GTP sensitive (Makman et al 1980a). DA agonists, however, showed about 100-fold greater affinity for ³H-ADTN binding than for stimulating retinal adenylate cyclase activity. Retina and striatum thus appear quite similar in the DARs that they contain. Both possess D-1, D-2 receptors, and high affinity binding sites for ³H-agonists that may represent a high affinity agonist state of the D-1 receptor, a separate D-3 site, or both. The retina should prove to be an advantageous system in which to study dopaminergic neurotransmission due to its relative accessibility to physiological study.

SOLUBILIZATION AND ISOLATION OF DOPAMINE RECEPTORS

Complete characterization of the various DARs will ultimately require the isolation of the purified proteins and other membrane components involved, followed by successful reconstitution. Substantial steps in this direction have already been taken. Several groups have now reported the solubilization of ³H-BUTY binding sites. Gorissen & Laduron (1979) employed 1% digitonin treatment of dog striatal membranes followed by ultracentrifugation. This results in a supernatant containing binding sites for ³H-SP, assayable using gel filtration, polyethylene glycol precipitation, or charcoal adsorption to separate bound from free ³H-ligand. Several criteria for identifying these sites as DARs, such as appropriate rank order potency of antagonists, regional distribution, and stereospecificity, have been met. Similar results have now been reported using rat (Gorissen et al 1980) and human (Madras et al 1980, Davis et al 1981) striatum. We have successfully followed a similar solubilization procedure and found that the guanine nucleotide sensitivity of agonist displacement of ³H-SP binding is lost following solubilization (Leff & Creese 1982). Agonist displacement is steep (pseudo-Hill slope $\simeq$ 1) and of low affinity ($IC_{50} \sim 10 \ \mu M$). This suggests that either the guanine nucleotide binding protein is not solubilized along with the ³H-BUTY site or that it can no longer couple functionally with the receptor after solubilization. Solubilization after DA preincubation, however, preserves guanine nucleotide sensitivity of the site.

Salt extraction with potassium chloride and chloroform/methanol extraction have also been employed to solubilize ³H-BUTY binding sites from calf caudate (Clement-Cormier & Kendrick 1980, Clement-Cormier et al

1980, Boyan-Salyers & Clement-Cormier 1980). Saturation and displacement studies, as well as gel filtration elution patterns, suggest the presence of several subtypes of ^{3}H-SP binding sites using this technique. At least one of these sites demonstrates affinity, specificity, and regional distribution comparable to D-2 membrane binding sites. However, in light of the previously reported solubilization of a high affinity "spirodecanone" (i.e. nondopaminergic) ^{3}H-SP binding site (Gorissen et al 1980), the relationship of these additional sites to multiple membrane bound ^{3}H-SP binding sites must be considered unknown, pending a more detailed characterization.

Recently, Kerdelhue et al (1981) have reported the presence of soluble binding sites in the cytosol from pituitary. A high speed supernatant fraction from homogenates of steer anterior pituitary was shown to contain a high affinity, saturable, and stereoselective binding site for ^{3}H-SP. Cytosolic binding sites represented 3% of the total anterior pituitary binding. Further studies are necessary to elucidate the possible significance of these soluble binding sites that may represent internalized receptors. Clement-Cormier and collaborators have also reported preliminary studies using the KCl and the detergent deoxycholate and octyl-beta-glycosyl pyranoside to solubilize high affinity binding sites for the agonist ligands ^{3}H-NPA, ^{3}H-APO, and ^{3}H-ADTN (Clement-Cormier et al 1980).

NEUROANATOMICAL LOCALIZATION OF DOPAMINE RECEPTORS IN THE CNS

Neostriatum

Studies in rat striatum have provided evidence for a differential localization of D-1, D-2, and D-3 receptor subtypes. Kainic acid-induced lesions of intrinsic striatal neurons almost completely eliminates striatal dopamine-sensitive adenylate cyclase activity, indicating that almost all of this enzyme, and thus presumably D-1 receptors, is present on these cells (Schwarcz et al 1978, Govoni et al 1978). We have observed a similar loss of ^{3}H-flupentixol binding, which labels D-1 receptors, after striatal kainic acid lesion (Leff et al 1981). Such lesions indicate that about 50% of ^{3}H-BUTY binding sites are also localized on intrinsic neurons (Govoni et al 1978, Schwarcz et al 1978, Fields et al 1979, Leff et al 1981). Most of the remaining BUTY binding sites in the striatum appear to be localized on cortico-striate terminals and are removable by cortical ablation (Schwarcz et al 1978, Garau et al 1978).

The localization of D-3 sites identified with ^{3}H-agonists has been more elusive. Several dopaminergic agonists have been used in such studies. Differing assay conditions have led to a lack of consensus concerning the pharmacological specificity, neuronal localizations, and functional roles of these sites labeled by ^{3}H-agonists (Creese & Sibley 1979, Titeler et al 1979, Leysen 1979, Sokoloff et al 1980a,b).

Initial studies comparing striatal kainate lesion-induced losses in the binding of ^{3}H-BUTYs and ^{3}H-agonists, such as ^{3}H-APO and ^{3}H-ADTN (Creese et al 1979c, Fujita et al 1980, Fuxe et al 1979), reported that high affinity agonist sites were decreased to a greater extent than ^{3}H-BUTY binding. Further studies using ^{3}H-NPA as an agonist ligand confirmed the original findings of a 70–80% loss of agonist binding in kainate lesioned striata (Leff et al 1981). Current studies indicate that this large decrease in ^{3}H-agonist binding may represent labeling of a high affinity agonist binding state of D-1 receptors as well as D-2 receptors (S. Leff, M. Hamblin, and I. Creese, in preparation).

Following nigral lesions with 6-OHDA, increased numbers of striatal D-2 receptors were identified by both ^{3}H-BUTY and ^{3}H-APO binding (Creese et al 1977a, Nagy et al 1978, Creese & Snyder 1979). Contrary to Creese's finding, Nagy et al (1978), using different assay conditions, reported a decrease in ^{3}H-APO binding in striata ipsilateral to the 6-OHDA lesion, which led them to hypothesize that ^{3}H-APO specifically labels presynaptic autoreceptors. Some groups failed to replicate these findings using similar assay conditions (Leysen 1979).

A recent study by Sokoloff et al (1980a) found a change in the distribution of pharmacologically differentiable ^{3}H-APO sites in striatum after 6-OHDA lesions of the nigrostriatal tract. ^{3}H-Apomorphine sites having nanomolar affinity for BUTYs and DOM (D-2 sites—termed by them as Class I) increased 25–30% in density, whereas sites having lower or micromolar affinity for BUTYs and DOM (D-3 sites—termed by them as Class II) showed 30–50% decreases in density. In this same study, striatal kainate lesions produced a 57% decrease in Class I sites and no change in Class II sites. This data suggests that a portion of ^{3}H-APO binding, i.e. a portion of Class II sites, could represent presynaptic autoreceptors. Using ^{3}H-NPA, ^{3}H-APO, and ^{3}H-DA and unlabeled DOM or SP to discriminate between these classes of agonist sites, we too have found a decrease in D-3 sites after unilateral 6-OHDA lesions, whereas ^{3}H-agonist-labeled D-2 sites increased 10–20%. However, it should be pointed out that the changes in relative amounts of such pharmacologically defined sites may not represent changes in receptor numbers per se but rather denervation-induced changes in their pharmacological characteristics. Furthermore, D-3 sites may not necessarily represent autoreceptors on nigro-striatal terminals as prior acute administration in vivo of reserpine may lead to similar decreases in ^{3}H-DA binding (Bacopolous 1981; S. Leff, M. Hamblin, and I. Creese, in preparation).

We have recently found that losses in D-3 binding produced by both 6-OHDA denervation and acute reserpine treatments can be reversed by preincubating the membranes in the presence of added DA (100 nM). We interpret this to indicate that the denervation-induced losses in D-3 binding are probably the result of the depletion of striatal DA and that D-3 binding sites are in fact located on cells post-synaptic to nigrostriatal terminals.

Future studies will be required to pursue the pharmacological and morphological identity of these putative subclasses of DARs in order to further comprehend their role on the physiological and behavioral function of the neostriatum.

Substantia Nigra

Several groups have reported the presence of D-1 receptors in rat substantia nigra (Kebabian & Saavedra 1976, Phillipson & Horn 1976, Spano et al 1976, Traficante et al 1976) that are localized largely on presynaptic terminals of descending afferents to the substantia nigra (Gale et al 1977, Phillipson et al 1977, Premont et al 1976, Schwarcz & Coyle 1977, Spano et al 1977) and not on pars compacta neurons. Quick et al (1979) found that a large portion (40%) of nigral ^{3}H-SP binding was lost after 6-OHDA lesioning of dopaminergic nigral neurons while striatal kainate lesions, which produced large losses in nigral D-1 receptors, did not affect levels of ^{3}H-SP binding. Murrin et al (1979) using light microscopic autoradiography found similar localizations of ^{3}H-SP sites in substantia nigra. These data thus provide evidence for a localization of D-2 receptors on dopaminergic cell bodies in the nigra.

CONCLUDING COMMENTS

The interpretation of early DAR binding studies with striatal membranes was difficult—the data did not describe a system containing a single set of homogenous receptors. A number of approaches have now allowed the clear division of CNS DARs into subtypes. Among the most important advances have been the examination of binding characteristics in other tissues such as anterior pituitary, the use of discrete CNS lesions to remove particular presynaptic or postsynaptic cellular elements, and the development of methods to irreversibly modify DARs. Such studies have now characterized three DAR subtypes known as D-1, D-2, and D-3, which have the characteristics shown in Table 2.

Table 2 Characteristics of dopaminergic binding sites

	D-1	D-2		D-3
		$R_H \rightleftharpoons R_L$		
Usable radioligands				
[3H] Thioxanthenes	+	+	+	?a
[3H] Butyrophenones	–	+	+	–
[3H] Agonists	?b	+	–	+
Agonist affinity	μM	nM	μM	nM
Butyrophenone affinity	μM	nM	nM	μM
Adenylate cyclase association	Stimulatory	Inhibitory or unassociated		?c
Guanine nucleotide sensitivity	+	+	–	+
Function	Parathyroid hormone release	Inhibition of pituitary hormone release		Autoregulation of DA neurons?
	Striatum: unknown	DA mediated behavioral responses and their antagonism by neuroleptics		
Striatal location	Intrinsic neurons	Intrinsic neurons		Intrinsic neurons?d
		Cortico-striate afferents?e		Nigro-striatal terminals
Pituitary location	–	+		–

a [3H] Flupentixol binding to D-3 receptors has yet to be investigated.

b [3H] Agonists may label a high affinity state of D-1 receptors.

c D-3 Autoreceptors are definitely not linked to stimulation of cAMP levels. However, their association with inhibition of adenylate cyclase has not been studied. Postsynaptic D-3 binding sites may be a high agonist-affinity state of D-1 receptors.

d D-3 binding sites appear to be found both on, and postsynaptic to, dopamine terminals in the striatum.

e [3H] Butyrophenone binding sites on cortico-striate terminals may be a distinct receptor subtype (D-4), as they have low agonist affinity and no guanine nucleotide sensitivity. Alternatively, their lack of guanine nucleotide regulation may be an artifact of the kainic acid lesion used to isolate them.

This classification is yet preliminary, as a number of questions remain. For instance, the possibility that a portion of the D-3 binding sites is merely a high affinity agonist binding state of the D-1 receptor has not yet been rigorously excluded. Also, other classes of DAR sites, such as those having high affinity for substituted benzamides, may be further elucidated in future studies. Additionally, the degree to which D-2 receptors on corticostriate terminals and intrinsic striatal neurons differ is as yet uncertain. Space has not allowed a discussion of dopamine receptors on the peripheral vasculature, which may also represent other subtypes (for review see Goldberg & Kohli 1979).

The availability of selective agonists and antagonists has been central to therapeutic and experimental advances in the field of dopaminergic transmission in the past. The advent of even more selective D-1, D-2, and D-3 agents may not only allow the resolution of the above questions, but also allow better pharmacological treatment of disorders involving these receptor subtypes.

Literature Cited

Ahn, H. M., Gardner, E., Makman, M. H. 1979. Anterior pituitary adenylate cyclase: Stimulation by dopamine and other monoamines. *Eur. J. Pharmacol.* 53:313–17

Attie, M. F., Brown, E. M., Gardner, D. G., Spiegel, A. M., Aurbach, G. D. 1980. Characterization of dopamine-responsive adenylate cyclase of bovine parathyroid cells and its relationship to parathyroid hormone secretion. *Endocrinology* 107:1776–81

Bacopoulos, N. G. 1981. Acute changes in the state of dopamine receptors: In vitro monitoring with ³H-dopamine. *Life Sci.* 29:2407–14

Baldessarini, R. J., Kula, N. S., Arana, G. W., Neumeyer, J. L., Law, S. J. 1980. Chloroethylnorapomorphine, a proposed long-acting dopamine antagonist: Interactions with dopamine receptors of mammalian forebrain in vitro. *Eur. J. Pharmacol.* 67:105–10

Bennett, J. P. Jr. 1978. Methods in binding studies. In *Neurotransmitter Receptor Binding*, ed. H. I. Yamamura, S. J. Enna, M. J. Kuhar, pp. 57–90. New York: Raven

Boeynaems, J. M., Dumont, J. E. 1977. The two-step model of ligand-receptor interaction. *Mol. Cell. Endocrinol.* 7: 33–47

Boyan-Salyers, B. D., Clement-Cormier, Y. C. 1980. Identification and partial purification of a hydrophobic protein component associated with [³H]spiroperidol-binding activity. *Biochim. Biophys. Acta* 617:274–81

Brown, E. M., Attie, M. F., Reen, S., Gardner, D. G., Kebabian, J., Aurbach, G. D. 1980. Characterization of dopaminergic receptors in dispersed bovine parathyroid cells. *Mol. Pharmacol.* 18: 335–40

Brown, E. M., Carrol, R. J., Aurbach, G. D. 1977. Dopaminergic stimulation of cyclic AMP accumulation and parathyroid hormone release from dispersed bovine parathyroid cells. *Proc. Natl. Acad. Sci. USA* 74:4210–13

Brown, J. H., Makman, M. H. 1972. Stimulation by dopamine of adenylate cyclase in retinal homogenates and of adenosine 3'-5'-cyclic monophosphate formation in intact retina. *Proc. Natl. Acad. Sci. USA* 69:539–43

Burt, D. R., Creese, I., Snyder, S. H. 1976. Properties of [³H]haloperidol and [³H]dopamine binding associated with dopamine receptors in calf brain membranes. *Mol. Pharmacol.* 12:800–12

Burt, D. R., Enna, S. J., Creese, I., Snyder, S. H. 1975. Dopamine receptor binding in the corpus striatum of mammalian brain. *Proc. Natl. Acad. Sci. USA* 72: 4655–59

Calabro, M. A., MacLeod, R. M. 1978. Binding of dopamine to bovine anterior pituitary gland membranes. *Neuroendocrinology* 25:32–46

Caron, M. C., Beaulieu, M., Raymond, V., Gagne, B., Drouin, J., Lefkowitz, R. J., Labrie, F. 1978. Dopaminergic receptors in the anterior pituitary gland. *J. Biol. Chem.* 253:2244–53

Clement-Cormier, Y. C., Heindel, J. J., Robison, G. A. 1977. Adenylyl cyclase from a prolactin producing tumour cell: The effect of phenothiazines. *Life Sci.* 21: 1357–64

Clement-Cormier, Y. C., Kebabian, J. W., Petzold, G. L., Greengard, P. 1974. Dopamine-sensitive adenylate cyclase in mammalian brain: A possible site of action of antipsychotic drugs. *Proc. Natl. Acad. Sci. USA* 71:1113–17

Clement-Cormier, Y. C., Kendrick, P. E. 1980. Solubilization and characterization of [³H]spiroperidol binding sites from subcellular fractions of the calf striatum. *Biochem. Pharmacol.* 29: 897–903

Clement-Cormier, Y. C., Meyerson, L. R., McIsaac, A. 1980. Solubilization of multiple binding sites for the dopamine receptor from calf striatal membranes. *Biochem. Pharmacol.* 29:2009–16

Clement-Cormier, Y. C., Parrish, R. A., Petzold, G. L., Kebabian, J. W., Greengard, P. 1975. Characterization of a dopamine-sensitive adenylate cyclase in the rat caudate nucleus. *J. Neurochem.* 25:143–49

Costall, B., Fortune, D. H., Granchelli, F. E., Law, S.-J., Naylor, R. J., Neumeyer, J. L., Nohria, V. 1980a. On the ability of N-chloroethyl apomorphine derivatives to cause irreversible inhibition of dopamine receptor mechanisms. *J. Pharm. Pharmacol.* 32:571–76

Costall, B., Fortune, D. H., Law, S.-J., Naylor, R. J., Neumeyer, J. L., Nohria, V. 1980b. (–) *N*-(Chloroethyl)norapomorphine inhibits striatal dopamine function via irreversible receptor binding. *Nature* 285:571–73

Cote, T. E., Grewe, C. W., Kebabian, J. W. 1981. Stimulation of the D-2 dopamine receptor in the intermediate lobe of the rat pituitary gland decreases the responsiveness of the beta-adrenoceptor: Biochemical mechanism. *Endocrinology* 108:420–26

Cote, T., Munemura, M., Eskay, R. L., Kebabian, J. W. 1980. Biochemical identification of the beta-adrenoceptor and evidence for the involvement of a cyclic AMP system in the beta-adrenergic-induced release of alpha-melanocyte stimulating hormone in the intermediate lobe of the rat pituitary gland. *Endocrinology* 107:108–16

Creese, I., Burt, D. R., Snyder, S. H. 1975. Dopamine receptor binding: Differentiation of agonist and antagonist states with ³H-dopamine and ³H-haloperidol. *Life Sci.* 17:993–1002

Creese, I., Burt, D. R., Snyder, S. H. 1976. Dopamine receptor binding predicts clinical and pharmacological potencies of antischizophrenic drugs. *Science* 192:481–83

Creese, I., Burt, D. R., Snyder, S. H. 1977a. Dopamine receptor binding enhancement accompanies lesion-induced behavioral supersensitivity. *Science* 197:596–98

Creese, I., Burt, D. R., Snyder, S. H. 1978a. Biochemical actions of neuroleptic drugs: Focus on the dopamine receptor. In *Handbook of Psychopharmacology,* ed. L. L. Iversen, S. D. Iversen, S. H. Snyder, 10:37–89. New York: Plenum

Creese, I., Padgett, L., Fazzini, E., Lopez, F. 1979a. ³H-*N*-propylnorapomorphine: A novel agonist ligand for central dopamine receptors. *Eur. J. Pharmacol.* 56:411–12

Creese, I., Prosser, T., Snyder, S. H. 1978b. Dopamine receptor binding: Specificity, localization and regulation by ions and guanyl nucleotides. *Life Sci.* 23:495–500

Creese, I., Schneider, R., Snyder, S. H. 1977b. ³H-Spiroperidol labels dopamine receptors in pituitary and brain. *Eur. J. Pharmacol.* 46:377–81

Creese, I., Sibley, D. R. 1979. Radioligand binding studies: Evidence for multiple dopamine receptors. *Commun. Psychopharmacol.* 3:385–95

Creese, I., Sibley, D. R. 1980. Regulation of dopamine receptors. In *Psychopharmacology and Biochemistry of Neurotransmitter Receptors,* ed. H. I. Yamamura, R. W. Olsen, E. Usdin, pp. 387–410. New York: Raven

Creese, I., Snyder, S. H. 1977. A novel, simple and sensitive radioreceptor assay for antischizophrenic drugs in blood. *Nature* 270:180–82

Creese, I., Snyder, S. H. 1978. Dopamine receptor binding of ³H-ADTN (2-amino-6,7-dihydroxy-1,2,3,4-tetrahydronaphthalene) regulated by guanyl nucleotides. *Eur. J. Pharmacol.* 50:459–61

Creese, I., Snyder, S. H. 1979. Nigrostriatal lesions enhance striatal ³H-apomorphine and ³H-spiroperidol binding. *Eur. J. Pharmacol.* 56:277–81

Creese, I., Stewart, K., Snyder, S. H. 1979b. Species variations in dopamine receptor binding. *Eur. J. Pharmacol.* 60:55–66

Creese, I., Usdin, T., Snyder, S. H. 1979c. Guanine nucleotides distinguish between two dopamine receptors. *Nature* 278:577–78

Cronin, M. J., Roberts, J. M., Weiner, R. I. 1978. Dopamine and dihydroergocryptine binding to the anterior pituitary and other brain areas of the rat and sheep. *Endocrinology* 103:302–9

Cronin, M. J., Weiner, R. I. 1979. [³H]Spiroperidol (spiperone) binding to a putative dopamine receptor in sheep and steer pituitary and stalk median eminence. *Endocrinology* 104:307–12

Cross, A. J., Owen, F. 1980. Characteristics of ³H-*cis*-flupenthixol binding to calf brain membranes. *Eur. J. Pharmacol.* 65:341–47

Crow, T. J., Owen, F., Cross, A. J., Lofthouse, R., Longden, A. 1978. Letters to the editor. *Lancet* 1:36–37

Davis, A., Madras, B., Seeman, P. 1981. Solubilization of neuroleptic/dopamine receptors of human brain striatum. *Eur. J. Pharmacol.* 70:321–29

De Camilli, P., Macconi, D., Spada, A. 1979. Dopamine inhibits adenylate cyclase in human prolactin-secreting pituitary adenomas. *Nature* 278:252–54

De Lean, A., Stadel, J. M., Lefkowitz, R. J. 1980. A ternary complex model explains the agonist-specific binding properties of the adenylate cyclase-coupled beta-adrenergic receptor. *J. Biol. Chem.* 255:7108–17

Dowling, J. E., Watling, K. J. 1981. Dopaminergic mechanisms in the Teleost retina. II. Factors affecting the accumulation of cyclic AMP in pieces of intact carp retina. *J. Neurochem.* 36:569–79

Ehinger, B. 1976. Biogenic amines as transmitters in the retina. In *Transmitters in the Visual Process,* ed. S. L. Bonting, pp. 145–63. Oxford: Pergamon

Fields, J. Z., Reisine, T. D., Yamamura, H. I. 1977. Biochemical demonstration of dopaminergic receptors in rat and human brain using [3H]spiroperidol. *Brain Res.* 136:578–84

Fields, J. Z., Reisine, T. D., Yamamura, H. I. 1979. Loss of striatal dopaminergic receptors after intrastriatal kainic acid injection. *Life Sci.* 23:569–74

Freedman, S. B., Poat, J. A., Woodruff, G. N. 1981. Effect of guanine nucleotides on dopaminergic agonist and antagonist affinity for [3H]sulpiride binding sites in rat striatal membrane preparations. *J. Neurochem.* 37:608–12

Fujita, N., Saito, K., Iwatsubo, K., Hirata, A., Noguchi, Y., Yoshida, H. 1980. Binding of [3H]apomorphine to striatal membranes prepared from rat brain after 6-hydroxydopamine and kainic acid lesions. *Brain Res.* 190:593–96

Fuxe, K., Hall, H., Kohler, C. 1979. Evidence for an exclusive localization of 3H-ADTN binding sites to postsynaptic nerve cells in the striatum of the rat. *Eur. J. Pharmacol.* 58:515–17

Gale, K., Guidotti, A., Costa, E. 1977. Dopamine-sensitive adenylate cyclase: Location in substantia nigra. *Science* 195:503–5

Garau, L., Govoni, S., Stefanini, E., Trabucchi, M., Spano, P. F. 1978. Dopamine receptors: Pharmacological and anatomical evidences indicate that two distinct dopamine receptor populations are present in rat striatum. *Life Sci.* 23:1745–50

Giannattasio, G., DeFerrari, M. E., Spada, A. 1981. Dopamine-inhibited adenylate cyclase in female rat adenohypophysis. *Life Sci.* 28:1605–12

Goldberg, L. I., Kohli, J. D. 1979. Peripheral pre- and post-synaptic dopamine receptors: Are they different from dopamine receptors in the central nervous system?, *Commun. Psychopharmacol.* 3:447–56

Goldsmith, P. C., Cronin, M. J., Weiner, R. I. 1979. Dopamine receptor sites in the anterior pituitary. *J. Histochem. Cytochem.* 27:1205–7

Gorissen, H., Ilien, B., Aerts, G., Laduron, P. 1980. Differentiation of solubilized dopamine receptors from spirodecanone binding sites in rat striatum. *FEBS Lett.* 121:133–38

Gorissen, H., Laduron, P. 1979. Solubilization of high affinity dopamine receptors. *Nature* 279:72–74

Govoni, S., Olgiati, V. R., Trabucchi, M., Garau, L., Stefanini, E., Spano, P. F. 1978. [3H]Haloperidol and [3H]spiroperidol receptor binding after striatal injection of kainic acid. *Neurosci. Lett.* 8:207–10

Greengard, P. 1976. Possible role for cyclic nucleotides and phosphorylated membrane proteins in postsynaptic actions of neurotransmitters. *Nature* 260:101–8

Hamblin, M., Creese, I. 1980. Phenoxybenzamine discriminates multiple dopamine receptors. *Eur. J. Pharmacol.* 65:119–21

Hamblin, M., Creese, I. 1982a. Phenoxybenzamine differentiates dopaminergic 3H-ligand binding to bovine caudate membranes. *Mol. Pharmacol.* 21:44–51

Hamblin, M., Creese, I. 1982b. Heat treatment mimics guanosine triphosphate effects on dopaminergic 3H-ligand binding to bovine caudate membranes. *Mol. Pharm.* 21:52–56

Hamblin, M., Creese, I. 1982c. 3H-Dopamine binding to rat striatal D-2 and D-3 sites: Enhancement by magnesium and inhibition by sodium. *Life Sci.* 30:1587–95

Hamblin, M. W., Leff, S. E., Creese, I. 1981. Regulation of 3H-dopamine binding to rat striatal membranes by guanine nucleotides and cations. *Soc. Neurosci. Abstr.* 7:428

Hoffman, B. B., Lefkowitz, R. J. 1980. Radioligand binding studies of adrenergic receptors: New insights into molecular and physiological regulation. *Ann. Rev. Pharmacol. Toxicol.* 20:581–68

Howlett, D. R., Nahorski, S. R. 1978. A comparative study of [3H]haloperidol and [3H]spiroperidol binding to receptors on rat cerebral membranes *FEBS Lett.* 87:152–56

Howlett, A. C., Van Arsdale, P. M., Gilman, A. G. 1978. Efficiency of coupling between the beta adrenergic receptor and adenylate cyclase. *Mol. Pharmacol.* 14:531–39

Huff, R. M., Molinoff, P. B. 1981. Effects of N-ethylmaleimide on 3H-spiroperidol binding to rat striatal membranes. *Soc. Neurosci. Abstr.* 7:125

Hyttel, J. 1978a. A comparison of the effect of neuroleptic drugs on the binding of ^{3}H-haloperidol and ^{3}H-*cis*(z)-flupenthixol and on adenylate cyclase activity in rat striatal tissue in vitro. *Prog. Neuro-Psychopharmacol.* 2:329–35

Hyttel, J. 1978b. Effects of neuroleptics on ^{3}H-haloperidol and ^{3}H-*cis*(Z)-flupenthixol binding and on adenylate cyclase activity in vitro. *Life Sci.* 23:551–56

Hyttel, J. 1980. Further evidence that ^{3}H-Cis(Z)flupenthixol binds to the adenylate cyclase-associated dopamine receptor (D-1) in rat corpus striatum. *Psychopharmacology* 67:107–9

Hyttel, J. 1981. Similarities between the binding of ^{3}H-piflutixol and ^{3}H-flupentixol to rat striatal dopamine receptors in vitro. *Life Sci.* 28:563–69

Iversen, L. L. 1975. Dopamine receptors in the brain. *Science* 188:1084–89

Iversen, L. L., Rogawski, M. A., Miller, R. J. 1976. Comparison of the effects of neuroleptic drugs on pre- and postsynaptic dopaminergic mechanisms in the rat striatum. *Mol. Pharmacol.* 12:251–62

Jacobs, S., Cuatrecasas, P. 1976. The mobile receptor hypothesis and "cooperativity" of hormone binding. *Biochim. Biophys. Acta* 433:482–95

Jenner, P., Marsden, C. D. 1979. The substituted benzamides—a novel class of dopamine antagonists. *Life Sci.* 25: 479–86

Kebabian, J. W., Calne, D. B. 1979. Multiple receptors for dopamine. *Nature* 277: 93–96

Kebabian, J. W., Petzold, G. L., Greengard, P. 1972. Dopamine-sensitive adenylate cyclase in caudate nucleus of rat brain and its similarity to the "dopamine receptor." *Proc. Natl. Acad. Sci. USA* 79:2145–49

Kebabian, J. W., Saavedra, J. M. 1976. Dopamine-sensitive adenylate cyclase occurs in a region of substantia nigra containing dopaminergic dendrites. *Science* 193:693–85

Kent, R. S., De Lean, A., Lefkowitz, R. J. 1980. A quantitative analysis of beta-adrenergic receptor interactions: Resolution of high and low affinity states of the receptor by computer modeling of ligand binding data. *Mol. Pharmacol.* 17:14–23

Kerdelhue, B., Weisman, A. S., Weiner, R. I. 1981. A dopaminergic binding site in the high speed supernatant of steer anterior pituitary homogenates. *Endocrinology* 109:307–9

Klawans, H. L. 1973. The pharmacology of tardive dyskinesias. *Am. J. Psychiatr.* 130:82–86

Komiskey, H. L., Bossart, J. F., Miller, D. D., Patil, P. N. 1978. Conformation of dopamine at the dopamine receptor. *Proc. Natl. Acad. Sci. USA* 75:2641–43

Kramer, S. G. 1976. Dopamine in retinal neurotransmission. See Ehinger 1976, pp. 165–98

LaBrie, F., Ferland, L., DiPaolo, T., Veilleux, R. 1980. Modulation of prolactin secretion by sex steroids and thyroid hormones. In *Central and Peripheral Regulation of Prolactin Function,* ed. R. M. MacLeod, U. Scapagnini, pp. 97–113. New York: Raven

Lautin, A., Wazer, D., Stanley, M., Rotrosen, J., Gershon, S. 1980. Chronic treatment with metoclopramide induces behavioral supersensitivity to apomorphine and enhances specific binding of ^{3}H-spiroperidol to rat striata. *Life Sci.* 27:305–16

Lee, T., Seeman, P. 1980. Elevation of brain neuroleptic/dopamine receptors in schizophrenia. *Am. J. Psychiatr.* 137: 191–97

Lee, T., Seeman, P., Tourtellotte, W., Farley, W. W., Hornykiewicz, O. 1978. Binding of ^{3}H-neuroleptics and ^{3}H-apomorphine in schizophrenic brains. *Nature* 274: 897–900

Leff, S., Adams, L., Hyttel, J., Creese, I. 1981. Kainate lesion dissociates striatal dopamine receptor radioligand binding sites. *Eur. J. Pharmacol.* 70:71–75

Leff, S., Creese, I. 1982. Solubilization of a guanine nucleotide sensitive form of the D-2 dopamine receptor from brain requires agonist occupancy. *Fed. Proc.* 41:1633

Lefkowitz, R. J. 1980. See Creese & Sibley 1980, pp. 155–70

Lefkowitz, R. J., De Lean, A., Hoffman, B., Stadel, J., Kent, R., Michel, T., Limbird, L. 1981. Molecular pharmacology of adenylate cyclase coupled α and β adrenergic receptors. *Adv. Cyclic Nucleotide Res.* 14:145–61

Leysen, J. E. 1979. Unitary dopaminergic receptor composed of cooperatively linked agonist and antagonist sub-unit binding sites. *Commun. Psychopharmacol.* 3:397–410

Leysen, J. E., Awouters, F., Kennis, L., Laduron, P. M., Vandenberk, J., Janssen, P. A. J. 1981. Receptor binding profile of R 41 468, a novel antagonist at 5-HT₂ receptors. *Life Sci.* 28: 1015–22

Leysen, J. E., Gommeren, W., Laduron, P. M. 1978a. Spiperone: A ligand of choice for neuroleptic receptors. 1. Kinetics and characteristics of *in vitro* binding. *Biochem. Pharmacol.* 27:307–16

Leysen, J. E., Niemegeers, C. J. E., Tollenaere, J. P., Laduron, P. M. 1978b. Serotonergic component of neuroleptic receptors. *Nature* 272:168–71

Limbird, L. E. 1981. Activation and attenuation of adenylate cyclase. *Biochem. J.* 195:1–13

Limbird, L. E., Gill, D. M., Lefkowitz, R. J. 1980a. Agonist-promoted coupling of the beta-adrenergic receptor with the guanine nucleotide regulatory protein of the adenylate cyclase system. *Proc. Natl. Acad. Sci. USA* 77:775–79

Limbird, L. E., Gill, D. M., Stadel, J. M., Hickey, A. R., Lefkowitz, R. J. 1980b. Loss of beta-adrenergic receptor-guanine nucleotide regulatory protein interactions accompanies decline in catecholamine responsiveness of adenylate cyclase in maturing rat erythrocytes, *J. Biol. Chem.* 255:1854–61

List, S. J., Seeman, P. 1981. Resolution of dopamine and serotonin receptor components of ³H-spiperone binding to rat brain regions. *Proc. Natl. Acad. Sci. USA* 78:2620–24

List, S., Titeler, M., Seeman, P. 1980. High-affinity ³H-dopamine receptors (D₃ sites) in human and rat brain. *Biochem. Pharmacol.* 29:1621–22

MacLeod, R. M., Nagy, I., Login, I. S., Kimura, H., Valdenegro, C. A., Thorner, M. D. 1980. See LaBrie et al 1980, pp. 27–41

Mackay, A. V. P., Bird, E. D., Iversen, L. L., Spokes, E. G., Creese, I., Snyder, S. H. 1980a. Dopaminergic abnormalities in postmortem schizophrenic brain. In *Long-Term Effects of Neuroleptics,* ed. F. Cattabeni, G. Racagni, P. F. Spano, E. Costa, pp. 325–33. New York: Raven

Mackay, A. V. P., Bird, E. D., Spokes, E. G., Rossor, M., Iversen, L. L., Creese, I., Snyder, S. H. 1980b. Dopamine receptors and schizophrenia: Drug effect or illness? *Lancet* 2:915–16

Madras, B. K., Davis, A., Kunashko, P., Seeman, P. 1980. See Creese & Sibley 1980, pp. 411–19

Magistretti, P. J., Schorderet, M. 1979. Dopamine receptors in bovine retina: Characterization of the ³H-spiroperidol binding and its use for screening dopamine receptor affinity of drugs. *Life Sci.* 25:1675–86

Makman, M. H., Dvorkin, B., Horowitz, S. G., Thal, L. J. 1980a. Properties of dopamine agonist and antagonist binding sites in mammalian retina. *Brain Res.* 194:403–18

Makman, M. H., Dvorkin, B., Horowitz, S. G., Thal, L. J. 1980b. Retina contains guanine nucleotide sensitive and insensitive classes of dopamine receptors. *Eur. J. Pharmacol.* 63:217–22

Marchais, D., Bockaert, J. 1980. Is there a connection between high affinity ³H-spiperone binding sites and DA-sensitive adenylate cyclase in corpus striatum? *Biochem. Pharmacol.* 29:1331–36

Martres, M.-P., Sokoloff, P., Schwartz, J. C. 1980. Three classes of dopaminergic receptors evidenced by two radioligands: ³H-apomorphine and ³H-domperidone. See Madras et al 1980, pp. 421–34

Memo, M., Spano, P. F., Trabucchi, M. 1980. Characterization and localization of dopamine-D₂ central receptors. *Br. J. Pharmacol.* 72:124P–25P

Miller, R. J., Horn, A. S., Iversen, L. L. 1974. The action of neuroleptic drugs on dopamine-stimulated adenosine cyclic 3',5'-monophosphate production in rat neostriatum and limbic forebrain. *Mol. Pharmacol.* 10:759–66

Miller, R. J., McDermed, J. 1979. Dopamine-sensitive adenylate cyclase. In *The Neurobiology of Dopamine,* ed. A. S. Horn, J. Korf, B. H. C. Westerink, pp. 159–77. New York: Academic

Mowles, T. F., Burghardt, B., Burghardt, C., Charneki, A., Sheppard, H. 1978. The dopamine receptor of the rat mammotroph in cell culture as a model for drug action. *Life Sci.* 22:2103–8

Muller, P., Seeman, P. 1978. Dopaminergic supersensitivity after neuroleptics: Timecourse and specificity. *Psychopharmacology* 60:1–11

Munemura, M., Cote, T. E., Tsuruta, K., Eskay, R. L., Kebabian, J. W. 1980a. The dopamine receptor in the intermediate lobe of the rat pituitary: Pharmacological characterization. *Endocrinology* 107:1683–86

Munemura, M., Eskay, R. L., Kebabian, J. W. 1980b. Release of alpha-melanocyte-stimulating hormone from dispersed cells of the intermediate lobe of the rat pituitary gland: involvement of catecholamines and adenosine 3',5'-monophosphate. *Endocrinology* 106:1795–1803

Munson, P. J., Rodbard, D. 1980. Ligand: A versatile computerized approach for characterization of ligand-binding systems. *Anal. Biochem.* 107:220–39

Murrin, L. C., Gale, K., Kuhar, M. J. 1979. Autoradiographic localization of neuroleptic and dopamine receptors in the caudate-putamen and substantia nigra: Effects of lesions. *Eur. J. Pharmacol.* 60:229–35

Nagy, J. I., Lee, T., Seeman, P., Fibiger, H. C. 1978. Direct evidence for presynaptic and postsynaptic dopamine receptors in brain. *Nature* 274:278–81

Neumeyer, J. L., Law, S. J., Baldessarini, R. J., Kula, N. S. 1980. (–)-*N*-(2-Chloroethyl)-10, 11-dihydroxynoraporphine (Chloroethylnorapomorphine), a novel irreversible dopamine receptor antagonist. *J. Med. Chem.* 23:595–99

Ogren, S. O., Hall, H., Kohler, C. 1978. Studies on the stereoselective dopamine receptor blockade in the rat brain by rigid spiro amines. *Life Sci.* 23:1769–74

Onali, P., Schwartz, J. P., Costa, E. 1981. Dopaminergic modulation of adenylate cyclase stimulation of vasoactive intestinal peptide (VIP) in anterior pituitary. *Proc. Natl. Acad. Sci. USA* 78:6531–34

Owen, F., Cross, A. J., Waddington, J. L., Poulter, M., Gamble, S. J., Crow, T. J. 1980. Dopamine-mediated behaviour and ³H-spiperone binding to striatal membranes in rats after nine months haloperidol administration. *Life Sci.* 26:55–59

Pawlikowski, M., Karasek, E., Kunert-Radek, J., Jaranowska, M. 1981. Effects of dopamine on cyclic AMP concentration in the anterior pituitary gland in vitro. *J. Neural Trans.* 50:179–84

Pawlikowski, M., Karasek, E., Kunert-Radek, J., Stepien, H. 1979. Dopamine blockade of the thyroliberin-induced cyclic AMP accumulation in rat anterior pituitary. *J. Neural Trans.* 45: 75–79

Peroutka, S. J., Snyder, S. H. 1979. Multiple serotonin receptors: Differential binding of [³H]5-hydroxytryptamine, [³H]lysergic acid diethylamide and [³H]spiroperidol. *Mol. Pharmacol.* 16: 687–99

Phillipson, O. T., Emson, P. C., Horn, A. S., Jessell, T. 1977. Evidence concerning the anatomical location of the dopamine-stimulated adenylate cyclase in the substantia nigra. *Brain Res.* 136: 45–58

Phillipson, O. T., Horn, A. S. 1976. Substantia nigra of the rat contains a dopamine-sensitive adenylate cyclase. *Nature* 261:418–20

Pike, L. J., Lefkowitz, R. J. 1980. Activation and desensitization of beta-adrenergic receptor-coupled GTPase and adenylate cyclase of frog and turkey erythrocyte membranes. *J. Biol. Chem.* 255:6860–67

Premont, J., Thierry, A. M., Taassin, J. P., Glowinski, J. G., Bockaert, J. 1976. Is the dopamine-sensitive adenylate cyclase in the rat substantia nigra coupled with autoreceptors? *FEBS Lett.* 68:99–104

Quik, M., Emson, P. C., Joyce, E. 1979. Dissociation between the presynaptic dopamine-sensitive adenylate cyclase and [³H]spiperone binding sites in rat substantia nigra. *Brain Res.* 167:355–75

Quik, M., Iversen, L. L. 1979. Regional study of ³H-spiperone binding and the dopamine-sensitive adenylate cyclase in rat brain. *Eur. J. Pharmacol.* 56:323–30

Quik, M., Iversen, L. L., Larder, A., Mackay, A. V. P. 1978. Use of ADTN to define specific ³H-spiperone binding to receptors in brain. *Nature* 274:513–14

Ray, K. P., Wallis, M. 1980. Is cyclic adenosine 3':5'-monophosphate involved in the dopamine-mediated inhibition of prolactin secretion? *J. Endocrin.* 85:59p

Redburn, D. A., Clement-Cormier, Y., Lam, D. M. K. 1980. Dopamine receptors in the goldfish retina: ³H-spiroperidol and ³H-domperidone binding; and dopamine-stimulated adenylate cyclase activity. *Life Sci.* 27:23–31

Rodbell, M. 1980. The role of hormone receptors and GTP-regulatory proteins in membrane transduction. *Nature* 284:17–22

Ross, E. M., Gilman, A. G. 1980. Biochemical properties of hormone-sensitive adenylate cyclase. *Ann. Rev. Biochem.* 49:533–64

Ross, E. M., Maguire, M. E., Sturgill, T. W., Biltonen, R. L., Gilman, A. G. 1977. Relationship between the beta-adrenergic receptor and adenylate cyclase—Studies of ligand binding and enzyme activity in purified membranes of S49 lymphoma cells. *J. Biol. Chem.* 252:5761–75

Sarthy, P. J., Lam, D. M. K. 1979. The uptake and release of [³H]-dopamine in the goldfish retina. *J. Neurochem.* 32: 1269–77

Schmidt, M. J. 1979. Perspectives on dopamine-sensitive adenylate cyclase in the brain. In *Neuropharmacology of Cyclic Nucleotides*, ed. G. C. Palmer, pp. 1–52. Baltimore: Urban & Schwarzenberg

Schmidt, M. J., Hill, L. E. 1977. Effects of ergots on adenylate cyclase activity in the corpus striatum and pituitary. *Life Sci.* 20:789–98

Schwarcz, R., Coyle, J. T. 1977. Neurochemical sequelae of kainate injections in corpus striatum and substantia nigra of the rat. *Life Sci.* 20:431–36

Schwarcz, R., Creese, I., Coyle, J. T., Snyder, S. H. 1978. Dopamine receptors localised on cerebral cortical afferents to rat corpus striatum. *Nature* 271:766–68

Seeman, P., Chau-Wong, M., Tedesco, J., Wong, K. 1975. Brain receptors for antipsychotic drugs and dopamine: Direct binding assays. *Proc. Natl. Acad. Sci. USA* 72:4376–80

Seeman, P., Lee, T., Chau-Wong, M., Wong, K. 1976. Antipsychotic drug doses and neuroleptic/dopamine receptors. *Nature* 261:717–19

Seeman, P., Tedesco, J. L., Lee, M., Chau-Wong, M., Muller, P., Bowles, J., Whitaker, P. M., McManus, C., Tittler, M., Weinreich, P., Friend, W. C., Brown, G. M. 1978. Dopamine receptors in the central nervous system. *Fed. Proc.* 37:130–36

Seeman, P., Woodruff, G. N., Poat, J. A. 1979. Similar binding of ³H-ADTN and ³H-apomorphine to calf brain dopamine receptors. *Eur. J. Pharmacol.* 55:137–42

Sibley, D. R., Creese, I. 1979. Multiple pituitary dopamine receptors: Effects of guanine nucleotides. *Soc. Neurosci. Abstr.* 5:352

Sibley, D. R., Creese, I. 1980a. Anterior pituitary dopamine receptors: Heterogeneity of agonist binding. *Fed. Proc.* 39:1098

Sibley, D. R., Creese, I. 1980b. Dopamine receptor binding in bovine intermediate lobe pituitary membranes. *Endocrinology* 107:1405–9

Sibley, D. R., De Lean, A., Creese, I. 1982. Anterior pituitary dopamine receptors: Demonstration of interconvertible high and low affinity states of the D-2 dopamine receptor. *J. Biol. Chem.* 257:6351–61

Sibley, D. R., Mahan, L. C., Creese, I. 1981. Dopamine receptor binding on intact cells: Absence of a high affinity agonist state. *Soc. Neurosci. Abstr.* 7:126

Snyder, S. H., Creese, I., Burt, D. R. 1975. The brain's dopamine receptor: Labeling with ³H]dopamine and ³H]haloperidol. *Psycho-pharmacol. Commun.* 1:663–673

Sokoloff, P., Martres, M.-P., Schwartz, J.-C. 1980a. ³H-Apomorphine labels both dopamine postsynaptic receptors and autoreceptors. *Nature* 288:283–86

Sokoloff, P., Martres, M. P., Schwartz, J. C. 1980b. Three classes of dopamine receptor (D-2, D-3, D-4) identified by binding studies with ³H-apomorphine and ³H-domperidone. *Naunyn-Schmiedebergs. Arch. Pharmacol.* 315:89–102

Spano, P. F., DiChiara, G., Tonon, G. C., Trabucchi, M. 1976. A dopamine-stimulated adenylate cyclase in rat substantia nigra, *J. Neurochem.* 27:1565–68

Spano, P. F., Trabucchi, M., DiChiara, G. 1977. Localization of nigral dopamine-sensitive adenylate cyclase on neurons originating from the corpus striatum. *Science* 196:1343–45

Stefanini, E., Devoto, P., Marchisio, A. M., Vernaleone, F., Collu, R. 1980. [³H]Spiroperidol binding to a putative dopaminergic receptor in rat pituitary gland. *Life Sci.* 26:583–87

Stefanini, E., Clement-Cormier, Y., Vernaleone, F., Devoto, P., Marchisio, A. M., Collu, R. 1981. Sodium-dependent interaction of benzamides with dopamine receptors in rat and dog anterior pituitary glands. *Neuroendocrinology* 32:103–7

Stoof, J. C., Kebabian, J. W. 1981. Opposing roles for D-1 and D-2 dopamine receptors in efflux of cyclic AMP from rat neostriatum. *Nature* 294:366–68

Thal, L., Creese, I., Snyder, S. H. 1978. ³H-Apomorphine interactions with dopamine receptors in calf brain. *Eur. J. Pharmacol.* 49:295–99

Theodorou, A., Crockett, M., Jenner, P., Marsden, C. D. 1979. Specific binding of [³H]sulpiride to rat striatal preparations. *J. Pharm. Pharmacol.* 31:424–26

Theodorou, A. E., Hall, M. D., Jenner, P., Marsden, C. D. 1980. Cation regulation differentiates specific binding of [³H]sulpiride and [³H]spiperone to rat striatal preparations. *J. Pharm. Pharmacol.* 32:441–44

Theodorou, A., Reavill, C., Jenner, P., Marsden, C. D. 1981. Kainic acid lesions of striatum and decortication reduce specific [³H]sulpiride binding in rats, so D-2 receptors exist postsynaptically on cortico-striate afferents and striatal neurons. *J. Pharm. Pharmacol.* 33:439–44

Titeler, M., List, S., Seeman, P. 1979. High affinity dopamine receptors (D₃) in rat brain. *Commun. Psychopharmacol.* 3:411–20

Titeler, M., Seeman, P. 1978. Antiparkinsonian drug doses and neuroleptic receptors. *Experientia* 34:1490–92

Titeler, M., Seeman, P. 1979. Selective labeling of different dopamine receptors by a new agonist ³H-ligand: ³H-*n*-propylnorapomorphine. *Eur. J. Pharmacol.* 56:291–92

Titeler, M., Weinreich, P., Sinclair, D., Seeman, P. 1978. Multiple receptors for brain dopamine. *Proc. Natl. Acad. Sci. USA* 75:1153–56

Traficante, L. J., Friedman, E., Oleshansky, M. A., Gershon, S. 1976. Dopamine-sensitive adenylate cyclase and cAMP phosphodiesterase in substantia nigra and corpus striatum of rat brain. *Life Sci.* 19:1061–66

Walton, K. G., Liepmann, P., Baldessarini, R. J. 1978. Inhibition of dopamine-stimulated adenylate cyclase activity by phenoxybenzamine. *Eur. J. Pharmacol.* 52:231–34

Watling, K. J., Dowling, J. E. 1981. Dopaminergic mechanisms in the Teleost retina. I. Dopamine-sensitive adenylate cyclase in homogenates of carp retina: Effects of agonists, antagonists and ergots. *J. Neurochem.* 36:559–68

Watling, K. J., Dowling, J. E., Iversen, L. L. 1979. Dopamine receptors in the retina may all be linked to adenylate cyclase.

Nature 281:578–80

Watling, K. J., Iversen, L. L. 1981. Comparison of the binding of [³H]spiperone and [³H]domperidone in homogenates of mammalian retina and caudate nucleus. *J. Neurochem.* 37:1130–43

Weiner, R. I., Cronin, M. J., Cheung, C. Y., Faure, N., Clark, B. R., Goldsmith, P. C. 1979. Anterior pituitary dopamine receptors and prolactin. In *Catecholamines: Basic and Clinical Frontiers,* ed. E. Usdin, I. J. Kopin, J. Barchas, pp. 1218–20. New York: Pergamon

Weiner, R. I., Ganong, W. F. 1978. Role of brain monoamines and histamine in regulation of anterior pituitary secretion. *Physiol. Rev.* 58:905–76

Withy, R. M., Mayer, R. J., Strange, P. G. 1980. [³H]Spiroperidol binding to brain neurotransmitter receptors. *FEBS Lett.* 112:293–95

Woodruff, G. N., Freedman, S. B. 1981. Binding of [³H]sulpiride to purified rat striatal synaptic membranes. *Neuroscience* 6:407–10

Zahniser, N. R., Molinoff, P. B. 1978. Effect of guanine nucleotides on striatal dopamine receptors. *Nature* 275:453–55

Ann. Rev. Neurosci. 1983. 6:73–94

SPECULATIONS ON THE FUNCTIONAL ANATOMY OF BASAL GANGLIA DISORDERS

J. B. Penney, Jr. and A. B. Young

Department of Neurology, The University of Michigan, Ann Arbor, Michigan 48109

Introduction

The disturbed movements seen in disorders of the basal ganglia, such as Huntington's disease and parkinsonism, emphasize the importance of the basal ganglia for normal motor control. Understanding of the pathophysiology of these disorders has been limited because most proposed models of basal ganglia function fail to explain adequately how disturbances in the structure and function of neuronal populations and alterations in neurochemistry result in abnormal motor behavior. This lack of adequate models, in turn, is related to deficiencies in our understanding of the neuronal interactions that mediate normal movement. A useful model for basal ganglia function should describe the circuitry involved in normal movement and provide a rationale for the signs produced by pathologic processes. The model should be consistent with current knowledge of the anatomy, physiology, and neurochemistry of the basal ganglia. The model should account also for the responses of patients with basal ganglia disorders to various pharmacological and surgical manipulations.

The anatomy, pharmacology, and physiology of the basal ganglia have been reviewed extensively (Baldessarini & Tarsy 1980, Barbeau 1979, Denny-Brown 1962, Fallon & Moore 1978, Graybiel & Ragsdale 1979, Hornykiewicz 1979, Iversen 1977, Jung & Hassler 1960, Oberg & Divac 1979). Recently, a very careful, detailed review by DeLong & Georgopoulos (1982) on the motor functions of the basal ganglia has appeared. In view of this we have not attempted to provide an extensive review but rather to present a model based on the accumulating evidence that suggests the importance of a cortico-striato-pallido-thalamocortical feedback circuit as the major extrapyramidal influence on the motor system in man. The anatomy and potential function of this circuit have been well described by Kemp &

73

Powell (1971). We propose that this circuit is organized anatomically and neurochemically so that the striatum can select and maintain motor (and probably also nonmotor) behaviors. The model proposes that behaviors generated from the cerebral cortex are focused and facilitated by projections through the basal ganglia. These behaviors vary from species to species. Furthermore, the basal ganglia function to suppress other conflicting activities while reinforcing ongoing behaviors. According to the hypothesis, the chorea of Huntington's disease (which is characterized pathologically by the loss of intrinsic and efferent striatal neurons), and the bradykinesia of Parkinson's disease (characterized pathologically by degeneration of the nigrostriatal pathway) represent opposite extremes in dysfunction of this system. Choreic movements, which were described by Wilson (1929) as fragments of normal movements, result from deficits in the ability to maintain ongoing activities and suppress unwanted ones. Bradykinesia results from overactivity in the primary facilitatory circuit, with inadequate inhibitory modulation of ongoing behaviors by the nigrostriatal dopamine pathway. In this model, we have placed emphasis on the primary facilitatory cortico-striato-pallido-thalamocortical circuit and its regulation by systems such as the nigrostriatal dopamine pathway. Although undoubtedly of importance, less emphasis is placed on pathways whose function in man is less certain (e.g. the peptidergic pathways).

Maintenance of Motor Behavior

In movement disorders, clinical observations indicate major alterations in the ability to regulate sustained motor activity. During normal motor behaviors in primates there is sustained activity of cortical output neurons to spinal segments (Evarts 1967). How do normal cortical neurons maintain their own activity? One mechanism by which a cortical output cell could do this is through a positive feedback circuit. For example, the cortical neuron could send collaterals to excite another cell or cells that would, in turn, reexcite it. This mutual excitation would maintain the activity of both cells once one cell becomes active. Such a positive feedback circuit potentially exists in the reciprocal connections between cortex and thalamus (Figure 1, Circuit A). Physiological studies show that the cells of the ventral anterior and the ventral lateral nuclei of the thalamus (VA/VL) that project to the cortex are monosynaptically excited by collaterals of corticospinal tract neurons (Endo et al 1973). At the same time, the cells of VA/VL project to the motor cortex and are there monosynaptically excitatory on the dendrites of corticospinal tract cells (Yoshida et al 1966). Steriade et al (1972) have provided evidence that the cortico-thalamocortical circuit acts as a positive feedback mechanism. In anatomical studies, synapses of thalamocortical cells on corticothalamic neurons have been demonstrated with electron microscopy (Hersch & White 1981). Also, some VA/VL cells send branches diffusely throughout the molecular layer of the frontal cortex (Herkenham 1979, 1980). Thus, the neuronal circuitry exists for active cortical output neurons to maintain their activity via this positive cortico-thalamocortical feedback circuit.

The neurotransmitter of the corticothalamic tract is probably the excitatory amino acid, glutamate. High affinity uptake and high glutamate levels are associated with putative glutamatergic presynaptic terminals. Fonnum et al (1981) demonstrated in rats that interruption of the cortical input to the anterior thalamus results in a decrease in high affinity glutamate uptake in the thalamus. We have shown recently that interruption of cortical input specifically to VA/VL in the rat and cat leads to a selective decrease in glutamate uptake and in glutamate levels (Bromberg et al 1981, Young et al 1981). In addition, glutamate may subserve a neurotransmitter function for corticostriatal (Kim et al 1977, McGeer et al 1977) and corticospinal fibers (Young et al 1981).

The cortico-thalamocortical circuit alone cannot regulate motor control

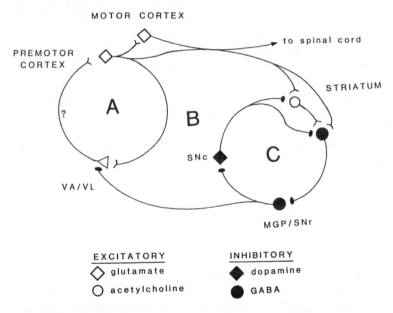

Figure 1 A schematic diagram of the basal ganglia pathways discussed in this paper. There are three feedback circuits: **A,** a cortex to thalamus to cortex positive feedback loop; **B,** a cortex to striatum to medial globus pallidus to thalamus to cortex positive feedback loop; **C,** a striatum to substantia nigra pars reticulata neuron to substantia nigra dopamine neuron to striatum negative feedback loop. Abbreviations used: GABA, gamma-aminobutyric acid; MGP, medial globus pallidus; SN$_r$, substantia nigra pars reticulata; SN$_c$, substantia nigra pars compacta; VA/VL, ventral anterior and ventral lateral nuclei of the thalamus.

because inhibitory mechanisms must function to modulate this feedback circuit. Which cells might provide this inhibition? One possibility is the cells of the medial segment of the globus pallidus (MGP) and of the substantia nigra pars reticulata (SN_r) (Figure 1). DeLong & Georgopoulos (1979) have proposed that the MGP and SN_r are actually one nucleus that has been divided by the internal capsule. The cells of the SN_r and MGP are morphologically and physiologically similar and constitute a major output of the basal ganglia (Carter & Fibiger 1978, Kemp & Powell 1971). They are inhibitory onto the cells of VA/VL (Uno & Yoshida 1975). Recent studies in the rat suggest that nigrothalamic neurons are GABAergic (DiChiara et al 1979, Kilpatrick et al 1980, MacLeod et al 1980). We have investigated the possibility that gamma-aminobutyric acid (GABA) is a pallidothalamic neurotransmitter (Penney & Young 1981). Three indicators of GABAergic function [amino acid levels, high affinity GABA uptake, and glutamic acid decarboxylase (GAD) levels] were measured in VA/VL after kainic acid lesions of the entopeduncular nucleus (the murine equivalent of the MGP). GABA levels, GABA uptake, and GAD activity were all reduced by approximately 50%, suggesting that GABA is a pallidothalamic transmitter. MGP projects to the rostral, ventral, lateral portion of VA/VL (Carter & Fibiger 1978), which in turn gives specific projections to premotor cortex (Herkenham 1980). The premotor cortex also receives an input from more dorsal and caudal portions of VA/VL, which in turn receive an excitatory input from the dentate nucleus of the cerebellum (Faull & Mehler 1976; P. L. Strick, personal communication). The SN_r projects to medial portions of VA/VL, which, in turn, project diffusely to the molecular layer of the entire frontal lobe (Carpenter et al 1976, Herkenham 1979). Furthermore, the cells of MGP/SN_r are tonically active (DeLong & Georgopoulos 1979). Thus, the normal action of the cells of MGP/SN_r may be to dampen the amount of positive feedback between the frontal cortex and VA/VL. The cells of the prefrontal cortex project to motor cortex (Muakkassa & Strick 1979). Thus, the output of the basal ganglia (MGP/SN_r) influences an area of cortex that in turn influences motor cortex.

Since the MGP inhibitory projection is tonically active (DeLong & Georgopoulos 1979), either an overwhelming excitation of the VA/VL cells or inhibition of MGP cells must occur in order for the positive feedback circuit A to become active. Facilitation of cortico-thalamocortical circuits could be achieved by inhibiting their inhibitory pallidothalamic and nigrothalamic afferents. Excitation via inhibition of inhibitors, so-called disinhibition, is a concept that has been proposed in other neuronal systems (Roberts 1976) and is well established in invertebrates (Maynard 1972). The inhibitory output cells of the neostriatum (putamen and caudate nucleus), which utilize GABA as their neurotransmitter, project to the MGP and the SN_r (DiChiara et al 1980, Fonnum et al 1978, Nagy et al 1978) (Figure 1).

There is considerable evidence to suggest that the striato-pallidothalamic pathway consists of two sequential GABAergic neurons. Somogyi et al (1979) have shown, in ultrastructural studies, that the striatonigral fibers synapse on nigrothalamic neurons. We have found similar contacts between striatoentopeduncular and entopeduculothalamic cells (J. B. Penney, unpublished observations). We have also performed a quantitative autoradiographic study of GABA receptors after unilateral striatal kainate lesions as shown by [^{3}H]-muscimol binding to frozen sections of rat brain (Pan et al 1981). We found increased numbers of GABA receptors in the globus pallidus, entopeduncular nucleus, and SN$_r$, but decreased numbers of receptors in VA/VL on the side of the striatal lesion. These findings suggest that decreased striatal output results in increased activity of thalamic GABAergic afferents and a decrease in the number of thalamic GABA receptors. Thus under normal conditions, GABAergic striatal output cells may disinhibit thalamocortical cells.

Furthermore, cells of the cortex excite GABAergic cells in the neostriatum and therefore may excite their own disinhibitors (Kitai et al 1976b, McGeer et al 1978, Somogyi et al 1981a) (Figure 1, Circuit B). Studies have shown that caudate neurons increase their activity during movements and during the maintenance of postures (Buser et al 1974). The cells that project from the cortex to the spinal cord have their cell bodies in the lower part of Layer V (Jones & Wise 1977), and many of these have collaterals to striatal neurons (Donoghue & Kitai 1981). In addition, there are corticostriatal cells that are located in the upper part of Layer V and possibly Layer III (Kitai et al 1976b).

The anatomical and physiological organization of the basal ganglia is very precise. Although there is overlap between areas, by and large the sensorimotor cortex projects somatotopically to the putamen (Kemp & Powell 1970, Künzle 1975, Liles 1979), the association cortex to the caudate (Kemp & Powell 1970), and the limbic cortex to the nucleus accumbens (Newman & Winans 1980). The caudate, putamen, and accumbens maintain these somatotopic relationships in their projections onto the MGP/SN$_r$ (Bunney & Aghajanian 1976, Grofova 1975, Johnson & Rosvold 1971) and the ventral tegmental area (Nauta et al 1978). MGP neurons fire in relation to limb movement whereas SN$_r$ neurons primarily respond to cranial and axial movements (DeLong & Georgopoulos 1979). These discrete somatotopic relationships are maintained in projections to the thalamus (Herkenham 1979) and thence to the cortex. Although the details of the circuitry are currently unknown, it is possible that a set of cortical neurons can maintain its own firing by activating a subset of corticostriatal neurons; this would result in disinhibition of the set's corticothalamic positive feedback circuit. Thus, the circuit described by Kemp & Powell (1971) that extends from the cortex to the striatum, MGP, VA/VL, and thence back to the cortex can function as another positive feedback system, and activation of striatal output cells by the cortex could facilitate maintenance of ongoing movements.

The nigrostriatonigral circuit (Figure 1, Circuit C) has been thought to have a major role in basal ganglia function. The GABAergic output cells of the striatum project to both the MGP and the SN_r (Yoshida et al 1972). Current evidence suggests that the striatonigral GABAergic cells make synapses on inhibitory nigral interneurons, which in turn make synapses on nigrostriatal dopamine neurons (Cheramy et al 1978, Waszczak et al 1979). The neurotransmitter of the interneuron is thought to be GABA (Waszczak et al 1979, Waszczak & Walters 1980) or glycine (Cheramy et al 1978, James and Starr 1979). This means that the striatonigrostriatal feedback loop (Figure 1, Circuit C) consists of three inhibitory neurons in sequence that respectively have GABA, GABA (or glycine), and dopamine as their transmitters. Such a circuit is a negative feedback loop. The role of this circuit in the regulation of nigral dopaminergic activity may be less important than has been suspected (Baring et al 1980, Waszczak et al 1979).

What, then, could be responsible for selecting which movement will be maintained? What determines which cortical functional units become active and which do not? Divac (1977) has proposed that the striatum makes this selection. Since there is a convergence of cortical output from many areas onto striatal cells, the latter might function to integrate different cortical outputs. The corticostriatal convergence puts striatal neurons in a position to be involved in the learning of motor behavior. In fact, striatal lesions lead to learning deficits in experimental animals (Divac et al 1978, Levine et al 1978, Thompson & Mettler 1963). A rich system of local axon collaterals interconnects the inhibitory striatal output neurons with one another (Park et al 1980, Somogyi et al 1981a). Activity of one output cell of the striatum will thus tend to suppress the activity of other output neurons (Figure 2). The extent and patterning of these axon collaterals may develop during the acquisition of new motor and perhaps cognitive capacities. Striatal output may be capable therefore of determining which cortical activities are maintained and which are suppressed. The basal ganglia may function to perform similar operations on inputs from various parts of the cortex, and the functional deficits resulting from lesions in specific regions of the basal ganglia may reflect the different somatotopic relationships. Divac's and our hypotheses are compatible if the cells of the striatum are responsible for selecting and maintaining certain desired movements or patterns of motor behavior while suppressing other antagonistic or unwanted activities. A group of striatal neurons firing together would constitute what Buchwald et al (1979) have called a "behavioral set," or the readiness to perform a certain movement. This selection of "behavioral sets" by the striatum would

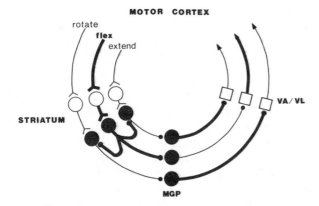

Figure 2 Schematic drawing of how activity in one corticostriatopallidothalamocortical loop (Figure 1, Circuit B) might sustain one movement and suppress other movements. A corticostriatal projection cell excites a striatal acetylcholine interneuron, which excites a GABAergic striatal output cell. This cell has many local axon collaterals that inhibit other striatal output cells. In the MGP, a tonically active pallidal output cell is inhibited. This pallidal cell normally inhibits a thalamocortical projection cell in VA/VL. Meanwhile, other pallidal output cells continue to suppress VA/VL cells because they do not get inhibitory input from the striatum. Thus, activity of a striatal output cell causes a VA/VL cell to be released from inhibition. This VA/VL cell, in turn, excites the cortex. Abbreviations used: GABA, gamma-aminobutyric acid; MGP, medial globus pallidus; SN_r, substantia nigra pars reticulata; SN_c, substantia nigra pars compacta; VA/VL, ventral anterior and ventral lateral nuclei of the thalamus.

also explain MacLean's (1978) finding that the basal ganglia select and maintain species-specific behaviors.

The Pathophysiology of Dyskinesias

How does disruption of the anatomy and physiology reviewed above relate to the movement disorders? One extreme of basal ganglia dysfunction is manifest in man by the dyskinesias, which include chorea, tardive dyskinesia, and athetosis. According to the model, these could result from varying degrees of inability to maintain circuit B for a particular behavioral set. Kinnier Wilson (1929) provided an insightful description of chorea:

> The spontaneous movements of chorea offer an intimate resemblance to those executed at the bidding of volition. They appear to be as complex (on occasion), as co-ordinated, and as purposeful as those of a frankly voluntary kind. Of their occasional or frequent elaborateness no doubt can be entertained; . . . and each fresh movement appears to be directed to an end—which is never attained. Choreic motility may be described with sufficient accuracy as subjectively purposeful but objectively purposeless, in appearance only are the movements purposive, but to describe them as purposeless without the qualification just given would be rather misleading.
>
> It is difficult to resist the conclusion that choreic motor disorder manifests itself through the pyramidal (corticospinal) tracts, which must be in a condition of relative integrity

for hyperkinetic symptoms to appear . . . Hence the voluntary movements of choreic
patients have some of the characters of the involuntary movements, . . . the latter take
the same route and usurp it, at the expense of the former. At the same time, choreic
motor derangement need not actually originate on the motor (effector) side of cortical
mechanisms.

Wilson went on to suggest that abnormal activity of the premotor cortex
(Area 6) must be responsible for the production of chorea since the primary
motor cortex is organized to control individual muscle contractions and not
movements that involve several muscles. This view of the organization of
the motor cortex has much more data to support it now than it did in
Wilson's day (Phillips & Porter 1977). The description of chorea as fragments of normal cortically generated movements that are superimposed on
other activities suggests that the striatum in these cases is unable to properly
inhibit unwanted movements during ongoing behaviors. Tardive dyskinesia,
the movement disorder induced after long-term neuroleptic therapy, resembles chorea and choreoathetosis clinically except that the movements are
generally most prominent in the oral-buccal-lingual regions and there is less
motor impersistence. Athetosis is characterized by slower, more writhing
movements and has been described by Denny-Brown (1962) as a release of
primitive cortical reflex movements. Chorea and athetosis could result ultimately from a combination of inadequate activity of GABAergic striatal
output neurons and inadequate collateral inhibition of output neurons subserving other behaviors. Dysfunction of this sort could arise from overactivity of dopaminergic nigrostriatal neurons or underactivity of the striatal
cholinergic interneurons and/or GABAergic output neurons.

Relative overactivity of the dopamine cells appears to decrease the activity of striatal cholinergic and GABAergic cells, but the mechanism of
dopamine's effects are controversial. In the superior cervical ganglion,
dopamine causes a slow inhibitory postsynaptic potential (Greengard 1976).
Extracellular recording studies (Bloom et al 1965, Connor 1970, McLennan
& York 1967) and pharmacological studies (McGeer et al 1974b, Sethy &
Van Woert 1974, Agid et al 1975, Mao et al 1977) also indicate that
dopamine decreases the activity of striatal cells. However, intracellular
recordings suggest that dopamine causes a depolarization of striatal output
neurons (Hull et al 1970, Kocsis & Kitai 1977). It has been suggested that
dopamine excites GABAergic striatal cells, which in turn inhibit cholinergic interneurons and GABAergic output cells (Park et al 1980). An alternative explanation is that depolarization results in blocked action potentials
and thereby functional inhibition of caudate output. Regardless of the
mechanism, drugs that increase dopaminergic activity result in choreiform
dyskinesias (Barbeau et al 1971). Furthermore, tardive dyskinesia, which
appears to result from drug-induced striatal dopamine receptor supersen-

sitivity after long-term high dose neuroleptic therapy, also manifests chorei-form movements (Baldessarini & Tarsy 1980).

In Huntington's disease there is a primary deficit in striatal output. Huntington's disease is the classic hereditary neurodegenerative disorder that results in choreoathetosis as well as dementia (Bruyn 1968). The most prominent feature of the pathology of Huntington's disease is a marked loss of striatal cells (Bruyn 1968). These cells are the cholinergic interneurons and peptidergic and GABAergic output neurons (Arregui et al 1979, Bird & Iversen 1974, Kanazawa et al 1979, McGeer et al 1973, Perry et al 1973). These cells are very important in maintaining a posture or motor behavior. Without them, the thalamus is tonically inhibited; this in turn results in the loss of ability to maintain a motor behavior. Hence, motor impersistence and cortically generated adventitious movements such as chorea develop. Thus, chorea may be a failure to select and maintain the appropriate move-ment and, at the same time, a failure to suppress inappropriate movements. The exact nature of the choreiform movement disorder depends on the extent of the striatal lesion. Mild lesions allow normal cortically generated movements to interrupt briefly ongoing activity (chorea). More severe le-sions, such as those seen in terminal Huntington's disease or in the status marmoratus of cerebral palsy, result in athetosis (Denny-Brown 1962).

The proposed connections of circuit B may explain a puzzling clinical observation about the pharmacology of Huntington's disease. The GABA-ergic striatopallidal cells (plus many others) degenerate in Huntington's disease (Barbeau 1979) and supersensitive GABA receptors develop in the globus pallidus (Enna et al 1976). In this respect Huntington's disease is analogous to Parkinson's disease, in which the dopaminergic nigrostriatal cells degenerate and supersensitive dopamine receptors develop in the stria-tum (Lee et al 1978). Dopamine agonists relieve the symptoms of Parkin-son's disease, but GABA mimetic agents do little to relieve the symptoms of Huntington's disease (Shoulson 1979). If, as appears to be the case, there are two sequential GABAergic inhibitory neurons, the striatopallidal and the pallidothalamic, then the actions of GABA mimetics on the globus pallidus would be counteracted by their effects on the thalamus. Perhaps lower doses of GABA mimetic drugs would be of benefit in Huntington's disease, by interacting preferentially with the supersensitive receptors in the pallidum.

Two other areas of brain when damaged result in dyskinetic movements. First, damage of the subthalamic nucleus in humans results in the hyperki-netic syndrome, hemiballismus. Surgical subthalamotomy in humans (Hassler et al 1965) and animals (Carpenter et al 1950) causes a hyperki-netic syndrome. The subthalamic nucleus receives inputs from the cortex (Hartmann-Von Monakow 1978, Kitai & Deniau 1981) and the lateral globus pallidus (LGP) (Nauta 1979) (Figure 3). The subthalamic nucleus projects to LGP and also has a major projection to MGP and SN$_r$ (Ricardo 1980). If subthalamic output is inhibitory as some recent evidence suggests (Perkins & Stone 1980), the subthalamic nucleus will have the same effect on the thalamus as does the striatum. Furthermore, the striatum inhibits the LGP which is in turn inhibitory onto the subthalamic nucleus (Rouzaire-Dubois et al 1980). Thus, both direct cortical influences on the subthalamic nucleus and indirect cortical influences on the subthalamic nucleus via the striatum would have the ultimate effect of disinhibiting the motor thalamus. If the above circuitry proves correct, subthalamic lesions would be expected to result in chorea and motor impersistence.

Degeneration of the centromedian nucleus of the thalamus also causes a hyperkinetic syndrome in man (Adams & Malamud 1971). The cells of the centromedian nucleus receive an input from the MGP (van der Kooy &

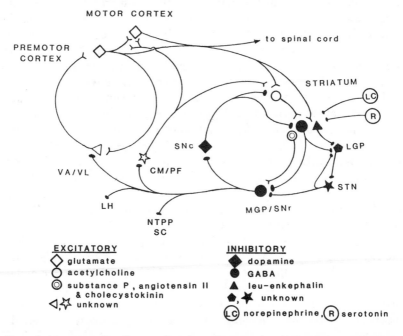

Figure 3 A more complete diagram of basal ganglia connections. Abbreviations used: GABA, gamma-aminobutyric acid; MGP, medial globus pallidus; SN$_r$, substantia nigra pars reticulata; SN$_c$, substantia nigra pars compacta; VA/VL, ventral anterior and ventral lateral nuclei of the thalamus; LGP, lateral globus pallidus; STN, subthalamic nucleus; LH, lateral habenula; CM, centromedian nucleus of thalamus; PF, parafascicular nucleus; NTPP, nucleus tegmenti pedunculopontinus; SC, superior colliculus; LC, locus coeruleus; R, raphe; Sub P, substance P; CCK, cholecystokinin; Ang II, angiotensin II; leu-ENK, leucine-enkephalin; ACh, acetylcholine.

Carter 1981) and are excitatory on the striatum (Kitai et al 1976a). A decrease in their function should give the same clinical syndrome as that of striatal dysfunction. Hassler (1978) has suggested that a major function of the basal ganglia is to transmit sensory information from the centromedian nucleus to the cortex. We agree with Marsden's (1980) conclusion that the centromedian nucleus is to be thought of as part of the ascending reticular activating system and that its main function is arousal of cortex and striatum.

The Pathophysiology of Bradykinesia, Rigidity, and Tremor

Circuit B is a positive feedback system. Activity in this circuit will continue to reinforce itself unless modulated by other pathways. The clinical manifestations of such unmodulated activity would be a tendency to maintain ongoing activity and a difficulty in establishing new behaviors. Thus, if one is walking, one will tend to keep walking (until an obstacle intervenes). If one is seated then one will tend to remain seated. Clinically, this describes the akinesia of Parkinson's disease.

This interpretation of the akinesia of parkinsonism has support pathologically and physiologically. The nigrostriatal dopamine projection degenerates in Parkinson's disease. Under normal circumstances, this dopamine system modifies the actions of striatal neurons and, thus, may modify the activity of circuit B. The nigrostriatal cells synapse on both the cholinergic interneurons and on the GABAergic projection neurons of the striatum (Kitai et al 1976a, McGeer et al 1975). These are the same type of cells as those contacted by the glutamatergic corticostriatal projection (Kitai et al 1976b, McGeer et al 1978, Somogyi et al 1981a). We suggest that the function of the nigrostriatal projection is to regulate the ease of transmission through circuit B so that ongoing activities can be maintained, but also so that new activities may be initiated relatively easily by overriding ongoing ones. Dopamine does not appear to function in an executive fashion as an on-off switch for specific motor behaviors. The nigral dopamine cells have extensive axon collaterals (Anden et al 1966), fire slowly, and have few phasic changes in firing rate (Guyenet & Aghajanian 1978). Thus they appear to have a tonic influence on striatal activity. Perhaps dopamine aids the initiation of new movements by modulating the activity (sensitivity or resistance) of transmission through circuit B. This can be accomplished by inhibiting striatal output neurons.

Our model would predict that an overactivity of thalamic neurons or an underactivity of MGP/SN_r neurons will produce a parkinsonian syndrome that is unresponsive to dopamine agonists. Either dysfunction will result in overactivity of circuit A and a loss of the ability to select specific behavioral sets. Thus, antagonistic sets of cortical neurons could be activated simultaneously and attempt to perform two movements at once. The limb would then become fixed at some position at which the force exerted by antagonist muscle equals that exerted by the agonist muscles, producing a rigid, akinetic and dystonic posture. Thalamotomy would relieve this problem regardless of whether the pathology is in the thalamus or the medial globus pallidus. Further support for this concept comes from the observation that patients with the rare syndrome of degeneration of medial pallidum have parkinsonian and dystonic symptoms (Jellinger 1968). [This is not the same disease as dystonia musculorum deformans, for which no pathology is known (Zeman 1970).] Other diseases that involve multiple neuronal systems but which have major pathology in the MGP and SN_r also result in akinesia, rigidity, and dystonia such as progressive supranuclear palsy, Hallervorden-Spatz, striatonigral degeneration, and carbon monoxide poisoning. The widespread pathology in these latter disorders makes it impossible to localize with certainty the areas responsible for the parkinsonian symptoms, but the common feature of severe MGP/SN_r degeneration makes these regions likely candidates.

Animal studies of pallidal lesions have not shown dramatic effects of bilateral pallidotomy on motor behavior. The most consistent finding after moderate sized pallidal lesions is a brief period of hypokinesia followed by complete improvement (Carpenter & Whittier 1952, Hore & Vilis 1980). Large bilateral lesions result in flexion dystonia and akinesia (Denny-Brown 1962). Little investigation has been done on animals with bilateral lesions of both the MGP and SN_r. Among the few studies reporting both lesions was one in which carbon disulfide was used to cause marked pathology in MGP/SN_r with sparing of nigral pigmented cells (Richter 1945). The lesions produced severely akinetic, rigid animals. Most of the animal studies have used indiscriminate destructive lesions that destroy fibers of passage as well as cell bodies. For this reason, some investigators have suggested that the transient akinesia resulting from MGP lesions is secondary to interruption of nigrostriatal dopaminergic fibers that pass through this region in man and primate (DeLong & Georgopoulos 1982). However, this hypothesis is unlikely to be valid since the symptoms fail to respond to dopamine agonists (Péchadre et al 1976).

Once the activity of circuit B reaches a steady state (i.e. a posture or action is maintained), the activity of each striatonigral and each nigrostriatal cell will also be in a steady state. Thus, by itself, the striatonigral feedback loop cannot be the source for the initiation of new movements. There are, however, other influences on the substantia nigra. Another projection onto the nigrostriatal cells comes from the nucleus accumbens (Nauta et al 1978, Somogyi et al 1981b). It has been suggested that there is also a pathway from the extreme prefrontal cortex directly onto nigral dopamine cells (Kimura & Maeda 1978).

Both the nucleus accumbens and the prefrontal cortex receive dopaminergic input from the dopamine cells in the ventral tegmental area (Carter & Fibiger 1977, Fallon & Moore 1978). They send projections back to the ventral tegmental area as well as to the substantia nigra (Kimura & Maeda 1978, Nauta et al 1978) where they make synaptic contact on nigrostriatal cells (Somogyi et al 1981b). The nucleus accumbens and prefrontal cortex are generally considered to be part of the limbic system, yet their projection to the substantia nigra suggests that they may have a direct influence on motor behavior. Evidence to support this suggestion comes from animal experiments. Lesioning the dopaminergic input to the nucleus accumbens and prefrontal cortex produces an animal model of parkinsonian akinesia. (Gaffori et al 1980). There is evidence for a descending GABAergic output from the accumbens to substantia nigra (Walaas & Fonnum 1980). If these GABAergic neurons are those which contact nigrostriatal cells directly, their activity will cause a decrease in nigrostriatal activity and akinesia. Dopaminergic output to the accumbens would thus prevent akinesia. If the prefrontonigral pathway is excitatory, activity in this pathway would increase the activity of the nigrostriatal dopamine system. Loss of this prefrontonigral excitatory pathway could explain the akinesia seen in patients with frontal lobe disease.

The model as presented would thus account for parkinsonian akinesia and rigidity, but the model would predict that once a movement begins it ought to be performed at a normal or even an increased rate. This is not the case. Draper & Johns (1964) have shown that parkinsonian movements are slower than normal because they are intermittent. The rate of acceleration is normal in a parkinsonian when acceleration occurs but the acceleration is interrupted by pauses at a rate of seven to eight per second (Draper & Johns 1964). This is the same as the rate of parkinsonian tremor.

The pathophysiology of parkinsonian tremor is unknown. One explanation may be that tremor is secondary to bursting of striatal output cells. In fact, bursting of both striatal and pallidal neurons has been observed after nigral lesions and treatment with dopamine antagonists (Hull et al 1974, Filion 1979). What might the cause of this bursting be? Since anticholinergic agents are effective against tremor, one might expect that the striatal cholinergic neurons play a role in the etiology of tremor. In sympathetic ganglia and mouse spinal cord neurons in culture, muscarinic cholinergic agonists cause a decrease in a voltage-dependent potassium conductance, which in turn results in a voltage-dependent excitation of the cholinoceptive cell (Brown & Adams 1980, Nowak & Macdonald 1981). Reduction of potassium current has been shown to promote bursting (Schwindt & Crill 1980, Heyer et al 1981). If similar effects of muscarinic cholinergic agonists occur in the striatum, loss of inhibitory dopaminergic striatal afferents in parkinsonism might cause membrane depolarization of striatal neurons, augment muscarinic cholinergic responses, and induce bursting. In fact, nigral lesions do induce bursting in striatal (Hull et al 1970) and pallidal neurons, and in monkeys the bursting correlates with tremor (Filion 1979). This may explain why tremor is very responsive to muscarinic cholinergic antagonists. Rigidity and akinesia would be predicted to be less responsive to muscarinic cholinergic antagonists because GABAergic output neurons would only decrease their overall activity to a limited extent after cholinergic blockade. Thus GABAergic output neurons would still be disinhibited secondarily due to loss of dopaminergic nigrostriatal input.

The mechanism of tremor proposed above would be compatible with the fact that pallidotomy relieves the tremor of Parkinson's disease (Carman 1968). Loss of pallidal neurons would eliminate the effects of striatal bursting on basal ganglia output. Similarly, tremor is unusual in diseases such as Hallervorden-Spatz and other pallidal degenerations presumably because pallidothalamic pathways are necessary for the manifestation of parkinsonian tremor. Tremor and rigidity can also be alleviated by surgery further along in circuit B at the level of the thalamus. The operation most often performed today is ventral lateral and ventral anterior thalamotomy. This operation has been performed with success for both parkinsonism (Carman 1968, Hassler et al 1965, Pagni et al 1965) and dystonia (Cooper 1969).

The dopamine-acetylcholine interaction whose physiology was described above also explains the clinical observation that the pharmacologic actions of the nigrostriatal dopaminergic pathway seem to be antagonistic to those of intrinsic striatal cholinergic neurons (for review see Baldessarini & Tarsy 1980, Chase 1976, Klawans 1970, Zeisel et al 1980). Thus, the movements of Huntington's disease are treated most successfully by dopamine receptor blocking agents such as the butyrophenones and phenothiazines and these same drugs cause parkinsonism as a side effect. Furthermore, cholingeric agents (physostigmine, choline, lecithin) improve chorea and worsen parkinsonism while anticholinergic agents improve parkinsonism and worsen chorea. L-Dopa produces choreiform movements as a side effect. Tardive dyskinesia (orofacial dyskinesia), a side effect of long-term use of antidopaminergic agents, is thought to result from increased dopamine receptor sensitivity and has been treated with dopaminergic antagonists and cholinergic agonists.

The McGeers and their collaborators (1975) have demonstrated that dopamine neurons make synaptic contacts on striatal cholinergic neurons. They have demonstrated also that the striatal cholinergic neurons function entirely as interneurons within the striatum (McGeer et al 1975, McGeer & McGeer 1976). Kitai et al (1976a) have demonstrated that dopamine neurons make synaptic contacts on striatal output neurons. Thus, presumably the dopamine neurons not only synapse on the cholinergic neurons but also on the striatal output neurons.

Other Aspects of The Model

This model is intentionally oversimplified. Our aim is to suggest a model that can account for much of the current clinical and experimental data. We have attempted to do this without ignoring data that are relevant to the common syndromes of chorea and parkinsonism. Our model does not deal with many neurotransmitters and pathways known to be present in the basal ganglia. We have included these pathways in Figure 3.

Norepinephrine (Moore & Bloom 1979, Ungerstedt 1971) and serotonin (Moore et al 1978) are both present in afferents to the striatum from the brainstem. Angiotensin II (Arregui et al 1979), cholecystokinin (Emson et al 1980), leu-enkephalin (Cuello & Paxinos 1978, Sar et al 1978), and substance P (Brownstein et al 1977, Jessell et al 1978) are all present in striatal efferents. All four peptides are decreased in Huntington's disease (Arregui et al 1979, Emson et al 1980, Kanazawa et al 1979). These neurotransmitters have not been excluded from the model arbitrarily. Norepinephrine and serotonin have not been discussed because their presence in the striatum is part of a diffuse projection to much of the forebrain (Moore & Bloom 1979, Moore et al 1978, Ungerstedt 1971) and because pharmacological manipulation of these neurotransmitters has not had any consistent effects on the movement disorders. The nature of the angiotensin II and cholecystokinin cells is unknown, as are angiotensin's and cholecystokinin's effects on the movement disorders. Enkephalinergic striatal efferents project largely to the lateral segment of the globus pallidus (LGP) (Cuello & Paxinos 1978, Sar et al 1978). Substance P neurons project to the substantia nigra and to the MGP (Brownstein et al 1977, Jessel et al 1978). These neurons, however, are located almost entirely rostral to the anterior commissure in the rat (Jessell et al 1978). This area of striatum receives input from prefrontal but not from motor areas of the cortex (Kemp & Powell 1970).

We have not dealt with a number of other basal ganglia projections because little is known of their clinical importance in man. These include the entopeduncular projection to the nucleus tegmenti pedunculopontinus (Garcia-Rill et al 1981, Jackson & Crossman 1981) and the habenula (Nauta 1974) and nigral projections to the superior colliculus (Vincent et al 1978), nucleus tegmenti pedunculopontinus (Beckstead et al 1979), and spinal cord (Commissiong et al 1978). The descending SN_r pathways to the tectum and midbrain tegmentum appear to be responsible for the rotatory behavior seen in models of parkinsonism in the rat (Morelli et al 1981) and probably in the cat (Garcia-Rill et al 1981). In fact, the rotatory behavior exhibited by rats after lesioning of nigral dopamine neurons continues even in the absence of the telencephalon and thalamus (Papadopolous & Huston 1981). It is, however, abolished by lesioning the colliculi (Imperato et al 1981). The relevance of this pathway to human disease is unclear, since in primates the motor cortex (Horsley 1909) and thalamus (Hassler et al 1965) are vital to the expression of abnormal movements.

Our model does not provide an explanation for dystonic reactions to phenothiazines, or the L-dopa induced on-off phenomenon of Parkinson's Disease.

Conclusions

We have attempted to provide a model of basal ganglia function to explain much of the known pharmacology and pathology of Huntington's and Parkinson's diseases. In this model, the role of the striatum is to select which behaviors will be carried out and to suppress unwanted ones. This concept is very similar to those recently proposed by Cools (1980), Divac (1977), and Marsden (1980). We differ from these authors by proposing a mechanism through which the striatum may be exerting its efforts. The model proposes a positive feedback circuit that travels from the motor cortex through the striatum, medial globus pallidus, and thalamus to return to the cortex. Failure to maintain the integrity of these pathways leads to chorea. Inability to suppress the activity of this loop leads to parkinsonism. Cholinergic neurons promote the positive feedback loop. Dopaminergic neurons inhibit the circuit. Thus, cholinergic and antidopaminergic agents relieve chorea and produce parkinsonism, whereas anticholinergic and dopaminergic agents relieve parkinsonism and produce chorea. Several predictions follow from this model.

1. The cells of the subthalamic nucleus are inhibitory on the cells of MGP/SN$_r$.
2. Direct positive feedback circuits exist for neurons of the cortex, particularly the premotor area, which go through the striatum, pallidum, and thalamus on their way back to the same cortical neurons.
3. The tremor and intermittency of movement in Parkinson's disease is due to bursting in striatal neurons induced by disinhibited cholinergic interneurons.

ACKNOWLEDGMENTS

We acknowledge the helpful comments of Sid Gilman, Michael Johnston, Mark Bromberg, and Robert Macdonald. We thank Kathy Lundquist for help in preparation of the manuscript. This work is derived in part from the concepts of Eugene Roberts and Donald MacRae. This work was supported by United States Public Health Service grants NS00464-01, NS00420-01, NS15140-02, the United Cerebral Palsy Research Foundation, the Committee to Combat Huntington's Disease, and the University of Michigan Memorial Phoenix Project.

Literature Cited

Adams, J. E., Malamud, N. 1971. Severe chorea with degeneration of nucleus centrum medianum. *Arch. Neurol.* 24: 101–5

Agid, Y., Guyenet, P., Glowinski, J., Beaujouan, J. C., Javoy, F. 1975. Inhibitory influence of the nigrostriatal dopamine system on the striatal cholinergic neurons in the rat. *Brain Res.* 86:488–92

Anden, N. E., Fuxe, K., Hamberger, B., Hökfelt, T. 1966. A quantitative study on the nigro-neostriatal dopamine neuron system in the rat. *Acta Physiol. Scand.* 67:306–12

Arregui, A., Iversen, L. L., Spokes, E. G. S., Emson, P. C. 1979. Alterations in postmortem brain angiotensin-converting enzyme activity and some neuropeptides in Huntington's disease. *Adv. Neurol.* 23:517–26

Baldessarini, R. J., Tarsy, D. 1980. Dopamine and the pathophysiology of dyskinesias induced by antipsychotic drugs. *Ann. Rev. Neurosci.* 3:23–41

Barbeau, A. 1979. Update on the biochemistry of Huntington's chorea. *Adv. Neurol.* 23:449–61

Barbeau, A., Mars, H., Gillo-Joffroy, L. 1971. Adverse clinical side effects of levodopa therapy. In *Recent Advances in Parkinson's Disease*, ed. F. H. MacDowell, C. H. Markham, pp. 203–8. Philadelphia: Davis

Baring, M. D., Walters, J. R., Eng, N. 1980. Action of systemic apomorphine on dopamine cell firing after neostriatal kainic acid lesion. *Brain Res.* 181: 214–18

Beckstead, R. M., Domesick, V. B., Nauta, W. J. H. 1979. Efferent connections of the substantia nigra and ventral tegmental area in the rat. *Brain Res.* 175:191–217

Bird, E. D., Iversen, L. L. 1974. Huntington's chorea. Postmortem measurement of glutamic acid decarboxylase, choline acetyltransferase and dopamine in basal ganglia. *Brain* 97:457–72

Bloom, F. E., Costa, E., Salmoiraghi, G. C. 1965. Anesthesia and the responsiveness of individual neurons of the caudate nuclei to acetylcholine, norephinephrine and dopamine administered by microelectrodephoresis. *J. Pharmacol. Exp. Ther.* 150:244–52

Bromberg, M. B., Penney, J. B., Stephenson, B. S., Young, A. B. 1981. Evidence for glutamate as the neurotransmitter of corticothalamic and corticorubral pathways. *Brain Res.* 215:369–74

Brown, D. A., Adams, P. R. 1980. Muscarinic suppression of a novel voltage-sensitive K^+ current in a vertebrate neurone. *Nature* 283:673–76

Brownstein, M. J., Mroz, E. A., Tappaz, M. L., Leeman, S. E. 1977. On the origin of substance P and glutamic acid decarboxylase (GAD) in the substantia nigra. *Brain Res.* 135:315–23

Bruyn, G. W. 1968. Huntington's chorea: Historical, clinical and laboratory synopsis. In *Handbook of Clinical Neurology*, ed. P. Vinken, G. W. Bruyn, 6:298–378. Amsterdam: North Holland

Buchwald, N. A., Hull, C. D., Levine, M. S. 1979. Basal ganglia neuronal activity and behavioral set. *Appl. Neurophys.* 42:109–12

Bunney, B. S., Aghajanian, G. K. 1976. The precise localization of nigral afferents in the rat as determined by a retrograde tracing technique. *Brain Res.* 117: 423–35

Buser, P., Pouderoux, G., Mereaux, J. 1974. Single-unit recording in the caudate nucleus during sessions with elaborate movements in the awake monkey. *Brain Res.* 71:337–44

Carman, J. B. 1968. Anatomic basis of surgical treatment of Parkinson's disease. *N. Engl. J. Med.* 279:919–29

Carpenter, M. B., Nakano, K., Kim, R. 1976. Nigrothalamic projections in the monkey demonstrated by autoradiographic techniques. *J. Comp. Neurol.* 165:401

Carpenter, M. B., Whittier, J. R. 1952. Study of methods for producing experimental lesions of the central nervous system with special reference to the stereotaxic technique. *J. Comp. Neurol.* 97:73–132

Carpenter, M. B., Whittier, J. R., Mettler, F. A. 1950. Analysis of choreoid hyperkinesia in the rhesus monkey. Surgical and pharmacological analysis of hyperkinesia resulting from lesions in the subthalamic nucleus of Luys. *J. Comp. Neurol.* 92:293–332

Carter, D. A., Fibiger, H. C. 1977. Ascending projections of presumed dopamine containing neurons in the ventral tegmentum of the rat as demonstrated by horseradish peroxidase. *Neuroscience* 2:569–76

Carter, D. A., Fibiger, H. C. 1978. The projections of the entopeduncular nucleus and globus pallidus in rat as demonstrated by autoradiography and horseradish peroxidase histochemistry. *J. Comp. Neurol.* 177:113–24

Chase, T. 1976. Rational approaches to the pharmacotherapy of chorea. In *The*

Basal Ganglia, ed. M. D. Yahr, pp. 337–50. New York: Raven

Cheramy, A., Nieoullon, A., Glowinski, J. 1978. Inhibition of dopamine release in the cat caudate nucleus by nigral applications of glycine. *Eur. J. Pharmacol.* 47:141–47

Commissiong, J. W., Galli, C. L., Neff, N. H. 1978. Differentiation of dopaminergic and noradrenergic neurons in rat spinal cord. *J. Neurochem.* 30:1095–99

Connor, J. D. 1970. Caudate nucleus neurones: Correlation of the effects of substantia nigra stimulation with iontophoretic dopamine. *J. Physiol.* 208:691–703

Cools, A. R. 1980. Role of the neostriatal dopaminergic activity in sequencing and selecting behavioural strategies: Facilitation of processes involved in selecting the best strategy in a stressful situation. *Behav. Brain Res.* 1:361–78

Cooper, I. S. 1969. *Involuntary Movement Disorders.* New York: Hoeber Med. Div./Harper & Row

Cuello, A. C., Paxinos, G. 1978. Evidence for a long leu-enkephalin striatopallidal pathway in rat brain. *Nature* 271: 178–80

DeLong, M. R., Georgopoulos, A. P. 1979. Motor functions of the basal ganglia as revealed by studies of single cell activity in the behaving primate. *Adv. Neurol.* 24:131–40

DeLong, M. R., Georgopoulos, A. P. 1982. Motor functions of the basal ganglia. In *Handbook of Physiology. The Nervous System II.* Washington DC: Am. Physiol. Soc. In press

Denny-Brown, D. 1962. *The Basal Ganglia and Their Relation to Disorders of Movement.* London: Oxford

Di Chiara, G., Morelli, M., Porceddu, M. L., Mulas, M., Del Fiacco, M. 1980. Effect of discrete kainic acid-induced lesions of corpus caudatus and globus pallidus on glutamic acid decarboxylase of rat substantia nigra. *Brain Res.* 189:193–208

Di Chiara, G., Porceddu, M. L., Morelli, M., Mulas, M., Gessa, G. L. 1979. Evidence for a GABA-ergic projection from the substantia nigra to the ventromedial thalamus and to the superior colliculus of the rat. *Brain Res.* 176:273–84

Divac, I. 1977. Does the neostriatum operate as a functional entity? In *Psychobiology of the Striatum,* ed. A. R. Cools, A. H. M. Lohman, J. H. L. Van den Bercken, pp. 21–30. Amsterdam: Elsevier

Divac, I., Markowitsch, H. J., Pritzel, M. 1978. Behavioral and anatomical consequences of small intrastriatal injections

of kainic acid in the rat. *Brain Res.* 151:523–32

Donoghue, J. P., Kitai, S. T. 1981. A collateral pathway to the neostriatum from corticofugal neurons of the rat sensory motor cortex: An intracellular HRP study. *J. Comp. Neurol.* 201:1–14

Draper, L. T., Johns, R. J. 1964. The disordered movement in parkinsonism and the effect of drug treatment. *Bull. Johns Hopkins Hosp.* 115:465–80

Emson, P. C., Rehfeld, J. F., Langevin, H., Rossor, M. 1980. Reduction in cholecystokinin-like immunoreactivity in the basal ganglia in Huntington's disease. *Brain Res.* 198:497–500

Endo, K., Araki, T., Yagi, N. 1973. The distribution and pattern of axon branching of pyramidal tract cells. *Brain Res.* 57:484–91

Enna, S. J., Bennett, J. P. Jr., Bylund, D. B., Snyder, S. H., Bird, E. D., Iversen, L. L. 1976. Alterations of brain neurotransmitter receptor binding in Huntington's chorea. *Brain Res.* 116:531–37

Evarts, E. V. 1967. Representation of movements and muscles by pyramidal tract neurons of the precentral motor cortex. In *Neurophysiological Basis of Normal and Abnormal Motor Activity,* ed. M. D. Yahr, D. P. Purpura, pp. 215–53. New York: Raven

Fallon, J. H., Moore, R. Y. 1978. Catecholamine innervation of the basal forebrain. IV. Topography of the dopamine projection to the basal forebrain and neostriatum. *J. Comp. Neurol.* 180:545–80

Faull, R. L. M., Mehler, W. R. 1976. Subdivision of the ventral tier nuclei in the rat thalamus based on their afferent connections. *Anat. Rec.* 184:400

Filion, M. 1979. Effects of interruption of the nigrostriatal pathway and of dopaminergic agents on the spontaneous activity of globus pallidus neurons in the awake monkey. *Brain Res.* 178:425–41

Fonnum, F., Gottesfeld, Z., Grofova, I. 1978. Distribution of glutamate decarboxylase, choline acetyltransferase and aromatic amino acid decarboxylase in the basal ganglia of normal and operated rats. Evidence for striatopallidal, striatoentopeduncular and striatonigral GABAergic fibres. *Brain Res.* 143: 125–38

Fonnum, F., Störm-Mathisen, J., Divac, I. 1981. Biochemical evidence for glutamate as neurotransmitter in corticostriatal and corticothalamic fibres in rat brain. *Neuroscience* 6:863–73

Gaffori, O., LeMoal, M., Stinus, L. 1980. Locomotor hyperactivity and hypoexploration after lesion of the dopaminergic-A10 area in the ventral mesencephalic tegmentum (VMT) of rats. *Behav. Brain Res.* 1:313–29

Garcia-Rill, E., Skinner, R. D., Gilmore, S. A. 1981. Pallidal projections to the mesencephalic locomotor region (MLR) in the cat. *Am. J. Anat.* 161:311–21

Graybiel, A. M., Ragsdale, C. W. Jr. 1979. Fiber connections of the basal ganglia. *Prog. Brain Res.* 51:239–83

Greengard, P. 1976. Possible role for cyclic nucleotides and phosphorylated membrane proteins in postsynaptic actions of neurotransmitters. *Nature* 260:101–8

Grofova, I. 1975. The identification of striatal and pallidal neurons projecting to substantia nigra. An experimental study by means of retrograde axonal transport of horseradish peroxidase. *Brain Res.* 91: 286–91

Guyenet, P. G., Aghajanian, G. T. 1978. Antidromic identification of dopaminergic and other output neurons of rat substantia nigra. *Brain Res.* 150:69–84

Hartmann-Von Monakow, K., Akert, K., Künzle, H. 1978. Projections of the precentral motor cortex and other cortical areas of the frontal lobe to the subthalamic nucleus in the monkey. *Exp. Brain Res.* 33:395–403

Hassler, R. 1978. Striatal control of locomotion, intentional actions and of integrating and perceptive activity. *J. Neurol. Sci.* 36:187–224

Hassler, R., Mundinger, F., Riechert, T. 1965. Correlations between clinical and autopic findings in stereotaxic operations of Parkinsonism. *Confin. Neurol.* 26:282–90

Herkenham, M. 1979. The afferent and efferent connections of the ventromedial thalamic nucleus in the rat. *J. Comp. Neurol.* 153:487–518

Herkenham, M. 1980. Laminar organization of thalamic projections to the rat cortex. *Science* 207:532–33

Hersch, S. M., White, E. L. 1981. Thalamo-cortical synapses with corticothalamic projection neurons in mouse SmI cortex: Electron microscopic demonstration of a monosynaptic feedback loop. *Neurosci. Lett.* 24:207–10

Heyer, E. J., Nowak, L. M., Macdonald, R. L. 1981. Bicuculline: A convulsant with synaptic and nonsynaptic actions. *Neurology* 31:1381–90

Hore, J., Vilis, T. 1980. Arm movement performance during reversible basal ganglia lesions in the monkey. *Exp. Brain Res.* 39:217–28

Hornykiewicz, O. 1979. Pharmacology of Huntington's disease. *Adv. Neurol.* 23: 679–86

Horsley, V. 1909. The Linacre lecture on the function of the so-called motor area of the brain. *Br. Med. J.* 2:125–32

Hull, C. D., Bernardi, G. A., Buchwald, N. A. 1970. Intracellular responses of caudate neurons to brain stem stimulation. *Brain Res.* 22:163–79

Hull, C. D., Levine, M. S., Buchwald, N. A., Heller, A., Browning, R. A. 1974. The spontaneous firing pattern of forebrain neurons. I. The effects of dopamine and nondopamine depleting lesions on caudate unit firing patterns. *Brain Res.* 73:241–62

Imperato, A., Porceddu, M. L., Morelli, M., Faa, G., Di Chiara, G. 1981. Role of dorsal mesencephalic reticular formation and deep layers of superior colliculus as out-put stations for turning behaviour elicited from the substantia nigra pars reticulata. *Brain Res.* 216: 437–43

Iversen, S. D. 1977. Striatal function and stereotyped behavior. See Divac 1977, pp. 333–84

Jackson, A., Crossman, A. R. 1981. Basal ganglia and other afferent projections to the peribrachial region in the rat: A study using retrograde and anterograde transport of horseradish peroxidase. *Neuroscience* 6:1537–49

James, T. A., Starr, M. S. 1979. Is glycine an inhibitory synaptic transmitter in the substantia nigra? *Eur. J. Pharm.* 57:15–125

Jellinger, K. 1968. Degenerations and exogenous lesions of the pallidum and striatum. In *Handbook of Clinical Neurology,* ed. P. J. Vinken, G. W. Bruyn 6:632–93 Amsterdam: North Holland

Jessell, T. M., Emson, P. C., Paxinos, G., Cuello, A. C. 1978. Topographical projections of substance P and GABA pathways in the striatal pallido-nigral system: A biochemical and immunohistochemical study. *Brain Res.* 152:487–98

Johnson, T. N., Rosvold, H. E. 1971. Topographic projections on the globus pallidus and the substantia nigra of selectively placed lesions in the precommissural caudate nucleus and putamen in the monkey. *Exp. Neurol.* 33:584–96

Jones, E. G., Wise, S. P. 1977. Laminar distribution of efferent cells in the sensory motor cortex of monkeys. *J. Comp. Neurol.* 175:391–438

Jung, R., Hassler, R. 1960. The extrapyramidal motor system. In *Handbook of*

Physiology-Neurophysiology, ed. J. Field, H. W. Magoun, V. E. Hall, 2:863–927, Sect. 1. Washington DC: Am. Physiol. Soc.

Kanazawa, I., Bird, E. D., Gale, J. S., Iversen, L. L., Jessell, T. M., Muramoto, O., Spokes, E. G., Sutoo, D. 1979. Substance P: Decrease in substantia nigra and globus pallidus in Huntington's disease. *Adv. Neurol.* 23:495–503

Kemp, J. M., Powell, T. P. S. 1970. The cortico-striate projection in the monkey. *Brain* 93:525–46

Kemp, J. M., Powell, T. P. S. 1971. The connections of the striatum and globus pallidus: Synthesis and speculation. *Philos. Trans. R. Soc. London Ser. B* 262: 441–57

Kilpatrick, I. C., Starr, M. S., Fletcher, A., James, T. A., MacLeod, N. K. 1980. Evidence for a GABAergic nigrothalamic pathway in the rat. I. Behavioural and biochemical studies. *Exp. Brain Res.* 40:45–54

Kim, J. S., Hassler, R., Haug, P., Paik, K. S. 1977. Effect of frontal cortex ablation on striatal glutamic acid level in rat. *Brain Res.* 132:370–74

Kimura, H., Maeda, T. 1978. Reciprocal connection between dopamine containing prefrontal cortex and deep mesencephalic region. *Soc. Neurosci. Abstr.* 4:382

Kitai, S. T., Deniau, J. M. 1981. Cortical inputs to the subthalamus: Intracellular analysis. *Brain Res.* 214:411–15

Kitai, S. T., Kocsis, J. D., Preston, R. J., Sugimori, M. 1976a. Monosynaptic inputs to caudate neurons identified by intracellular injection of horseradish peroxidase. *Brain Res.* 109:601–6

Kitai, S. T., Kocsis, J. D., Wood, J. 1976b. Origin and characteristics of the cortico-caudate afferents: An anatomical and electrophysiological study. *Brain Res.* 118:137–41

Klawans, H. L. 1970. A pharmacologic analysis of Huntington's chorea. *Eur. Neurol.* 4:148–63

Kocsis, J. D., Kitai, S. T. 1977. Dual excitatory inputs to caudate spiny neurons from substantia nigra stimulation. *Brain Res.* 138:271–83

Künzle, H. 1975. Bilateral projections from precentral motor cortex to the putamen and other parts of the basal ganglia: An autoradiographic study in Macacca fascicularis. *Brain Res.* 88:195–209

Lee, T., Seeman, P., Rajput, A., Farley, I. J., Hornykiewicz, O. 1978. Receptor basis for dopaminergic supersensitivity in Parkinson's disease. *Nature* 273:59–61

Levine, M. S., Hull, C. D., Buchwald, N. A., Villablanca, J. R. 1978. Effects of caudate nuclei or frontal cortical ablations in kittens: Motor activity and visual discrimination performance in neonatal and juvenile kittens. *Exp. Neurol.* 62:555–69

Liles, S. L. 1979. Topographic organization of neurons related to arm movement in the putamen. *Adv. Neurol.* 23:155–62

MacLean, P. D. 1978. Effects of lesions of globus pallidus on species-typical display behavior of squirrel monkeys. *Brain Res.* 149:175–96

MacLeod, N. K., James, T. A., Kilpatrick, I. C., Starr, M. S. 1980. Evidence for a GABAergic nigrothalamic pathway in the rat. II. Electrophysiological studies. *Exp. Brain Res.* 40:55–61

Mao, C. C., Cheney, D. L., Marco, E., Revuelta, A., Costa, E. 1977. Turnover times of gamma-aminobutyric acid and acetylcholine in nucleus caudatus, nucleus accumbens, globus pallidus and substantia nigra: Effects of repeated administration of haloperidol. *Brain Res.* 132:375–79

Marsden, C. D. 1980. The enigma of the basal ganglia and movement. *Trends Neurosci.* 3:284–87

Maynard, D. M. 1972. Simpler networks. *Ann. NY Acad. Sci.* 193:59–72

McGeer, E. G., McGeer, P. L., Grewaal, D. S., Singh, V. K. 1975. Striatal cholinergic interneurons and their relation to dopaminergic nerve endings. *J. Pharmacol.* 6:143–52

McGeer, E. G., McGeer, P. L., Singh, K. 1978. Kainate induced degeneration of neostriatal neurons: Dependency upon corticostriatal tract. *Brain Res.* 139: 381–83

McGeer, P. L., Grewaal, D. S., McGeer, E. G. 1974b. Influence of noncholinergic drugs on rat striatal acetylcholine levels. *Brain Res.* 80:211–17

McGeer, P. L., McGeer, E. G. 1976. The GABA system and function of the basal ganglia: Huntington's disease. In *GABA in Nervous System Function,* ed. E. Roberts, T. N. Chase, D. B. Tower, pp. 487–95. New York: Raven

McGeer, P. L., McGeer, E. G., Fibiger, H. C. 1973. Choline acetylase and glutamic acid decarboxylase in Huntington's chorea. A preliminary study. *Neurology* 23:912–17

McGeer, P. L., McGeer, E. G., Scherer, U., Singh, K. 1977. A glutamatergic corticostriatal path? *Brain Res.* 128: 369–73

McLennan, H., York, D. H. 1967. The action of dopamine on neurons of the caudate nucleus. *J. Physiol. London* 189:393–402

Morelli, M., Imperato, A., Porceddu, M. L., Di Chiara, G. 1981. Role of dorsal mesencephalic reticular formation and deep layers of superior collicus in turning behaviour elicited from the striatum. *Brain Res.* 215:337–41

Moore, R. Y., Bloom, F. E. 1979. Central catecholamine neuron systems: Anatomy and physiology of the norepinephrine and epinephrine systems. *Ann. Rev. Neurosci.* 2:113–68

Moore, R. Y., Halaris, A. E., Jones, B. E. 1978. Serotonin neurons of the midbrain raphe: Ascending projections. *J. Comp. Neurol.* 180:417–38

Muakkassa, K. F., Strick, P. L. 1979. Frontal lobe input to primate motor cortex. *Soc. Neurosci. Abstr.* 5:379

Nagy, J. I., Carter, D. A., Fibiger, H. C. 1978. Anterior striatal projections to the globus pallidus, entopeduncular nucleus and substantia nigra in the rat: The GABA connection. *Brain Res.* 158:15–29

Nauta, H. J. W. 1974. Evidence of a pallidohabenular pathway in the cat. *J. Comp. Neurol.* 156:19–28

Nauta, H. J. W. 1979. Projections of the pallidal complex: An autoradiographic study in the cat. *Neuroscience* 4: 1853–71

Nauta, H. J. W., Smith, G. P., Faull, R. L. M., Domesick, V. B. 1978. Efferent connections and nigral afferents of the nucleus accumbens septi in the rat. *Neuroscience* 3:385–401

Newman, R., Winans, S. S. 1980. An experimental study of the ventral striatum of the golden hamster I. Neuronal connections of the nucleus accumbens. *J. Comp. Neurol.* 191:167–92

Nowak, L. M., Macdonald, R. L. 1981. DL-Muscarine decreases a potassium conductance to depolarize mammalian spinal cord neurons in cell culture. *Soc. Neurosci. Abstr.* 7:725

Oberg, R. G. E., Divac, I. 1979. "Cognitive" functions of the neostriatum. In *The Neostriatum*, ed. I. Divac, R. G. W. Oberg, pp. 291–314. Oxford: Pergamon

Pagni, C. A., Wildi, E., Ettore, G., Infuso, L., Marossero, F., Cabrini, G. P. 1965. Anatomic verification of lesions which abolished tremor and rigor in Parkinsonism. *Confin. Neurol.* 26:291–94

Pan, H. S., Young, A. B., Penney, J. B. Jr. 1981. Changes in [³H]-muscimol binding in substantia nigra, entopeduncular nucleus, and thalamus after striatal lesions as demonstrated by quantitative receptor autoradiography. *Soc. Neurosci. Abstr.* 7:500

Papadopoulos, G., Huston, J. P. 1981. Removal of the telencephalon spares turning induced by injection of GABA agonists and antagonists into the substantia nigra. *Behav. Brain Res.* 1:25–38

Park, M. R., Lighthall, J. W., Kitai, S. T. 1980. Recurrent inhibition in the rat neostriatum. *Brain Res.* 194:359–69

Péchadre, J. C., Larochelle, L., Poirier, L. J. 1976. Parkinsonian akinesia, rigidity and tremor in the monkey. *J. Neurol. Sci.* 28:147–57

Penney, J. B., Young, A. B. 1981. GABA as the pallidothalamic neurotransmitter: Implications for basal ganglia function. *Brain Res.* 207:195–99

Perkins, M. N., Stone, T. W. 1980. Subthalamic projections to the globus pallidus: An electrophysiological study in the rat. *Exp. Neurol.* 68:500–11

Perry, T. L., Hansen, S., Kloster, M. 1973. Huntington's chorea: Deficiency of gamma-aminobutyric acid in brain. *N. Engl. J. Med.* 288:337–42

Phillips, C. G., Porter, R. 1977. *Corticospinal Neurons: Their Role in Movement.* New York: Academic

Ricardo, J. A. 1980. Efferent connections of the subthalamic region in the rat. I. The subthalamic nucleus of Luys. *Brain Res.* 202:257–71

Richter, R. 1945. Degeneration of the basal ganglia in monkeys from chronic carbon disulfide poisoning. *J. Neuropathol. Exp. Neurol.* 4:324–53

Roberts, E. 1976. Some thoughts about GABA and the basal ganglia. See Chase 1976, pp. 191–203

Rouzaire-Dubois, B., Hammond, C., Hamon, B., Feger, J. 1980. Pharmacological blockade of the globus pallidus-induced inhibitory response of subthalamic cells in the rat. *Brain Res.* 200:321–29

Sar, M., Stumpf, W. E., Miller, R. J., Chang, K. J., Cuatrecasas, P. 1978. Immunohistochemical localization of enkephalin in rat brain and spinal cord. *J. Comp. Neurol.* 182:17–38

Schwindt, P., Crill, W. 1980. Role of a persistent inward current in motoneuron bursting during spinal seizures. *J. Neurophysiol.* 43:1296–318

Sethy, V. H., Van Woert, M. H. 1974. Modification of striatal acetylcholine concentration by dopamine receptor agonists and antagonists. *Res. Commun. Chem. Pathol. Pharmacol.* 8:13–28

Shoulson, I. 1979. Huntington's disease: Overview of experimental therapeutics. *Adv. Neurol.* 23:751–57

Somogyi, P., Bolam, J. P., Smith, A. D. 1981a. Monosynaptic cortical input and local axon collaterals of identified striatonigral neurons. A light and electron microscopic study using the Golgi-peroxidase transport-degeneration procedure. *J. Comp. Neurol.* 195:567–84

Somogyi, P., Bolam, J. P., Totterdell, S., Smith, A. D. 1981b. Monosynaptic input from the nucleus accumbens—ventral striatum region to retrogradely labelled nigrostriatal neurons. *Brain Res.* 217:245–63

Somogyi, P., Hodgson, A. J., Smith, A. D. 1979. An approach to tracing neuron networks in the cerebral cortex and basal ganglia, combination of Golgi staining, retrograde transport of horseradish peroxidase and anterograde degeneration of synaptic boutons in the same material. *Neuroscience* 4:1805–52

Steriade, M., Wyzinski, P., Apostol, V. 1972. Corticofugal projections governing rhythmic thalamic activity. In *Corticothalamic Projections and Sensorimotor Activities,* ed. T. L. Frigyesi, E. Rinvik, M. D. Yahr, pp. 221–72. New York: Raven

Thompson, R. L., Mettler, F. A. 1963. Permanent learning deficit associated with lesions in the caudate nuclei. *Am. J. Ment. Defic.* 67:526–35

Ungerstedt, U. 1971. Stereotaxic mapping of the monamine pathways in the rat brain. *Acta Physiol. Scand. Suppl.* 367: 1–48

Uno, M., Yoshida, M. 1975. Monosynaptic inhibition of thalamic neurons produced by stimulation of the pallidal nucleus in cats. *Brain Res.* 99:377–80

van der Kooy, D., Carter, D. A. 1981. The organization of the efferent projections and striatal afferents of the entopedun-cular nucleus and adjacent areas in the rat. *Brain Res.* 211:15–36

Vincent, S. R., Hattori, T., McGeer, E. G. 1978. The nigrotectal connection: A biochemical and ultrastructural characterization. *Brain Res.* 151:159–64

Walaas, I., Fonnum, F. 1980. Biochemical evidence for gamma-aminobutyrate containing fibers from the nucleus accumbens to the substantia nigra and ventral tegmental area in the rat. *Neuroscience* 5:63–73

Waszczak, B. L., Eng, N., Walters, J. R. 1979. Effects of picrotoxin upon single unit activity of substantia nigra neurons. *Soc. Neurosci. Abstr.* 5:81

Waszczak, B. L., Walters, J. R. 1980. Intravenous GABA agonist administration stimulates firing of A_{10} dopaminergic neurons. *Eur. J. Pharmacol.* 66: 141–44

Wilson, S. A. K. 1929. *Modern Problems in Neurology,* p. 211–22. New York: Wood

Yoshida, M., Rabin, A., Anderson, M. 1972. Monosynaptic inhibition of pallidal neurons by axon collaterals of caudate-nigral fibers. *Exp. Brain Res.* 15:333–47

Yoshida, M., Yajima, K., Uno, M. 1966. Differential activation of the two types of pyramidal tract neurons through the cerebellothalamocortical pathway. *Experientia* 22:331–32

Young, A. B., Bromberg, M. B., Penney, J. B. Jr. 1981. Decreased glutamate uptake in subcortical areas deafferented by sensorimotor cortical ablation in the cat. *J. Neurosci.* 1:241–49

Zeisel, S. H., Gelenberg, A. J., Growdon, J. H., Wurtman, R. J. 1980. Use of choline and lecithin in the treatment of tardive dyskinesia. *Adv. Biochem. Psychopharmacol.* 24:463–70

Zeman, W. 1970. Pathology of the torsion dystonias (dystonia musculorum deformans). *Neurology* 20(2):79–88

Ann. Rev. Neurosci. 1983. 6:95–120

ORGANIZATION OF THE THALAMOCORTICAL AUDITORY SYSTEM IN THE CAT

Thomas J. Imig and Anne Morel

Department of Physiology, Kansas University Medical Center, Kansas City, Kansas 66103

INTRODUCTION

The demonstration of multiple orderly representations of the cochlear partition in the cat's auditory cortex by Woolsey & Walzl (1942) laid the foundation for much of the current approach for studying the organization of the thalamocortical auditory system. Instead of relying solely upon cytoarchitectonic criteria to differentiate cortical fields (Campbell 1905, von Economo 1929), an auditory field could be physiologically defined as a cochlear representation. Assuming that a physiologically defined cortical field represented a fundamental unit of cortical structure and function, anatomists began to describe the organization of thalamocortical, corticothalamic, and corticocortical pathways in relation to maps produced by Woolsey and his collaborators (e.g. Rose & Woolsey 1949). Recent revisions of the tonotopic map and the discovery of a binaural map in cat auditory cortex have been paralleled by anatomical studies of cortical and thalamic auditory pathways. The purpose of this report is to describe some aspects of the topographic organization of these pathways, especially as they relate to physiologically derived cortical maps.

AUDITORY CORTEX

Tonotopic Organization

Multiple cortical foci are activated by electrical or acoustical stimulation of a limited length of the cochlear partition. Woolsey (1960) divided the ectosylvian region into four fields (AI, AII, Ep, and SF). Each field contains a representation of the cochlear base and apex (Figure 1A), and Woolsey hypothesized an orderly topographic representation of the cochlear parti-

95

0147-006X/83/0301-0095$02.00

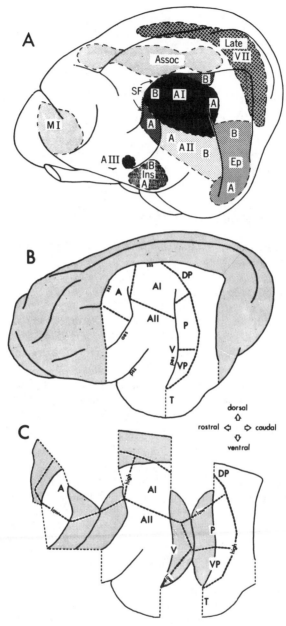

Figure 1 Auditory cortical fields in the cat. *A.* Woolsey's summary diagram showing four cortical areas with cochlea represented anteroposteriorly from apex (A) to base (B) in the suprasylvian fringe sector (SF), from base to apex in AI; from apex to base in AII. In Ep, representation is base above, apex below. In insula (Ins), evidence suggests that the base is represented above, apex below. AIII is Tunturi's third auditory area. "Association" cortex (Assoc) and precentral motor cortex (MI) gave responses to click with 15-msec latencies under chloralose. Visual area II (late VII) gave responses with 100 ms latency, also under chloralose

tion between these basal and apical extremes. During the past few years, a somewhat different view of cortical tonotopic organization has emerged as a result of microelectrode studies in which best-frequency[1] gradients were mapped in detail (Knight 1977, Merzenich et al 1975, Reale & Imig 1980). In Figures 1B and C, the primary (AI), anterior (A), posterior (P), and ventroposterior (VP) areas each contains a complete and orderly tonotopic representation. Moving from rostral to caudoventral along the arc of tissue occupied by these fields, best frequencies progress from low to high in field A, from high to low in field AI, from low to high in field P, and from high to low in field VP (Figure 1C). Borders between the four fields are located in regions in which best-frequency gradients show reversals. In spite of the differences between the two maps, inspection of the locations of the low and high extremes of the frequency representation in Figure 1C reveals a re-markable correspondence with similarly located representations of the cochlear apex (A) and base (B) respectively, in Figure 1A.

Neurons with similar best frequencies within each of fields AI, A, P, and VP define isofrequency contours that are oriented orthogonal to the low-to-high best-frequency gradients. Woolsey & Walzl (1942) described the pro-jection of the cochlear nerve upon AI as "a system in which functionally related groups of elements are oriented approximately in a dorso-ventral direction" (p. 323). These findings have been confirmed in microelectrode mapping experiments (Merzenich et al 1975, Reale & Imig 1980). Within field A (Knight 1977, Merzenich et al 1975, Reale & Imig, 1980, Woolsey 1961) and fields P and VP (Reale & Imig 1980), isofrequency contours are also evident. Thus, each point along the cochlear partition is represented upon a strip of tissue in each of the four fields.

Although maps obtained in individual animals demonstrate an orderly tonotopic organization in AI (Merzenich et al 1975, Reale & Imig 1980, Woolsey 1961, Woolsey & Walzl 1942), other studies in which data from several animals were pooled using sulci for reference (Evans et al 1965,

---

(from Woolsey 1961). *B.* A lateral view of the left cerebral hemisphere showing positions of fields. *C.* Unfolded cortical surface forming the gyral surfaces and sulcal banks in the unshaded region in *B;* cortical surfaces forming sulcal banks are shaded, while those forming gyral surfaces are not. *Heavy interrupted lines* delimit four tonotopically organized fields (A, AI, P, and VP). The locations of the lowest and highest best-frequencies in these fields are indicated in *c* by "low" and "high," respectively. Surrounding these four fields is a belt of cortex containing neurons responsive to acoustic stimulation. (*B* and *C* from Imig & Reale 1980).

---

[1]Best frequency is that frequency of sound stimulation for which a neuron's response threshold is the lowest.

Hind 1953, Hind et al 1960, Goldstein et al 1970) present a much less orderly picture. A key to resolving this apparent discrepancy lies in the observation that the location of AI can vary considerably with respect to cortical sulci (Merzenich et al 1975). Consequently, one would expect considerable blurring of tonotopy in maps produced from pooled data.

Under conditions of general anesthesia, the vast majority (98%) of single units sampled within the mapped borders of AI exhibit narrow tuning curves (Phillips & Irvine 1981). Within single electrode penetrations oriented perpendicular to the cortical surface, narrowly tuned units generally have best frequencies in close registration (Phillips & Irvine 1981, Abeles & Goldstein 1970). The higher proportion of broadly tuned units observed in unanesthetized animals may in part reflect inclusion in the sample of neurons located in surrounding cortical areas (Abeles & Goldstein 1970, Goldstein et al 1970, Hind et al 1960, Oonishi & Katsuki 1965).

Rostral to AI is a tonotopic representation that Knight (1977) named the anterior auditory field. Woolsey (1960) included this region in the SF area, the organization of which he proposed on the basis of partial maps obtained in earlier evoked potential studies (Hind 1953, Woolsey & Walzl 1942). Woolsey and his colleagues (Woolsey 1961) later produced the first and most complete map of this field in the cat, which is in accord with more recent microelectrode studies (Knight 1977, Merzenich et al 1975, Reale & Imig 1980), although his summary map (Figure 1A) was not changed to reflect these results. Single neurons in field A exhibit frequency selectivities similar to neurons in AI. In anesthetized animals, the vast majority of neurons appear to be narrowly tuned (Phillips & Irvine 1982a).

Fields P and VP are located largely within the posterior ectosylvian sulcus. In evoked potential studies, a representation of the basal end of the cochlea that is seen in some maps (Downman et al 1960, Woolsey 1961, Woolsey & Walzl 1942) closely corresponds in location to the adjacent high-frequency representations of fields P and VP. Furthermore, Sindberg & Thompson (1962) described a representation of the apical end of the cochlea near the ventral end of the posterior ectosylvian sulcus and gyrus that may in part correspond to the low-frequency representation of field VP. Preliminary reports suggest that the majority of single neurons in field P are as narrowly tuned as those in fields AI and A, although field P neurons respond at somewhat longer minimum latencies (20–45 ms) and a greater proportion of them exhibit nonmonotonic spike count-intensity functions (Phillips & Orman 1982). Responses of clusters of neurons often exhibit a short-latency component that is narrowly tuned and tonotopically organized, and a long-latency component (50 msec or greater) that is broadly tuned (Reale & Imig 1980). Long-latency, broadly-tuned responses are rarely encountered in fields A and AI under similar conditions (Phillips & Irvine 1981, 1982a, Reale & Imig 1980).

Surrounding fields A, AI, P, and VP is a peripheral belt of auditory responsive cortex that has been partially subdivided into the second (AII), ventral (V), temporal (T), and dorsoposterior (DP) areas. In V there is evidence of a tonotopic representation (Reale & Imig 1980), although it has not been worked out in detail. Results of evoked potential studies suggest the existence of one (Woolsey 1960, 1961) or two (Volkov 1980) tonotopic representations in AII although these findings currently receive little support from microelectrode mapping studies (Reale & Imig 1980, Schreiner & Cynader 1981). Clusters of neurons in AII generally exhibit broader tuning and respond at longer latencies than those in AI (Schreiner & Cynader 1981). It appears that tonotopy in AII is considerably less precise than that present in fields A, AI, P, and VP. The remainder of the peripheral auditory belt has not been studied in detail but the limited data available do not suggest a precise tonotopic representation. Thus, in auditory cortex, fields A and AI exhibit specific characteristics (predominance of narrowly tuned neurons in a precise tonotopic arrangement), field AII exhibits diffuse characteristics (predominance of broadly tuned neurons and lack of precise tonotopy), and areas P and VP exhibit both specific and diffuse characteristics.

Corticocortical Connections Related to Tonotopic Maps

Cortical neurons whose axons interconnect auditory cortical fields form a complex neural network. Anatomists have described topographic patterns of corticocortical connections in relationship to fissural patterns of cat auditory cortex (Diamond et al 1968a,b, Heath & Jones 1971, Kawamura 1973, Paula-Barbosa et al 1975), but due to variation of the positions of tonotopic maps in relation to sulci it has not been possible to relate closely the results of anatomical studies to current physiological maps. More recent experiments combining anatomical tracing of corticocortical pathways and electrophysiological mapping of cortical fields in the same brain have overcome this difficulty (Brugge & Imig 1978, Imig & Reale 1980).

A summary diagram of the known connections between fields A, AI, P, and VP is shown in the left panel of Figure 2. Each of the four fields is reciprocally connected with the other three in the same hemisphere. Additionally, neurons in each field project to the contralateral hemisphere. Projections have been found from field A to contralateral fields A and AI; from field AI to contralateral fields A, AI, and P; and from each of fields P and VP to both fields in the opposite hemisphere. Finally, each of four fields projects to several areas within the peripheral auditory belt. Projections from auditory belt fields have not been extensively studied, but it is clear that AII projects upon both specific and diffuse cortical areas (Imig & Reale 1980). Thus, at the level of the cortex, specific and diffuse regions are mutually interconnected.

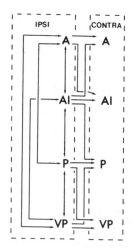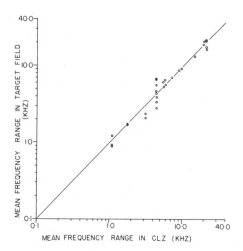

Figure 2 Pathways interconnecting tonotopic fields and their relationship to the frequency representation. The panel on the left diagramatically illustrates pathways connecting fields A, AI, P, and VP in both hemispheres. All known pathways originating in the fields of one hemisphere (IPSI) are indicated. A pathway connecting two fields is represented as a *line* originating in one field and terminating in another with an *arrow*. This diagram summarizes findings of Imig & Reale (1980) and unpublished results of A. Morel and T. J. Imig (interhemispheric reciprocal connections between fields P and VP). The panel on the right graphically relates the geometric mean of best frequencies represented in the dense central labeled zone (CLZ) at the site of tritiated amino acid injection and the geometric mean of best frequencies represented in labeled regions of target fields for projections among fields A, AI, P, and VP (from Imig & Reale 1980).

Neurons whose best frequencies lie within a common frequency range are found near the sources and terminations of corticocortical projections in fields A, AI, P, and VP of the same and opposite hemispheres. This relationship is graphically illustrated in the right panel of Figure 2. The abscissa shows the geometric mean of the range of best frequencies represented in the heavily labeled region surrounding the injection site, and the ordinate shows the geometric mean of the range of best frequencies represented in labeled areas of a target field. Points cluster about the diagonal of the graph, indicating that corticocortical projections interconnect similar portions of the best-frequency representations of the fields in both hemispheres.

These findings provide an anatomical substrate for the results of physiological experiments reported by Downman et al (1960). In one experiment, Downman et al defined areas on the middle ectosylvian and posterior ectosylvian gyri in which responses to electrical stimulation of the basal end of the cochlea were recorded. These regions correspond to the high-frequency representations located near the borders between fields AI and A, and fields P and VP, respectively. Direct electrical stimulation of the cortical surface in each of these regions resulted in a focus of evoked responses in the other, and Downman et al concluded that "some areas of cortex, which are activated by stimulation of a limited portion of the cochlear nerve, are rather specifically interrelated" (p. 139).

A single injection of tritiated amino acid into one of the four fields generally produces multiple dense patches of labeled axon terminals within each of several cortical fields. These patches of labeling are separated from each other by regions of less dense labeling. Patchy patterns of corticocortical terminations have been described in the somatic-motor (Jones et al 1978, Künzle 1978), visual (Montero 1981, Wong-Riley 1979), auditory (Brugge & Imig 1978, Ebner & Myers 1965, Fitzpatrick & Imig 1980, Imig & Brugge 1978, Imig & Reale 1980), and other cortical areas (Goldman & Nauta 1977) in monkeys and carnivores, suggesting that this pattern may be a rather general property of corticocortical connectivity.

There appears to be a considerable amount of convergence and divergence in the projections between cortical fields. Axons from neurons located along a limited portion of an isofrequency contour in one field project upon relatively longer portions of isofrequency contours located in other fields. Nevertheless, the divergence is limited to the isofrequency dimension and does not occur in the frequency gradient dimension (Imig & Reale 1980).

Binaural Organization of Auditory Cortex

Neurons within the auditory brainstem nuclei receive bilateral input and their projection upon higher centers ultimately carries binaural information to the cortex. Within AI, most neurons appear to be sensitive to binaural stimulation (Abeles & Goldstein 1970, Brugge et al 1969, Hall & Goldstein 1968, Phillips & Irvine 1979a, 1982b) and two classes of physiologically defined columns have been recognized in the high-frequency representation on the basis of neurons' responses to monaural and binaural stimulation.

Although some neurons in AI are sensitive only to binaural stimulation (Hall & Goldstein 1968, Imig & Adrian 1977, Kitzes et al 1980, Middlebrooks et al 1980, Phillips & Irvine 1982b), most may be excited by monaural stimulation as well. For some neurons, stimulation to either the contralateral or ipsilateral ear may elicit the strongest excitatory monaural response (contralateral or ipsilateral ear dominance), while for others, stimulation to either ear may be equally effective. Studies of both single units (Phillips & Irvine 1982b) and unit clusters (Imig & Adrian 1977) in AI suggest that neurons within a vertical column exhibit similar ear dominance.

Another type of columnar organization found in AI is based on the binaural interactions displayed by single neurons and clusters of neurons (Imig & Adrian 1977, Middlebrooks et al 1980). Binaural interactions may be defined using various stimulation paradigms, but in several mapping studies, stimulus intensity to the dominant ear was held constant while stimulus intensity to the opposite ear was varied to produce maximal interaction. For most neurons, the binaural responses are either stronger (summation) or weaker (suppression) than responses to monaural stimulation to the dominant ear. In a few of their electrode penetrations into AI, Abeles & Goldstein (1970) found that each neuron exhibited the same binaural interaction as every other neuron in the penetration. Later studies of responses of single neurons and neuron clusters suggest that the vast majority of neurons in a vertical column display similar binaural interactions (Imig & Adrian 1977, Middlebrooks et al 1980). Phillips & Irvine (1979a, 1982b) have recently questioned the columnar organization of binaural interactions within AI, as single neurons displaying different binaural interactions were encountered in 39% of the individual penetrations they analyzed. They proposed that the heterogeneity of binaural interactions within a column was masked in the response of a cluster of neurons. Nevertheless, several lines of evidence argue for a high degree of homogeneity of binaural interactions within a cluster as well as within a column:

1. Neuron clusters with ambiguous or mixed binaural properties are only rarely encountered and these tend to be located on or near column borders (Imig & Adrian 1977, Middlebrooks et al 1980).
2. Different binaural properties were observed in only 5% of the cases involving simultaneous recording of single neuron and neuron cluster responses (Imig & Adrian 1977).
3. Middlebrooks et al (1980) observed that all units in a given cluster of suppression neurons were inhibited by adequate ipsilateral stimulation and stated that any neurons showing summation would have been easily detected.

Differences in results relating to the homogeneity of binaural interactions within a column may be due to the following:

1. The use of different stimulus conditions to define binaural interactions [zero interaural intensity differences (Phillips & Irvine 1982b) versus variable interaural intensity differences (Imig & Adrian 1977, Middlebrooks et al 1980)].
2. The sampling of neurons in the frequency representation below 3 kHz (Phillips & Irvine 1982b), in which binaural columns have not been described.

Topographically, binaural interaction columns occupy elongated patches or bands of cortex running orthogonal to isofrequency contours (Imig & Adrian 1977, Imig & Brugge 1978, Imig & Reale 1981b, Middlebrooks et al 1980). Bands of summation and suppression neurons alternate across the dorsoventral extent of AI and at least three bands of each type can be recognized in the most complete maps. Bands are somewhat irregular in shape, running for various distances across the high-frequency representation, occasionally giving rise to branches or joining with other bands.

Corticocortical Connections Related to Binaural Maps

Ipsilateral and contralateral projections to field AI terminate in patches that are often elongated in the direction of the low-to-high best-frequency gradient (Ebner & Myers 1965, Heath & Jones 1971, Imig & Brugge 1978, Imig & Reale 1980, 1981b). In tissue sections cut parallel to the flattened cortical surface, this pattern of labeling is particularly apparent. Figure 3 illustrates patterns of terminal labeling in AI as a result of injection of tritiated amino acids into ipsilateral fields A (Figure 3A) and P (Figure 3B), and contralateral field A (Figure 3C). Elongated patches of terminal labeling are also evident in AI as a result of an HRP injection in AI of the opposite hemisphere, but in addition, the distribution of neurons giving rise to callosal axons can be seen (Figure 3D).

Elongated patches of terminal labeling are reminiscent of bands formed by binaural interaction columns. In brains in which labeling of corticocortical projections is combined with mapping of binaural interaction columns, a close correspondence between the two patterns is often seen (Imig & Brugge 1978, Imig & Reale 1981b). Both the terminals of callosal axons and the perikarya giving rise to callosal axons are more densely aggregated in summation columns than in suppression columns. Furthermore, in experiments in which both callosal axon terminals and callosal cell bodies are labeled with HRP, the two distributions appear in register in tissue sections cut through layer III parallel to the cortical surface (Figure 3D). The covariation of the densities of the two elements is less apparent in tissue sections cut in the transverse plane (Imig & Brugge 1978, Kelly & Wong 1981). In contrast to the callosal projection pattern, terminals of ipsilateral projections from fields A and P are more densely aggregated in suppression columns than in summation columns. These findings indicate that binaural interaction columns differ in their corticocortical connectivities.

Corticocortical connections between the tonotopic fields in both hemispheres are organized with respect to both tonotopic and binaural maps. Distributed among the four tonotopic representations are neurons that receive ascending input from the same limited portion of the basilar membrane. These same elements are linked via corticocortical connections into a network that allows neurons located in different fields, and linked to the same portion of the basilar membrane via ascending pathways, to influence each others' activities. Thus, neurons that share a common frequency selectivity appear to be interconnected.

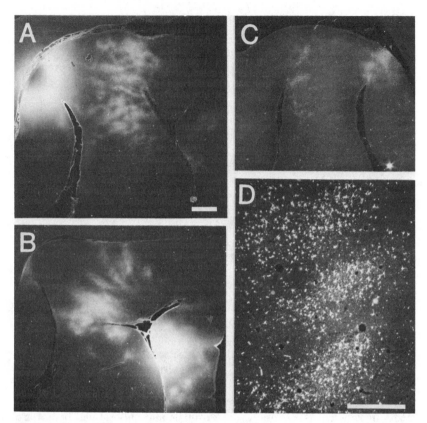

Figure 3 Topographic patterns of projections to AI from tonotopic auditory fields. Each panel illustrates a tissue section cut through layer III parallel to the flattened cortical surface. Projections were labeled by injection of tritiated proline into ipsilateral field A (*A*), ipsilateral field P (*B*), and contralateral field A (*C*), and by injection of HRP into contralateral field AI (*D*). In panels *A, B,* and *C,* axon terminals are seen to aggregate in patches or bands elongated in the direction of the low-to-high frequency gradient in AI. In panel *D,* both callosal axon terminals (fine grains) and perikarya giving rise to callosal axons are labeled with HRP. The two distributions appear coextensive in layer III and form patches elongated along the low-to-high frequency gradient in AI. Bar represents 2 mm in *A, B,* and *C,* and 0.5 mm in *D.*

Whether the correspondence between corticocortical connections and the binaural map in AI represents a linking of neurons with common functions is less clear. Middlebrooks & Pettigrew (1981), using free field stimulation, have described receptive fields of AI neurons and have suggested that neurons displaying summation respond to sounds regardless of location, whereas neurons displaying contralateral ear dominance and suppression respond to sounds located only on the contralateral side of the head. A particular region of sound space may be represented within a binaural column. Lacking more detailed knowledge of the interconnections between individual binaural columns, it is not possible to determine whether the relation between patterns of corticocortical connections and binaural columns reflects interconnections of areas representing common regions of sound space.

AUDITORY THALAMUS

Cytoarchitecture and Physiology

A major auditory thalamic structure, the medial geniculate body (MGB), is composed of several anatomically and physiologically distinct nuclei, each of which projects upon the auditory cortex. In tissue sections stained for Nissl substance, the MGB can be divided into a lateral *pars principalis* composed of small, densely packed cells, and a medial *pars magnocellularis* composed of larger, less densely packed cells (Rioch 1929). Based largely upon dendritic morphology of principal neurons viewed in Golgi stained sections, Morest (1964, 1965) recognized ventral, medial, and dorsal divisions within the MGB. Pars principalis corresponds to the ventral division and the caudolateral part of the dorsal division, whereas pars magnocellularis corresponds to the medial division and the medial part of the dorsal division. Many investigators relate results of their electrophysiological or anatomical pathway tracing experiments to Morest's divisions of the MGB. Nevertheless, some uncertainty exists in establishing such relationships, as the borders of the divisions are difficult to define unequivocally in tissue sections that are not stained by the Golgi technique.

Throughout most of the ventral nucleus, neurons have disc-shaped dendritic fields that are oriented to form laminae. Morest (1964, 1965) described two subdivisions, a *pars lateralis* (LV) in which laminae are arranged more or less parallel to the lateral margin of the geniculate, and a *pars ovoidea* (OV), in which laminae are arranged in a more complex coiled configuration (Figure 4).

The majority of neurons in pars lateralis have narrow tuning curves and respond at minimum latencies of 30 msec or less. A somewhat greater proportion of broadly tuned neurons is found in animals anesthetized with nitrous oxide (Morel 1980, Toros-Morel et al 1981) than with barbiturates and ketamine (Aitkin & Webster 1972, Calford & Webster 1981, Phillips & Irvine, 1979b). Neurons in the caudal part of the nucleus tend to exhibit somewhat longer latencies and broader tuning than those located more rostrally (Toros et al 1979).

The general pattern of tonotopy within pars lateralis was initially described by Rose & Woolsey (1958). Electrical stimulation of the cochlear apex evoked potentials laterally in the nucleus whereas electrical stimula-

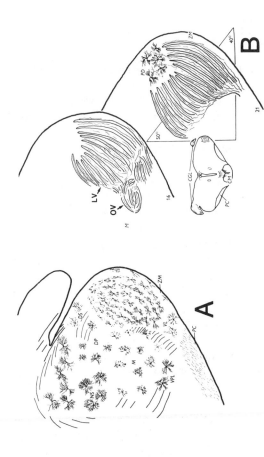

Figure 4 Dendritic morphology of principal neurons and laminar structure of the MGB. *A.* Typical distribution of principal neuron types at the junction of the anterior and middle thirds of the medial geniculate body. Transverse section, Golgi-Cox, 15-day-old cat (from Morest 1964). *B.* Transverse reconstructions of the fibro-dendritic layers of the ventral nucleus of a 15-day-old cat. Only half of the dendritic layers are shown. The ends of every other *heavy line* are connected for clarity and to indicate the probable points at which dendritic layers end. In section 21 are indicated the angles formed with the parasagittal (50°) and horizontal (40°) planes by the best straight line through the point of least flexure of each layer. The typical appearances of the principal neurons of the dorsal cap of the ventral nucleus (PD) and some of their afferent fibers are shown in section 21. Section 16 corresponds approximately to Figure 4A. The *inset,* an outline drawing of section 16, indicates the plane of sectioning through the supramamillary decussation and the posterior pole of the lateral geniculate (*B* adapted from Morest 1965). CGL, lateral geniculate body; DP, deep dorsal nucleus; DS, superficial dorsal nucleus; LV, pars lateralis; M, medial division; NS, suprageniculate nucleus; OV, pars ovoidea; PC, cerebral peduncle; PD, dorsal cap; VL, ventrolateral nucleus; ZM, marginal zone. Reprinted with permission of Cambridge University Press.

tion of the base evoked potentials at more medial loci. The mediolateral tonotopic gradient has been confirmed in microelectrode studies (Aitkin et al 1981, Aitkin & Webster 1971, 1972, Calford & Webster 1981, de Ribaupierre & Toros 1976, Morel 1980, Woolsey 1972, Morel & Imig 1982). Representative sequences of units' best frequencies encountered along electrode tracks oriented in a dorsolateral-to-ventromedial direction are illustrated in Figure 5. From sequences of units' best frequencies obtained in a number of electrode penetrations from several experiments (Morel 1980), isooctave laminae have been reconstructed at different anterposterior levels of the pars lateralis (Figure 6). The laminae form arcs curving from dorsolateral to ventromedial in the caudal half of the nucleus, whereas more rostrally, they are vertically oriented. A similar pattern is found in maps of pars lateralis obtained in individual animals (Morel & Imig 1982). The orientations of the isooctave laminae bear a striking resemblance to the laminae drawn by Morest on the basis of his anatomical studies (Morest 1965; Figure 4B).

The amount of scatter of units' best frequencies sequentially encountered along a track is variable (Figure 6). When measured on a large number of electrode penetrations distributed throughout the structure, the scatter tends to be lower in the dorsoanterior part of pars lateralis (A. Morel, Y. de Ribaupierre, E. Rouiller, F. de Ribaupierre, in preparation).

Neurons in pars ovoidea appear to be similar to those in pars lateralis with respect to response latencies and frequency selectivity (Aitkin & Webster 1972, Morel 1980). Sequences of best frequencies encountered in individual electrode penetrations through this nucleus are generally less regular than those in pars lateralis. However, there does appear to be a low-to-high gradient of best frequencies that runs from lateral to medial (Morel 1980, Morel & Imig 1982).

The dorsal cap of the ventral nucleus (PD in Figure 4B) and the ventrolateral nucleus (VL in Figure 4A) are nonlaminated regions within the ventral division. Neurons in the ventrolateral nucleus generally have broader tuning curves and respond at longer latencies than neurons in pars lateralis and pars ovoidea (Toros-Morel et al 1981). Electrophysiological studies have not been specifically directed toward describing the properties of neurons in the dorsal cap, but Calford & Webster (1981) found neurons that responded to broad ranges of frequencies, generally at long latencies, as well as neurons that responded to narrow ranges of frequencies at short latencies which may be located in this region.

Anatomical (Andersen et al 1980a,b, Jones & Powell 1971, Moore & Goldberg 1963) and electrophysiological studies (Phillips & Irvine 1979b) show that lateral division of the posterior group of thalamic nuclei (posterior group) is a part of the auditory thalamus. The majority of neurons in

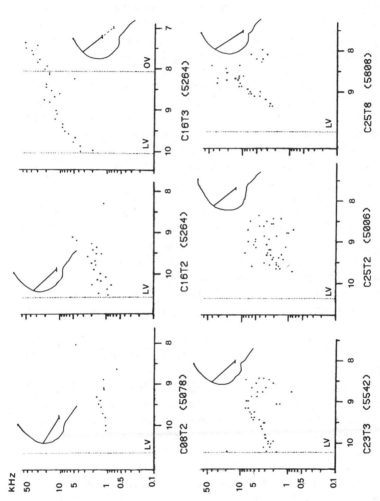

Figure 5 Sequences of units' best frequencies obtained along six electrode penetrations in the MGB. Each *dot* represents the best frequency of a unit (in kilohertz) according to its lateral coordinate (in millimeters). The lateral border of LV and its boundary with OV (in one example) are indicated by *vertical dotted lines*. *Insets* represent the orientation and location of the electrode penetrations in frontal sections of the MGB. The anteroposterior Horsley-Clark coordinate (in micrometers) is indicated below each diagram in parenthesis (adapted from Morel 1980).

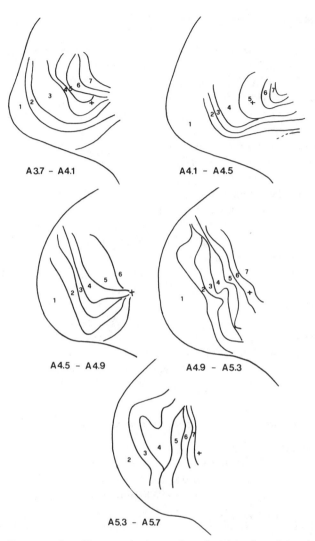

A 3.7 - A 4.1   A 4.1 - A 4.5

A 4.5 - A 4.9   A 4.9 - A 5.3

A 5.3 - A 5.7

Figure 6 Representation of isooctave laminae on frontal sections through pars lateralis. The laminae were reconstructed from sequences of units' best frequencies obtained in individual electrode tracks, such as illustrated in Figure 5. The tracks were divided into segments of linear progressions of best frequencies, upon which regression lines were computed. For all tracks included within a 400 μm frontal section, segments of the regression lines corresponding to a given octave were projected onto a frontal section of a standard MGB. This procedure was repeated for each octave and each 400 μm frontal section. The contours of isooctave laminae were obtained by drawing the limits including most of the segments of tracks corresponding to each octave. Numbers in isooctave laminae indicate ranges of best frequencies: (1) 0.1–0.8 kHz; (2) 0.8–1.6 kHz; (3) 1.6–3.2 kHz; (4) 3.2–6.4 kHz; (5) 6.4–12.8 kHz; (6) 12.8–25.6 kHz; (7) 25.6–51.2 kHz. Anteroposterior Horsley-Clark coordinates (in milimeters) of the caudal and rostral limits of the 400 μm thick section are indicated below each panel (from Morel 1980).

this region are narrowly tuned and respond at short latencies, similar to neurons in the laminated part of the ventral nucleus of the MGB (Phillips & Irvine 1979b). Within the posterior group, a low-to-high gradient of best frequencies is observed in a rostromedial-to-caudolateral direction. The caudolateral border of the high-frequency representation in the posterior group adjoins the rostromedial border of the high-frequency representation of pars lateralis (Morel & Imig 1982).

The dorsal division of the MGB contains neurons with spherical dendritic fields. Morest (1964) recognized several subdivisions (e.g. dorsal, deep dorsal, and suprageniculate nuclei) based on differences in perikaryal diameters and dendritic branching patterns (Figure 4A). The medial division is characterized by neurons with both radiate and tufted dendritic arbors. Radiate neurons are more numerous posteriorly, whereas tufted neurons are the major component anteriorly.

In contrast to the high degree of frequency selectivity displayed by the majority of neurons in pars lateralis, pars ovoidea, and the posterior group, the majority of neurons in the dorsal and medial divisions respond to broader ranges of frequencies (Aitken 1973, Aitkin et al 1981, Calford & Webster 1981, Phillips & Irvine 1979b, Toros-Morel et al 1981). Latencies of responses may be shorter (medial division and suprageniculate nucleus) or longer (dorsal nucleus) than latencies in pars lateralis (Aitkin et al 1981, Calford & Webster 1981, Toros-Morel et al 1981).

Ascending Input to the Auditory Thalamus

The central nucleus of the inferior colliculus (ICC) projects upon the posterior group and the MGB (Andersen et al 1980b, Kudo & Niimi 1978, Moore & Goldberg 1963). Projections to the MGB take the form of spatially segregated lateral and medial arrays. Afferents from restricted portions of the ICC terminate in laminae passing through pars ovoidea and pars lateralis, paralleling the dendritic and isooctave laminae in the latter structure (Figures 4 and 6). The tonotopic organization of the pars lateralis and pars ovoidea reflects the topography of projections from the ICC, as fibers from low-frequency loci project laterally and fibers from high-frequency loci project medially. The ICC also projects upon a caudorostral column of tissue that passes continuously through the medial division and deep dorsal nucleus of the MGB (Andersen et al 1980b).

Neurons in the dorsal and medial divisions receive auditory input from other brainstem nuclei, including nucleus sagulum, dorsal and ventral nuclei of the lateral lemniscus, and pericentral and external nuclei of the inferior colliculus (Aitkin et al 1981, Andersen et al 1980b, Kudo & Niimi 1978).

Organization of Thalamocortical Connections

Each auditory cortical field receives input from several auditory thalamic nuclei. The results of recent experimental studies relate to two questions. First, which thalamic nuclei project to each of fields A, AI, P, VP, and AII? Second, how are projections topographically related to cortical maps?

THALAMIC NUCLEI PROJECTING TO DIFFERENT AUDITORY FIELDS Thalamic nuclei projecting upon each of fields A, AI, P, VP, and AII are summarized in Table 1. These data come from experiments in which retrogradely transported pathway tracers were injected into physiologically mapped fields. Three distinctly different patterns of projections are seen in Table 1.

1. Similar portions of the frequency representations in fields A and AI receive projections from coextensive regions of the MGB (laminated ventral, deep dorsal and medial nuclei) and the posterior group (Andersen et al 1980a). Thalamic regions projecting to fields A and AI also receive projections from the central nucleus of the inferior colliculus (Andersen et al 1980b). The description of the connections of the central nucleus of the inferior colliculus and cortical fields A and AI with the thalamus by Andersen and his colleagues represents a major advance in understanding the organization of the central auditory system. By injecting tritiated amino acids and HRP into different sites in the frequency representations of cortical fields A and AI, they identified five separate thalamic nuclei projecting to each field and accurately described the topographic organization of several of these projection systems. Neurons that project to both fields A and AI via axon collaterals are found within each of these thalamic zones. Figure 7A shows a tissue section in which neurons projecting to field A (*open circles*) or to field AI (*dots*) were labeled with different retrogradely transported substances. *Squares* indicate double-labeled cells that project upon both fields. Within portions of the ventral nucleus (Figure 7A), the majority of cells that project to field A also project to field AI (Imig & Reale 1981, Imig et al 1981). Although the spatial distributions of neu-

Table 1 Thalamic nuclei projecting upon auditory cortical fields

Field	Laminated ventral nucleus	Posterior group	Periventral nuclei	Deep dorsal nucleus	Medial division
A	X	X		X	X
AI	X	X		X	X
P	X		X	X	X
VP	X		X	X	X
AII			X		X

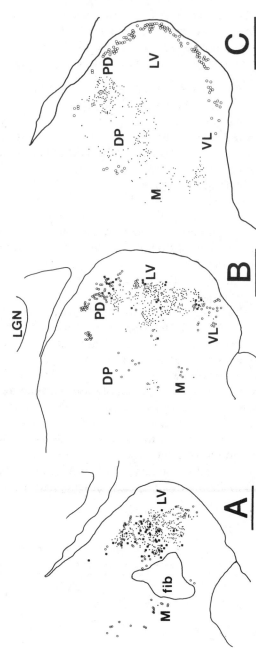

Figure 7 Distributions of MGB neurons projecting upon auditory cortex. Each panel represents a transverse tissue section cut through the medial geniculate of a brain in which two retrogradely transported substances (HRP and tritiated bovine serum albumin, or ³[H]-BSA) were injected at two cortical sites. Cells labeled with HRP are represented by *dots*, cells labeled with ³[H]-BSA are represented by *open circles*, and cells labeled with both substances are represented by *squares*. *A*. HRP was injected into field AI, and ³[H]-BSA was injected into field A, both near 10 kHz best-frequency sites. *B*. HRP was injected into field AI, and ³[H]-BSA was injected into field P, both near 6 kHz best-frequency sites. *C*. Both tracers were injected into field P, HRP into a 14 kHz best-frequency site and ³[H]-BSA into a 1.7 kHz best-frequency site. *Bars* represent 1 mm. LGN, lateral geniculate nucleus; DP, deep dorsal nucleus; fib, fibers; LV, pars lateralis; M, medial division; PD, dorsal cap of the ventral nucleus; VL, ventrolateral nucleus.

rons projecting to the two fields appear coextensive, the patterns do differ. Field AI receives a relatively stronger input from the ventral nucleus (Andersen et al 1980a) and field A receives a stronger input from the posterior group (A. Morel, R. A. Reale, and T. J. Imig, in preparation).

2. Fields P and VP receive projections from a second group of MGB regions (laminated ventral, "periventral," and deep dorsal nuclei, and the medial division), but this group is not identical to the group projecting to fields A and AI (Fitzpatrick et al 1977, Imig & Reale 1981, Imig et al 1981, Sousa-Pinto 1973). We use the term *periventral nuclei* to refer to regions surrounding the dorsal, caudal, and ventral borders of that portion of the laminated ventral nucleus which contains neurons projecting upon fields A and AI. This region probably includes the ventrolateral nucleus, the dorsal cap of the ventral nucleus, and a portion of the dorsal nucleus. Both fields P and VP receive a projection from the laminated portion of the ventral nucleus, but it is sparser than the projection to AI. Furthermore, neurons projecting to similar portions of the frequency representations of the two fields do not necessarily occupy coextensive thalamic regions. Clusters of neurons in the caudal portion of the periventral region projecting to field P are spatially segregated from those projecting to field VP (A. Morel, R. A. Reale, and T. J. Imig, in preparation). Double-labeling experiments show that some cells located within the laminated ventral nucleus project via axon collaterals to both fields AI and P (e.g. Figure 7B) and others to both fields A and P (Imig and Reale 1981b, Imig et al 1981).

3. Field AII receives projections from a group of geniculate regions [periventral nuclei, caudal tip of the MGB (dorsal division), and the medial division] that differs from the two groups described above (Andersen et al 1980a, Niimi & Matsuoka 1979, Sousa-Pinto 1973). The caudal tip of the dorsal division projects upon field AII but not upon fields P and VP, whereas all three fields receive projections from the periventral nuclei.

The specific or diffuse characteristics of a cortical auditory field appear to correspond with physiological characteristics of a subset of auditory thalamic nuclei (laminated ventral nucleus, periventral nuclei, and posterior group) from which it receives input. Fields AI and A have specific characteristics and receive input only from nuclei of the group with specific characteristics (laminated ventral nucleus, posterior group). On the other hand, field AII has diffuse characteristics and receives input only from areas with diffuse characteristics (periventral nuclei). Fields P and VP have both specific and diffuse characteristics and receive inputs from regions with both specific (laminated ventral nucleus) and diffuse (periventral nuclei) characteristics. It is tempting to speculate that the periventral nuclei might be one source of broadly tuned, long-latency responses found in fields AII, P, and VP. Finally, the medial division and deep dorsal nucleus of the MGB project upon cortical areas with specific and diffuse characteristics.

TOPOGRAPHIC ORGANIZATION OF THALAMOCORTICAL PROJEC-
TIONS   A single locus in either field AI or field A receives projections from
at least four topographically organized thalamic nuclei including pars lateralis and ovoidea, the deep dorsal nucleus and posterior group (Andersen et
al 1980a). Projections to high frequency sites lie dorsal within the deep
dorsal nucleus. A second array occupies a folded sheet of tissue passing
through pars lateralis and pars ovoidea of the ventral nucleus (Andersen et
al 1980a, A. Morel, R. A. Reale, and T. J. Imig, in preparation). The sheet
of labeled neurons passing through the pars lateralis takes the same form
as the isooctave laminae in this structure (Figure 6). Projections to sites of
successively higher frequency representation in either field A or AI arise
from arrays of neurons displaced toward a more medial location within the
ventral nucleus (Andersen et al 1980a), and toward a more caudolateral
location within the tonotopic representation of the posterior group (A.
Morel, R. A. Reale, and T. J. Imig, in preparation).

The projection from the laminated ventral nucleus to fields A and AI is
a part of a larger topographic projection upon fields A, AI, P, and VP. The
population of neurons projecting to the same limited portion of the frequency representation in all four fields occupies a folded sheet of tissue
extending across the laminated ventral nucleus and into adjacent portions
of the periventral nuclei. While neurons projecting to fields A and AI appear
confined to the central portion of the sheet that passes through the ventral
nucleus (Figure 7A, *dots* and *squares* in Figure 7B), neurons projecting to
fields P and VP are distributed throughout the entire sheet and are most
densely aggregated in the periventral nuclei that occupy the dorsal, caudal,
and ventral perimeter of the sheet (e.g. *open circles* and *squares* in Figure
7B, Imig & Reale 1981b). Injections into sites of low and high frequency
representation in field P label sheets of neurons located laterally and medially, respectively, which extend across the ventral and periventral nuclei
(Figure 7C). Thus, the mediolateral location of the entire array, not just the
portion in the laminated ventral nucleus projecting upon fields A and AI,
appears related to the frequency representation of its cortical target.

Single loci in fields A and AI receive convergent input from sheets and
columns of cells in the MGB. Andersen et al (1980a) suggested that all
sectors along an isofrequency contour in the cortex receive convergent input
from the same set of MGB neurons. Thus, as it is the case for corticocortical
connections, there appears to be an orderly topographic segregation of
connections within the dimension of the frequency gradient, and convergence and divergence of connections within the isofrequency dimension.

The binaural organization of the MGB has not been worked out in detail.
A binaural organization of the ventral nucleus, topographically comparable
to that in AI, would consist of slabs of tissue containing summation neurons

that are oriented orthogonally to the isofrequency laminae and parallel to the frequency gradient. These slabs would alternate with similarly oriented slabs containing suppression neurons. A regional segregation of neurons with similar binaural properties exists in the ICC (Roth et al 1978). Projections from limited regions of the ICC terminate in patches distributed along isofrequency laminae (Andersen et al 1980b) and small injections of HRP into AI (perhaps confined to a single binaural column) label several clusters of cells distributed along an isofrequency lamina (Andersen et al 1980a, Middlebrooks & Zook 1981). Both patterns suggest a slab-like organization. There is currently no electrophysiological support for such a binaural organization in the ventral nucleus, although ear dominance does appear to be differentially represented within the pars lateralis (de Ribaupierre et al 1979, Toros et al 1979).

AUDITORY CORTICOFUGAL PROJECTIONS

Three major corticofugal projection systems are topographically related to the cortical best-frequency map:

1. Corticothalamic and thalamocortical connections are reciprocally organized such that thalamic areas projecting to a cortical location receive a projection from the same location (Andersen et al 1980a, Colwell 1975). Thus, corticothalamic projections exhibit the same patterns of organization as do thalamocortical projections.
2. Field AI projects upon the pericentral nucleus and dorsomedial division of the inferior colliculus. The topography of these projections is in register with respect to the tonotopic organization of AI and both targets in the inferior colliculus (Andersen et al 1980c).
3. Projections from fields AI and P to the basal ganglia are topographically organized, as projections from low-frequency cortical loci terminate medially with respect to projections from high-frequency cortical loci (Reale & Imig 1983). In general, all corticofugal projections from fields A, AI, P, and VP that have been studied exhibit a topographic organization in relation to the cortical frequency representation.

SUMMARY

Cat auditory cortex can be divided into a number of fields, some of which exhibit a precise tonotopic organization (A, AI, P, and VP), and others which do not (e.g. AII). Based on differences in frequency selectivity of constituent neurons and tonotopic organization, fields A and AI have predominantly specific characteristics (narrowly tuned neurons and precise tonotopy), field AII has diffuse characteristics (broadly tuned neurons and absence of tonotopy), and fields P and VP have both specific and diffuse characteristics.

Fields A and AI appear to be the cortical representation of the central nucleus of the inferior colliculus (ICC), as the ICC projects upon the same nuclei of the auditory thalamus that project upon these two fields (laminated ventral and deep dorsal nuclei and medial division of the MGB, posterior group). Fields P and VP receive projections from the same thalamic regions that project to A and AI except for the posterior group, but in addition these fields receive inputs from the periventral nuclei, which include nonlaminated portions of the ventral division as well as portions of the dorsal division. Field AII receives projections from the medial division, the periventral nuclei, and the caudal tip of the MGB. Physiological studies suggest that input from the laminated portion of the ventral nucleus and the posterior group may be responsible for the specific characteristics of fields A, AI, P, and VP, while input from the periventral nuclei may be in part responsible for the diffuse characteristics of fields P, VP, and AII.

The thalamocortical pathway is divided into several topographically organized arrays of projections. A lateral array in the MGB is composed of the laminated portion of the ventral nucleus and the periventral nuclei. Neurons within a sheet of tissue extending across the ventral and periventral nuclei project upon the same limited portion of the frequency representation of fields A, AI, P, and VP. The higher the frequency representation at the cortical target, the more medial is the location of the projecting array. Within a sheet, neurons projecting to the four fields are differentially distributed; those projecting to fields A and AI are located only within the laminated portion of the ventral nucleus, while those projecting to fields P and VP are more densely aggregated in the periventral nuclei. A rostromedial array of projections from the tonotopically organized portion of the posterior group to fields A and AI is also topographically organized. The higher the frequency representation at their cortical target, the more caudolateral is the location of the projection neurons in the posterior group.

The tonotopic organization of portions of the colliculo-geniculo-cortical pathway has been characterized electrophysiologically. The ascending colliculo-geniculo-cortical pathways, the corticocortical pathways, and the corticofugal pathways are organized such that projections from one tonotopic array terminate in register with the tonotopic organization of the target array. Although there is a clear segregation of projections to representations of different frequencies, there is a considerable overlap of projections within the isofrequency dimension.

In addition to a tonotopic organization, the ICC-laminated ventral nucleus-AI pathway also contains a binaural organization. Although many of the details remain to be worked out, the topography of projections may be as precise with respect to binaural organization as it is with respect to the tonotopic organization.

ACKNOWLEDGMENTS

The authors wish to express their gratitude to Linda Carr and Helen Knefel for typing this manuscript. Our work was supported by NIH Grant NS 17220 and Biomedical Research Support Program, NIH BRSG S07 RR05373.

Literature Cited

Abeles, M., Goldstein, M. H. 1970. Functional architecture in cat auditory cortex: Columnar organization and organization according to depth. *J. Neurophysiol.* 33:172–87

Aitkin, L. M. 1973. Medial geniculate body of cat: Responses to tonal stimuli of neurons in medial division. *J. Neurophysiol.* 36:275–83

Aitkin, L. M., Calford, M. B., Kenyon, C. E., Webster, W. R. 1981. Some facets of the organization of the principal division of the cat medial geniculate body. In *Neuronal Mechanisms of Hearing*, ed. J. Syka, L. Aitkin, pp. 163–82. New York/London: Plenum. 540 pp.

Aitkin, L. M., Webster, W. R. 1971. Tonotopic organization in the medial geniculate body of the cat. *Brain Res.* 26:402–5

Aitkin, L. M., Webster, W. R. 1972. Medial geniculate body of the cat: Organization and responses to tonal stimuli in ventral division. *J. Neurophysiol.* 35:365–80

Andersen, R. A., Knight, P. L., Merzenich, M. M. 1980a. The thalamocortical and corticothalamic connections of AI, AII, and the anterior auditory field (AAF) in the cat: Evidence for two largely segregated systems of connections. *J. Comp. Neurol.* 194:663–701

Andersen, R. A., Roth, G. L., Aitkin, L. M., Merzenich, M. M. 1980b. The efferent projections of the central nucleus and the pericentral nucleus of the inferior colliculus in the cat. *J. Comp. Neurol.* 194:649–62

Andersen, R. A., Snyder, R. L., Merzenich, M. M. 1980c. The topographic organization of corticocollicular projection from physiologically identified loci in the AI, AII, and anterior auditory cortical fields of the cat. *J. Comp. Neurol.* 191:479–94

Brugge, J. F., Dubrovsky, N. A., Aitkin, L. M., Andersen, D. J. 1969. Sensitivity of single neurons in auditory cortex of cat to binaural tonal stimulation: Effects of varying interaural time and intensity. *J. Neurophysiol.* 32:1005–24

Brugge, J. F., Imig, T. J. 1978. Some relationships of binaural response patterns of single neurons to cortical columns and interhemispheric connections of auditory areas AI of cat cerebral cortex. In *Evoked Electrical Activity in the Auditory Nervous System*, ed. R. F. Naunton, C. Fernandez, pp. 487–503. New York: Academic

Calford, M. B., Webster, W. R. 1981. Auditory representation within principal division of cat medial geniculate body: An electrophysiological study. *J. Neurophysiol.* 45:1013–28

Campbell, A. W. 1905. *Histological studies on the localisation of cerebral function.* Cambridge: Cambridge Univ. Press

Colwell, S. A. 1975. Thalamocortical-corticothalamic reciprocity: A combined anterograde-retrograde tracer study. *Brain Res.* 92:443–49

de Ribaupierre, F., Toros, A. 1976. Single unit properties related to the laminar structure of the MGN. *Exp. Brain Res.* Suppl. 1:503–5

de Ribaupierre, F., Toros, A., Rouiller, E. 1979. Binaural representation in the medial geniculate body. *Experientia* 35:926

Diamond, I. T., Jones, E. G., Powell, T. P. S. 1968a. Interhemispheric fiber connections of the auditory cortex of the cat. *Brain Res.* 11:177–93

Diamond, I. T., Jones, E. G., Powell, T. P. S. 1968b. The association connections of the auditory cortex of the cat. *Brain Res.* 560–79

Downman, C. B. B., Woolsey, C. N., Lende, R. A. 1960. Auditory areas I, II and EP: Cochlear representation, afferent paths and interconnections. *Bull. Johns Hopkins Hosp.* 106:127–42

Ebner, F. F., Myers, R. E. 1965. Distribution of corpus callosum and anterior commissure in cat and raccoon. *J. Comp. Neurol.* 124:353–66

Evans, E. F., Ross, H. F., Whitfield, I. C. 1965. The spatial distribution of unit characteristic frequency in the primary auditory cortex of the cat. *J. Physiol.* 179:238–47

Fitzpatrick, K. A., Imig, T. J. 1980. Auditory cortico-cortical connections in the owl monkey. *J. Comp. Neurol.* 192:589–610

Fitzpatrick, K. A., Imig, T. J., Reale, R. A. 1977. Thalamic projections to the posterior auditory cortical field in cat. *Soc. Neurosci. Abstr.* 3:6

Goldman, P. S., Nauta, W. J. H. 1977. Columnar distribution of cortico-cortical fibers in the frontal association, limbic and motor cortex of the developing rhesus monkey. *Brain Res.* 122:393–413

Goldstein, M. H., Abeles, M., Daly, R. L., McIntosh, J. 1970. Functional architecture in cat primary auditory cortex: Tonotopic organization. *J. Neurophysiol.* 33:188–97

Hall, J. L., Goldstein, M. H. 1968. Representation of binaural stimuli by single units in primary auditory cortex of unanesthetized cats. *J. Acoust. Soc. Am.* 43:456–61

Heath, C. J., Jones, E. G. 1971. The anatomical organization of the suprasylvian gyrus of the cat. *Ergeb. Anat. Entwicklung gesch.* 45:1–64

Hind, J. E. 1953. An electrophysiological determination of tonotopic organization in auditory cortex of cat. *J. Neurophysiol.* 16:475–89

Hind, J. E., Rose, J. E., Davies, P. W., Woolsey, C. N., Benjamin, R. M., Welker, W. I., Thompson, R. F. 1960. Unit activity in the auditory cortex. In *Neural Mechanisms of the Auditory and Vestibular Systems,* ed. G. L. Rasmussen, W. F. Windle, pp. 201–10. Springfield, Ill: Thomas

Imig, T. J., Adrian, H. O. 1977. Binaural columns in the primary field (AI) of cat auditory cortex. *Brain Res.* 138:241–57

Imig, T. J., Brugge, J. F. 1978. Sources and terminations of callosal axons related to binaural and frequency maps in primary auditory cortex of the cat. *J. Comp. Neurol.* 182:637–60

Imig, T. J., Reale, R. A. 1980. Patterns of cortico-cortical connections related to tonotopic maps in cat auditory cortex. *J. Comp. Neurol.* 192:293–332

Imig, T. J., Morel, A., Reale, R. A. 1981. Organization of thalamic neurons projecting to auditory cortical fields in the cat. *Presented at the Int. Sem. Neurosci., Algiers, Algeria*

Imig, T. J., Reale, R. A. 1981a. Medial geniculate projections to auditory cortical fields A, AI and P in the cat. *Soc. Neurosci. Abstr.* 7:230

Imig, T. J., Reale, R. A. 1981b. Ipsilateral cortico-cortical projections related to binaural columns in cat primary auditory cortex. *J. Comp. Neurol.* 203:1–14

Jones, E. G., Coulter, J. D., Hendry, S. H. C. 1978. Intracortical connectivity of architectonic fields in the somatic sensory, motor and parietal cortex of monkeys. *J. Comp. Neurol.* 118:291–348

Jones, E. G., Powell, T. P. S. 1971. An analysis of the posterior group of thalamic nuclei on the basis of its afferent connections. *J. Comp. Neurol.* 143:185–216

Kawamura, K. 1973. Corticocortical fiber connections of the cat cerebrum. I. The temporal region. *Brain Res.* 51:1–21

Kelly, J. B., Wong, D. 1981. Laminar connections of the cat's auditory cortex. *Brain Res.* 212:1–15

Kitzes, L. M., Wrege, K. S., Cassady, J. M. 1980. Patterns of responses of cortical cells to binaural stimulation. *J. Comp. Neurol.* 192:455–72

Knight, P. L. 1977. Representation of the cochlea within the anterior auditory field (AAF) of the cat. *Brain Res.* 130:447–67

Kudo, M., Niimi, K. 1978. Ascending projections of the inferior colliculus onto the medial geniculate body in the cat studied by anterograde and retrograde tracing techniques. *Brain Res.* 155:113–17

Künzle, H. 1978. Cortico-cortical efferents of primary motor and somatosensory regions of the cerebral cortex in *Macaca fascicularis. Neuroscience* 3:25–39

Merzenich, M. M., Knight, P. L., Roth, G. L. 1975. Representation of cochlea within primary auditory cortex in the cat. *J. Neurophysiol.* 28:231–49

Middlebrooks, J. C., Dykes, R. W., Merzenich, M. M. 1980. Binaural response-specific bands in primary auditory cortex (AI) of the cat: Topographical organization orthogonal to isofrequency contours. *Brain Res.* 181:31–48

Middlebrooks, J. C., Pettigrew, J. D. 1981. Functional classes of neurons in primary auditory cortex of the cat distinguished by sensitivity to sound location. *J. Neurosci.* 1:107–20

Middlebrooks, J. C., Zook, J. M. 1981. Thalamic connections to and from binaural interaction bands in AI of the cat. *Soc. Neurosci. Abstr.* 7:230

Montero, V. M. 1981. Topography of the cortico-cortical connections from the striate cortex in the cat. *Brain Behav. Evol.* 18:194–218

Moore, R. Y., Goldberg, J. M. 1963. Ascending projections of the inferior colliculus in the cat. *J. Comp. Neurol.* 121:109–36

Morel, A. 1980. *Codage des sons dans le corps genouillé médian du chat: évaluation de l'organisation tonotopique de ses différents noyaux.* PhD thesis. Univ. de Lausanne. Zurich: Juris Druck Verlag. 154 pp.

Morel, A., Imig, T. J. 1982. Frequency representation in the cat's medial geniculate body (MGB) and posterior complex (PO). *Soc. Neurosci. Abstr.* 8:In press

Morest, D. K. 1964. The neuronal architecture of the medial geniculate body of the cat. *J. Anat. London* 98:611–30

Morest, D. K. 1965. The laminar structure of the medial geniculate body of the cat. *J. Anat. London* 99:143–60

Niimi, K., Matsuoka, H. 1979. Thalamocortical organization of the auditory system in the cat studied by retrograde axonal transport of horseradish peroxidase. *Adv. Anat. Embryol. Cell Biol.* 57:1–56

Oonishi, S., Katsuki, Y. 1965. Functional organization and integrative mechanism of the auditory cortex of the cat. *Jpn. J. Physiol.* 15:342–65

Paula-Barbosa, M. M., Feyo, P. B., Sousa-Pinto A. 1975. The association connections of the suprasylvian fringe (SF) and other areas of the cat auditory cortex. *Exp. Brain Res.* 23:535–54

Phillips, D. P., Irvine, D. R. F. 1979a. Methodological consideration in mapping auditory cortex: binaural columns in AI of cat. *Brain Res.* 161:342–46

Phillips, D. P., Irvine, D. R. F. 1979b. Acoustic input to single neurons in pulvinar-posterior complex of cat thalamus. *J. Neurophysiol.* 42:123–36

Phillips, D. P., Irvine, D. R. F. 1981. Responses of single neurons in physiologically defined primary auditory cortex (AI) of the cat: Frequency tuning and responses to intensity. *J. Neurophysiol.* 45:48–58

Phillips, D. P., Irvine, D. R. F. 1982a. Properties of neurons in AAF. *Brain Res.* In press

Phillips, D. P., Irvine, D. R. F. 1982b. Binaural neurons in cat AI. *J. Neurophysiol.* In press

Phillips, D. P., Orman, S. S. 1982. Field P of cat auditory cortex: Coding of frequency and intensity by single neurons. *Soc. Neurosci. Abstr.* In press

Reale, R. A., Imig, T. J. 1980. Tonotopic organization of auditory cortex in the cat. *J. Comp. Neurol.* 192:265–91

Reale, R. A., Imig, T. J. 1983. Auditory cortical field projections to the basal ganglia of the cat. *Neuroscience.* In press

Rioch, D. M. 1929. Studies on the diencephalon of carnivora. 1. The nuclear configuration of the thalamus, epithalamus, and hypothalamus of the dog and cat. *J. Comp. Neurol.* 49:1–119

Rose, J. E., Woolsey, C. N. 1949. The relations of thalamic connections, cellular structure and evocable electrical activity in the auditory region of the cat. *J. Comp. Neurol.* 91:441–46

Rose, J. E., Woolsey, C. N. 1958. Cortical connections and functional organization of the thalamic auditory system of the cat. In *Biological and Biochemical Bases of Behavior,* ed. H. F. Harlow, C. N. Woolsey, pp. 127–50. Madison: Univ. Wisc. Press

Roth, G. L., Aitkin, L. M., Andersen, R. A., Merzenich, M. M. 1978. Some features of the spatial organization of the central nucleus of the inferior colliculus of the cat. *J. Comp. Neurol.* 182:661–80

Schreiner, C., Cynader, M. 1981. Response properties and topographic maps of cat AI and AII auditory cortex. *Soc. Neurosci. Abstr.* 7:231

Sindberg, R. M., Thompson, R. F. 1962. Auditory response fields in ventral temporal and insular cortex in cat. *J. Neurophysiol.* 25:21–28

Sousa-Pinto, A. 1973. Cortical projections of the medial geniculate body in the cat. *Adv. Anat. Embryol. Cell Biol.* 48:1–42

Toros, A., Rouiller, E., de Ribaupierre, Y., Ivarsson, C., Holden, M., de Ribaupierre, F. 1979. Changes of functional properties of medial geniculate body neurons along the rostrocaudal axis. *Neurosci. Lett. Suppl.* 3:S5

Toros-Morel, A., de Ribaupierre, F., Rouiller, E. 1981. Coding properties of the different nuclei of the cat's medial geniculate body. In *Neuronal Mechanisms of Hearing,* ed. J. Syka, L. Aitkin, pp. 239–43. New York/London: Plenum. 540 pp.

Volkov, I. O. 1980. The cochleotopic organization of the cat second auditory cortex. *Neirofiziologiya* 12:18–26. (in Russian)

von Economo, C. 1929. *The Cytoarchitectonics of the Human Cerebral Cortex,* pp. 110–28. London: Oxford Univ. Press

Wong-Riley, M. 1979. Columnar cortico-cortical interconnections within the visual system of the squirrel and macaque monkeys. *Brain Res.* 162:201–17

Woolsey, C. N. 1960. Organization of cortical auditory system: A review and a synthesis. See Hind et al 1960, pp. 165–80

Woolsey, C. N. 1961. Organization of cortical auditory system. In *Sensory Communication,* ed. W. A. Rosenblith, pp. 235–57. Cambridge: MIT Press

Woolsey, C. N. 1972. Designated discussion. *Brain Behav. Evol.* 6:323–25

Woolsey, C. N., Walzl, E. M. 1942. Topical projections of nerve fibers from local regions of the cochlea to the cerebral cortex of the cat. *Bull. Johns Hopkins Hosp.* 71:315–44

Ann. Rev. Neurosci. 1983. 6:121–48

CLINICAL IMPLICATIONS OF RECEPTOR SENSITIVITY MODIFICATION

Arnold J. Friedhoff and Jeannette C. Miller

Department of Psychiatry, New York University School of Medicine, New York, New York 10016

Introduction

Many drugs are known to act on specific cell membrane receptors. Most drugs acting in this manner have been designed to interact with a receptor to trigger a desired effect at the instant of the interaction; however, membrane receptors are not only transducers of specific ligand-dependent effects, but many receptors are adaptive structures as well, responsive to various long-term changes in the receptor environment. In seminal studies of denervation supersensitivity, Cannon & Rosenbleuth (1949) observed adaptive changes resulting from the loss of the neurotransmitter secondarily to degeneration of axon terminals. Although the mechanism was not clear at the time of these studies, it is now known that certain receptors can adjust to changes in specific ligand supply by regulation of their sensitivity or efficiency. It appears that these adaptive changes in receptors can have important clinical implications in the pathophysiology of various disorders, and are important in the mediation of certain types of drug effects. Through deliberate manipulation of these receptor systems, it is possible to produce therapeutic effects by changing receptor sensitivity. These issues, in relation to central biogenic amine systems, are the subject of this review.

Changes in the sensitivity of a receptor system involve a change in the response to a given concentration of agonist (Iversen 1975). The relationship between the sensitivity of synaptic receptors and the supply of available neurotransmitter has been the subject of many studies (Von Voigtlander et al 1973, Von Voigtlander & Moore 1973, Ungerstedt et al 1975). Hornykiewicz (1975) reported that increased activity of dopaminergic neurons may occur in Parkinsonism and in syndromes resulting from prolonged administration of dopamine (DA) blockers, as a compensation for the de-

121

0147-006X/83/0301-0121$02.00

crease of available DA at the receptor. Chronic administration of blockers or chronic inhibition of DA synthesis produces an increase in the number of central DA receptors, so that a greater than normal response occurs when the interfering agent is discontinued and the neurotransmitter can again interact with the receptor (Moore & Thornburg 1975).

Dopamine blockade is believed to be important in antipsychotic drug action, and the extent of DA blockade has been used as a measure of the potency of neuroleptics as antipsychotic agents (Creese et al 1976). Carlsson & Lindqvist (1963), in a classic study, demonstrated that antipsychotic drugs increase DA turnover, which probably serves as a compensatory response to the blockade of central DA receptors. Their work was the stimulus for subsequent efforts (Anden et al 1964, Van Rossum 1966). The DA receptor, when blocked for long periods, becomes chemically "denervated," and this is followed by the development of supersensitivity (Sharpless 1964, Trendelenberg 1966, Fleming et al 1973, Von Voigtlander 1974, Moore & Thornburg 1975, Friedhoff et al 1977, Muller & Seeman 1978) in a manner reminiscent of the supersensitivity described by Cannon & Rosenbleuth. Supersensitivity may (a) provide a means by which the receptor can overcome the blocking agent and thus return to normal functioning, and (b) be an end point of a normal regulatory system of the receptor, which modifies receptor sensitivity to correspond to transmitter availability. In such a regulatory system, subsensitivity of the receptor might also be produced under appropriate conditions. Mukherjee et al (1975) and Mukherjee & Lefkowitz (1976) have shown that specific norepinephrine (NE) sites on nucleated frog erythrocytes decrease in number after continued exposure to NE or other β-adrenergic agonists in culture. Comparable findings have been reported in the rat pineal β-adrenergic system (Kebabian et al 1975) and in the striatal dopaminergic systems after chronic in vivo treatment (Friedhoff et al 1977, Mishra et al 1978). More recently, it has been found that repeated administration of antidepressant drugs to animals results in down-regulation of noradrenergic receptors (Vetulani et al 1976). Thus, some cells appear to be capable of "tuning" their sensitivity up or down to compensate for chronic alterations in agonist supply.

In viewing the receptor as a regulated and regulatory unit, it is necessary to consider effects at the cell surface recognition or binding sites as well as effects on coupling and response mechanisms. In general, significant changes in the regulation of specific membrane receptors should produce changes in the integrated action of the cell. In the case of certain central nervous system (CNS) biogenic amine systems, these receptor changes are also reflected in behavioral changes. It is thus possible to study the effects of a specific receptor-ligand interaction at the molecular, cellular, and the behavioral level. By changing the sensitivity of the various elements through pharmacological manipulation, further insight can be gained into the relationship of the various elements of an integrated behavioral response.

The therapeutic effects of certain psychotropic drugs may be mediated by their long-term adaptive effects, rather than by their immediate effects, at target receptors. In a sense, this is ironic, inasmuch as these drugs have been largely developed via animal screening tests that are based on their acute effects. If, however, it is the compensatory effect that is desired, it would clearly be desirable to test agents for their ability to produce the desired compensatory changes. For this reason, an approach to treatment has been developed that attempts, directly, to produce the desired compensatory or adaptive effect using the principles of receptor sensitivity modification (RSM) (Friedhoff 1977, Friedhoff et al 1977, Friedhoff & Alpert 1978, Friedhoff 1979, Alpert & Friedhoff 1980). This approach involves a novel use of drugs: the pharmacological treatment initially worsens the condition somewhat; the therapeutic benefit occurs after the drug treatment is terminated.

Many types of adaptive changes of receptors are of interest; however, in this chapter we limit the discussion to those relatively long-term regulatory changes in biogenic amine receptor systems that appear to have importance in the pathophysiology or clinical management of neuropsychiatric disorders.

Possible Involvement of Receptor Regulation in the Pathophysiology of Several Conditions

Various drugs developed as treatments for psychotic disorders have been found to be antagonists of central DA receptors. Therefore, the DA system is believed to be involved in the production of psychotic symptoms. Of particular interest is the curious relationship between Parkinsonism and psychosis. Parkinsonian-like extrapyramidal symptoms are often caused by the action of antipsychotic drugs (Klein & Davis 1969) and are believed to result from the DA blocking action of these drugs, inasmuch as naturally occurring Parkinsonism is known to be associated with striatal hypodopaminergia.

Two other DA-related syndromes, tardive dyskinesia and withdrawal dyskinesia, are also of interest. The former occurs in some patients who have been taking neuroleptics for long periods, and is characterized by involuntary mouth movements, tongue fibrillation and thrusting, and tic-like movements of the extremities, which often persist long after the medication has been discontinued. Withdrawal dyskinesia occurs in some patients after abrupt withdrawal of neuroleptics, and is characterized by the same symptoms as tardive dyskinesia, but usually disappears after a short time. These two conditions, appearing after prolonged blockade of DA receptors by antipsychotic drugs, have the characteristics of a compensatory response. These motoric disturbances are believed to result from compensatory supersensitivity of striatal DA receptors (Klawans 1973, Klawans et al 1975, Hornykiewicz 1975), inasmuch as laboratory animals demonstrate a characteristic increase in the number of striatal DA receptors, without change in affinity, after administration of antipsychotic drugs for several weeks or more (see Seeman 1980 for a review). An associated increase in DA-stimulated adenylate cyclase and an increase in apomorphine-induced behavioral stereotypy also occurs. All of these changes can be seen only after the antipsychotic drug is withdrawn, whereas tardive dyskinesia often emerges while the patient is still taking antipsychotic medication. A better model for this latter type of dyskinesia is that of Clow et al (1979, 1980), who administered antipsychotic drugs to rats for periods of up to one year and found that supersensitivity to apomorphine occurred even while the animal was receiving neuroleptics. In these animals the compensatory process apparently had gone beyond simply compensating for the DA receptor blockade by antipsychotic agents, and had overcompensated, producing increased domaminergic activity beyond that necessary to offset the original blockade. As expected, L-dopa, administered for brief periods, markedly increases symptoms of tardive dyskinesia (Barbeau 1969, Markham 1971, Klawans 1973, Klawans et al 1975), although as is discussed below, prolonged administration and abrupt withdrawal may have remedial effects by down-regulating the sensitivity of DA systems.

Several other psychiatric syndromes also appear to involve hyperdopaminergic states, possibly involving supersensitivity of the DA systems. It has been presumed that disorders that respond to antipsychotic drugs that are DA receptor blocking agents may be caused initially by excessive dopaminergic activity. One such condition is schizophrenia. Hyperdopaminergia in schizophrenia could occur through several mechanisms as, for instance, increased synthesis and release of DA, or through blocked degradation; however, evidence of these metabolic abnormalities has not been forthcoming despite intensive investigation (see Lipton et al 1978 for an extensive review).

The most popular, and probably most useful, theory of the pathophysiology of depressive disorders has been the so-called "catecholamine hypothesis." In its original version it was believed that decreased noradrenergic activity was responsible for depressive symptoms, although Schildkraut, one of the principal proponents of this hypothesis, has been careful to state that changes in NE metabolism, observed in depressives, may be part of the pathophysiology of depression, although not necessarily of primary etiological significance (Schildkraut 1965). This hypothesis was developed initially from observations that reserpine, which depletes biogenic amine stores, sometimes causes depression in humans after chronic use (Rech & Moore 1971), and the idea was strongly reinforced when it was discovered that tricyclic antidepressants (TCAs) blocked the reuptake of biogenic amines and thus initially increased central aminergic activity. The proposal was undermined by Vetulani et al (1976), who demonstrated that the acute increase in synaptic NE, secondary to reuptake blockade, resulted in a later decrease in postsynaptic β-receptors, and thus presumably a later decrease, rather than an increase, in noradrenergic activity. Thus, it appears that the delayed decrease in noradrenergic activity, rather than the immediate increase, may be involved in the therapeutic action of the TCAs and other antidepressant treatments, because of its better temporal correspondence with the delayed therapeutic action of these drugs. Using the same rationale as in the earlier hypothesis, one might conclude that depression results from overactivity of the noradrenergic system, rather than underactivity.

A similar logic was used in developing the DA hypothesis of schizophrenia (Friedhoff & Van Winkle 1964, Friedhoff 1969), that is, that schizophrenia involved hyperdopaminergia because antipsychotic drugs blocked DA receptors. In the case of the antipsychotic drugs, the relationship of their therapeutic effects and changes in receptor sensitivity is more complex. Maximal blockade can be produced rapidly (van Rossum 1967); however, the chronic effect of antipsychotic drug administration is up-regulation of postsynaptic DA receptors, which should result in increased dopaminergic activity. Increasing dopaminergic activity acutely by other means can result in exacerbation, rather than amelioration, of psychiatric symptoms. Thus, the role of receptor sensitivity changes in mediating the therapeutic action of conventional antipsychotic drugs is not clear.

It is tempting, in the absence of other leads, to presume that effective psychotropic agents normalize the primary pathophysiological aspects of psychiatric illness; however, these drugs may affect aspects of brain function not intimately involved in the disease process. Direct evidence for the etiological role of NE in depression or of DA in psychosis has been scant. In regard to schizophrenia, of greatest interest have been post mortem studies, in which Seeman (1980) and Cross et al (1980) have shown that brains of schizophrenics have a greater number of specific DA binding sites than brains of nonpsychotic individuals (Crow et al 1979, Cross et al 1980, Seeman 1980). An obvious danger in interpreting these findings is that a subject's exposure to antipsychotic drugs during life could produce a compensatory increase in specific DA sites. Both Seeman and Crow report that they have found an increase in DA binding sites in the brains of some subjects with no history of antipsychotic drug use, but this may be difficult to validate in post mortem subjects.

Tourette syndrome is another disorder in which hyperdopaminergia has been proposed as an etiological factor, again because of the therapeutic response of many Tourette patients to haloperidol. Attempts to demonstrate a hyperdopaminergic state in this condition have also been largely unsuccessful. In fact, there is evidence of a small reduction in concentration of the DA metabolite, homovanillic acid, in the spinal fluid of these patients (Koslow 1982), which has been interpreted as a compensatory response to a hypothesized supersensitivity of DA receptors.

Apart from the validity of these specific hypotheses, which is not of concern in this chapter, is there a basis for concluding that the primary disturbance in psychosis and depression resides in the brain function that is normalized by a therapeutic agent? In the case of both antipsychotic and antidepressant drugs there are decided reasons for concluding that the drugs do not correct a primary etiological disturbance. Perhaps the most compelling evidence against this notion is that both antipsychotic and antidepressant drugs are active against a wide range of psychiatric entities, all of which could not have the same etiology. Thus, antipsychotic drugs are active against symptoms of organic psychosis, amphetamine psychosis, manic depressive psychosis, schizophrenia, and the nonpsychotic illness, Tourette syndrome; similarly, antidepressant drugs are active against a number of types of depression that are almost certainly of different etiologies.

Biogenic amine receptor systems have adaptive properties that contribute to the plasticity of the brain, particularly its ability to accommodate to alien substances or unusual environmental contingencies. We have proposed that these receptor systems may be involved in mechanisms for maintaining mental stability. From this perspective, psychotropic drugs could have beneficial effects in disorders of varying etiology by enhancing the ability of these adaptive systems to accommodate to different pathological processes that upset some aspect of mental function. For instance, Stone (1979) has shown that antidepressants produce the same kind of biochemical adaptation as occurs in response to stress, and has proposed that antidepressant drugs increase resistance to stress, much as prior exposure to a stress can reduce the reaction resulting from subsequent exposure to that stress. Thus, there may be central restitutive systems, each of which can be activated by specific agents, that protect against insults to the CNS of varying etiology.

Despite certain logical objections and the failure to find direct evidence of abnormalities in biogenic amine function in disorders responding to antipsychotic and antidepressant drugs, the possibility remains that there are primary abnormalities in psychosis and depression that involve biogenic amine receptors rather than transmitters. Abnormalities such as those described in autopsied brains of schizophrenics (Crow et al 1979, Cross et al 1980, Seeman 1980) would be very difficult to detect in living human subjects.

What Is Receptor Sensitivity Modification Treatment?

Deliberate receptor sensitivity modification (RSM) was developed as a treatment for conditions believed to be responsive to a reduction in dopaminergic activity, i.e. tardive dyskinesia, schizophrenia, and Gilles de la Tourette syndrome (Friedhoff 1977, Friedhoff & Alpert 1978). This treatment involves the production of increased dopaminergic activity by the administration of a dopaminergic agonist or agonist precursor, which in turn results in a secondary down-regulation of dopamine receptors, and a net reduction in dopaminergic activity that occurs when the agonist or precursor is withdrawn. This decreased activity of the dopaminergic system has been investigated as a treatment for the three conditions.

Deliberate modification of the sensitivity of other transmitter systems, using specific agents, might also have therapeutic effects in the treatment of other conditions, although no such studies have yet been performed to our knowledge. The down-regulating effect of TCAs, although it may be involved in the therapeutic action of these drugs, was discovered after the fact. These agents are, however, now being evaluated to determine how their RSM properties may be involved in the mediation of their therapeutic action.

Because a receptor is a functional system rather than a simple structure, the net effect of an alteration of one of its components is not easy to determine. A decrease in the number of surface recognition sites, for example, may involve a series of adjustments in synaptic function, each of which can serve to offset or to enhance the decreased activity of the system that would be expected from the reduced number of specific binding sites. Alterations can occur in coupling of the receptor, in ion flux across synaptic membranes, in the electrical activity of the neuron, and in its metabolic function. Thus, the way in which the alteration of one component of the system will affect more complex functions, such as behavior, is not always predictable and may be different, initially, from longer term effects.

One neuronal system in which these complex interactions have been intensively investigated is the striatal DA system. In the corpus striatum, DA terminals originating from cell bodies in the zona compacta region of the substantia nigra either terminate directly on cholinergic cells or act on them via short interneurons (Hattori et al 1976). These postsynaptic DA receptors are believed to mediate, primarily, inhibitory actions (see Siggins 1978 for review). Thus, release of DA from presynaptic terminals results in the inhibition of acetylcholine (ACh) release from the axon terminal of the postsynaptic cholinergic cell (Bartholini et al 1975).

A number of dopaminergic agonists and antagonists have been identified that produce behavioral effects, and specific binding assays have been developed (Creese et al 1975) to study changes in the concentration or affinity of the ligand recognition sites of the receptor. In addition, certain DA receptors (D_1) have been found to be coupled to adenylate cyclase (Kebabian & Greengard 1971, Kebabian et al 1972), an enzyme activated as a result of the receptor-ligand interaction. This cyclic AMP-forming system is itself regulated by the calcium binding protein, calmodulin, and probably by GTP (Gnegy et al 1977). The phosphodiesterase responsible for cyclic AMP degradation is also regulated (Cheung 1970). Polypeptides, so called co-transmitters, are also released with dopamine (Hökfelt et al 1980a,b), although their role in synaptic function is not known. Adjustments in transmitter release can also occur through modification of presynaptic autoreceptors (D_3) (Muller & Seeman 1979) and through long loop feedback systems (Meller et al 1980).

There is evidence, albeit circumstantial, that postsynaptic DA receptors that are not coupled to adenylate cyclase (D_2) mediate the action of neuroleptics. The clinical potency of neuroleptics generally parallels their ability to block striatal D_2 receptors (Seeman 1980). Most of these drugs also weakly block DA stimulation of adenylate cyclase; however, their ability to inhibit this enzyme does not correlate with their clinical potency (Clement-Cormier et al 1974, Karobath & Leitich 1974, Iversen 1975). When DA receptors are deprived of DA—through lesioning, by chemical blockade, or by synthesis inhibition—supersensitivity of both dopamine-stimulated adenylate cyclase and D_2 receptors results (Burt et al 1976, Friedhoff et al 1977, Muller & Seeman 1978).

The study of supersensitivity of the DA receptor has been greatly facilitated by the behavioral model developed by Ungerstedt & Arbuthnott (1970), and through the use of apomorphine as a direct DA agonist (Ernst 1967, Anden et al 1967). Tarsy & Baldessarini (1974) and others (Christensen & Moller-Nielsen 1974, Sayers et al 1975) showed that apomorphine-induced stereotyped behavior was increased in rats, following relatively prolonged pretreatment with the dopamine synthesis inhibitor, α-methyl-p-tyrosine (AMPT). Chronic administration of antipsychotic drugs, followed by wash-out, also produced this effect. They concluded that this behavioral stereotypy was probably specifically related to compensatory supersensitivity of the DA receptor. Supersensitivity did not occur when non-DA receptor blockers like phenobarbital were used as a pretreatment, even though behavioral depression was produced as with the antipsychotic drugs.

Long-term effects of antipsychotic drugs may be involved in the delayed therapeutic action of these drugs via mechanisms that are not yet clear, and also in some of their adverse effects. The relationships between acute and longer term effects are not simple. Although quieting of psychotic patients occurs very soon after administration of a sizeable dose, possibly because of nonspecific sedative effects of these drugs, specific effects on psychotic symptoms do not occur for days to weeks after initiation of treatment. Interestingly, the sedative effects believed to be mediated by the α-noradrenergic and possibly histamine$_1$ receptor blocking effects of antipsychotic drugs tend to disappear over time, possibly because of compensatory effects occurring in these systems (Hill & Young 1978, Quach et al 1979, Rehavi et al 1980).

The RSM treatment was developed specifically to reduce postsynaptic DA receptor sensitivity, based on the supposition that supersensitivity of the dopaminergic system could be reversed by administration of L-dopa to animals. When access of DA to the receptor is limited, either through blockade with antipsychotic drugs or by interference with presynaptic function, an increase in the number of specific postsynaptic DA binding sites results. Therefore, overstimulation of these sites with DA might be expected to return the system toward normal.

The results of an experiment carried out to test this hypothesis can be found in Figure 1. Neuroleptic drugs produced an increase in the concentration of DA binding sites, an increase in DA-stimulated adenylate cyclase activity, and in apomorphine-induced stereotypy. Administration of L-dopa, following the neuroleptic, produced a reversal of these three manifestations of supersensitivity, presumably by increasing the supply of DA at the receptor site. These changes all persisted for substantial periods of time (Friedhoff et al 1977).

Seeman (1980) and J. W. Schweitzer (unpublished observations) have shown that the duration of persistent drug-induced DA supersensitivity is relatively independent of drug dose, but is a function of the duration of treatment, lasting, after the drug is withdrawn, for about 60% of the initial treatment period. Pharmacological manipulation of striatal DA receptors also has effects on related neurons. A type of reciprocal balance is believed to exist between striatal dopmainergic and cholinergic neurons (Friedhoff & Alpert 1973, Bartholini et al 1975, Siggins 1978). For instance, the symptoms of Parkinsonism can frequently be relieved not only by dopaminergic agonists such as L-dopa (Birkmayer & Hornykiewicz 1961, Friedhoff et al 1963, Cotzias et al 1969) but also by anticholinergic agents (Duvoisin 1967). DA has been found to produce a dose-related inhibition of K+-stimulated release of ^{3}H-ACh from rat striatal slices (Miller & Friedhoff 1979). Chronic pretreatment with haloperidol for 28 days, followed by

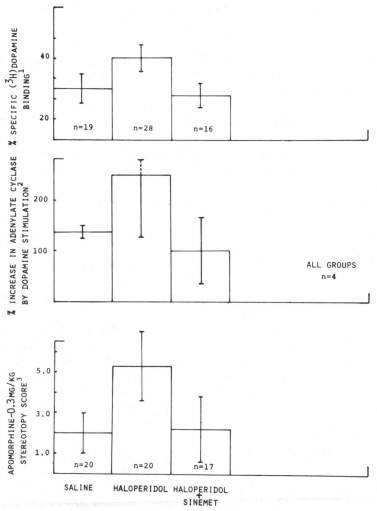

Figure 1 Reversal of behavioral and biochemical manifestations of dopamine receptor super-sensitivity by administration of Sinemet®. Rats in the saline group were treated with saline for 28 days. Rats in the haloperidol group and the haloperidol plus Sinemet®(L-dopa-carbidopa) group were treated with haloperidol 2.0 mg/kg during that same period. At the end of that time haloperidol treatment was terminated. For the next ten days the saline group continued to receive saline, the haloperidol group also received only saline, and the haloperidol plus Sinemet® group received Sinemet® in a dose equivalent to 200 mg/kg L-dopa alone. All treatments were stopped at the end of the thirty-eighth day. Rats were taken for behavioral or biochemical assays five days later. Specific ³H-dopamine binding is defined as the percentage of total ³H-binding displaced by (+)-butaclamol. Adenylate cyclase activity is defined as the amount of cAMP formed in 0.3 mm striatal slices in the presence of the phosphodiesterase inhibitor, RO 20-1724. Stereotypy observations were made 30 min after treatment with apomorphine.

treatment with AMPT for ten days, resulted in a significant shift to the left in the dose-dependent inhibition of K^+-stimulated release of ^{3}H-ACh by DA. This shift to the left in the dose response curve is apparently the result of an increase in the number of striatal postsynaptic DA receptors produced by chronic DA receptor blockade (Gianutsos et al 1974, Friedhoff et al 1977, Friedhoff & Alpert 1978) and inhibition of DA synthesis by AMPT. Thus, in addition to the increase in DA receptor number and the increase in DA-stimulated adenylate cyclase activity in striatum following chronic haloperidol, there is also an increased sensitivity to dopaminergic inhibition of cholinergic release.

The molecular mechanisms involved in up- and down-regulation of receptors are not clear. Mukherjee & Lefkowitz (1976) showed that increased β-receptor number, induced by culture of frog erythrocytes with the β-blocker, propanolol, as well as the decreased number resulting from culture with the agonist, isoproterenol, could not be prevented by protein synthesis inhibitors. Thus, regulation of receptor number, at least for β-receptors in nucleated frog erythrocytes, apparently does not involve *de novo* protein synthesis. In support of this view, Chuang & Costa (1979) have shown that β-adrenergic desensitization produced by prolonged exposure to isoproterenol [reflected by decreased ^{3}H-dihydroalprenolol (^{3}H-DHA) binding to surface membranes] also resulted in increased ^{3}H-DHA binding in the cytosol of frog erythrocyte preparations. These changes were independent of RNA and protein synthesis. These authors suggest that the disappearance of surface binding sites and the increase in binding sites in the cytosol may result from internalization of membrane receptors.

These are interesting leads, but the molecular mechanisms involved in the regulation of receptor sensitivity are still poorly understood. It is, however, well established that the sensitivity of biogenic amine systems can be modified by manipulation of agonist supply. This principle provides the basis for the clinical application of RSM.

Possible Role of RSM In The Mechanism of Action of Antidepressant Drugs

Antidepressant drugs were not developed for their long-term effects on receptors; however, it now appears that RSM may also be involved in the action of these drugs. In this review we emphasize recent biochemical findings regarding the possible involvement of receptor sensitivity changes in the action of antidepressant drugs.

Antidepressant drugs include monoamine oxidase inhibitors (MAOIs) such as pargyline, tertiary and secondary tricyclics such as imipramine and desmethylimipramine (DMI), intermediate structures such as amoxapine, and novel structures such as iprindole and mianserin. The acute action shared by all of these agents is to enhance biogenic amine activity in the CNS. Tricyclic antidepressants block the reuptake mechanisms for NE and serotonin and also reduce their turnover (Axelrod et al 1961, Glowinski et al 1966, Carlsson et al 1966, Rosloff & Davis 1978, Svensson, 1978) but do not significantly affect DA reuptake or inhibit DA turnover (Glowinski et al 1966, Carlsson et al 1966). In vivo NE reuptake is inhibited equally by secondary and tertiary TCAs, whereas the tertiary compounds are two to three times more potent in inhibiting reuptake of serotonin (Ross & Renyi 1967, Carlsson et al 1969). The atypical antidepressant, iprindole, has been shown to be clinically effective (Davemen 1967, Rickels et al 1973) but does not inhibit the reuptake of biogenic amines (Gluckman & Baum 1969, Ross et al 1971, Zis & Goodwin 1979) or inhibit either NE or serotonin turnover (Sanghvi & Gershon 1975).

An additional property of most antidepressants is their ability to bind to several neurotransmitter receptor binding sites. The tertiary and secondary TCAs are very potent inhibitors (nanomolar concentrations) of histamine$_1$ receptors (Green & Maayani 1977, Kanof & Greengard 1978, Richelson 1978, Tran et al 1978) and fairly potent inhibitors of muscarinic cholinergic (Snyder & Yamamura 1977) and serotonin receptors (Fuxe et al 1977). Recent studies point to two physically distinct serotonin receptors (Peroutka et al 1981): an S_1 receptor, associated with serotonin-stimulated adenylate cyclase (Peroutka & Snyder, 1980), and an S_2 receptor, which is not coupled to the enzyme. Apparently, the TCAs are potent only against the S_2 binding sites of the cerebral cortex, whereas iprindole is a very poor inhibitor of S_2 (^{3}H-spiroperidol) receptor binding (Peroutka & Snyder 1980). Neither DA-receptor binding sites nor β-receptor sites are inhibited to any significant extent by any of the antidepressant agents (Peroutka & Snyder 1980). Thus those sites that would be immediately blocked by TCA treatment include histamine$_1$, muscarinic cholinergic, α_1, and serotonin$_2$ receptors.

All TCAs, MAOIs, ECS, and some atypical antidepressants reduce NE-stimulated adenylate cyclase and decrease the concentration but not affinity of β-receptors (Vetulani et al 1976) and possibly S_2 receptors (Peroutka & Snyder 1980) secondarily to increasing synaptic NE or serotonin (see Creese & Silbley 1981 for a detailed review). This view is supported by the inability of TCAs to produce β-receptor desensitization in animals in which the presynaptic terminals are destroyed by 6-OH-DA (Wolfe et al 1978, Schweitzer et al 1979, Janowsky et al 1981). The atypical antidepressant, iprindole, which does not inhibit NE reuptake, also can induce β-receptor subsensitivity (Wolfe et al 1978) and also induces S_2 receptor subsensitivity (Peroutka & Snyder 1980). Mianserin, whose effect on NE reuptake is unclear, was found to down-regulate β-receptors (Clement-Jewery 1978), an effect that may be mediated by presynaptic α_2 receptor blockade, which would also increase synaptic NE levels (see Green 1981 for a review). These results could not be confirmed by Brunello et al (1981), who suggest that ^{3}H-mianserin binds to postsynaptic serotonin receptors inasmuch as lesioning of serotonin terminals results in an increase in ^{3}H-mianserin binding.

The down-regulation is a delayed response and coincides temporally with the time at which improvement occurs in most depressives. Electroconvulsive shock has been reported to desensitize β-receptors in rats more rapidly than the TCAs (Sulser et al 1978), and may even have a greater effect in down-regulating the NE-sensitive adenylate cyclase. Amoxapine has been reported to be a rapidly acting antidepressant (Hekimian et al 1978); however, no information is available as to its speed of down-regulation of receptors.

Peroutka & Snyder (1980) have reported that long-term antidepressant treatments did not effect changes in the number or affinity of S_1, muscarinic cholinergic, α_1, or postsynaptic DA receptors; however, Rehavi et al (1980) found that α_1 receptors were increased after treatment with antidepressants. DA autoreceptor sites in the substantia nigra, however, may be desensitized by chronic antidepressant treatment (Chiodo & Antelman 1980a,b, Green 1981). The mechanism by which this effect on the DA autoreceptor is produced is not completely clear.

Neurophysiological studies have shown that repeated injections of TCAs in animals produces increased baseline firing of hippocampal pyramidal cells that receive locus coeruleus efferent fibers (Huang 1979). This increased firing could result from desensitization of postsynaptic receptors, inasmuch as NE inhibits hippocampal pyramidal cells. Serotonergic neuronal firing in midbrain raphe cells is reduced by acute administration of TCAs; however, CA3 hippocampal cells that are innervated by raphe serotonergic fibers have been found to have an increased response to serotonin after chronic antidepressant treatment (de Montigny & Aghajanian 1978). Thus, supersensitivity, not subsensitivity, of some serotonin receptors results from chronic antidepressant treatment. Further study is needed of sensitivity changes in serotonin receptors as a result of chronic treatment. Some antidepressants are blockers of S_2 binding and are also serotonin uptake blockers. The net result of these competing actions is not clear (Peroutka & Snyder 1980, deMontigny & Aghajanian 1978).

The therapeutic efficacy of antidepressant drugs does not appear to result from a simple action at a single locus, especially considering the multiple interactions of biogenic amine systems. Increasingly, however, it appears that down-regulation of the β-system is important in mediating the therapeutic action of antidepressant drugs.

Deliberate RSM As a Novel Treatment Approach

In investigating the possibility that RSM might be useful as a treatment, the assumption has been made that a change in postsynaptic receptor density or affinity has an effect similar to that resulting from a change in agonist supply. It is well established that antipsychotic drugs (DA receptor blockers) acutely reduce the concentration of DA at the receptor by blocking its access. Thus, all other factors being equal, in the presence of antipsychotic drugs the inhibition of neural activity by DA would be decreased, and the release of ACh from the axon terminal would be increased. It seemed reasonable that the same result could be achieved by chronic administration of L-dopa, which would produce an initial increase in DA concentration, but an ultimate adaptive decrease in receptor density. Therefore, when L-dopa was withdrawn, there would be a decrease in the inhibitory action of DA and an increase in ACh release. Conversely, the acute effects of DA agonists might be duplicated by chronic administration and then withdrawal of antipsychotic DA blocking drugs.

Adaptive changes in receptors, once induced, appear to last for a long time after termination of drug treatment. Thus, the basis exists for using the adaptive changes in the DA system and other similarly responding receptor systems as a novel means of pharmacological treatment—the desired therapeutic effect occurring when the treatment is stopped, and persisting for some time after that.

TREATMENT OF TARDIVE DYSKINESIA   As has been pointed out above, tardive dyskinesia is one condition in which there is strong circumstantial evidence in support of the idea that chronic DA receptor blockade by antipsychotic drugs results in an excess of DA receptors in the caudate. In an attempt to restore the receptor number to normal, victims of tardive dyskinesia have been treated with the RSM approach, using L-dopa as the agent for enhancing dopaminergic activity. In the original studies, L-dopa was administered in gradually increasing doses, over several weeks, beginning with 0.5 g and increasing to 6.0 g. Once having achieved the maximum, the dose was held constant for two to three weeks and then abruptly terminated. The step limiting the rate of increase of L-dopa was the severity of side effects, principally nausea, and worsening of tardive and psychotic symptoms.

The first treatment approach involved termination of antipsychotic treatment before initiation of L-dopa treatment and was, therefore, limited to those patients not requiring continued antipsychotic medication for control of their psychiatric symptoms. Since these clinical studies were initiated, co-administration of L-dopa and haloperidol to rats in a ratio of 320/1 has been found to prevent the development of striatal DA supersensitivity (H. Rosengarten and A. J. Friedhoff, unpublished observations). This ratio is achievable in the clinical situation, using moderate but effective doses of haloperidol.

In the studies described below (Alpert & Friedhoff 1980, Alpert et al 1982), tardive dyskinesia patients were treated with L-dopa-carbidopa alone (Sinemet®), or in combination with antipsychotic medication. When antipsychotic drugs were co-administered, doses were adjusted to achieve ratios roughly equivalent to those found to block the development of supersensitivity in rats. Before entry into the study, patients were observed for six weeks, either on a constant dose of antipsychotic medication, or without medication. If tardive symptoms showed evidence of regression, patients were dropped from the study. The remaining patients were then randomly assigned to either L-dopa-carbidopa treatment or no L-dopa-carbidopa. Thus, four groups resulted: (*a*) antipsychotic only, (*b*) antipsychotic and L-dopa-carbidopa, (*c*) L-dopa-carbidopa, (*d*) no treatment. Treatment was carried out according to the schedule in Figure 2.

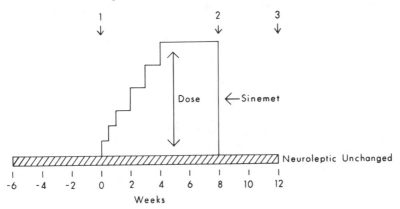

Figure 2   Schedule of RSM treatment of tardive dyskinesia. Subjects were patient volunteers hospitalized for chronic schizophrenia, who also had a confirmed diagnosis of tardive dyskinesia. Most were receiving antipsychotic medication, but a few were receiving no medication. At time minus six weeks, antipsychotic medication was terminated in additional subjects so as to provide more untreated subjects. Tardive dyskinesia symptoms were assessed over the next six weeks. During this period, subjects who showed changes in severity of tardive dyskinesia, as assessed by the Abnormal Involuntary Movement Scale (AIMS), were dropped from the study. At time zero, fifteen subjects, selected without knowledge of their AIMS score, were entered into the Sinemet® (L-dopa-carbidopa) group. The remaining ten subjects were followed, as were the Sinemet® group, but did not receive Sinemet®. Both the Sinemet and no Sinemet® group included subjects receiving or not receiving concommittant antipsychotic medication. The Sinemet dose was increased stepwise to a maximum equivalent of from 3 to 6 gm of L-dopa alone, and maintained at that level of approximately four weeks before treatment was abruptly terminated. The evaluator was blind as to treatment or treatment phase.

In Figure 3, it can be seen that L-dopa-carbidopa produced a significant decrease in tardive dyskinesia symptoms after discontinuation of this treatment (week 12) as compared to their level before treatment. No such change was seen in the patients who did not receive the L-dopa-carbidopa, regardless of whether their antipsychotic medication was discontinued. Although it is not shown in the figure, it is of interest that the L-dopa-carbidopa treatment produced transient worsening of tardive dyskinesia symptoms during the period of drug administration.

These results are compatible with the hypothesis that DA receptor supersensitivity is involved in tardive dyskinesia, that increased synthesis of DA, stimulated by the L-dopa carbidopa treatment, initially worsens the condition, and that improvement results after discontinuation of L-dopa-carbidopa because of compensatory down-regulation of DA receptors. Although evidence from rat studies supports this conclusion, in the absence of direct pathological studies of humans and validating studies by additional investigators, this proposal must be considered to be tentative.

Several investigators have reported that depression is a risk factor in tardive dyskinesia (Alpert et al 1976, Davis et al 1976, Rosenbaum et al

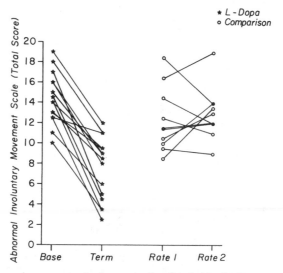

Figure 3 Abnormal movements of L-dopa treated tardive dyskinetics ($N = 15$) and comparison group of equivalent baseline pathology ($N = 10$). Second rating is four months after baseline. Subjects who received Sinemet® (L-dopa-carbidopa) were treated according to the schedule shown in Figure 2. Only subjects whose tardive symptoms remained constant during the six weeks preceding the trial were entered into the treatment phase. The baseline assessment of tardive dykinesia symptoms was made just before the initiation of the Sinemet® phase, and the termination assessment was made four weeks after discontinuation of Sinemet®. Subjects not receiving Sinemet® were assessed at the same time points.

1977, Cutler et al 1981). Alpert has suggested that this association might result from a tendency of depressed patients to abnormally up-regulate catecholamine receptors (see Rush et al 1981), thus extending the proposals of Vetulani et al (1976) about the action of antidepressant drugs (see section on Antidepressants, above). Up-regulation of NE receptors could lead to depression, while excessive up-regulation of DA receptors by antipsychotic medication could produce tardive dyskinesia.

TREATMENT OF TOURETTE SYNDROME   Tourette syndrome, which is characterized by both motoric and behavioral symptoms, is responsive to treatment with moderate doses of the DA blocker, haloperidol. A number of patients are unable to tolerate this treatment, because it reduces motivation and has other unpleasant side effects. To date, six patients who responded to haloperidol, but were unable to tolerate its side effects, have been treated with the same RSM paradigm as described for tardive dyskinesia. As with the tardive dyskinesia patients, some worsening occurred during the L-dopa administration followed by a general remission of symptoms after termination of L-dopa treatment. In all six cases, a recurrence occurred within four months to two years. Several patients have been successfully retreated.

TREATMENT OF SCHIZOPHRENIA   Several studies of the treatment of patients with schizophrenia have been carried out, using the RSM approach. Alpert et al (1978) studied a small number of patients treated either with chlorpromazine or L-dopa. The L-dopa group became somewhat worse during treatment, as has been reported earlier (Yaryura-Tobias et al 1970); however, in contrast to findings of the earlier study, much of the persistent worsening appeared to occur from an overlying toxic psychosis rather than an exacerbation of primary symptoms of schizophrenia. In fact, after an initial worsening, some symptoms diminished in severity while the patients were receiving the L-dopa. Patients were only observed for four days after termination of treatment, and significant overall improvement would not have been expected, and was not observed, at that time.

In a second study, Beramendi et al (1980) treated 12 patients with L-dopa-carbidopa, nine with haloperidol and ten with placebo. Approximately 50% of the L-dopa-carbidopa group showed clinically significant improvement four weeks after termination of treatment, although overall the haloperidol group fared better. As a corollary study, Gutierrez et al (1979) studied muscle rigidity in this group. They found, as expected, that some patients became more rigid during the haloperidol treatment, presumably from the extrapyramidal effects of this drug secondary to DA receptor blockade. Surprisingly, some of the L-dopa group also became more rigid.

Upon analysis of the data, they found that those patients who became most rigid with haloperidol had the least improvement, while those who became most rigid with L-dopa had the most improvement.

In a related study, Alpert et al (1977) measured resting finger tremor as an index of extrapyramidal function in patients receiving antipsychotic medication. These investigators found a negative correlation between tremor index of DA receptor blockade, measured four days after initiation of treatment, and a reduction in schizophrenic symptomatology after three weeks of treatment, i.e. those individuals who showed evidence of the most effective DA receptor blockade tended to respond least well to the antipsychotic medication.

The findings of these two studies are unexpected indeed, considering that antipsychotic drugs are believed to produce their therapeutic effect via their ability to block DA receptors. It should be noted, however, that blockade of DA receptors by antipsychotics produces a compensatory release of DA, which can compete with antipsychotic drugs for the receptor sites, particularly for the D_1 cyclase linked receptors for which antipsychotic drugs have a low affinity. Thus, those individuals in whom haloperidol provoked the greatest release of DA would have the greatest reduction in blockade and therefore the least extrapyramidal deficit, particularly at the onset of treatment when blood levels of the antipsychotic drug were low and blockade was incomplete. Chase et al (1970) have provided direct evidence that this occurs, showing that a greater release, induced by antipsychotics, of DA metabolites into spinal fluid was associated with a lesser degree of antipsychotic-induced extrapyramidal rigidity. It is thus tempting to speculate that the release of DA, provoked by blockade of DA receptors by antipsychotics, is somehow involved in the therapeutic action of these drugs.

Although the number of studies of the clinical efficacy of RSM treatment are still few in number, the approach is promising, and it provides a useful tool for the exploration of the behavioral consequences of the adaptation of various receptor systems.

Effect of Fetal Exposure to Psychotropic Drugs on Receptor Development And Sensitivity

Alterations in the fetal environment through exposure to certain drugs can alter the regulation of specific receptor systems on which the drugs act (Rosengarten & Friedhoff 1979, Kellogg et al 1980). These changes can be associated with permanent alterations in brain function, in the absence of gross morphological changes (Butcher et al 1975, Vorhees et al 1978). Prenatal drug exposure can also have profound and enduring effects on postnatal behavior, in human and in nonhuman species. In humans, postnatal functional deficits are found in infants exposed prenatally to alcohol

(Jones & Smith 1973) and to barbiturates and anticonvulsants (Smith 1977, Feldman et al 1977). The literature on the effects of drug exposure prenatally have been reviewed by Joffe (1969), Kornetsky (1970), Coyle et al (1976), and Hutchings (1978).

Learning deficits have been consistently reported in animal offspring exposed to neuroleptic drugs during gestation (Hoffeld & Webster 1965, Ordy et al 1966, Clark et al 1970, Lundborg 1972, Ahleneius et al 1973, Engel & Lundborg 1974, Lundborg & Engel 1976). Lundborg (1972) reported that rabbits born to haloperidol-treated mothers had severe motor dysfunctions such as head tremors, problems with head lifting, and abnormal gait. Chronic haloperidol treatment during pre- and postnatal development has also been reported to produce hyperactivity and an attenuated response to amphetamine in young adult rats (Spear et al 1980, Shalaby & Spear 1980). Anticonvulsants given prenatally also produce motor disturbances in offspring (Ata & Sullivan 1977). Prenatal exposure to the ataractic agent, diazepam, also alters postnatal behavior (Kellogg et al 1980), producing depressed locomotor and acoustic startle responses in the rat.

There are few published studies of the effects of psychotropic agents on the differentiation and maturation of transmitter receptors. Rosengarten & Friedhoff (1979) have demonstrated that certain drugs that are administered during the fetal and early postnatal periods have pronounced prolonged effects on central DA receptors and on a DA-mediated behavior. Pregnant female rats were treated with either haloperidol, AMPT, or saline for 16 days beginning on day 4 or 5 of gestation. DA receptor binding (^{3}H-spiroperidol) and apomorphine-induced stereotyped behavior were subsequently assessed in the offspring of the treated mothers at 14, 21, 28, 35, and 60 days of age. Both receptor blockade and synthesis inhibition treatments resulted in a marked decrement in the concentration of DA receptors in the striatum of offspring, up to weaning age. The haloperidol effect persisted well into adulthood. Apomorphine-induced stereotypy was also significantly reduced in these animals. When pups were exposed to haloperidol immediately after birth, through haloperidol given to their nursing mothers and transmitted via their milk, they demonstrated a significant increase in the concentration of striatal DA receptors and enhanced apomorphine-induced stereotypy, effects similar to those observed in adult rats. In contrast, L-dopa administered during a short period, toward the end of the second and beginning of the third gestational week, greatly increased the number of DA sites. Because haloperidol prevents the access of DA to the receptor and AMPT blocks DA synthesis, whereas L-dopa increases the turnover or release of DA, it appears that DA receptor number is influenced by exposure to DA during appropriate periods of fetal development. It is suggested that at least some neurotransmitters play an important role in programming the number of postsynaptic receptors during development, and that adult compensatory mechanisms are not yet operative during fetal and possibly very early postnatal receptor ontogeny.

In the rat, significant differentiation and functional maturation of central transmitter systems occurs during gestational days 13–16 (Lauder & Bloom 1974, Olson & Seiger 1972, Schultze et al 1974, Johnston et al 1981, Beaulieu & Coyle 1981). It is of considerable interest that a similar period has been found to be the critical or sensitive period for the effects produced by antipsychotic drugs or L-dopa. Administration of these substances before this gestational period has no effect on receptor number, whereas later administration produces the adult compensatory response (H. Rosengarten and A. J. Friedhoff, unpublished observations).

Recently, Deskin et al (1981) suggested that presynaptic dopaminergic activity may be necessary for the maturation of postsynaptic DA receptors. They found that elimination of dopaminergic presynaptic terminal development by intracisternal 6-OH DA on days 1 and 6 after birth resulted in the retardation of the ontogeny of dopamine receptors in the whole brain, an effect consistent with that of *in utero* haloperidol treatment but in contrast to that of haloperidol given to the early postnatal animal via mothers milk (Rosengarten & Friedhoff, 1979, Madsen et al 1981) or by direct intraperitoneal injection (J. C. Miller and A. J. Friedhoff, unpublished observations), which produced an increase in the number of specific DA binding sites and not the decrease seen after prenatal exposure.

It is not clear why haloperidol administered immediately after birth produces the expected increase in DA sites seen in mature animals, while 6-OH DA treatment during the same period produces a decrease, i.e. the same result as that produced by haloperidol prenatally. Perhaps there are regional differences in response, inasmuch as the 6-OH DA studies were carried out using whole brain assays, whereas the haloperidol effect was assessed in the striatum. Also, feedback mechanisms in different DA systems may develop differentially and, therefore, be more or less sensitive to either 6-OH DA or antipsychotic drug treatment. Alteratively, depriving the receptor of DA by receptor blockade with an antipsychotic drug and by presynaptic denervation with 6-OH DA may have differential effects in the developing animal.

In a study carried out with a somewhat different receptor binding assay, Madsen et al (1981) reported a decrease in sensitivity to apomorphine-induced stereotypy in 10 and 15 day old offspring of mothers treated during pregnancy with antipsychotic drugs. These investigators also found a reduction in the affinity of ^{3}H-spiroperidol for striatal receptors at day 7 of postnatal life, but not at older ages. They concluded that the changes following prenatal haloperidol exposure resulted from residual haloperidol in the newborn pups. The issue of residual antipsychotic medication in pups born to mothers exposed to these drugs has been investigated directly in other studies. H. Rosengarten and A. J. Friedhoff (unpublished observations) have found persistent reduction in the density of striatal dopamine receptors in pups fostered to drug-free mothers. Similar defects have also been found after prenatal DA synthesis inhibition with AMPT, which does not block DA receptors.

In preliminary studies it has been found that prenatal exposure to haloperidol produces a marked increase in the number of muscarinic cholinergic receptors. (J. C. Miller and A. J. Friedhoff, unpublished observations). Since differentiation of the dopaminergic system apparently precedes differentiation of the cholinergic system in the rat (Hattori & McGeer 1973), a change in the sensitivity of striatal DA receptors at an early critical period may alter the normal development of the functional relationship between these two systems in the striatum. Such changes may have important implications for later development of normal extrapyramidal function because of a shift in the balance between dopaminergic and cholinergic activity.

Although many humans have been exposed to psychotropic agents in utero, the behavioral consequences of this exposure have been little studied (Vorhees et al 1978). Fetal exposure to antipsychotic drugs has occurred in offspring of psychotic women maintained on neuroleptics for management of psychotic symptoms during pregnancy. In addition, many women have received neuroleptics such as prochlorperazine for control of nausea and vomiting during pregnancy. Although no evidence of anatomical teratology has been found, if human exposure produces decrements similar to those found in rats, these decrements could have significant behavioral effects. It is not known for instance, what the implications of diminished dopaminergic activity would be for offspring of schizophrenics, who are at higher risk for schizophrenia (Cancro 1973, Rosenthal & Kety 1968), a disorder that responds to reduction in dopaminergic activity. Persistent behavioral consequences of exposure to other types of drugs must also be suspected (Vorhees et al 1978, Kellogg et al 1980).

Conclusions

Psychobiological investigations into the cause and treatment of psychiatric and behavioral disorders have largely followed a model successful in identifying the etiology of a number of mental deficiencies and medical disorders. This strategy has involved a search for deviant metabolites, presumably secondary to genetic abnormalities, but has been generally unsuccessful. Psychological and behavioral functions involve continuous, long-term adaptations to environmental contingencies that are almost certainly mediated by adaptations at the cellular level. One significant regulatory aspect of biogenic amine function occurs at the cell membrane receptor. Thus, a better understanding of the long-term adaptive properties of specific receptor systems is necessary to understand the pathophysiology and treatment of mental and behavioral disorders.

Increasingly, it appears that existing psychotropic agents, particularly those involved in the treatment of psychosis and depression, affect the sensitivity of specific receptor systems. Also, overcompensation of DA receptor sensitivity seems to be an important factor in the pathogenesis of tardive dyskinesia. There is also other evidence that supersensitivity of the dopaminergic system might be associated with Tourette syndrome and schizophrenia, even if in a tangential manner. The failure to have found clear evidence of disturbances in receptor regulation until now is not surprising, inasmuch as this has not been investigated in these disorders until recently. It is known that fetal experience with drugs can modify receptor regulatory properties for long periods after birth in nonhuman species, and it seems likely that variations in the human fetal environment during critical periods can influence the adaptive properties of brain receptors, and perhaps affect adaptability in later life.

The use of deliberate strategies for modifying CNS receptors provides an opportunity to study these adaptive mechanisms and to attempt new treatment approaches. The use of deliberate RSM as a treatment for tardive dyskinesia and Tourette syndrome seems promising; however, the present strategy for RSM, through the modification of agonist supply, has an obvious deficiency: patients must get worse before they get better. The discovery of methods to provoke up- or down-regulation directly, by manipulating the as yet poorly understood molecular mechanisms for regulating receptor sensitivity, would be an important step in the direction of inducing RSM without the attendant worsening. This is an important area for future research.

Literature Cited

Ahlenius, S., Brown, R., Engel, J., Lundborg, P. 1973. Learning deficits in 4 week old offspring of the nursing mothers treated with the neuroleptic drug penfluridol. *Naunyn-Schmiedebergs Arch. Pharmacol.* 279:31–37

Alpert, M., Diamond, F., Friedhoff, A. J. 1976. Tremographic studies in tardive dyskinesia. *Psychopharmacol. Bull.* 12(2):5

Alpert, M., Diamond, F., Kesselman, M. 1977. Correlation between extrapyramidal and therapeutic effects of neuroleptics. *Comp. Psychiatr.* 18:333–36

Albert, M., Friedhoff, A. J. 1980. Clinical application of receptor sensitivity modification treatment. In *Tardive Dyskinesia: Research and Treatment,* ed. W. E. Fann, J. M. Davis, R. C. Smith, E. F. Domino, pp. 471–74. New York: Spectrum

Alpert, M., Friedhoff, A. J., Diamond, F. 1982. Use of dopamine receptor agonists to reduce dopamine receptor number as a treatment for tardive dyskinesia. In *Experimental Therapeutics of Movement Disorders,* ed. S. Faln, D. Calne, I. Shoulson. New York: Raven

Alpert, M., Friedhoff, A. J., Marcos, L. R., Diamond, F. 1978. Paradoxical reaction to L-dopa in schizophrenic patients. *Am. J. Psychiatr.* 135:1329–32

Anden, N. E., Roos, B. E., Wedinus, B. 1964. Effects of chlorpromazine, haloperidol and reserpine on the levels of phenolic acids in rabbit corpus striatum. *Life Sci.* 3:149–58

Anden, N. E., Rubenson, A., Fuxe, K., Hökfelt, T. 1967. Evidence for dopamine receptor stimulation by apomorphine. *J. Pharm. Pharmacol.* 19:627–29

Ata, M. M., Sullivan, F. M. 1977. Effects of prenatal phenytoin treatment on postnatal development. *Proc. Br. Pharmacol. Soc.* 59:494

Axelrod, J., Whitby, L. G., Hertting, G. 1961. Effect of psychotropic drugs on the uptake of ^{3}H-norepinephrine by tissues. *Science* 133:383–84

Barbeau, A. 1969. L-Dopa therapy in Parkinson's disease: Critical review of nine years experience. *Can. Med. Assoc. J.* 101:59–68

Bartholini, G., Stadler, H., Lloyd, K. G. 1975. Cholinergic-dopa-minergic interregulation within the extrapyramidal system. In *Cholinergic Mechanisms*, ed. P. G. Waser, pp. 411–18. New York: Raven

Beaulieu, M., Coyle, J. T. 1981. Effects of fetal methylazoxymethanol acetate lesion on the synaptic neurochemistry of the adult rat striatum. *J. Neurochem.* 37:878–87

Beramendi, V., Alpert, M., Guimon, J., Friedhoff, A. J., Gutierrez, M. 1980. Estudio controlado de la L-dopa enesquizofrenia con referencia a la teroria de la hipersesibilidad de los receptores dopaminergicos. *Arch. Neurobiol.* 43: 107–24

Birkmayer, W., Hornykiewicz, O. 1961. Der L-3,4-dioxyphenylalanin (-DOPA)-effekt bei der Parkinson-Akinese. *Wien. Klin. Wochenschr.* 73:787–88

Brunello, N., Chuang, D. M., Costa, E. 1981. Differences in brain recognition sites for typical and atypical antidepressants. *Soc. Neurosci. Abstr.* 7:934

Burt, D. R., Creese, I., Snyder, S. H. 1976. Antischizophrenic drugs: Chronic treatment elevates dopamine receptor binding in brain. *Science* 196:326–28

Butcher, R. E., Hawver, K., Kazmaier, K., Scott, W. 1975. Postnatal behavioral effects from prenatal exposure to teratogens. In *Basic and Therapeutic Aspects of Perinatal Pharmacology*, ed. P. L. Morselli, S. Garattini, F. Sereni, pp. 171–76. New York: Raven

Cancro, R., ed. 1973. *Annual Review of the Schizophrenic Syndrome*, Vol. 3, pp. 33–57. New York: Brunner/Mazel

Cannon, W. B., Rosenbleuth, A. 1949. *The Supersensitivity of Denervated Structures: A Law of Denervation*. New York: MacMillan

Carlsson, A., Corrodi, H., Fuxe, H., Hökfelt, T. 1969. Effect of antidepressant drugs on the depletion of intraneuronal brain 5-hydroxy-tryptamine stores caused by 4-methyl-α-ethyl-meta-tyramine. *Eur. J. Pharmacol.* 5:357–66

Carlsson, A., Fuxe, K., Hamburger, B., Lindqvist, M. 1966. Biochemical and histochemical studies on the effects of imipramine-like drugs and (+)-amphetamine on central and peripheral catecholamine neurons. *Acta Physiol. Scand.* 67:481–97

Carlsson, A., Lindqvist, M. 1963. Effect of chlorpromazine or haloperidol on formation of 3-methoxytyramine and normetanephrine in mouse brain. *Acta Pharmacol. Toxicol.* 20:140–44

Chase, T. N., Schnur, J. A., Gordon, E. K. 1970. Cerebrospinal fluid monoamine metabolites in drug-induced extrapyramidal syndrome. *Neuropharmacology* 9:265–68

Cheung, W. Y. 1970. Cyclic 3'5'-nucleotide phosphodiesterase, a demonstration of an activator. *Biochem. Biophys. Res. Comm.* 38:533–38

Chiodo, L. A., Antelman, S. M. 1980a. Electroconvulsive shock: Progressive dopamine autoreceptor subsensitivity independent of repeated treatment. *Science* 210:799–801

Chiodo, L. A., Antelman, S. M. 1980b. Repeated tricyclics induce a progressive dopamine autoreceptor subsensitivity independent of daily drug treatment. *Nature* 287:451–54

Christensen, A. V., Moller-Nielsen, I. 1974. Influences of flupenthixol and fluphenthixol-deconate on methylphenidat-and apomorphine-induced compulsive gnawing in mice. *Psychopharmacologica* 34:119–26

Chuang, D. M., Costa, E. 1979. Evidence for internalization of the recognition site of β-adrenergic receptors during receptor sensitivity induced by (−)-isoproterenol. *Proc. Natl. Acad. Sci. USA* 76:3024–28

Clark, C. V. H., Gorman, D., Vernadakis, A. 1970. Effects of prenatal administration of psychotropic drugs on behavior of developing rats. *Dev. Psychobiol.* 3: 225–35

Clement-Cormier, Y. C., Kebabian, J. W., Petzold, G. L., Greengard, P. L. 1974.

Dopamine-sensitivite adenylate cyclase in mammalian brain: A possible site of action of antipsychotic drugs. *Proc. Natl. Acad. Sci. USA* 71:1113–17

Clement-Jewery, S. 1978. The development of cortical β-adrenoreceptor subsensitivity in the rat by chronic treatment with trazodone, doxepin and mianserine. *Neuropharmacology* 17:799–81

Clow, A., Jenner, P., Theodorou, A., Marsden, C. D. 1979. Striatal dopamine receptors become supersensitive while rats are given trifluoperzine for six months. *Nature* 278:60–61

Clow, A., Theodorou, A., Jenner, P., Marsden, C. D. 1980. Changes in rat striatal dopamine turnover and receptor activity during one years neuroleptic administration. *Eur. J. Pharmacol.* 63:135–44

Cotzias, G. C., Papavasilious, P. S., Gellene, R. 1969. Modification of Parkinsonism by chronic treatment with L-dopa. *N. Engl. J. Med.* 280:331–45

Coyle, I., Wayner, M. J., Singer, G. 1976. Behavioral teratogenesis: A critical evaluation. *Pharmacol. Biochem. Behav.* 4:191–200

Creese, I., Burt, D. R., Snyder, S. H. 1975. Dopamine receptor binding: Differentiation of agonists and antagonist states with ³H-haloperidol. *Life Sci.* 17:993–1022

Creese, I., Burt, D. R., Snyder, S. H. 1976. Dopamine receptor binding predicts clinical and pharmacological potencies of antischizophrenic drugs *Science* 192:481–83

Creese, I., Sibley, D. R. 1981. Receptor adaptation to centrally acting drugs. *Ann. Rev. Pharmacol. Toxicol.* 21:357–91

Cross, A., Crow., T. J., Owen, F. 1980. ³H-cis-flupenthixol (³H-FPT) binding in postmortem brains of schizophrenics, evidence for a selective increase in dopamine (D-2) receptors. *Proc. 12th Coll. Int. Neuropsychopharmacol., Goteberg. Progr. Neuropsychopharmacol. Suppl.* 4:147

Crow, T. J., Baker, H. F., Cross, A. J., Joseph, M. H., Lofthouse, R., Longden, A., Owen, F., Riley, G. J., Glover, V., Killpack, W. 1979. Monoamine mechanisms in chronic schizophrenia: Postmortem neurochemical findings. *Br. J. Psychiatr.* 134:249–56

Cutler, N. R., Post, R. M., Rey, A. C., Bunney, W. E. 1981. Depression-dependent dyskinesias in two cases of manic depressive illness. *N. Engl. J. Med.* 304:1088–89

Davemen, E. 1967. Treatment of depressed patients with iprindole. *Psychosomatics* 8:216–21

Davis, K. L., Berger, P. A., Hollister, L. E. 1976. Tardive dyskinesia and depressive illness. *Psychopharmacol. Commun.* 2:125–30

deMontigny, C., Aghajanian, G. K. 1978. Tricyclic antidepressants: Long-term treatment increases responsivity of rat forebrain neurons to serotonin. *Science* 202:1303–6

Deskin, R., Seidler, F. J., Whitmore, W. L., Slotkin, T. A. 1981. Development of alpha-noradrenergic and dopaminergic receptor systems depends on maturation of their presynaptic nerve terminals in rat brain. *J. Neurochem.* 36:1683–90

Duvoisin, R. C. 1967. Cholinergic anticholinergic antagonism in Parkinsonism. *Arch. Neurol.* 17:124–36

Engel, J., Lundborg, P. 1974. Regional changes in monoamine levels and in the rate of tyrosine hydroxylation in 4 week old offspring of nursing mothers treated with the neuroleptic drug penfluridol. *Nauyn-Schmiedebergs Arch. Pharmacol.* 282:327–32

Ernst, A. M. 1967. Mode of action of apomorphine and dexamphetamine on gnawing compulsion in rats. *Psychopharmacologia* 10:316–23

Feldman, G. L., Weaver, D. D., Lovrien, E. W. 1977. The fetal trimethadione syndrome. *Am. J. Dis. Child.* 131:1389–92

Fleming, W. W., McPhillips, J. W., Westfall, D. P. 1973. Postjunctional supersensitivity and subsensitivity of excitable tissues to drugs. *Ergeb. Physiol.* 68:55–119

Friedhoff, A. J. 1969. Methylation processes in schizophrenia. In *Schizophrenia: Current Concepts and Research*, ed. D. V. Siva Sankar, pp. 552–56. Hicksville, N.Y.: PJD Publ. 944 pp.

Friedhoff, A. J. 1977. Receptor sensitivity modification (RSM)—a new paradigm for potential treatment of some hormonal and transmitter disturbances. *Comp. Psychiatr.* 18:309–17

Friedhoff, A. J. 1979. Receptor sensitivity modification (RSM) produced by chronic administration of psychotropic agents. In *New Clinical Frontiers in Psychotropic Drug Research*, ed S. Fielding, R. Effland, pp. 105–14. New York: Futura

Friedhoff, A. J., Alpert, M. 1973. Dopaminergic-cholinergic mechanism in the production of psychotic symptoms. *Biol. Psychiatr.* 6:165–69

Friedhoff, A. J., Alpert, M. 1978. Receptor sensitivity modification as a potential treatment. See Lipton et al, 1978, pp. 797–801

Friedhoff, A. J., VanWinkle, E. 1964. Biological-0-methylation and schizophrenia. *Psychiatr. Res. Rep., Am. Psychiatr. Assoc.* 19:149–53

Friedhoff, A. J., Bonnet, K. A., Rosengarten, H. 1977. Reversal of two manifestations of dopamine receptor supersensitivity by administration of L-dopa. *Res. Commun. Chem. Pathol. Pharmacol.* 16: 411–23

Friedhoff, A. J., Hekimian, L., Alpert, M., Tobach, E. 1963. Dihydroxyphenylalanine in extrapyramidal disease. *J. Am. Med. Assoc.* 184:285–86

Fuxe, K., Ogren, S., Agnati, L., Gustafsson, J. A., Jonsson, G. 1977. On the mechanism of action of the antidepressant drugs amitriptyline and nortriptyline, evidence for 5-hydroxtryptamine receptor. *Neurosci. Lett.* 6:339–43

Gianutsos, G., Drawbaugh, R. B., Hynes, M. D., Lal, H. 1974. Behavioral evidence for dopaminergic supersensitivity after chronic haloperidol. *Life Sci.* 14: 887–98

Glowinski, J., Axelrod, J., Iversen, L. L. 1966. Regional studies of catecholamines in the rat brain. IV. Effects of drugs on the disposition and metabolism of ³H-norepinephrine and ³H-dopamine. *J. Pharmacol. Exp. Ther.* 153:30–41

Gluckman, M. I., Baum, T. 1969. The pharmacology of iprindole, a new antidepressant. *Psychopharmacologia* 15: 169–85

Gnegy, M., Uzunov, P., Costa, E. 1977. Participation of an endogenous Ca²⁺ binding protein activator in the development of drug-induced supersensitivity of striatal dopamine receptors. *J. Pharmacol. Exp. Ther.* 202(3):558–64

Green, J. P. 1981. Typical and atypical antidepressants. *Trends Pharm. Sci.* 2:7–9

Green, J. P., Maayani, S. 1977. Tricyclic antidepressant drugs block histamine H₂ receptors in brain. *Nature* 269:163–65

Gutierrez, M., Alpert, M., Guimon, M., Friedhoff, A. J., Beramendi, V. 1979. Evolucion de la rigidez como medida de la funcion extrapirimidal en un grup de pacientes esquizofrencios cronicos. *Informaciones Psiquiatricas* 2°-Trimestre, 99–107. Barcelona

Hattori, T., McGeer, P. L. 1973. Synaptogenesis in the corpus striatum of infant rat. *Exp. Neurol.* 38:70–79

Hattori, T., Singh, V. K., McGeer, E. G., McGeer, P. L. 1976. Immunocytohistochemical localization of choline acetyl-

transferase containing neostriatal neurons and their relationship with dopaminergic synapses. *Brain Res.* 102: 164–73

Hekimian, L. J., Friedhoff, A. J., Deever, E. 1978. A comparison of the onset of action and therapeutic efficacy of amoxapine and amitriptyline. *J. Comp. Psychiatr.* 39:633–37

Hill, S. J., Young, M. 1978. Antagonism of central histamine H₁ receptors by antipsychotic drugs. *Eur. J. Pharmacol.* 52:397–99

Hoffeld, D. R., Webster, R. L. 1965. Effect of injection of tranquilizing drugs during pregnancy on offspring. *Nature* 205: 1070–72

Hökfelt, T., Johansson, O., Ljungdahl, A., Lundbert, J. M., Schultzberg, M. 1980a. Peptidergic neurons. *Nature* 284:515–21

Hökfelt, T., Lundberg, J. M., Schultzberg, M., Johansson, O., Ljungdahl, A., Rehfeld, J. 1980b. Coexistence of peptides and putative neurotransmitters in neurons. In *Neural Peptides and Neuronal Communication*, ed. E. Costa, M. Trabucchi, pp. 1–23. New York: Raven

Hornykiewicz, O. 1975. Parkinsonism induced by dopaminergic antagonists. *Adv. Neurol.* 9:155–64

Huang, Y. H. 1979. Chronic desipramine treatment increases activity of noradrenergic postsynaptic cells. *Life Sci.* 25:709–16

Hutchings, D. E. 1978. Behavioral teratology: Embryopathic and behavioral effects of drugs during pregnancy. In *Studies on the Development of Behavior and the Neurons System*, Vol. 4, *Early Influences*, ed. G. Gottlieb. London: Academic

Iversen, L. L. 1975. Summing up. *Adv. Neurol.* 9:415–18

Janowsky, A., Steranka, L. R., Gillespie, D. D., Sulser, F. 1981. Role of neuronal signal input in the regulation of central noradrenergic receptor function by antidepressant drugs. *Soc. Neurosci. Abstr.* 7:817

Joffe, J. M. 1969. *Prenatal Determinants of Behavior.* London: Pergamon

Johnston, M. V., Carman, A. B., Coyle, J. T. 1981. Effects of fetal treatment with methylazoxymethanol acetate at various gestational dates on the neurochemistry of the adult neocortex of the rat. *J. Neurochem.* 36:124–28

Jones, K. L., Smith, D. W. 1973. Recognition of the fetal alcohol syndrome in early infancy. *Lancet* 2:999–1001

Kanof, P. D., Greengard, P. 1978. Brain histamine receptors as targets for antidepressant drugs. *Nature* 272:329–33

Karobath, M., Leitich, H. 1974. Antipsychotic drugs and dopamine-stimulated adenylate cyclase prepared from corpus striatum of rat brain. *Proc. Natl. Acad. Sci. USA* 71:2915–18

Kebabian, J. W., Greengard, P. 1971. Dopamine sensitive adenylate cyclase: Possible role in synaptic transmission. *Science* 174:1346

Kebabian, J. W., Petzold, G. L., Greengard, P. 1972. Dopamine sensitive adenylate cyclase in caudate nucleus of rat brain and its similarity to the "dopamine receptor." *Proc. Natl. Acad. Sci. USA* 69:2145–49

Kebabian, J. W., Zatz, J., Romero, J., Axelrod, J. 1975. Rapid changes in pineal β-adrenergic receptors: Alterations in L-(³H-alprenolol binding and adenylate cyclase. *Proc. Natl. Acad. Sci. USA* 72:3735–39

Kellogg, C., Tervo, D., Ison, J., Parisi, T., Miller, R. K. 1980. Prenatal exposure to diazepam alters behavioral development in rats. *Science* 207:205–7

Klawans, H. L. 1973. The pharmacology of tardive dyskinesias. *Am. J. Psychiatr.* 130:82–86

Klawans, H. L., Crossett, P., Dana, N. 1975. Effect of chronic amphetamine exposure on sterotyped behavior: Implications for pathogenesis of L-dopainduced dyskinesias. *Adv. Neurol.* 9:105–12

Klein, D. F., Davis, J. M., eds. 1969. *Diagnosis and Drug Treatment of Psychiatric Disorders*, pp. 187–299. Baltimore: William & Wilkins. 480 pp.

Kornetsky, C. 1970. Psychoactive drugs in the immature organism. *Psychopharmacologia* 17:105–36

Koslow, S. H. 1982. Monoamine metabolites in Gilles de la Tourette syndrome. In *Tourette Syndrome*, ed. A. J. Friedhoff, T. N. Chase. New York: Raven

Lauder, J. M., Bloom, F. E. 1974. Ontogeny of monoamine neurons in the locus coeruleus raphe nuclei and substantia nigra of the rat. *J. Comp. Neurol.* 155:469–81

Lipton, M., DiMascio, A., Killam, K. F., eds. 1978. *Psychopharmacology: A Generation of Progress*, New York: Raven. 1731 pp.

Lundborg, P. 1972. Abnormal ontogeny in young rabbits after chronic administration of haloperidol to nursing mothers. *Brain Res.* 44:684–87

Lundborg, P., Engel, J. 1976. Learning deficits and selective biochemical brain changes in 4-week old offspring of nursing rat mothers treated with neuroleptics. In *Antipsychotic Drugs: Pharmacodynamics and Pharmacokinetics*, ed. G. Sedvall, B. Uvnas, Y. Zitterman, p. 261. Oxford: Pergamon. 459 pp.

Madsen, J. R., Campbell, A., Baldessarini, R. J. 1981. Effects of prenatal treatment of rats with haloperidol due to altered drug distribution in neonatal brain. *Neuropharmacology*, 20:931–39

Markham, C. H. 1971. The choreathetoid movement disorder induced by levodopa. *Clin. Pharmacol. Ther.* 12:340–43

Meller, E., Friedhoff, A. J., Friedman, E. 1980. Differential effects of acute and chronic haloperidol treatment on striatal and nigral 3,4-dihydroxyphenylacetic acid (DOPAC) levels. *Life Sci.* 26:541–47

Miller, J. C., Friedhoff, A. J. 1979. Dopamine receptor-coupled modulation of the K⁺-depolarized overflow of ³H-acetylcholine from rat striatal slices: Alteration after chronic haloperidol and α-methylparatyrosine pretreatment. *Life Sci.* 25:1249–56

Mishra, R. K., Wong, Y. W., Varmuza, S. L., Tuff, L. 1978. Chemical lesion and drug induced supersensitivity of caudate dopamine receptors. *Life Sci.* 23:443–46

Moore, K. E., Thornburg, J. E. 1975. Drug-induced dopaminergic supersensitivity. *Adv. Neurol.* 9:93–104

Mukherjee, C., Caron, M. G., Lefkowitz, R. J. 1975. Catecholamine-induced subsensitivity of adenylate cyclase associated with loss of β-adrenergic receptor binding sites. *Proc. Natl. Acad. Sci. USA* 72:1945–49

Mukherjee, C., Lefkowitz, R. J. 1976. Desensitization of β-adrenergic agonists in a cell-free system: Resensitization by guanosine 5"-(β-4-imino)triphosphate and other purine nucleotides. *Proc. Natl. Acad. Sci. USA* 73:1494–98

Muller, P., Seeman, P. 1978. Dopaminergic supersensitivity after neuroleptics. Time-course and specificity. *Psychopharmacology* 60:1–11

Muller, P., Seeman, P. 1979. Pre-synaptic subsensitivity as a possible basis for sensitization by long-term dopamine mimetics. *Eur. J. Pharmacol.* 55:149–57

Olson, L., Seiger, A. 1972. Early prenatal ontogeny of central monoamine neurons in the rat: Fluorescence histochemical observations. *Z. Anat. Entwicklungsgesch.* 137:301–16

Ordy, J. M., Samorajski, T., Collings, R. L., Rolsten, C. 1966. Prenatal chlorpromazine effects on liver survival and behavior of mice offspring. *J. Pharmacol.* 151:110–25

Peroutka, S. J., Lebovitz, R. M., Snyder, S. H. 1981. Two distinct serotonin receptors with different physiological functions. *Science* 212:827–29

Peroutka, S. J., Snyder, S. H. 1980. Long term antidepressant treatment decreases spiroperidol-labeled serotonin receptor binding. *Science* 210:88–90

Quach, T. T., Duchemin, A. M., Rose, C., Schwartz, J. C. 1979. In vivo occupation of cerebral histamine H_1-receptors evaluated with ^3H-mepyramine may predict sedative properties of psychotropic drugs. *Eur. J. Pharmacol.* 60: 391–92

Rech, R. H., Moore, K. E., eds. 1971. *An Introduction to Psychopharmacology,* p. 120. New York: Raven. 353 pp.

Rehavi, M., Ramot, O., Yavetz, B., Sokolovsky, M. 1980. Amitriptyline: Long term treatment elevates α-adrenergic and muscarinic receptor binding in mouse brain. *Brain Res.* 194:43–53

Richelson, E. 1978. Tricyclic antidepressants block histamine H_2 receptors of mouse neuroblastoma cells. *Nature* 274: 176–77

Rickels, K., Chung, H. R., Csanalosi, I., Sablosky, W., Simon, J. H. 1973. Iprindole and imipramine in non-psychotic depressed outpatients. *Br. J. Psychiatr.* 123:329–39

Rosenbaum, A. H., Niven, R. G., Hanson, N. P., Swanson, D. W. 1977. Tardive dyskinesia: Relationship with a primary affective disorder. *Dis. Nerv. Syst.* 38: 423–27

Rosengarten, H., Friedhoff, A. J. 1979. Enduring changes in dopamine receptor cells of pups from drug administration to pregnant and nursing rats. *Science* 203:1133–35

Rosenthal, D., Kety, S. S., eds. 1968. *Transmission of Schizophrenia,* pp. 345–62, 379–91. New York: Pergamon

Rosloff, B. N., Davis, J. M. 1978. Decrease in brain NE turnover after chronic DMI treatment: No effect with iprindole. *Psychopharmacology* 56:335–41

Ross, S. B., Reyni, A. L. 1967. Inhibition of the uptake of tritiated catecholamines by antidepressant and related agents. *Eur. J. Pharmacol.* 2:181–86

Ross, S. B., Renyi, A. L., Ogren, S. O. 1971. A comparison of the inhibitory activities of iprindole and imipramine on the uptake of 5-hydroxytryptamine and noradrenaline in brain slices. *Life Sci.* 10:1267–77

Rush, M., Diamond, F., Alpert, M. 1981. Depression as a risk factor in tardive dyskinesia. *Biol. Psychiatr.* 17:387–92

Sanghvi, I., Gershon, S. 1975. Effect of acute and chronic iprindole on serotonin turnover in mouse brain. *Biochem. Pharmacol.* 24:2103–4

Sayers, A. C., Burki, H. R., Ruch, W., Asper, H. 1975. Neuroleptic-induced hypersensitivity of striatal dopamine receptors in the rat as a model of tardive dyskinesias. Effects of clozapine, haloperidol, loxapine and chlorpromazine. *Psychopharmacologia* 41:97–104

Schildkraut, J. 1965. The catecholamine hypothesis of affective disorders. A review of supportive evidence. *Am. J. Psychiatry.* 5:509–22

Schultze, B., Nowak, B., Maurer, W. 1974. Cycle times of the neural epithelial cells of various types of neuron in the rat. An autoradiographic study. *J. Comp. Neurol.* 158:207–18

Schweitzer, J. W., Schwartz, R., Friedhoff, A. J. 1979. Intact presynaptic terminals required for beta-adrenergic receptor regulation by desipramine. *J. Neurochem.* 33:377–79

Seeman, P. 1980. Brain dopamine receptors. *Pharmacological Rev.* 32:229–313

Shalaby, I. A., Spear, L. P. 1980. Chronic administration of halo-peridol during development: Later psychopharmacological responses to apomorphine and arecoline. *Pharmacol. Biochem. Behav.* 13:685–90

Sharpless, S. K. 1964. Reorganization of function in the nervous system—use and disuse. *Ann. Rev. Physiol.* 26: 357–88

Siggins, G. R. 1978. Electrophysiological role of dopamine in striatum: Excitatory inhibitory? See Lipton et al 1978, pp. 143–57

Smith, D. W. 1977. Teratogenicity of anticonvulsive medications. *Am. J. Dis. Child.* 131:1337–39

Snyder, S. H., Yamamura, H. I. 1977. Antidepressants and the muscarinic acetylcholine receptor. *Arch. Gen. Psychiatr.* 34:236–39

Spear, L. P., Shalaby, I. A., Rick, J. 1980. Chronic administration of haloperidol during development: Behavioral and psychopharmacological effects. *Psychopharmacology* 70:47–56

Stone, E. A. 1979. Subsensitivity to norepinephrine as a link between adaptation to stress and antidepressant therapy: An hypothesis. *Res. Commun. Psychol. Psychiatr. Behav.* 4:241–55

Sulser, F., Vetulani, J., Mobley, P. L. 1978. Mode of action of anti-depressant drugs. *Trends Pharm. Sci.* 1:92–94

Svensson, T. H. 1978. Attenuated feed-back inhibition of brain serotonin synthesis following chronic administration of imipramine. *Nauyn-Schmeidbergs Arch. Pharmacol.* 302:115–18

Tarsy, D., Baldessarini, R. J. 1974. Behavioral supersensitivity to apomorphine following chronic treatment with drugs which interfere with synptic function of catecholamines. *Neuropharmacology* 13:927–40

Tran, V. T., Chang, R. S. L., Snyder, S. H. 1978. Histamine HI receptors identified in mammalian brain membranes with (³H) mepyramine. *Proc. Natl. Acad. Sci. USA* 75:6290–94

Trendelenberg, U. 1966. Mechanisms of supersensitivity and subsensitivity to sympathomimetic amines. *Pharmacol. Rev.* 18:629–40

Ungerstedt, U., Arbuthnott, G. W. 1970. Quantitative recordings of rotational behavior in rats after 6-hydroxydopamine lesions of the nigrostriatal dopamine system. *Brain Res.* 24:485–93

Ungerstedt, U., Ljungbert, T., Hoffer, B., Siggins, G. 1975. Dopaminergic supersensitivity in the striatum. *Adv. Neurol.* 9:57–65

van Rossum, J. M. 1966. The significance of dopamine-receptor blockade for the mechanism of the action of neuroleptic drugs. *Arch. Int. Pharmacodyn. Ther.* 160:492–94

van Rossum, J. M. 1967. The significance of dopamine-receptor blockade for the action of neuroleptic drugs. In *Proc. 5th Int. Congr. Coll. Int. Neuro-Psycho-Pharmacologicum*, pp. 321–29. Amsterdam: Excerpta Medica

Vetulani, J., Stawarz, R. J., Dingell, J. V., Sulser, F. 1976. A possible common mechanism of action of antidepressant treatment. *Nauyn-Schmiedebergs Arch. Pharmacol.* 293:109–14

Von Voigtlander, P. F. 1974. Behavioral and biochemical investigation of dopamine supersensitivity induced by chronic neuroleptic treatment. *Fed. Proc.* 33:578

Von Voigtlander, P. F., Boukam, S. Y., Johnson, G. A. 1973. Dopaminergic denervation supersensitivity and dopamine stimulated adenylate cyclase activity. *Neuropharmacology* 12:1081–86

Von Voigtlander, P. F., Moore, K. E. 1973. Turning behavior of mice with unilateral 6-hydroxydopamine lesions in the striatum: Effects of apomorphine, L-dopa, amanthadine, amphetamine and other psychomotor stimulants. *Neuropharmacology* 12:451–62

Vorhees, C. V., Brunner, R. L., McDaniel, C., Butcher, R. E. 1978. The relationship of gestational age to vitamin A induced postnatal dysfunction. *Teratology* 18:379–84

Wolfe, B. B., Harden, T. K., Sporn, J. R., Molinoff, P. B. 1978. Presynaptic modulation of beta-adrenergic receptors in rat cerebral cortex after treatment with antidepressants. *J. Pharmacol. Exp. Ther.* 207:446–57

Yaryura-Tobias, J. A., Wolpert, A., Dana, L., Merlis, S. 1970. Action of L-dopa in drug-induced extrapyramidalism. *Dis. Nerv. Syst.* 31:60–63

Zis, A. P., Goodwin, F. K. 1979. Novel antidepressants and the biogenic amine hypothesis of depression. The case for iprindole and mianserin. *Arch. Gen. Psychiatr.* 36:1097–1107

Ann. Rev. Neurosci. 1983. 6:149–85

MICROCIRCUITRY OF THE CAT RETINA

Peter Sterling

Department of Anatomy, University of Pennsylvania, Philadelphia,
Pennsylvania 19104

INTRODUCTION

As a device for extracting information from a visual image, the vertebrate
retina is unparalleled in its range, reliability, and compactness. Signaling in
the retina is slower by six orders of magnitude than in an integrated digital
circuit. The advantage of the biological structure must therefore derive from
the variety of its fundamental elements and from the subtlety of their
connections. Each of the five major classes of retinal neuron, whose synaptic
contacts were first described systematically by Dowling & Boycott (1966),
is now known to have multiple types, totaling in the cat about 60. Specific
local circuits involving about one-third of these neurons have been recognized in the electron microscope. Physiological responses have also been
documented for about one-third of the types, and evidence regarding the
neural transmitter, or at least the sign of the synapse, has accumulated also
for about one-third.

These discoveries have abundantly supported certain concepts of retinal
function developed in the 1960s by Lettvin & Maturana. The function of
the retina, they proposed, "is not to transmit information about the point-
to-point distribution of light and dark in the image, but to analyze this
image at every point in terms of . . . arbitrary contexts . . ." (Maturana et
al 1960). Each of these "contexts," they suggested, corresponds to some
operation on the local image performed by a ganglion cell of particular size
and shape (Lettvin et al 1961). This idea, based on studies of the frog,
seemed for a time inapplicable to the cat, which was thought to have a
"simple" retina with only center-surround type ganglion cells. Subsequent
studies to be reviewed here have firmly established for the cat the validity
of this idea.

149

0147-006X/83/0301-0149$02.00

Lettvin & Maturana also paid special attention to the stratification of processes in the frog's inner plexiform layer, believing that the operation performed by a ganglion cell is determined by specific bipolar inputs delivered to the strata of its dendritic arbor. This idea, too, was thought to be inapplicable to the cat, whose inner plexiform layer is less obviously stratified than the frog's. Studies to be reviewed here now strongly support this concept for the cat.

Nothing of the actual circuits between particular neuron types was known to Lettvin & Maturana, but fragments of such knowledge accumulated for the cat during the 1970s. Some of the first observations were extremely puzzling. It turned out, for example, that rods and cones have separate bipolars and thus apparently separate pathways to ganglion cells. But rod signals are also transmitted directly to cones, so why the separate bipolars? Further, the rod bipolar does not contact most ganglion cells, as one might have anticipated, but contacts an amacrine, which in turn contacts, not ganglion cells, but cone bipolar axons! What could be the meaning of this second convergence of rod and cone pathways, and why send the rod signal through such a tortuous route?

The functions of such apparently bizarre paths have been difficult to comprehend for the same reasons that fragments of an integrated circuit cannot be grasped except in the context of its larger diagram. Now, however, with links established between about one-third of the neuron types, broad pathways can be identified and specific hypotheses can be suggested regarding their function. The outline of a detailed mechanistic account of retinal function emerges, and many of the necessary strategies and techniques for achieving it seem at hand. In reviewing this subject now, when many puzzling findings of the last 15 years begin to fit, one is deeply impressed with the intelligence and care of the many individual studies from laboratories that literally girdle the globe.

In this article the first section reviews our knowledge of particular cell types comprising each major class of retinal neuron. Information is presented, where available, for each type regarding morphology, circuitry, distribution, transmitter, and physiology. Such details constitute the primary evidence that the retina is composed of many discrete cell types arranged in a regular mosaic. This section also serves in effect as a "parts list" for the second section, which describes a complex circuit involving thirteen types of neuron and suggests how the circuit might function.[1]

---

[1]Many features of the cat retina reviewed here were described first in other species. Were careful reference made in each case to this prior work, the bibliography would mushroom and the special focus on cat retina would be lost. Therefore, I have omitted reference to other species except where data is unavailable for the cat. Access to the larger literature may be found, of course, in the original papers I have cited.

CELL TYPES

In describing a cell type I refer, where the data permit, to its "coverage factor." The *physiological* coverage factor is the number of neurons of a particular type whose receptive fields overlap a particular point on the retina. It is calculated for a given point as the receptive field area (mm^2) X the cell density (cells/mm^2) (Cleland et al 1975). The *anatomical* coverage factor is similar: the number of cells of a given type whose dendritic fields overlap a particular point. It is calculated similarly, as the dendritic field area (mm^2) X the cell density (cells/mm^2) (Wässle & Riemann 1978). The coverage factor for a given eccentricity appears to be characteristic for each cell type. Another useful analytic expression introduced by Wässle & Riemann (1978) as a measure of regularity in a pattern is the ratio, mean/standard deviation. The higher this ratio, the greater the regularity. This value, too, is apparently characteristic for a given feature of a specific cell type.

Receptors

RODS   Only two types of photoreceptor have been identified in the cat retina, rods, and cones. The rod has an extremely fine outer segment, which increases in diameter (1.0 μm to 1.6 μm) and length (25 μm to 50 μm) from the periphery to the central area (A. Laties, cited by Barlow et al 1971; Steinberg et al 1973). The rod soma is small (4.5 by 6 μm) and the inner segment is narrow (2 μm). The axon is extremely fine (0.25 μm), ending in a "spherule" (3 μm diameter) in the outer half of the outer plexiform layer. The spherule, which contains mitochondria and 1–2 synaptic ribbons, is invaginated by 2–3 laterally placed processes from horizontal cells and two centrally placed processes from rod bipolars (Figure 1; Boycott & Kolb 1973, Kolb 1974, Kolb 1977). The rod spherule receives 4–6 punctate gap junction contacts from the basal processes of neighboring cone pedicles (Kolb 1977).

The response of the cat rod to light must be inferred from the responses of rods in other species and from responses of neurons postsynaptic to rods in cat. In darkness, according to Penn & Hagins (1972), a steady depolarizing electric current flows in the interstitial space and enters the rod outer segment (rat). This "dark current" is suppressed transiently by a flash, the hyperpolarizing change being termed the "photocurrent." The amplitude of the rod photocurrent increases with light intensity until the dark current is just balanced. Beyond this point, which corresponds psychophysically to rod saturation, increases in light intensity increase the photocurrent's rate of rise and greatly prolong its duration (to more than 50 min; Penn &

A   B

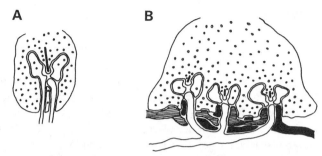

Figure 1 A. Rod spherule (with synaptic ribbon and vesicles) invaginated by two lateral processes (B horizontal cell axon) and a central process (rod bipolar dendrite). B. Cone pedicle (with vesicles and three synaptic ribbons) invaginated by four types of process: lateral elements, A and B horizontal cell dendrites; pale process opposite ribbon, "invaginating" cone bipolar dendrite; dark and shaded processes, "flat" cone bipolar dendrites. Reprinted from Boycott & Kolb (1973).

Hagins 1972). This prolonged photocurrent is caused apparently by the stimulus-induced accumulation of a "cytoplasmic transmitter" that controls the rod's sodium permeability (see Penn & Hagins 1972).

When a similarly prolonged photocurrent is observed in a neuron postsynaptic to rods, it is called the "rod after-effect" (Steinberg 1969c). The cat rod then is presumed to depolarize in the dark and hyperpolarize in response to light. Above rod saturation its light response is greatly prolonged. The rod's transmitter molecule (released in the dark) is unknown but its effect on horizontal and rod bipolar cells is to depolarize, probably by increasing sodium permeability (reviewed by Kaneko 1979). Neurons postsynaptic to rods, therefore, hyperpolarize to light.

The rod receptive field in lower vertebrates is extremely large by virtue of extensive rod-rod gap junctions (Fain et al 1976, Schwartz 1973). The rod receptive field in cat is inferred to be relatively small from the small receptive fields of cones to which the rod is electrically connected (Nelson 1977). The cat rod must be extremely efficient at collecting light and also at signaling its presence because at the wave length of its maximum sensitivity (500 nm; reviewed by Daw & Pearlman 1969) three quanta at the cornea or one quantum absorbed can produce at least one extra impulse in a ganglion cell (Barlow et al 1971).

CONES   The cone has a thicker outer segment (about 5 μm diameter) than the rod and a larger cell body (7 by 8.5 μm, Wässle & Rieman 1978, Kolb 1977). The inner segment (4 μm diameter), and the axon (1.5 μm diameter), are also larger, as is the flattened cone "pedicle" (5–8 μm diameter) by which the cone terminates in the inner half of the outer plexiform layer (Kolb 1977). The pedicle contains a large mitochondrion and 10–12 synaptic ribbons, each of which points to an invagination formed by a triad of processes (Figure 1). The lateral elements of the cone triad are dendrites of A or B horizontal cells (Kolb 1974) and the central element is the dendrite of an "invaginating" cone bipolar (Boycott & Kolb 1973). "Flat" cone bipolar dendrites contact the base of the pedicle and enter part way into the invaginations (Boycott & Kolb 1973). The pedicle emits fine "basal processes," which are short in the central area but up to ten microns long further peripherally. These contact neighboring cone pedicles and rod spherules by small gap junctions (Raviola & Gilula 1975, Kolb 1977).

The cone resembles the rod physiologically in hyperpolarizing to light and in having a dynamic range for a given adaptation level of about three log units. The cone differs from the rod in having a faster response and a threshold about three log units higher (Figure 2; Nelson 1977, Rodieck & Rushton 1976). The spectral sensitivity maximum of the cones studied individually is 556 nm (Nelson 1977). Behavioral studies and recordings from single ganglion and lateral geniculate neurons provide solid evidence for a second cone type with maximal sensitivity at about 450 nm (Daw & Pearlman 1970, Pearlman & Daw 1970). Ganglion cell recordings have recently given evidence for a third cone type with maximal sensitivity at around 500 nm (Ringo et al 1977, Ringo & Wolbarsht 1981, Zrenner & Wienrich 1981). Evidence of only the 556 nm cone is detectable in recordings from horizontal cells (Nelson 1977). In certain vertebrates, fish, for

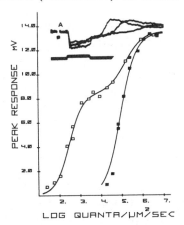

Figure 2 Response characteristics of cat cone. *Inset A:* Responses to rod-saturating, blue (441 nm) stimuli increasing in intensity in steps of 0.6 log units. Note increases in peak response and in "rod after-effect." Calibration pulse, 4.1 mv; flash length 540 msec. *Curve on right* (647 nm stimuli) shows cone threshold, smooth intensity/response relation, and saturation at about three log units above threshold. *Curve on left* (441 nm stimuli) shows "rod response" recorded in same cone. The rod threshold is 3 log units below cone threshold with a sharp inflection marking rod saturation. Reprinted from Nelson et al (1976).

example, chromatic differences between cones are associated with morphological differences (Scholes 1975, Stell & Lightfoot 1975). Such correlates have not been observed in the cat, nor has the mosaic distribution of the chromatic types been determined as in goldfish (Marc & Sperling 1976) and monkey (Marc & Sperling 1977, DeMonasterio et al 1981). The antagonistic surround demonstrated for cones in the turtle (Baylor et al 1971) has not been found in cat, possibly for technical reasons (Nelson 1977).

Among the more astonishing of recent findings is the discovery that cones have input from rods (Nelson 1977). The main lines of evidence, illustrated in Figure 2, are as follows: The threshold of the 556 nm cone to a rod stimulus (blue, 441 nm) is the same as the rod's threshold and about 3 log units below the cone's threshold to a cone stimulus (red, 647 nm). The cone's response function to blue light of increasing intensity shows a break at rod-saturation, whereas its response function to red light is smooth. Finally, the rod after-effect is recorded in cones. This rod input is presumably mediated by the rod-cone gap junctions. Evidence of the reverse pathway, passage of the cone signal to rods, has not been obtained from rod bipolars, where it might have been anticipated (Nelson et al 1976). Whether this reflects rectification at the gap junction or simply dilution of the cone signal among the rods, which are at least ten times more numerous, is unknown. It seems clear, however, that rod and cone signals mix in the cone pedicle before transmission to the horizontal cells.

MOSAIC DISTRIBUTION OF RODS AND CONES   The rods are arranged in well-defined rows; each rod is surrounded by six others (Figure 3). Their density reaches a maximum of 460,000/mm² at an eccentricity of 10–15° and is high (275,000/mm²) even in the central area (Steinberg et al 1973). The cone density reaches a maximum of about 26,000/mm² in the central area, where the rod/cone ratio is 10.6, and falls to 3100/mm² in the periphery where the rod/cone ratio is 90 (Steinberg et al 1973; see also Höllander & Stone 1972, Wässle & Riemann 1978).

The distribution of cones appears upon simple inspection to form a regular pattern (Figure 3). The reason for this impression, expressed by Wässle & Riemann (1978), is that the distances of each cone to its nearest neighbors are regular. The distribution of these distances is Gaussian and the ratio, mean standard/deviation, for this distribution is relatively high. A standard of comparison is the distribution of nearest neighbor distances for a random dot pattern of the same density. This distribution is skewed and the ratio, mean/standard deviation, is relatively low. Wässle & Riemann made their measurements on what is almost certainly a heterogeneous population. When these measurements can be repeated on distributions for individual cone types, the degree of regularity may be even greater. This

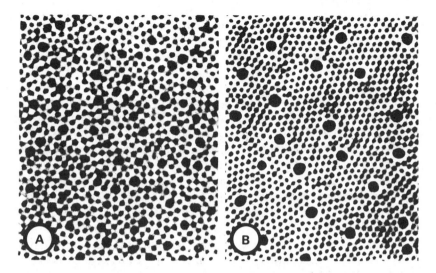

Figure 3   Rod and cone mosaic at level of inner segments. A. Central area. Density for cones, $32 \times 10^3/mm^2$; for rods, $338 \times 10^3/mm^2$. B. About 1 mm from central area. Density for cones, $11 \times 10^3/mm^2$; for rods, $537 \times 10^3/mm^2$. Reprinted from Steinberg et al (1973).

has been found to be strikingly so for blue cones in the monkey (DeMonasterio et al 1981).

Horizontal Cells

TYPES A AND B   There are two types of horizontal cell in the cat retina (Dowling et al 1966, Fisher & Boycott 1974, Kolb 1977, Boycott et al 1978). Type A has a large soma (12 by 15 μm), an indented nucleus, and 4–6 thick, filament-packed dendrites radiating to form a circular or oval field about 80 μm in diameter.[2] Type B has a smaller soma, an unindented nucleus, and densely branched, relatively slender dendrites radiating to form a circular field about 60 μm in diameter. Clusters of terminals from the finer dendritic branches of both cell types lie in register with the overlying cone pedicles to form the lateral elements of the triad. In each triad one lateral element is from type A and the other from type B (Kolb 1974, Boycott et al 1978). Neither type of horizontal cell makes or receives typical chemical synaptic contacts. The type A cells, however, commonly form gap junctions with each other and, less commonly, "basal junction"-type contacts with cone bipolars (Kolb 1977), at least one of which is the type CBb_2 of McGuire et al (1983b). The type B horizontal cell emits a fine (0.5 μm) axon

[2]Cell sizes and dendritic diameters, unless otherwise noted, are given for the central area.

that meanders for 300–500 μm before producing an elaborate arborization whose terminals penetrate the rod spherule to form the lateral elements of the triad. Each triad receives from two different axons (Kolb 1974, Nelson et al 1975, Boycott et al 1978).

MOSAIC DISTRIBUTION   The mosaic distribution of each horizontal cell type is regular and independent of the other. The densities of both types peak centrally and decline peripherally, though their ratio (2.7 B/A) remains constant. This decline is offset, however, by a corresponding increase in dendritic field diameter so that the coverage factor (See above) for each type remains constant at about 4 (Figure 5), except in the central area where the coverage by type B increases to 7 (Wässle et al 1978b). In a tour de force of quantitative light microscopy, Wässle et al (1978a) estimated that each type A cell contacts 120–170 cones and each type B cell contacts 60–90. (Every cone is connected with several horizontal cells of each type.) They argued that since each horizontal cell contacts at least 80% of the cones in its field, at least 60% of the cones must be common to both types and there can be no strict selectivity of cone type. The axon of type B contacts 2000–3000 rods (Wässle et al 1978a).

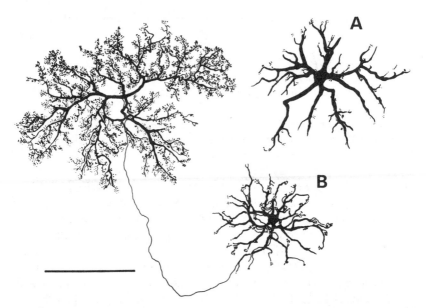

Figure 4   Type A and B horizontal cells drawn from Golgi preparations. Dendrites from the A and B somas contact only cones; axon terminal of type B (*upper left*) contacts only rods. Reprinted from Nelson et al (1976).

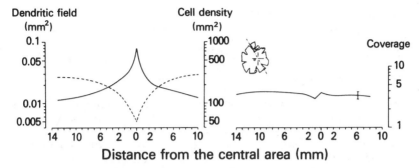

Dendritic field
(mm²)

Cell density
(mm²)

Coverage

Distance from the central area (mm)

Figure 5 Coverage of retina by A-horizontal cells. A. *Left solid curve,* cell density; *dotted curve,* dendritic field area. Reciprocity of these curves holds coverage *(right)* constant across retina. Schematic retinal map indicates sampled strip. Reprinted from Wässle et al (1978b).

PHYSIOLOGY   Intracellular recordings followed by dye injection (Nelson et al 1975, Nelson 1977) show that the A and B somas and also the B axon respond to a flash of light with a sharp, sustained hyperpolarization, called the "S-potential" (Steinberg 1969a–c, 1971, Steinberg & Schmidt 1970, Niemyer & Gouras 1973). To a dim flash the response is rod-generated with a spectral sensitivity peaking at 500 nm, coincident with the absorption peak of rhodopsin (Steinberg 1969a). The amplitude of the S-potential increases with intensity to about 2.5 log units above threshold. At higher intensities (3 log units or more above threshold) the peak amplitude is saturated but the S-potential's return to baseline is greatly prolonged, and this is the rod after-effect whose basis is described above.

Rod input to the type B axon terminal probably comes directly from the rod spherule in which it is housed. This, however, is probably not the source of rod input to the type B soma because the connecting axon does not spike and seems too thin to conduct much of the rod signal passively (Nelson et al 1975). Further, were the B axon to convey rod signals, the soma's rod receptive field would be offset from its cone receptive field, but in fact they are almost exactly superimposed (Nelson 1977). The rod signal to the A and B somas is believed to be conveyed from rod spherule to cone pedicle and thence to the dendrites of these horizontal cells.

To a bright flash, three log units or more above rod threshold, the S-potential in the type A and B somas is cone-driven (Figure 6). Experiments with chromatic adaptation indicate the input to be overwhelmingly from the 556 nm cone with no evidence of contributions from the 450 or 500 nm types (Nelson 1977). The axon terminal of type B shows relatively little response to cone stimuli (Nelson et al 1975).

The horizontal cell receptive fields show spatial summation over 1.0–1.7 mm for rod stimuli and over 0.8–1.0 mm for cone stimuli (Steinberg 1969a)

A-TYPE CELL   B-TYPE CELL   B-TYPE AXON TERMINAL

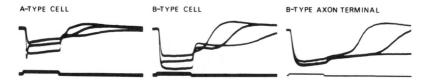

Figure 6 Horizontal cell responses to rod-saturating blue (400 nm) stimuli of increasing intensity (roughly half log unit steps). Note for A and B, somas increase in peak amplitude of hyperpolarization (cone response); this is absent for B axon. For all three elements, prolonged hyperpolarization following the stimulus (rod after-effect) increases with intensity. Reprinted from Nelson et al (1976).

with space constants ranging from 210–410 μm (Nelson 1977). The larger space constants probably belong to the type A cells with their larger dendritic fields and gap-junction interconnections (Nelson 1977). Intracellular recordings in vivo provide evidence for three classes of response based on the critical flicker frequency, but their assignment to the three anatomical types of horizontal cell process is not yet established (Foerster et al 1977a,b).

In species other than cat the horizontal cell hyperpolarization induced by a broad field stimulus feeds back to the cone as a depolarization, thereby providing the cone with an antagonistic surround (Baylor et al 1971). Cone bipolars, which share the cone triad with horizontal cell dendrites, also have antagonistic surrounds (Werblin & Dowling 1969, Kaneko 1970, Schwartz 1974). Whether the cone bipolar surrounds arise directly from horizontal cell input or indirectly via the cone's surround is unknown (reviewed by Kaneko 1979).

To assign distinct functions to the A and B horizontal cells in the cat is at this point purely speculative. One might suggest that the type A horizontal cell, activated by either rod or cone input, provides the antagonistic surrounds of the cones and/or cone bipolars. The axonal terminals of the type B horizontal cell might conceivably perform a corresponding function for rods and/or rod bipolars, but there is no evidence from any species that rods or rod bipolars have surrounds. On the contrary, rods show broad spatial summation (See above), apparently maximizing sensitivity at the expense of contrast-enhancing mechanisms. The dendritic terminals of type B might act in similar fashion to those of type A, but this would not explain why the two types should inhabit the same cone, nor would it offer a role for the axonal pathway connecting the axonal and dendritic terminals of type B.

One function for the type B horizontal cell might be to regulate the direct rod → cone pathway. This pathway might be open under mesopic illumination, giving rods access to high-acuity, cone bipolar pathways, and closed at the end of dark adaptation when rod bipolars are thought to take over and acuity is sacrificed for the last degree of sensitivity (See Microcircuitry of the Beta Ganglion Cell Receptive Field, below; Sterling & Megill 1983). The rod after-effect recorded in the type B axon terminal, when the rods are saturated, might represent the state of adaptation. This prolonged hyperpolarization might be transduced to a chemical signal that might be conveyed by axoplasmic transport to the type B dendritic terminals in the cone pedicle. There, the hypothetical substance would hold the rod → cone pathway open. This signal would decay during dark adaptation, consequent to the decay of the rod after-effect, allowing the rod-cone pathway to close after about an hour in the dark. Such chemical signaling would require an axoplasmic transport rate of about 12 mm/day which is well within the known rates. This speculation is unsupported by evidence but links several physiological and anatomical observations for which at present there is no other interpretation.

Bipolar Cells

Cajal (1892) observed that certain bipolars in mammals are associated exclusively with rods (rod bipolars) and others exclusively with cones (cone bipolars). Boycott & Kolb (1973), examining Golgi-impregnated bipolars by electron microscopy, showed that in the cat rod bipolar dendrites do indeed form the central elements of the triad in rod spherules (Figure 1A). Rod bipolar axon terminals end in sublamina *b* of the inner plexiform layer where they receive input from reciprocal and nonreciprocal amacrines and send output to reciprocal and AII amacrines (Kolb & Famiglietti 1974, Kolb 1979, McGuire et al 1983b).

Cone bipolar dendrites were shown by Boycott & Kolb (1973) to be associated only with cones (Figure 1B). They found the dendrites of certain cone bipolars, called "invaginating," to form the central element of the cone pedicle triad and the dendrites of other cone bipolars, called "flat," to form superficial contacts with the cone pedicle at some distance from the synaptic ribbon. This distinction proved fundamental when evidence from other species showed that invaginating cone bipolars depolarize to illumination of their receptive field centers, whereas flat cone bipolars hyperpolarize (reviewed by Kaneko 1979). Surprisingly, the rod bipolar, despite its invaginating, ribbon-related dendrite, is hyperpolarizing (Figure 7A; Nelson et al 1976).

It was believed at first that axons of flat (hyperpolarizing) bipolars terminate exclusively in the outer third of the inner plexiform layer (sublamina *a*) and those of invaginating (depolarizing) bipolars exclusively in its inner two-thirds (sublamina *b*) (Famiglietti & Kolb 1976). This fit the observation that ganglion cells branching exclusively in sublamina *a* or *b* are, respec-

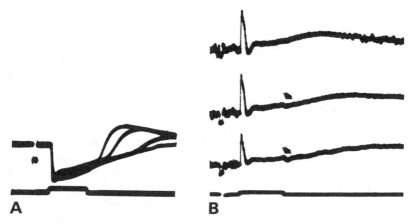

Figure 7 A. Hyperpolarizing responses of rod bipolar to rod-saturating, blue (441 nm) stimuli of increasing intensity (roughly half log unit steps). Note similarity to cone response (Figure 2A). B. Responses of AII amacrine to similar stimuli (intensity increasing from top to bottom). Note all-or-none spike followed by prolonged depolarization. Arrows mark "cone notch" seen at stimulus "off." Calibrating pulse, 2 mv; flash length 520 msec. Reprinted from Nelson et al (1976).

tively, off- or on-center (Famiglietti & Kolb 1976). In turn this implied that both flat and invaginating bipolars are excitatory, and this fit the fragmentary (though hard won) evidence from other species (see Kaneko 1979).

The situation, however, proved more complicated. Nelson (1980) found an hyperpolarizing bipolar ending in sublamina *b,* and McGuire et al (1983b) described an invaginating cone bipolar (CBa$_2$) with an axon terminal in sublamina *a.* There is reason to believe, furthermore, that at least one cone bipolar type in each sublamina is inhibitory (McGuire et al 1983b, Sterling et al 1983).

Renewed study of cone bipolars with the Golgi method now suggests at least eight types based on the morphology of the axon and its stratification and tangential spread in the inner plexiform layer (Kolb et al 1981, Famiglietti 1981). Strong support for such a robust number of types is provided by detailed reconstructions, (Stevens et al 1980a,b) performed on rod and cone bipolars from electron micrographs of serial sections (McGuire et al 1983b). These reconstructions, all taken from a small region near the central area of a single retina, provide morphological detail comparable to that of the Golgi method and simultaneously provide information on cytology and patterns of synaptic connection. This approach has so far revealed at least seven types of cone bipolar, four terminating in sublamina *a* and three terminating in *b.* Their detailed description and comparisons with the Golgi types are presented in McGuire et al (1983b). The description is restricted here to four types (Figure 8): two from each sublamina, whose connections with other types of retinal neuron suggests their specific functions (p. 31).

In sublamina *a*, type CBa_1 is pale with flat cone contacts, while type CBa_2 is dark with invaginating contacts (Table 1). Both types get input from reciprocal and nonreciprocal amacrines, though in different proportions, but only CBa_1 has extensive input from the AII amacrine. Both types have output to reciprocal and nonreciprocal amacrines, again in different proportions, but only CBa_1 has a strong output to the AII. Both types are presynaptic to dark (off-beta) and pale (some, probably off-alpha) ganglion cell dendrites. CBa_2 appears to accumulate exogenous ^{3}H-glycine.

In sublamina *b*, type CBb_1 is pale and is believed to be invaginating, while type CBb_2 is darker and believed to be flat (Table 1). Both receive inputs from reciprocal and nonreciprocal amacrines in different proportions, but only CBb_1 makes multiple, extensive gap junctions with the AII amacrine.

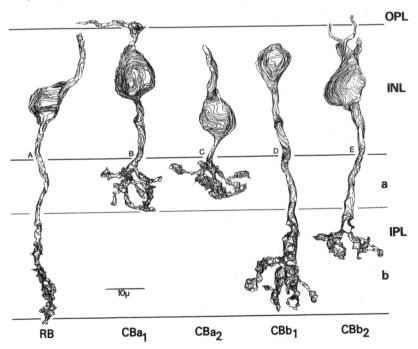

Figure 8 Five types of bipolar partially reconstructed from electron micrographs of 189 serial sections by the method of Stevens et al (1980a). RB, rod bipolar; CBa_1, flat cone bipolar ending in sublamina *a;* CBa_2, invaginating cone bipolar also ending in *a;* CBb_1, cone bipolar believed invaginating, ending in sublamina *b;* CBb_2, cone bipolar believed flat, ending in *b*. Note morphological differences in their axonal arborizations. See Table 1 for differences in input and output. IPL, inner plexiform layer; OPL, outer plexiform layer; *a, b,* sublaminae of IPL. Reprinted from McGuire et al (1983b).

Table 1 Characteristics of four types of cone bipolar[a]

Bipolar type	Cone contact	Cyto-plasm	IPL stratum	Inputs			Outputs					Transmitter
				AII	NR%	R%	AII%	NR%	R%	DG%	PG%	
CBa$_1$	F	pale	1, 2	36% (chem)	57	7	24	12	2	28	18	E (?)
CBa$_2$	I	dark	1, 2	3% (chem)	70	28	4	21	15	23	11	glycine uptake −In (?)
CBb$_1$	I(?)	pale	3, 5	4.5 gj	73	27	0	21	10	26	22	E (?)
CBb$_2$	F(?)	dark	3, 4	1 gj	44	56	0	17	20	34	4	glycine uptake −In (?)

[a]Calculated from Tables 1–3, McGuire et al (1983b). Percentages, based on small sample (1–4) of each type, indicate differences between types. AII, AII amacrine; DG, PG, dark and pale ganglion cell dendrite; chem., chemical synapse; gj, gap junction; E, excitatory; F, flat; I, invaginating; In, inhibitory; NR, nonreciprocal; R, reciprocal.

Both types have outputs, in different proportions, to reciprocal and nonreciprocal amacrines and to dark (on-beta) and pale (some, probably on-alpha) ganglion cell dendrites. CBb$_2$ may accumulate exogenous ^{3}H-glycine (McGuire et al 1983b, Sterling et al 1983).

Amacrine Cells

Golgi-impregnated amacrine neurons in the cat retina have been sorted into at least 22 types based on differences in soma size and in the length, stratification, and morphology of their processes in the inner plexiform layer (Kolb et al 1981). The expectation by now is strong that each morphological type will also have a specific wiring pattern and physiology. Anatomical support for this belief comes from partial EM reconstructions of amacrine processes that show consistent correlations between fine morphology and synaptic patterns (Kolb 1979, Sterling & Megill 1983). Simultaneous physiological and anatomical support for the view comes from intracellular recordings in three amacrine types followed by electron microscopy of their HRP-injected processes (Kolb & Nelson 1981).

Amacrine types have also been identified by their association with a particular neural transmitter (Table 2). Such data can not yet be perfectly integrated with the Golgi and physiological studies but we shall see (see section below on Microcircuitry of the Beta ganglion Receptive Field) that where integration is possible, rather specific hypotheses emerge regarding function.

BIOGENIC AMINES, NEUROPEPTIDES, AND ACETYLCHOLINE   One type of amacrine in the cat retina fluoresces for catecholamines (Ehinger 1966, Boycott et al 1975) and is apparently dopaminergic (Kramer 1971, Kramer et al 1971). Törk & Stone (1979) described this type as a large cell distributed at a density of 40–50/mm^2, whose intertwining dendrites form rings of catecholamine fluorescence at the base of another amacrine type. Kolb et al (1981) argue convincingly that this dopamine amacrine corresponds to their wide-field (about 500 μm) amacrine, A_{18}. The coverage factor at a density of 40/mm^2 and a field diameter of 500 μm is 7.9. Pourcho (1981) has shown that the cell encircled by the dopaminergic rings and postsynaptic to them is the AII amacrine. An indoleamine amacrine was described along with its synaptic contacts by Holmgren et al (1981), and amacrines immunoreactive for substance P, vasoactive intestinal polypeptide, and possibly cholecystokinin have been observed by H. Karten and N. Brecha (unpublished observations). Choline acetyltransferase activity in the cat retina is less than a tenth that in the rabbit (Ross & McDougal 1976), where Masland & Mills (1979) have identified cholinergic amacrines. The enzyme in the cat retina peaks, however, in the amacrine and inner plexiform layers, thus suggesting that at least one amacrine type is cholinergic.

Table 2 Amacrine transmitters

Transmitter	Cell density	Field diam.	Coverage	Percentage of total amt. layer
Dopamine[a, b, c]	40/mm^2[c]	500 μm[d]	7.9	1
GABA-accumulating:				
interplexiform[e]	90/mm^2[e]	250[f]	4.4	2[e]
GABA I[g]				12[g]
GABA II[g]				5[g]
GABA III[g]				6[g]
GABA IV[g]				1[g]
Glycine-accumulating:				
AII[h, i]	5000/mm^2[j]	35 μm[j, k]	4.8	25[j]
unidentified[h, i]				15[h]
				67
Indoleamine[l]				
Substance P[m]				
Vasoactive intestinal polypeptide[m]				
Cholecystokinin[m]				
Acetylcholine[n] (no morphological identification)				

[a] Ehinger (1966).
[b] Kramer et al (1971).
[c] Tork & Stone (1979).
[d] Kolb et al (1981).
[e] Nakamura et al (1980).
[f] Boycott et al (1975).
[g] Freed et al (1983).
[h] Nakamura et al (1978).
[i] Pourcho (1980).
[j] Sterling & Megill (1983).
[k] Famiglietti & Kolb (1975).
[l] Holmgren et al (1981).
[m] H. Karten and N. Brecha (unpublished).
[n] Ross & McDougal (1976).

AMINO ACIDS   Gamma aminobutyric acid (GABA) and glycine are widely considered to be transmitters in the mammalian retina because of their high endogenous levels, the presence of the GABA synthetic enzyme glutamic acid decarboxylase (Graham 1972, Wood et al 1976, Brandon et al 1979), the light-evoked, calcium-dependent release of glycine (Ehinger & Lindberg-Bauer 1976), and the potent physiological effects of their antagonists (Kirby & Enroth-Cugell 1976, Kirby 1979, Saito 1981, Caldwell & Daw 1978). Many neurons in the cat amacrine layer accumulate ^{3}H-GABA or ^{3}H-glycine following intravitreal injections (Bruun & Ehinger 1974, Marshall & Voaden 1975), and the problem since this discovery has been to determine which specific amacrine types are involved and the nature of their circuitry.

Five cell types accumulating GABA have been described by partial reconstructions from serial, electron microscope autoradiograms (Freed et al 1983), and one of these has been identified more specifically as the interplexiform cell (Nakamura et al 1980). This type is known from Golgi studies (Gallego 1971, Boycott et al 1975) to ramify in the outer and inner plexiform layers. The dendritic field diameter of the best Golgi-impregnated interplexiform cell is about 250 μm (Figure 8; Boycott et al 1975), and the type is distributed near the central area at a density of about 90/mm^2. The coverage factor calculated from these figures is about 4.4. The cat interplexiform cell receives amacrine input and has output to amacrines. The cell also provides output in both the outer and inner plexiform layers to rod bipolars and cone bipolars of both the flat and invaginating types (Kolb & West 1977, Nakamura et al 1980, McGuire et al 1983b). One might well wonder what the function of such a cell could be that innervates, and perhaps affects simultaneously, all bipolars, including those with opposite responses to light. One possibility is that the interplexiform cell might function in dim light as part of a gain-control system for bipolars.

AII AMACRINE   The best characterized amacrine in the cat retina is the AII, first described from Golgi and electron microscope studies (Kolb & Famiglietti 1974, Famiglietti & Kolb 1975) and later from detailed reconstructions (Sterling & Megill 1983). The AII cell, with a medium soma (about 9 μm) and narrow dendritic field (about 35 μm near the central area), sends a distinctly different set of processes to each sublamina. To sublamina b it sends radially a single stout process that branches to at least the eighth order and contains large mitochondria. To sublamina a it sends laterally fine processes that swell into large varicosities ("lobular appendages") and contain large mitochondria and synaptic vesicles.

The base of the AII soma bears roughly a dozen synaptic contacts (Sterling & Megill 1983) that are apparently from the dopamine amacrine (Pour-

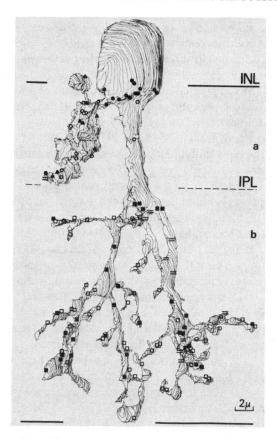

Figure 9 AII amacrine and its synaptic connections partially reconstructed from electron micrographs of 189 serial sections. Radial process branches in sublamina *b;* receives input from rod bipolars (□) and unidentified amacrine (■) and has gap junctions (=) with cone bipolars (primarily type CBb_1). Lobular appendage in sublamina *a* and soma receive amacrine (probably dopamine) input (•); lobular appendage receives input also from cone bipolar CBa_1 (o) and has outputs to cone bipolars CBa_1 and ganglion cell dendrities (Δ). Reprinted from Sterling & Megill (1983).

cho 1981). The AII arborization in sublamina *b* collects numerous synaptic contacts from rod bipolars (Famiglietti & Kolb 1975). One rod bipolar axon can contribute at least nine contacts to a single AII, and the amacrine as a whole may collect from as many as 30 rod bipolars, representing the convergence of an estimated 420 rods (Sterling & Megill 1983). The AII cells near their rod bipolar inputs make small gap junction contacts with each other (Kolb 1979). Large gap junctions are established between the AII processes and certain cone bipolar axons (Famiglietti & Kolb 1975, Kolb 1979). Detailed reconstructions show these to be principally with cone bipolar CBb_1 and to a lesser extent with types CBb_2 and CBb_3 (McGuire et al 1983b). Amacrine contacts, unidentified as to origin, distribute sparsely over the AII dendrites in sublamina b (Sterling & Megill 1983). The AII lobular appendages in sublamina a appear to receive input from cone bipolar CBa_1 (McGuire et al 1983b) and the dopamine amacrine (Pourcho 1981). The chemical synaptic output of the lobular appendage is directed to other amacrines, ganglion cells, and cone bipolars, specifically type CBa_1 (Famiglietti & Kolb 1975, Kolb 1979, McGuire et al 1983b).

The AIIs response to light is an all-or-none, depolarizing transient followed by a sustained depolarization (Figure 7b; Nelson et al 1976). This response, as anticipated from the circuitry, is strongly rod-driven. There is also evidence at higher stimulus intensities, for a small cone signal, which Nelson et al (1976) suggest may enter the AII via its gap junctions with cone bipolars. The AII accumulates ^{3}H-glycine (Nakamura et al 1978, Pourcho 1980, Sterling et al 1983). Its lobular appendages may, therefore, be glycinergic (inhibitory). The AII is the most numerous of the amacrine types, representing almost one quarter of all cells in the amacrine layer (Sterling & Megill 1983). Its distribution near the central area is quite regular, with a density of about 5000 cells/mm^2 and a coverage factor about 4.8. Possible roles for this cell in states of light and dark adaptation are suggested below (Microcircuitry of Beta Ganglion Cell Receptive Field).

In summary, the amacrines as a class are extremely heterogeneous. There is evidence so far for at least 22 types and eight transmitters. Such types as are already linked to a particular transmitter account numerically for over two-thirds of the amacrine layer cells (Table 2). Much of the evidence linking particular cells and transmitters derives from morphological studies following in vivo transmitter accumulation. That a correspondence exists between a cell's morphology and its accumulation of a particular transmitter has been firmly established. It remains to be determined whether such accumulation invariably reflects a cell's *use* of that transmitter. Evidence on this point is badly needed and can be gathered by localizing antibodies to transmitter-synthesizing enzymes.

The distribution in the retinal mosaic has been determined for three cell types in the amacrine layer (Table 2). In each case the distribution is quite regular, but the density can vary between types by a factor of more than 100. The dendritic field area varies inversely with density so that every point on the retina falls within the field of at least one member of each type. The range of coverage factors (4.4–7.9) is, so far, surprisingly narrow. For only one type, the AII, is there detailed knowledge of circuitry. One reason is that electron microscopy of Golgi-impregnated amacrines has been hampered by the difficulty of obtaining good ultrastructural preservation. Serial reconstruction has also proved difficult because the large amacrine profiles, so obvious in a single thin section, mostly represent varicosities that are interconnected by extremely fine cytoplasmic strands. These strands, being in thickness the same order of magnitude as the section, may be followed when orthogonal to the plane of section but are easily lost when they turn tangentially (Ellias & Stevens 1980). Newer marking methods such as intracellular staining with HRP (Kolb & Nelson 1981) and immunocyto-chemical staining with monoclonal antibodies Sterling & Lampson 1983) hold considerable promise.

Ganglion Cells

Ganglion cells in the cat retina, according to the recent Golgi study of Kolb et al (1981), are of 23 distinct morphological types. At this point in the present review such a figure may no longer seem astonishing. But it is sobering to realize that a decade ago this conclusion would probably have been judged to be unpublishable. To grasp why this claim now seems reasonable requires some historical perspective. From 1953, when Kuffler first described the on- and off-center ganglion cell receptive fields, until 1966, only these two physiological types were distinguished. Evidence sug-gesting additional types (Wiesel 1960) went unappreciated. As to morpho-logical types, many had been illustrated by Cajal (1892), but whether these represented discrete types or a continuum was unclear (Brown & Major 1966).

The number of recognized physiological types doubled in 1966 when Enroth-Cugell & Robson distinguished on- and off-center X and Y type ganglion cells. These were later described also by Cleland et al (1971) as "brisk-sustained" and "brisk-transient." By 1974 fully a dozen physiologi-cal types had been described (reviewed by Rodieck 1979, Rowe & Stone 1977, Stone et al 1979), and in that year as well the problem of morphologi-cal typing was dramatically clarified.

The crucial step in defining discrete morphological types of ganglion cell was to compare neurons at equivalent retinal positions (Boycott & Wässle 1974). With this approach, two fundamental types, termed alpha and beta, suddenly became obvious (Table 3). The alpha has a large soma and axon and a wide-field (180–1000 μm diameter), sparsely branched, dendritic tree. The beta has a medium soma and axon, and a narrow-field (20–300 μm diameter), densely branched, dendritic tree (Figure 10). Each type is repre-sented at *all* eccentricities and increases in size from center to periphery. Evidence that alpha cells are physiologically brisk-transient (Y) while beta cells are brisk-sustained (X) was obtained from comparisons at a given eccentricity between soma and dendritic field size on the one hand, and axonal conduction velocity and receptive field center size on the other (Cleland & Levick 1974).

Table 3 Characteristics of alpha and beta ganglion cells in central area

Cell type	Soma size	Sub-lamina of dend. strat.	Dend. field diam.	R.F. center diam.	Density	Physiol. coverage
on-alpha	17 μm[a]	b	180 μm[b]	175 μm[d]	77/mm^2[c]	1.35[a]
off-alpha	17 μm[a]	a	180 μm[b]	175 μm[d]	94/mm^2[c]	1.7[a]
on-beta	13 μm[e]	b	20 μm[b]	66 μm[d]	3250/mm^2[d]	11[d]
off-beta	13 μm[e]	a	20 μm[b]	66 μm[d]	3250/mm^2[d]	11[d]

Cell type	Inputs[k] from:				Output[j] to:
	CBa$_{1-4}$[f,g]	CBb$_{1-3}$[f,g]	AII lob. append.[h,i]	AII via CBb$_1$[h,i]	
on-alpha	−	+	−	+	Lgn, Sc
off-alpha	+	−	+	−	Lgn, Sc
on-beta	−	+	−	+	Lgn
off-beta	+	−	+	−	Lgn

[a] Wassle et al 1981a.
[b] Boycott & Wassle 1974.
[c] Calculated from Wassle et al 1981a.
[d] Peichl & Wassle 1979.
[e] Wassle et al 1981b.
[f] McGuire et al 1983b.
[g] McGuire et al 1983c.
[h] Sterling & Megill 1983.
[i] Kolb 1979.
[j] Reviewed by Rodieck 1979.
[k] Known only for region 1–2° from central area.

Cleland et al (1975) mapped the position of every brisk-transient cell in a small patch of retina. Subsequently they identified all the alpha cells in the same patch and showed the correspondence to be essentially complete. They showed further that over a wide range of eccentricities every retinal locus is covered by the receptive fields of 3–7 alpha cells. Although the distribution density of alpha cells declines from the central area outward, at each locus they are a constant fraction (about 4%) of the total ganglion cell population (Wässle et al 1975). Their increase in dendritic field diameter with eccentricity offsets their decline in numbers; this is the reason the coverage factor is constant.

When Boycott & Wässle (1974) pointed out the resemblance between a central alpha cell and a peripheral beta cell, the root of much doubt and confusion was exposed. Since then, anatomical observations have been reported in the context of their eccentricity; hence the conclusion of Kolb et al (1981) that there are several dozen types of ganglion cell is convincing. Boycott & Wässle (1974) also described several types of smaller ganglion cells with rather wide dendritic trees, termed gamma and delta, and to this Leventhal et al (1980) have added the epsilon cell. Kolb et al (1981) have

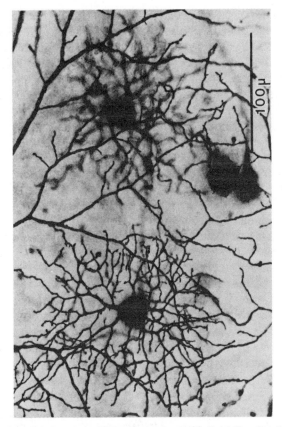

Figure 10 Micrograph from retinal flat-mount stained with Golgi-Cox. Focal plane through sublamina *b* of inner plexiform layer. Narrow-field cell with bushy dendrites on left is an on-beta ganglion cell. Matching cell on right with dendrites out of focus in sublamina *a* is off-beta. Sparsely-branched dendrites, belonging to two on-alpha cells, overlap the whole field. Reprinted from Wässle et al (1981b).

also recognized the gamma and delta types, terming them, respectively, G_3 and G_{19}. These types probably have physiological counterparts among the heterogeneous W category (see Rodieck 1979). Because so little is known of their circuitry, these types are omitted here from further discussion.

The next crucial observation in developing a typology for ganglion cells was made by Famiglietti & Kolb (1976). They noticed that the dendrites of both alpha and beta cells stratify either in the outer third of the inner plexiform layer (sublamina *a*) or in the inner two-thirds (sublamina *b*) (Figure 10). Believing that invaginating (depolarizing) cone bipolars end exclusively in sublamina *b* and flat (hyperpolarizing) cone bipolars end exclusively in *a,* they proposed that ganglion cells branching in *b* are on-center, while those branching in *a* are off-center. Intracellular recordings from ganglion cells followed by dye-injection confirmed this conjecture (Nelson et al 1978). Only later was it learned that flat and invaginating cone bipolars innervate *both* sublaminae (see Section above on Bipolar Cells). In summary, dendrites of off- and on-center ganglion cells stratify, respectively, in sublaminae *a* and *b* (see also Peichl & Wässle 1981, Wässle et al 1981b), and appear in each sublamina to receive input from both depolarizing and hyperpolarizing bipolars (McGuire et al 1983b,c,).

ALPHA CELLS   Knowledge of the on- and off- alpha cells' dendritic stratification permitted Wässle et al (1981a) to work out, using neurofibrillar stains selective for alpha cells, their detailed distributions in the retinal mosaic. The on-alpha was found to be slightly smaller and less numerous than the off-alpha, and the coverage factor for each type, respectively, was 1.4 and 1.7. The distribution of somas formed a regular lattice independently for each alpha type. For such a lattice the ratio, mean/s.d., for the distribution of nearest neighbor distances is about 4.5 whereas this ratio for a random distribution of the same density is 2.4. By this measure (mean/s.d.) the distribution of on-alpha dendritic field diameters for a small patch of retina is also extremely regular, the numerical value being 10.9 (Wässle et al 1981a). Peichl & Wässle (1979) found similar regularity in the size of receptive field centers within a small patch of retina.

BETA CELLS   Across most of the retina the beta cells constitute about 55% of the ganglion cells, the on-betas being slightly less numerous than the off-betas (Peichl & Wässle 1979, Wässle et al 1981b). Their density is highest in the central area (about 3250 cells/mm^2 for each type; Peichl & Wässle 1979) and their dendritic fields the smallest (about 20 μm diameter; Boycott & Wässle 1974). Each beta type forms a regular lattice (Figure 11) that is independent of the other, and provides an anatomical coverage factor in the central area of about one. The minimum diameter of the physiological receptive field center is about 66 μm, providing a *physiological coverage* for each type of about 11 (Peichl & Wässle 1979). This spacing and field size matches closely, according to Hughes (1981), the optimum predicted from sampling theory for a system that can resolve about 6 cycles/deg. using elements with the cut-off frequencies of the beta cells (Cleland et al 1979). Thus the evidence grows that the maximum spatial resolution demonstrated behaviorally in cat (6 cycles/deg; reviewed by Hughes 1981) depends on the simultaneous reports from these two types of beta ganglion cells. This is consistent with the observation of Berkeley & Sprague (1979) that the major effect of removing area 17 (to which beta cells project via the lateral geniculate nucleus) is a loss of spatial resolution.

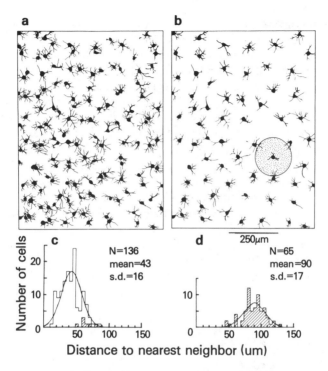

Figure 11   (*a*) Retrograde filling by HRP of all beta cells in a small field. (*b*) Same field with only on-betas drawn. Note regularity of mosaic. Stippling indicates dendritic field of a beta cell from Golgi material at same eccentricity. Below, distributions of distances to nearest neighbor for drawings above (*c*) Mean/sd for all betas is 2.7. (*d*) Mean/sd for on-betas only is 5.3. Reprinted from Wässle et al (1981b).

CIRCUITRY OF ALPHA AND BETA GANGLION CELLS   Once the alpha and beta types of ganglion cell were defined by light microscopy, they could be recognized in serial sections at the electron microscope level and a start could be made in defining their circuitry. The first observations showed that the inputs to each type are not on the soma but are concentrated on the dendrites in the sublamina of the cell's major arborization (Kolb 1979, Stevens et al 1980b). Beta cells receive about 70% of their input from cone bipolars, whereas the alpha cells get relatively more of their input from amacrines (Kolb 1979). McGuire et al (1938b) found that all four cone bipolar types in sublamina *a* (CBa$_1$–CBa$_4$) and all three types in sublamina *b* (CBb$_1$–CBb$_3$) contact dark ganglion cell dendrites (presumed beta) and pale ganglion cell dendrites (some probably alpha). Thus, whatever information is conveyed to the inner plexiform layer by this heterogeneous group of bipolar neurons, it all appears to reach both the alpha and the beta ganglion cells.

The convergence onto on- and off-beta cells from two pairs of cone bipolars was reconstructed in more detail (McGuire et al 1983c). The partially reconstructed beta cells had adjacent somas near the central area and overlapping dendritic fields estimated to be 30–40 μm diameter. The on-beta reconstructed to its sixth order branches received a total of 69 synaptic contacts from three CBb$_1$ cells and 34 contacts from a single CBb$_2$ (Figure 12). The contacts from CBb$_1$ were concentrated on the proximal dendrites in strata 4–5, while those from CBb$_2$ were concentrated more distally in strata 3–4. The off-beta, reconstructed to its seventh order branches, received 23 contacts from a single CBa$_1$ and 14 from a CBa$_2$. The actual proportions of input from the different bipolar types cannot be read from these data because the reconstructions were incomplete. It was estimated, for example, that the on-beta might actually collect as many as 400 contacts from up to six cone bipolars. Clearly, however, each type of beta cell receives substantial input from at least two types of cone bipolar. the possible functions of these arrangements are discussed below (See Microcircuitry of the Beta Ganglion Cell Receptive Field).

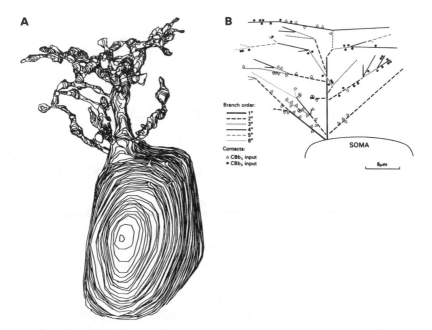

Figure 12   A. On-beta ganglion cell partially reconstructed from electron micrographs of 189 serial sections. B. Branch schematic of same cell, showing branches to the sixth order and distribution of inputs from cone bipolars CBb$_1$ (Δ) more proximally and CBb$_2$ (•) more distally. Reprinted from McGuire et al (1983c).

The diameter of the receptive field center of an alpha cell corresponds almost exactly to the diameter of its dendritic field. In contrast, the receptive field center of the beta cell is about three times larger than its dendritic field (Peichl & Wässle 1979). According to Hochstein & Shapley (1976), the alpha's receptive field center contains nonlinear subunits corresponding roughly in size to the beta's receptive field center. These authors suggested that both the subunit of the alpha cell and the full receptive field center of the beta cell might correspond to the receptive field of a single bipolar.

Although it is now clear that each beta cell receives from several bipolars, this basic idea may still have merit. The relation between a single cone bipolar and its target beta cell is strong, the narrow-field bipolar richly enveloping the beta's dendritic tree and investing it with multiple contacts (Kolb 1979, McGuire et al 1983c). These contacts distribute widely over the beta's dendritic tree, which might contribute to the linear summation of their postsynaptic effects (Rall 1967). The relation of the same cone bipolar and its target alpha cell is both weaker and more punctate, the bipolar contributing a smaller number of contacts to a relatively short segment of the alpha dendrite. The postsynaptic effects of these contacts, which are locally concentrated, might tend to sum nonlinearly (Rall 1967). Obviously, more detail must be gathered on the alpha cells' circuitry before this argument can be pursued.

Summary of Retinal Cell Types

The crucial insight emerging from the work reviewed here is that the neurons in the cat retina belong to discrete types. A "type" has come to be defined by the regular association of particular morphological, cytological, connectional, chemical, and physiological features. Thus, among ganglion cells, neither soma size nor dendritic branching pattern alone defines a fundamental type. The strict association, however, of medium size and "bushy" dendritic branching restricted to sublamina *b,* and the physiological properties, "X-ness," brisk-sustained, and on-center, define what is almost certainly a fundamental type, the on-center beta/X cell. Similarly, the accumulation of a particular neural transmitter does not define a type because neurons of several different morphologies accumulate the same transmitter. But a strict association between accumulation of a particular transmitter, such as glycine, and a distinctive morphology, connectivity, and physiology does define a fundamental type, the glycine-accumulating, depolarizing, AII amacrine. Each type, so defined, turns out to have a characteristic stoichiometry and distribution in the retinal mosaic; thus further supporting the idea that the type is fundamental.

In certain respects, the definition of a type is not absolute but relational. Thus, the on-alpha cell cannot be defined by its absolute dendritic field diameter because this increases about five-fold from center to periphery. Nor can the on-alpha be defined by its absolute distribution density, because this falls from center to periphery by about sixteen-fold. The relation *between* these two features is strong, however, so that coverage factor at every retinal locus is the same and this becomes a defining feature of the type. Furthermore, although the on-alpha's absolute density changes, its fractional relationship to the other ganglion cells remains constant across the retina at about 4%, and this, too, is a defining feature. The same is true physiologically. Absolute receptive field size rises from center to periphery, but the fundamental properties, Y-ness, brisk-transient, and on-center, can be recognized at every eccentricity. This is probably because the synaptic relations are constant between the on-alpha and the various cell types that provide its input. Direct evidence on this point is badly needed.

For a particular cell type at a given retinal locus there is evidence, though still fragmentary, for a surprising degree of regularity of feature. Thus, the distances between neighboring cells of a particular type form a relatively narrow, Gaussian distribution, the ratio, mean/sd, (see Cell Types, above) ranging from 3.5–6.5 (Table 4). The soma diameters for a particular type at a particular locus are similarly regular, the ratio, mean/sd, for each of five types ranging from 5.4 to 7.7 (Table 4). Similar regularity has been observed for the dendritic field diameters of on- and off-alpha ganglion cells; the mean/sd for their distributions is about seven (Table 4). The synaptic relationships between particular cell types may be even more highly regular (Table 4; McGuire et al 1983b). Thus, inputs to adjacent rod bipolars from reciprocal and nonreciprocal amacrines form rather constant fractions of the total input, and rod bipolar outputs to each of these amacrines form rather constant fractions of the total output (mean/sd, 10.3–46.4). Inputs and outputs of the cone bipolar CBb_1 are similarly regular (mean/sd, 6.-0–16.3).

Considered simply as nervous tissue, a small patch of cat retina resembles in some ways an invertebrate ganglion. Both tissues are composed of many discrete cell types, each highly regular in size, distribution, synaptic connections, and neural transmitter. Just as the developmental rules for forming an invertebrate ganglion are repeated for each segment with appropriate modification, so the rules for forming a patch of the mosaic seem to be repeated with appropriate modification across the retina. Possibly, mammalian neurons are wired with a precision similar to that for which the invertebrates are famous (Sterling 1982). If so, perhaps there exist wide similarities across phyla in the principles, if not the detailed instructions, by which nervous tissue is formed. Such speculation aside, an appreciation of retinal tissue as composed of many definite types has sharpened the confidence of microscopists that even subtle differences between cells can have meaning and has led to a renewed and truly inspired microscopy.

Table 4 Quantitative evidence for regularity of feature in certain cell types

						Total input from (%):			Total output to (%):			
	Coverage factor[a]	Distance nearest neighbor	Soma size	Axon diam.	Dend. field diam.	Recip. Am	Non Recip.	All	Recip. Am	Non Recip.	Dark Gang.	Pale Gang.
cone		6.5^g										
A-horizontal	4^c	6.7^h										
B-horizontal	7^c	6.0^h										
rod bipolar			7.7^k	11.5^k		46.4^k	16.9^k	10.3^k	17.9^k			
cone bipolar CBb$_1$						6.0^k	16.3^k		7.4^k	12^k	13^k	9.9^k
interplexiform	4.4^d	3.5^i										
on-alpha	1.4^e	4.5^e	5.9^e		7.5^e							
off-alpha	1.7^e	4.7^e	6.6^e		6.7^e							
on-beta	11^f	5.3^j	5.4^j									
off-beta	11^f	5.3^j	5.4^j									

[a] Horizontal and ganglion cell values are for central area.
[b] Ratio introduced by Wassle & Riemann (1978) as a measure of regularity in a pattern.
[c] Anatomical coverage, Wassle et al 1978b.
[d] Anatomical coverage, calculated from Nakamura et al 1980 and Boycott et al 1975.
[e] Wassle et al 1981a.
[f] Physiological coverage, Peichl & Wassle 1979.
[g] Wassle & Riemann 1978.
[h] Wassle et al 1978b.
[i] Calculated by M. Freed from data in Nakamura et al 1980.
[j] Wassle et al 1981b.
[k] McGuire et al 1983b.

COVERAGE FACTOR   The coverage factor (see Cell Types, above) has been determined for nine cell types (Tables 2–4). In every case coverage is complete. It has been determined for six types that the coverage is largely constant across the retina from center to periphery. Because cell density falls continuously across the retina while dendritic field diameter rises, a constant coverage factor implies a close matching of these two parameters at every eccentricity. This appears to be a large scale regularity that should perhaps be distinguished in our thinking from local regularities such as dendritic field diameter, cell spacing, and synaptic connections.

Certain cell types, the wide-field, sparsely distributed, alpha ganglion cells, for example, have small coverage factors (1.4–1.7), close to the minimum required for completeness. Other types, such as the narrow field, densely distributed, beta ganglion cells, have much larger coverage factors (as large as 11, in the central area). It is believed that the beta cell's large coverage is not redundant. Rather, the particular combination of receptive field size and density that produces such a coverage appears matched to the optimum predicted by sampling theory for a system resolving about 6 cycles/deg (Wässle et al 1981b, Hughes 1981). The distribution density of the narrow-field, AII amacrine resembles that of the beta ganglion cells, and this is probably because the AII forms the beta cell's dark-adapted receptive field (see below). On the other hand, the dopamine amacrine, a cell that provides input to the AII, is wide-field and sparse, suggesting that the information it conveys to the AII is poor in spatial detail. Such observations as a whole suggest that the coverage factor for each cell type reflects the specific role of that type and that coverage beyond the minimum has little to do with abstract notions of "redundancy" (Sterling 1982).

MICROCIRCUITRY OF THE BETA GANGLION CELL RECEPTIVE FIELD

The preceding sections described individual elements and anatomical circuits linking photoreceptors to ganglion cells. An obvious next step is to infer which of these circuits form the actual physiological pathways that generate the ganglion cell receptive fields. With this aim I consider for the beta cell a circuit involving 13 neuron types. Connections between eight of these types, reconstructed directly from a series of 189 thin sections, are shown in Figure 13. Two ganglion cell somas, an on- and an off-beta, abut each other in the series, and their dendritic fields overlap, but in Figure 13 they are teased apart for clarity.

Major input to each beta cell is from a pair of cone bipolars, each of which is presumed to have a center and an antagonistic surround. The bipolars seem to be arranged in "push-pull" fashion; that is, excitation delivered from one bipolar to the beta cell is accompanied by withdrawal of inhibition delivered by the other, and vice-versa (Figure 14). For the on-beta cell, one

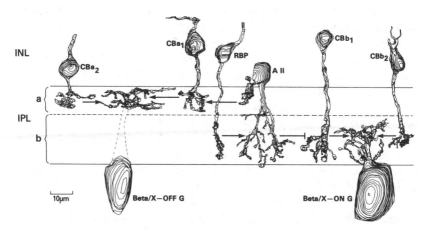

Figure 13 Eight neurons with synaptic interconnections partially reconstructed from electron micrographs of 189 serial sections. Chemical synapse (→); gap junctions (⊣). Dendritic fields of all these neurons overlap extensively and are separated here only for clarity. [Reprinted from Sterling & Megill (1983).] INL, inner nuclear layer; CB, cone bipolar; RBP, rod bipolar; G, ganglion cell.

member of the cone bipolar pair (CBb$_1$) is believed to be depolarizing and excitatory, and the other (CBb$_2$) is believed to be hyperpolarizing and inhibitory. Turning on a spot in the on-beta cell's receptive field center should depolarize CBb$_1$, delivering excitation, and simultaneously hyperpolarize CBb$_2$, withdrawing inhibition. Turning off the spot or turning on an annulus in the surround should cause the opposite response: withdrawal of excitation by CBb$_1$ and delivery of inhibition from CBb$_2$. For the off-beta cell, one member of the cone bipolar pair (CBa$_1$) is believed to be hyperpolarizing and excitatory, and the other (CBa$_2$ is believed to be depolarizing and inhibitory. Turning off a spot in the center should cause excitation from CBa$_1$ and withdrawal of inhibition from CBa$_2$.

This "push-pull" hypothesis, which stems directly from the microcircuitry illustrated in Figure 13, might have several advantages for a cell designed to detect local contrast at high spatial resolution. First, it would allow a cell to fire at high frequencies under some conditions and be totally supressed under others. This may be what enables ganglion cells such as the on-beta (detecting local brightness) and the off-beta (detecting local darkness) to operate over such a wide range of spike frequencies (0–700hz; Kuffler 1953). The push-pull mechanism should also quicken the beta's

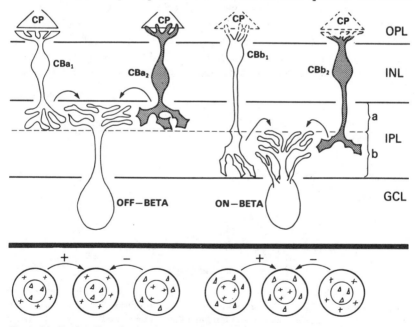

Figure 14   Push-pull mechanism for generating photopic receptive fields of on- and off-center beta ganglion cells. CP, cone pedicle. Further explanation in text. Reprinted from McGuire et al (1983c).

response to *changes* in contrast and thus may contribute to one of the beta's notable physiological attributes, "briskness" of response.

Figure 15 shows that under mesopic illumination the beta ganglion cell receptive field still has a center and an antagonistic surround, both of which are rod-driven (Barlow et al 1957, Kaplan et al 1979). The operative pathway suggested in Figure 16 is the one from the rod spherules to adjacent cone pedicles (Kolb 1979, Nelson 1977). A dim spot (below the cone threshold) causes a photocurrent to flow from rods into the cone pedicle. This current may activate the horizontal cell mechanism contained in the pedicle for generating the center-surround fields of cone bipolars and thence the center-surround receptive fields of beta cells. This might be the mechanism by which the center-surround organization, so crucial to the beta cell's spatial resolution, is maintained even in dim illumination.

After an hour or so in total darkness, the beta cell's sensitivity increases still further (Figure 15), to the point where a single photon can cause several extra impulses (Barlow et al 1971). But now spatial resolving power is sacrificed as the receptive field center enlarges by two-fold and the antagonistic surround drops out (Figure 15; Barlow et al 1957, Enroth-Cugell & Robson 1966, Kaplan et al 1979). The primary pathway suggested for this condition is no longer rod → cone → cone bipolar, but rod → rod bipolar → AII amacrine (Figure 16). The depolarization evoked in the AII by a spot (Nelson et al 1976) may spread via gap junctions to the CBb_1 axon terminals, causing them to excite the on-beta cell. The on-beta cell's enlarged receptive field center would correspond in diameter to the aggregate field

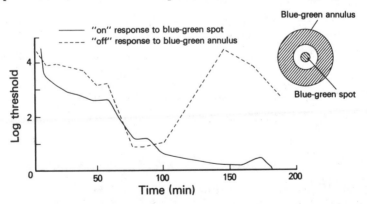

Figure 15   Thresholds of on-center ganglion cell to flashes of blue-green light during dark adaptation. Stimuli 380 msec in duration focused on *hatched areas* of inset diagram. Cones are responsible for the early sharp fall in threshold (first 10 min). Rods take over at between 10 and 30 min in both center and surround (see Barlow et al 1957). Threshold falls for both center and surround until about 100 min; thereafter, center threshold continues falling while surrounded threshold rises sharply. [Condensed from Figure 2 of Barlow et al (1957).]

of all the AII cells that feed it (Sterling & Megill 1983). It could, perhaps, be even a little larger than this, because the AIIs are interconnected by gap junctions. The antagonistic surround is lost from the beta cell apparently because it is conveyed from the outer plexiform layer by the cone bipolar dendrites, which under complete dark adaptation may carry no signal.

The same depolarization of the AII that excites the on-beta ganglion cell may suppress the off-beta cell by two mechanisms. The depolarized lobular appendage of the AII, as it is probably glycinergic, may cause postsynaptic inhibition via its contact with off-beta dendrites (Kolb 1979) and presynaptic inhibition via its contacts with the axon terminal of CBa_1 (Figure 16; McGuire et al 1983b, Sterling & Megill 1983). Because for both types of beta cell the push-pull arrangement is sacrificed in the dark, one would anticipate a loss in temporal as well as spatial resolution, and this is in accord with the physiological observations of Kaplan et al (1979). The circuitry in Figure 16 is also in accord with the reported effects of strychnine, a glycine blocker, on beta cell receptive fields (Kirby 1979, Saito 1981).

The two rod pathways discovered by Famiglietti, Kolb, and Nelson thus may serve different functions, and one wonders whether there are mecha-

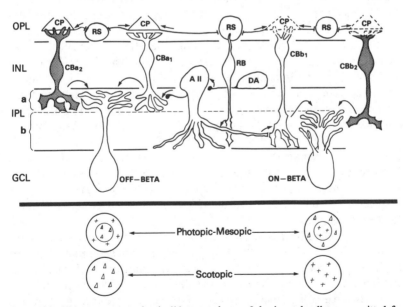

Figure 16 Thirteen neuron circuit (11 types shown; 2 horizontal cell types omitted for clarity) that may generate the receptive fields of on- and off-center beta ganglion cells in states of light and dark adaptation. Further explanation in text. CP, cone pedicle; RS, rod spherule; OPL, outer plexiform layer; INL, inner nuclear layer; CB, cone bipolar; RB, rod bipolar; DA, dopamine amacrine; AII, AII amacrine; GCL, ganglion cell layer.

nisms to regulate which pathway is active. It would not be surprising if the rod → cone pathway serving mesopic vision were switched off at the end of dark adaptation and the rod → AII pathway were switched on. The patency of the rod → cone pathway might be regulated by the B-horizontal cell as described above (Horizontal Cells, Physiology). The rod bipolar → AII pathway might be regulated by the dopamine amacrine. One would not expect the AII to be totally suppressed by dopamine at photopic levels because the AII can carry a cone signal (Figure 7). However, the rod bipolar pathway into or out of the AII might be regulated by dopamine in a more specific fashion (Sterling & Megill 1983).

CONCLUSION

Hubel & Wiesel proposed in 1962 a hierarchical structure for the cat visual system. A particular property, they argued, is established at one level, sharpened at successively higher levels, and generalized to a larger region of visual field. Their idea developed from studies of single units at the third through sixth levels (ganglion cells to complex cells in area 17). The studies reviewed here strongly support their concept as applied to the first through third levels (receptors to ganglion cells). In regard to simple and complex cortical cells, the hierarchical model has been strongly criticized (Stone et al 1979). However, if area 17 is anything like the retina, it probably contains many discrete cell types connected in specific ways. These probably do not form a single simple-to-complex hierarchy but many parallel hierarchies to further abstract and generalize the "qualitative contexts" relayed there from the many types of retinal ganglion cells. Anatomical and physiological evidence grows that a single sublayer of area 17 does have many discrete types (Davis & Sterling 1979, Solnick et al 1983, Hamos et al 1983), and new "contexts" apparently are created in the domains of orientation (Hubel & Wiesel 1962) and spatial frequency (Movshon et al 1978). What is needed to identify the continuation of the retinal hierarchies into area 17 is more detailed knowledge of the cortical cell types and their circuitry (Gilbert 1983, this volume).

Strong concepts have emerged to guide research on microcircuitry of the cat retina. It is now believed that cell types are discrete and number roughly 60. Each type is believed to have a particular transmitter, set of connections, and mosaic distribution. As knowledge of circuitry becomes very detailed, hypotheses regarding function emerge. These hypotheses are quite specific and testable; whether they are correct seems less important than that they can be read directly from the circuitry. Powerful technologies to extend our knowledge of circuitry in cat retina exist and are still developing. We may expect, in addition to the approaches already noted, to have monoclonal antibodies specific for particular cell types (Sterling & Lampson 1983). Further, since neural activity is strongly reflected in oxidative metabolism and this can be recognized at the electron microscope level (Wong-Riley et al 1978), it may be possible to determine directly by electron microscopy which pathways are active under particular conditions. Where concepts and methods are strong, one may expect progress to be rapid. At a recent meeting of the Society for Neuroscience, D. H. Hubel declared, "The good old days are right now." Those who work on the cat retina will not argue.

ACKNOWLEDGMENTS

I thank M. Freed, B. McGuire, and L. Palmer for critical comments on this manuscript. Thanks also to T. Davis, M. Freed, B. McGuire, J. Megill, Y. Nakamura, R. Smith, and J. Stevens for their collaborative efforts supported by grants EY00828, EY01583, and Research Career Development Award EY00080 from the National Eye Institute. I am also grateful to D. Heany and J. Michaud for assistance, to H. Abriss for preparing the manuscript, and to J. and B. Woolsey for the illustrations.

Literature Cited

Barlow, H. B., Fitzhugh, R., Kuffler, S. W. 1957. Change of organization in the receptive fields of the cat's retina during dark adaptation. *J. Physiol.* 137:338–54

Barlow, H. B., Levick, W. R., Yoon, M. 1971. Responses to single quanta of light in retinal ganglion cells of the cat. *Vision Res.* 3:87–101

Baylor, D. A., Fuortes, M. G. F., O'Bryan, P. M. 1971. Receptive fields of cones in the retina of the turtle. *J. Physiol.* 214:265–71

Berkeley, M. A., Sprague, J. M. 1979. Striate cortex and visual acuity functions in the cat. *J. Comp. Neurol.* 187:679–702

Boycott, B. B., Dowling, J. E., Fisher, S. K., Kolb, H., Laties, A. M. 1975. Interplexiform cells of the mammalian retina and their comparison with catecholamine-containing retinal cells. *Proc. R. Soc. London Ser. B* 191:353–68

Boycott, B. B., Kolb, H. 1973. The connections between bipolar cells and photoreceptors in the retina of the domestic cat. *J. Comp. Neurol.* 148:91–114

Boycott, B. B., Peichl, L., Wässle, H. 1978. Morphological types of horizontal cell in the retina of the domestic cat. *Proc. R. Soc. London Ser. B* 203:229–45

Boycott, B. B., Wässle, H. 1974. The morphological types of ganglion cells of the domestic cat retina. *J. Physiol.* 240:397–419

Brandon, C., Lam, D. M. K., Wu, J-Y. 1979. The γ-aminobutyric acid system in rabbit retina: Localization by immunocytochemistry and autoradiography. *Proc. Natl. Acad. Sci. USA* 76:3557–61

Brown, J. E., Major, D. 1966. Cat retinal ganglion cell dendritic fields. *Exp. Neurol.* 15:70–78

Bruun, A., Ehinger, B. 1974. Uptake of certain possible neurotransmitters into retinal neurons of some mammals. *Exp. Eye Res.* 19:435–47

Cajal, S. Ramon y, 1892. La retine des vertebres. In *La Cellule* 9:119–257. Engl. trans., *The Structure of the Retina*, Compiled and transl. S. A. Thorpe, M. Glickstein. Springfield, Ill.: Thomas

Caldwell, J. H., Daw, N. W. 1978. Effects of picrotoxin and strychnine on rabbit retinal ganglion cells: Changes in center surround receptive fields. *J. Physiol.* 276:299–310

Cleland, B. G., Dubin, M. W., Levick, W. R. 1971. Sustained and transient neurones in the cat's retina and lateral geniculate nucleus. *J. Physiol.* 273:473–96

Cleland, B. G., Harding, T. H., Tulunay, U. 1979. Visual resolution and receptive field size: Examination of two kinds of cat retinal ganglion cell. *Science* 205:1015–17

Cleland, B. G., Levick, W. R. 1974. Brisk and sluggish concentrically organized ganglion cells in the cat's retina. *J. Physiol.* 240:421–56

Cleland, B. G., Levick, W. R., Wässle, H. 1975. Physiological identification of a morphological class of cat retinal ganglion cells. *J. Physiol.* 248:151–71

Davis, T. L., Sterling, P. 1979. Microcircuitry of cat visual cortex: Classification of neurons in layer IV of area 17, and identification of the patterns of lateral geniculate input. *J. Comp. Neurol.* 188:599–628

Daw, N. W., Pearlman, A. L. 1969. Cat color vision: One cone process or several. *J. Physiol.* 201:745–64

Daw, N. W., Pearlman, A. L. 1970. Cat color vision: Evidence for more than one cone process. *J. Physiol.* 211:125–37

DeMonasterio, F. M., Schein, S. J., McCrane, E. P. 1981. Staining of blue-sensitive cones of the Macaque retina by a fluorescent dye. *Science* 213:1278–81

Dowling, J. E., Boycott, B. B. 1966. Organization of the primate retina: Electron microscopy. *Proc. R. Soc. London Ser. B* 166:80–111

Dowling, J. E., Brown, J. E., Major, D. 1966. Synapses of horizontal cells in rabbit and cat retinas. *Science* 153:1639–41

Ehinger, B. 1966. Adrenergic junctional neurons. *Z. Zellforsch. Mikrosk. Anat.* 71:146–52

Ehinger, B., Lindberg-Bauer, B. 1976. Light evoked release of glycine from cat and rabbit retina. *Brain Res.* 113:535–49

Ellias, S. A., Stevens, J. K. 1980. The dendritic varicosity: A mechanism for electrically isolating the dendrites of cat retinal amacrine cells? *Brain Res.* 196:365–72

Enroth-Cugell, C., Robson, J. G. 1966. The contrast sensitivity of retinal ganglion cells of the cat. *J. Physiol.* 187:517–52

Fain, G. L., Gold, G. H., Dowling, J. E. 1976. Receptor coupling in the toad. *Cold Spring Harbor Symp. Quant. Biol.* 40:547–61

Famiglietti, E. V. 1981. Functional architecture of cone bipolar cells in mammalian retina. *Vision Res.* 21:1559–63

Famiglietti, E. V., Kolb, H. 1975. A bistratified amacrine cell and synaptic circuitry in the inner plexiform layer of the retina. *Brain Res.* 84:293–300

Famiglietti, E. V., Kolb, H. 1976. Structural basis for "ON" and "OFF"-center responses in retinal ganglion cells. *Science* 194:193–95

Fisher, S. K., Boycott, B. B. 1974. Synaptic connexions made by horizontal cells within the outer plexiform layer of the

retina of the cat and the rabbit. *Proc. R. Soc. London Ser. B* 186:317–31

Foerster, M. H., van de Grind, W. A., Grüsser, O-J. 1977a. Frequency transfer properties of three distinct types of cat horizontal cells. *Exp. Brain Res.* 29:347–66

Foerster, M. H., van de Grind, W. A., Grüsser, O-J. 1977b. The response of cat horizontal cells to flicker stimuli of different area, intensity and frequency. *Exp. Brain Res.* 29:367–85

Freed, M., Nakamura, Y., Sterling, P. 1983. Four types of amacrine cell in cat retina accumulate ^{3}H-gamma-aminobutyric acid. Manuscript in preparation

Gallego, A. 1971. Horizontal and amacrine cells in the mammal's retina. *Vision Res.* 3:33–50

Gilbert, C. D. 1983. Microcircuitry of visual cortex. *Ann. Rev. Neurosci.* 6:217–47

Graham, L. T. Jr. 1972. Intraretinal distribution of GABA content and GAD activity. *Brain Res.* 36:476–79

Hamos, J., Davis, T. D., Sterling, P. 1983. Four types of neurons in layer IVab of cortical area 17 accumulate ^{3}H-GABA. Manuscript in preparation

Hochstein, S., Shapley, R. M. 1976. Linear and nonlinear spatial subunits in Y cat retinal ganglion cells. *J. Physiol.* 262:265–84

Holländer, H., Stone, J. 1972. Receptor pedicle density in the cat's retina. *Brain Res.* 42:497–502

Holmgren, I. T., Ehinger, E., Dowling, J. E. 1981. Synaptic organization of the indoleamine-accumulating neurons in the cat retina. *ARVO Abstr. Suppl. Invest. Ophthal. Visual Sci.* 20:203

Hubel, D. H., Wiesel, T. N. 1962. Receptive fields, binocular interaction and functional architecture in the cat's visual cortex. *J. Physiol.* 160:106–54

Hughes, A. 1981. Cat retina and the sampling theorem: The relation of transient and sustained brisk-unit cut-off frequency to α and β-mode cell density. *Exp. Brain Res.* 42:196–202

Kaneko, A. 1970. Physiological and morphological identification of horizontal, bipolar and amacrine cells in goldfish retina. *J. Physiol.* 207:623–33

Kaneko, A. 1979. Physiology of the retina. *Ann. Rev. Neurosci.* 2:169–92

Kaplan, E., Marcus, S., So, Y. T. 1979. Effects of dark adaptation on spatial and temporal properties of receptive fields in cat lateral geniculate nucleus. *J. Physiol.* 294:561–80

Kirby, A. W. 1979. The effect of bicuculline and picrotoxin on X and Y cells in the cat retina. *J. Gen. Physiol.* 74:71–84

Kirby, A. W., Enroth-Cugell, C. 1976. The involvement of gamma-aminobutyric acid in the organization of cat retinal ganglion cell receptive fields. *J. Gen. Physiol.* 68:465–84

Kolb, H. 1974. The connections between horizontal cells and photoreceptors in the retina of the cat: Electron microscopy of Golgi preparations. *J. Comp. Neurol.* 155:1–14

Kolb, H. 1977. The organization of the outer plexiform layer in the retina of the cat: Electron microscopic observations. *J. Neurocytol.* 6:131–53

Kolb, H. 1979. The inner plexiform layer in the retina of the cat: Electron microscopic observations. *J. Neurocytol.* 8:295–329

Kolb, H., Famiglietti, E. V. 1974. Rod and cone pathways in the inner plexiform layer of cat retina. *Science* 186:47–49

Kolb, H., Nelson, R. 1981. Three amacrine cells of the cat retina: Morphology and intracellular responses. *ARVO Abstr. Suppl. Invest. Ophthal. Visual Sci.* 20:184

Kolb, H., Nelson, R., Mariani, A. 1981. Amacrine cells, bipolar cells, and ganglion cells of the cat retina: A Golgi study. *Vision Res.* 21:1081–1114

Kolb, H., West, R. W. 1977. Synaptic connections of the interplexiform cell in the retina of the cat. *J. Neurocytol.* 6:155–70

Kramer, S. G. 1971. Dopamine: A retinal neurotransmitter. I. Retinal uptake, storage, and light-stimulated release of H³-dopamine in vivo. *Invest. Ophthalmol.* 10:438–52

Kramer, S. G., Potts, A. M., Mangnall, Y. 1971. Dopamine: A retinal neurotransmitter. II. Autoradiographic localization of H³-dopamine in the retina. *Invest. Ophthalmol.* 10:617–24

Kuffler, S. W. 1953. Discharge patterns and functional organization of mammalian retina. *J. Neurophysiol.* 16:37–68

Leventhal, A., Keens, J., Törk, I. 1980. The afferent ganglion cells and cortical projections of the retinal recipient zone of the cat's pulvinar complex. *J. Comp. Neurol.* 194:535–54

Lettvin, J. Y., Maturana, H. R., Pitts, W. H., McCulloch, W. S. 1961. Two remarks on the visual system of the frog. In *Sensory Communication,* ed. W. Rosenblith, pp. 757–76. Cambridge: MIT Press

Marc, R., Sperling, H. G. 1976. The chromatic organization of the goldfish cone mosaic. *Vision Res.* 16:1211–24

Marc, R., Sperling, H. G. 1977. Chromatic organization of primate cones. *Science* 196:454–56

Marshall, J., Voaden, M. 1975. Autoradiographic identification of the cells accumulating ³H-γ-aminobutyric acid in mammalian retina: A species comparison. *Vision Res.* 15:459–61

Masland, R. H., Mills, F. W. 1979. Autoradiographic identification of acetylcholine in the rabbit retina. *J. Cell Biol.* 83:159–78

Maturana, H. R., Lettvin, J. Y., McCulloch, W. S., Pitts, W. H. 1960. Anatomy and physiology of vision in the frog (*rana pipiens*). *J. Gen. Physiol.* 43(2):129–76

McGuire, B. A., Goldman, S., Sterling, P. 1983a. Distribution of ganglion cells in cat area centralis. Manuscript in preparation

McGuire, B. A., Stevens, J. K., Sterling, P. 1980. *Soc. Neurosci. Abstr.* 6:347 (Abstr.)

McGuire, B. A., Stevens, J. K., Sterling, P. 1983b. Microcircuitry of bipolar cells in cat retina. Manuscript in preparation

McGuire, B. A., Stevens, J. K., Sterling, P. 1983c. Microcircuitry of the beta ganglion cell. Manuscript in preparation

Movshon, J. A., Thompson, I. D., Tolhurst, D. J. 1978. Spatial and temporal contrast sensitivity of neurons in areas 17 and 18 of the cat's visual cortex. *J. Physiol.* 283:101–20

Nakamura, Y., McGuire, B. A., Sterling, P. 1978. Selective uptake of ³H-gamma aminobutyric acid (GABA) and ³H-glycine by neurons of the amacrine layer of the cat retina. *Soc. Neurosci. Abstr.* 4:639

Nakamura, Y., McGuire, B. A., Sterling, P. 1980. Interplexiform cell in cat retina: Identification by selective uptake of [³H]-GABA. *Proc. Natl. Acad. Sci. USA* 77:658–61

Nelson, R. 1977. Cat cones have rod input: A comparison of response properties of cones and horizontal cell bodies in the retina of the cat. *J. Comp. Neurol.* 172:109–36

Nelson, R. 1980. Functional stratification of cone bipolar cell axons in the cat retina. *ARVO Abstr. Suppl. Invest. Ophthal. Visual Sci.* p. 130

Nelson, R. E., Famiglietti, E. V., Kolb, H. 1978. Intracellular staining reveals different levels of stratification for ON-

and OFF-center ganglion cells in cat retina. *J. Neurophysiol.* 41:472–83

Nelson, R., Kolb, H., Famiglietti, E. V., Gouras, P. 1976. Neural responses in rod and cone systems of the cat retina: Intracellular records and Procion stains. *Invest. Ophthal. Visual Sci.* 15: 946–53

Nelson, R., Lutzow, A. V., Kolb, H., Gouras, P. 1975. Horizontal cells in cat retina with independent dendritic systems. *Science* 189:137–39

Niemeyer, G., Gouras, P. 1973. Rod and cone signals in S-potentials of the isolated perfused cat eye. *Vision Res.* 13: 1603–12

Pearlman, A. L., Daw, N. W. 1970. Opponent color cells in the cat lateral geniculate nucleus. *Science* 167:84–86

Peichl, L., Wässle, H. 1979. Size, scatter, and coverage of ganglion cell receptive field centers in the cat retina. *J. Physiol.* 291:117–41

Peichl, L., Wässle, H. 1981. Morphological identification of on- and off-center brisk transient(Y) cells in the cat retina. *Proc. R. Soc. London Ser. B* 212:139–56

Penn, R. D., Hagins, W. A. 1972. Kinetics of the photocurrent of retinal rods. *Biophys. J.* 12:1073–94

Pourcho, R. G. 1980. Uptake of [³H]-glycine and [³H]-GABA by amacrine cells in the cat retina. *Brain Res.* 198:333–46

Pourcho, R. C. 1981. Dopaminergic amacrine cells in the cat retina. *ARVO Abstr. Suppl. Invest. Ophthal. Visual Sci.,* 20:203

Rall, W. 1967. Distinguishing theoretical synaptic potentials computed for different soma-dendritic distributions of synaptic input. *J. Neurophysiol.* 30: 1138–68

Raviola, E., Gilula, N. B. 1975. Intramembrane organization of specialized contacts in the outer plexiform layer of the retina. A freeze-fracture study in monkeys and rabbits. *J. Cell Biol.* 65:192–222

Ringo, J., Wolbarsht, M. L. 1981. Color coding in cat retinal ganglion cell receptive fields. *ARVO Abstr. Suppl. Invest. Ophthal. Visual Sci.,* 20:185

Ringo, J., Wolbarsht, M. L., Wagner, H. G., Crocker, R., Amthor, F. 1977. Trichromatic vision in the cat. *Science* 198:753–54

Rodieck, R. W. 1979. Visual pathways. *Ann. Rev. Neurosci.* 2:193–226

Rodieck, R. W., Rushton, W. A. H. 1976. Isolation of rod and cone contributions to cat ganglion cells by a method of light

exchange. *J. Physiol. London Ser. B* 254:759–73

Ross, C. D., McDougal, D. B. Jr. 1976. The distribution of choline acetyltransferase activity in vertebrate retina. *J. Neurochem.* 26:521–26

Rowe, M. H., Stone, J. 1977. Naming of neurons. Classification and naming of cat retinal ganglion cells. *Brain Behav. Evol.* 14:185–216

Saito, H. 1981. The effects of strychnine and bicuculline on the responses of X- and Y-cells of the isolated eye-cup preparation of the cat. *Brain Res.* 212:243–48

Scholes, J. H. 1975. Colour receptors, and their synaptic connexions, in the retina of cyprinid fish. *Philos. Trans R. Soc. London Ser. B* 270:61–118

Schwartz, E. A. 1973. Responses of single rods in the retina of the turtle. *J. Physiol.* 232:503–14

Schwartz, E. A. 1974. Responses of bipolar cells in the retina of the turtle. *J. Physiol.* 236:211–24

Solnick, B., Davis, T. L., Sterling, P. 1983. Number of neurons in layer IVab of a cortical module. Manuscript in preparation

Steinberg, R. H. 1969a. Rod and cone contributions to S-potentials from the cat retina. *Vision Res.* 9:1319–29

Steinberg, R. H. 1969b. Rod-cone interaction in S-potentials from the cat retina. *Vision Res.* 9:1331–44

Steinberg, R. H. 1969c. The rod after-effect in S-potentials from the cat retina. *Vision Res.* 9:1345–55

Steinberg, R. H. 1971. The evidence that horizontal cells generate S-potentials in the cat retina. *Vision Res.* 11:1029–31

Steinberg, R. H., Reid, M., Lacy, P. L. 1973. The distribution of rods and cones in the retina of the cat (*Felis domesticus*). *J. Comp. Neurol.* 148:229–48

Steinberg, R. H., Schmidt, R. 1970. Identification of horizontal cells as S-potential generators in the cat retina by intracellular dye injections. *Vision Res.* 10: 817–20

Stell, W. K., Lightfoot, D. O. 1975. Color-specific interconnection of cones and horizontal cells in the retina of the goldfish. *J. Comp. Neurol.* 159:473–502

Sterling, P. 1982. Identified neurons in the cat retina. In *Changing Concepts of the Nervous System, Proc.* 1st *Inst. Neurol. Sci. Symp. Neurobiol.,* pp. 281–93. New York: Academic.

Sterling, P., Lampson, L. A. 1983. Monoclonal antibodies stain specific types of amacrine in cat retina. Manuscript in preparation

Sterling, P., Freed, M., McGuire, B. A., Nakamura, Y. 1983. Accumulation of glycine by certain types of amacrine and bipolar neuron in cat retina. Manuscript in preparation

Sterling, P., Megill, J. R. 1983. Microcircuitry of the AII amacrine in the cat retina and its possible roles in states of light and dark adaptation. Manuscript in preparation

Stevens, J. K., Davis, T. L., Friedman, N., Sterling, P. 1980a. A systematic approach to reconstructing microcuitry by electron microscopy of serial sections. Brain Res. Rev. 2:265–93

Stevens, J. K., McGuire, B. A., Sterling, P. 1980b. Toward a functional architecture of the retina: Serial reconstruction of adjacent ganglion cells. Science 207:317–19

Stone, J., Dreher, B., Leventhal, A. 1979. Hierarchical and parallel mechanisms in the organization of visual cortex. Brain Res. Rev. 1:345–94

Törk, I., Stone, J. 1979. Morphology of catecholamine-containing amacrine cells in the cat's retina, as seen in retinal whole mounts. Brain Res. 169:261–73

Wässle, H., Boycott, B. B., Illing, R. B. 1981b. Morphology and mosaic of on- and off-beta cells in the cat retina and some functional considerations. Proc. R. Soc. London Ser. B 212:177–95

Wässle, H., Boycott, B. B., Peichl, L. 1978a. Receptor contacts of horizontal cells in the retina of the domestic cat. Proc. R. Soc. London Ser. B 203:247–67

Wässle, H., Levick, W. R., Cleland, B. G. 1975. The distribution of the alpha type of ganglion cells in the cat's retina. J. Comp. Neurol. 159:419–38

Wässle, H., Peichl, L., Boycott, B. B. 1978b. Topography of horizontal cells in the retina of the domestic cat. Proc. R. Soc. London Ser. B 203:269–91

Wässle, H., Peichl, L., Boycott, B. B. 1981a. Morphology and topography of on- and off-alpha cells in the cat retina. Proc. R. Soc. London Ser. B 212:157–75

Wässle, H., Riemann, H. J. 1978. The mosaic of nerve cells in the mammalian retina. Proc. R. Soc. London Ser. B. 200: 441–61

Werblin, F. S., Dowling, J. E. 1969. Organization of the retina of the mudpuppy, Necturus maculosus. II. Intracellular recording. J. Neurophysiol. 32:339–55

Wiesel, T. N. 1960. Receptive fields of ganglion cells in the cat's retina. J. Physiol. 153:583–94

Wong-Riley, M. T. T., Merzenich, M. M., Leake, P. A. 1978. Changes in endogenous enzymatic reactivity to DAB induced by neuronal inactivity. Brain Res. 141:185–92

Wood, J. G., McLaughlin, B. J., Vaughn, J. E. 1976. Immunocytochemical localization of GAD in electron microscopic preparations of rodent CNS. In GABA in Nervous System Function, ed. E. Roberts, T. N. Chase, D. B. Tower, pp. 133–48. New York: Raven

Zrenner, E., Wienrich, J. 1981. Chromatic signals in the retina of cat and monkey. ARVO Abstr. Suppl. Invest. Ophthalmol. Vis Sci. 20:185

Ann. Rev. Neurosci. 1983. 6:187–215

MECHANOELECTRICAL TRANSDUCTION BY HAIR CELLS IN THE ACOUSTICOLATERALIS SENSORY SYSTEM

A. J. Hudspeth [1]

Division of Biology, California Institute of Technology, Pasadena, California 91125

INTRODUCTION

Hair cells are sensory receptors that occur ubiquitously in the vertebrate sensory organs, collectively constituting the acousticolateralis system, responsible for the detection of sound, linear and angular accelerations, water motion, and substrate vibration. In subserving these various sensitivities, hair cells occur in organs of widely differing structures and of frequency sensitivities covering the range 0–100 kHz. Despite this variability of environment and function, however, all hair cells share a sensory apparatus of similar structure and all seem to operate in fundamentally the same manner. This paper describes the structure of the sensory apparatus of hair cells and what is known of their transduction process.

In the context of the acousticolateralis system, the term "transduction" is employed with both a broad and a narrow meaning. In the general sense, transduction is taken to include the whole complex of events—acoustical, mechanical, hydrodynamic, and physiological—that occur between the application of stimuli and genesis of a neural response. The present review focuses on transduction in its more restricted, cellular sense, as the transformation of a mechanical stimulus into a conductance change and an electrical response of the hair cell's membrane. Because the number of publications addressing this point is fairly restricted, and since there have

[1] Present address: Departments of Physiology and Otolaryngology, University of California School of Medicine, San Francisco, California 94143

0147-006X/83/0301-0187$02.00

been few reviews oriented toward the biophysical nature of transduction by hair cells (Wiederhold 1976, Russell 1979), this communication will cover not only contemporary research, but also some work dating back as much as one century. Readers with an interest in the broader question of transduction in the auditory system are referred to a recent review by Dallos (1981).

STRUCTURE OF HAIR CELLS

The organs of the acousticolateralis system originate from an ectodermal thickening, the otic placode, during the neurulation stage of ontogeny. Despite the complex series of structural rearrangements that ensue during the development of the labyrinth and of the lateral-line system, the organs maintain their epithelial organization. The essential cells, in other words, are epithelial cells lying in a continuous sheet and separating two dissimilar fluids. The sensory receptors, or hair cells, are themselves among the epithelial cells derived from the otic placode.

An individual hair cell is generally cylindrical or flask-shaped and lacks extensions such as dendrites or an axon (Figure 1). Its one striking morpho-

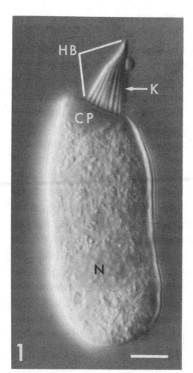

Figure 1 Light micrograph of a living hair cell enzymatically isolated from the bullfrog's sacculus. Like those in other acousticolateralis organs, this hair cell has no axon or dendrites protruding from its smooth basolateral surface. The hair bundle (HB) extends from the cuticular plate (CP) at the apical surface of the cell. K, Kinocilium; N, nucleus; calibration bar, 5 μm.

logical specialization is the hair bundle which protrudes from the apical cellular surface (Figure 2).

Hair Bundle

The morphology of hair bundles varies considerably from species to species, organ to organ, and even cell to cell within a given organ. Nevertheless, three organizational features of the hair bundle are universal (Figure 3).

1. The predominant constituent of every hair bundle is a specialized derivative of the microvillus termed the stereocilium (Engström & Engström 1978). From as few as about 30 (Iurato 1967, Miller 1978b) to as many as 150 (Kimura 1966, Takasaka & Smith 1971, von Düring et al 1974) of these cylindrical or clavate organelles protrude almost perpendicularly from the apical surface of the hair cell. Most hair bundles also have a single true, axonemal cilium, or kinocilium, although this organelle may degenerate during ontogeny (Kimura 1966, Lindeman et al 1971).

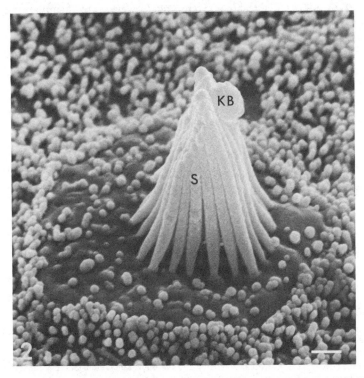

Figure 2 Scanning electron micrograph of the apical surface of a hair cell in a bullfrog's saccular epithelium from which the otolithic membrane has been dissected. The hair cell's perimeter is delineated by a palisade of microvilli projecting from the adjacent supporting cells. As in Figures 1 and 3, the plane of bilateral symmetry of the hair bundle lies approximately in the plane of the figure. S, stereocilium; KB, kinociliary bulb; calibration bar, 1 μm.

2. The stereocilia extend from the cellular surface in a hexagonal array with relatively constant center-to-center spacing. This array is present whether the cross-sectional shape of the hair bundle is round, as is characteristic of hair bundles in vestibular and lateral-line organs (Flock 1965 a,b, Hillman 1976), rectangular, as with inner hair cells of the mammalian cochlea (Engström & Engström 1978), or W-shaped, as with the cochlea's outer hair cells (Kimura 1966, Takasaka 1981).

3. Hair bundles display morphological polarization: regardless of the gross configuration of the hair bundle, the stereocilia are shortest at one edge of the hair bundle and grow monotonically longer toward the opposite edge. The kinocilium, when present, always occurs eccentrically at the tall edge of the bundle. With the exception of some saurian hair cells and vestibular hair cells whose longest stereocilia are exaggerated (Hillman 1976, Miller 1978b, Tilney et al 1980), the increments in stereociliary length from rank to rank are approximately equal (Lewis & Li 1975, Hillman 1976, Lim 1976, 1980). Cilia within a given rank, however, are of similar lengths. Hair bundles thus possess a plane of bilateral symmetry that lies perpendicular to the apical cellular surface and extends from the short edge of the hair bundle to the long edge (Shotwell et al 1981). In outer hair cells of the

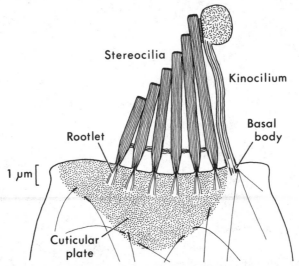

Figure 3 Schematic diagram of the apical end of a vertebrate hair cell, showing the important structural features in the hair bundle's plane of bilateral symmetry. The continuous plasma membrane envelops the individual stereocilia and the single kinocilium. Some of the microfilaments of the stereociliary cores extend as rootlets into the filamentous cuticular plate. The kinocilium's axoneme terminates at a basal body. Note the regular progression of stereociliary lengths, with the longest stereocilium at the edge of the hair bundle adjacent to the kinocilium.

apical portion of the mammalian cochlea, the symmetrical plan of the hair bundle is apparent despite some torsional distortion of the W-shaped hair bundles (Lim 1980). One of the three axes of hexagonal symmetry in the array of ciliary insertions appears always to parallel the plane of bilateral symmetry (Kimura 1966, Engström & Ades 1973, Miller 1973, 1978b, Wever 1978, Hama & Saito 1981, Takasaka 1981).

Kinocilium

The kinocilium of a vertebrate hair cell is cylindrical, membrane-bounded, and endowed with an axoneme, or 9 + 2 array of microtubules. Its diameter is 0.3 μm and its length in some organs exceeds 50 μm. Electron micrographs reveal all the major features of motile cilia: nine doublet outer tubules, a central pair of tubules, dynein arms (Hamilton 1969), radial spokes (Hama & Saito 1981), a ciliary necklace (A. J. Hudspeth, unpublished observations), and a basal body with accessory structures (Flock 1965b, Flock & Duvall 1965). The kinocilia of hair cells in some organs are defective, showing, for example, a 9 + 0 tubular array (Hamilton 1969). This appearance also results on occasion from sectioning immediately above the basal body, at a site proximal to the origin of the central pair of microtubules (Takasaka & Smith 1971). As might be expected from their structure, there are reports of relatively slow, spontaneous, oscillatory movement of kinocilia (Bowen 1932). There are no observations to date of receptor potentials or other physiological effects of kinociliary motion; Tumarkin (1980) has speculated that the cilium could serve to position the hair bundle in its most sensitive position.

The presence of a kinocilium in the hair bundle has attracted attention because true cilia or their derivatives have been implicated in the transduction of mechanical stimuli in numerous sensory organs of invertebrates (Wiederhold 1976). The receptive organelles in sensilla of arthropods, for example, are elongate or swollen cilia protruding from the cell body to contact the cuticle of the body surface or of a sensory hair. Transduction by such sensilla may involve sliding of the axonemal microtubules (Moran et al 1977), although responsiveness persists in some organs after lysis of microtubules by *Vinca* alkaloids (Erler 1980). The hair cells of invertebrate statocysts are endowed with motile cilia, which, however, need not be active to mediate transduction (Gallin & Wiederhold 1977, Stommel et al 1980). While the mechanosensitivity of protists was formerly thought to reside in cilia, it now appears that the mechanosensitive channels are situated in the membrane of the cell body (Machemer & Ogura 1979), whereas voltage-sensitive calcium channels, which amplify depolarizing receptor potentials, occur in the ciliary membrane (de Peyer & Deitmer 1980).

Stereocilium

By comparison to the elaborate structure of the kinocilium, that of the stereocilium is perhaps deceptively simple. The stereocilium is a cylindrical organelle, 0.2 to 0.8 μm in diameter, covered by plasma membrane and containing a prominent, longitudinally oriented bundle of parallel microfilaments (Engström & Engström 1978, DeRosier et al 1980). The substructure of the stereocilium suggests that many of its protein constituents will prove to be similar to or identical with those of the microvillus, to which it is developmentally related.

The microfilaments of stereocilia contain actin identifiable by its heavy meromyosin-binding capacity (Flock & Cheung 1977, DeRosier et al 1980, Tilney et al 1980, Flock et al 1981a) and immunoreactivity (Flock et al 1981b, Zenner 1981). The microfilaments are held in longitudinal register and are rather regularly spaced with respect to one another by cross-bridges of unknown molecular nature (De Rosier et al 1980). The comparable links in microvilli are presently believed to be formed of the proteins villin (Bretscher & Weber 1979, Matsudaira & Burgess 1979) and fimbrin (Bretscher 1981). An additional set of filaments connects the stereociliary core to the surface membrane; this role is played in microvilli by a protein of 110 kilodaltons molecular weight (Matsudaira & Burgess 1979). At its distal end, the filamentous core of the stereocilium is surmounted at its contact with the membrane by an electron-dense structure (Kimura 1966, Engström & Engström 1978) whose molecular nature is unknown. While the comparable density in microvilli was initially thought to represent α-actinin, this identification now appears to be artifactual (Bretscher & Weber 1978, Geiger et al 1979).

A single report (Macartney et al 1980) indicates that stereocilia of the guinea pig cochlea contain a myosin-like protein detectable by immunofluorescence. Contractile interactions between myosin and the actin in the stereocilia and cuticular plate could be important in controlling the mechanical properties of the hair bundle. It is therefore important that this result be confirmed in additional species and with other anti-myosin antibodies.

Filamentous Attachments of the Hair Bundle

CONTACTS BETWEEN THE HAIR BUNDLE AND ACCESSORY STRUCTURES   It is essential to the role of hair cells in transduction that mechanical displacements representing various stimuli reach the hair bundle. In some instances, the hair bundle is free of accessory structures (Bagger-Sjöbäck 1974, Miller 1974, 1978b, Turner et al 1981) and is presumably deflected by hydrodynamic coupling to moving endolymph. In other cases, an accessory structure such as a cupula, otolithic membrane, or tectorial membrane is present and is bound to the hair bundle by filamentous connections.

In otolithic organs of the vestibular system, for example, the tip of the kinocilium is attached to the filamentous matrix of the otolithic membrane by an assemblage of fine filaments (Hillman 1969, Hillman & Lewis 1971). The distal tip of the kinocilium is often swollen at the site of contact to form a spherical or ellipsoidal bulb about 1 μm in diameter; this bulb is filled with filamentous material of unknown composition (Hillman 1969). In semicircular canals, the kinocilium and adjacent long stereocilia extend across the subcupular space; the tip of the kinocilium is embedded a few micrometers into the cupula (Hillman 1977).

In the mammalian cochlea, whose hair cells lose their kinocilia shortly after birth (Lindeman et al 1971), the tectorial membrane contacts the longest stereocilia of the outer hair cells directly. It appears that the filaments of the tectorial membrane insert on the membrane of the stereociliary tips (Kimura 1966, Engström & Engström 1978, Takasaka 1981); the contact at this site is sufficiently close that the dissected tectorial membrane bears imprints of the attached cilia (Kimura 1966, Lim 1972).

CONNECTIONS AMONG ELEMENTS OF THE HAIR BUNDLE   Three types of filamentous connections have been described among elements of the hair bundle.

1. The kinocilia of some hair cells in otolithic organs have striking attachments to the adjacent stereocilia (Hillman 1969, Bagger-Sjöbäck & Wersäll 1973, Bagger-Sjöbäck 1974). Since mechanical stimuli are delivered to the hair bundle via connections of the kinocilium to the otolithic membrane, it seems very probable that the bonds between the kinocilium and stereocilia transmit displacements to the latter.

2. The stereocilia of a variety of hair cells have filamentous attachments to nearest neighbors near their basal insertions (Bagger-Sjöbäck & Wersäll 1973, Bagger-Sjöbäck 1974, Hama & Saito 1981). These interciliary strands usually insert onto the membranes within zone approximately 0.3 μm wide beginning distal to the tapered region of the stereocilium, roughly 1 μm above the cuticular plate. The role of these connections is unknown. Most likely they simply aid in holding the stereocilia of a hair bundle in an appropriate array—resisting, for example, the tendency of an unrestrained bundle to undergo torsion—without significantly restricting motions at the stereocilliary tips.

3. Another set of filamentous attachments links stereocilia near their distal tips. In published micrographs (Kimura 1966, Flock 1977, Hama & Saito 1981), the attachments appear as tenuous strands with a region of

increased density midway between the apposed stereocilia. Because the distal tips of stereocilia probably slide past one another when a hair bundle is deflected, these attachments occur at one of the most plausible sites for an involvement in transduction.

Cuticular Plate

A final area of considerable anatomical interest is the region at which the stereocilia insert into the cellular apex. From each stereocilium, a rootlet consisting of some of its longitudinal actin filaments (Tilney et al 1980) penetrates into the cuticular plate, a feltlike network of filaments lying within the cytoplasm at the cell's apex. The actin microfilaments of each rootlet enter the cell in a circular bundle, then fan out to a degree that varies among preparations as they descend into the cuticular plate (Itoh & Nakashima 1980, Hama & Saito 1981, Takasaka 1981). Fine filaments 3 nm in diameter appear to run radially from each rootlet into the filaments of the plate (Flock 1965b, Itoh & Nakashima 1980, Tilney et al 1980).

The cuticular plate itself is heterogeneous in composition. Heavy meromyosin decoration and immunofluorescence labeling indicate that many of its filaments are actin (Tilney et al 1980, Flock et al 1981a,b, Zenner 1981). In addition to these and the fine filaments noted above, there are decorated microtubules inserted into its basal aspect (Tilney et al 1980) and immunofluorescence evidence for tubulin within the plate itself (Zenner 1981). A striated organelle of unknown composition and function lies at the base of the cuticular plate in some mammalian hair cells (Slepecky et al 1980, 1981a).

The site of insertion of the stereocilia into the cuticular plate, which appears to be where they pivot, is an obvious candidate for the transducer's location. In addition, the insertion site could be important in regulating the apposition of the stereocilia. Although it is often depicted as flat and appears so in some OsO_4-fixed preparations (Kimura 1966), the apical surface of the cuticular plate is generally curved in living, mechanically sensitive hair cells of the bullfrog's sacculus (A. J. Hudspeth and R. Jacobs, unpublished observations) and in many published micrographs of aldehyde-fixed material (Spoendlin 1970, Engström & Ades 1973, Engström & Engström 1978, Slepecky et al 1981a). This curvature, moreover, is directional: the surface is concave in sections parallel with the cell's axis of symmetry, but relatively flat in the orthogonal direction. If it is normally present in living hair cells, this curvature would have the effect of forcing stereocilia together along the hair bundle's axis of mechanosensitivity, and might thereby bring transduction sites along the stereocilia into contact or provide appropriate stress on the stereocilia to activate transduction. Changes in the radius of curvature of the cuticular plate, mediated for example by the actin there, could thereby be involved in processes such as adaptation of the transduction system (Eatock et al 1979).

Mechanics of Hair-Bundle Motion

The elaborate structure of the hair bundle very probably has developed in order to convert gross displacement of the hair bundle's tip into the application of force at the site of the transducer molecules. An understanding of the micromechanics of hair bundle motion is accordingly essential to our description of the transduction process.

Although fixed stereocilia sometimes appear flaccid in micrographs, cilia observed in living preparations are quite stiff (Flock 1977; A. J. Hudspeth and R. Jacobs, unpublished observations). Retzius (1881, 1884) noted that fresh stereocilia are straight and stiff enough to fracture when excessively stressed. Both cochlear (Engström et al 1962) and vestibular (Flock 1977) stereocilia behave as rigid bodies. If they are inextensible and firmly attached, the extensive cross-bridges between microfilaments in the cores of stereocilia would account for this rigidity (DeRosier et al 1980, Tilney et al 1982). The observation that acoustical trauma can render stereocilia limp (Hunter-Duvar 1977, Lim 1980, Robertson & Johnstone 1980, Engström & Berg 1981, Slepecky et al 1981b) might be explained by scission of actin-crosslink bonds (Tilney et al 1982), caused perhaps by elevation of the intracellular Ca^{2+} concentration and its effect on filament-associated proteins (Glenney et al 1980, Mooseker et al 1980).

The pivot for motion of a stereocilium as a stiff lever may be provided by either or both of two specializations at its basal insertion. First, stereocilia taper to a diameter of 0.1 μm over a segment roughly 1 μm in length immediately above their bases (Kimura 1966, Spoendlin 1970, Engström & Engström 1978). In this region, many microfilaments terminate by inserting on the plasma membrane (Tilney et al 1980) and the remaining filaments, evidently without the usual cross-bridges, come to lie closely packed together as a rootlet (Itoh & Nakashima 1980). Second, the actin microfilaments of this rootlet do not fuse with those of the cuticular plate immediately upon penetrating the plate, but instead make contacts with fine, radially-directed, 3-nm filaments (Flock 1965b, Itoh & Nakashima 1980, Tilney et al 1980). This linkage, which appears tenuous in electron microscopic preparations, may allow some back-and-forth motion and flexion of stereociliary rootlets even within the cuticular plate. These regional differences in stereociliary structure suggest that the cilia may be more compliant near their insertions than elsewhere along their lengths.

If stereocilia move as rigid rods with basal pivots at fixed positions on the cuticular plate, their distal tips must undergo shear when the hair bundle's tip is displaced by a stimulus. A horizontal mechanical displacement of the hair bundle's tip results in a principally vertical shear between stereocilia in successive rows along the axis of stimulation. This region of shear is a candidate for the location of transduction molecules.

It remains obscure how stereocilia interact mechanically during deflections of the hair bundle. When a hair bundle is manipulated with a probe in an in vitro preparation under light microscopic observation, the bundle consistently moves as a unit, without visible separation of the constituent stereocilia (Flock et al 1977, Hudspeth & Corey 1977). This behavior persists after removal of the plasmalemma with detergent (Flock et al 1977), suggesting that membrane surface charges and proteins are not essential in keeping stereocilia together. Since it is not known whether the two types of filamentous, inter-stereociliary connections (see above) are left intact after such treatment, it is uncertain whether these strands couple stereociliary movements. As an alternative, it is possible that the cuticular plate holds stereocilia against one another by exerting forces at their basal insertions.

GATING OF THE TRANSDUCTION CHANNEL

The sensitivity of a hair cell to sound, acceleration, or some other stimulus is not an intrinsic property of the cell. Instead, an appropriate stimulus is converted by a mechanical and hydrodynamic linkage into a force applied near the distal end of the hair bundle (Pumphrey 1950). Each hair cell is thus essentially a mechano-receptor and the crucial event of transduction is that between the receipt of a stimulus by the hair bundle and the onset of an electrical response by the hair cell.

Molecular Event at the Transducer

Regardless of the precise subcellular location of the transduction apparatus, the mechanical action of the hair bundle must culminate in the application of force to some molecule or multimolecular assembly which then, directly or indirectly, opens transduction channels. By analogy with voltage-dependent and postsynaptic channels, which are controlled or gated by membrane potential and synaptic transmitter, respectively, we may refer to this process as the gating of transduction channels in hair cells. Unfortunately very little information is available on this process, which is at the heart of mechanoelectrical transduction in hair cells and elsewhere.

One may imagine that the molecular event in transduction is analogous to an allosteric transition of an enzyme. The transduction molecule might alternate, under thermal excitation, between its active and its inactive forms. The application of force would alter the free energy difference between the forms, stabilizing one relative to the other and inducing thereby an increase in the fraction of transducers in that state (Corey & Hudspeth 1980).

In the simplest scheme, the transducer protein might be a transmembrane channel. Were this anchored to the cytoskeleton—for example, to the microfilaments of the stereocilium—stress applied to the filaments could distort the transducer protein and favor or oppose channel opening. Similarly, an ion channel situated in the membrane might have two states that differ in their molecular volumes. Stretch or compression applied within the plane of the membrane itself would therefore influence the relative population of the open or closed channels.

The hypothesized molecular event need not occur within the membrane. The production of stress in the filamentous cytoskeleton by stimulation could affect the binding of some small molecule. This species would then act as a second messenger, diffusing to the surface membrane and opening or closing ion channels. The diffusion distance permitted this model is restricted by the short delay, a few hundredths of a millisecond, that is observed between mechanical stimulation and the onset of a response that signals the operation of the transduction process (Corey & Hudspeth 1979a). Diffusion of a small molecule in the time available is certainly possible, however, over a range such as the radius of a stereocilium.

One noteworthy feature of stochastic models of transduction, in which channels open and close randomly under thermal bombardment, is that they have no absolute threshold for transduction. An arbitrarily small stimulus will, if averaged for a sufficiently long time, produce some change in the average fraction of open channels. Such models are therefore in accordance with estimates that the human auditory threshold involves motions of the basilar membrane, and presumably of the hair bundles, through submolecular dimensions (see below).

Relationship Between Hair-Bundle Deflection and Response

SENSITIVITY OF THE TRANSDUCER   Perhaps the most impressive property of the transduction apparatus of vertebrate hair cells is its sensitivity. Receptor potentials have not yet been recorded from hair cells during calibrated stimulation of their hair bundles at very small amplitudes. It is possible in several experimental systems, however, to correlate behavioral thresholds, an indication of cellular response, with stimulus intensity, a gauge of hair-bundle deflection. Both of these measures present problems. Behavioral thresholds depend upon the amount of neural time-averaging and upon the extent of convergence of inputs from hair cells onto nerve fibers, both of which affect the capacity of a given system to extract periodic signals from noise. While stimulus-intensity measurements can be made with great accuracy, we are limited in our ability to assess how much hair-bundle deflection ensues from a given strength of stimulation, particularly at very low stimulus intensities. Despite these difficulties, one can make order-of-magnitude estimates for the extent of hair-bundle motion required to produce a behaviorally significant response in various organs.

Extensive psychophysical experimentation suggests a human threshold for the detection of angular accelerations near 0.1 deg/s^2 (Guedry 1974). Both theoretical calculations for humans (Oman & Young 1972) and experimental measurements in frogs (McLaren & Hillman 1979) indicate that such a stimulus produces a peak bowing of the central portion of the cupula in a semicircular canal through ± 3–10 nm. The deflection of the tips of the hair bundles roughly 50 μm in length (Bowen 1932, Hillman 1972) is probably about 40% smaller (Hillman & McLaren 1979). If the stereocilia pivot rigidly at their bases, their angular motion at threshold is ±2–7 X 10^{-3} deg. Even very robust stimuli produce cupular and hair-bundle displacements of less than 3–5 μm in semicircular canals (Oman et al 1979).

A brief report (Crawford & Fettiplace 1980a) suggests that the displacement of the turtle's basilar membrane is about ± 170 pm at the behavioral threshold, a sound-pressure level of about 40 dB relative to 20 μN/m^2. Unfortunately no available information permits estimating the amplitude of the hair-bundle motion caused by such minute stimuli. Another preliminary communication (Khan et al 1979) mentions the range of hair-bundle deflections over which various reptilian organs respond, but fails to explain how these values are measured.

The amplitude of motion by the basilar membrane in the mammalian cochlea in response to moderate to high intensity sounds has been investigated extensively by stroboscopic, capacitance-probe, and Mössbauer techniques (von Békésy 1960, Johnston et al 1970, Rhode 1971, Wilson & Johnstone 1975). The data suggest that the motion of the basilar membrane is close to ± 1 pm for stimulation at 0 dB, the approximate threshold for human hearing. This estimate is dependent upon the assumption that basilar-membrane-motion data acquired at moderate to high sound pressure levels may be extrapolated linearly to 0 dB. Evidence is now accumulating, however, that this assumption is erroneous and that basilar membrane motion is quite nonlinear (Rhode 1971, Rhode & Robles 1974). Capacitance-probe (LePage & Johnstone 1980), Mössbauer (Sellick et al 1981), and laser-interferometric (Khanna & Leonard 1981) studies of basilar membrane motion have recently been undertaken in several species and at sound pressure levels as low as 30 dB. If confirmed, the preliminary results of these experiments will increase the estimate of basilar membrane motion at 0 dB by over an order of magnitude (Rhode 1971, 1978), perhaps to ± 100 pm.

While the mechanical description of the organ of Corti is incomplete, shear of the tectorial membrane with respect to the reticular lamina—in other words, deflection of hair-bundle tips—is probably comparable to basilar membrane motion. In the basal portion of the cochlea (Rhode & Geisler 1967), where the frequency range of 3–20 kHz is represented and where most of the measurements of basilar membrane motion occur, the tallest ranks of stereocilia are about 3 μm in length (Iurato 1967, Lim 1980). The threshold deflection of these stereocilia at 0 dB is therefore ± 2–200×10^{-5} deg.

The results imply that hair cells in situ are sensitive to hair-bundle deflections on the order of thousandths of a degree, and perhaps much less. This extrapolation may well be inexact; at the level of very small stimulation, stereocilia cannot be considered as rigid bodies that move as units. In addition, the Brownian motion of stereocilia is probably of a size comparable to the calculated displacement of threshold (de Vries 1949, Naftalin 1981). Transduction must nevertheless occur when stimuli produce average displacements of stereocilia well below the dimensions of protein molecules (von Békésy 1960).

It remains unclear whether various hair cells, despite the structural differences in their hair bundles, are able to transduce very small stimuli with similar sensitivities. Although hair cells of the mammalian cochlea appear to be as much as two orders of magnitude more sensitive than other known hair cells, this estimate rests heavily on the assumption of linear basilar-membrane motion and hence may be inaccurate. If so, the absolute sensitivities of the mammalian cochlea, turtle basilar papilla, and mammalian semicircular canal may lie within about one order of magnitude of one another. One may accordingly entertain the notion that various hair cells are actually quite comparable in their sensitivities, presumably by virtue of a common transduction apparatus.

DISPLACEMENT-RESPONSE CURVE   A direct, calibrated measurement of the relationship between hair bundle deflection and receptor potential is thus far available only for hair cells of the bullfrog's sacculus (Hudspeth & Corey 1977, Shotwell et al 1981). The displacement-response curve produced by micromanipulation of the hair bundle's tip in vitro has an apparent width of about 1 μm, corresponding to a total angular displacement approximating 6°. Larger stimuli transiently saturate the transducer, activating an adaptation process that restores sensitivity (Eatock et al 1979). The displacement-response relationship demonstrates an asymmetry and saturating nonlinearities that can account for such phenomena as the double-frequency (2f) responses and summating potentials found in numerous acousticolateralis organs (Flock 1965a, Johnstone & Johnstone 1966,

Furukawa & Ishii 1967, Engebretson & Eldredge 1968, Furukawa et al 1972, Hubbard et al 1979, Boston 1980).

The precise relationship between hair bundle deflection and the ensuing conductance of transduction channels is unknown. The receptor-potential data taken with stimuli of 10 Hz (Hudspeth & Corey 1977, Shotwell et al 1981) include the effects of the membrane time constant, time- and voltage-dependent conductances (Corey & Hudspeth 1979b), and the adaptation process associated with the transducer (Eatock et al 1979). Voltage-clamp experiments in an appropriate ionic environment should allow direct measurement of the transducer conductance uncontaminated by these effects. In order to compare the in vitro data with those obtained in intact preparations, it is also necessary to control for the altered ionic environment, for enzymatic removal of the otolithic membrane from hair cells, and for imperfections in coupling between the stimulus probe and the hair bundle.

If measurements can be suitably refined, displacement-response relationships may provide a window through which the workings of the hair bundle can be discerned. It seems probable that the geometrical relationships among elements of the hair bundle control the number of open transducer channels associated with the stereocilia. Study of the displacement-response curves for various hair cells should therefore suggest which geometrical relationships are crucial. It would be interesting, for example, to compare the responses of otherwise similar hair cells whose hair bundles differ in the numbers of ranks or files of stereocilia, in stereociliary lengths, or in the length increments between contiguous ranks. Changes might even be observed in the form of the displacement-response curve for an individual hair cell as successive ranks of stereocilia are extirpated (Hudspeth & Jacobs 1979).

Hypotheses for Transduction

While there is no strong evidence that the fundamental transduction process is identical in all vertebrate hair cells, several points favor this assumption.

1. The organization of the hair bundle and the structure of stereocilia are conserved throughout the vertebrates.
2. The special ionic milieu facing the hair cell's apex—and the site of transduction—seems common to most species.
3. The transduction mechanism of hair cells from fish (Wersäll & Flock 1964) to mammals (Wersäll et al 1973) is blocked by aminoglycoside antibiotics.

The analysis below is based in part on the assumption that there is a general and probably a universal transduction mechanism in vertebrate hair cells, and hence that evidence from any system is applicable to all.

Several theories of transduction ascribe the basic event to structures exogenous to the hair cells, that is, to the tectorial membrane or cupula. For example, O'Leary (1970) posits that motion of the cupula establishes an electric field that hair cells, acting as electroreceptors, then detect. Naftalin (1965) has suggested the stress of the tectorial membrane produces a potential by a piezoelectric effect; he now proposes that the stimulated tectorial membrane releases magnesium or hydrogen ions, which diffuse to hair cells and excite them (Naftalin 1977, 1981). These theories, which ascribe to hair cells a passive electro- or chemoreceptive role, fail to explain why the elaborate mechanical arrangement of the hair bundle has arisen and why it has been retained throughout phylogeny. Moreover, they cannot account for transduction in organs in which no tectorial structure is ordinarily present or in the bullfrog's sacculus, where transduction persists after removal of the otolithic membrane (Hudspeth & Corey 1977, Corey & Hudspeth 1979b, Shotwell et al 1981). Finally, were a potential change produced in the hair cell's membrane by a piezoelectric mechanism, it should dissipate at a rate set by the cell's time constant. This type of transduction process accordingly cannot account for the steady-state or low-frequency transduction currents that some hair cells can produce (von Békésy 1960, Hudspeth & Corey 1977, Corey & Hudspeth 1979b).

The role of cilia in transduction elsewhere has prompted the suggestion that the transducers of vertebrate hair cells are also associated with their kinocilia (Hillman 1969, Hillman & Lewis 1971, Hillman 1972). In this model, the stereocilia pivoting on the cuticular plate exert alternately tensile and compressional forces on the distal end of the kinocilium. Acting along the kinocilium, these forces are supposed to distort the membrane and thus to control a transducer at or near the kinocilium's basal insertion. This hypothesis has the attractive feature of explaining the functional polarity of the hair bundle. Were the hair bundle deflected toward the kinocilium, the tension along the cilium would be reduced, the region of the ciliary insertion depressed, and the cell thereby somehow depolarized. Conversely, motion in the opposite direction would increase tension on the cilium, elevate its insertion, and yield a hyperpolarization. Stimulus components along the perpendicular axis would yield little or no response, as is observed experimentally (Flock 1965a,b, Shotwell et al 1981).

The kinociliary model for transduction suffers from two weaknesses. First, it is applicable neither to those hair cells of the mammalian cochlea which lack a cilium in their mature form (Kimura 1966, Lindeman et al 1971), nor to those reptilian (von Düring et al 1974) and avian (Jahnke et al 1969) hair cells whose kinocilia appear to be mechanically decoupled from the stereocilia and from accessory structures such as tectorial and otolithic membranes. Second, even in hair cells whose mechanical arrangement conforms with the models, decoupling or detachment of the kinocilium by microdissection does not prevent transduction in vitro (Hudspeth & Jacobs 1979).

In sum, it presently appears that the kinocilium is neither itself the site of transduction nor essential in conducting forces from the stereociliary bundle to the site of transduction. The kinocilium may, however, be involved in the establishment of polarity in the hair cell: a kinocilium is universally present during the development of hair cells, its emergence precedes that of mature stereocilia (Lewis & Li 1973, Li & Lewis 1974), and its eccentric position and internal asymmetry both predict the functional polarization of the transducer (Flock & Duvall 1965). Kinocilia certainly serve to transmit stimuli from extracellular accessory structures to the stereocilia in some hair cells, and are in many cases modified in length (Lewis & Li 1975, Hillman 1976, Lim 1976) or in their terminal specializations (Miller 1973, 1974, 1978a,b) to conform with the accessory structures which convey mechanical stimuli.

Another model that ascribes an important role in transduction to the micromechanics of the hair bundle is that of Malcolm (1974). Here the hypothesis is that the stereociliary membrane forms a passive resistance across which current flows. Mechanical stimulation is thought to modulate the current by varying the amount of membrane surface that is unobstructed by close contact with adjacent cilia. This model, while plausible, suffers from three possible flaws.

1. Whereas the model is based upon the notion that stimuli along the axis of cellular symmetry act by increasing the separations between successive ranks of stereocilia, the experimental evidence suggests that hair bundles move as units, without such spreading (Hillman 1972, Flock 1977, Hudspeth & Corey 1977).

2. Unless the spaces between stereocilia are filled with material of unusually high resistivity, rather than with endolymph, the contribution of interciliary spaces to the resistance to current flow is small by comparison to a specific membrane resistance of 1–12 $k\Omega \cdot cm^2$ (Corey & Hudspeth 1979b, Crawford & Fettiplace 1981, Hudspeth 1982). Electron microscopic views of hair bundles (Kimura 1966, Takasaka & Smith 1971, Bagger-Sjöbäck & Wersäll 1973, Engström & Engström 1978) suggest that, below a region of contact at their distal tips, stereocilia are separated to the extent that about half the cross-sectional area of the hair bundle is occupied by extracellular fluid. Only if the cilia were close-packed in an hexagonal array, so that as little as 9% of the area were extracellular, would the resistance of the spaces between the cilia approach that expected across their membranes.

3. The model predicts a substantial conductance—proportional to the unobstructed membrane surface area—for hair cells at rest. Malcolm

(1974) estimates the maximal modulation of exposed membrane as $\pm 30\%$ and predicts that transduction-channel current should vary over a similar range. The limited experimental observations available (Hudspeth & Corey 1977, Corey & Hudspeth 1979b, Crawford & Fettiplace 1981) suggest that the modulation of the transduction conductance is much greater, at least several times its resting value.

NATURE OF THE TRANSDUCTION CHANNEL

It is unknown whether the channel through which transduction current enters hair cells is a hydrated protein pore, a transmembrane pump, or a diffusile ion carrier. It is even conceivable that the channel is not formed of protein, but consists of a permeability pathway through the lipid bilayer when this is suitably distorted by stimulation. If the transduction channel is a protein pore, it may well be detected in the near future on the basis of its substantial unitary conductance, either by membrane-noise analysis or by single-channel recording from membrane patches or whole cells.

Ionic Selectivity

Consistent with their epithelial origin and character, the hair cells of vertebrates separate two fluid compartments of differing composition. The basolateral cellular surfaces are bathed in ordinary extracellular fluid, termed perilymph in those organs of the inner ear in which substantial volumes are present. The apical pole of the hair cell—the site of the hair bundle and of the transduction process—faces a special fluid. In the inner ear, this is termed endolymph; in lateral-line organs, it is the fluid within the subcupular space. By comparison to perilymph and other extracellular fluids, endolymph is relatively enriched with K^+ and correspondingly depleted in Na^+ (Citron et al 1956, Bosher & Warren 1968, Peterson et al 1978). Although only one measurement with ion-selective electrodes is available (Bosher & Warren 1978), it also seems that the endolymphatic Ca^{2+} activity is unusually low (23 μM). It is unknown whether the higher Ca^{2+} concentrations measured by electron-probe analysis (Peterson et al 1978) reflect species differences or the presence of Ca^{2+} in a complexed form.

The consistency of these findings in fishes, amphibians, reptiles, and mammals suggests that the ionic content of endolymph and related fluids is somehow important for the operation of hair cells, presumably in the transduction process. In particular, the bulk of the transduction current in hair cells of the inner ear probably is carried by K^+. This is potentially advantageous in the energetic economy of hair cells, for K^+ can both enter their apical surfaces and exit their basolateral surfaces passively (Matsuura et al 1971a, Sellick & Johnstone 1975). Unlike photoreceptors, which must expend energy maintaining their dark Na^+ currents (Hagins et al 1970), hair cells can transfer the burden of ionic pumping to other cells of the epithelium.

Whether ions other than K^+ can pass through the transduction channel, however, is less clear. Experimental tests involving replacement of normal inner ear or lateral-line fluids with other solutions do not give consistent results in all preparations. In the sacculus of the goldfish, quantitative replacement of endolymphatic K^+ with Na^+ produces a negligible change in the microphonic response (Matsuura et al 1971a). Potassium ion may likewise be wholly replaced in endolymph by Li^+, Na^+, Rb^+, or Cs^+ without significant effect on the microphonic potentials of the frog's sacculus (Corey & Hudspeth 1979b). In the latter preparation, substitution of NH_4^+ yields an enhanced microphonic, while isotonic Ca^{2+} gives a diminished response. Weak but significant responses persist when K^+ is replaced by small organic cations such as tetramethylammonium; acetylcholine, carbamylcholine, choline, and tris(hydroxymethyl)aminomethane also support a response (D. P. Corey and A. J. Hudspeth, unpublished observations). The results from these two experimental models suggest that the transduction channel of hair cells is a relatively poor discriminator of cations and will admit many ions, probably in their hydrated state, up to a molecular diameter of about 0.6 nm (Corey & Hudspeth 1979b).

Substantial replacement of K^+ with Na^+ in the frog's semicircular canal results in reduction of the extracellularly recorded receptor potential and of afferent nerve activity (Valli et al 1977). A more pronounced effect is a diminution in the inhibitory (presumably hyperpolarizing) phase of the response relative to the excitatory (depolarizing) component. Complete removal of endolymphatic K^+ blocks all responses (Valli et al 1979).

Transduction in the lateral-line organ of *Necturus*, assayed by afferent-axon recordings, is sensitive to the presence of Ca^{2+} in the solution at the apical epithelial surface (Sand 1975). As occurs in many processes involving Ca^{2+} channels (Hagiwara & Byerly 1981), Sr^{2+} replaces Ca^{2+} in supporting transduction, while Co^{2+}, La^{3+}, Mg^{2+}, and the organic compound D-600 block the response (Sand 1975, Jørgensen 1979).

It is possible that transduction currents in lateral-line hair cells, whose reported responses are no larger than a few millivolts (Harris et al 1970, Flock et al 1973, Sand et al 1975, Russell 1979), are carried by the small amounts of Ca^{2+} present in most fresh water. The interpretation of the results from the semicircular canal and lateral-line organ is complicated, however, by uncertainty as to what extent alterations in the fluid bathing such organs, particularly those with intact cupulae, are effective in changing the ionic environment at the transduction site. While Ca^{2+} evidently equilibrates promptly between the bathing fluid and the subcupular volume (McGlone et al 1979), the existences of an intracupular potential and a high

K^+ concentration within the cupula (Russell & Sellick 1976) suggest that some permeability barrier regulates the ionic milieu at the apical hair-cell surface.

Finally, exchange of Na^+ for K^+ in the scala media of the cochlea irreversibly abolishes the microphonic response (Konishi et al 1966). That high-Na^+ solutions grossly shrink the tectorial membrane (Kronester-Frei 1979) complicates the interpretation of this result. It is uncertain whether Na^+ cannot carry transduction current in the cochlea or whether Na^+-rich endolymph simply disturbs the propagation of stimuli to hair bundles by distorting the tectorial membrane. An effect of Na^+ on cupular mechanics might also account for the reduction in responses in the semicircular canal exposed to high-Na^+ solutions (Valli et al 1977, 1979).

It is worth noting that one other technical problem affects several of these experiments. Substitution of test solutions may alter the resting potentials of hair cells to an extent that is unknown unless intracellular recordings are made. The apical surfaces of hair cells have a resting K^+ conductance (Bracho & Budelli 1978), so replacement of endolymph with other solutions may affect the resting potential. A reduction in the stimulus-induced response measured as a microphonic potential or in eighth-nerve fibers might then be due to changes in the driving force on the ion(s) carrying transduction current or to biassing of synaptic discharge by the hair cell, rather than to a reduced conductance of the transduction channel to the test ion.

The substitution experiments cited above deal almost exclusively with cations, although there is a report that exchange of endolymphatic Cl^- with SO_4^{2-} is without effect on transduction (Valli et al 1979). In the light of the apparent nonselectivity of the transduction channel, and of the probably substantial HCO_3^- concentration in endolymph, it would be valuable to perform systematic exchanges of anions in the presence of a fixed K^+ concentration.

The most serious uncertainty in assessing which ions can carry transduction current in hair cells—and accordingly in beginning to characterize the chemical nature of the channel—results from the Ca^{2+} requirement of the transduction process (see below). So long as Ca^{2+} must be present for transduction to occur, it may be argued that this ion actually carries the transduction current, while K^+ and other ions somehow facilitate this process. A strong test of the idea that other ions can carry transduction current would be to demonstrate their entry into hair cells' cytoplasm as a result of stimulation. One such test has occurred fortuitously (D. P. Corey, personal communication). When Cs^+ is used in the solution bathing the apical surface of the saccular epithelium, the hair cells' voltage-dependent K^+ conductance gradually diminishes during repeated mechanical stimulation. Since Cs^+ blocks similar K^+ channels of squid axons from the inner membrane surface (Bezanilla & Armstrong 1972), this finding suggests that Cs^+ enters through transduction channels, then blocks voltage-sensitive K^+ channels upon attempting to exit.

A more satisfactory test would be to measure stimulus-dependent accumulation in a hair cell of a labeled compound that can be chemically fixed. An analogous procedure has already confirmed the nonselectivity of the acetylcholine-gated synaptic channel in a sympathetic ganglion (Yoshikami 1981).

Role of Calcium Ion

While there is a diversity of opinion on the question of which ion(s) can carry transduction current, there is virtual unanimity on the requirement for Ca^{2+} in transduction. Removal of Ca^{2+} or addition of an appropriate chelator blocks transduction in the *Necturus* lateral-line organ, the sacculus of the bullfrog, the frog semicircular canal, and the mammalian cochlea (Sand 1975, Corey & Hudspeth 1979b, Valli et al 1979, Tanaka et al 1979, 1980). The minimal concentration of Ca^{2+} required for transduction is around 10–100 μM (Sand 1975, Corey & Hudspeth 1979b); as noted earlier, Ca^{2+} may be replaced by Sr^{2+} (Sand 1975), but not by Ba^{2+} or Mg^{2+} (Corey & Hudspeth 1979b).

To date data are insufficient to decide among four plausible roles for Ca^{2+} in transduction. Evidence to the contrary notwithstanding, Ca^{2+} might itself carry all transduction currents. It might operate from within the cytoplasm, for example through interaction with the proteins that cross-link actin in the stereocilia (Mooseker et al 1980, Bretscher 1981). The ion might act nonspecifically by screening surface charges on the hair cell's membrane. Finally, Ca^{2+} might act specifically at the transduction site, for example by binding to the transduction molecule and causing it to assume an active configuration. Consistent with the last two hypotheses, fixation in the presence of high Ca^{2+} concentrations produces calcium-containing densities on the surfaces of stereocilia and at the base of the kinocilium (Moran et al 1981).

Action of Ototoxic Drugs

Many drugs in the contemporary pharmacological armamentarium have deleterious side effects on the inner ear (Porter & Jick 1977). Some, such as the diuretics ethacrynic acid and furosemide, act against ion-pumping epithelia rather than directly against hair cells (Bosher 1980). Several other types of drugs, notoriously aminoglycoside antibiotics such as streptomycin, gentamicin, and tobramycin, exert their toxicity on hair cells, quite possibly at the level of the transduction apparatus.

Aminoglycosides are without effect on the function of hair cells when experimentally administered into the perilymphatic compartment at modest concentrations, but block microphonic responses when added to the endolymph (Matsuura et al 1971b, Konishi 1979). The gradual onset and decline of clinical ototoxicity therefore probably reflect the pharmacokinetics of drug entry into and loss from the endolymphatic compartment (Tran-Ba-Huy et al 1979, 1981). Prolonged exposure to aminoglycosides causes abnormalities of hair-bundle structure, notably progressive fusion of the plasma membranes of contiguous stereocilia (Wersäll et al 1973), and eventually cell death.

The application of aminoglycosides to lateral-line organs, sacculi, semicircular canals, and cochleas causes a prompt reduction in or abolition of microphonic responses and afferent-nerve activity (Wersäll & Flock 1964, Matsuura et al 1971b, Gallais 1979, Konishi 1979, Kroese & van den Bercken 1980, Yanagisawa 1981). Removal of the agents after a brief exposure restores at least partial activity, indicating that permanent ototoxic effects take some time to develop. Taking into account the atomic masses of the counterions and bound CO_2 and H_2O in commercial preparations of aminoglycosides, the half-blocking concentrations of the drugs are in the range 10–250 μM. This low affinity of the drugs for hair cells unfortunately renders them inadequate as ligands for location of their site of binding to hair cells. Moreover, since aminoglycosides are strongly cationic, they may be expected to bind to the negatively charged phospholipids found in biological membranes. Labeled aminoglycosides injected into the cochlea in fact bind to membrane surfaces of a wide variety of cell types (Balough et al 1970).

The limited experimental evidence available suggests three ways in which aminoglycosides might block transduction.

1. As a consequence of their cationic nature, the drugs may compete with Ca^{2+} for a binding site that activates transduction (see above). It is interesting in this regard that increasing the Ca^{2+} concentration in the fluid bathing the apices of hair cells counters the effects of aminoglycosides (Kroese & van den Bercken 1980, Yanagisawa 1981).

2. The drugs might interfere with transduction by blocking the turnover of polyphosphoinositides in the membranes of hair cells (Schacht 1974, 1976, 1979). This hypothesis is strengthened by (a) a correspondence between the ototoxicities of various aminoglycosides and their strengths of binding to lipid films (Lohdi et al 1979, 1980) and (b) by the fact that polyphosphoinositide turnover is stimulated by sound exposure in an insect's auditory organ (Kilian & Schacht 1980). The hypothesis leaves unanswered, however, the critical question of whether lipid phosphorylation is central to the transduction process or is a secondary effect of cellular activity.

3. The poor ionic selectivity that the transduction channel seems to possess (see above) suggests a third possible ototoxic action for aminoglycosides: The cationic functional groups on these drugs enter the transduction channel and block it by "plugging". The interference of tetraethylammonium ion with transduction (Katsuki et al 1966, Salt & Konishi 1979) is also consistent with this type of blockage mechanism. This hypothesis may be tested; blockade of the transduction channel should be very rapidly established upon introduction of a drug and should be dependent upon the transmembrane potential (Woodhull 1973, Lewis 1979).

Regardless of the mechanism by which aminoglycosides interfere with transduction, it remains obscure why this effect should also be lethal to hair cells. The prompt, reversible blockage of transduction may proceed by a mechanism different from the insidious, lethal effect. Alternatively, simple transduction-channel blockage might conceivably stimulate continuous production of new channels by an up-regulation process and cause cellular death by this metabolic stress.

Number of Active Transduction Channels

The total conductances of transduction channels in hair cells of various preparations during maximal stimulation are remarkably similar in magnitude. In the bullfrog's sacculus, a direct measurement of transduction current by voltage clamp reveals a peak-to-peak response of over 200 pA in healthy cells (Corey & Hudspeth 1979b). The driving force for transduction current in such hair cells at a resting potential of -60 mV is about 58 mV (Corey & Hudspeth 1979b), implying a peak conductance at the transduction channels of at least 3.5 nS. Current-voltage relationships for two hair cells of the turtle's basilar papilla (Crawford & Fettiplace 1979, 1981) yield a mean total transduction conductance of 3.7 nS; the largest responses in this preparation, 30–45 mV (Crawford & Fettiplace 1980b), presumably correspond to slightly larger conductances. Data from the guinea pig (Russell & Sellick 1978) indicate peak responses of 13 mV in two inner hair cells of average input resistances 55 MΩ. In the cochlea, both the endocochlear potential of $+90$ mV and the resting potential of -40 mV contribute to the driving force for K$^+$. If the hair-cell membrane behaves fairly linearly, these values suggest a maximal transduction current of 262 pA and a transduction conductance near 2.0 nS. Even lateral line organ hair cells, with responses no larger than 500 μV and input impedances of 1–3 MΩ (Sand et al 1975), may have total transduction conductances in the same range.

In conjunction with the observations on the ionic selectivity of the transduction channel, these estimates for peak transduction conductances of 2–5 nS permit an inference about the minimal number of transduction channels that may be open in a stimulated hair cell. If the transduction channel were a hydrated pore about 0.6 nm in diameter (Corey & Hudspeth 1979b), it might be expected to have a conductance on the order of 50 pS, similar to that of the acetylcholine-activated synaptic channel of comparable ionic selectivity (Lewis 1979, Horn & Patlak 1980). It follows that the peak transduction conductance of 2–5 nS could be due to the simultaneous opening of as few as 40–100 channels, or about one per stereocilium.

It should be noted that this value depends critically on the estimate of unitary conductance for a transduction channel. If the channel has a lower unitary conductance than suggested above, additional transduction channels would be active. There may also be many transduction channels that, on average, are not open. In any event the possibility that there are few transduction channels bears on two issues.

1. Freeze-fracture electron micrographs of stereocilia (Jahnke 1975, Bagger-Sjöbäck & Flock 1977, Hama & Saito 1981) demonstrate only small numbers of intramembrane particles in desultory array; among these may nevertheless lie the full complement of transduction channels.
2. The possible dearth of channels may considerably impede the isolation and biochemical characterization of a transducer molecule from vertebrate hair cells.

CONCLUSION

The present article emphasizes the considerable recent progress toward a cell biological understanding of the transduction process in the hair cells of vertebrates. Because of its focus on the cellular and subcellular bases of transduction, this report has not conveyed the parallel advances in the investigation of the auditory and vestibular systems in situ. The next decade should yield a satisfying synthesis of what is known about the hair cell as a discrete entity with an understanding of the inner ear as a whole. In particular, progress may be anticipated in electrical modeling of the responses of hair cells and of the organs that contain them, in definition of the electrical and perhaps the mechanical role of hair cells in cochlear frequency tuning, and in understanding the effects of efferent innervation of hair cells.

The prospects are somewhat less bright for prompt advances in the biochemistry of transduction by the acousticolateralis system. The initial chemical steps in some transduction processes are obvious: Vision must commence with an interaction between light and an absorptive molecule, taste and smell with the binding of appropriate ligands to their receptors. There is no comparably straightforward interaction in the operation of mechanoreceptors, no known ligand whose binding is obligatorily associated with the transduction site. The challenge for the near future lies in identifying the crucial molecules of the transduction mechanism and in learning how they are activated by mechanical stimulation of hair cells.

ACKNOWLEDGMENTS

The author's original research cited herein has been supported by Grant NS-13154 from the National Institutes of Health and by grants from the William Randolph Hearst, Ann Peppers, Pew, and Alfred P. Sloan Foundations. I thank P. Brown and particularly C. Hochenedel for preparing the manuscript, R. Jacobs for the scanning electron micrograph and for technical assistance, and R. A. Eatock, R. Jacobs, R. Lewis, J. H. R. Maunsell, and especially T. Holton for perspicacious critiques.

Literature Cited

Bagger-Sjöbäck, D. 1974. The sensory hairs and their attachments in the lizard basilar papilla. *Brain Behav. Evol.* 10:88–94

Bagger-Sjöbäck, D., Flock, Å. 1977. Freeze-fracturing of the auditory basilar papilla in the lizard *Calotes versicolor. Cell Tissue Res.* 177:431–43

Bagger-Sjöbäck, D., Wersäll, J. 1973. The sensory hairs and tectorial membrane of the basilar papilla in the lizard *Calotes versicolor. J. Neurocytol.* 2:329–50

Balough, K., Hiraide, F., Ishii, D. 1970. Distribution of radioactive dihydrostreptomycin in the cochlea. *Ann. Otol. Rhinol. Laryngol.* 79:641–52

Bezanilla, F., Armstrong, C. M. 1972. Negative conductance caused by entry of sodium and cesium ions into the potassium channels of squid axons. *J. Gen. Physiol.* 60:588–608

Bosher, S. K. 1980. The nature of the ototoxic actions of ethacrynic acid upon the mammalian endolymph system. II. Structural-functional correlates in the stria vascularis. *Acta Otolaryngol.* 90:40–54

Bosher, S. K., Warren, R. L. 1968. Observations on the electrochemistry of the cochlear endolymph of the rat: A quantitative study of its electrical potential and ionic composition as determined by means of flame spectrophotometry. *Proc. R. Soc. London Ser. B* 171:227–47

Bosher, S. K., Warren, R. L. 1978. Very low calcium content of cochlear endolymph, an extracellular fluid. *Nature* 273:377–78

Boston, J. R. 1980. A model of lateral line microphonic response to high-level stimuli. *J. Acoust. Soc. Am.* 67:875–81

Bowen, R. E. 1932. The ampullar organs of the ear. *J. Comp. Neurol.* 55:273–313

Bracho, H., Budelli, R. 1978. The generation of resting membrane potentials in an inner ear hair cell system. *J. Physiol.* 281:445–65

Bretscher, A. 1981. Fimbrin is a cytoskeletal protein that crosslinks F-actin *in vitro. Proc. Natl. Acad. Sci. USA* 78:6849–53

Bretscher, A., Weber, K. 1978. Localization of actin and microfilament-associated proteins in the microvilli and terminal web of the intestinal brush border by immunofluorescence microscopy. *J. Cell Biol.* 79:839–45

Bretscher, A., Weber, K. 1979. Villin: The major microfilament-associated protein of the intestinal microvillus. *Proc. Natl. Acad. Sci. USA* 76:2321–25

Citron, L., Exley, D., Hallpike, C. S. 1956. Formation, circulation and chemical properties of the labyrinthine fluids. *Br. Med. Bull.* 12:101–4

Corey, D. P., Hudspeth, A. J. 1979a. Response latency of vertebrate hair cells. *Biophys. J.* 26:499–506

Corey, D. P., Hudspeth, A. J. 1979b. Ionic basis of the receptor potential in a vertebrate hair cell. *Nature* 281:675–77

Corey, D. P., Hudspeth, A. J. 1980. Gating kinetics of the transduction element in a vertebrate hair cell: Evidence for a three-state model. *Soc. Neurosci. Abstr.* 6:96 (Abstr.)

Crawford, A. C., Fettiplace, R. 1979. Reversal of hair cell responses by current. *J. Physiol.* 295:66P (Abstr.)

Crawford, A. C., Fettiplace, R. 1980a. Auditory nerve responses to mechanical deflections of the basilar membrane. *J. Physiol.* 308:86P (Abstr.)

Crawford, A. C., Fettiplace, R. 1980b. The frequency selectivity of auditory nerve fibres and hair cells in the cochlea of the turtle. *J. Physiol.* 306:79–125

Crawford, A. C., Fettiplace, R. 1981. Nonlinearities in the responses of turtle hair cells. *J. Physiol.* 315:317–38

Dallos, P. 1981. Cochlear physiology. *Ann. Rev. Psychol.* 32:153–90

de Peyer, J. E., Deitmer, J. W. 1980. Divalent cations as charge carriers during two functionally different membrane currents in the ciliate *Stylonychia. J. Exp. Biol.* 88:73–89

DeRosier, D. J., Tilney, L. G., Egelman, E. 1980. Actin in the inner ear: The remarkable structure of the stereocilium. *Nature* 287:291–96

de Vries, H. 1949. The minimum perceptible angular acceleration under various conditions. *Acta Otolaryngol.* 37:218–29

Eatock, R. A., Corey, D. P., Hudspeth, A. J. 1979. Adaptation in a vertebrate hair cell: Stimulus-induced shift of the operating range. *Soc. Neurosci. Abstr.* 5:19 (Abstr.)

Engebretson, A. M., Eldredge, D. H. 1968. Model for the nonlinear characteristics of cochlear potentials. *J. Acoust. Soc. Am.* 44:548–54

Engström, B., Borg, E. 1981. Lesions to cochlear inner hair cells induced by noise. *Arch. Otorhinolaryngol.* 230:279–84

Engström, H., Ades, H. W. 1973. The ultrastructure of the organ of Corti. In *The Ultrastructure of Sensory Organs,* ed. I. Friedmann, pp. 83–151. New York: American Elsevier

Engström, H., Ades, H. W., Hawkins, J. E. 1962. Structure and functions of the sensory hairs of the inner ear. *J. Acoust. Soc. Am.* 34:1356–63

Engström, H., Engström, B. 1978. Structure of the hairs on cochlear sensory cells. *Hearing Res.* 1:49–66

Erler, G. 1980. Are microtubules involved in mechanosensory transduction? *Abstr. Cold Spring Harbor Lab. Mtg. Forms Microtubule Org. Cells,* p. 14 (Abstr.)

Flock, Å. 1965a. Transducing mechanisms in the lateral line canal organ receptors. *Cold Spring Harbor Symp. Quant. Biol.* 30:133–45

Flock, Å. 1965b. Electron microscopic and electrophysiological studies on the lateral line canal organ. *Acta Otolaryngol. Suppl.* 199:1–90

Flock, Å. 1977. Physiological properties of sensory hairs in the ear. In *Psychophysics and Physiology of Hearing,* ed. E. F. Evans, J. P. Wilson, pp. 15–25. London: Academic

Flock, Å., Cheung, H. C. 1977. Actin filaments in sensory hairs of inner ear receptor cells. *J. Cell Biol.* 75:339–43

Flock, Å., Cheung, H. C., Flock, B., Utter, G. 1981a. Three sets of actin filaments in sensory cells of the inner ear. Identification and functional orientation determined by gel electrophoresis, immunofluorescence and electron microscopy. *J. Neurocytol.* 10:133–47

Flock, Å., Duvall, A. J. 1965. The ultrastructure of the kinocilium of the sensory cells in the inner ear and lateral line organs. *J. Cell Biol.* 25:1–8

Flock, Å., Flock, B., Murray, E. 1977. Studies on the sensory hairs of receptor cells in the inner ear. *Acta Otolaryngol.* 83:85–91

Flock, Å., Hoppe, Y., Wei, X. 1981b. Immunofluorescence localization of proteins in semithin 0.2–1 μm frozen sections of the ear. *Arch. Otorhinolaryngol.* 233:55–66

Flock, Å., Jørgensen, M., Russell, I. 1973. The physiology of individual hair cells and their synapses. In *Basic Mechanisms in Hearing,* ed. A. R. Møller, pp. 273–306. New York: Academic

Furukawa, T., Ishii, Y. 1967. Effects of static bending of sensory hairs on sound reception in the goldfish. *Jpn. J. Physiol.* 17:572–88

Furukawa, T., Ishii, Y., Matsuura, S. 1972. An analysis of microphonic potentials of the sacculus of goldfish. *Jpn. J. Physiol.* 22:603–16

Gallais, A. 1979. Comparative study of the influence of aminoglycoside antibiotics on the activity of the horizontal semicircular canal in the frog. *Acta Otolaryngol.* 88:88–96

Gallin, E. K., Wiederhold, M. L. 1977. Response of *Aplysia* statocyst receptor cells to physiologic stimulation. *J. Physiol.* 266:123–37

Geiger, B., Tokuyasu, K. T., Singer, S. J. 1979. Immunocytochemical localization of α-actinin in intestinal epithelial cells. *Proc. Natl. Acad. Sci. USA* 76:2833–37

Glenney, J. R., Bretscher, A., Weber, K. 1980. Calcium control of the intestinal microvillus cytoskeleton: Its implications for the regulation of microfilament organizations. *Proc. Natl. Acad. Sci. USA* 77:6458–62

Guedry, F. E. 1974. Psychophysics of vestibular sensation. In *Handbook of Sensory Physiology,* Vol. 6/2, ed. H. H. Kornhuber, pp. 3–154. Berlin: Springer-Verlag

Hagins, W. A., Penn, R. D., Yoshikami, S. 1970. Dark current and photocurrent in retinal rods. *Biophys. J.* 10:380–412

Hagiwara, S., Byerly, L. 1981. Calcium channel. *Ann. Rev. Neurosci.* 4:69–125

Hama, K., Saito, K. 1981. The fine structure of the sensory epithelium in the acustico-lateralis system. *Adv. Neurol. Sci.* 25:756–76 (In Japanese)

Hamilton, D. W. 1969. The cilium on mammalian vestibular hair cells. *Anat. Rec.* 164:253–58

Harris, G. G., Frishkopf, L. S., Flock, Å. 1970. Receptor potentials from hair cells of the lateral line. *Science* 167:76–79

Hillman, D. E. 1969. New ultrastructural findings regarding a vestibular ciliary apparatus and its possible functional significance. *Brain Res.* 13:407–12

Hillman, D. E. 1972. Observations on morphological features and mechanical properties of the peripheral vestibular receptor system in the frog. *Prog. Brain Res.* 37:69–75

Hillman, D. E. 1976. Vestibular and lateral line system. Morphology of peripheral and central vestibular systems. In *Frog Neurobiology,* ed. R. Llinas, W. Precht, pp. 452–80. Berlin: Springer-Verlag

Hillman, D. E. 1977. Relationship of the sensory cell cilia to the cupula. In *Scanning Electron Microscopy/1977,* ed. O. Johari, R. P. Becker, 2:415–20. Chicago: IIT Res. Inst.

Hillman, D. E., Lewis, E. R. 1971. Morphological basis for a mechanical linkage in otolithic receptor transduction in the frog. *Science* 174:416–19

Hillman, D. E., McLaren, J. W. 1979. Displacement configuration of semicircular canal cupulae. *Neuroscience* 4:1989–2000

Horn, R. Patlak, J. 1980. Single channel currents from excised patches of muscle membrane. *Proc. Natl. Acad. Sci. USA* 77:6930–34

Hubbard, A. E., Mountain, D. C., Geisler, C. D. 1979. The spectral content of the cochlear microphonic measured in scala media of the guinea pig cochlea. *J. Acoust. Soc. Am.* 66:415–30

Hudspeth, A. J. 1982. Extracellular current flow and the site of transduction by vertebrate hair cells. *J. Neurosci.* 2:1–10

Hudspeth, A. J., Corey, D. P. 1977. Sensitivity, polarity, and conductance change in the response of vertebrate hair cells to controlled mechanical stimuli. *Proc. Natl. Acad. Sci. USA* 74:2407–11

Hudspeth, A. J., Jacobs, R. 1979. Stereocilia mediate transduction in vertebrate hair cells. *Proc. Natl. Acad. Sci. USA* 76:1506–9

Hunter-Duvar, I. M. 1977. Morphology of the normal and the acoustically damaged cochlea. See Hillman 1977, pp. 421–28, 300

Itoh, M., Nakashima, T. 1980. Structure of the hair rootlets on cochlear sensory cells by tannic acid fixation. *Acta Otolaryngol.* 90:385–90

Iurato, S. 1967. *Submicroscopic Structure of the Inner Ear.* Oxford: Pergamon Press. 367 pp.

Jahnke, K. 1975. The fine structure of freeze-fractured intercellular junctions in the guinea pig inner ear. *Acta Otolaryngol. Suppl.* 336:1–40

Jahnke, V., Lundquist, P.-G., Wersäll, J. 1969. Some morphological aspects of sound perception in birds. *Acta Otolaryngol.* 67:583–601

Johnstone, B. M., Taylor, K. J., Boyle, A. J. 1970. Mechanics of the guinea pig cochlea. *J. Acoust. Soc. Am.* 47:504–9

Johnstone, J. R., Johnstone, B. M. 1966. Origin of summating potential. *J. Acoust. Soc. Am.* 4:1405–13

Jørgensen, F. 1979. The influence of calcium and D 600 on the mechanosensitivity of the hair cells of the lateral line system of Necturus. *16th Scand. Congr. Physiol. Pharmacol.,* p. 59 (Abstr.)

Katsuki, Y., Yanagisawa, K., Kanzaki, J. 1966. Tetraethylammonium and tetrodotoxin: Effects on cochlear potentials. *Science* 151:1544–45

Khan, N. S., Müller-Arnecke, H., Röskenbleck, H., Trinckner, D. E. W. 1979. Functions of different receptor systems in the reptilian labyrinth. *Arch. Otorhinolaryngol.* 224:31–35

Khanna, S. M., Leonard, D. G. B. 1981. Laser interferometric measurements of basilar membrane vibrations in cats using a round window approach. *J. Acoust. Soc. Am.* 69:S51 (Abstr.)

Kilian, P. L., Schacht, J. 1980. Sound stimulates labeling of polyphosphoinositides in the auditory organ of the Noctuid moth. *J. Neurochem.* 34:709–12

Kimura, R. S. 1966. Hairs of the cochlear sensory cells and their attachment to the tectorial membrane. *Acta Otolaryngol.* 61:55–72

Konishi, T. 1979. Effects of local application of ototoxic antibiotics on cochlear potentials in guinea pigs. *Acta Otolaryngol.* 88:41–46

Konishi, T., Kelsey, E., Singleton, G. T. 1966. Effects of chemical alteration in the endolymph on the cochlear potentials. *Acta Otolaryngol.* 62:393–404

Kroese, A. B. A., van den Bercken, J. 1980. Dual action of ototoxic antibiotics on sensory hair cells. *Nature* 283:395–97

Kronester-Frei, A. 1979. The effect of changes in endolymphatic ion concentrations on the tectorial membrane. *Hearing Res.* 1:81–94

LePage, E. L., Johnstone, B. M. 1980. Nonlinear mechanical behaviour of the basilar membrane in the basal turn of the guinea pig cochlea. *Hearing Res.* 2:183–89

Lewis, C. A. 1979. Ion-concentration dependence of the reversal potential and the single channel conductance of ion channels at the frog neuromuscular junction. *J. Physiol.* 286:417–45

Lewis, E. R., Li, C. W. 1973. Evidence concerning the morphogenesis of saccular receptors in the bullfrog (*Rana catesbeiana*). *J. Morphol.* 139:351–62

Lewis, E. R., Li, C. W. 1975. Hair cell types and distributions in the otolithic and auditory organs of the bullfrog. *Brain Res.* 83:35–50

Li, C. W., Lewis, E. R. 1974. Morphogenesis of auditory receptor epithelia in the bullfrog. In *Scanning Electron Microscopy/1974*, ed. O. Johari, I. Corvin, pp. 792–98. Chicago: IIT Res. Inst.

Lim, D. J. 1972. Fine morphology of the tectorial membrane: Its relationship to the organ of Corti. *Arch. Otolaryngol.* 96:199–215

Lim, D. J. 1976. Morphological and physiological correlates in cochlear and vestibular sensory epithelia. In *Scanning Electron Microscopy/1976*, Vol. 2, ed. O. Johari, R. P. Becker, pp. 269–76, 366. Chicago: IIT Res. Inst.

Lim, D. J. 1980. Cochlear anatomy related to cochlear micromechanics. A review. *J. Acoust. Soc. Am.* 67:1686–95

Lindeman, H. H., Ades, H. W., Bredberg, G., Engström, H. 1971. The sensory hairs and the tectorial membrane in the development of the cat's organ of Corti. *Acta Otolaryngol.* 72:229–42

Lohdi, S., Weiner, N. D., Mechigian, I., Schacht, J. 1980. Ototoxicity of aminoglycosides correlated with their action on monomolecular films of polyphosphoinositides. *Biochem. Pharmacol.* 29:597–601

Lohdi, S., Weiner, N. D., Schacht, J. 1979. Interactions of neomycin with monomolecular films of polyphosphoinositides and other lipids. *Biochem. Biophys. Acta* 557:1–8

Macartney, J. C., Comis, S. D., Pickles, J. O. 1980. Is myosin in the cochlea a basis for active motility? *Nature* 288:491–92

Machemer, H., Ogura, A. 1979. Ionic conductances of membranes in ciliated and deciliated *Paramecium. J. Physiol.* 296:49–60

Malcolm, R. 1974. A mechanism by which the hair cells of the inner ear transduce mechanical energy into a modulated train of action potentials. *J. Gen. Physiol.* 63:757–72

Matsudaira, P. T., Burgess, D. R. 1979. Identification and organization of the components in the isolated microvillus cytoskeleton. *J. Cell Biol.* 83:667–73

Matsuura, S., Ikeda, K., Furukawa, T. 1971a. Effects of Na^+, K^+, and ouabain on microphonic potentials of the goldfish inner ear. *Jpn. J. Physiol.* 21:563–78

Matsuura, S., Ikeda, K., Furukawa, T. 1971b. Effects of streptomycin, kanamycin, quinine, and other drugs on the microphonic potentials of goldfish sacculus. *Jpn. J. Physiol.* 21:579–90

McGlone, F. P., Russell, I. J., Sand, O. 1979. Measurement of calcium ion concentrations in the lateral line cupulae of *Xenopus laevis. J. Exp. Biol.* 83:123–30

McLaren, J. W., Hillman, D. E. 1979. Displacement of the semicircular canal cupula during sinusoidal rotation. *Neuroscience* 4:2001–8

Miller, M. R. 1973. A scanning electron microscope study of the papilla basilaris of *Gekko gecko. Z. Zellforsch.* 136:307–28

Miller, M. R. 1974. Scanning electron microsopy of the lizard papilla basilaris. *Brain Behav. Evol.* 10:95–112

Miller, M. R. 1978a. Scanning electron microscope studies of the papilla basilaris of some turtles and snakes. *Am. J. Anat.* 151:409–36

Miller, M. R. 1978b. Further scanning electron microscope studies of lizard auditory papillae. *J. Morphol.* 156:381–418

Mooseker, M. S., Graves, T. A., Wharton, K. A., Falco, N., Howe, C. L. 1980. Regulation of microvillus structure: Calcium-dependent solation and cross-linking of actin filaments in the microvilli of intestinal epithelia cells. *J. Cell Biol.* 87:809–22

Moran, D. T., Rowley, J. C., Asher, D. L. 1981. Calcium-binding sites on sensory processes in vertebrate hair cells. *Proc. Natl. Acad. Sci. USA* 78:3954–58

Moran, D. T., Varela, F. J., Rowley, J. C. 1977. Evidence for active role of cilia in sensory transduction. *Proc. Natl. Acad. Sci. USA* 74:793–97

Naftalin, L. 1965. Some new proposals regarding acoustic transmission and transduction. *Cold Spring Harbor Symp. Quant. Biol.* 30:169–80

Naftalin, L. 1977. The peripheral hearing mechanism: New biophysical concepts for transduction of the acoustic signal to an electrochemical event. *Physiol. Chem. Phys.* 9:337–82

Naftalin, L. 1981. Energy transduction in the cochlea. *Hearing Res.* 5:307–15

O'Leary, D. P. 1970. An electrokinetic model of transduction in the semicircular canal. *Biophys. J.* 10:859–75

Oman, C. M., Frishkopf, L. S., Goldstein, M. H. 1979. Cupula motion in the semicircular canal of the skate, *Raja erinacea. Acta Otolaryngol.* 87:528–38

Oman, C. M., Young, L. R. 1972. The physiological range of pressure difference and cupula deflections in the human semicircular canal. *Acta Otolaryngol.* 74:324–31

Peterson, S. K., Frishkopf, L. S., Lechène, C., Oman, C. M., Weiss, T. F. 1978. Element composition of inner ear lymphs in cats, lizards, and skates determined by electron probe microanalysis of liquid samples. *J. Comp. Physiol.* 126:1–14

Porter, J., Jick, H. 1977. Drug-induced anaphylaxis, convulsions, deafness, and extrapyramidal symptoms. *Lancet* 1:587–88

Pumphrey, R. J. 1950. Hearing. In *Physiological Mechanisms in Animal Behaviour*, ed. J. F. Danielli, R. Brown, pp. 3–18. Cambridge: Cambridge Univ. Press

Retzius, G. 1881. *Das Gehörorgan der Wirbeltiere. I. Das Gehörorgan der Fische und Amphibien.* Stockholm: Samson & Wallin. 222 pp.

Retzius, G. 1884. *Das Gehörorgan der Wirbeltiere. II. Das Gehörorgan der Reptilen, der Vögel und der Säugetiere.* Stockholm: Samson & Wallin. 368 pp.

Rhode, W. S. 1971. Observations of the vibration of the basilar membrane in squirrel monkeys using the Mössbauer technique. *J. Acoust. Soc. Am.* 49:1218–31

Rhode, W. S. 1978. Some observations on cochlear mechanics. *J. Acoust. Soc. Am.* 64:158–76

Rhode, W. S., Geisler, C. D. 1967. Model of the displacement between opposing points on the tectorial membrane and reticular lamina. *J. Acoust. Soc. Am.* 42:185–90

Rhode, W. S., Robles, L. 1974. Evidence from Mössbauer experiments for nonlinear vibration in the cochlea. *J. Acoust. Soc. Am.* 55:588–96

Robertson, D., Johnstone, B. M. 1980. Acoustic trauma in the guinea pig cochlea: Early changes in ultrastructure and neural threshold. *Hearing Res.* 3:167–79

Russell, I. J. 1979. The responses of vertebrate hair cells to mechanical stimulation. In *Neurones without Impulses*, ed. A. Roberts, B. M. H. Bush, pp. 117–45. Cambridge: Cambridge Univ. Press

Russell, I. J., Sellick, P. M. 1976. Measurement of potassium and chloride ion concentrations in the cupulae of the lateral lines of *Xenopus laevis. J. Physiol.* 257:245–55

Russell, I. J., Sellick, P. M. 1978. Intracellular studies of hair cells in the mammalian cochlea. *J. Physiol.* 284:261–90

Salt, A. N., Konishi, T. 1979. The role of potassium and sodium in cochlear transduction: A study with amiloride and tetraethylammonium. *J. Acoust. Soc. Am.* 66:S47 (Abstr.)

Sand, O. 1975. Effects of different ionic environments on the mechano-sensitivity of lateral line organs in the mudpuppy. *J. Comp. Physiol. Ser. A.* 102:27–42

Sand, O., Ozawa, S., Hagiwara, S. 1975. Electrical and mechanical stimulation of hair cells in the mudpuppy. *J. Comp. Physiol. Ser. A* 102:13–26

Schacht, J. 1974. Interaction of neomycin with phosphoinositide metabolism in guinea pig inner ear and brain tissues. *Ann. Otol.* 83:613–18

Schacht, J. 1976. Biochemistry of neomycin ototoxicity. *J. Acoust. Soc. Am.* 59:940–44

Schacht, J. 1979. Isolation of an aminoglycoside receptor from guinea pig inner ear tissues and kidney. *Arch. Otorhinolaryngol.* 224:129–34

Sellick, P. M., Johnstone, B. M. 1975. Production and role of inner ear fluid. *Prog. Neurobiol.* 5:337–62

Sellick, P. M., Russell, I. J., Patuzzi, R., Johnstone, B. M. 1981. Generation of hair cell receptor potentials and basilar membrane tuning. *J. Acoust. Soc. Am.* 70:S51–S52 (Abstr.)

Shotwell, S. L., Jacobs, R., Hudspeth, A. J. 1981. Directional sensitivity of individual vertebrate hair cells to controlled deflection of their hair bundles. *Ann. NY Acad. Sci.* 374:1–10

Slepecky, N., Hamernik, R. P., Henderson, D. 1980. A re-examination of a hair cell organelle in the cuticular plate region and its possible relation to active processes in the cochlea. *Hearing Res.* 2:413–21

Slepecky, N., Hamernik, R., Henderson, D. 1981a. The consistent occurrence of a striated organelle (Friedmann body) in the inner hair cells of the normal chinchilla. *Acta Otolaryngol.* 91:189–98

Slepecky, N., Hamernik, R., Henderson, D., Coling, D. 1981b. Ultrastructural changes to the cochlea resulting from impulse noise. *Arch. Otorhinolaryngol.* 230:273–78

Spoendlin, H. 1970. Auditory, vestibular, olfactory and gustatory systems. In *Ultrastructure of the Peripheral Nervous System and Sense Organs*, ed. A. Bischoff, pp. 173–337. St. Louis: Mosby

Stommel, E. W., Stephens, R. E., Alkon, D. L. 1980. Motile statocyst cilia transmit rather than directly transduce mechanical stimuli. *J. Cell Biol.* 87:652–62

Takasaka, T. 1981. Inner ear ultrastructure in a high voltage electron microscope. *Adv. Neurol. Sci.* 25:777–91 (In Japanese)

Takasaka, T., Smith, C. A. 1971. The structure and innervation of the pigeon's basilar papilla. *J. Ultrastruct. Res.* 35:20–65

Tanaka, Y., Asanuma, A., Yanagisawa, K. 1979. Effect of EDTA in the scala media on cochlear potentials. *Proc. Jpn. Acad. Ser. B* 55:31–36

Tanaka, Y., Asanuma, A., Yanagisawa, K. 1980. Potentials of outer hair cells and their membrane properties in cationic environments. *Hearing Res.* 2:431–38

Tilney, L. G., DeRosier, D. J., Mulroy, M. J. 1980. The organization of actin filaments in the stereocilia of cochlear hair cells. *J. Cell Biol.* 86:244–59

Tilney, L. G., Saunders, J. C., Egelman, E., DeRosier, D. J. 1982. Changes in the organization of actin filaments in the stereocilia of noise damaged lizard cochleae. *Hearing Res.* 7:181–97

Tran-Ba-Huy, P., Manuel, C., Meulemans, A. 1979. Pharmacokinetics of gentamicin in perilymph and endolymph, studied in the rat by radioimmunoassay. *Arch. Otorhinolaryngol.* 224:135–36

Tran-Ba-Huy, P., Manuel, C., Meulemans, A., Sterkers, O., Amiel, C. 1981. Pharmacokinetics of gentamicin in perilymph and endolymph of the rat as determined by radioimmunoassay. *J. Infect. Dis.* 143:476–86

Tumarkin, A. 1980. The controversial cupula and the enigmatic kinocilium. *J. Laryngol. Otol.* 94:917–27

Turner, R. G., Muraski, A. A., Nielsen, D. W. 1981. Cilium length: Influence on neural tonotopic organization. *Science* 213:1519–21

Valli, P., Zucca, G., Casella, C. 1977. The importance of potassium in the function of frog semicircular canals. *Acta Otolaryngol.* 84:344–51

Valli, P., Zucca, G., Casella, C. 1979. Ionic composition of the endolymph and sensory transduction in labyrinthine organs. *Acta Otolaryngol.* 87:466–71

von Békésy, G. 1960. *Experiments in Hearing.* New York: McGraw-Hill. 745 pp.

von Düring, M., Karduck, A., Richter, H.-G. 1974. The fine structure of the inner ear in *Caiman crocodilus*. *Z. Anat. Entwicklungsgesch.* 145:41–65

Wersäll, J., Björkroth, B., Flock, Å., Lundquist, P.-G. 1973. Experiments on ototoxic effects of antibiotics. *Adv. Oto. Rhino. Laryngol.* 20:14–41

Wersäll, J., Flock, Å. 1964. Suppression and restoration of the microphonic output from the lateral line organ after local application of streptomycin. *Life Sci.* 3:1151–55

Wever, E. G. 1978. *The Reptile Ear.* Princeton: Princeton Univ. Press. 1024 pp.

Wiederhold, M. L. 1976. Mechanosensory transduction in "sensory" and "motile" cilia. *Ann. Rev. Biophys. Bioengin.* 5: 39–62

Wilson, J. P., Johnstone, J. R. 1975. Basilar membrane and middle-ear vibration in guinea pig measured by capacitive probe. *J. Acoust. Soc. Am.* 57:705–23

Woodhull, A. M. 1973. Ionic blockage of sodium channels in nerve. *J. Gen. Physiol.* 61:687–708

Yanagisawa, K. 1981. Molecular mechanism of neomycin actions on the acousticolateralis receptors. *Adv. Neurol. Sci.* 25:849–856 (In Japanese)

Yoshikami, D. 1981. Transmitter sensitivity of neurons assayed by autoradiography. *Science* 212:929–30

Zenner, H. P. 1981. Cytoskeletal and musclelike elements in cochlear hair cells. *Arch. Otorhinolaryngol.* 230:81–92

Ann. Rev. Neurosci. 1983. 6:217–47
Copyright © 1983 by Annual Reviews Inc. All rights reserved

MICROCIRCUITRY OF THE VISUAL CORTEX

Charles D. Gilbert

Department of Neurobiology, Harvard Medical School, Boston,
Massachusetts 02115

Introduction

Compared to the detailed information available on the microcircuitry of the retina (see review by Sterling in this volume), our understanding of the microcircuitry of the cortex is still in its infancy. Nonetheless, the cortex has increasingly become perceived as a highly ordered structure, and the way in which this order contributes to cortical function is becoming better understood. In the middle of the nineteenth century it was known, through Nissl and myelin stains, that the cortex was a multilayered structure comprised of a multiplicity of cell and fiber types. It was not until the end of the nineteenth century, with the advent of the Golgi technique and the work of Golgi and Cajal, that a more concrete idea of the structural elements of the cortical circuit was obtained. In the middle of this century, further advances were made by the development of electron microscopy, which enabled investigators to identify points of synaptic contact between neurons (for review see Colonnier 1981), and of a technique for tracing neuronal pathways, the Nauta stain (Nauta & Gygax 1954, Fink & Heimer 1967). Over the past two decades, electrophysiological studies have contributed additional information concerning the function of the cortex at the single cell level, and during the last few years the field has become considerably enriched by the further addition of a number of anatomical, physiological, and chemical techniques for studying how the cortical circuit acts to produce the functional properties of cortical cells.

In a few pages it is of course impossible to give a complete account of the work in this area and it is therefore necessary to limit the scope of this chapter. The visual cortex as a whole is divided into a number of areas, each with a more or less precise representation of the visual field on its surface (for review see Van Essen 1979). The area that has been studied in the greatest detail with the most techniques is area 17 of cat and monkey visual

217

cortex. Using this area as the primary focus, I also draw upon information obtained in other mammalian species, other visual cortical areas, and areas serving other modalities where studies have contributed additional insights to cortical organization.

While the term "microcircuitry" most aptly applies to the local connections made within an individual cortical area, it is difficult to think of the role of the intrinsic cortical connections without also discussing the various sources of input to the cortex, the laminar distribution of these inputs, and the distant targets to which different classes of cortical cells provide the product of cortical processing.

Inputs and Outputs of the Striate Cortex

The most striking aspect of the cerebral cortex is its lamination. Originally seen in unstained sections of brain tissue by Gennari and by Baillarger, the layers have since been distinguished on the basis of cell and fiber stains (Campbell 1905, Brodmann 1909, von Economo & Koskinas 1925, Otsuka & Hassler 1962). These show that the cells and fibers in each layer have a characteristic size, shape, and packing density. As is described below, the layers have specific functional roles, as reflected in the afferents they receive, the sites to which they project, and the receptive field properties of the cells within them.

Thalamocortical afferents were originally described on the basis of Golgi stains by Cajal (1911). Golgi and degeneration studies showed that the afferents terminate principally in layer 4 (Cajal 1911, Garey & Powell 1971, Wilson & Cragg 1967, Colonnier & Rossignol 1969, Szentagothai 1973, Peters & Feldman 1976). More recent degeneration and autoradiographic studies have shown the involvement of other layers in receiving thalamic input.

Each cortical area receives input from a number of thalamic nuclei, and each nucleus contains more than one class of cortically projecting (or principal) cells. Area 17 receives input from the lateral geniculate nucleus (LGN), medial interlaminar nucleus (in the cat), and lateral posterior (LP) nucleus/pulvinar complex. Within the LGN there exist at least three classes of principal cell, each projecting to area 17. Geniculate cells, which can be subdivided into a number of morphological classes (Guillery 1966), have been classified physiologically either as on- and off-center (Hubel & Wiesel 1961) or as X, Y, and W (also known as brisk sustained and transient and sluggish; Enroth-Cugell & Robson 1966, Cleland et al 1971, Cleland & Levick 1974, Stone & Fukuda 1974, Wilson et al 1976), and there is a relationship between the morphological and physiological classification schemes (Friedlander et al 1981, Stanford et al 1981). The X, Y, and W cells have been differentiated on the basis of linearity of spatial summation, firing properties, conduction velocities, and so on. While there has been some speculation as to the functional significance of these differences, more work needs to be done to clarify their relative roles in cortical function. Each thalamic cell class forms a projection to the cortex with a unique laminar distribution. This has been demonstrated with a number of methods: (*a*) tracing the projections of individual thalamic nuclei or of the layers within a nucleus using degeneration techniques (Hubel & Wiesel 1972, Peters & Feldman 1976, Hornung & Garey 1981a) or using the autoradiographic technique (Rosenquist et al 1975, LeVay & Gilbert 1976, Ogren & Hendrickson 1977, Hendrickson et al 1978, Rezak & Benevento 1979, Miller et al 1980); (*b*) making focal injections of HRP restricted to individual laminae in the cortex (Leventhal 1979); (*c*) filling of thalamocortical fibers by large extracellular injections of HRP (Ferster & LeVay 1978); (*d*) intracellular injection of physiologically characterized axons (Gilbert & Wiesel 1979); (*e*) current source density analysis (Mitzdorf & Singer 1978, 1979); and (*f*) electrical stimulation in the optic radiation (Hoffmann & Stone 1971, Singer et al 1975, Toyama et al 1977, Henry et al 1978, Bullier & Henry 1979a–c). The last method may not have sufficient resolution to differentiate accurately between the inputs coming from different afferent types, since there tends to be a substantial overlap in their conduction velocities (Ferster & Lindstrom 1982).

The results from these studies show that each cell class in the thalamus has a unique laminar pattern of termination in the cortex. This is summarized for area 17 of the cat and monkey in Figure 1. In the cat, X-afferents terminate in layer 4c and the upper half of layer 6; Y-afferents terminate in layer 4ab and also in the upper half of layer 6. The cells giving rise to these terminals lie in the dorsal layers of the LGN. The C-laminae, which

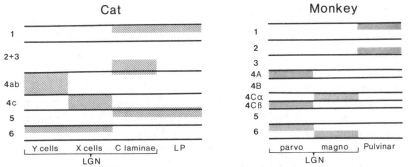

Figure 1 Laminar distribution of thalamic afferents in cat and monkey. Projections of X- and Y-afferents (Ferster & LeVay 1978, Gilbert & Wiesel 1979), C laminae (LeVay & Gilbert 1976), and latero-posterior nucleus (LP; Miller et al 1980) in the cat; magno- and parvocellular layers of the monkey's lateral geniculate nucleus (Hubel & Wiesel 1972, Hendrickson et al 1978) and pulvinar (Ogren & Hendrickson 1977, Rezak & Benevento 1979).

contain Y- and W-cells, project in two bands bracketing layer 4 and also to the upper half of layer 1. The LP/pulvinar complex projects to layers 1 and 5. In the monkey, one can draw similar distinctions, with magnocellular geniculate laminae projecting to layer $4C\alpha$ and parvocellular laminae projecting to layer $4C\beta$ and 4A. The LP/pulvinar complex projects to layers 1 and 2. Thus, the cells lying in many of the cortical layers combine some form of thalamic input with the intrinsic connections coming from other cortical layers.

Although there is a considerable segregation of the various afferent types into different cortical layers, it is not known to what extent this segregation is maintained at subsequent stages in the cortical circuit. There is, however, a convergence of a different sort by thalamic afferents onto cortical cells: Though cells in the LGN are driven exclusively by stimulation of one eye or the other, most cells in the cortex are binocular, and therefore receive convergent input from LGN cells driven by the ipsilateral and contralateral eyes (Hubel & Wiesel 1962). Also, pharmacological studies have demonstrated the integration of on- and off- inputs by individual cortical cells (Schiller 1982).

Within the cortical layers receiving thalamic input, evidence indicates that many cell classes in a given layer synapse with thalamic terminals, though they vary substantially in the number of terminals received (Davis & Sterling 1979; for review, see White 1979). In addition to the layer 4 spiny and smooth stellate cells that receive thalamic input, it is also possible for the apical dendrites of pyramidal cells whose somata are in deeper layers to receive thalamic input (Peters et al 1979) and for the pyramids in the bottom of layer 3 to receive thalamic input onto their basal dendrites, as shown by anatomical (Davis & Sterling 1979) and physiological (Ferster & Lindstrom 1982) experiments. Terminals from the lateral geniculate nucleus are of the asymmetric type, and most commonly end on synaptic spines (Colonnier 1968, Jones 1969, Colonnier & Rossignol 1969; Garey & Powell 1971, LeVay & Gilbert 1976; Peters & Feldman 1976). Approximately 30% of the asymmetric terminals in layer 4 come from the lateral geniculate nucleus (LeVay & Gilbert 1976), so that the remaining terminals are formed by intrinsic connections from other layer 4 cells, input from layer 6 cells (Lund & Boothe 1975, Gilbert & Wiesel 1979), and input from the claustrum (Carey et al 1980, LeVay & Sherk 1981).

The information above has important implications for understanding the role of the thalamic input in producing the functional properties of cortical cells. Many cortical cells receive direct geniculate input in combination with input from other cortical cells, while others (particularly those in layer 2) do not receive direct input from the thalamus. Among the neurons receiving direct thalamic input, one can imagine that their properties would reflect the properties of the particular thalamic cell class that contacts them. It has been found, for example, that only those cells receiving direct input from dorsal layers of the LGN (i.e. from X- and Y-cells) have simple receptive field properties, so that the distribution of simple cells coincides with the distribution of the X- and Y-afferents (Hubel & Wiesel 1962, Gilbert 1977, Gilbert & Wiesel 1979).

There is much that needs to be clarified in this regard, since there is evidence indicating that some complex cells may also receive thalamic input. In their studies of the receptive field properties of cortical cells, Hubel & Wiesel (1962) suggested that, because of their receptive field structure, simple cells were the "first order" cells in the cortex, and that complex cells received their input in turn from simple cells. Physiological studies in which cortical cells were studied while the optic radiation was electrically stimulated (Hoffman & Stone 1971) suggested instead that complex cells receive Y-input, whereas simple cells receive X-input, and these investigators proposed that "parallel processing," rather than the Hubel & Wiesel model of "serial processing," accounted for the distinction between simple and complex cells. It is now clear that simple cells can receive either X- or Y-input (Gilbert & Wiesel 1979, Henry et al 1979, Ferster & LeVay 1978), with one variety of simple cells lying in the Y-afferent zone (layer 4ab) and the another class lying in the X-afferent zone (layer 4c). Reflecting the receptive field properties and patterns of arborization of their inputs, simple cells in the X-afferent zone have smaller receptive fields than those in the Y-afferent zone (Bullier & Henry 1979b, Gilbert & Wiesel 1979). There is, however, some evidence for direct thalamic input onto complex cells. This does not apply equally to all complex cells in all layers. Some physiological studies suggest that layer 5 complex cells receive thalamic input (Toyama et al 1977, Bullier & Henry 1979a–c). Ferster & Lindstrom (1982) have suggested that these findings may be artifactual, and that they were due to antidromic activation of layer 2 + 3 cells, followed by the activation of the synapse between these and layer 5 cells. However, they do find evidence for monosynaptic activation of a few layer 5 cells. These cells, lying at the border between layers 4 and 5, do not project to the superior colliculus, and are likely to be the pyramidal cells described by O'Leary (1941) which do not project into the white matter. Also, layer 3 complex cells can be activated monosynaptically from the LGN. In the monkey the problem is somewhat different, since the first-order cells of the monkey striate cortex have center-surround receptive field organization (Hubel & Wiesel 1974a), and the issue of simple versus complex cell input is even less well resolved.

The anatomical substrate for direct activation of complex cells by geniculate afferents is not well established. It has been suggested that lower layer pyramidal cells can receive thalamic input upon their apical dendrites as they pass through layer 4, but there is some disagreement as to the extent of this input. Golgi studies in the monkey show that apical dendrites of lower layer pyramidal neurons have fewer spines as they course through the geniculate afferent layer 4C (Lund 1973), suggesting that they do not receive input from the lateral geniculate nucleus. In cat striate cortex, ultrastructural studies combining degeneration techniques with serial reconstruction of apical dendrites indicate that apical dendrites do not receive thalamic input (Davis & Sterling 1979), but combined Golgi/degeneration studies show that there can be direct connections between afferents and layer 5 apical dendrites (Hornung & Garey 1981a). Also, Golgi/EM studies of rat visual and mouse somatosensory cortex show that apical dendrites can receive direct thalamic input (Peters et al 1979, White 1978). Due to limitations in these techniques it is difficult to evaluate the importance of the connection, because on any process only a minute proportion of the synapses are formed by degenerating terminals. The discrepancies between different studies may, to some extent, reflect species or areal differences. Interestingly, there appears to be considerable variability in the extent to which different types of layer 5 pyramidal cells receive thalamic input (White & Hersch 1982). This is consistent with the findings of physiological studies that certain classes of layer 5 complex cells receive thalamic input while others do not. The result has important implications for the cortical circuit as a whole, since it shows that cells within a given layer can be very selective for the input they receive and that proximity between a dendrite and a set of axon terminals does not ensure that there will be contact between them. Given that a cell can receive an input on its apical dendrite that it does not receive on its basal dendrite, one must then determine the relative functional effects of inputs on the different sets of dendrites. As yet no studies have dealt with this issue.

In the superficial layers, complex cells can receive thalamic input in one of two ways: Layer 3 "border pyramids" (near layer 4) extend their basal dendrites into layer 4. Also, the C laminae of the cat's geniculate projects in tufts that extend well into layer 3 (LeVay & Gilbert 1976). There is direct evidence for thalamic terminals on layer 3 pyramids (Hornung & Garey 1981a). Similar superficial layer thalamic inputs are seen in the monkey (Ogren & Hendrickson 1977, Miller et al 1980, Hubel & Livingstone 1981). For complex cells in any layer, the pivotal issue, which is dealt with below, is the functional role played by each of the thalamic and intrinsic inputs.

There are afferents to the cortex other than those coming from the thalamus. A given cortical area can receive input from several other cortical areas. Corticocortical afferents can come from the opposite hemisphere via the corpus collosum or from other cortical areas in the same hemisphere.

Commissural afferents to the striate cortex are limited to a narrow band at the 17/18 border, corresponding to the representation of the vertical meridian (Daniel & Whitteridge 1961). The corticocortical afferents generally ramify in the superficial layers (Garey et al 1968, Jones & Powell 1969, Colonnier & Rossignol 1969, Tigges et al 1974, Fisken et al 1975, Shatz 1977). Subcortical sources of input include the claustrum (Levay & Sherk 1981, Olson & Graybiel 1980, Carey et al 1981), which projects most densely to layer 4. There is also a diffuse projection from the raphe and locus coeruleus, which carry the serotonergic and noradrenergic innervation to the cortex (Fuxe 1965, Freedman et al 1975, Morrison et al 1978). Each of the aminergic afferent systems has distinctive laminar distributions (Morrison et al 1982).

In thinking of the function of the cortical microcircuit, one should be aware that each cortical layer has a specific function, as shown by its distant projections. The superficial layers (layers 2 and 3) are responsible for cortico-cortical connections (Toyama et al 1974, Maciewicz 1974, Gilbert & Kelly 1975, Lund et al 1975, Kawamura & Naito 1980), though reciprocal cortico-cortical connections may also arise from deeper layers (Gilbert & Kelly 1975, Rockland & Pandya 1981). Cells in layer 5 project to the superior colliculus (Toyama et al 1974, Palmer & Rosenquist 1974, Gilbert & Kelly 1975, Lund et al 1975, Magalhaes-Castro et al 1975, Kawamura & Konno 1979), to the pulvinar (Lund et al 1975, Ogren & Hendrickson 1977), and to the pons (Albus and Donate-Oliver 1977, Brodal 1978, Gibson et al 1978, Kawamura & Chiba 1979, Glickstein et al 1980). Layer 6 is responsible for the recurrent projection from the cortex to the thalamus (Gilbert & Kelly 1975, Lund et al 1975, Tombol et al 1975) and for the projection to the claustrum (Carey et al 1980, LeVay & Sherk 1981). The cortex shows considerable uniformity in this division of labor by the different cortical layers, with analogous projections seen in a number of visual cortical areas (Gilbert & Kelly 1975) as well as in parts of the cortex serving other sensory modalities (Jones & Wise 1977, Kelly & Wong 1981). Within a given layer the projecting cell populations can be further subdivided: Distinct populations of layer 2 and 3 cells are responsible for projections to different cortical areas (Gilbert & Wiesel 1981b), and in layer 6 different cells are responsible for projections to geniculate and claustrum (LeVay & Sherk 1981).

The laminar segregation of the various cortical afferents and the specialization in the projections of the different cortical layers is reflected in the elaboration of a variety of functional cell classes in the cortex; particular classes are restricted to particular layers (Hubel & Wiesel 1962, Palmer & Rosenquist 1974, Camarda & Rizzolatti 1976, Gilbert 1977, Leventhal &

Hirsch 1978, Henry et al 1979). These specialized properties are generated though the interlaminar and intercolumnar intrinsic cortical connections, which are considered below.

Cell Types and Synaptic Interaction

Because it is composed of an extraordinary variety of cells, the cortical circuit can seem hopelessly intricate (Cajal 1911, 1922, O'Leary 1941, Lorente de Nó 1944, Poljakov 1953, Polyak 1957, Shkolnik-Yarros 1960, Globus & Scheibel 1967, Jones & Powell 1970, Garey 1971, Valverde 1971, LeVay 1973, Lund 1973, Jones 1975, Peters & Fairen 1978). The number of neuronal types depends on the criteria upon which the classification is based. Many features have been used to differentiate cortical neurons: the presence or absence of spines, the projection pattern of the axon and the nature of the postsynaptic structure, the shape of the dendritic tree, the pattern of synapses on the cell soma, and the ultrastructural appearance of the cytoplasm. Most classification schemes that have been presented to date are subjective, and whether a given set of morphological criteria can be used to differentiate classes of cells that are functionally distinct remains to be demonstrated. One difficulty with the existing schemes is that two cells that appear to be different may belong to distinct classes or may simply reflect a continuum of variation within a single neuronal class. In this review, no attempt is made to present an exhaustive list of cell types, but instead a few features that are most relevant to cortical circuitry are discussed.

First, one can divide cortical cells into two principal classes, one with spiny dendrites and the other with smooth or sparsely spinous dendrites. Both classes are found in all cortical layers (Cajal 1911, Lorente de Nó 1933, O'Leary 1941, Sholl 1955, Polyak 1957, Lund 1973, Jones 1975, Valverde 1976, Parnavelas et al 1977, Feldman & Peters 1978). For the most part, each class has a characteristic synaptic morphology: the axons of spiny cells forming asymmetric (Gray's type I) synapses and most smooth cells forming symmetric (Gray's type II) synapses (LeVay 1973, Peters & Fairen 1978). Indirect evidence indicates that there are approximately ten times as many spiny cells in the cortex as smooth cells (Lund 1981). It has been suggested that, because of their synaptic morphology [by analogy to studies of the cerebellum (Uchizono 1965)], the cells whose axons form asymmetric synapses are responsible for excitatory connections and cells forming symmetric synapses are inhibitory. Although still quite speculative, there is mounting evidence from studies of the visual cortex in support of this hypothesis (see below).

Spiny cells include pyramidal cells, which are present in all layers, though rarely in layer 4, and spiny stellate cells, which are found exclusively in layer 4. Spiny cells form a series of interlaminar and horizontal connections, which are described below. They tend to receive asymmetric input onto their spines and symmetric input onto their somata, dendritic shafts and axon initial segments (Peters & Kaiserman-Abramof 1970, LeVay 1973, Hornung & Garey 1981b).

Smooth stellate cells have been subclassified by many investigators. Lorente de Nó identified more than 60 varieties (Lorente de Nó, 1922, 1933). Interestingly, the axons of some varieties are quite specific as to the structures with which they synapse. Most specific are the chandelier cells, which synapse exclusively with axon initial segments, and basket cells, which are thought to form synapses with cell somata, though this has not been demonstrated conclusively. Intermediate in specificity are bipolar or double bouquet cells, which synapse upon dendritic shafts and spines, but not upon somata or axon initial segments. There has been some speculation that these cells preferentially contact the apical dendrites of pyramidal cells, though this has been disputed (see below). Least specific are smooth multipolar stellates, which can form symmetric synapses with apical or basal dendritic shafts, cell somata, and axon hillocks (Peters & Proskauer 1980). At the ultrastructural level, smooth stellate cells receive a mixture of symmetric and asymmetric synapses on their somata and dendrites (Colonnier 1968, LeVay 1973, Somogyi 1977, Peters & Fairen 1978).

Chandelier cells can be found in any layer, though they most commonly occur in the superficial layers. They form short, vertically oriented axon collaterals consisting of strings of boutons, and the boutons form numerous symmetric synaptic contacts with individual pyramidal cell axon initial segments (Somogyi 1977, Fairen & Valverde 1980, Peters et al 1982). The chandelier complex formed by a single cell crosses several layers and can extend for 200 to 300 μm in the horizontal direction. From serial EM reconstruction of terminal complexes stained with antibodies to GAD, these cells have been shown to be GABAergic (Peters et al 1982). This indicates that they are inhibitory, and from their structure one can infer that they are capable of lowering the activity of relatively large regions of cortex.

The specificity of the basket cell contacts is at present only presumptive, as their terminals have not been examined at the ultrastructural level. Basket cells and pericellular baskets are found in layers 3 to 5 (Marin-Padilla 1969). Their axons tend to form long, horizontally running arcades, giving off branches which form the baskets all along their length (Marin-Padilla 1970, 1974, Jones 1975, Peters & Regidor 1981). They may provide a basis for long-range horizontal inhibitory interactions across the cortex, and it has been suggested that the distribution of basket cell axons may coincide with the "slab-shaped" orientation columns of Hubel & Wiesel (Marin-Padilla 1970).

Another population of cells is distinguished by having very narrow axonal fields, crossing many layers. Their axonal fields were sometimes referred to as having a "horse-tail" shape (Szentagothai 1973), whereas others, including the "double-bouquet cells," have a somewhat wider axonal field (Cajal 1911, Colonnier 1966, Szentagothai 1969, Valverde & Ruiz-Marcos 1969). Because of the distinctive distribution of the axons of these cells, they were originally thought to synapse preferentially with apical dendrites (Colonnier 1968, Szentagothai 1969), but have recently been shown to synapse with other dendrites as well (Somogyi & Cowey 1981). The bipolar cells in the rat and a subset of monkey double-bouquet cells are somewhat exceptional among smooth stellate cells, in that they form asymmetric synapses (Somogyi & Cowey 1981, Peters & Kimerer 1981). One monkey double-bouquet cell that was observed to form asymmetric synapses was also distinguished by having spines; other double bouquet cells in cats and monkeys were sparsely spinous and formed symmetric synapses. In the rat, some of the bipolar cells show vasoactive intestinal polypeptide (VIP)-like immunoreactivity (Emson & Hunt 1981). VIP is a putative excitatory transmitter (Phillis et al 1978), so its presence in these cells is consistent with the hypothetical association between excitatory cells and asymmetric synapses. Because of their very narrow axonal fields, crossing many cortical layers, it has been suggested that the bipolar cells may serve to synchronize the output of cells residing within a cortical column.

Despite the presence of numerous anatomical cell classes, there has been no rigorous relationship drawn between a given morphological cell class and a particular functional class. While there is a correlation of simple cells with stellate cells and of pyramidal cells with complex cells (Kelly & VanEssen 1974), there are numerous exceptions (Gilbert & Wiesel 1979, Lin et al 1979). It seems, rather, that the laminar position of a cell, and consequently the inputs it receives, is more determinative of its functional properties (Gilbert & Wiesel 1979). Though a cell could be responsible for producing a given functional property, it need not be distinguishable from other cells in its receptive field properties.

The fact that particular inputs, and therefore particular functional classes of neurons, are associated with laminar position, does not necessarily imply that lamination is responsible for the specificity in connections between afferents and cell types. Rather, chemical specificity may play an important role. Within the axonal field of a given cell there may be tens of thousands of other cells, only a fraction of which are contacted by the axon of that cell, and it is intriguing that axons appear to be able to contact the appropriate sites even in the face of a disruption of the normal cortical structure. This is seen in the reeler mutant mouse, which manages to maintain normal receptive field properties for the cells in the visual cortex despite a disorder of the cortical lamination (Caviness 1977, Drager 1981).

Interlaminar Connections

A principal feature of cortical structure around which many of the intrinsic connections are organized is cortical lamination. It has been established, using a number of techniques, that there is a systematic flow of information from one cortical layer to another, starting with the layers that receive thalamic input.

Using the Golgi technique, Lorente de Nó suggested that cortical connections ran predominantly in a vertical direction (across cortical layers) with very little horizontal transfer of information (Lorente de Nó 1922, 1938). From the earliest studies it was clear that the cortical layers were richly interconnected (Cajal 1911, Lorente de Nó 1922, 1938, Szentagothai 1973). By examining a number of Golgi impregnated cells, Lund & Boothe (1975) found that the intracortical projections of the spiny cells in each layer are precisely laminated, and they worked out the details of a circuit that transfers information from layer to layer. One drawback of this technique is that it is difficult to get a full picture of the axonal arbor of a cell because of partial impregnation. Also, from Golgi studies done at the light level it is not possible to identify the cells that are presynaptic or postsynaptic to the labeled cell. That one can identify the layer(s) to which a labeled cell projects, however, does supply one with a considerable amount of information in trying to relate interlaminar connections to the functional properties of cortical cells. By combining the Golgi technique with electron microscopy, one can sometimes determine the identity of the postsynaptic cell. The Golgi/EM technique has been refined by a technique known as "gold toning" (Fairen et al 1977), which enables one to see the intracellular detail of Golgi impregnated neurons, and consequently to see synaptic specializations. In some instances it then becomes possible to identify the types of cells or processes that are postsynaptic to the impregnated cell (Levay 1973, Somogyi 1977, Fairen & Valverde 1980, Peters & Proskauer 1980, Somogyi & Cowey 1981, Peters & Kimerer 1981).

Another anatomical technique that reveals some of the principal interlaminar connections in the cortex is the stain for degenerating terminals, namely the Nauta technique and its modifications. By making tiny lesions that are restricted to individual layers of the cortex, one can determine the projections of cells in the lesioned layer. This brought out the importance of a projection from the superficial cortical layers to layer 5 (Szenthagothai 1965, Spatz et al 1970, Nauta et al 1973), which is formed by the same pyramidal cells that project from area 17 to areas 18 and 19 (Gilbert & Wiesel 1981a).

A variety of physiological techniques have been employed as well. One technique involves activating cortical cells by stimulating fibers in the optic radiation and optic tract while recording, either intracellularly or extracellularly, from a cortical cell (Hoffmann & Stone 1971, Toyama et al 1977, Bullier & Henry 1979a–c, Ferster & Lindstrom 1982). By measuring the time delay between stimulation in the radiation and the appearance of the first synaptic potentials in a given cortical cell, it is sometimes possible to determine whether that cell is monosynaptically or disynaptically activated from the radiation. The difficulty with this type of study is in interpreting the layers of origin of polysynaptic excitation. It is also possible to mistake various complicated forms of interaction (such as antidromic activation followed by collateral activation) for monosynaptic activation (see above).

The current source density technique identifies the current sources and sinks occurring at different depths in the cortex and at different times after the delivery of electrical or visual stimuli (Mitzdorf & Singer 1978, 1979). This produces a profile of depth in the cortex versus time after stimulus delivery. At any given depth one sees a series of current peaks, with the earliest peaks arriving in the layers receiving input from the LGN. Later peaks in other layers are interpreted as reflecting polysynaptic activity, but the pathways mediating this activity are subject to different interpretations. The technique is also subject to the difficulties inherent in the single-cell recording/optic radiation stimulation experiments.

The cross-correlation technique involves recording from two cells in the cortex and correlating the appearance of spikes in one cell with spikes in the other. If the two cells are monosynaptically related, then there is a high probability of the appearance of a spike in one cell at a short delay after the spike in the other, and this results in a characteristic cross-correlogram. The advantage of this technique is that it allows one to establish the functional properties of both the pre- and postsynaptic cells, as well as the nature of the interaction between them (excitatory vs inhibitory). Originally developed by Gerstein and co-workers (Perkel et al 1967, Dickson & Gerstein 1974, Stevens & Gerstein, 1976), it has been employed in the investigation of interlaminar connections in visual cortex (Toyama et al 1981a,b). One of the pitfalls of the technique is that it requires precise alignment of the recording electrodes to determine reliably the connectivity of two layers. There are presently some discrepancies that may be partially due to this difficulty between the findings of the cross-correlation technique and other physiological and anatomical techniques.

Finally, intracellular recording and injection of cells with horseradish peroxidase enables one to combine information on the functional properties of a cell with its dendritic and axonal morphology. By comparing the receptive field properties of the injected cell with the cells lying in the layers to which it projects, it is possible to develop hypotheses on the functional role of individual cortical connections.

Figure 2 provides a summary diagram of the principal intrinsic cortical connections. Although primarily based on findings in the cat, it bears great resemblance to the monkey, as described by Lund and co-workers (Lund 1973, Lund & Boothe 1975, Lund et al 1979, Lund 1981) on the basis of Golgi studies and by Hubel & Wiesel (1972) on the basis of degeneration studies. The Golgi work in cat and monkey has been confirmed and extended by the intracellular injection technique (Gilbert & Wiesel 1979) and by electrical stimulation experiments (Bullier & Henry 1979a–c, Ferster & Lindstrom 1982). Though there are clear differences in the appearance of the layers in different cortical areas, there is a remarkable similarity in the general pattern of interlaminar connections between areas 17 and 18 (Lund et al 1981). As mentioned above, in area 17 the thalamic input to the cortex arrives in different laminae, depending on the cell class in question, but the strongest input is to layer 4. Cells in layer 4 project to the superficial layers. From layers 2 and 3 there is a substantial projection to layer 5, and from layer 5 to layer 6. The layer 6 cells then form a closed loop by projecting back to layer 4. Although there are other connections, these constitute the most substantial connections formed by spiny cells in the cortex. The smooth stellate cells show a variety of connections, though not necessarily associated with particular layers. Their projections can be quite restricted, such as those formed by the Golgi II cells, which are contained to the same area as covered by their dendrites, and usually restricted to an individual layer. Other smooth stellate cells can project across many layers, as do the double bouquet cells, or they can project for fairly long distances horizontally, as do the basket cells.

The functional role of each of the elements of this circuit is now undergoing considerable scrutiny. The first model of a "functional microcircuitry" in the visual cortex was advanced by Hubel & Wiesel (1962), who suggested that simple cells represented the first order cortical cells, and that complex cells receive input from simple cells. The circuit described above lends some support to this scheme, in that simple cells are found in the layers receiving X- and Y-cell input, and that complex cells are found in layers 2 and 3, which receive a substantial input from the simple cells in layer 4 (Hubel & Wiesel 1962, Gilbert 1977, Leventhal & Hirsch 1978, Henry et al 1979, Gilbert & Wiesel 1979, 1981a, Lin et al 1979). Other receptive field properties may be ascribed to other elements in the circuit, relying on the tangential distribution of intracortical connections (see next section).

Because an individual cortical layer can receive input from the thalamus as well as from other cortical layers, it is difficult to know which inputs are responsible for particular functional properties. A given cell in layer 4, for example, receives comparable projections from the lateral geniculate nucleus and from cells in layer 6. Another example is layer 3. Whereas most of the cells in layer 4 project up into layers 2 and 3, cells in layer 3 also receive input from the C laminae of the geniculate (LeVay & Gilbert 1975, Leventhal 1976, Hornung & Garey 1981a). It has been shown that certain

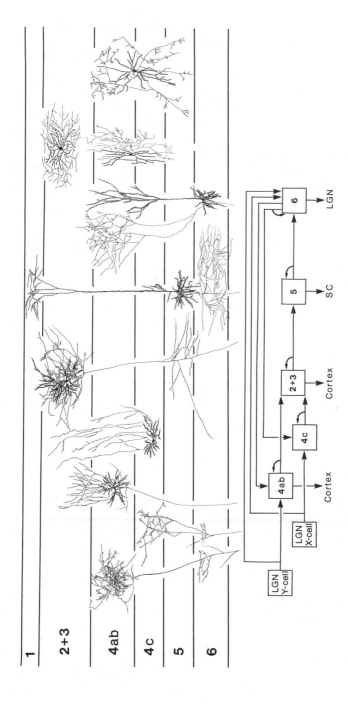

Figure 2 Schematic diagram of the intracortical connections of the cat's striate cortex. The spiny stellate and pyramidal cells are presumably responsible for most of the excitatory connections, and their intracortical and efferent connections are summarized in the block diagram. Smooth stellate cells, several types shown at upper right, are presumed to mediate inhibitory interactions in the cortex.

complex cells (probably those in layer 5) can be activated by textured visual stimuli that are incapable of activating simple cells, although other complex cells (for the most part in the superficial layers) are less responsive to such stimuli (Hammond & MacKay 1977). The finding that superficial layer complex cells have properties that are similar to layer 4 simple cells is consistent with the anatomical observation of a powerful projection from layer 4 to layer 2 + 3. The finding that layer 5 cells can be activated without activating layer 4 is somewhat at variance with the electrical stimulation experiments. There is evidence, however, that layer 4 can be bypassed in activating cells in the superficial layers. One can readily demonstrate monosynaptic activation of layer 3 cells by stimulation in the optic radiation (Ferster & Lindstrom 1982). It has been suggested that this pathway is sufficient to endow layer 3 cells with some of their normal functional properties. This was demonstrated by inactivating the dorsal geniculate layers by injections of cobalt (Malpeli 1981), which had the effect of inactivating layer 4 cells, yet left layer 3 cells with little change in the size of response and with normal orientation specificity. In view of the rich input layer 3 cells receive from layer 4, however, one would expect that eliminating this input should substantially alter their properties. The respective roles of the thalamic and intrinsic cortical inputs for the function of layer 3 cells is therefore an issue that is still unresolved.

Horizontal Connections and Functional Organization

Although the cortex had for many years been viewed as consisting of many independent columnar units, with most of the intercellular interaction occurring across the cortical layers, it has become increasingly evident that there is also a significant amount of horizontal interaction, covering quite large distances.

The original work investigating horizontal interaction in the cortex was done with the use of degeneration techniques (Szentagothai 1973, Fisken et al 1975, Creutzfeldt et al 1975). These studies, done in cat and monkey striate cortex, showed that the degree of horizontal interaction depends on the cortical layer. From a narrow lesion, spanning all of the layers in the monkey's striate cortex, the maximum extent of degeneration is 5–6 mm (Fisken et al 1975). The layer with the shortest ranging degeneration is 4c, where it extends for only 200 μm on either side of the lesion, and the layer with the most is 4b (or the stria of Gennari), where it extends for a total of 5–8 mm. Other layers are intermediate, with horizontal connections extending over a distance of 2–3 mm. At the ultrastructural level, the degenerating terminals form asymmetric synapses, 90% of them with spines.

Contemporary neuroanatomical methods have revealed additional features concerning horizontal cortical connections. Using HRP as an extracellular tracer, it is possible to examine the nature of these connections. Because of its tendency to diffuse over long distances, one cannot use the technique to quantify the extent of the horizontal connections. It does, however, reveal that these connections tend to be patchy in nature, so that the HRP label tends to break up into discrete patches of cells and terminals as one moves away from the injection site (Rockland & Lund 1982). Another method that demonstrates this phenomenon, and one that allows one to investigate it at the single cell level, is that of intracellular injection of HRP. This has shown that the connections formed by individual cells can span considerable distances, sometimes extending up to 6 mm, and that these projections are clustered in nature. If one views the distribution of axon collaterals formed by a single cell in a view parallel to the cortical surface, one often sees a number of discrete clumps of collaterals (Gilbert & Wiesel 1979, 1982).

Widespread connections formed by individual axons had been ascribed to the diffuse system of noradrenergic input from the locus coeruleus, which is thought to serve a "modulatory" function rather than that of being responsible for specific receptive field properties. It was therefore something of a surprise to see the extent to which the specific cortical afferents spread within the cortex. An individual axon coming from the lateral geniculate nucleus can cover more than one ocular dominance column of a given eye, skipping over the columns of the other eye (Ferster & LeVay 1978, Gilbert & Wiesel 1979). Knowing that the receptive fields of the cells that receive these afferents are reasonably well restricted, one wonders whether the horizontal spread of the geniculate afferents goes beyond what one would expect from the size of these cells' fields. This question awaits a quantitative study.

While the functional significance of the clustering is unknown, one might speculate that cells may tend to project to other cells having certain functional properties in common, and that these cells are nonrandomly distributed within the cortex. One reason the collaterals are clustered may be that a cell is connected with cells in distant orientation columns having the same orientation specificity as the injected cells, and that this cell skips over cells with the "wrong" orientation specificity. It has been suggested that the clustering of labeled cells and terminals seen near the focal extracellular HRP injection sites is due to this tendency of cells to connect with other cells of like orientation (Mitchison & Crick 1982). In some instances the shape and extent of the axonal field of a given cell provides some information on the likely role of the connections formed by that cell. For example, cells in layer 6 have very long receptive fields, much longer than those of cells in other layers of the same column. Cells in layer 5 project to layer 6, forming axonal arbors that extend many millimeters. This allows individual layer 6 cells to combine input from layer 5 cells over a wide area of cortex, and therefore may account for the manner in which they construct such long receptive fields (Gilbert & Wiesel 1979). Other long-ranging connections may be used for generating inhibitory flanks around the excitatory portion of a cell's receptive field (Gilbert & Wiesel 1982).

Although dendrites do not extend over nearly so wide an area as axons, their shape may also influence these interactions. The reason for this is that the dendritic fields of cortical cells are frequently elongated, having a preferred orientation (Colonnier 1964, Jones 1975). This would enable a cell to receive more input from axons oriented along the long axis of the dendritic field than from those oriented along the short axis. It would allow for input selection based on geometric relationships between dendrites and axons, rather than on chemical specificity. As yet no one has determined the precise role of dendritic orientation. It is not known whether there is any consistent relationship between dendritic orientation and either the visuotopic or columnar organization of the cortex. The intracellular recording and injection techniques have provided an opportunity to add another dimension to this study, since it will be possible to examine the relationship between dendritic orientation, receptive field structure, and the cortical functional architecture.

Though perhaps not directly relevant to cortical microcircuitry, there is a pattern in the horizontal distribution of the connections between different cortical areas that is reminiscent of the clustering of intrinsic connections within an individual cortical area. This pattern was first observed in examining the distribution of cells that give rise to cortico-cortical connections (Gilbert & Kelly 1975). Using retrograde tracing techniques, that study showed that cells projecting from one cortical area to another, in addition to being predominantly located in the superficial cortical layers, are distributed in numerous discrete clusters. The clustering pattern has been seen in the monkey as well as in the cat (Gilbert & Wiesel 1980, Maunsell et al 1980, Tigges et al 1981). Other studies, which use techniques that trace connections in the anterograde direction, have shown that a particular site in one cortical area gives rise to patches of terminals in other cortical areas (Zeki 1976, Tigges et al 1977, Wong-Riley 1979a, Montero 1980). The patchy pattern of connectivity appears to be universal; it is seen in cortical sensory areas serving all modalities, including visual, somatosensory (Jones et al 1978), and auditory (Imig & Reale 1980), and also in the frontal cortex (Goldman & Nauta 1977). For some areas, callosal projections as well as ipsilateral associational cortical projections form a banded pattern of termination (Imig & Brugge 1978, Jones et al 1978, 1979, Kelly & Wong 1981).

For the most part, the functional significance of this pattern remains a mystery. In the auditory cortex it has been demonstrated that the cortical projection bands are associated with aural interaction columns, such that callosal projection bands are aligned with binaural summation response columns (Imig & Brugge 1978), and ipsilateral projection bands are associated with contralateral dominant suppression responses (Imig & Reale 1981). Although this relationship holds for certain parts of the projection from area A to area A1, it breaks down in other parts. In the visual cortex the functional role of the clusters has thus far eluded discovery. It is difficult to conceive of an analogy between the visual and auditory cortex, because in striate cortex the callosal projections are restricted to the 17/18 border, with no clear clustering, and the ipsilateral projection bands arise from all parts of area 17.

Increasingly the horizontal domain is recognized as being of crucial importance in cortical microcircuitry, and will undoubtedly be the subject of further studies in the near future. From a functional standpoint, the horizontal organization of the cortex has been well studied, though there has recently been a resurgence in interest in this topic due to the discovery of new methods for investigating the columnar architecture of the cortex. At a macroscopic level, it is known that visual cortical areas are mapped retinotopically (Talbot 1940, Talbot & Marshall 1941, Woolsey et al 1955, Daniel & Whitteridge 1961, Hubel & Wiesel 1962, 1965, 1974b, Bilge et al 1967, Allmann & Kaas 1975, Palmer et al 1978, Tusa et al 1978; Van Essen & Zeki 1978, Zeki 1978; see review by Van Essen 1979).

At a finer level, the cortical map is broken up into functional units, extending from the pia to the white matter, which have been called "columns" (Hubel & Wiesel 1962, 1974a, 1977, Hubel et al 1978). The columns consist of groups of cells that share the same property, such as ocular dominance or orientation specificity. When mapped onto the cortical surface, they form a fingerprint-like pattern. The morphological substrate for the ocular dominance columns is well understood. These columns are formed by the segregation of the geniculate afferents serving the two eyes (Hubel & Wiesel 1972, LeVay & Gilbert 1975, Wiesel et al 1974). The morphological substrate for the orientation columns is not at all well understood. Hubel & Wiesel (1962) have suggested that orientation specificity is produced by the simple cells selecting input from geniculate afferents whose receptive fields fall along a straight line. The way in which layer 4 stellate cells could show this degree of specificity in their connection is not known. It may not require a highly programmed chemical specificity, however, but could be obtained by "geometric" factors, such as oriented dendritic fields. An alternative hypothesis is that orientation specificity is produced by mutual inhibition of the oriented cortical cells (see section on inhibition).

A set of columns having all orientation specificities or a set of columns of cells having all ocular preferences is termed a "hypercolumn" (Hubel & Wiesel 1974b). Because there is a linear relationship between magnification factor, receptive field size, and the scatter in receptive field positions, Hubel & Wiesel (1974b) suggested that the cortex could be thought of as a uniform structure, consisting of repeating units performing a stereotyped analysis of different parts of the visual field. It is now apparent that these units are not independent of one another because of the extensive interconnections between them.

Histochemical techniques have revealed further aspects of horizontal cortical organization. Wong-Riley (1979b) examined the distribution of the enzyme, cytochrome oxidase. In the cat this enzyme appears to be distributed uniformly in any given layer, but in the superficial layers of the monkey striate cortex it has a distinct patchy distribution (Humphrey & Hendrickson 1980, Horton & Hubel 1981, Tootell & Silverman 1981). These patches have a fixed relationship with ocular dominance and orientation columns (Horton & Hubel 1981, Tootell & Silverman 1981). The GABA synthetic enzyme, glutamic acid decarboxylase, also has a patchy distribution, which coincides with the distribution of cytochrome oxidase (Hendrickson et al 1981). This suggests a relationship between the intrinsic cortical circuitry and cortical columns. While the purpose of the various sets of patches is not known, they appear to coincide with patches of thalamic input to the superficial cortical layers, and with zones of cortex where cells lack orientation specificity (Hubel & Livingstone 1981). At present these phenomena are undergoing intensive investigation.

Inhibitory Connections

As mentioned above, there is a broad class of cells, the smooth stellate cells, that is associated with cortical inhibition. It has been possible to investigate the role of these connections by pharmacological means, because they constitute one of the few instances in which the transmitter employed by a given set of cortical cells is known. Many of the inhibitory connections in the cortex may be mediated by one transmitter, gamma-amino butyric acid (GABA).

The way in which GABA has come to be associated with smooth stellate cells has been through two techniques: uptake of 3H-GABA applied locally (Hökfelt & Ljundahl 1972, Chronwall & Wolff 1978, Hamos et al 1981, Hendry & Jones 1981, Somogyi et al 1981a,b), and immunocytochemical techniques, employing antibodies to the GABA synthetic enzyme, glutamic acid decarboxylase (Ribak 1978, Hendrickson et al 1981). The result of these experiments is that GABA uptake and GAD immunoreactivity have been associated with smooth stellate cells. Attempts are now being made to relate specific subclasses of smooth stellate cells to GABA uptake. One approach is to use the Golgi technique in conjunction with autoradiography (Somogyi et al 1981b). This technique has not yet led to a determination of which smooth stellate subclasses are GABAergic, but indirect evidence has led to speculation that double bouquet (Somogyi et al 1981a) and basket cells (Hendry & Jones 1981) are among the cells responsible for GABA uptake.

Whatever specific cell classes are responsible for GABA transmission, one can ascertain their functional role in the visual cortex by investigating how a specific GABA antagonist, bicuculline, influences the functional properties of cortical cells. Cells are recorded from before, during, and after a period of bicuculline iontophoresis (Sillito 1975), and their receptive field properties during iontophoresis are compared with those before and after. These studies have shown that properties such as orientation specificity and directionality can be reduced or eliminated during GABA blockade (Rose & Blakemore 1974, Sillito 1977a, 1979, Sillito et al 1980, Tsumoto et al 1979). Other properties, such as end-inhibition, remain relatively unaffected (Sillito 1977b). This leads to a model for orientation specificity involving inhibition between cells in orientation columns having opposing orientation preferences (Sillito 1979). The idea that neighboring columns interact with one another was originally suggested by Blakemore & Tobin (1972), who studied interaction between stimuli using psychophysical tests. Although the experiments with GABA antagonists offer an intriguing model for the generation of specific receptive field properties, it is not sufficiently complete to accept as fact. It is not clear whether the necessary connections exist for the sort of inhibitory interactions that are postulated in this model. Also, if orientation specificity is mediated by inhibition, it is necessary to demonstrate how it is possible to generate oriented cells from unoriented inputs; some form of "bootstrapping" operation must take place. In addition to the pharmacological experiments, one can demonstrate with multiple visual stimuli that certain fundamental properties of cortical cells, such as the side flanks of simple cells, are generated by inhibitory interactions within the cortex (Heggelund 1981).

Other Transmitters

The studies on the role of GABA in cortical function show the importance of transmitter chemistry and pharmacology in drawing the relationship between the cortical circuitry and the function of cortical cells. In this review I make no attempt to cover the present state of knowledge concerning cortical transmitters, except to mention the general classes that are known to exist and where they may be localized in the cortex. For reviews on this subject, see Emson & Lindvall (1979) and Emson & Hunt (1981).

In addition to GABA, acetylcholine (ACh) and norepinephrine are well established as transmitters. These are both probably associated with diffuse pathways: norepinephrine with input from the locus coeruleus, and ACh from basal forebrain nuclei. Presently, there is some controversy as to whether ACh may be associated with a particular population of cortical cells. The functional role of ACh and norepinephrine in the adult cortex has not been established. One set of studies indicates that norepinephrine may be involved in the process of cortical development (Kasamatsu & Pettigrew 1976, 1979, Kasamatsu et al 1979).

For the most part, however, the transmitters associated with the specific thalamic afferents or with the cortical cells themselves are unknown. One class of putative transmitters that have become associated with a particular class of cortical cells are the acidic amino acids, aspartate and glutamate. These have been shown to be released from the cortex in a calcium-dependent manner, and localization studies have associated them with pyramidal cells in layer 6 (Baughman & Gilbert 1981). For the corticostriatal pathway in rat layer 5, pyramids may be involved as well (Streit 1980). Other classes of pyramidal and spiny stellate cells have not been associated with particular transmitters as yet.

Another class of transmitters is the peptides, such as substance P, neurotensin, somatostatin, enkephalin, vasoactive intestinal polypeptide (VIP; Fuxe et al 1977), and so on. Individual transmitters tend to be distributed nonuniformly throughout the cortex, and are found more commonly in some cortical areas than in others. Immunohistochemical studies associate them with smooth stellate cells. VIP, for example, is thought to be associated with bipolar cells (see above).

Most of the work done presently is involved with transmitter identification and localization. It is necessary for future work also to develop pharmacological tools for each class of neurotransmitters to determine the functional role of the transmitters and to aid in identifying the cells in which they reside.

Other Forms of Interaction between Cortical Neurons

Much of the discussion to this point has been presented in classic Sherringtonian terms, in which processing at a cellular level consists of a straightforward addition of excitatory and inhibitory inputs; however, there are other ways in which cortical cells may interact. While there is presently little or no experimental evidence to show how these nonstandard interactions may contribute to cortical function, they should be taken into account in constructing models of cortical processing.

One of the most prominent features of cortical cell morphology is the apical dendrite of pyramidal cells, a remarkable specialization whose function is unknown. Little has been done so far to ascertain whether different inputs are segregated onto different sets of dendrites of pyramidal cells. However, as mentioned above, many pyramids receive thalamic input only on their apical dendrites. In addition to providing a functional architecture on a subcellular scale, apical dendrites may modulate the firing properties of pyramidal cells. It is known that pyramidal cells are the predominant class of output neurons of the cortex (though spiny stellate cells in layer 4 also participate in cortico-cortical connections). It is also known that apical dendrites tend to bundle together (Peters & Walsh 1972, Fleischauer et al 1972, Fleischauer 1974: see Roney et al 1979 for review). Apical dendritic bundles consist of 6 to 30 dendrites in close proximity (sometimes separated only by extracellular space) at intervals of 50 to 100 μm. The bundles have different relationships in different parts of the cortex. In visual cortex apical dendrites are relatively unbranched, and apical dendrites of cells in the superficial layers join those of cells in layer 5; however, in somatosensory cortex the dendrites may branch and join in an adjacent bundle (Fleischauer 1974). It has been suggested that these interactions serve to synchronize the activity of all pyramidal cells in a given cortical column (Fleischauer et al 1972).

Whereas it has been traditional to think of synapses as being one of two types, symmetric or asymmetric, excitatory or inhibitory, there is evidence suggesting a greater variety. Though not present in any great number, gap junctions have been found in some areas of cortex, suggesting the possibility for electrical synapses. Sloper & Powell (1978) have described infrequently occurring dendrosomatic and dendrodendritic gap junctions between spiny stellate cells of layers 4 and 5 in sensory-motor cortex. Experiments in tissue slices have also suggested the possibility of electrotonic coupling between superficial layer pyramidal cells (Gutnick & Prince 1981), but this may have been an artifact of the preparation. Within the category of chemical transmission, there is room for further differentiation of synaptic function than simply excitation or inhibition. Taking the sympathetic ganglion as a model, one must also be aware of the possibility that excitatory and inhibitory interactions can be of short, medium, and long duration (Hartzell 1981). All these factors allow for a complexity of interaction in the cortical circuit that make a functional understanding of the cortical microcircuitry a rather distant objective.

Two of the most obvious features of cortical cells are the presence of numerous branched dendrites and of dendritic spines. A simplistic explanation for this morphology is the need for neurons to have a large surface area to receive a large number of synaptic inputs. The most telling argument against this is that cell somata and dendrites receive input on a relatively small proportion of their surface. Thus, one must develop other hypotheses. While there is very little experimental evidence available on the details of synaptic interactions on an individual neuron, mathematical models have provided some insight into this issue. Because the primary effect of a given input is a conductance change rather than an injection of a given amount of current, nearby inputs sum nonlinearly. Dendritic branching provides a way of bringing about linear summation, if the two inputs are on different positions of the dendritic tree (Rall 1964). Dendritic branching also allows for electrical isolation of patches of membrane, so that not all interactions are governed by the dendrites' length constant, and dendrites are consequently allowed to act independently of one another. The relative location of excitatory and inhibitory inputs greatly affects the effectiveness of the inhibitory input. The best arrangement is to have the inhibitory input proximal to the excitatory input, on a direct line between the excitatory input and soma (Rall 1964, Poggio 1982). Dendritic spines insulate excitatory inputs against the shunting effect of inhibitory inputs. The proper placement of excitatory and inhibitory inputs, and the use of timing differences in synaptic inputs, can theoretically account for some of the functional properties of neurons, such as directional sensitivity of visual cells (Barlow & Levick 1965, Poggio et al 1982). Finally, nonlinearities in the membrane properties of cortical neurons, such as voltage-sensitive conductances and rectification, can allow these neurons to perform more complex operations on their inputs. It might, for example, be desirable to have neurons acting as "and gates," firing only in the presence of two different inputs; this cannot be accomplished by a simple linear summation of inputs.

Summary

In the last ten years there has been an extraordinary amount of progress in our understanding of the microcircuitry of the visual cortex, owing in great part to the development of a number of anatomical, physiological, and chemical techniques. Many studies have furthered our understanding of the functions of the cortical layers, the connections existing between identified cell classes, interlaminar and horizontal interactions, the cortical inputs, and the source and function of intracortical inhibition and the cortical columns. Even with these advances, investigators have only begun to scratch the surface. Research in this area continues to be very active, however, and in the next decade there will undoubtedly be advances equivalent in scope to those of the last one.

ACKNOWLEDGMENTS

I would like to thank Dr. Torsten Wiesel, who has provided invaluable suggestions and criticisms throughout the preparation of the manuscript. I also thank Drs. Alan Peters and Jean-Pierre Hornung for their expert advice.

240   GILBERT

Literature Cited

Albus, K., Donate-Oliver, F. 1977. Cells of origin of the occipitopontine projection in the cat: Functional properties and intracortical location. *Exp. Brain Res.* 28:167–74

Allman, J. M., Kaas, J. H. 1975. The dorsomedial cortical visual area: A third tier area in the occipital lobe of the owl monkey (*Aotus Trvirgatus.*) *Brain Res.* 100:473–87

Barlow, H. B., Levick, W. R. 1965. The mechanism of directionally selective units in rabbit's retina. *J. Physiol.* 178:477–504

Baughman, R. W., Gilbert, C. D. 1981. Aspartate and glutamate as possible neurotransmitters in the visual cortex. *J. Neurosci.* 1:427–39

Bilge, M., Bingle, A., Seneviratne, K. N., Whitteridge, D. 1967. The primary visual receptive area of the cerebral cortex of the cat. *J. Physiol.* 17:39–58

Blakemore, C., Tobin, E. 1972. Lateral inhibition between orientation detectors in the cat's visual cortex. *Exp. Brain Res.* 15:439–40

Brodal, P. 1978. The corticopontine projections in the rhesus monkey. Origin and principles of organization. *Brain* 101:251–83

Brodmann, K. 1909. *Vergleichende Lokalisationslehte der Groshirnrinde in ihren Prinzipien dargestelit auf Grund des Zellenhanes.* Leipzig: Barth

Bullier, J., Henry, G. H. 1979a. Ordinal position of neurons in cat striate cortex. *J. Neurophysiol.* 42:1251–63

Bullier, J., Henry, G. H. 1979b. Neural path taken by afferent streams in striate cortex of the cat. *J. Neurophysiol.* 42:1264–70

Bullier, J., Henry, G.H. 1979c. Laminar distribution of first order neurons and afferent terminals in cat striate cortex. *J. Neurophysiol.* 42:1271–81

Cajal, S. R. 1911. *Histologie du Systeme Nerveux de l'Homme et des Vertebres.* Madrid: CSIC, 1972 ed.

Cajal, S. R. 1922. Studien uber die Sehrinde der Katze *J. Psychol. Neurol.* 29:161–81

Camarda, R., Rizzolatti, G. 1976. Receptive fields of cells in the superficial layers of the cat's area 17. *Exp. Brain Res.* 24:423–27

Campbell, A. W. 1905. *Histological Studies on the Localisation of Cerebral Function.* Cambridge: Univ. Press

Carey, R. G., Bear, M. F., Diamond, I. T. 1980. The laminar organization of the reciprocal projections between the claustrum and striate cortex in the tree shrew, *Tupaia glis. Brain Res.* 184:193–98

Caviness, V. S. Jr. 1977. Reeler mutant mouse: A genetic experiment in developing mammalian cortex. *Neurosci. Symp.* 2:27–46

Chronwall, B., Wolff, J. B. 1978. Classification and location of neurons taking up [3H]-GABA in the visual cortex of rats. In *Amino Acids as Chemical Transmitters*, ed. F. Fonnum, pp. 297–303. New York: Plenum

Cleland, B. G., Dubin, M. W., Levick, W. R. 1971. Sustained and transient neurons in the cat's retinae and lateral geniculate nucleus. *J. Physiol.* 267:473–96

Cleland, B. G., Levick, W. R. 1974. Brisk and sluggish concentrically organized ganglion cells in the cat's retina. *J. Physiol.* 240:421–56

Colonnier, M. 1964. The tangential organization of the visual cortex. *J. Anat.* 98:327–44

Colonnier, M. 1966. The structural design of the neocortex. In *Brain and Conscious Experience*, ed J. C. Eccles, pp. 1–23. New York: Springer Verlag

Colonnier, M. 1968. Synaptic patterns on different cell types in the different laminae of the cat visual cortex. An electron microscopic study. *Brain Res.* 9:268–87

Colonnier, M. 1981. The electron-microscopic analysis of the neuronal organization of the cerebral cortex. In *The Organization of the Cerebral Cortex*, ed. F. O. Schmitt, F. G. Worden, G. Adelman, S. G. Dennis, pp. 125–52. Cambridge: MIT Press

Colonnier, M., Rossignol, S. 1969. Heterogeneity of the cerebral cortex. In *Basic Mechanisms of the Epilepsies*, ed. H. Jasper, A. Ward, A. Pope, pp. 29–40. Boston: Little Brown

Creutzfeldt, O. D., Garey, L. J., Kuroda, R., Wolff, J. -R. 1975. The distribution of degenerating axons after small lesions in the intact and isolated visual cortex of the cat. *Exp. Brain Res.* 27:419–40

Daniel, P.M., Whitteridge, D. 1961. The representation of the visual field on the cerebral cortex in monkeys. *J. Physiol.* 159:203–21

Davis, T. L., Sterling, P. 1979. Microcircuitry of cat visual cortex: Classification of neurons in layer 4 of area 17, and identification of the patterns of lateral geniculate input. *J. Comp. Neurol.* 188:599–628

Dickson, J. W., Gerstein, G. L. 1974. Interactions between neurons in the auditory cortex of cat. *J. Neurophysiol.* 37: 1239–61

Drager, U. C. 1981. Observations on the organization of the visual cortex in the reeler mouse. *J. Comp. Neurol.* 201: 555–70

Emson, P. C., Hunt, M. C. 1981. Anatomical chemistry of the cerebral cortex. See Colonnier 1981, pp.325–45

Emson, P. C., Lindvall, O. 1979. Distribution of putative neurotransmitters in the neocortex. *Neuroscience* 4:1–30

Enroth-Cugell, C., Robson, J. G. 1966. The contrast sensitivity of retinal ganglion cells in the cat. *J. Physiol.* 187:517–52

Fairen, A., Peters, A., Saldanha, J. 1977. A new procedure for examining Golgi impregnated neurons by light and electron microscopy. *J. Neurocytol.* 6:311–37

Fairen, A., Valverde, F. 1980. A type of neuron in the visual cortex of cat. A Golgi and electron microscope study of chandelier cells. *J. Comp. Neurol.* 194: 761–79

Feldman, M. L., Peters, A. 1978. The forms of non-pyramidal neurons in the visual cortex of the rat. *J. Comp. Neurol.* 179:761–94

Ferster, D., LeVay, S. 1978. The axonal arborization of lateral geniculate neurons in the striate cortex of the cat. *J. Comp. Neurol.* 182:923–44

Ferster, D., Lindstrom, S. 1982. An intracellular analysis of geniculocortical and intracortical connectivity in area 17 of the cat. *J. Physiol.* In press

Fink, R. P., Heimer, L. 1967. Two methods for selective silver impregnation of degenerating axons and their synaptic endings in the central nervous system. *Brain Res.* 4:369–74

Fisken, R. A., Garey, L. J., Powell, T. P. S. 1975. The intrinsic association and commissural connections of area 17 of the visual cortex. *Philos. Trans. R. Soc. London Ser. B* 272:487–536

Fleischhauer, K. 1974. On different patterns of dendritic bundling in the cerebral cortex of the cat. *Z. Anat. Entwicklungsgesch.* 143:115–26

Fleischhauer, K., Petsche, H., Wittkowski, W. 1972. Vertical bundles of dendrites in the neocortex. *Z. Anat. Entwicklungsgesch.* 136:213–23

Freedman, R., Foote, S. L., Bloom, F. E. 1975. Histochemical characterization of a neocortical projection of the nucleus locus coeruleus in the squirrel monkey. *J. Comp. Neurol.* 164:209–32

Friedlander, M. J., Lin, C. -S., Stanford, L. R., Sherman, S. M. 1981. Morphology of functionally identified neurons in lateral geniculate nucleus of cat. *J. Neurophysiol.* 46:80–129

Fuxe, K. 1965. Evidence for the existence of monoamine terminals in the central nervous system. IV. Distribution of monoamine terminals in the central nervous system. *Acta Physiol. Scand.* 64 (Suppl. 247):37–85

Fuxe, K., Hökfelt, T., Said, S. I., Mutt, V. 1977. Vasoactive intestinal polypeptide and the nervous system: Immunohistochemical evidence for localization in central and peripheral neurons, particularly intracortical neurons of the cerebral cortex. *Neurosci. Lett.* 5:241–46

Garey, L. J. 1971. A light and electron microscopic study of the visual cortex of the cat and monkey. *Proc. R. Soc. London Ser. B* 179:21–40

Garey, L. J., Jones, E. G., Powell, T. P. S. 1968. Interrelationships of striate and extrastriate cortex with the primary relay sites of the visual pathway. *J. Neurol. Neurosurg. Psychiatr.* 31:135–57

Garey, L. H., Powell, T. P. S. 1971. An experimental study of the termination of the lateral geniculo-cortical pathway in the cat and monkey. *Proc. R. Soc. London Ser. B* 179:1–63

Gibson, A., Baker, J., Mower, G., Glickstein, M. 1978. Corticopontine cells in area 18 of the cat. *J. Neurophysiol.* 41:484–95

Gilbert, C. D. 1977. Laminar differences in receptive field properties in cat primary visual cortex. *J. Physiol.* 268:391–421

Gilbert, C. D., Kelly, J. P. 1975. The projections of cells in different layers of the cat's visual cortex. *J. Comp. Neurol.* 163:81–106

Gilbert, C. D., Wiesel, T. N. 1979. Morphology and intracortical projections of functionally identified neurons in cat visual cortex. *Nature* 280:120–25

Gilbert, C. D., Wiesel, T. N. 1980. Interleaving projection bands in cortico-cortical connections. *Neurosci. Abstr.* 6:315

Gilbert, C. D., Wiesel, T. N. 1981a. Laminar specialization and intracortical connections in cat primary visual cortex. See Colonnier 1981, pp. 163–94

Gilbert, C. D., Wiesel, T. N. 1981b. Projection bands in visual cortex. *Neurosci. Abstr.* 7:356

Gilbert, C. D., Wiesel, T. N. 1982. Projection bands in visual cortex. *J. Neurosci.* In press

Glickstein, M., Cohen, J. L., Dixon, B., Gibson, A., Hollins, M., Labossiere, E., Robinson, F. 1980. Corticopontine visual projections in Macaque monkeys. *J. Comp. Neurol.* 190:209–29

Globus, A., Scheibel, A. B. 1967. Pattern and field in cortical structure: The rabbit. *J. Comp. Neurol.* 131:155–72

Goldman, P. S., Nauta, W. J. H. 1977. Columnar distribution of corticocortical fibers in the frontal association, limbic and motor cortex of the developing rhesus monkey. *Brain Res.* 122:393–413

Guillery, R. W. 1966. A study of Golgi preparations from the dorsal lateral geniculate nucleus of the adult cat. *J. Comp. Neurol.* 128:21–50

Gutnick, M. J., Prince, D. A. 1981. Dye coupling and possible electrotonic coupling in the guinea pig neocortical slice. *Science* 211:67–70

Hammond, P., MacKay, D. M. 1977. Differential responsiveness of simple and complex cells in cat striate cortex to visual texture. *Exp. Brain Res.* 30:275–96

Hamos, J. E., Davis, T. L., Sterling, P. 1981. Several groups of neurons in layer IVab of cat area 17 accumulate 3H-gamma-aminobutyric acid (GABA). *Neurosci. Abstr.* 7:173

Hartzell, H. C. 1981. Mechanisms of slow postsynaptic potentials. *Nature* 291:539–44

Heggelund, P. 1981. Receptive field organization of simple cells in cat striate cortex. *Exp. Brain Res.* 42:89–98

Hendrickson, A. E., Hunt, S. P., Wu, J.-Y. 1981. Immunocytochemical localization of glutamic acid decarboxylase in monkey striate cortex. *Nature* 292:605–7

Hendrickson, A. E., Wilson, J. R., Ogren, M. P. 1978. The neuroanatomical organization of pathways between the dorsal lateral geniculate nucleus and visual cortex in old world and new world primates. *J. Comp. Neurol.* 182:123–36

Hendry, S. H. C., Jones, E. G. 1981. Sizes and distributions of intrinsic neurons incorporating tritiated GABA in monkey sensory-motor cortex. *J. Neurosci.* 1:390–408

Henry, G. H., Harvey, A. R., Lund, J. S. 1979. The afferent connections and laminar distribution of cells in the cat striate cortex. *J. Comp. Neurol.* 187:725–44

Henry, G. H., Lund, J. S., Harvey, A. R. 1978. Cells of the striate cortex projecting to the Clare-Bishop area of the cat. *Brain Res.* 151:154–58

Hoffmann, K.-P., Stone, J. 1971. Conduction velocity of afferents to cat visual cortex: A correlation with cortical receptive field properties. *Brain Res.* 32:460–66

Hökfelt, T., Ljundahl, A. 1972. Autoradiographic identification of cerebral and cerebellar cortical neurons accumulating labeled gamma-aminobutyric acid (3H-GABA). *Exp. Brain Res.* 14:354–62

Hornung, J. P., Garey, L. J. 1981a. The thalamic projection to cat visual cortex: Ultrastructure of neurons identified by Golgi impregnation or retrograde horseradish peroxidase transport. *Neuroscience* 6:1053–68

Hornung, J. P., Garey, L. J. 1981b. Ultrastructure of visual callosal neurons in cat identified by retrograde axonal transport of horseradish peroxidase. *J. Neurocytol.* 10:297–314

Horton, J. C., Hubel, D. H. 1981. Regular patchy distribution of cytochrome oxidase staining in primary visual cortex of macaque monkeys. *Nature* 292:762–64

Hubel, D. H., Livingstone, M. S. 1981. Regions of poor orientation tuning coincide with patches of cytochrome oxidase staining in monkey striate cortex. *Neurosci. Abstr.* 7:357

Hubel, D. H., Wiesel, T. N. 1961. Integrative action in the cat's lateral geniculate body. *J. Physiol.* 155:385–98

Hubel, D. H., Wiesel, T. N. 1962. Receptive fields, binocular interaction and functional architecture in the cat's visual cortex. *J. Physiol.* 160:106–54

Hubel, D. H., Wiesel, T. N. 1965. Receptive fields and functional architecture in two non-striate visual areas (18 and 19) of the cat. *J. Neurophysiol.* 28:229–89

Hubel, D. H., Wiesel, T. N. 1972. Laminar and columnar distribution of geniculocortical fibers in the macaque monkey. *J. Comp. Neurol.* 146:421–50

Hubel, D. H., Wiesel, T. N. 1974a. Sequence regularity and geometry of orientation columns in the monkey striate cortex. *J. Comp. Neurol.* 158:267–94

Hubel, D. H., Wiesel, T. N. 1974b. Uniformity of monkey striate cortex: A parallel relationship between field size, scatter, and magnification factor. *J. Comp. Neurol.* 158:295–306

Hubel, D. H., Wiesel, T. N. 1977. Functional architecture of monkey visual cortex. *Proc. R. Soc. London Ser. B* 198:1–59

Hubel, D. H., Wiesel, T. N., Stryker, M. S. 1978. Anatomical demonstration of orientation columns in macaque monkey. *J. Comp. Neurol.* 177:361–80

Humphrey, A. L., Hendrickson, A. E. 1980. Radial zones of high metabolic activity in monkey striate cortex. *Neurosci. Abstr.* 6:315

Imig, T. J., Brugge, J. F. 1978. Sources and terminations of callosal axons related to binaural and frequency maps in primary auditory cortex of the cat. *J. Comp. Neurol.* 182:637–60

Imig, T. J., Reale, R. A. 1980. Patterns of cortico-cortical connections related to tonotopic maps in cat auditory cortex. *J. Comp. Neurol.* 192:293–32

Imig, T. J., Reale, R. A. 1981. Ipsilateral corticocortical projections related to binaural columns in cat primary auditory cortex. *J. Comp. Neurol.* 203:1–14

Jones, E. G. 1969. An electron microscopic study of the terminations of afferent fibre systems within the somatic sensory cortex of the cat. *J. Anat.* 103:595–97

Jones, E. G. 1975. Varieties and distribution of non-pyramidal cells in the somatic sensory cortex of the squirrel monkey. *J. Comp. Neurol.* 160:205–68

Jones, E. G., Coulter, J. D., Hendry, S. H. C. 1978. Intracortical connectivity of architectonic fields in the somatic sensory, motor and parietal cortex of monkeys. *J. Comp. Neurol.* 181:291–348

Jones, E. G., Coulter, J. D., Wise, S. P. 1979. Commissural columns in the sensory-motor cortex of monkeys. *J. Comp. Neurol.* 188:113–36

Jones, E. G., Powell, T. P. S. 1969. Connexions of the somatic sensory cortex of the rhesus monkey. I. Ipsilateral cortical connexions. *Brain* 92:477–502

Jones, E. G., Powell, T. P. S. 1970. Electron microscopy of the somatic sensory cortex of the cat. I. Cell types and synaptic organization. *Philos. Trans. R. Soc. London Ser. B* 257:1–11

Jones, E. G., Wise, S. P. 1977. Size, laminar and columnar distribution of efferent cells in the sensory-motor cortex of monkeys. *J. Comp. Neurol.* 175:391–438

Kasamatsu, T., Pettigrew, J. D. 1976. Depletion of brain catecholamines: failure of ocular dominance shift after monocular occlusion in kittens. *Science* 194:206–9

Kasamatsu, T., Pettigrew, J. D. 1979. Preservation of binocularity after monocular deprivation in the striate cortex of kittens treated with 6-hydroxydopamine. *J. Comp. Neurol.* 185:139–62

Kasamatsu, T., Pettigrew, J. D., Ary, M. 1979. Restoration of visual cortical plasticity by local microperfusion of norepinephrine. *J. Comp. Neurol.* 185:163–82

Kawamura, K., Chiba, M. 1979. Cortical neurons projecting to the pontine nuclei in the cat. An experimental study with the horseradish peroxidase technique. *Exp. Brain Res.* 35:269–85

Kawamura, K., Konno, T. 1979. Various types of corticotectal neurons of cats as demonstrated by means of retrograde axonal transport of horseradish peroxidase. *Exp. Brain Res.* 35:161–75

Kawamura, K., Naito, J. 1980. Corticocortical neurons projecting to the medial and lateral banks of the middle suprasylvian sulcus in the cat: An experimental study with the horseradish peroxidase method. *J. Comp. Neurol.* 193:1009–22

Kelly, J. P., Van Essen, D. C. 1974. Cell structure and function in the visual cortex of the cat. *J. Physiol.* 328:515–47

Kelly, J. P., Wong, D. 1981. Laminar connections of the cat's auditory cortex. *Brain Res.* 212:1–15

LeVay, S. 1973. Synaptic patterns in the visual cortex of the cat and monkey. Electron microscopy of Golgi preparations. *J. Comp. Neurol.* 150:53–86

LeVay, S., Gilbert, C. D. 1976. Laminar patterns of geniculocortical projection in the cat. *Brain Res.* 113:1–19

LeVay, S., Sherk, H. 1981. The visual claustrum of the cat. I. Structure and connections. *J. Neurosci.* 1:956–80

Leventhal, A. G. 1979. Evidence that the different classes of relay cells of the cat's lateral geniculate nucleus terminate in different layers of the striate cortex. *Exp. Brain Res.* 37:349–72

Leventhal, A. G., Hirsch, H. V. B. 1978. Receptive field properties of neurons in different laminae of visual cortex of the cat. *J. Neurophysiol.* 41:948–62

Lin, C.-S., Friedlander, M. J., Sherman, S. M. 1979. Morphology of physiologically identified neurons in the visual cortex of the cat. *Brain Res.* 172:344–48

Lorente de Nó, R. 1922. La Corteza cerebral del raton. *Trab. Lab. Invest. Biol.* (*Univ. Madrid*) 20:41–78

Lorente de Nó, R. 1933. Studies on the structure of the cerebral cortex. *J. Psychol. Neurol.* 45:382–438

Lorente de Nó, R. 1944. Cerebral cortex: Architecture, intracortical connections, motor projections. In *Physiology of the Nervous System,* ed. J. F. Fulton, pp. 291–325. London: Oxford Univ. Press

Lund, J. S. 1973. Organization of neurons in the visual cortex, area 17, of the monkey (*Macaca mulatta*). *J. Comp. Neurol.* 147:455–96

Lund, J. S. 1981. Intrinsic organization of the primate visual cortex, area 17, as seen in Golgi preparations. See Colonnier 1981, pp. 325–45

Lund, J. S., Boothe, R. G. 1975. Interlaminar connections and pyramidal neuron organization in the visual cortex, area 17, of the Macaque monkey. *J. Comp. Neurol.* 159:305–34

Lund, J. S., Hendrickson, A. E., Ogren, M. P., Tobin, K. A. 1981. Anatomical organization of primate visual cortex area VII. *J. Comp. Neurol.* 202:19–45

Lund, J. S., Henry, G. H., Macqueen, C. L., Harvey, A. R. 1979. Anatomical organization of the primary visual cortex (area 17) of the cat. A comparison with area 17 of the macaque monkey. *J. Comp. Neurol.* 184:599–618

Lund, J. S., Lund, R. D., Hendrickson, A. E., Bunt, A. H., Fuchs, A. F. 1975. The origin of efferent pathways from the primary visual cortex, area 17, of the macaque monkey as shown by retrograde transport of horseradish peroxidase. *J. Comp. Neurol.* 164:287–304

Maciewicz, R. J. 1974. Afferents to the lateral suprasylvian gyrus of the cat traced with horseradish peroxidase. *Brain Res.* 78:139–43

Magalhaes-Castro, H. H., Saraiva, P. E. S., Magalhaes-Castro, B. 1975. Identification of corticotectal cells of the visual cortex of the cat by means of horseradish peroxidase. *Brain Res.* 83:474–79

Malpeli, J. 1981. Effects of blocking A-layer geniculate input on cat area 17. *Soc. Neurosci. Abstr.* 7:355

Marin-Padilla, M. 1969. Origin of the pericellular baskets of the pyramidal cells of the human motor cortex: A Golgi study. *Brain Res.* 14:633–46

Marin-Padilla, M. 1970. Prenatal and postnatal ontogenesis of the human motor cortex II. The basket-pyramidal system. *Brain Res.* 23:185–91

Marin-Padilla, M. 1974. Three-dimensional reconstruction of the pericellular nests (baskets of the motor (area 4) and visual (area 17), areas of the human cerebral cortex. A Golgi Study. *Z. Anat. Entwicklungsgesch.* 144:123–35

Maunsell, J. H. R., Newsome, W. T., Van Essen, D. C. 1980. The spatial organization of connections between V1 and V2 in the macaque: Patchy and non-patchy projections. *Soc. Neurosci. Abstr.* 6:580

Miller, J. W., Buschmann, M. B. T., Benevento, L. A. 1980. Extrageniculate thalamic projections to the primary visual cortex. *Brain Res.* 189:221–27

Mitchison, G., Crick, F. 1982. Long axons within the striate cortex: Their distribution, orientation, and patterns of connection. *Proc. Natl. Acad. Sci. USA* 79:3661–5

Mitzdorf, U., Singer, W. 1978. Prominent excitatory pathways in the cat visual cortex (A 17 and A 18): A current source density analysis of electrically evoked potentials. *Exp. Brain Res.* 33:371–94

Mitzdorf, U., Singer, W. 1979. Excitatory synaptic ensemble properties in the visual cortex of the Macaque monkey: A current source density analysis of electrically evoked potentials. *J. Comp. Neurol.* 187:71–84

Montero, V. M. 1980. Patterns of connections from the striate cortex to cortical visual areas in superior temporal sulcus of macaque and middle temporal gyrus of owl monkey. *J. Comp. Neurol.* 189:45–59

Morrison, J. H., Foote, S. L., Molliver, M. E., Lidov, H. G. W. 1982. Noradrenergic and serotonergic fibers innervate complementary layers in monkey primary visual cortex: An immunohistochemical study. *Proc. Natl. Acad. Sci. USA* 79:2401–5

Morrison, J. H., Grzanna, R., Molliver, M. E., Coyle, J. T. 1978. The distribution and orientation of noradrenergic fibers in neocortex of the rat: An immunofluorescence study. *J. Comp. Neurol.* 181:17–40

Nauta, H. J. W., Butler, A. B., Jane, J. A. 1973. Some observations on axonal degeneration resulting from superficial lesions of the cerebral cortex. *J. Comp. Neurol.* 150:349–60

Nauta, W. J. H., Gygax, P. A. 1954. Silver impregnation of degenerating axons in the central nervous system. *Stain Technol.* 29:91–3

Ogren, M. P., Hendrickson, A. E. 1977. The distribution of pulvinar terminals in visual areas 17 and 18 of the monkey. *Brain Res.* 137:343–50

O'Leary, J. L. 1941. Structure of the area striata of the cat. *J. Comp. Neurol.* 75:131–61

Olson, C. R., Graybiel, A. M. 1980. Sensory maps in the claustrum of the cat. *Nature* 288:479–81

Otsuka, R., Hassler, H. 1962. Uber aufbau und Gliederung der corticalen Sehsphare bei der Katze. *Arch. Psych. Nervenkr.* 203:212–34

Palmer, L. A., Rosenquist, A. C. 1974. Visual receptive fields of single striate cortical units projecting to the superior colliculus in the cat. *Brain Res.* 67:27–42

Palmer, L. A., Rosenquist, A. C., Tusa, R. J. 1978. The retinotopic organization of lateral suprasylvian visual areas in the cat. *J. Comp. Neurol.* 177:237–56

Parnavelas, J. G., Lieberman, A. R., Webster, K. E. 1977. Organization of neurons in the visual cortex, area 17, of the rat. *J. Anat.* 124:305–22

Perkel, D. H., Gerstein, G. L., Moore, G. P. 1967. Neuronal spike trains and stochastic point processes. II. Simultaneous spike trains. *Biophys. J.* 7:419–40

Peters, A., Feldman, M. L. 1976. The projection of the lateral geniculate nucleus to area 17 of rat cerebral cortex. IV. Terminations upon spiny dendrites. *J. Neurocytol.* 6:669–89

Peters, A., Fairen, A. 1978. Smooth and sparsely-spined stellate cells in the visual cortex of the rat: A study using a combined Golgi-electron microscope technique. *J. Comp. Neurol.* 181: 129–72

Peters, A., Kaiserman-Abramof, J. R. 1970. The small pyramidal neuron of the rat cerebral cortex. The perikaryon, dendrites and spines. *Am. J. Anat.* 127: 321–56

Peters, A., Kimerer, L. 1981. Bipolar neurons in rat visual cortex: A combined Golgi-electron microscopic study. *J. Neurocytol.* 10:921–46

Peters, A., Proskauer, C. C. 1980. Synaptic relationships between a multipolar stellate cell and a pyramidal neuron in the rat visual cortex. A combined Golgi-electron microscope study. *J. Neurocytol.* 9:163–83

Peters, A., Proskauer, C. C., Feldman, M. L., Kimerer, L. 1979. The projection of the lateral geniculate nucleus to area 17 of the rat cerebral cortex. V. Degenerating axon terminals synapsing with Golgi impregnated neurons. *J. Neurocytol.* 8:331–57

Peters, A., Proskauer, C. C., Ribak, C. E. 1982. Chandelier cells in rat visual cortex. *J. Comp. Neurol.* 206:397–416

Peters, A., Regidor, J. 1981. A reassessment of the forms of nonpyramidal neurons in area 17 of cat visual cortex. *J. Comp. Neurol.* 203:685–716

Peters, A., Walsh, T. M. 1972. A study of the organization of apical dendrites in the somatic sensory cortex of the rat. *J. Comp. Neurol.* 144:253–68

Phillis, J. W., Kirkpatrick, J. R., Said, S. I. 1978. Vasoactive intestinal polypeptide excitation of central neurons. *Can. J. Physiol. Pharmacol.* 57:337–40

Poggio, T., Koch, C., Torre, V. 1982. Microelectronics in nerve cells: Dendritic morphology and information processing. *Philos. Trans. R. Soc. London Ser. B* In press

Poljakov, G. L. 1953. On the fine structural characteristics of the human cerebral cortex and on interneuronal functional interaction. *Arh. Anat. Gistol. Embriol.* 30:48–60

Polyak, S. 1957. *The Vertebrate Visual System.* Chicago: Univ. Chicago Press

Rall, W. 1964. Theoretical significance of dendritic trees for neuronal input-output relations. In *Neural Theory and Modeling,* ed R. F. Reiss. Stanford: Stanford Univ. Press

Rezak, M., Benevento, L. A. 1979. A comparison of the organization of the projections of the dorsal lateral geniculate nucleus, the inferior pulvinar and adjacent lateral pulvinar to primary visual cortex (area 17) in the macaque monkey. *Brain Res.* 169:19–40

Ribak, C. E. 1978. Aspinous and sparsely-spinous stellate neurons in the visual cortex rats contain glutamic acid decarboxylase. *J. Neurocytol.* 7:461–78

Rockland, K. S., Lund, J. S. 1982. Widespread periodic intrinsic connections in the tree shrew visual cortex. *Science* 215:1532–34

Rockland, K. S., Pandya, D. N. 1981. Cortical connections of the occipital lobe in the Rhesus monkey: Interconnnections between areas 17, 18, 19 and the superior temporal sulcus. *Brain Res.* 5:249–70

Roney, K. J., Scheibel, A. B., Shaw, G. L. 1979. Dendritic bundles. Survey of anatomical experiments and physiological theories. *Brain Res. Rev.* 1:225–71

Rose, D., Blakemore, C. 1974. Effects of bicuculline on functions of inhibition in visual cortex. *Nature* 249:375–77

Rosenquist, A. C., Edwards, S. B., Palmer, L. A. 1975. An autoradiographic study of the projections of the dorsal lateral geniculate nucleus and the posterior nucleus in the cat. *Brain Res.* 80:71–93

Schiller, P. H. 1982. Central connections of the retinal ON and OFF pathways. *Nature* 297:580–83

Shatz, C. J. 1977. Anatomy of interhemispheric connections in the visual system of Boston Siamese and ordinary cats. *J. Comp. Neurol.* 173:497–518

Shkolnik-Yarros, E. G. 1960. Neurons of visual cortex in man. *Arh. Anat. Gistol. Embriol.* 38:24–38

Sholl, D. A. 1955. The organization of the visual cortex in the cat. *J. Anat.* 89:33–46

Sillito, A. M. 1975. The effectiveness of bicuculline as an antagonist of GABA and visually evoked inhibition in the cat's striate cortex. *J. Physiol.* 250:287–304

Sillito, A. M. 1977a. Inhibitory processes underlying the directional specificity of simple, complex and hypercomplex cells in the cat's visual cortex. *J. Physiol.* 271:699–720

Sillito, A. M. 1977b. The contribution of excitatory and inhibitory inputs to the length preference of hypercomplex cells in layers II and III of the cat's striate cortex. *J. Physiol.* 273:775–90

Sillito, A. M. 1979. Inhibitory mechanisms influencing complex cell orientation selectivity and their modification at high resting discharge levels. *J. Physiol.* 289:33–53

Sillito, A. M., Kemp, J. A., Milson, J. A., Berardi, N. 1980. A re-evaluation of the mechanisms underlying simple cell orientation selectivity. *Brain Res.* 194:517–20

Sloper, J. J., Powell, T. P. S. 1978. Gap junctions between dendrites and somata of neurons in the primate sensori-motor cortes. *Proc. R. Soc. London Ser. B* 203:39–47

Somogyi, P. 1977. A specific 'axo-axonal' interneuron in the visual cortex of the rat. *Brain Res.* 136:345–50

Somogyi, P., Cowey, A. 1981. Combined Golgi and electron microscopic study on the synapses formed by double bouquet cells in the visual cortex of the cat and monkey. *J. Comp. Neurol.* 195:547–66

Somogyi, P., Cowey, A., Halasz, N., Freund, T. F. 1981a. Vertical organization of neurons accumulating 3H-GABA in visual cortex of rhesus monkey. *Nature* 294:761–63

Somogyi, P., Freund, T. F., Halasz, N., Kisvarday, Z. F. 1981b. Selectivity of neuronal 3H-GABA accumulation in the visual cortex as revealed by Golgi staining of the labeled neurons. *Brain Res.* 225:431–36

Spatz, W. B., Tigges, J., Tigges, M. 1970. Subcortical projections, cortical associations and some intrinsic interlaminar connections of the striate cortex in the squirrel monkey. (*Saimiri*). *J. Comp. Neurol.* 140:155–74

Stanford, L. R., Friedlander, M. J., Sherman, S. M. 1981. Morphology of physiologically identified W-cells in the C laminae of the cat's lateral geniculate nucleus. *J. Neurosci.* 1:578–84

Sterling, P. 1983. Microcircuitry of the cat retina. *Ann. Rev. Neurosci.* 6:149–85

Stevens, J. K., Gerstein, G. L. 1976. Interactions between cat lateral geniculate neurons. *J. Neurophysiol.* 39:239–56

Stone, J., Fukuda, Y. 1974. Properties of cat retinal ganglion cells. A comparison of W-cells with X- and Y-cells. *J. Neurophysiol.* 37:722–48

Streit, P. 1980. Selective retrograde labeling indicating the transmitter of neuronal pathways. *J. Comp. Neurol.* 191:429–63

Szentagothai, J. 1965. The use of degeneration methods in the investigation of short neuronal connections. *Progr. Brain Res.* 14:1–22

Szentagothai, J. 1969. Architecture of the cerebral cortex. In *Basic Mechanisms of the Epilepsies*, ed. H. H. Jasper, A. A. Ward, A. Pope, pp. 13–28. Boston: Little, Brown

Szentagothai, J. 1973. Synaptology of the visual cortex. In *Handbook of Sensory Physiology Vol. 7, Central Visual Information*, ed. R. Jung, pp. 269–324. Berlin: Springer-Verlag

Talbot, S. A. 1940. Arrangement of visual area of cat's cortex. *Am. J. Physiol.* 129:477P–78P

Talbot, S. A., Marshall, W. H. 1941. Physiological studies on neural mechanisms of visual localization and discrimination. *Am. J. Ophthalmol.* 24:1255–64

Tigges, J., Spatz, W. B., Tigges, M. 1974. Reciprocal point-to-point connections between parastriate and striate cortex in the squirrel monkey (*Saimiri*). *J. Comp. Neurol.* 148:481–90

Tigges, J., Tigges, M., Anschel, S., Cross, N. A., Letbetter, W. D., McBride, R. L. 1981. Areal and laminar distribution of neurons interconnecting the central visual cortical areas 17, 18, 19 and MT in Squirrel monkey (*Saimiri*). *J. Comp. Neurol.* 202:539–60

Tigges, J., Tigges, M., Perachio, A. 1977. Complementary laminar terminations of afferents to area 17 originating in area 18 and in the lateral geniculate nucleus in squirrel monkey. *J. Comp. Neurol.* 176:87–100

Tombol, T., Hajdu, F., Somogyi, G. 1975. Identification of the Golgi picture of the layer VI cortico-geniculate projection neurons. *Exp. Brain Res.* 24:107–10

Tootell, R. B. H., Silverman, M. S. 1981. A comparison of cytochrome oxidase and deoxyglucose patterns in macaque visual cortex. *Neurosci. Abstr.* 7:356

Toyama, K., Matsunami, K., Ohno, T., Tokashiki, S. 1974. An intracellular study of neuronal organization in the visual cortex. *Exp. Brain Res.* 21:45–66

Toyama, K., Maekawa, K., Takeda, T. 1977. Convergence of retinal inputs onto visual cortical cells. I. A study of the cells monosynaptically excited from the lateral geniculate body. *Brain Res.* 137: 207–20

Toyama, K., Kimura, M., Tanaka, K. 1981a. Cross-correlation analysis of interneuronal connectivity in cat visual cortex. *J. Neurophysiol.* 46:191–201

Toyama, K., Kimura, M., Tanaka, K. 1981b. Organization of cat visual cortex as investigated by cross-correlation technique. *J. Neurophysiol.* 46:202–14

Tsumoto, T., Eckart, W., Creutzfeldt, O. D. 1979. Modification of orientation sensitivity of cat visual cortex neurons by removal of GABA-mediated inhibition. *Exp. Brain Res.* 34:351–63

Tusa, R. J., Palmer, L. A., Rosenquist, A. C. 1978. The retinotopic organization of area 17 (striate cortex) in the cat. *J. Comp. Neurol.* 177:213–36

Uchizono, K. 1965. Characteristics of excitatory and inhibitory synapses in the central nervous system of the cat. *Nature* 207:642–43

Valverde, F. 1971. Short axon neuronal subsystems in the visual cortex of the monkey. *Int. J. Neurosci.* 1:181–97

Valverde, F. 1976. Aspects of cortical organization related to the geometry of neurons with intra-cortical axons. *J. Neurocytol.* 5:509–29

Valverde, F., Ruiz-Marcos, A. 1969. Dendritic spines in the visual cortex of the mouse: Introduction to a mathematical model. *Exp. Brain Res.* 8:269–83

Van Essen, D. C. 1979. Visual areas of the mammalian cerebral cortex. *Ann. Rev. Neurosci.* 2:227–63

Van Essen, D. C., Zeki, S. M. 1978. The topographic organization of rhesus monkey prestriate cortex. *J. Physiol.* 277:273–90

von Economo, C., Koskinas, G. N. 1925. *Die Cytoarchitecktonik der Hirnrinde des erwachsenen Menschen.* Wien: Springer

White, E. L. 1978. Identified neurons in mouse Smi cortex which are postsynaptic to thalamocortical axon terminals: A combined Golgi-electron microscopic and degeneration study. *J. Comp. Neurol.* 181:627–62

White, E. L. 1979. Thalamocortical synaptic relations: A review with emphasis on the projections of specific thalamic nuclei to the primary sensory areas of the neocortex. *Brain Res. Rev.* 1:275–311

White, E. L., Hersch, S. M. 1982. A quantitative study of thalamocortical and other synapses involving the apical dendrites of corticothalamic projection cells in mouse Smi cortex. *J. Neurocytol.* 11: 137–57

Wiesel, T. N., Hubel, D. H., Lam, D. 1974. Autoradiographic demonstration of ocular dominance columns in the monkey striate cortex by means of trans-synaptic transport. *Brain Res.* 79: 273–79

Wilson, M. E., Cragg, B. G. 1967. Projection from the lateral geniculate nucleus in the cat and monkey. *J. Anat.* 101: 677–92

Wilson, P. D., Rowe, M. H., Stone, J. 1976. Properties of relay cells in cat's lateral geniculate nucleus: A comparison of W-cells with X- and Y-cells. *J. Neurophysiol.* 39:1193–1209

Wong-Riley, M. 1979a. Columnar cortico-cortical interconnections within the visual system of the squirrel and macaque monkeys. *Brain Res.* 162:201–17

Wong-Riley, M. 1979b. Changes in the visual system of monocularly sutured or enucleated cats demonstrable with cytochrome oxidase histochemistry. *Brain Res.* 171:11–28

Woolsey, C. N., Akert, K., Benjamin, R. M., Liebowitz, H., Welker, W. I. 1955. Visual cortex of the marmoset. *Fed. Proc.* 14:166

Zeki, S. M. 1976. The projections to the superior temporal sulcus from areas 17 and 18 in the rhesus monkey. *Proc. R. Soc. London Ser. B* 193:199–207

Zeki, S. M. 1978. Uniformity and diversity of structure and function in rhesus monkey prestriate visual cortex. *J. Physiol.* 277:273–90

Reference added in proof:

Singer, W., Tretter, F., Cynader, M. 1975. Organization of cat striate cortex: A correlation of receptive-field properties with afferent and efferent connections. *J. Neurophysiol.* 38:1080–98

Ann. Rev. Neurosci. 1983. 6:249–67

POSITRON EMISSION TOMOGRAPHY

Marcus E. Raichle

The Edward Mallinckrodt Institute of Radiology, and Department of Neurology and Neurological Surgery, Washington University School of Medicine, St. Louis, Missouri 63110

INTRODUCTION

An understanding of most processes occurring in the human brain in health and disease must ultimately be based on a precise knowledge of the dynamic underlying biochemical events. Progress toward the development of a satisfactory means of obtaining dynamic biochemical information on the human brain began with the pioneering work of Kety & Schmidt (1948). Their nitrous oxide technique for the measurement of cerebral blood flow (Kety & Schmidt 1948), when combined with arteriovenous measurements of various substrates and metabolites, provided our first clear information on the relationship between substrate delivery and utilization in the human brain. Although this technique was widely used and provided much information, it suffered from the obvious flaw of not yielding regional information. Dynamic regional differences in both circulation and metabolism were obscured, if not lost completely, by this approach. The introduction of a method for the measurement of regional cerebral blood flow based on the external detection of the clearance of freely diffusible radioactive gases such as xenon-133 from the brain (Hoedt-Rasmussen et al 1966) provided much additional information on regional hemodynamics in the human brain, and demonstrated for the first time the dynamic interplay between localized functional events in the brain and its circulation (Olesen 1971). This technique did not, however, provide information about regional metabolism and therefore was of limited value for investigating problems related to specific brain areas.

The introduction into the medical environment of cyclotron-produced, positron-emitting radiopharmaceuticals for the measurement of brain hemodynamics and metabolism partially fulfilled this need for regional

249

0147-006X/83/0301-0249$02.00

metabolic and biochemical information. In vivo techniques were developed for the measurement, not only of regional blood flow (Ter-Pogossian et al 1969), glucose (Raichle et al 1975), and oxygen utilization (Ter-Pogossian et al 1970, Raichle et al 1976b,c) in humans, but also regional cerebral blood volume (Eichling et al 1975) and vascular permeability (Raichle et al 1976a). These techniques, however, have received limited application because they require the intracarotid injection of the radiopharmaceutical as well as immediate access to a cyclotron. Thus, until recently we have been unable to obtain more than a quasi-regional assessment of cerebral blood flow in the human brain and, when circumstances permit, only limited metabolic information.

Three significant developments move us closer to the capability of safely acquiring regional in vivo biochemical information in the human brain. First, the appearance within the medical environment of apparatus for nuclear bombardment, such as cyclotrons and linear accelerators, coupled with ingenious techniques for rapid synthesis of radiopharmaceuticals, has provided many radiopharmaceuticals suitable for in vivo regional hemodynamic and metabolic studies (Welch 1977). Second, the parallel development of appropriate mathematical models has provided practical algorithms that allow parameters of biochemical and physiological significance to be estimated from the data. Finally, the development of positron emission tomography (PET) permits us to detect safely these radiopharmaceuticals in vivo in a truly regional and quantitative manner.

Emission tomography is a visualization technique in nuclear medicine that yields an image of the distribution of a positron-emitting radionuclide in any desired section of the body. Positron emission tomography (PET) utilizes the unique properties of the radiation generated when positions, produced by radioactive decay, are annihilated by interacting with electrons. A PET image reconstructed from the radioactive counting data is an accurate and quantitative representation of the spatial distribution of a radionuclide in the chosen section. This approach is analogous to quantitative autoradiography but has the added advantage of allowing in vivo studies. This review presents an introduction to the radionuclides, radiopharmaceuticals, detection systems, and tracer techniques used with PET, and some of the current applications of this technique in human neurobiology.

RADIONUCLIDES AND RADIOPHARMACEUTICALS

A small number of radionuclides, especially ^{15}O, ^{13}N, ^{11}C, and ^{18}F, possess the following chemical and physical properties that make them uniquely suitable for obtaining quantitative, regional, in vivo, biochemical, and physiological information:

1. They decay by positron emission (Figure 1). Positrons, which are positively charged electrons, are emitted from the nucleus of these radionuclides because they have too few neutrons to be stable. Positrons quite rapidly lose their kinetic energy in matter and interact with an electron. The positron and electron annihilate one another and give rise to two annihilation photons traveling in approximately opposite directions with an energy of 511 keV. This is known as annihilation radiation. The high energy of this annihilation radiation (511 keV) gives it greater tissue penetration and, thus, better detectability than, for example, the 140 keV gamma-ray photons of technetium-99m more commonly used in nuclear medicine imaging.

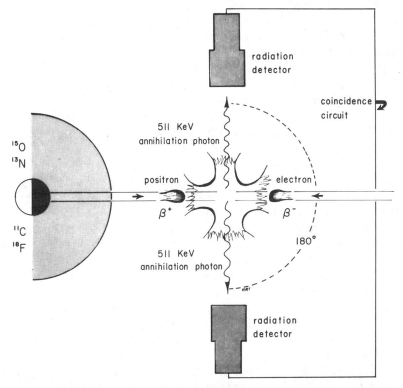

Figure 1 Radionuclides employed in positron emission tomography (PET) decay by the emission of positrons (β^+) or positive electrons from a nucleus unstable because of a deficiency of neutrons. Positrons lose their kinetic energy in matter after traveling a finite distance ($\sim$ 1–6 mm; Phelps et al 1975a) and, when they are brought to rest, interact with electrons (β^-). The two particles undergo annihilation and their mass is converted to two annihilation photons traveling at approximately 180° from each other with an energy of 511 keV. It is these annihilation photons that are detected by the imaging device, using opposing radiation detectors connected by electronic circuits (coincidence circuits) that record an event only when these two photons arrive simultaneously. (Adapted from Raichle 1979).

The annihilation radiation can be differentiated from background by two radiation detectors placed well outside a region of interest and connected to a coincidence circuit (Figure 1) that records an event from the tissue only if both detectors sense an annihilation photon simultaneously. This coincidence detection of annihilation radiation forms the basis for detection systems used in PET (see below).

2. The chemical nature of ^{15}O, ^{13}N, and ^{11}C so resembles the normal constituents of molecules that compose living matter that they can be incorporated in substances that will be included in most metabolic processes (Table 1).

3. These radionuclides decay with short half lives ($^{15}O = 2.05$ min; $^{13}N = 9.96$ min; $^{11}C = 20.34$ min; $^{18}F = 110$ min). Initially these short half lives might appear to be a major disadvantage, because they not only require a dedicated cyclotron and facilities for rapid radiopharmaceutical synthesis, but also metabolic processes sufficiently short to be studied. Indeed, the need for a dedicated cyclotron has, in the past, limited the use of these radionuclides to a few major medical centers worldwide. This will continue to be a requirement for this type of research, but the number of medical cyclotrons is gradually increasing and their operation is becoming simplified. Further, many ingenious, fast labeling techniques have been developed and have already provided a surprisingly large number of metabolic substrates and other molecules of potential importance in the study of the brain (Table 1). An increased awareness of this area of research, and stimulation by interested neuroscientists, will undoubtedly lead to larger numbers of highly specific radiopharmaceuticals.

4. Fortunately the overwhelming majority of brain metabolic processes studied over the past two decades have proved to be fast on the time scale of radionuclide lifetimes. A number of studies of importance to our understanding of brain biochemistry and physiology have been carried out with molecules labeled with these radionuclides. These include studies in man and animal brain of oxygen consumption; glucose transport, kinetics, and metabolism; permeability; and blood volume.

Two additional advantages accrue from the short half life of these radionuclides:

5. Because of their short half life, the dose of radiation to the subject as well as to those handling the radionuclides is significantly reduced. Examples of typical radiation doses to subjects undergoing PET studies are shown in Table 2. These doses are expressed as doses to "critical organs," which represent those organs receiving the greatest radiation exposure relative to doses permitted by the Food and Drug Administration (Code of Federal Regulation, Title 21, page 161, April 1, 1981).

6. The short physical half life of the radionuclides, particularly of ^{15}O,

Table 1 Substances labeled with positron-emitting radionuclides

Label	Half life	Labeled substance	Use
^{15}O	2 min	oxygen	metabolism
		water	blood flow
		carbon monoxide	blood volume
		carbon dioxide	blood flow
^{13}N	10 min	ammonia	organ perfusion
			metabolism
		various amino acids	metabolism
			permeability (brain)
		nitrous oxide	blood flow
		1,3 bis-(2-chloroethyl)	tumor drug levels
		1-nitrosourea (BCNU)	
^{11}C	20 min	carbon monoxide	blood volume
		carbon dioxide	tissue pH
		various alcohols	permeability (brain)
			blood flow
		various ethers	blood flow
			tissue lipid content
		acetate	metabolism (heart)
		palmitate	metabolism (heart)
		methylalbumin	tissue hematocrit (ratio of cells to plasma)
		octylamine	mapping of monoamine oxidase receptors (lung)
		glucose	metabolism
		2-deoxy-D-glucose	metabolism
		methionine	metabolism
		phenytoin	tissue drug levels (epilepsy)
		thymidine	metabolism (tumors)
		dopamine ⎫	neurotransmitter pharmacology
		norepinephrine ⎭	(brain)
		flunitrazepam ⎫	neuroreceptor pharmacology
		entorphine ⎬	(brain)
		pimozide ⎭	
^{18}F	110 min	2-deoxy-D-glucose	metabolism
		3-deoxy-D-glucose	metabolism
		haloperidol ⎫	neuroreceptor pharmacology
		spiroperidol ⎬	(brain)
		fluorodopa ⎭	

permits repeated studies in the same subject, not only because of the low radiation dose but also because the amount of background activity from one study to the next remains minimal. This minimizes the need for cumbersome schemes that would otherwise be necessary to account for complex decaying background radioactivity.

Table 2 Radiation doses to critical organs

Labeled compound (route of administration)	Critical organ	Administered activity (mCi)	Total dose (rads)	Permissible dose[a] (rads)
[18F] fluorodeoxyglucose (intravenous)	bladder	5	1.5	5
[11C] glucose (intravenous)	kidney	30	1.0	5
[15O] water (intravenous)	high flow organs[b]	100	0.7	5
[15O] oxygen (single breath)	trachea	50	1.0	5
[15O] carbon monoxide (single breath)	trachea	50	1.0	5
[15O] oxygen (continuous inhalation)	lung	—	1.9	5
[15O] carbon dioxide (continuous inhalation)	lung			

[a] *Code of Federal Regulation*, Title 21, page 161, April 1, 1981.
[b] Heart, lungs, brain, kidney, gastrointestinal tract.

DETECTION SYSTEMS AND IMAGES

The coincidence detection of the annihilation radiation provides a method of "electronic" collimation, since the two detectors can record coincidence events only in a volume of space established by straight lines joining the exposed surfaces of the two detectors (Figure 2). Thus, coincidence detection provides a nearly uniform field of view (or sensitivity) in the region between the two detectors. Coincidence detection also permits an easy, accurate correction for the attenuation or absorption of the radiation by the tissue (Phelps et al 1975a). Thus, it can be shown that the attenuation suffered by the annihilation radiation detected by coincidence is independent of the position of the source of the positrons within the tissue between the two detectors (Phelps et al 1975b).

A number of detection systems have been built and are currently in use, some of which have been designed especially for studies of the human brain (Ter-Pogossian et al 1978, Ter-Pogossian et al 1982, Hoffman et al 1981). In its simplest form, such a detection system consists of a pair of scintillation detectors scanning the imaged object at different angles. However, all systems in current use incorporate multiple detectors placed around the imaged object (usually arranged in banks to form a hexagon, or in a full circle) to improve the efficiency of the radiation collection. Each detector is then operated in coincidence with multiple opposing detectors, thus creating many coincidence lines through the imaged object. In addition, the

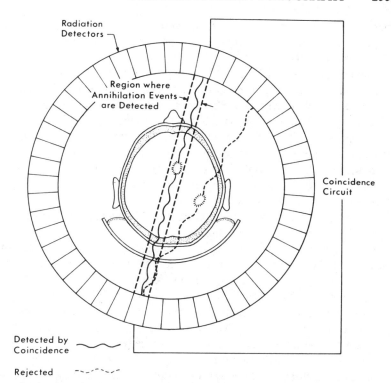

Figure 2 A schematic representation of a radiation detector arrangement for positron emission tomography (PET). Because each radioactive decay of a positron-emitting radionulcide
results in two annihilation photons traveling at approximately 180° from each other (Figure
1), coincidence detection is employed to localize the event in the tissue. A coincidence circuit
between pairs of detectors arrayed about the imaged object records an event only if both
detectors in the circuit record an event simultaneously. In order to increase the number of such
coincidence lines through the imaged object and, hence, the information gathering ability of
the imaging device, each radiation detector is in coincidence with many opposing radiation
detectors.

detectors are usually moved about the imaged object by a combination of
rotation, translation, or more complex movements. These lines or rays of
data, representing the summation of radioactivity measured between many
opposing detectors, are sorted into parallel lines and stored in a digital
computer for processing into an image of the distribution of radioactivity
in slices of the brain. Before these data can be used for the construction of
such an image, however, they must be corrected for the attenuation or
absorption of annihilation radiation by the object being imaged (i.e. the
human head). Appreciable attenuation occurs despite the relatively high
energy of the annihilation radiation (Phelps et al 1977a) but can be quite
satisfactorily compensated because the attenuation is independent of the

position of the positrons between the detector pairs. Various procedures have been devised to accomplish a correction for attenuation (e.g. see Huang et al 1979).

The actual construction of a PET image of the in vivo distribution of a radiopharmaceutical from the sorted and attenuation-corrected radioactive counting data obtained by the PET imaging device is a multi-step, computer-assisted procedure involving mathematical techniques developed many years ago for such diverse applications as gravitational theory, radio astronomy, and electron microscopy. More recently, these mathematical techniques have been applied to both transmission (x-ray) and emission tomography. A discussion of these techniques is well beyond the scope of this review. Interested readers will find both concise (Brooks & DiChiro 1975) and comprehensive reviews (Gordon & Herman 1974, Herman 1980) available describing these techniques.

It is important to note that the computer reconstruction of the distribution of a radionuclide in a section of the imaged object is a quantitative representation of its spatial distribution. Actual calibration of the instrument is achieved by imaging appropriately designed phantoms containing known amounts of radioactivity (Eichling et at 1977). This is an especially important feature of these detection systems because it forms the basis of the quantitative tracer methods employed (see below).

The resolution of detection systems has steadily improved. Early instruments had resolution elements approximately 1.5–2.0 cm in the plane of the tomographic section. This has now been reduced to less than 1.0 cm in the plane of the section, while the thickness remains about 1.4 cm for most tomographs. In considering ultimate resolution capabilities of PET one must remember that the positron itself travels a non-zero distance in tissue before the annihilation event occurs (Figure 1) (Phelps et al 1975a). This distance—which varies from 1 to 6 mm, depending in part on the energy of the particular positron—obviously limits the ultimate resolution of the method. It is especially important to keep the resolution limits of PET images in mind when interpreting measurements attributed to specific structures within the human brain. Attempts to measure quantitative metabolic rates in structures smaller than the measured resolution of PET will lead to serious errors (Hoffman et al 1979, Mazziotta et al 1981).

The proper imaging of an organ such as the brain obviously requires several tomographic sections. Detection systems capable of yielding only one section at a time must be operated sequentially with movement of the imaged object with respect to the tomograph between measurements. Refinements in technology have now led to detection systems that are capable of obtaining multiple tomographic sections simultaneously (Ter-Pogossian et al 1982, Hoffman et al 1981). The ability to obtain multiple tomographic sections simultaneously makes it possible to reconstruct the data in sagittal and coronal slices in addition to the standard horizontal slices.

A typical set of PET images of the human brain are shown in Figure 3.

TRACER TECHNIQUES

Blood Volume

The determination of regional cerebral blood volume (rCBV) in transverse sections of the human brain was one of the first measurements accomplished with PET. The early introduction of this measurement was, in part, the

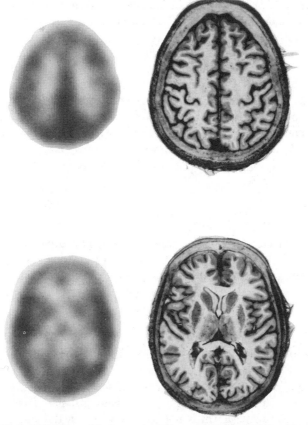

Figure 3   Typical PET scans of the human brain (upper and lower left) showing quantitative maps of local cerebral blood flow. These studies were performed following the intravenous injection of a bolus of water labeled with ^{15}O. The darkest areas represent the highest blood flow and the light areas the lowest blood flow. Corresponding horizontal sections of the human head are shown on the right for comparison.

result of its simplicity and accuracy. The most satisfactory way to measure rCBV using PET is to label the circulating blood pool of the body with the vascular tracer ^{11}C-labeled or ^{15}O-labeled carboxyhemoglobin. This is achieved when the subject inhales air that contains trace amounts of carbon monoxide labeled with either of these radionuclides. The rCBV is then determined from the equilibrium tomographic image of [^{11}C]carboxyhemoglobin in the brain (counts/second/milliliter of tissue) and venous blood samples (counts/second/gram of blood) obtained during the time of the emission scan.

Measurements of rCBV in man by PET not only reveal the expected regional differences in vascular density between gray and white matter, but also provide a clear delineation of the major vascular structures, primarily venous, surrounding the brain. Furthermore, the responsiveness of the cerebral blood volume to changes in arterial carbon dioxide tension observed with emission tomography agree with a variety of other such measurements of cerebral blood volume (Greenberg et al 1978). The measurement of rCBV is important, not only because it permits an accurate assessment of brain hemodynamics, but also because it provides the only means of determining the concentration of a radiopharmaceutical in the vascular compartment as compared to the extravascular compartment of tissue. The ability to make this distinction accurately is a unique feature of PET, and is of importance in the measurement of metabolism and tissue chemical composition.

Substrate Utilization

Quantitative measurement of regional substrate utilization in vivo is possible with PET, but is rather complex. Such measurements are exemplified by the quantitative studies of cerebral glucose utilization using ^{11}C-labeled glucose (Raichle et al 1978) and ^{18}F-labeled 2-deoxy-D-glucose (Reivich et al 1979, Phelps et al 1979, Huang et al 1980).

Use of these tracers for the measurement of local metabolism requires several important assumptions.

1. We assume that the tracer is transported and metabolized in the same manner and at the same rate as the compound being traced (i.e. tracee). This requirement is clearly met when a substrate such as glucose is labeled with ^{11}C, as ^{11}C-Glucose is biologically indistinguishable from unlabeled D-glucose. The requirement is not met, however, in the case of substrate analog tracers such as ^{18}F-2-deoxy-D-glucose. These analog tracers therefore require appropriate correction factors [i.e. so-called "lumped constant" (Sokoloff et al 1977)]; that must be determined in circumstances often different from the actual experimental situation.

2. We assume that the metabolized tracer is retained within the area of interest during the measurement period. In the case of a radiolabeled substrate such as ^{11}C-glucose, this assumption is justified by the following reasoning. Immediately upon entry into brain cells ^{11}C-glucose is phosphorylated to ^{11}C-glucose-6-phosphate. The glucose-6-phosphate is trapped within brain cells until it has been metabolized, because the brain is thought not to contain sufficient quantities of glucose-6-phosphatase necessary for dephosphorylation [this assumption has recently been challenged (Phelps et al 1979)] and brain cells are impermeable to glucose-6-phosphate. As ^{11}C-glucose is metabolized, it enters several large metabolic pools, the specific activity of the labeled metabolites remains low, and the egress of tracer from the brain is minimal for a period of about 5 min (Raichle et al 1975). Thus, satisfactory measurement of regional cerebral glucose metabolism can be achieved using ^{11}C-glucose when the measurement is performed within the first few minutes, or when a satisfactory correction can be achieved for the egress of the metabolized tracer. The requirement for metabolic trapping within the tissue is more easily achieved when analogs such as ^{18}F-labeled 2-deoxy-D-glucose are employed. In the case of such a tracer, metabolism does not proceed past the phosphorylation step (Sokoloff et al 1977) because of the structure of the molecule. Thus, ^{18}F-labeled 2-deoxy-D-glucose-6-phosphate is theoretically retained within the cells of the central nervous system for the period of the measurement.

3. We assume that the amount of tracer not metabolized, i.e. free tracer in blood and extracellular fluid, is either negligible or accounted for at the time of the measurement. This requirement is most easily met when radiolabeled analogs are employed. Because they are irreversibly trapped within the cells of the region of interest, measurements can be delayed until free tracer in tissue and blood has fallen to insignificant levels, thus minimizing the errors in estimating the free tracer in the tissue. In the case of glucose analogs, this usually requires 30 to 60 min. Such a delay is not possible when radiolabeled substrates such as ^{11}C-glucose are employed, because, as pointed out above, significant egress of radiolabeled metabolites from the tissue begins within 5 min of the time of injection. Thus, in the case of radiolabeled substrates, the tracer strategy employed must explicitly account for the free tracer in blood and tissue. This is accomplished, in part, by measuring the regional cerebral blood volume in association with the measurement of regional metabolism. By knowing the regional cerebral blood volume, one may compute not only the free tracer in blood, but also the free tracer within the tissue itself (Raichle et al 1978).

It should be clear from the foregoing that tracers in the form of radiolabeled substrates (e.g. ^{11}C-glucose) and radiolabeled substrate analogs (e.g. ^{18}F-2-deoxy-D-glucose) each have their special advantages and disadvantages. These are summarized in Table 3. The various advantages & disadvantages must be carefully considered when planning a specific experiment.

Oxygen Utilization

The strategies outlined above do not permit, in their present form, the measurement of tissue oxygen consumption. The difficulty is not tracer labeling—the positron-emitting radionuclide ^{15}O is readily available from most cyclotrons in the form of a gas that can then be used to label the subject's blood as ^{15}O-oxyhemoglobin—but rather, the prompt conversion of ^{15}O-labeled oxygen into ^{15}O-labeled water that rapidly leaves cells. Quite simply, the egress of metabolized tracer from the tissue occurs immediately and in significant quantities.

Because it is important to measure tissue oxygen consumption by PET, Jones and his colleagues (Jones et al 1976, Lenzi et al 1978, Frackowiak et al 1980, Lammertsma et al 1981), as well as others (Subramanyam et al 1978), have developed a technique for the measurement of the regional oxygen utilization by the brain. This measurement is accomplished by the sequential inhalation to equilibrium of trace amounts of ^{15}O-oxygen and ^{15}O-labeled carbon dioxide.

An alternative approach to the measurement of brain oxygen utilization, made possible by the development of fast, multislice PET imaging devices (Ter-Pogossian et al 1982), employs a modification of a tracer model previously described for the measurement of local glucose utilization (Raichle et

Table 3 Comparison of labeled substrates (e.g. ^{11}C-glucose) and substrate analogs (e.g. ^{18}F-2-deoxy-D-glucose) for measurement of metabolism with positron emission tomography

Compound	Advantage	Disadvantage
Labeled substrate	1. tracer and tracee are identical 2. short measurement time (< 5 min) 3. many compounds available	1. inefficiently trapped in tissue 2. blood volume must be measured 3. model parameters must be known accurately
Labeled substrate analogs	1. efficiently trapped in tissue 2. model parameters need not be known 3. blood volume need not be measured	1. tracer and tracee are not identical 2. long measurement time (> 30 min)[a] 3. relatively large radiation dose[b] 4. potentially toxic carrier 5. few compounds available

[a] Errors resulting from this long measurement time have recently been evaluated (Huang et al 1981).

[b] The absorbed dose of radiation (whole body as well as critical organ) is directly related to the half life of the radionuclide used. ^{18}F has a half life of 110 min compared with 20 min for ^{11}C. Administration of equivalent millicurie quantities of ^{18}F and ^{11}C, therefore, leads to a substantially higher dose when the tracer is labeled with ^{18}F.

al 1981). This technique, employing a single breath of ^{15}O-labeled air for the measurement of local tissue oxygen consumption, appears to provide a direct means of measuring local oxygen consumption with PET.

Blood Flow

Satisfactory techniques for measuring regional cerebral blood flow (rCBF) employing PET have developed more slowly than other areas of PET tracer methodology. The reason is that rCBF measurements have, to date, depended upon the analysis of dynamic tracer data obtained with sampling times of 1 sec or less. Most PET units currently in operation have not been able to collect sufficient data for an adequate reconstruction of a tissue slice in less than 1 min. As a result, methods for the measurement of blood flow have focused primarily on techniques that require equilibrium imaging.

One approach relates the equilibrium image of ^{15}O-labeled water, obtained by the continuous inhalation of ^{15}O-labeled carbon dioxide, to rCBF. As Lenzi and his colleagues have shown (Lenzi et al 1978, Frackowiak et al 1980, Lammertsma et al 1981), the amount of the tracer ^{15}O-water in a region of the brain or other organ under these circumstances is quantitatively related to rCBF. Thus, cerebral blood flow (CBF) can be calculated from:

$$CBF = K_D/(C_a/C_T - \lambda^{-1})$$

where (K_D) is the physical constant of oxygen-15 (0.335 min^{-1}), (λ) is the tissue-to-blood equilibrium partition coefficient for water (grey matter = 1.04; white matter = 0.88; whole brain = 0.95), and C_a and C_t are the arterial blood and tissue concentration of circulating $H_2{}^{15}O$, respectively. Although superficially appealing, this approach has two significant problems:

1. The relationship between blood flow and the measured quantities (C_a) and (C_t) is nonlinear.
2. The estimate of blood flow by this method is very sensitive to errors in the measurement of (C_a) and (C_t). A more detailed analysis of such errors and strategies for minimizing them is presented elsewhere (Lammertsma et al 1981, Jones et al 1982).

A second approach to the estimation of rCBF with PET is analogous to the radiolabeled microsphere method employed in laboratory animals. While the microsphere method employed in animals depends upon mechanical trapping of the tracer in the microcirculation of the organ of interest, PET studies have employed the concept of metabolic trapping. Specifically, ^{13}N-labeled ammonia has been used as a tracer to map perfusion in the brain

(Phelps et al 1977b, Phelps et al 1981a), because of its ability to be efficiently trapped, presumably by immediate incorporation into amino acids. Although conceptually sound, this approach is not likely to succeed in the case of ^{13}N-ammonia because this tracer is not freely permeable (Raichle & Larson 1981); some of the tracer in blood always exists in the impermeable form of ammonium ion (^{13}NH^{+4}).

Finally, it is now possible to measure rCBF with PET in a manner analogous to the tissue autoradiographic method originally proposed by Landau and his colleagues (1955) and later refined by Reivich et al (1969) and Sakurada et al (1978). The method is an application of the principles of inert gas exchange between blood and tissue developed by Kety. In this technique, as originally applied to animals, a freely diffusible radiolabeled tracer is infused intravenously for one minute, followed immediately by decapitation. Quantitative autoradiograms of brain slices and the time activity curve of arterial blood radioactivity form the data base form which rCBF is calculated. The PET image can substitute for the tissue autoradiogram provided that the image is obtained in a relatively short period of time ($\sim$ 40 sec). Recent developments in PET technology (see above) now permit such a measurement to be accomplished, as illustrated by the study in Figure 3 using ^{15}O-labeled water as the diffusible tracer. A number of freely diffusible tracers labeled with positron-emitting radionuclides could serve satisfactorily as the tracer for such measurements, provided, of course, that they are freely diffusible in the range of blood flow being studied.

Additional Tracer Techniques

A variety of additional tracer techniques are currently under development in several laboratories throughout the world. These include techniques: for the measurement of brain protein synthesis using ^{11}C-labeled leucine (Smith et al 1980); for the measurement of brain tissue pH using ^{11}C-labeled carbon dioxide (Raichle et al 1979), as well as ^{11}C-labeled DMO (5-dimethyl-2,4-oxazolidinedione; Ginos et al 1982); for the measurement of blood brain barrier permeability, using a variety of radiopharmaceuticals (Blasberg et al 1980, Raichle 1980, McCulloch & Angerson 1981); to study neuroreceptor pharmacology, using several radiolabeled ligands (Comar et al 1981); using very lipid soluble agents for the in vivo staining of myelin (Frey et al 1981); and for measuring neurotransmitter metabolism using ^{18}F-labeled L-DOPA (Garnett et at 1980). Of further interest is the very recent observation (Scheinberg et al 1982) that cell specific monoclonal antibodies can be labeled with bifunctional radioactive metal chelates, thus permitting their distribution to be determined in vivo.

There are insufficient published data or experience with these techniques at the time of this writing to comment further on them. Their diversity suggests exciting developments in PET in the near future.

CLINICAL APPLICATIONS OF PET

We are currently on the threshold of a wide variety of important biochemical, metabolic, and hemodynamic measurements on the human brain in vivo. These studies should provide us with new and unique information on important clinical and theoretical issues in human neurobiology. As a preface to a survey of the potential application of PET, one must remember that brain function depends on an extremely complex interaction between biochemical and physiological processes, heterogeneously distributed throughout the organ. Furthermore, these processes are continually changing in accordance with the normal functional activity of the brain. In the normal brain, metabolism and blood flow are clearly altered in discrete regions of the brain as a result of highly specific activity (Raichle et al 1976c, Yarowsky & Ingvar 1981). This observation provides the basis for a detailed exploration of the relationship of various structures within the brain to specific tasks performed by the central nervous system of man (Greenberg et al 1981, Phelps et al 1981b). Such studies will not only evaluate the organization of motor activity, but will also extend our understanding of how the brain of man processes information, stores data, and organizes such complex functions as language. Not only will such data be of immense theoretical interest in furthering our understanding of the functioning of the central nervous system of man, but it should also prove invaluable in understanding the manner in which the nervous system recovers, as it does in some instances, from head injury and stroke.

The evaluation of patients with stroke has always presented a most difficult challenge to clinicians. It is particularly difficult, for example, to distinguish areas of the brain permanently damaged from those rendered nonfunctional but not dead, due to severe reduction in oxygen supply. Failure to make such distinctions in individual patients makes it extremely difficult to plan appropriate therapy. With tracer techniques now available with PET, it is possible to make such distinctions in individual patients, choose appropriate therapy, and evaluate the effectiveness of the therapy (Kuhl et al 1980b, Baron et al 1981). Such an approach will ultimately allow us to state conclusively whether a given treatment in a patient with stroke accomplished its intended purpose—be it metabolic, biochemical, or hemodynamic—and, further, whether in accomplishing this, the patient was actually benefited. Studies of this type have already commenced in a number of centers with PET facilities.

Epilepsy is one of the most common diseases to afflict the nervous system of man. It is well known from experimental work in animals that the abnormal electrical activity is associated with striking changes in metabolism and blood flow. On the basis of this observation, studies are now being used to define more clearly the origin of epileptic seizures in humans

(Kuhl et al 1980a). The ability to make such a determination is important, not only for understanding more clearly the clinical manifestations of seizures in an individual patient, but also in determining the appropriate therapy. For example, some patients fail to respond to medical treatment for their seizures and then become candidates for surgery. Surgery is only performed when a very localized area of abnormally functioning brain can be found. Finding this area is often difficult using electrical recording techniques. Such abnormal areas of brain can now be clearly shown with in vivo studies of metabolism. The ability to localize these abnormal areas in patients with epilepsy is not only of importance in patients considered appropriate for surgery but may well be of importance in evaluating the effect of medical therapy. Thus, once knowing the area responsible for the seizure, one may, by appropriately labeling various anti-epileptic drugs determine whether therapeutic concentrations are achieved in the seizure focus. Low concentrations of anti-epileptic drugs in a seizure focus may well explain why certain drugs fail in a given individual despite adequate blood levels.

The evaluation of patients with tumors of the brain is also an important area for research. The ability to determine the biochemical and metabolic characteristics of an individual tumor may well prove valuable in the diagnosis as well as the treatment of specific tumor types. Various chemotherapeutic agents and x-ray treatments, for example, induce changes in tumors well before there is any obvious change in their size. Clearly, in vivo tracer techniques now offer a direct means for detecting such changes. Furthermore, as in the case of epilepsy, it should be possible to determine local tissue concentrations of various chemotherapeutic agents in tumors and know with certainty whether sufficient drugs reach the tumor to effect a cure.

A variety of other diseases affecting the central nervous system of man also challenge our new technical capabilities. For example, it should be possible to determine not only the location but the activity of diseased processes in multiple sclerosis by measuring changes in local tissue lipid content. Because this disease follows such a variable course, evaluating any form of therapy is often very difficult. The ability to not only localize but characterize the activity of a specific lesion in relationship to clinical symptoms and treatment should be a major advance in the management of multiple sclerosis.

Finally, various diseases of the central nervous system of man are thought to be related to disorders of neurotransmitter function in the brain. These include such well-known diseases as Parkinson's disease and Huntington's chorea, as well as mania, depression, and schizophrenia. It should be possible to explore for the first time the pathophysiology of these important diseases as a prelude to much more effective management.

In summary, we now have an extremely flexible and powerful tool in PET for the in vivo evaluation of the brain of man. If wisely used in the exploration of new and previously inaccessible frontiers in human neurobiology, this new approach should substantially further our understanding of the human central nervous system in health and disease.

ACKNOWLEDGMENTS

This work was supported by NIH Grants NS06833, HL13851, and NS14834, and the McDonnell Center for Studies of Higher Brain Function.

Literature Cited

Baron, J. C., Bousset, M. G., Rey, A., Guillard, A., Comar, D., Castaigne, P. 1981. Reversal of focal "misery perfusion syndrome" by extra-intracranial bypass in hemodynamic cerebral ischemia. A case study with ^{15}O positron emission tomography. *Stroke* 12:454–59

Blasberg, R. G., Gazendam, J., Patlak, C. S., Fenstermacher, J. D. 1980. Quantitative autoradiographic studies of brain edema and a comparison of multiisotope autoradiographic techniques. *Adv. Neurol.* 28:255–70

Brooks, R. A., DiChiro, G. 1975. Theory of image reconstruction in computed tomography. *Radiology* 117:561–72

Comar, D., Maziere, M., Cepeda, C., Godot, J. M., Menini, C., Naquet, R. 1981. The kinetics and displacement of [^{11}C]flunitrazepam in the brain of the living baboon. *Eur. J. Pharmacol.* 75:21–26

Eichling, J. O., Raichle, M. E., Grubb, R. L. Jr., Larson, K. B., Ter-Pogossian, M. M. 1975. In vivo determination of cerebral blood volume with radioactive oxygen-15 in the monkey. *Circ. Res.* 37:707–14

Eichling, J. O., Higgins, C. S., Ter-Pogossian, M. M. 1977. Determination of radionuclide concentrations with positron CT scanning (PETT). *J. Nucl. Med.* 18:845–47

Frackowiak, R. S. J., Lenzi, G-L., Jones, T., Heather, J. D. 1980. Quantitative measurement of regional cerebral blood flow and oxygen metabolism using ^{15}O and positron emission tomography: Theory, procedures and normal values. *J. Comput. Assist. Tomogr.* 4:727–36

Frey, K. A., Wieland, D. M., Brown, L. E., Rogers, W. L., Agranoff, B. W. 1981. Development of a tomographic myelin scan. *Ann. Neurol.* 10:214–21

Garnett, E. S., Firnau, G., Nahmias, C., Sood, S., Belbeck, L. 1980. Blood brain barrier transport and cerebral utilization of dopa in living monkeys. *Am. J. Physiol.* 238:R318–27

Ginos, J. Z., Tilbury, R. S., Haber, M. T., Rottenberg, D. A. 1982. Synthesis of [2-^{11}C]5,5-dimethyl-2,4-oxazolidimedione for studies with positron tomography. *J. Nucl. Med.* 23:255–58

Gordon, R., Herman, G. T. 1974. Three dimensional reconstruction from projections: A review of algorithms. *Int. Rev. Cytol.* 38:111–51

Greenberg, J. H., Alavi, A., Reivich, M., Kuhl, D., Uzzell, B. 1978. Local cerebral blood volume response to carbon dioxide in man. *Cir. Res.* 43:324–30

Greenberg, J. H., Reivich, M., Alavi, A., Haud, P., Rosenquist, A., Rintelmann, W., Stein, A., Tusa, R., Dann, R., Christman, D., Fowler, J., MacGregor, B., Wolf, A. 1981. Metabolic mapping of functional activity in human subjects with the [^{18}F] fluorodeoxyglucose technique. *Science* 212:675–80

Herman, G. T. 1980. *Image Reconstruction from Projections,* p. 316. New York: Academic

Hoedt-Rasmussen, K., Sveinsdottir, E., Lassen, N. A. 1966. Regional cerebral blood flow in man determined by intra-arterial injection of radioactive inert gas. *Circ. Res.* 18:237–47

Hoffman, E. J., Huang, S. C., Phelps, M. E. 1979. Quantitation in positron computed tomography: 1. Effect of object size. *J. Comput. Assist. Tomogr.* 3:299–308

Hoffman, E. J., Phelps, M. E., Huang, S. C., Kuhl, D. E. 1981. A new tomograph for quantitative positron emission computed tomography of the brain. *IEEE Trans. Nucl. Sci.* 28:99–103

Huang, S. C., Hoffman, E. J., Phelps, M. E., Kuhl, D. E. 1979. Quantitation in positron emission computed tomography: 2.

Effects of inaccurate attenuation correction. *J. Comput. Assist. Tomogr.* 3:804–14

Huang, S. C., Phelps, M. E., Hoffman, E. J., Sideris, K., Selin, C. J., Kuhl, D. E. 1980. Noninvasive determination of local cerebral metabolic rate of glucose in man. *Am. J. Physiol.* 238:E69–82

Huang, S. C., Phelps, M. E., Hoffman, E. J., Kuhl, D. E. 1981. Error sensitivity of fluorodeoxyglucose method for measurement of cerebral metabolic rate of glucose. *J. Cereb. Blood Flow Metab.* 1:391–401

Jones, S. C., Greenberg, J. H., Reivich, M. 1982. Error analysis for the determination of cerebral blood flow with continuous inhalation of ^{15}O-labeled carbon dioxide and positron emission tomography. *J. Comput. Assist. Tomogr.* 6:116–24

Jones, T., Chesler, D. A., Ter-Pogossian, M. M. 1976. The continuous inhalation of oxygen-15 for assessing regional oxygen extraction in the brain of man. *Br. J. Radiol.* 49:339–43

Kety, S. S., Schmidt, C. F. 1948. The nitrous oxide method for the quantitative determination of cerebral blood flow in man: Theory procedure and normal values. *J. Clin. Invest.* 27:476–83

Kuhl, D. E., Engle, J. Jr., Phelps, M. E., Selin, C. 1980a. Epileptic patterns of local cerebral metabolism and perfusion in humans determined by emission computed tomography of ^{18}FDG and ^{13}NH$_3$. *Ann. Neurol.* 8:348–60

Kuhl, D. E., Phelps, M. E., Kowell, A. P., Metter, J., Selin, C., Winter, J. 1980b. Effects of stroke on local cerebral metabolism and perfusion: Mapping by emission computed tomography of ^{18}FDG and ^{13}NH$_3$. *Ann. Neurol.* 8:47–60

Lammertsma, A. A., Jones, T., Frackowiak, R. S. J., Lenzi, G-L. 1981. A theoretical study of the steady-state model for measuring regional cerebral blood flow and oxygen utilization using oxygen-15. *J. Comput. Assist. Tomogr.* 5:544–50

Landau, W. M., Freygang, W. H. Jr., Rowland, L. P., Sokoloff, L., Kety, S. S. 1955. The local circulation of the living brain: Values in the anesthetized and unanesthetized cat. *Trans. Am. Neurol. Assoc.* 80:125–29

Lenzi, G. L., Jones, T., McKenzie, C. G., Buckingham, P. D., Clark, J. C., Moss, S. 1978. Study of regional cerebral metabolism and blood flow relationships in man using the method of continuously inhaling oxygen-15 and oxygen-15-labeled carbon dioxide. *J. Neurol. Neurosurg. Psychiatr.* 41:1–10

Mazziotta, J. C., Phelps, M. E., Plummer, D., Kuhl, D. E. 1981. Quantitation in positron emission computed tomography: 5. Physical-anatomical effects. *J. Comput. Assist. Tomogr.* 5:734–43

McCulloch, J., Angerson, W. 1981. Regional water permeability in the CNS of conscious rate. *J. Cereb. Blood Flow Metab. Suppl.* 1:S377–78

Olesen, J. 1971. Contralateral focal increase in cerebral blood flow in man during arm work. *Brain* 94:635–646

Phelps, M. E., Hoffman, E. J., Huang, S. C., Ter-Pogossian, M. M. 1975a. Effect of positron range on spatial resolution. *J. Nucl. Med.* 16:649–652

Phelps, M. E., Hoffman, E. J., Mullani, N. A., Ter-Pogossian, M. M. 1975b. Application of annihilation coincidence detection to transaxial reconstruction tomography. *J. Nucl. Med.* 16:210–224

Phelps, M. E., Hoffman, E. J., Mullani, N. A., Higgins, C. S., Ter-Pogossian, M. M. 1977a. Some performance and design characteristics of PETT III. In *Reconstruction Tomography In Diagnostic Radiology and Nuclear Medicine,* ed. M. M. Ter-Pogossian, M. E. Phelps, G. L. Brownell, J. R. Cox, Jr., D. O. Davis, et al, pp. 371–92. Baltimore: Univ. Park Press. 573 pp.

Phelps, M. E., Hoffman, E. J., Raybaud, C. 1977b. Factors which affect cerebral uptake and retention of ^{13}NH$_3$. *Stroke* 8:694–702

Phelps, M. E., Huang, S. C., Hoffman, E. J., Selin, C., Sokoloff, L., Kuhl, D. E. 1979. Tomographic measurement of local cerebral glucose metabolic rate in humans with (F-18)-2-fluoro-2-deoxy-D-glucose: Validation of method. *Ann. Neurol.* 6:371–88

Phelps, M. E., Huang, S. C., Hoffman, E. J., Selin, C., Kuhl, D. E. 1981a. Cerebral extraction of N-13 ammonia: Its dependence on cerebral blood flow and capillary permeability-surface area product. *Stroke* 12:607–19

Phelps, M. E., Mazziotta, J. C., Kuhl, D. E., Nuwer, M., Packwood, J., Melter, J., Engle, J. Jr. 1981b. Tomographic mapping of human cerebral metabolism: Visual stimulation and deprivation. *Neurology* 31:517–29

Raichle, M. E. 1979. Quantitative in vivo autoradiography with positron emission tomography. *Brain Res. Rev.* 1:47–68

Raichle, M. E. 1980. Brain edema: Evaluation in vivo with positron emission tomography. *Adv. Neurol.* 28:423–27

Raichle, M. E., Larson, K. B. 1981. The significance of the NH₃-NH₄+ equilibrium on the passage of ¹³N-ammonia from blood to brain. *Circ. Res.* 48:913–37

Raichle, M. E., Larson, K. B., Phelps, M. E., Grubb, R. L. Jr., Welch, M. J., Ter-Pogossian, M. M. 1975. In vivo measurement of brain glucose transport and metabolism employing glucose −¹¹C. *Am. J. Physiol.* 228:1936–48

Raichle, M. E., Eichling, J. O., Straatman, M. G., Welch, M. J., Larson, K. B., Ter-Pogossian, M. M. 1976a. Blood-brain barrier permeability of ¹¹C-labeled alcohols and ¹⁵O-labeled water. *Am. J. Physiol.* 230:543–52

Raichle, M. E., Grubb, R. J. Jr., Eichling, J. O., Ter-Pogossian, M. M. 1976b. Measurement of brain oxygen utilization with radioactive oxygen-15: Experimental verification. *J. Appl. Physiol.* 40:638–40

Raichle, M. E., Grubb, R. L. Jr., Gado, M. H., Eichling, J. O., Ter-Pogossian, M. M. 1976c. Correlation between regional cerebral blood flow and oxidative metabolism. *Arch. Neurol. Chicago* 33: 523–26

Raichle, M. E., Welch, M. J., Grubb, R. L. Jr., Higgins, C. S., Ter-Pogossian, M. M., Larson, K. B. 1978. Measurement of regional substrate utilization rates by emission tomography. *Science* 199: 986–87

Raichle, M. E., Grubb, R. L. Jr., Higgins, C. S. 1979. Measurement of brain tissue carbon dioxide content in vivo by emission tomography. *Brain Res.* 166: 413–17

Raichle, M. E., Larson, K. B., Markham, J., Depresseux, J-C., Grubb, R. L. Jr., Ter-Pogossian, M. M. 1981. Measurement of regional oxygen consumption by positron emission tomography. *J. Cereb. Blood Flow Metab. Suppl.* 1:S57–S58

Reivich, M., Jehle, J., Sokoloff, L., Kety, S. S. 1969. Measurement of regional cerebral blood flow with antipyrine-¹⁴C in awake cats. *J. Appl. Physiol.* 27:296–300

Reivich, M., Kuhl, D., Wolf, A., Greenberg, J., Phelps, M., Ido, T., Cosella, N., Fowler, J., Hoffman, E., Alavi, A., Som, P., Sokoloff, L. 1979. The [¹⁸F]fluorodeoxyglucose method for the measurement of local cerebral glucose metabolism in man. *Circ. Res.* 44:127–37

Sakurada, O., Kennedy, C., Jehle, J., Brown, J. D., Carbin, G. L., Sokoloff, L. 1978.

Measurement of local cerebral blood flow with iodo[¹⁴C]antipyrine. *Am. J. Physiol.* 234:H59–66

Scheinberg, D. A., Strand, M., Gansow, O. A. 1982. Tumor imaging with radioactive metal chelates conjugated to monoclonal antibodies. *Science* 215:1511–13

Smith, C. B., Davidsen, L., Deibler, G., Patlak, C., Pettigrew, K., Sokoloff, L. 1980. A method for the determination of local rates of protein synthesis in brain. *Trans. Am. Soc. Neurochem.* 11:94. (Abstr.)

Sokoloff, L., Reivich, M., Kennedy, C., DeRosiers, M. H., Pattak, C. S., Pettigrew, K. C., Sakurada, O., Shinohara, M. 1977. The [¹⁴C]deoxyglucose method for the measurement of local cerebral glucose utilization: Theory, procedure, and normal values in the conscious and anesthetized albino rat. *J. Neurochem.* 28:897–916

Subramanyam, R., Alpert, N. M., Hoop, B. Jr., Brownell, G. L., Taveras, J. M. 1978. A model for regional cerebral oxygen distribution during continuous inhalation of ¹⁵O₂, C¹⁵O and C¹⁵O₂. *J. Nucl. Med.* 19:48–53

Ter-Pogossian, M. M., Eichling, J. O., Davis, D. O., Welch, M. J., Metzger, J. M. 1969. The determination of regional blood flow by means of water labeled with radioactive oxygen-15. *Radiology* 93:31–40

Ter-Pogossian, M. M., Eichling, J. O., Davis, D. O., Welch, M. J. 1970. The measure in vivo of regional cerebral oxygen utilization by means of oxyhemoglobin labeled with radioactive oxygen-15. *J. Clin. Invest.* 49:381–91

Ter-Pogossian, M. M., Mullani, N. A., Hood, J. T. Sr., Higgins, C. S., Ficke, D. C. 1978. Design consideration for a positron emission transverse tomograph (PETT V) for imaging of the brain. *J. Comput. Assist. Tomogr.* 2:539–44

Ter-Pogossian, M. M., Ficke, D. C., Hood, J. T. Sr., Yamamoto, M., Mullani, N. A. 1982. PETT VI: A positron emission tomograph utilizing cesium fluoride scintillation detectors. *J. Comput. Assist. Tomogr.* 6:125–33

Welch, M. J. 1977. *Radiopharmaceuticals and Other Compounds Labelled with Short-lived Radionuclides.* New York: Pergamon

Yarowsky, P. J., Ingvar, D. H. 1981. Neuronal activity and energy metabolism. *Fed. Proc.* 40:2353–3262

Ann. Rev. Neurosci. 1983. 6:269–324

HYPOTHALAMIC INTEGRATION: Organization of the Paraventricular and Supraoptic Nuclei

L. W. Swanson and P. E. Sawchenko

The Salk Institute for Biological Studies, and The Clayton Foundation for Research, California Division, San Diego, California 92138

INTRODUCTION

Living organisms display a remarkable capacity to maintain a relatively constant internal milieu in the face of unrelenting challenges posed by the requirements of metabolism, and by an often hostile environment. This stability is assured by a series of biochemical, physiological, and behavioral mechanisms that have become ever more complex during the course of evolution. In vertebrates, the central nervous system mediates a broad range of autonomic, endocrine, and behavioral responses that maintain homeostasis, and a great deal of work indicates that the hypothalamus, in particular, plays a critical role in the coordination of such responses (for recent reviews see Morgane & Panksepp 1980). In mammals, the hypothalamus is thought to regulate body temperature, the cardiovascular system, and the abdominal viscera, as well as ingestive behaviors that replenish nutrients and water, and sexual and maternal behaviors that assure survival of the species.

The neural mechanisms underlying such adaptive responses have been difficult to unravel, and remain largely undefined in terms of specific antomical circuitry (Swanson & Mogenson 1981). This is due primarily to the fact that the hypothalamus, which is commonly regarded as the rostralmost part of the reticular formation, is a poorly differentiated region, with the exception of the magnocellular neurosecretory system. In addition, the hypothalamus gives rise to, and is traversed by, a large number of fibers in the medial forebrain bundle and the periventricular system, two of the most

269

complex fiber systems in the brain. Because of this, the organization of hypothalamic pathways involved in the modulation of autonomic centers, of the anterior pituitary, and of somatomotor control systems were poorly understood until quite recently (Nauta & Haymaker 1969). However, considerable advances in neuroanatomical methodology have been made in the last decade, based largely on ideas adopted from cell biology and immunology. As reviewed in the first volume of this series (Jones & Hartman 1978), these sensitive new methods rely on the anterograde and retrograde axonal transport of locally injected markers, and on the immunohistochemical identification of neural systems that contain specific antigens such as neurotransmitters and their synthetic enzymes.

In this review we focus on the morphology and functions of the paraventricular (PVH) and supraoptic (SO) nuclei of the hypothalamus, which give rise to the magnocellular neurosecretory system, and thus contain cells that synthesize either oxytocin or vasopressin, and release them into the blood stream in the posterior pituitary. The full battery of modern neuroanatomical techniques has been used to study the connections of these nuclei, and the surprising conclusion is emerging that the PVH may also be involved in the autonomic mode of cardiovascular regulation, and in the visceral and behavioral responses associated with appetitive behaviors such as eating and drinking. The PVH may thus come to serve as a convenient model system in which to study the hypothalamic integrative mechanisms outlined above. Along with these functional insights, it has become clear that the PVH is much more complex morphologically than previously thought. In the rat, for example, the PVH occupies less than a third of a mm^3 of tissue on either side of the third ventricle, and yet it can be divided into at least eight clearly distinguishable subdivisions, within which some 30 different putative neurotransmitters have been identified either in cell bodies or in presumed terminals. The job of unraveling this circuitry will indeed be difficult, but from work done thus far, several general principles that appear to underly the organization of biochemically distinct pathways into and out of the nucleus are beginning to emerge. Because the literature on the PVH and SO is quite extensive, we concentrate here on what is known, and what remains to be learned, about the connections of these nuclei, and their possible functional significance, in the rat.

CELLULAR ORGANIZATION

The PVH and SO are easily recognized in standard Nissl preparations as discrete aggregates of large, intensely stained perikarya. Although often treated as rather homogeneous cell groups, recent morphological studies have established that they are composed of a number of different cell types, which show a hitherto unexpected degree of topographic organization.

Cytoarchitectonics

SUPRAOPTIC NUCLEUS   The SO (of Lenhossék 1887), which was also referred to in the older literature as the tangential or perichiasmatic nucleus (Cajal 1894, 1911), or as the basal optic ganglion (Meynert 1869), has the appearance in Nissl preparations of a uniform mass of large neurons that straddles the lateral border of the optic chiasm. Magnocellular neurons that lie along blood vessels penetrating the optic chiasm and nerve (pars intraoptica of Léránth et al 1975), and that lie medial to the optic tract along the base of the brain at the level of the arcuate nucleus (the pars tuberalis of Bodian 1939, or the retrochiasmatic magnocellular cell group of Peterson 1966), are sometimes included within the SO, but here we limit the borders of the SO to the compact, or principal, cell group described above. Cell counts indicate that each SO in the rat, as thus defined, consists of between 4400 (Léránth et al 1975) and 7000 cells (Rasmussen 1940, Bodian & Maren 1951, Olivecrona 1957), although the higher estimates appear to have been inadequately corrected for double-counting errors.

PARAVENTRICULAR NUCLEUS   The PVH (of Malone 1910), which has also been referred to as the filiform (Fortuyn 1912) or subventricular (Cajal 1911) nucleus, has often been thought of strictly as a compact group of magnocellular neurons (e.g. Malone 1916, Loo 1931, Le Gros Clark 1938, Christ 1969). It is now conventional, however, to view the nucleus as composed of distinct magnocellular and parvocellular divisions (e.g. Gurdjian 1927, Grunthal 1931, Krieg 1932, Rose 1935, Bodian 1939). Recently, several different parcellations of each major division, which are based on cytoarchitectonic, connectional, and immunohistochemical criteria, have been proposed (Armstrong et al 1980, Swanson & Kuypers 1980, Koh & Ricardo 1980). Fortunately, the differences between these schemes are more semantic than substantive, and they are quite easily reconciled. In the present paper we use the terminology proposed by Swanson & Kuypers (1980), which has been broadened somewhat here to incorporate recent experimental findings.

Eight distinct subdivisions, three of which are magnocellular, can be recognized in standard Nissl preparations. The *anterior magnocellular* part of the PVH lies ventromedial to the descending column of the fornix as it enters the hypothalamus from the septum. Immediately caudal to this level, the *medial magnocellular* subdivision is embedded within the anterior periventricular nucleus of Krieg (1932). Together, the anterior and medial magnocellular parts of the PVH correspond to the anterior commissural cell group described by Peterson (1966) in material stained with the Gomori method. The *posterior magnocellular* part of the nucleus is the largest of the magnocellular subdivisions, and corresponds to the magnocellular division of the PVH of most authors. In frontal sections this cell group first appears just caudal to the medial magnocellular part of the nucleus, in ventromedial regions of the PVH. It then expands dorsolaterally, and comes to form a compact cell mass at the lateral margin of the nucleus. Hatton et al (1976) used cytoarchitectonic criteria to define two distinct populations of large cells in the anteroventromedial (or "medial") and posterodorsolateral (or "lateral") halves of this region. Cells in the lateral group tend to be larger, to have multiple nucleoli, and to have a more uniform orientation than those of the medial group. Subsequent immunohistochemical studies have established that this distinction also has a neurochemical basis (see below). Cell counts in Nissl material of the rat indicate that the posterior magnocellular part of the PVH contains 1300 (Olivecrona 1957) to 2000 (Bodian & Maren 1951) cells on either side of the brain. These values correspond well to the 1700 cells counted by Bandaranayake (1971) in Gomori-stained preparations.

The parvocellular division consists of five distinct parts. The *periventricular part* is composed mostly of small, vertically oriented fusiform cells, and corresponds to the region of the anterior periventricular nucleus of Krieg (1932) that lies medial to the rest of the PVH. The *anterior parvocellular* part is a relatively poorly defined group of small- to medium-sized cells at the level of the anterior and medial magnocellular parts of the nucleus. In the older literature, only Rioch (1929, cat and dog) and Rose (1935, rabbit) considered this cell group to be a rostral extension of the PVH. The *medial parvocellular* part of the PVH is a heterogeneous collection of densely packed cells that extends throughout the caudal half of the nucleus. Koh & Ricardo (1980) have subdivided this region into dorsal and ventral components on the basis of distinct efferent projections (see below). The *dorsal parvocellular* part of the nucleus has been recognized by most contemporary authors, and is an elliptical group of horizontally-oriented, medium-sized cells in the caudal half of the nucleus. Finally, the *lateral parvocellular* part of the PVH forms a lateral expansion of the parvocellular division in the caudal third of the nucleus. This part contains medium-sized cells that are characteristically fusiform in shape and horizontally oriented; it corresponds to part of the "posterior" subdivision of Armstrong et al (1980). At its maximum extent, the lateral parvocellular part of the PVH extends laterally over the descending column of the fornix (about 750—1000 μm from the midline), where it lies next to another "accessory" group of magnocellular neurosecretory neurons, the posterior forniceal group of Peterson (1966). Only Koh & Ricardo (1980) have included the latter cell group within the PVH proper.

Counts of nucleoli, corrected for double-counting errors, indicate that the parvocellular division of the PVH contains about 7000 neurons (Sawchenko

& Swanson 1981a; L. W. Swanson, unpublished observations). Each part contains approximately the following number of cells: anterior, 1650; dorsal 600; lateral 1300, medial 2175; periventricular 1275.

Golgi Studies

The refractoriness of magnocellular neurosecretory neurons to Golgi impregnation has been known since the time of Cajal (1894). Thus, only a few detailed morphological descriptions of cells in the PVH and SO are available, and these have been carried out in a variety of species, and in animals at different stages of development. Nevertheless, some consistent findings can be gleaned from the available literature, and are important for two reasons.

1. A knowledge of the branching pattern and extent of the dendritic trees of these neurons is prerequisite to a determination of the particular cell types that are influencd by specific inputs to the nuclei.
2. A considerable body of electrophysiological evidence suggests that the activity of magnocellular neurosecretory neurons is modulated by recurrent inhibition (or excitation) that may involve axon collaterals and interneurons (see Barker 1977 and Renaud 1978 for reviews).

SUPRAOPTIC NUCLEUS   Large bipolar and multipolar neurons in the SO have been observed in the adult monkey (LuQui & Fox 1976) and rabbit (Felten & Cashner 1979), and in young dogs (Leontovich 1969/1970), cats, and guinea pigs (Lefranc 1966). Such cells have one to five primary dendrites that do not branch extensively and that are confined largely within the morphological boundaries of the nucleus. The dendrites tend to be oriented vertically (i.e. perpendicular to the optic tract), and prominent ventrally directed processes have been described in the dog (Leontovich 1969/1970) and rat (Armstrong et al 1982, McNeill & Sladek 1980). At least some ventrally directed dendrites of large neurons in the SO aggregate along the pial surface in what has been called a ventral glial lamina; these processes have been shown to receive both symmetric and asymmetric synaptic contacts at the ultrastructural level (Armstrong et al 1982).

There is general agreement that the axons of most, if not all, of the large neurons in the SO contribute to the supraoptico-hypophysial tract. Felten & Cashner (1979) described collaterals that emerge at right angles from the axons of large bipolar neurons in the rabbit, although Leontovich (1969/1970) did not observe such branching in the puppy. In the primate, collateral branching was seen only after axons had entered the supraoptico-hypophysial tract (LuQui & Fox 1976).

Smaller multipolar somata have also been described in the SO of several species (Felten & Cashner 1979, LuQui & Fox 1976). Although these may be interneurons, it is important to point out that locally ramifying axonal plexuses have not been demonstrated, and recent immunohistochemical evidence suggests that cell size is not a sufficient criterion for distinguishing cells that contain neurosecretory material (Sawchenko & Swanson 1982a). That some degree of intranuclear interaction occurs in the SO is supported by a report that surgical deafferentation results in the degeneration of only about one-third of the synapses in the nucleus (Záborszky et al 1975).

PARAVENTRICULAR NUCLEUS   The available Golgi or Golgi-like impregnations of magnocellular neurons in the PVH indicate that they are quite similar to those in the SO. Bipolar and multipolar neurons with two or three simple dendrites have been described in the anterior and posterior magnocellular parts of the PVH in the rat (Armstrong et al 1980, Van Den Pol 1982), guinea pig (Lefranc 1966), dog (Leontovich 1969/1970), and mouse (Barry 1975). As in the SO, the dendrites of these large neurons stay, for the most part, within the morphological boundaries of the nucleus. In the rat, magnocellular neurons often have a prominent dendrite that is directed medially or ventromedially (Armstrong et al 1980, Van Den Pol 1982), suggesting that most of the dendritic mass of these neurons is confined to the subdivision in which they lie. The anterior commissural nucleus may be an exception, since the medially oriented dendrites of cells here appear to traverse the anterior parvocellular part of the nucleus (Armstrong et al 1980).

The axons of magnocellular neurons in the PVH leave the nucleus laterally or ventrolaterally, course above or below the fornix, and then arch posteroventromedially toward the median eminence (Krieg 1932, Lefranc 1966, Leontovich 1969/1970, Van Den Pol 1982). There are, however, no unambiguous descriptions in Golgi material of local collaterals arising from the axons of magnocellular neurons within the PVH (Van Den Pol 1982).

Sofroniew & Glassman (1981) have used a modified immunoperoxidase method for thick sections stained with antisera directed against oxytocin, vasopressin, or neurophysin to obtain "Golgi-like" impregnations of magnocellular neurosecretory neurons in the rat. Short collaterals, typically with a smaller diameter than their parent axons, were observed, and were seen most frequently in posterodorsal regions of the PVH. Because cells in the compact magnocellular parts of the PVH and SO were intensely stained, most of the cells they described were either in accessory groups or along the margins of the PVH. Thus, the extent to which magnocellular neurosecretory neurons in the PVH, proper, give rise to axon collaterals remains unresolved. Detailed light- and electron microscopic analyses of physiologically identified neurons injected intracellularly with either HRP or a fluorochrome are needed to clarify this issue. The feasibility of such an approach has recently been established by Reaves & Hayward (1981), who were able to fill a small number of antidromically identified neurosecretory neurons in the PVH of the cat.

Interestingly, the few available morphological descriptions of *parvocellular* neurons in the PVH suggest that their dendritic trees are also quite simple. Small cells in the PVH commonly have two or three short primary dendrites that branch once or twice, but usually show no preferential orientation within the nucleus (Krieg 1932, Barry 1975, Millhouse 1979). Bipolar cells in the periventricular part of the PVH give rise to short, vertically oriented dendrites that appear to stay within this part of the nucleus (Krieg 1932, Millhouse 1979, Van Den Pol 1982). Long primary dendrites of fusiform cells in the lateral parvocellular part of the PVH tend to be oriented horizontally (Armstrong et al 1980, Van Den Pol 1982). In a comprehensive Golgi study of neurons in the PVH, Van Den Pol (1982) has described axon collaterals of *parvocellular* neurons that ramify locally and appear to contact the dendrites of cells in both the magnocellular and parvocellular divisions. This result provides the first clear evidence for local interactions between cells of the two major divisions of the PVH, and suggests one possible means through which their outputs may be integrated.

Although a review of the ultrastructure of magnocellular neurosecretory neurons is unnecessary here, one morphological specialization is relevant to the present discussion. Direct soma-soma appositions are observed quite frequently between magnocellular neurons in the PVH and SO (Tweedle & Hatton 1976, 1977, Van den Pol 1982). In addition, the use of freeze-fracture replicas has revealed the presence of apparent gap junctions between magnocellular neurons, and intracellular injections of small molecular weight dyes, in vitro, have provided evidence for dye-coupling between magnocellular neurons (Andrew et al 1981). Stimuli that release oxytocin or vasopressin, such as suckling or dehydration, respectively, have been shown to increase the extent of membrane involved in soma-soma contacts (Tweedle & Hatton 1976, 1977, Gregory et al 1980), and to increase the frequency of dye-coupled magnocellular neurons (Hatton & Tweedle 1980). The possibility that labile electrotonic interactions occur between adjacent magnocellular neurosecretory neurons may help to explain how the pulsatile release of oxytocin is achieved during suckling, and how vasopressinergic neurons are recruited into a phasic bursting pattern during periods of dehydration or hemorrhage.

In summary, the available morphological studies of neurons in the PVH and SO suggest that the cells in each nucleus are rather simple morphologically. The simplicity of the dendritic trees in particular should facilitate detailed analyses of specific inputs to each cell type. Surprisingly, anatomical evidence for the existence of axon collaterals and interneurons, presumed substrates for the recurrent modulation of magnocellular neurosecretory neurons, is quite sparse. Intranuclear interactions between adjacent magnocellular neurons may occur by way of local collaterals of parvocellular neurons or by way of gap junctions, the extent of which may vary with the physiological state of the animal.

Immunohistochemistry

The role of magnocellular neurosecretory neurons in the synthesis, transport, and release of the posterior pituitary hormones oxytocin and vasopressin, and their respective "carrier" proteins (the neurophysins), is well-documented (Brownstein et al 1980). Apart from a small population of parvocellular vasopressinergic cells in the suprachiasmatic nucleus (Vandesande et al 1975), these peptides have been localized exclusively in neurons of the PVH, the SO, and the scattered accessory cell groups of the magnocellular neurosecretory system, and thus provide a biochemical "signature" for these nuclei. Recent immunohistochemical studies, using well-characterized antisera, have clarified the organization of oxytocin- and vasopressin-immunoreactive cells in a number of species (see Dierickx 1980).

A second distinguishing feature of the PVH and the SO that has emerged from immunohistochemical studies is that they contain an exceptional variety of neuroactive substances. More than 30 putative transmitter substances, transmitter-related enzymes, and biologically active peptides have been identified in cells or fibers within the SO and the PVH (Table 1). Indeed, of all the central neuropeptides that have been characterized at all thoroughly, only carnosine has not been identified in these nuclei.

OXYTOCIN, VASOPRESSIN   Before the application of imunohistochemical methods to the magnocellular neurosecretory system in the mid 1970s, it was not clear whether oxytocin and vasopressin are localized within the same, or within separate, populations of neurons, and whether one or the other of these peptides is localized preferentially in either the PVH or the SO. In fact, it was held for a time that oxytocin is synthesized primarily in the PVH, and vasopressin in the SO (e.g. Olivecrona 1957, Lederis 1961). However, the use of antisera against oxytocin and vasopressin, cross-adsorbed in the solid phase against the heterologous antigen, showed convincingly that the two nonapeptides occur in separate populations of cells in a variety of species (see Dierickx 1980), including the rat (Vandesande & Dierickx 1975, Vandesande et al 1977). This observation is supported by the fact that the sum of the oxytocin- and vasopressin-stained cells in either the PVH or the SO corresponds to the total number of magnocellular neurons in each nucleus (Swaab et al 1975a,b, Rhodes et al 1981a, Saw-

Table 1 Neuroactive substances identified in the paraventricular and supraoptic nuclei

Antigen	Comments	References
A. Cell Bodies		
Oxytocin (also neurophysin I) Vasopressin (also neurophysin II)	Topographically segregated in SO and magno-cellular division of PVH; also some in parvo-cellular division of PVH.	Swaab et al 1975a,b, Vandesande & Dierickx 1975, Rhodes et al 1981a, Sawchenko & Swanson 1982a
Somatostatin	Periventricular part of parvocellular division of PVH.	Dierickx & Vandesande 1979, Fisher et al 1979, Sawchenko & Swanson 1982a
Dopamine	Periventricular and adjoining regions of parvo-cellular division of PVH.	Hökfelt et al 1973, Swanson et al 1981
Methionine-enkephalin Leucine-enkephalin	SO, magnocellular and parvocellular divisions of PVH. More met- than leu-enkephalin-stained cells in PVH.	Rossier et al 1979, Finley et al 1981a, Sawchenko & Swanson 1982a
Neurotensin	Medial and periventricular parts of parvocellu-lar division of PVH.	Kahn et al 1980
Gastrin-, cholecystokinin-like peptide(s)	Distribution incompletely characterized; some in dorsal (oxytocinergic) part of SO.	Innis et al 1979, Lorén et al 1979a, Vanderhaegen et al 1980
Dynorphin	Coexists with vasopressin in magnocellular division of PVH, SO.	Watson et al 1981, 1982
Substance P	Distribution incompletely characterized; ap-parently localized primarily in cells of parvo-cellular division of PVH.	Ljungdahl et al 1978
Glucagon	Vasopréssinergic regions of SO, magnocellular division of PVH.	Tager et al 1980

Table 1 *(Continued)*

Antigen	Comments	References
Renin	Oxytocinergic parts of magnocellular division of PVH, SO; absorption controls incomplete.	Fuxe et al 1980
Corticotropin releasing factor	Anterior and medial parts of parvocellular division of PVH.	Bloom et al 1982; Swanson et al 1982
Pro-opiocortin-derived peptides (ACTH 1–39, β-endorphin)	Magnocellular division of PVH, SO; absorption controls inconclusive.	Joseph & Sternberger 1979, Watkins 1980

Antigen	Distribution	Source	References
B. Fibers and Terminals			
Norepinephrine	Most parts of parvocellular division of PVH; vasopressinergic areas of magnocellular division of PVH and SO.	A1, A2, A6 catecholamine cell groups.	Swanson et al 1981, Sawchenko & Swanson 1981b, 1982b
Epinephrine	Most parts of parvocellular division of PVH.	C1, C2 catecholamine cell groups; not proven directly.	Hökfelt et al 1974, Swanson et al 1981, Sawchenko & Swanson 1981b, 1982b
Serotonin	Sparse inputs to parvocellular division of PVH; oxytocinergic part of magnocellular division, and SO.	Dorsal, median nuclei of the raphe; nucleus raphe magnus and ventral medullary serotonergic cell groups may also contribute.	Bobillier et al 1976, Loewy et al 1981, Steinbusch 1981, Sawchenko & Swanson 1981c
Acetylcholine	"Cholinoceptive" neurons described in PVH and SO (of cat); details of distribution not yet available.	Unknown.	Kimura et al 1981

ACTH (1–39)	Periventricular, medial and dorsal parts of parvocellular division of PVH; oxytocinergic regions of magnocellular division of PVH, and SO.	Arcuate nucleus of hypothalamus.	Joseph 1980, Sawchenko et al 1982
β-Endorphin	Presumably similar to that of ACTH (1–39).	Arcuate nucleus of hypothalamus.	Bloom et al 1978, Finley et al 1981b
α-MSH	Presumably similar to that of ACTH (1–39).	Arcuate nucleus of hypothalamus.	Jacobowitz & O'Donohue 1978, O'Donohue et al 1979
VIP	Periventricular part of parvocellular division of PVH.	Unknown. Perhaps suprachiasmatic nucleus of hypothalamus.	Sims et al 1980, Card et al 1981
GABA	Present in both PVH and SO; detailed distribution unknown.	Unknown	Pérez de la Mora et al 1981
TRH	Both divisions (primarily parvocellular) of PVH.	Unknown.	Hökfelt et al 1975, Johansson & Hökfelt 1979
LHRH	Sparse input to PVH (hamster).	Unknown.	Jennes & Stumpf 1980
Pancreatic Polypeptide	Fibers in PVH.	Unknown.	Lorén et al 1979b
Brandykinin	Few fibers in PVH.	Unknown.	Corrêa et al 1979
Molluscan Cardioexcitatory Peptide	Few fibers in PVH and SO.	Unknown.	Weber et al 1981
Prolactin	Fibers in PVH and SO.	Unknown.	Fuxe et al 1977
Angiotensin II	Scattered punctate varicosities in PVH and SO.	Unknown	Fixe et al 1976, Changaris et al 1978

chenko & Swanson 1982a) and by the fact that both the PVH and the SO contain substantial numbers of oxytocin- and vasopressin-stained cells, which are topographically organized in a way that varies from species to species (see Dierickx 1980 for a review).

In the SO of the rat, a clear but incomplete topographic segregation of oxytocin- and vasopressin-stained cells has been found, with oxytocin-stained cells being concentrated anterodorsally, and vasopressin-stained cells concentrated posteroventrally (Swaab et al 1975b, Vandesande & Dierickx 1975, Sokol et al 1976, McNeill & Sladek 1980, Rhodes et al 1981a, Swanson et al 1981). Estimates of the ratio of vasopressin- to oxytocin-stained cells in the SO range from about 1:1 (Vandesande & Dierickx 1975) to 1.6:1 (Swaab et al 1975a,b).

The organization of oxytocin- and vasopressin-immunoreactive neurons in the PVH is no less distinctive. The anterior and medial magnocellular parts of the nucleus are composed almost exclusively of some 400–600 oxytocin-stained cells (Rhodes et al 1981a, Sawchenko & Swanson 1982a). In the posterior magnocellular part of the PVH, oxytocin-stained cells are concentrated anteroventromedially, and vasopressin-stained cells are concentrated posterodorsolaterally, in a way that corresponds quite precisely to the medial and lateral subdivisions recognized by Hatton et al (1976) in Nissl material. This correspondence is underscored by the observations that cells in the lateral subdivision are significantly larger than those of the medial subdivision (Hatton et al 1976), and that vasopressin-stained cells throughout the PVH are larger, on the average, than oxytocin-stained cells (Sawchenko & Swanson 1982a). There is general agreement that the posterior magnocellular part of the PVH in the rat consists of roughly equal numbers of oxytocin- and vasopressin-stained cells (about 1000 of each type; Swaab et al 1975a,b, Vandesande & Dierickx 1975, Rhodes et al 1981a, Sawchenko & Swanson 1982a).

In a recent analysis of the distribution of oxytocin- and vasopressin-stained cells in each of the cytoarchitectonically defined subdivisions of the PVH (Sawchenko & Swanson 1982a), it was found that 31% and 20% of the respective total numbers of cells were localized in the *parvocellular* division of the nucleus. Measurements of the cross-sectional areas of each cell type indicated that although some of these immunoreactive cells may be considered, on the basis of size, to be displaced magnocellular neurons, the oxytocin- and vasopressin-stained cells in the parvocellular division are, on the average, significantly smaller than those in the magnocellular division. This result, which suggests the existence of a separate population of parvocellular neurohypophysial peptide-containing neurons, is not without precedent, since a discrete population of small vasopressin-containing cells in the suprachiasmatic nucleus is well-known (Vandesande & Dierickx

1975). It would appear, then, that neither size nor precise location can be inferred from the presence of oxytocin or vasopressin within a particular neuron.

SOMATOSTATIN   The tetradecapeptide somatostatin inhibits the release of growth hormone from the anterior pituitary, and a substantial number (nearly 800) of somatostatin-stained cells is found in the PVH (Sawchenko & Swanson 1982a), primarily in the periventricular part of the parvocellular division of the nucleus (Dierickx & Vandesande 1979, Fisher et al 1979, Sawchenko & Swanson 1982a). There are no reports of somatostatin-stained cells in the SO. Somatostatin-immunoreactive cells are typically small and bipolar, with primary dendrites that are oriented vertically, and are clearly contiguous with a larger group of cells that occupies the anterior periventricular nucleus of Krieg (1932) throughout much of the anterior hypothalamus. Although widely scattered somatostatin-stained cells are found in the magnocellular division of the PVH, Dierickx & Vandesande (1979) used a double-immunoperoxidase staining method to establish that somatostatin and the neurohypophysial peptides are contained within separate populations of cells.

DOPAMINE   Tyrosine hydroxylase (TH) is involved in catecholamine biosynthesis, and may, with appropriate controls, be used as a marker for dopaminergic cells in immunohistochemical studies. Confirming earlier suggestions by Hökfelt and his colleagues (1973, 1978), it has recently been shown that some 500 TH-stained cells are found in the periventricular part, and in adjacent regions of the anterior and medial parvocellular parts, of the PVH (Swanson et al 1981). Because no cells in the PVH are stained with antisera against dopamine-β-hydroxylase or phenylethanolamine-N-methyltransferase (markers for noradrenergic and adrenergic cells, respectively), the TH-containing neurons are probably dopaminergic. Although the distribution of TH-stained cells is somewhat similar to that of somatostatin-stained cells, they are not identical. The zone containing TH-stained cells in the medial region of the PVH is somewhat wider than that containing somatostatin-stained cells, and, in addition, some TH-stained cells are found in the lateral parvocellular part of the nucleus, where they appear to form an anteromedial extension of the A13 group of dopaminergic neurons (Bjorklund & Nobin 1973). It has not yet been determined whether dopamine and somatostatin coexist within some individual hypothalamic neurons.

OPIATE PEPTIDES   Methionine- and leucine-enkephalin, endogeneous pentapeptides with opioid activity, have been localized in somata throughout the central nervous system, including the PVH and SO. The distribution of such cells is somewhat controversial, and depends largely on whether experimental animals have been pretreated with colchicine, which disrupts axonal transport by interfering with microtubule aggregation. Some evidence suggests, however, that cell body staining with antisera against enkephalin may be in part artifactual in animals pretreated with colchicine (Bayon et al 1979, Guillemin 1979).

In the non-pretreated rat, some 100 met-enkephalin-, and some 40 leu-enkephalin-immunoreactive cells were counted in the PVH, and were localized primarily in the posterior magnocellular and in the medial parvocellular parts of the nucleus (Sawchenko & Swanson 1982a). A far greater number of enkephalin-immunoreactive cells, with a distribution that overlaps partially with that of the neurohypophysial peptide-containing cell bodies, have been described in the non-pretreated cat (Micevytch & Elde 1980).

In rats pretreated with colchicine, many enkephalin-stained cells have been visualized in the PVH and SO (Rossier et al 1979, Sar et al 1979, Finley et al 1981a), but descriptions of the immunoreactive cells in these studies are not sufficiently detailed to allow comparisons with the distribution of oxytocin- and vasopressin-stained cells. Although it has not yet been shown that enkephalin and oxytocin or vasopressin are found within individual hypothalamic neurons, it has recently been demonstrated that single terminals in the posterior pituitary can be stained with antisera to enkephalin, and to oxytocin, vasopressin, or neurophysin (Coulter et al 1981, Martin & Voigt 1981). The co-localization experiments of Martin & Voigt (1981) suggest that met-enkephalin staining is usually, but not always, found in vasopressinergic endings.

Dynorphin is an extremely potent opiate heptadecapeptide that has an NH_2-terminal pentapeptide sequence corresponding to that of leu-enkephalin (Goldstein & Ghazarossian 1980, Goldstein et al 1981). Many dynorphin immunoreactive cells have been found in the PVH and SO, along with stained terminals in the posterior lobe (Watson et al 1981). Dynorphin and vasopressin have recently been shown to coexist within magnocellular neurons in both the PVH and the SO (Watson et al 1982).

Finally, isolated reports suggest that magnocellular neurons in the PVH and SO may be stained with antisera directed against various pro-opiocortin-derived peptide fragments (e.g. Larsson 1977, Bugnon et al 1979, Joseph & Sternberger 1979, Watkins 1980). These results are at odds with the prevailing view that the cells which give rise to the intracerebral opiocortin system are localized exclusively in the arcuate nucleus and immediately adjoining regions of the basomedial hypothalamus (Joseph et al 1980, Knigge et al 1980). Finley et al (1981b) reported β-endorphin-stained magnocellular neurons in the PVH, but absorption of the antiserum with vasopressin eliminated the staining. Similarly, Watkins (1980) stained many magnocellular cells with an antiserum against ACTH (1–39). Absorption with neurophysin abolished staining in some, but not all, experiments; absorption controls with oxytocin or vasopressin were not performed. Although results such as these suggest that the detection of opiocortin-derived peptides in the hypothalamo-neurohypophysial system is artifactual, the recent isolation from bovine neurohypophysial extracts of an 80,000 dalton protein, from which neurophysin-, vasopressin-, β-endorphin-, and ACTH (1–39)-immunoreactive peptides could be obtained (Lauber et al 1981), raises the possibility that a large common precursor for both the classical neurohypophysial peptides and various pro opiocortin-derived peptides may exist in the PVH and the SO.

OTHER PEPTIDES   Many other neuropeptides have been localized in the PVH and the SO, although their precise distributions have not been described. Several appear to be found primarily in the parvocellular division of the PVH. Substance P- (Ljungdahl et al 1978) and neurotensin-stained (Kahn et al 1980) cells fall in this category. The distribution of neurotensin is particularly interesting since, based on the available illustrations (Kahn et al 1980), a substantial population of cells in the medial parvocellular and periventricular parts of the PVH appear to be stained. Preliminary immunohistochemical localization of the recently isolated corticotropin-releasing factor (CRF; Vale et al 1981) indicate that this peptide may be distributed similarly within the parvocellular division of the PVH (Bloom et al 1982; Swanson et al 1983).

An as yet uncharacterized peptide, or class of peptides, in the gastrin-cholecystokinin family has been localized in cells of the PVH (Innis et al 1979, Lorén et al 1979a, Vanderhaegen et al 1980) and the SO (Vanderhaegen et al 1980). The distribution of such immunoreactive cells in the PVH is unclear, but, using an antiserum against the terminal octapeptide of cholecytokinin that was cross-adsorbed in the solid phase against oxytocin and neurophysin, Vanderhaegen et al (1980) found staining of many cells in the dorsal (oxytocinergic) part of the SO. Similarly, cells that appear to lie in oxytocinergic regions of the PVH and SO have been stained with antisera against renin (Fuxe et al 1980), but detailed absorption controls were not included in this study. Finally, antisera against pancreatic glucagon have been shown to stain magnocellular neurons in vasopressinergic parts of the PVH and the SO in a specific manner (Tager et al 1980); pre-incubation of the antiserum with vasopressin did not eliminate this staining.

It must be pointed out that until the peptides that have been reported in magnocellular parts of the PVH and SO are completely characterized, and their distributions in the hypothalamus and pituitary studied thoroughly, the possibility that they coexist with oxytocin or vasopressin in single cells must be viewed with caution. Thus far, convincing evidence for colocalization has been marshalled only for enkephalins and the neurohypophysial peptides, and this, too, is incomplete (see above). It may be no coincidence that many of the peptides reported in cells of the PVH and SO are also found in the blood. This, coupled with the fact that the PVH and SO are the most highly vascularized parts of the hypothalamus (Ambach & Palkovits 1979), suggests the possibility that neuroactive substances localized in these nuclei may arise from the periphery, or from other parts of the brain.

These caveats notwithstanding, it is clear that a number of biochemically distinct cell types are found in these nuclei. As discussed in the next section, the possible significance of this diversity has just begun to be explored.

PROJECTIONS

Supraoptic Nucleus

Cajal (1894) first demonstrated convincingly that the SO projects massively to the neural lobe of the pituitary (in the mouse); no additional pathways from the nucleus have been firmly established since. In fact, Rasmussen (1940) reported that "practically all" of the cells in the SO disappear after extensive ablations of the pituitary and median eminence, suggesting in addition that the SO contains few, if any, interneurons. For many years it was thought that the supraoptico-hypophysial tract is involved in altering the release of pressor, anti-diuretic, and oxytocic factors synthesized by pituicytes in the neural (or intermediate) lobe of the pituitary (Ranson & Magoun 1939), a view that was dramatically changed by Bargmann's histological confirmation, with the Gomori stain for colloidal secretory material, of Scharrer's hypothesis that the neurohypophysial factors are synthesized in the SO (and the PVH) and are carried down the hypothalamo-hypophysial tract to the pituitary (Bargmann & Scharrer 1951). This observation, when viewed in the light of Weiss & Hiscoe's (1948) almost simultaneous demonstration of anterograde axonal transport, marked the beginning of modern neuroendocrinology, and the magnocellular neurosecretory system has since become the best understood peptidergic system in the CNS. Oxytocin and vasopressin were isolated and synthesized almost 30 years ago (du Vigneaud 1954–1955), and, as discussed above, specific immunohistochemical methods have since shown that separate neurons in the SO (and in the PVH) synthesize these two nonapeptides. The extent to which oxytocin- and vasopressin-containing fibers in the hypothalamo-neurohypophysial tract may be segregated topographically is still not known. It has been shown autoradiographically that fibers from the SO (and the PVH) on one side of the brain course through the internal lamina of the median eminence on both sides of the midline, and thus presumably end throughout much of the posterior lobe on both sides of the midline (Saper et al 1978). The functional significance of such a bilateral projection, first described by Laruelle (1934) in normal material, is not clear. It is also worth pointing out that there have been many reports of Gomori-stained fibers passing between the neural and the intermediate lobes (Christ 1966), although the existence of such fibers has not been studied in detail in more recent immunohistochemical studies.

It is important to stress that the cells of the SO are neurons despite the fact that their principal, and perhaps only, projection is involved in the delivery of hormones to capillaries, rather than to other neurons. The predominantly unmyelinated axons of the supraoptico-hypophysial tract are known to conduct action potentials at a velocity of somewhat less than 1 m/sec (Yagi et al 1966), and the release of hormones is a calcium-dependent process involving exocytosis (Douglas 1974), which appears to be closely related to the amount of electrical activity in the hypothalamo-neurohypophysial tract (see Lincoln & Wakerly 1975). In addition, since the magnocellular neurosecretory system is so well-defined anatomically, it has provided a useful model system for studying the biosynthesis and post-translational processing of peptide neurohormones (see Brownstein et al 1980).

Paraventricular Nucleus

Despite the obvious morphological similarities between the large cells in the SO and the PVH, the paraventriculo-hypophysial tract was not demonstrated until 1936, when Ingram and colleagues found retrograde degeneration in the PVH of the cat after hypophysectomy. The full extent of this tract was difficult to trace in normal material because it joins the supraoptico-hypophysial tract in a region just dorsal to the SO itself. It had been suspected for many years that there are major differences in the projections of the SO and the PVH, since retrograde cell loss always appears considerably less severe in the PVH after hypophysectomy. Christ (1966) suggested that such changes are less obvious in the PVH simply because the axons of cells in the nucleus are longer than those from the SO. An alternative explanation was offered by Rasmussen (1940), and later in more detail by Palay (1953), both of whom suggested that the Gomori-stained projections from the PVH end at varying levels within the hypothalamus and the median eminence, and that few fibers may actually reach the posterior lobe itself (Rasmussen 1940). Somewhat surprisingly, the existence of such intrahypothalamic endings has not been examined in detail at the ultrastructural level, although an input from the PVH to the external lamina of the median eminence has been confirmed (see below).

Bodian & Maren (1951), in an infrequently cited paper, clearly resolved this problem, however. Following an earlier suggestion by Harris (1948), they distinguished between magnocellular and parvocellular parts of the PVH in the rat, and showed that cells in the former undergo retrograde changes that are fully comparable to those found in the SO after hypophysectomy, while cells in the latter remain largely unaffected. They also made the striking observation, since largely forgotten, that the extent of retrograde changes in the PVH is also influenced by damage to the *anterior* lobe, which led them to postulate the existence of a "trophic factor" in the anterior lobe that acts on the magnocellular neurosecretory system (see Baertschi et al 1982 for recent evidence supporting this view). Recent experiments, in which retrograde tracers such as HRP (Sherlock et al 1975, Kelly & Swanson 1980) and true blue (Swanson & Kuypers 1980) have been injected into the pituitary, confirm that at least a large majority of the cells in all parts of the magnocellular division of the PVH project to the posterior lobe. In addition, it is clear that scattered cells in all parts of the parvocellular division of the nucleus project to the posterior lobe as well (Swanson & Kuypers 1980). The course of the paraventriculo-hypophysial tract has already been mentioned, and a more detailed account can be found in Vandesande & Dierickx (1975); it is not known whether there is a partial topographic segregation of oxytocinergic and vasopressinergic fibers in the tract, which consists almost entirely (although not exclusively) of unmyelinated axons (see Christ 1966).

In addition to oxytocin, vasopressin, and the associated neurophysins, the paraventriculo-hypophysial tract may also contain enkephalins, because (*a*) they are present in high concentrations in the posterior lobe (Bloom et al 1977), (*b*) lesions of the PVH reduce these concentrations (Rossier et al 1979), (*c*) enkephalin-stained cells are found in the PVH (see above), and (*d*) enkephalin- and vasopressin-(or oxytocin-)immunoreactivities have been detected within individual terminals in the posterior lobe (Coulter et al 1981, Martin & Voigt 1981).

A direct input to the external lamina of the median eminence from the magnocellular neurosecretory system, has, as mentioned earlier, been inferred on the basis of indirect evidence for some time (Rasmussen 1940). More recently, Gomori-positive fibers have been described in the external lamina of a wide range of vertebrates (see Benoit & Assenmacher 1955, Dierickx & Van den Abeele 1959), including the rat (Rinne 1960). These observations are of interest because they suggest that oxytocin and vasopressin may be released into the capillary plexus of the external lamina that is part of the "neuro-vascular" link between the hypothalamus and the anterior lobe of the pituitary (Harris 1948), and thus may influence the release of one or another of the adenohypophysial hormones. This hypothesis gained support from the observation that the number of Gomori-positive fibers, and the size of individual stained granules, in the external lamina is greatly increased by adrenalectomy or hypophysectomy (Bock 1970, Rinne 1970, Wittkowski & Bock 1972). In addition, treatment with glucocorticoids or corticotropin, respectively, prevents these effects (see Brinkmann & Bock 1973). In recent years, the origin and biochemical identity of this input has been clarified using immunohistochemical methods. With differentially cross-adsorbed antisera to vasopressin, oxytocin, and their respective neurophysins, Dierickx et al (1976) have shown that most of these fibers are vasopressinergic, although a small number are oxytocinergic, and that both types disappear from the external lamina after lesions of the PVH (Vandesande et al 1977, see also Antunes et al 1977). Evidence that this pathway influences the release of ACTH from the anterior lobe has been reviewed recently by Maraka et al (1981).

After the injection of retrograde tracers such as HRP and true blue into the external part of the median eminence, heavily labeled cells are concentrated in the periventricular, medial (dorsal half), and anterior parts of the parvocellular division of the PVH, areas that project only lightly to the posterior lobe (Weigand & Price 1980, Lechan et al 1980). Based on the number and the distribution of retrogradely labeled cells found in these experiments, it is apparent that neuroactive substances other than oxytocin and vasopressin must also be contained within this pathway. Quite variable numbers of lightly labeled cells were also reported in the magnocellular division of the PVH, and in the SO (which does not project to the external lamina; see above), suggesting either that some injected tracer spread to label fibers (in adjacent parts of the internal lamina) destined for the posterior lobe, or that some magnocellular neurons project both to the posterior lobe and to the external lamina. Similarly, electrophysiological studies have shown that a small number of individual cells in the PVH can be antidromically activated by electrical stimulation of both the external lamina and the posterior lobe (Pittman et al 1981), although the possibility that current spread from the external to the internal lamina of the median eminence in these experiments cannot be excluded. In summary, the evidence clearly indicates that neurons projecting to the median eminence are concentrated in medial regions of the parvocellular division of the PVH, and that neurons projecting to the posterior lobe are concentrated in the magnocellular division, although it is still not clear whether the PVH contains a small population of cells that projects to both areas. However, the suggestion that the external lamina is innervated by a separate group of parvocellular neurons is supported by the observation that Gomori-positive (Wittkowski & Bock 1972) and neurophysin-stained (Silverman & Zimmerman 1975) secretory granules in this region are distinctly smaller than those in the posterior lobe.

The existence of extrahypothalamic pathways associated with the magnocellular neurosecretory system was inferred in a variety of mammals by workers using the Gomori stain. In the most elegant studies of this kind, Barry (1958) described in detail the course of such fibers from the hypothalamus to a variety of sites in the forebrain and midbrain including the septum, the amygdala, the habenula, the tectum, the pineal gland, and the reticular formation. The significance of these observations was unclear, however, because the specificity of the Gomori stain is uncertain, and because such fiber systems are difficult to demonstrate under even optimal conditions (Dierickx 1980). Nevertheless, many of Barry's findings have been confirmed immunohistochemically, particularly by Buijs (1978) and by Sofroniew & Weindl (Weindl & Sofroniew 1976, Sofroniew & Weindl 1978, Sofroniew 1980), although the precise origins of these pathways in the forebrain have not yet been determined experimentally.

In contrast, long, descending pathways from the PVH have attracted considerable attention in the last five years. The seminal work of Kuypers & Maiskey (1975) with the HRP method demonstrated convincingly for the first time that cells in medial and lateral parts of the hypothalamus project directly to the spinal cord in the cat. This observation was quickly confirmed with the same method in the rat, cat, and monkey by Saper et al (1976), who showed that cells in the paraventricular nucleus, the lateral hypothalamic area, the zona incerta, and the posterior hypothalamic area contribute to this projection. The paraventriculo-spinal pathway has now been firmly established with the HRP method (Hancock 1976, Cratcher et al 1978, Kneisley et al 1978, Ono et al 1978, Blessing & Chalmers 1979, Hosoya & Matsushita 1979); however, it is important to point out that retrogradely labeled neurons in the SO have not been described in these studies. The functional significance of the descending projection from the region of the PVH to the spinal cord was also clarified by Saper et al (1976), who demonstrated autoradiographically a pathway to the intermediolateral column in thoracic levels of the cord after large ^{3}H-amino acid injections that involved the PVH and adjacent regions of the zona incerta and the lateral hypothalamic area. In the same material, a projection to the dorsal motor nucleus of the vagus nerve and to most parts of the nucleus of the solitary tract (NTS) was identified, confirming and extending an earlier report of an input to the rostral, taste-related, part of the NTS from the region of the PVH (Conrad & Pfaff 1976). The contribution of cells in the PVH itself to this input to the dorsal vagal complex was established with the HRP method (Saper et al 1976). This evidence strongly suggested that cells in the PVH, but not in the SO, project directly to preganglionic cell groups of both the parasympathetic and sympathetic divisions of the autonomic nervous system, and to the principal sensory nucleus of the vagus and glossopharyngeal nerves (the NTS), which is concerned with the relay of visceral information to a variety of sites in the CNS, including the PVH itself (see below).

Further studies have clarified the distribution and morphology of cells in the PVH that give rise to the descending autonomic projections, have examined the possibility that single cells in PVH give rise to collaterals that innervate more than one major terminal field, and have identified a surprising diversity of biochemically distinct cell types that project to autonomic centers. Hosoya & Matsushita (1979) first emphasized that, in the rat, cells in the PVH that are retrogradely labeled with HRP after injections in the spinal cord are smaller than, and are topographically separate from, cells in magnocellular neurosecretory parts of the nucleus. They also showed that the HRP-labeled cells tend to differ from magnocellular neurosecretory neurons at the ultrastructural level, and suggested that different populations of cells in the PVH project to the spinal cord and to the posterior pituitary. Retrograde transport studies with true blue (Swanson & Kuypers 1980) confirmed that most cells projecting to the spinal cord are confined to the parvocellular division of the PVH, although scattered cells are also found in the posterior part of the magnocellular division. Thus, because some cells that project to the spinal cord lie within the magnocellular division of the PVH, and because some cells that project to the posterior lobe lie within the parvocellular division (Swanson & Kuypers 1980), it was still possible that some cells might project to both areas. This possibility was tested directly with a double retrograde labeling method, and the results indeed showed that essentially separate groups of cells in the PVH project to the cord and to the posterior lobe, although a very small number of small, doubly-labeled neurons in the medial and lateral parts of the parvocellular division were identified (Swanson & Kuypers 1980). The latter observation has recently been confirmed electrophysiologically by Zerihun & Harris (1981), who also suggest that descending fibers from the PVH to the medulla are unmyelinated, because they have a conduction velocity of less than 1m/sec.

The cells in the PVH that project to the spinal cord are concentrated in the dorsal, lateral, and medial (ventral half) parts of the parvocellular division of the nucleus, although scattered cells are also found in all other parts of the parvocellular division, and in the posterior part of the magnocellular division (Swanson & Kuypers 1980). The distribution of cells that project to the dorsal vagal complex is quite similar, with the notable exception that very few cells in the dorsal parvocellular group contribute to this pathway (Swanson & Kuypers 1980). Direct cell counts confirm that injections of true blue in the spinal cord retrogradely label more cells in the PVH than do injections in the dorsal vagal complex (about 950 compared to about 650, respectively; Swanson & Kuypers 1980, see also Hosoya

1980), and the functional significance of the dorsal parvocellular group in particular is underscored by the finding that at least 88% of its neurons project to the spinal cord (Sawchenko & Swanson 1981a).

Double retrograde labeling studies indicate that at least 15% of the cells in the PVH that project to thoracic levels of the spinal cord also send a collateral to the dorsal vagal complex (Swanson & Kuypers 1980). It appears, therefore, that a significant number of individual cells in the ventral half of the medial parvocellular part, and in the lateral parvocellular part, of the PVH innervate both sympathetic and parasympathetic cell groups. And while the estimate of the proportion of cells with such collaterals may be conservative for a variety of technical reasons, it is clear that most of the cells in the dorsal parvocellular group project to the spinal cord and not to the dorsal vagal complex. In addition, similar double-labeling experiments show clearly that cells in the PVH projecting to the median eminence do not send axon collaterals to the dorsal vagal complex or to the spinal cord (Swanson et al 1980).

Taken together, then, the evidence suggests that essentially separate, and topographically distinct, groups of cells in the PVH project to the posterior pituitary, to the median eminence, and to autonomic centers in the brain stem and spinal cord, although some cells in the latter group appear to innervate both sympathetic and parasympathetic preganglionic cell groups. The functional significance of this organization, which is illustrated in Figures 1 and 2, is discussed more fully below.

The demonstration of a projection from the PVH to the dorsal medulla and spinal cord suggested immediately that this pathway contains either oxytocin or vasopressin. In the first immunohistochemical examination of this problem, differentially cross-adsorbed antisera to bovine neurophysin I and neurophysin II were used to show that neurophysin-I-stained (presumably oxytocinergic) fibers could be traced from the hypothalamus to the dorsal vagal complex and to the spinal cord in both the cow and the rat (Swanson 1977). Such fibers descend initially through the medial forebrain bundle, and then, after passing between the substantia nigra and the red nucleus, continue through ventrolateral parts of the reticular formation to enter the dorsolateral funiculus of the spinal cord; in the pons, fibers leave this pathway to innervate the parabrachial nucleus and the locus coeruleus, while in the medulla some fibers arch dorsally to innervate the dorsal motor nucleus of the vagus and the NTS. In addition, a second labeled pathway descends through the central gray and appears to innervate the Edinger-Westphal nucleus, the locus coeruleus, and perhaps the central gray as well. Pathways to the central gray, NTS, and spinal cord (central gray) were also stained with an antiserum to neurophysin by Sofroniew & Weindl (1978). Essentially similar results were obtained with antisera to oxytocin by Buijs

(1978), who also reported that a much smaller proportion of the fibers in the brainstem and spinal cord are stained with anti-vasopressin, and by Swanson & Hartman (1980).

Detailed comparisons of the distribution of oxytocin- and vasopressin-stained fibers in the medulla and spinal cord have not been carried out, although Nilaver et al (1980) reported that presumed vasopressinergic fibers predominate in the dorsal vagal complex whereas presumed oxytocinergic fibers predominate in the spinal cord. In this study, antisera to rat neurophysins, and to rat neurophysins cross-adsorbed with hypothalamic extracts from Brattleboro rats (which do not synthesize vasopressin and its associated neurophysin), were used to stain fibers in the normal and Brattleboro rat. However, a careful reexamination of this problem with differentially cross-adsorbed antisera to oxytocin and vasopressin (P. E. Sawchenko and L. W. Swanson, in preparation) clearly shows that oxytocin-stained fibers predominate in both areas. It is also clear from this work that whereas oxytocin-stained fibers course throughout virtually all parts of the NTS, they are concentrated in the medial division of the nucleus at the level of the area postrema. In addition, such fibers are not distributed uniformly in the dorsal motor nucleus; rather, they are concentrated in the medial and lateral tips of the nucleus.

The functional significance of the oxytocin-stained input to the dorsal vagal complex is not yet clear, although recent evidence suggests that oxytocin may be released as a neurotransmitter in this area, rather than as a neurohormone as it is in the posterior pituitary. Thus, some conventional asymmetrical synapses with round, clear vesicles in the NTS have been stained with an antiserum to oxytocin (R. M. Buijs, personal communication); iontophoretically applied oxytocin changes the firing rate of some neurons in the dorsal medulla (Morris et al 1980); and oxytocin is released from tissue slices of the medulla by potassium, in a calcium-dependent manner (R. M. Buijs, personal communication). In addition, degenerating terminals have been demonstrated at the ultrastructural level in the dorsal motor nucleus of the vagus after lesions that involve the PVH (Akmayev et al 1981).

The distribution of oxytocinergic fibers in the spinal cord has been studied in detail (Swanson & McKellar 1979). In the rat, such fibers course the length of the cord in the dorsolateral funiculus, and extend into the filum terminale. Within the gray matter three distinct regions contain labeled fibers: the intermediolateral column, the central gray, and the marginal zone of the dorsal horn. The latter two terminal fields extend the length of the cord, and all parts of the intermediolateral column contain at least a few fibers. However, it is clear that the column is particularly densely innervated at some levels (e.g. T_{1-3}, T_9, and L_1), and that even within a particular segment of the cord some groups of preganglionic neurons are more densely innervated than others. The evidence indicates, therefore, that certain descending hypothalamic inputs differentially innervate specific subpopulations of preganglionic autonomic neurons in the spinal cord, and, as mentioned above, in the dorsal motor nucleus of the vagus as well. It is also possible that oxytocinergic fibers modulate the processing of nociceptive, and perhaps thermal, information in the marginal zone of the dorsal horn, and in the spinal nucleus of the trigeminal nerve (Swanson & Hartman 1980).

The descending pathways discussed above were studied in normal material. However, their origin from the PVH could only be inferred because, although an origin from the SO seemed unlikely on the basis of retrograde transport studies, they could also arise from the accessory neurosecretory cells scattered outside the limits of the PVH and SO in other parts of the hypothalamus, in the zona incerta, and in the septum (Peterson 1966, Kelly & Swanson 1980). Direct evidence that these descending oxytocin- and vasopressin-stained pathways arise in the PVH has been obtained with the use of a combined retrograde tracer-immunofluorescence method (Sawchenko & Swanson 1980, 1982a). Injections of true blue were made into either the dorsal vagal complex or into upper thoracic segments of the spinal cord, and alternate sections were counterstained with differentially cross-adsorbed antisera to oxytocin and vasopressin. The results showed that all doubly-labeled cells for each antiserum were confined to the PVH, and that three to four times more oxytocin- than vasopressin-stained cells were retrogradely labeled after injections into either site. In addition, both types of doubly labeled neuron were found in parvocellular as well as magnocellular parts of the PVH, although they were more frequent in the medial and lateral parts of the parvocellular division. Sofroniew & Schrell (1981), using HRP as a retrograde tracer and using immunoperoxidase histochemistry on adjacent thin sections, have also reported that about four times more oxytocin- than vasopressin-stained cells in the PVH project to the medulla.

Interestingly, oxytocin- and vasopressin-stained cells account for only about 20% of the cells in the PVH that are retrogradely labeled after injections of true blue in the dorsal vagal complex and the spinal cord (Sawchenko & Swanson 1980, 1982a), thus suggesting that additional cell types may contribute to these projections. This has been found to be the case; with the retrograde tracer-immunofluorescence method, small populations of cells in the PVH that cross-react with antisera to (a) tyrosine hydroxylase (and are thus presumably dopaminergic) (Swanson et al 1981), (b) somatostatin, and (c) met-enkephalin have been found to project to the dorsal vagal complex and to the spinal cord (Sawchenko & Swanson 1980, 1982a). Each of the five cell types thus far identified have a unique distribution within the PVH, and together they still account for only about a quarter of the total population of retrogradely labeled cells in the nucleus.

In summary, the evidence suggests that at least five cell types, each with a unique distribution within the PVH, give rise to descending inputs to autonomic centers in the medulla and spinal cord. In addition, the PVH contains at least three distinct groups of oxytocin- and vasopressin-stained neurons: one group consists of large neurons that project to the pituitary; another group consists of smaller cells that give rise to descending projections to the brain stem and spinal cord; a third group of smaller cells projects to the external lamina of the median eminence.

NEURAL INPUTS

Less than ten years ago, Cross & Dyball (1974) wrote, "The sad truth is ... that no well defined and documented system of afferent fibers has been revealed by morphological procedures to either the paraventricular or the supraoptic nucleus" (p. 270). This was due in part to the inability of degen-

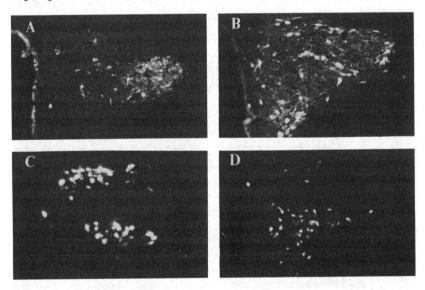

Figure 1 Photomicrographs to indicate different cell types at approximately the same level through the PVH. A. Cell bodies in the posterior magnocellular part, stained with anti-vasopressin. B. Cell bodies stained with anti-oxytocin. C. Retrogradely labeled neurons in the dorsal (*top*) and medial (*bottom*) parvocellular parts of the PVH after an injection of true blue in upper thoracic segments of the spinal cord. D. Retrogradely labeled neurons concentrated in the medial parvocellular part of the nucleus after an injection of true blue in the dorsal vagal complex. The third ventricle is to the left in each photomicrograph, and the sections were cut in the frontal plane. X 60 (reproduced with permission from Sawchenko & Swanson 1982a).

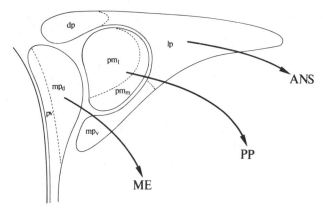

Figure 2 A summary of the major functional and morphological subdivisions of the PVH in the rat. Abbreviations: ANS, autonomic nervous system; ME, median eminence (external lamina); PP, posterior pituitary; dp, dorsal parvocellular part; lp, lateral parvocellular part; $mp_{d,v}$, dorsal and ventral subdivisions of the medial parvocellular part; $pm_{l,m}$, lateral (vasopressinergic) and medial (oxytocinergic) subdivisions of the posterior magnocellular part; pv, periventricular part.

eration techniques to impregnate small-diameter fibers destined for the hypothalamus, and in part to problems of interpretation since unrelated fibers-of-passage may have been damaged by lesions of presumed afferent cell groups or pathways. The introduction of sensitive neuroanatomical tracing methods that rely on the uptake and axonal transport of various marker molecules, and of immunohistochemical methods for identifying transmitter-specific cells and terminals, has circumvented many of these difficulties, and a flurry of recent studies has greatly clarified the distribution and cells of origin of afferent projections to the PVH and SO at the light microscopic level.

Three general features of the inputs to the PVH and SO have emerged, and are worthy of mention at the outset.

1. In line with morphological evidence that the dendrites of cells in the PVH and SO are restricted to the morphological boundaries of the nuclei, most inputs to the PVH, and, to a lesser extent to the SO, are quite discrete; the neuropil surrounding the nuclei are typically far less densely innervated.

2. Most inputs are restricted to specific parts of the nuclei, and they respect in a general way the boundaries between adjacent cytoarchitectonically defined subdivisions.

3. All known inputs to the magnocellular neurosecretory system display a similar distribution, with respect to oxytocinergic and vasopressinergic regions, in both the SO and the magnocellular division of the PVH.

In this section we summarize what is known about the distribution, cells of origin, and biochemical specificity of neuronal inputs from various levels of the CNS to the PVH and SO. Emphasis is placed on those projections whose distribution has been charted directly or can reasonably be inferred on the basis of immunohistochemical studies. A list of the neuroactive substances that have been localized in fibers or terminals in the PVH and the SO, and the cells of origin of these projections (if known), is given in Table 1.

Brain Stem

NORADRENERGIC AND ADRENERGIC   Noradrenergic projections are the best known, and from a functional standpoint the most thoroughly characterized, set of inputs to the PVH and SO. Shortly after the introduction of histofluorescence methods for the demonstration of catecholamine-containing neurons, the PVH and SO were shown to contain one of the most dense catecholaminergic terminal fields in the brain (Carlsson et al 1962, Fuxe 1965, Ungerstedt 1971, Lindvall & Bjorklund 1974). Nevertheless, until recently the origins of these pathways have remained obscure, and virtually every noradrenergic and adrenergic cell group described by Dahlström & Fuxe (1964) and by Hökfelt et al (1974) has been said to contribute to them (e.g. Fuxe et al 1970, Ungerstedt 1971, Brownstein et al 1976, Leibowitz & Brown 1980, Palkovits 1981). Similarly, the distribution within the PVH and SO of terminals from each of the contributing aminergic cell groups has only recently been clarified. This issue is important in view of evidence that specific subdivisions of the nuclei are innervated preferentially by adrenergic and by noradrenergic fibers (McNeill & Sladek 1980, Swanson et al 1981).

Recent studies with a variety of neuroanatomical methods (Sakumoto et al 1978, Swanson & Hartman 1980, Palkovits et al 1980a,b, Takagi et al 1980, Berk & Finkelstein 1981a, Tribollet & Dreifuss 1981) indicate that, in fact, three cell groups provide almost all of the noradrenergic and adrenergic fibers to the hypothalamus, including the PVH and SO. These include the A1 (and C1) cell group of the ventral medulla, which lies just dorsal to, but not within, the lateral reticular nucleus; the A2 (and C2) cell group centered in the medial part of the NTS; and the locus coeruleus (A6 cell group) in the lateral part of the pontine central gray. However, the distribution and number of aminergic cells that project to the hypothalamus were not determined in these studies, and experiments based on lesions may well have involved fibers-of-passage. The problem of the origin of the noradrenergic innervation of the PVH and SO has recently been reexamined with a combined retrograde transport-immunofluorescence method; injections of true blue were placed in the PVH or the SO, and the tissue was counter-stained with an antiserum against DBH to identify noradrenergic and adrenergic neurons (Sawchenko & Swanson 1981b, 1982b).

Injections of true blue that were centered in the PVH resulted in extensive labeling of DBH-stained neurons in the A1, A2, and A6 regions of the brain stem, and an average of 516, 197, and 43 cells were doubly labeled, respectively. In addition, the topography of doubly labeled cells in the A1 and A2 regions suggests that cells of the C1 and C2 adrenergic cell groups, which lie near the rostral poles of the A1 and A2 regions, also contribute to the catecholaminergic innervation of the PVH. This is consistent with the observation that the PVH (Hökfelt et al 1974), and particularly its parvocellular division (Swanson et al 1981), contains a relatively dense adrenergic (i.e. PNMT-stained) terminal field. Injections of true blue that were centered in the SO yielded double labeling in the same noradrenergic cell groups, but here the total number of doubly labeled neurons was less than that found after injections in the PVH, and an even greater portion of all doubly labeled cells were found in the A1 region. After injections centered in either the PVH or the SO, over 80% of all retrogradely labeled cells in the immediate vicinity of the A1, A2, and A6 cell groups were also found to be DBH-positive, indicating that the projection from each region is *primarily* noradrenergic. It is reasonable, therefore, to interpret the distribution of silver grains found over the PVH and SO after ^{3}H-amino acid injections centered in any one of these regions as representative of the pattern of DBH-stained inputs from the injected cell group.

Injections of ^{3}H-amino acids that involve the locus coeruleus label a moderately dense projection to the most medial, primarily periventricular, parts of the parvocellular division of the PVH (Jones & Moore 1977, Loewy & Saper 1978, Sawchenko & Swanson 1981b, 1982b). Only Jones & Moore (1977) reported a direct projection from the locus coeruleus to the SO.

Ricardo & Koh (1978) were the first to describe a direct projection from the NTS to the PVH. More detailed analyses have shown that injections centered in the A2 region of the NTS at the level of the obex label a moderately dense projection to most parts of the parvocellular division of the PVH that is most dense over the dorsal part, and the dorsal half of the medial parvocellular part, of the nucleus (Koh & Ricardo 1979, 1980, Sawchenko & Swanson 1981b, 1982b, McKellar & Loewy 1981). Again, there is no clear evidence that the A2 region projects to magnocellular parts of the PVH, or to the SO. It is important to point out that in the autoradiograms of Sawchenko & Swanson (1982b), evidence for a substantial projection from the NTS to the region of the A1 cell group was found. Double-labeling experiments with anti-DBH and true blue, after injections of the tracer into the A1 region, indicated that this pathway is primarily non-noradrenergic (Sawchenko & Swanson 1982b).

Finally, after injections centered in the A1 cell group, a dense projection to the parvocellular division of the PVH was labeled (Sawchenko & Swanson 1981b, 1982b); the distribution of this input was quite similar to that found after injections placed in the A2 region, but the density was considerably greater. In addition, and in contrast to the inputs from the A2 and A6 regions, a dense projection was traced from the A1 region to the magnocellular division of the PVH and to the SO, and was found to be distributed preferentially to those regions of both nuclei in which vasopressin-stained cells were concentrated (Sawchenko & Swanson 1981b, 1982b). It was also noted that en route to the hypothalamus, a clear projection from the A1 region diverged to end discretely and selectively in the locus coeruleus (Sawchenko & Swanson 1982b). Double-labeling experiments confirmed that this pathway is primarily noradrenergic.

These results indicate that the noradrenergic innervation of the PVH and the SO arises from three interrelated cell groups in the brain stem; the projection from each displays a unique distribution and relative density within the nuclei. These pathways probably convey visceroceptive information to the various functional classes of effector neurons in the PVH and SO—a conclusion that is based on a number of lines of evidence.

1. The NTS, which contains the A2 cell group, is the principal recipient of first order visceral afferent information carried by the vagus and glossopharyngeal nerves (Allen 1923, Torvik 1956), and projects to the parvocellular division of the PVH by a pathway that is primarily noradrenergic.
2. Non-noradrenergic cells in the NTS project to the region of the A1 cell group, which, in turn, projects massively to the parvocellular division of the PVH and to vasopressinergic regions in both the magnocellular division of the PVH and the SO.
3. Although physiological studies of the termination of primary and secondary visceral afferents have not examined the region of the A1 cell group in detail, Calaresu and his colleagues (Ciriello & Calaresu 1977, Thomas et al 1977) have found that single units in the vicinity of the lateral reticular nucleus respond to a variety of cardiovascular stimuli, including inputs from the carotid sinus nerve.
4. As discussed elsewhere (Sawchenko & Swanson 1982b), a substantial body of physiological evidence is congruent with the anatomical data in suggesting that the A1 region provides the route through which the control of vasopressin secretion by visceral afferent stimuli is achieved.
5. The locus coeruleus receives a substantial projection from the A1 region, and projects to the most medial parts of the parvocellular division of the PVH, and it has been shown that visceral afferent information from the

cardiovascular system reaches the locus coeruleus (Svensson & Thorén 1979, Ward et al 1980).

SEROTONERGIC   Studies of the normal distribution of serotonergic fibers in the hypothalamus with histofluorescence methods are in general agreement that such inputs to the PVH and SO are quite sparse (Fuxe 1965, Aghajanian et al 1973, Kent & Sladek 1978). In fact, recent immunohistochemical studies have shown that the neuropil surrounding the PVH and SO contains far more serotonergic fibers and varicosities than do the nuclei themselves (Steinbusch 1981). This result indicates that retrograde transport studies [which have demonstrated labeled cells in the region of serotonergic cell groups in the dorsal and median nuclei of the raphe and in the nucleus raphe magnus, after injections of tracers that were centered in the PVH (Tribollet & Dreifuss 1981, Berk & Finkelstein 1981a, Sawchenko & Swanson 1981c)] must be viewed with caution since the retrograde tracers may have been taken up and transported by fibers adjacent to the PVH. Nevertheless, autoradiographic studies have shown light projections from the midbrain nuclei of the raphe to the PVH and the SO in the rat (Conrad et al 1974, Azmitia & Segal 1978, Moore et al 1978). In the cat, Bobillier et al (1976) found evidence for sparse inputs to the PVH from the dorsal and median nuclei of the raphe, and from the region of the nucleus raphe magnus; inputs to the SO were seen only after injections into the dorsal raphe.

Loewy and his colleagues (1981) have placed injections of ^{3}H-amino acids into the region of serotonergic cells in the ventral medulla, and have described projections to both the PVH and SO. This pathway was unaffected in animals that were pretreated with 6-hydroxydopamine (a catecholamine neurotoxin), but was abolished in rats pretreated with 5,6-, or 5,7-dihydroxytryptamine (indoleamine neurotoxins), thus raising the possibility that the relatively few non-catecholaminergic cells in the ventral medulla that project to the PVH may be serotonergic. However, the neurotoxins that were used to establish the biochemical specificity of these projections are not entirely specific (Jonsson 1980), and the distribution of the postulated serotonergic projections to the PVH and SO, which were described as paralleling those from the adjoining A1 region (Loewy et al 1981), is inconsistent with the normal distribution of serotonin in these nuclei (Kent & Sladek 1978, Sawchenko & Swanson 1981c). It appears most likely, therefore, that the sparse serotonergic inputs to the PVH and SO arise primarily from the mesencephalic nuclei of the raphe, although additional evidence is needed to confirm this conclusion.

In the magnocellular division of the PVH, and in the SO, serotonergic varicosities are found primarily in regions in which oxytocinergic cells are concentrated (Sawchenko & Swanson 1981c). Microiontrophoresis of serotonin onto identified neurosecretory cells has been shown to inhibit the finding of some cells, and to have no effect on others (Barker et al 1971, Moss et al 1972).

OTHER BRAINSTEM AREAS The parabrachial nucleus surrounds the middle cerebellar peduncle in dorsolateral parts of the pons. Many cells, primarily in the lateral (or dorsal) part of the nucleus, are retrogradely labeled after injections of HRP centered in the PVH (Berk & Finkelstein 1981a, Tribollet & Dreifuss 1981), a result confirmed in autoradiographic studies, which indicate that the lateral part of the parabrachial nucleus projects primarily to medial parts of the parvocellular division of the PVH (Saper & Loewy 1980, Koh & Ricardo 1980, Sawchenko & Swanson 1981c). Injections of ^{3}H-amino acids centered in the medial part of the nucleus label the PVH much more lightly (Saper & Loewy 1980), and neither the magnocellular division of the PVH nor the SO receive a substantial input. The biochemical specificity of the projection from the parabrachial nucleus to the PVH is unknown, although both enkephalin- (Hökfelt et al 1977) and neurotensin-stained (Uhl et al 1979) cells have been localized in this region.

Based largely on the work of Norgren and colleagues (Norgren & Leonard 1973, Norgren 1978), it is known that both parts of the parabrachial nucleus receive a massive input from the rostral, primarily gustatory, part of the NTS. Interestingly, the more caudal, visceroceptive, part of the NTS projects heavily to the lateral part of the parabrachial nucleus. Thus, projections from the parabrachial nucleus appear to be complementary to the ascending noradrenergic projections described above in the sense that both serve to relay visceral sensory information to the forebrain.

Retrograde transport studies also suggest that the laterodorsal tegmental nucleus, which lies just rostral and ventral to the locus coeruleus, also projects to the PVH (Berk & Finkelstein 1981a, Tribollet & Dreifuss 1981). This nucleus has been considered a micturition center, and projects directly to sacral parasympathetic preganglionic neurons (Loewy et al 1979). However, a projection from the laterodorsal nucleus to the PVH has not been confirmed autoradiographically.

Hypothalamus

Autoradiographic studies have documented the existence of projections to the PVH from a number of hypothalamic cell groups, including various parts of the preoptic area (Swanson 1976), the ventromedial nucleus (Saper et al 1976), the anterior hypothalamic area (Saper et al 1978), the lateral hypothalamic area (Saper et al 1979), and the suprachiasmatic nucleus (Swanson & Cowan 1975). These results have, by and large, been confirmed by subsequent retrograde transport studies (Silverman et al 1981, Tribollet & Dreifuss 1981), and some additional sources of local inputs have been suggested. Recently, much of the autoradiographic material has been re-analyzed to describe in greater detail the distribution of projections from the hypothalamus to the PVH and SO (Sawchenko & Swanson 1981c). Inputs from the anterior hypothalamic area, from parts of the lateral hypo-thalamic area, and from the ventromedial and dorsomedial nuclei end primarily in medial parts of the parvocellular division of the PVH, as do light projections from the lateral, median, and periventricular parts of the preoptic area, and a more substantial input from the medial preoptic area (Sawchenko & Swanson 1981c). Surprisingly, only the projections from the dorsomedial nucleus and the median preoptic nucleus appear to end in the magnocellular division of the PVH, or in the SO.

The suprachiasmatic nucleus, the ventral part of which receives a direct input from the retina, projects to the periventricular part, and to the dorsal parvocellular part of the PVH, and to a region just ventral to the ventromedial edge of the nucleus (Berk & Finkelstein 1981b, Sawchenko & Swanson 1981c). Interestingly, the ventral part of the suprachiasmatic nucleus contains a dense concentration of vasoactive intestinal polypeptide (VIP)-stained cell bodies, and the distribution of VIP-positive fibers in and around the PVH (Card et al 1981) is quite similar to the projection described above. It seems likely that this pathway, which may contain VIP, plays a role in the mediation of circadian rhythms in the neuroendocrine and autonomic functions that are influenced by the PVH (Swanson & Sawchenko 1980).

Recent descriptions of a discrete intracerebral system of ACTH (1–39)-stained cells and fibers suggest that a substantial number of such fibers innervate the PVH (e.g. Watson et al 1978, Joseph 1980). ACTH-immunoreactive fibers are found in most parts of the parvocellular division of the PVH, and in those regions of the magnocellular division of the PVH, and of the SO, in which oxytocinergic cells are concentrated (Sawchenko et al 1982). In agreement with most localization studies of ACTH-stained cell bodies, double-labeling experiments indicate that these projections arise from immunoreactive neurons in and near the arcuate nucleus of the hypo-thalamus (Sawchenko et al 1982). Because β-endorphin and other pro-opiocortin-derived peptides have been co-localized within many, if not all, ACTH (1–39)-stained neurons (e.g. Bugnon et al 1979, Nilaver et al 1979), it is not surprising that β-endorphin-containing fibers also reach the PVH (Sawchenko et al 1982). In view of the well-established role of the arcuate nucleus in the control of anterior pituitary function, and considering electrophysiological evidence indicating that some cells in the arcuate nucleus project both to the median eminence and to the PVH, it is possible that the projection to the PVH serves to coordinate the endocrine regulatory influences mediated by the arcuate nucleus with more widespread visceral regulatory influences mediated by the PVH.

Projections to the PVH from the supraoptic nucleus and the accessory magnocellular cell groups have been inferred on the basis of experiments with HRP (Silverman et al 1981, Tribollet & Dreifuss 1981). In the absence of corroborating autoradiographic evidence, this observation must be viewed with caution. In both studies, the extent of injection sites was estimated with DAB, whereas a more sensitive substrate was used to demonstrate retrogradely labeled cells; the effective size of the injection sites was thus probably underestimated.

Although not strictly a part of the hypothalamus, the subfornical organ has been shown with both HRP (Silverman et al 1981, Tribollet & Dreifuss 1981) and autoradiographic methods (Miselis et al 1979, Sawchenko & Swanson 1981c, Miselis 1981) to project to both the PVH and the SO. The projection from the subfornical organ is distributed broadly to most parts of both nuclei; labeling in the magnocellular division of the PVH, and in the SO, is somewhat more dense over vasopressinergic, than over oxytocinergic neurons, although both cell types appear to receive an input. Accordingly, Renaud and his colleagues (1981) have recently shown electrophysiologically that stimulation of the subfornical organ most often excites phasically-discharging (presumably vasopressinergic; Poulain et al 1977) neurons in the SO.

In summary, projections to the PVH from nearby hypothalamic regions, though typically quite sparse, may provide a substrate for local integration of hypothalamic neuroendocrine and autonomic responses. Most of these short projections reach only the PVH, and tend to end preferentially in regions of the nucleus whose principal output is directed toward the median eminence or, to a lesser extent, toward autonomic centers in the brain stem and spinal cord. The subfornical organ and the arcuate nucleus, structures which themselves mediate neuroendocrine responses, project heavily to most regions of the PVH, including the magnocellular division, and to the SO.

Telencephalon

That higher forebrain structures influence cells in the PVH and SO is suggested by the observations that emotional factors can disrupt vasopressin secretion (Verney 1947), and that the secretion of oxytocin can be conditioned (Cleverly & Folley 1970). Ample physiological evidence indicates that various telencephalic structures, primarily within the limbic region, can affect the activity of magnocellular neurosecretory neurons, although the extent to which such influences are mediated by way of monosynaptic projections is not yet clear. Retrograde transport studies of afferents to the PVH in the rat suggest that parts of the septal region (primarily the lateral nucleus), the amygdaloid complex (primarily the medial nucleus), and the hippocampal formation (specifically the ventral part of the subicular complex) all project substantially to the PVH (Silverman et al 1981, Berk & Finkelstein 1981a, Tribollet & Dreifuss 1981). In one study (Silverman et al 1981), over 500 retrogradely labeled cells were counted in both the lateral septal nucleus and the ventral part of the subiculum in one experiment. Nevertheless, detailed autoradiographic studies of these regions have failed to corroborate such results (septal region: Swanson & Cowan 1979; amygdala: Krettek & Price 1978, Post & Mai 1980; hippocampal formation: Swanson & Cowan 1977). These discrepancies are probably due to problems of interpretation associated with the HRP method.

As noted above, the effective extent of HRP injection sites is difficult to determine, especially if different substrates are used to localize injection sites and retrogradely labeled neurons. Because both the medial nucleus of the amygdala (Krettek & Price 1978) and the lateral septal nucleus (Swanson & Cowan 1977) project to, or through, structures that lie immediately adjacent to the PVH, such as the anterior hypothalamic area, or the posterior part of the bed nucleus of the stria terminalis, it is likely that much of the labeling seen in these limbic structures is attributable to uptake of the marker by fibers that lie outside the PVH proper. In the case of retrograde labeling in the subiculum, it should be noted that the medial corticohypothalamic tract, a pathway through which the subiculum projects to a number of basal forebrain regions, traverses the anterior part of the PVH, and that the subiculum projects to the nucleus reuniens, which lies just dorsal to the PVH. It is likely, therefore, that uptake of HRP by fibers in the medial corticohypothalamic tract or in the nucleus reuniens accounts for most of the retrograde labeling of neurons in the subiculum.

One likely route through which the limbic region of the telencephalon may influence cells in the PVH and SO is by way of the bed nucleus of the stria terminalis. This area receives a massive input from the amygdala (Krettek & Price 1978), and a smaller, yet substantial, projection from the ventral part of the subiculum (Swanson & Cowan 1977). Autoradiographic studies have shown that the bed nucleus projects to the PVH (Swanson & Cowan 1979), and that this projection appears to innervate all parts of the parvocellular division and primarily oxytocinergic parts of the magnocellular division (Sawchenko & Swanson 1981c). Whether this projection influences, by way of intranuclear interactions, vasopressinergic neurons in the PVH and SO is not yet known.

Although the pathways taken by many inputs from limbic regions to the PVH and SO have not yet been determined, it is clear that stimulation of a number of sites within it can affect the release of oxytocin or vasopressin (see Cross & Dyball 1974 for a review). Electrophysiological studies, in which the responses of antidromically identified magnocellular neurosecretory neurons to stimulation of remote structures were monitored, have provided valuable information on the ways in which limbic regions can modulate the activity of cells in the PVH and SO. For example, stimulation of the septum or of the amygdala can have either excitatory or, more commonly, inhibitory effects on the activity of magnocellular neurons (Koizumi & Yamashita 1972, Negoro et al 1973, Pittman et al 1981). Using a characteristic phasic or bursting pattern of activity to identify vasopressinergic neurons, Renaud & Arnauld (1979) have shown that stimulation of the amygdala inhibits the activity of "phasic" cells in the PVH that project to either the median eminence or to the posterior pituitary. Poulain et al (1980) obtained similar results with phasic SO cells in lactating female rats after stimulation of the septal region, and also showed that the response was dependent on stimulation frequency. High-frequency stimulation abolished the phasic pattern altogether, whereas low-frequency stimulation decreased the firing rate of presumed oxytocinergic neurons. However, neither low- nor high-frequency stimulation interfered with the high-frequency discharge pattern that such cells display immediately prior to milk ejection. These results suggest that inputs from the septum to the SO do not play a major role in the release of oxytocin during suckling.

There is as yet no evidence for direct projections from neocortical areas to the PVH or the SO, although evoked potential studies suggest that indirect inputs to the SO may exist (Koizumi et al 1964).

In summary, it appears that a rather small number of cell groups in the brain stem, the hypothalamus, and limbic regions of the telencephalon provide most of the neural inputs to the PVH and SO. Projections from each of these levels have been shown to reach or to influence cells in both divisions of the PVH, and in the SO, and each projection ends in a distinctive way with respect to the cytoarchitectonic subdivisions of these nuclei. Thus it is now clear that a number of anatomically defined pathways are involved in the differential afferent control of the various functional subdivisions of the PVH and the SO. This applies to the magnocellular neurosecretory system in particular, since recent studies suggest that essentially separate pathways modulate the release of oxytocin and vasopressin.

Neurotransmitter Receptor Localization

Recent methods for the localization of neurotransmitter receptors in tissue sections, which utilize in vitro labeling of high-affinity receptors with radiolabeled ligands (Young & Kuhar 1979, Herkenham & Pert 1980), have provided an additional way to identify substances that modulate the activity of particular cell types in the central nervous system. Although the distribution of a given class of biochemically specified terminals cannot necessarily be inferred from the localization of receptors of the same specificity, the apparently widespread existence of multiple receptors for single transmitter substances, each of which may mediate quite different physiological responses (Goodman et al 1980), indicates that the localization of various receptor subtypes can provide a basis for selective functional and pharmacological manipulation of a system. Although the resolution provided by the newer methods of receptor localization appears to be sufficient to allow the description of receptor distributions in the various subdivisions of the PVH and SO, such analyses have not yet appeared. Nevertheless, a number of receptor types have been localized within these nuclei.

Alpha-bungarotoxin is known to bind selectively and irreversibly to cholinergic receptors in skeletal muscle (e.g. Miledi & Potter 1971) and has been used as a probe for nicotinic cholinergic receptors in the CNS. Although the specificity of bungarotoxin binding to nicotinic receptors in the brain is in doubt (see Oswald & Freeman 1981), binding in the SO (Polz-Tereja et al 1975, Silver & Billar 1976, Hunt & Schmidt 1978, Segal et al 1978) and in the PVH (Hunt & Schmidt 1978, Segal et al 1978) has been described in the rat, but not in the mouse (Arimatsu et al 1981). The presence of nicotinic receptors in the PVH and SO could provide a basis for the long-standing observation that nicotine stimulates the release of vasopressin from the posterior pituitary (e.g. Robinson 1975). Biochemical (Kobayashi et al 1978) and anatomical (Rotter et al 1979) localization studies have provided evidence for moderate to high levels of specific muscarinic receptor ligand binding in the SO, and, to a lesser extent, in the PVH, but as is the case for the presumed nicotinic receptors, nothing is known of their regional distribution.

Relative to other brain areas, the hypothalamus of the rat, and the PVH and SO in particular, have been shown to contain only low to moderate concentrations of opiate receptors (e.g. Atweh & Kuhar 1977). A recent comparison of the distribution of delta- and mu-opiate receptors, the postulated physiological receptors for met- and leu-enkephalin, respectively (Snyder & Goodman 1980), indicates that the hypothalamus contains a relatively high proportion of mu-receptors (Goodman et al 1980).

Suitable ligands for the localization of noradrenergic receptors have been identified only recently (Young & Kuhar 1979, Palacios & Kuhar 1980). Preliminary studies suggest that noradrenergic receptor subtypes are distributed preferentially in the rat brain. For example, especially high levels of specific $\alpha 2$ receptor binding have been found in the SO, whereas $\alpha 1$ and $\alpha 2$ receptors appear to occupy different parts of the periventricular region of the hypothalamus (Young & Kuhar 1980a).

Additional neurotransmitter receptors that have been localized in the PVH and the SO include the benzodiazipine receptor (Young & Kuhar 1980b) and the histamine H1 receptor (Palacios et al 1981).

In summary, although no detailed anatomical studies of neurotransmitter localization in the PVH and the SO are available, preliminary evidence suggests the presence of multiple receptors for a single transmitter substance (acetylcholine), and of topographically segregated receptor subtypes (α-noradrenergic, opiatergic) within these nuclei. Finer grained analyses of receptor distributions in the PVH and SO promise to broaden substantially our knowledge of functional neural circuits within these structures.

HUMORAL INFLUENCES

It has been known for many years that the PVH and the SO are the most richly vascularized nuclei in the CNS, and early workers suggested that this unusual feature might be related to the presence of Scharrer's "neurosecretory" cells (see Craigie 1940). It may well be that magnocellular neurons have an unusually high metabolic rate, in view of the large quantities of hormones that they synthesize and secrete in the posterior pituitary, and that this is reflected in the density of capillaries surrounding the cells. However, it has been suggested over the years that humoral factors, in addition to the neural inputs described in the last section, may influence the electrical activity or metabolism of cells in the PVH and the SO.

This notion was stimulated by Verney's (1947) demonstration that central osmoreceptors are involved in the regulation of vasopressin (antidiuretic hormone) release, and it is now clear that such receptors are exquisitely sensitive; in conscious sheep, an increase of only 1.2% in plasma osmolality is followed by a doubling of plasma vasopressin levels (Johnson et al 1970). Such osmoreceptors clearly are located in basomedial parts of the forebrain (Jewell & Verney 1957), and some have claimed, on the basis of rather indirect evidence, that magnocellular neurosecretory neurons themselves may be osmosensitive (see Cross & Green 1959, Brimble & Dyball 1977, Leng 1980). Other evidence indicates, however, that osmosensitive cells in the vicinity of, but not within, the PVH and the SO are osmosensitive and that projections from these cells are responsible for the activation of vasopressin-secreting neurons. The firing rate of magnocellular neurons in tissue slices of the hypothalamus is not increased when the slice is bathed in solutions of hypertonic NaCl within the physiological range, although it is increased by glutamate (Haller & Wakerly 1980). In addition, characteristic increases in vasopressin levels that accompany dehydration are abolished by lesions in the preoptic region that do not encroach upon the PVH or the SO (Van Gemert et al 1975). In summary, it has been quite difficult to show unequivocally that magnocellular neurosecretory neurons themselves are osmosensitive under physiological conditions, and there is now good reason to believe that they are not. However, it may only be possible to settle this issue directly when suitable primary cultures of magnocellular vasopressinergic cells are developed.

In the last several years it has become clear that injections of angiotensin II into either the ventricular system (see Andersson & Westbye 1970, Yamamoto et al 1978), or into preoptic, septal, and supraoptic sites (Simonnet et al 1979) act to release vasopressin. It is well known that conditions that deplete body water, such as hemorrhage and dehydration, are accompanied by a rise in circulating levels of angiotensin II, and it has been suggested that under such circumstances the hormone acts centrally to release vasopressin (antidiuretic hormone) (e.g. Malvin et al 1977). Angiotensin II increases the firing rate of neurons in the SO when applied either iontophoretically (Nicoll & Barker 1971) or to organ cultures that contain the nucleus (Sakai et al 1974). There is no evidence at this time, however, that angiotensin II crosses the blood-brain barrier to gain direct access to the PVH and the SO, or that it enters the third ventricle by way of the choroid plexus (Schelling et al 1977). It has therefore been proposed that circulating angiotensin II may act instead upon cells in or near regions like the subfornical organ, which lack a blood-brain barrier (Simpson 1981). This now seems likely in view of evidence that the subfornical organ has a high concentration of angiotensin II receptors (Van Houten et al 1980), is sensitive to local injections of angiotensin II, and projects directly to the PVH and the SO (Miselis 1981). On the other hand, it is also possible that the brain itself contains a complete renin-angiotensin system (Ganten et al 1981). Fibers (Fuxe et al 1976) and cells (Phillips et al 1980) that cross-react with antisera to angiotensin II have been described in the PVH and SO, as have cells that cross-react with an antiserum to renin (Fuxe et al 1980). However, some have questioned whether authentic angiotensin II, as opposed, for example, to an immunologically similar peptide, is synthesized in the brain, and to settle this controversy it may be necessary to isolate biochemically angiotensin II, or related peptides, directly from brain tissue. In summary, it seems likely that circulating angiotensin II influences vasopressin release by acting on tissue in or near certain circumventricular organs that lack a blood-brain barrier to peptides, but it is still not clear whether circulating angiotensin II acts directly on magnocellular neurosecretory cells, or whether authentic angiotensin II is synthesized by neurons in the CNS itself.

Steroid hormones, unlike peptide hormones, can readily cross the blood-brain barrier, and detailed autoradiographic studies have shown that the nuclei of cells in a variety of specific sites in the CNS, including the PVH and the SO, concentrate ^{3}H-estrogens (Stumpf 1970, Pfaff & Keiner 1973) and ^{3}H-androgens (Sar & Stumpf 1977) that have been injected peripherally. Double-labeling methods have also been used to show that neurophy-sin-stained cells in the PVH and the SO concentrate peripherally administered estradiol (Sar & Stumpf 1978, Rhodes et al 1981b), and that vasopressinergic (Sar & Stumpf 1980) and oxytocinergic (Rhodes et al 1982) cells in the PVH concentrate estradiol. The distribution of such oxytociner-gic cells in the PVH suggests that they project mostly to autonomic centers rather than to the pituitary (Rhodes et al 1982). The functional significance of this binding in the PVH and SO is not yet clear, however, although it is known that sex hormones influence reproductive behavior by way of central effects (see Pfaff 1980), and that estrogens can stimulate the release of neurophysin (Robinson et al 1976), oxytocin (Yamaguchi et al 1979), and vasopressin (Skowsky et al 1979). It is also of interest that each of the catecholamine cell groups that projects to the PVH and to the SO has also been shown to concentrate estrogen (Heritage et al 1980), and that drugs that interfere with noradrenergic neurotransmission can affect the concen-tration of steroid receptors in the hypothalamus (Nock et al 1981). Thus, the interaction between gonadal steroids and the circuitry under consider-ation here appears to be bidirectional, and extremely complex.

Taken as a whole, the evidence that humoral factors directly influence cells in the PVH and the SO is not strong, except in the case of the sex steroids. This is not to say, however, that important humoral influences will not be found, and it should be pointed out that part of the noradrenergic input to the PVH ends directly on capillaries (Swanson et al 1977). The potential significance of this observation lies in the fact that the central noradrenergic system may change the permeability of the blood-brain bar-rier, in particular to water (Raichle 1981).

As an extension of the idea that humoral factors may influence the PVH and the SO, it is worth noting that, in the rat, vasopressin is found in higher concentrations in the CSF than in plasma (Dogterom et al 1977). It is possible, therefore, that vasopressin (and oxytocin) may act as a central "hormone," distributed through the CSF, on periventricular tissue and on tissue adjacent to the subarachnoid space. This suggestion is supported by evidence that iontophoretically applied vasopressin affects the firing rate of neurons in a variety of sites (e.g. Olpe & Baltzer 1981), and that intraven-tricularly applied vasopressin dramatically alters the blood-brain barrier to water (Raichle & Grubb 1978).

FUNCTIONAL IMPLICATIONS

Since much of the circuitry outlined above for the PVH has only recently been clarified, its functional significance has not yet been subjected to de-tailed experimental analysis. This is clearly a most interesting topic, though space limitations do not permit an evaluation of recent hypotheses here. We have reviewed elsewhere evidence that the PVH is involved in the integration of autonomic and endocrine responses that regulate the cardiovascular system (Sawchenko & Swanson 1982b), and in the regulation of feeding behavior (Sawchenko 1982) and thirst (Swanson & Mogenson 1981). For the sake of completeness, it should also be mentioned that vasopressin and oxytocin have from time to time been implicated in a variety of other functions, based largely on responses obtained by intraventricular injections, although the physiological significance, and the neural mechanisms that underlie these responses, are not yet clear. Thus oxytocin has been implicated in the initiation of maternal behavior (Pedersen & Prange 1979) and in certain learning and memory processes (see de Weid 1980). Vasopressin may be involved in the mediation of febrile responses (Kasting et al 1981), in the processing of painful stimuli (Berntson & Berson 1980), and in learning and memory (see de Weid 1980).

CONCLUDING REMARKS

The evidence reviewed here indicates that the PVH, which contains some 10,000 neurons in the rat, is an unusually complex nucleus, and it may be useful in closing to provide a synthetic overview of current knowledge about its connections, in the hope that conceptually useful organizing principles will emerge. In general, the function of any discrete part of the CNS can be approached by considering (a) the function of the structures that it innervates, (b) the types of information that it receives from other parts of the nervous system, and (c) the nature of information processing within the structure itself, which may, for example, involve the participation of interneurons and recurrent or local axon collaterals from projection neurons.

It now seems clear that the PVH can be viewed most appropriately as a *visceral* effector nucleus in a rather broad sense. In contrast to the SO, which appears to project only to the pituitary, the PVH also projects to the portal capillary zone of the median eminence, and to a number of cell groups in the brain stem and spinal cord, many of which are related to the autonomic nervous system. Projections from the PVH can, therefore, influence the secretion of hormones from the anterior as well as from the posterior lobe of the pituitary and can influence the activity of preganglionic neurons associated with the sympathetic and parasympathetic divisions of the autonomic nervous system.

The first, and possibly the most important, generalization that can be made about the neural organization of the PVH is that essentially separate groups of neurons appear to project to the posterior pituitary, to the external lamina of the median eminence, and to autonomic centers in the brain stem and spinal cord. This point is of fundamental significance because it implies that, in principle, the nucleus can influence each of its major functionally distinct target zones independently. However, physiological responses that serve to maintain homeostasis normally involve coordinated autonomic and endocrine responses.

What, then, is the significance of separate output channels from the PVH, and how might appropriately integrated responses be effected? Clearly, the most straightforward mechanism for the integration of responses would involve axon collaterals, from individual cells, that innervate different functional regions, as indeed appears to be the case for a significant proportion of the cells in the PVH that innervate both sympathetic and parasympathetic centers. However, this mechanism has the disadvantage that it greatly reduces the variety of responses that can be elicited from a cell group. For example, if all of the cells in the PVH gave rise to collaterals that end in the pituitary, the median eminence, and autonomic centers, then any input to the nucleus would only serve to activate (or inhibit) a stereotyped pattern of endocrine and autonomic responses.

The physiological evidence suggests, however, that the PVH is involved in a number of different responses, from the release of oxytocin in the pituitary during lactation to the modulation of insulin secretion by way of the vagus nerve. In addition, as a consideration of the role of the PVH in cardiovascular regulation makes clear, a number of PVH-mediated responses to specific stimuli such as baroreceptor activation appear to be appropriately coordinated. There are several ways that coordinated responses might be elicited from separate projection systems. The mechanism for which there is now direct experimental support involves the innervation of two or more cell types in the PVH by a single afferent projection. In fact, all of the evidence to date supports a second generalization about the neural organization of the PVH, namely, that each input to the nucleus has a unique distribution within the nucleus. This principle is best illustrated by the fibers of the central noradrenergic system, which are found in all parts of the PVH (and the SO), except those in which oxytocinergic cell bodies are concentrated. In addition, these noradrenergic fibers originate in three different cell groups in the lower brain stem, and each group gives rise to a different pattern of fibers in the PVH. The locus coeruleus innervates only medial parts of the nucleus associated with the median eminence, the A2 group in the dorsal vagal complex innervates parvocellular parts of the nucleus associated with the median eminence and descending autonomic projections, and only the A1 group innervates vasopressinergic regions of the PVH and the SO. In contrast, ascending serotonergic fibers preferentially innervate oxytocinergic parts of the PVH and the SO, while inputs from the subfornical organ end among both oxytocinergic and vasopressin-ergic cell bodies. When all of the inputs to the PVH are considered in relation to the organization of its projection neurons, it seems clear that the circuitry within the nucleus is highly differentiated.

The full extent of this differentiation can only be understood after detailed immunocytochemical studies are carried out to determine specific input-output relationships at the ultrastructural level, and after these pathways are subjected to detailed functional analysis. However, it is tempting to speculate about the extent to which synaptic and functional specificity may occur in the PVH. It has already been stressed that the nucleus consists of at least three functionally and cytoarchitectonically distinct regions, but it should also be emphasized that each of these regions consists of several distinct cell types, each of which has a distinct topographic distribution. For example, the magnocellular division is composed of relatively discrete zones that contain predominantly either oxytocinergic or vasopressinergic cells; the median eminence part of the nucleus contains oxytocinergic, vasopres-sinergic, somatostatinergic, CRF-containing, and dopaminergic neurons; and the autonomic part of the nucleus contains at least a half a dozen different cell types with descending projections. The observation that nora-drenergic fibers end only in relation to vasopressin cell bodies, whereas serotonergic and ACTH-stained fibers end preferentially in relation to ox-ytocinergic cell bodies in the magnocellular division of the nucleus, indi-cates how specific the synaptic relationships between specific cell types in the PVH may be. It will be of particular interest to determine whether functionally defined inputs to the parvocellular division of the nucleus innervate one or another of the cell types that project to the median emi-nence or to autonomic centers. For example, although the full range of visceral information that is relayed to the PVH from the vagus nerve, by way of the NTS, is not known, it does include inputs from the aortic depressor, carotid sinus, and renal nerves. Do functionally distinct cells in the NTS project to biochemically distinct cells in the PVH, which in turn send fibers to the median eminence, or to functionally specific parts of the dorsal motor nucleus of the vagus? In view of the exquisite performance of the autonomic and endocrine systems in the maintenance of homeostasis, the answer to this type of question must almost certainly be yes.

There are several other mechanisms that may play a role in the integra-tion of responses from the PVH. One involves the coupling of different classes of projection neurons by recurrent or local axon collaterals. Unfortu-nately, while there is considerable electrophysiological evidence that some projection neurons in the PVH give rise to local collaterals, nothing is known with certainty about the distribution of such collaterals. The use of intracellular labeling methods can be expected to provide important information about this problem. Another involves the activity of interneurons, but as yet there is no clear indication for their existence in the PVH and the SO. Still another possibility involves humoral factors, such as the steroid hormone estradiol, that may bind to more than one cell type in the PVH, but again, this possibility requires further study. And finally, gap junctions between functionally similar cell types may occur under some conditions, as has been shown for magnocellular neurosecretory neurons (Andrew et al 1981). The extent of such coupling remains to be determined, however, and it is not known whether it may occur between functionally distinct classes of cells.

In summary, the primary functions of the PVH and the SO are probably best viewed in relation to the magnocellular neurosecretory projections to the posterior lobe of the pituitary. In addition, however, the PVH has a second component, which is part of the parvocellular neurosecretory system, that regulates the output of the anterior pituitary, and a third component that modulates autonomic reflexes, which are mediated by circuitry in the brain stem and spinal cord. It seems reasonable to suggest that these functionally distinct components are concentrated within a single, well-defined nucleus because inputs to the magnocellular neurosecretory system can easily affect other cell groups that influence adenohypophysial and autonomic responses that act in concert with the released neurohypophysial hormone to produce integrated visceral responses.

ACKNOWLEDGMENTS

Work described from our laboratory was supported in part by Grant NS-16686 from the NIH, and by Grant DA-00259 from ADAMHA. Dr. Swanson is a Clayton Foundation Investigator. We especially thank Drs. K. Dierickx, F. Vandesande, K. B. Helle, W. Vale, M. Brown, H. W. M. Steinbusch, and B. K. Hartman for generous supplies of antisera. We also thank P. Thomas for secretarial help, and C. Trulock for photographic assistance.

Literature Cited

Aghajanian, G. K., Kuhar, M. J., Roth, R. H. 1973. Serotonin-containing neuronal perikarya and terminals: Differential effects of p-chlorophenylalanine. *Brain Res.* 54:85–101

Akmayev, I. G., Vikhreva, O. V., Konovalova, L. K. 1981. Ultrastructural evidence for the existence of a direct hypothalamic-vagal descending pathway. *Brain Res.* 209:205–9

Allen, W. F. 1923. Origin and distribution of

the tractus solitarius in the guinea pig. *J. Comp. Neurol.* 35:171–204

Ambach, G., Palkovits, M. 1979. The blood supply of the hypothalamus in the rat. In *Handbook of the Hypothalamus,* ed. P. J. Morgane, J. Panksepp, 1:267–377. New York: Dekker

Andersson, B., Westbye, O. 1970. Synergistic action of sodium and angiotensin on brain mechanisms controlling water and salt balance. *Nature* 228:75

Andrew, R. D., MacVicar, B. A., Dudek, F. E., Hatton, G. I. 1981. Dye transfer through gap junctions between neuroendocrine cells of rat hypothalamus. *Science* 211:1187–89

Antunes, J. L., Carmel, P. W., Zimmerman, E. A. 1977. Projections from the paraventricular nucleus to the zona externa of the median eminence of the rhesus monkey: An immunohistochemical study. *Brain Res.* 137:1–10

Arimatsu, Y., Seto, A., Amano, T. 1981. An atlas of α-bungarotoxin binding sites and structures containing acetylcholinesterase in the mouse central nervous system. *J. Comp. Neurol.* 198: 603–31

Armstrong, W. E., Schöler, J., McNeill, T. H. 1982. Immunocytochemical, Golgi and electron microscopic characterization of putative dendrites in the ventral glial lamina of the rat supraoptic nucleus. *Neuroscience* 7:679–94

Armstrong, W. E., Warach, S., Hatton, G. I., McNeill, T. H. 1980. Subnuclei in the rat hypothalamic paraventricular nucleus: A cytoarchitectural, horseradish peroxidase and immunocytochemical analysis. *Neuroscience* 5:1931–58

Atweh, S. F., Kuhar, M. J. 1977. Autoradiographic localization of opiate receptors in rat brain. II. The brain stem. *Brain Res.* 129:1–12

Azmitia, E. C., Segal, M. 1978. An autoradiographic analysis of the differential ascending projections of the dorsal and median raphe nuclei in the rat. *J. Comp. Neurol.* 179:641–68

Baertschi, A. J., Bény, J.-L., Gahwiler, B. 1982. Hypothalamic paraventricular nucleus is a privileged site for brain-pituitary interaction in long-term tissue culture. *Nature* 295:145–47

Bandaranayake, R. C. 1971. Morphology of the accessory neurosecretory nuclei and of the retrochiasmatic part of the supraoptic nucleus of the rat. *Acta Anat.* 80:14–22

Bargmann, W., Scharrer, E. 1951. The site of origin of the hormones of the posterior pituitary. *Am. Sci.* 39:255–59

Barker, J. L. 1977. Physiological role of peptides in the nervous system. In *Peptides in Neurobiology*, ed. H. L. Gainer, pp. 295–343. New York: Plenum

Barker, J. L., Crayton, J. W., Nicoll, R. 1971. Noradrenaline and acetylcholine responses of supra-optic neurosecretory cells. *J. Physiol. London* 218:19–32

Barry, J. 1958. Les voies neurosécrétoires extra-hypophysaires et le probleme de l'action nerveuse centrale des hormones posthypophysaires. *J. Med. Lyon* 935: 1065–73

Barry, J. 1975. Essai de classification, en technique de Golgi, des diverses catégories de neurones de noyau paraventriculaire chez la souris. *C. R. Seances Soc. Biol.* 169:978–80

Bayon, A., Koda, L., Battenberg, E., Bloom, F. E. 1979. Redistribution of endorphin and enkephalin immunoreactivity in the rat brain and pituitary after in vivo treatment with colchicine or cytochalasin B. *Brain Res.* 183:103–11

Benoit, J., Assenmacher, I. 1955. Le côntrole hypothalamique de l'activité préhypophysaire gonadotrope. *J. Physiol. Paris* 47:427–567

Berk, M. L., Finkelstein, J. A. 1981a. Afferent projections to the preoptic area and hypothalamic regions in the rat brain. *Neuroscience* 6:1601–24

Berk, M. L., Finkelstein, J. A. 1981b. An autoradiographic determination of the efferent projections of the suprachiasmatic nucleus of the hypothalamus. *Brain Res.* 226:1–13

Berntson, G. G., Berson, B. S. 1980. Antinociceptive effects of intraventricular or systemic administration of vasopressin in the rat. *Life Sci.* 261:455–59

Bjorklund, A., Nobin, A. 1973. Fluorescence histochemical and microspectrofluorometric mapping of dopamine and noradrenaline cell groups in the rat diencephalon. *Brain Res.* 51:193–205

Blessing, W. W., Chalmers, J. P. 1979. Direct projection of catecholamine (presumably dopamine)-containing neurons from hypothalamus to spinal cord. *Neurosci. Lett.* 11:35–40

Bloom, F. E., Battenberg, E. L. F., Rivier, J., Vale, W. 1982. Corticotropin releasing factor (CRF): Immunoreactive neurones and fibers in rat hypothalamus. *Regul. Peptides* 4:43–48

Bloom, F., Battenberg, E., Rossier, J., Ling, N., Guillemin, R. 1978. Neurons containing β-endorphin exist separately from those containing enkephalin: Immunocytochemical studies. *Proc. Natl. Acad. Sci. USA* 75:1591–95

Bloom, F., Battenberg, E., Rossier, J., Ling, N., Leppaluoto, J., Vargo, T. M., Guillemin, R. 1977. Endorphins are located in the intermediate and anterior lobes of the pituitary gland, not in the neurohypophysis. *Life Sci.* 20:43–48

Bobillier, P., Seguin, F., Petitjean, D., Salvert, M., Touret, M., Jouvet, M. 1976. The raphe nuclei of the cat brain stem: A topographic atlas of their efferent projections as revealed by autoradiography. *Brain Res.* 113:449–86

Bock, R. 1970. Hypophysenhinterlappen des Rhesusaffen. *Z. Zellforsch.* 107:499–507

Bodian, D. 1939. Studies on the diencephalon of the Virginia opossum. Part I. The nuclear pattern in the adult. *J. Comp. Neurol.* 94:259–323

Bodian, D., Maren, T. H. 1951. Effect of neuro- and adenohypophysectomy on retrograde degeneration in the hypothalamic nuclei of the rat. *J. Comp. Neurol.* 94:485–511

Brimble, M. J., Dyball, R. E. J. 1977. Characterization of the responses of oxytocin- and vasopressin-secreting neurons in the supraoptic nucleus to osmotic stimulation. *J. Physiol. London* 271:253–71

Brinkmann, H., Bock, R. 1973. Influence of various corticoids on the augmentation of "Gomori-positive" granules in the median eminence of the rat following adrenalectomy. *Naunyn-Schmied. Arch. Pharmacol.* 280:49–62

Brownstein, M. J., Palkovits, M., Saavedra, J. M., Kizer, J. S. 1976. Distribution of hypothalamic hormones and neurotransmitters within the diencephalon. In *Front. Neuroendocrinol.* 4:1–23

Brownstein, M. J., Russell, J. T., Gainer, H. 1980. Synthesis, transport, and release of posterior pituitary hormones. *Science* 207:373–78

Bugnon, C., Block, B., Lenys, D., Gouget, A., Fellman, D. 1979. Comparative study of the neuronal populations containing β-endorphin, corticotropin and dopamine in the arcuate nucleus of the rat hypothalamus. *Neurosci. Lett.* 14:43–48

Buijs, R. M. 1978. Intra- and extrahypothalamic vasopressin and oxytocin pathways in the rat. Pathways to the limbic system, medulla oblongata and spinal cord. *Cell Tiss. Res.* 192:423–35

Cajal, S. Ramón y. 1894. Algunas contribuciones al concimento de los ganglios del encefalo. III. Hipofisis. *An Soc. Esp. Hist. Rat. 2 Ser.* 23:195–237

Cajal, S. Ramón y. 1911. *Histologie du Système Nerveux de l'homme et des Vertébrés,* Vol. II. Paris: Maloine

Card, J. P., Brecha, N., Karten, H. J., Moore, R. Y. 1981. Immunocytochemical localization of vasoactive intestinal polypeptide-containing cells and processes in the suprachiasmatic nucleus of the rat: Light and electron microscopic analysis. *J. Neurosci.* 1:1289–1303

Carlsson, A., Falck, B., Hillarp, N.-Å. 1962. Cellular localization of brain monoamines. *Acta Physiol. Scand.* 56(Suppl. 196):1–27

Changaris, D. G., Keil, L. C., Severs, W. B. 1978. Angiotensin II immunohistochemistry of the rat brain. *Neuroendocrinology* 25:257–74

Christ, J. F. 1966. Nerve supply, blood supply and cytology of the neurohypophysis. In *The Pituitary Gland, Vol. 3, Pars Intermedia and Neurohypophysis,* ed. G. W. Harris, B. T. Donovan, pp. 62–130. Berkeley/Los Angeles: Univ. Calif. Press

Christ, J. F. 1969. Derivation and boundaries of the hypothalamus, with atlas of hypothalamic grisea. In *The Hypothalamus,* ed. W. Haymaker, E. Anderson, W. J. H. Nauta, pp. 13–60. Springfield: Thomas

Ciriello, J., Calaresu, F. R. 1977. Lateral reticular nucleus: A site of somatic and cardiovascular integration in the cat. *Am. J. Physiol.* 233:R100–9

Cleverley, J. D., Folley, S. J. 1970. The blood levels of oxytocin during machine milking in cows with some observations on its half life in the circulation. *J. Endocrinol.* 46:347–61

Conrad, L. C. A., Leonard, C. M., Pfaff, D. W. 1974. Connections of the median and dorsal raphe nuclei in the rat: An autoradiographic and degeneration study. *J. Comp. Neurol.* 156:179–206

Conrad, L. C. A., Pfaff, D. W. 1976. Efferents from medial basal forebrain and hypothalamus in the rat. II. An autoradiographic study of the anterior hypothalamus. *J. Comp. Neurol.* 169:221–62

Corrêa, F. M., Innis, R. B., Uhl, G. R., Synder, S. H. 1979. Bradykinin-like immunoreactive neuronal systems localized histochemically in rat brain. *Proc. Natl. Acad. Sci. USA* 76:1489–93

Coulter, H. D., Elde, R. P., Unverzagt, S. L. 1981. Co-localization of neurophysin- and enkephalin-like immunoreactivity in cat pituitary. *Peptides* 2 (Suppl. 1):51–55

Craigie, E. H. 1940. Measurements of vascularity in some hypothalamic nuclei of the albino rat. *Assn. Res. Nerv. Ment. Dis.* 20:310–19

Cratcher, K. A., Humbertson, A. O., Martin, G. F. 1978. The origin of brainstem spinal pathways in the North American Opossum (*Didelphis virginiana*). Studies using the horseradish peroxidase method. *J. Comp. Neurol.* 179:169–94

Cross, B. A., Dyball, R. E. J. 1974. Central pathways for neurohypophysial hormone release. *Handb. Physiol.* 4(7): 269–85

Cross, B. A., Green, J. D. 1959. Activity of single neurones in the hypothalamus: Effect of osmotic and other stimuli. *J. Physiol. London* 148:554–69

Dahlström, A., Fuxe, K. 1964. Evidence for the existence of monoamine-containing neurons in the central nervous system. I. Demonstration of monoamines in the cell bodies of brain stem neurons. *Acta Physiol. Scand.* 62(Suppl. 232):1–80

de Weid, D. 1980. Behavioral actions of neurohypophysial peptides. *Proc. R. Soc. London Ser. B* 210:183–95

Dierickx, K. 1980. Immunocytochemical localization of the vertebrate cyclic nonapeptide neurohypophyseal hormones and neurophysins. *Int. Rev. Cytol.* 62:119–85

Dierickx, K., Van den Abeele, A. 1959. On the relations between the hypothalamus and the anterior pituitary in *Rana temporaria. Z. Zellforsch.* 51:78–87

Dierickx, K., Vandesande, F. 1979. Immunochemical localization of somatostatin neurons in the rat hypothalamus. *Cell Tiss. Res.* 201:349–59

Dierickx, K., Vandesande, F., DeMey, J. 1976. Identification, in the external region of the rat median eminence, of separate neurophysin-vasopressin and neurophysin-oxytocin containing nerve fibers. *Cell Tiss. Res.* 168:141–51

Dogterom, J., Van Wimersma Greidanus, T. B., Swaab, D. F. 1977. Evidence for the release of vasopressin and oxytocin into cerebrospinal fluid: Measurements in plasma and CSF of intact and hypophysectomized rats. *Neuroendocrinology* 24:108–18

Douglas, W. W. 1974. Mechanism of release of neurohypophysial hormones: Stimulus-secretion coupling. *Handb. Physiol.* 4(7):31–102

Du Vigneaud, V. 1954–1955. Hormones of the posterior pituitary glands: Oxytocin and vasopressin. *Harvey Lect. Ser. L*, pp. 1–26

Felten, D. L., Cashner, K. A. 1979. Cytoarchitecture of the supraoptic nucleus: A Golgi study. *Neuroendocrinology* 29: 221–30

Finley, J. C. W., Maderdrut, J. L., Petrusz, P. 1981a. The immunocytochemical localization of enkephalin in the central nervous system of the rat. *J. Comp. Neurol.* 198:541–65

Finley, J. C. W., Lindstrom, P., Petrusz, P. 1981b. Immunocytochemical localization of β-endorphin-containing neurons in the rat brain. *Neuroendocrinology* 33:28–42

Fisher, A. W. F., Price, P. G., Buford, G. D., Lederis, K. 1979. A 3-dimensional reconstruction of the hypothalamo-neurohypophyseal system of the rat. The neurons projecting to the neuro/intermediate lobe and those containing vasopressin and somatostatin. *Cell Tiss. Res.* 204:343–54

Fortuyn, Ae. B. D. 1912. Die Ontogenie der kerne des Zwischenhirns beim Kaninchen. *Arch. Anat. Physiol.* 302–52

Fuxe, K. 1965. Evidence for the existence of monoamine neurons in the central nervous system. IV: The distribution of monoamine nerve terminals in the central nervous system. *Acta Physiol. Scand.* 64(Suppl. 247):37–85

Fuxe, K., Ganten, D., Hökfelt, T., Bohne, P. 1976. Immunohistochemical evidence for the existence of angiotensin II containing nerve terminals in the brain and spinal cord of the rat. *Neurosci. Lett.* 2:229–34

Fuxe, K., Ganten, D., Hökfelt, T., Locatelli, V., Poulsen, K., Stock, G., Rix, E., Taugner, R. 1980. Renin-like immunocytochemical activity in the rat and mouse brain. *Neurosci. Lett.* 18: 245–50

Fuxe, K., Hökfelt, T., Eneroth, P., Gustafsson, J.-Å., Skett, P. 1977. Prolactin: Localization in nerve terminals of the rat hypothalamus. *Science* 196:899–900

Fuxe, K., Hökfelt, T., Ungerstedt, U. 1970. Morphological and functional aspects of central monoamine neurons. *Int. Rev. Neurobiol.* 13:93–126

Ganten, D., Speck, G., Schelling, P., Unger, Th. 1981. The brain renin-angiotensin system. *Adv. Biochem. Psychopharmacol.* 28:359–72

Goldstein, A., Fischli, W., Lowrey, L. I., Hunkapiller, M., Hood, L. 1981. Porcine pituitary dynorphin: Complete amino acid sequence of the biologically active heptadecapeptide. *Proc. Natl. Acad. Sci. USA* 78:7219–23

Goldstein, A., Ghazarossian, V. E. 1980. Immunoreactive dynorphin in pituitary and brain. *Proc. Natl. Acad. Sci. USA* 77:6207–10

Goodman, R. R., Snyder, S. H., Kuhar, M. J., Young, W. S. III. 1980. Differentiation of delta and mu opiate receptor localizations by light microscopic autoradiography. *Proc. Natl. Acad. Sci. USA* 77:6239–43

Gregory, W. A., Tweedle, C. D., Hatton, G. I. 1980. Ultrastructure of neurons in the paraventricular nucleus of normal, dehydrated and rehydrated rats. *Brain Res. Bull.* 5:301–6

Grunthal, E. 1931. Der Zellaufbau im Hypothalamus des Kanischens und des Macacus Rhesus nebst einigen allgemeinen Bemerkungen über dieses Organ. *J. Psychol. Neurol.* 42:425–64

Guillemin, R. 1979. Discussion of session chaired by R. Guillemin. In *Brain Peptides: A New Endocrinoogy,* ed. A. M. Gotto Jr., E. J. Peck Jr., A. E. Boyd III, pp. 67–70 Amsterdam: Elsevier

Gurdjian, E. S. 1927. The diencephalon of the albino rat. *J. Comp. Neurol.* 43:1–114

Haller, E. W., Wakerly, J. B. 1980. Electrophysiological studies of paraventricular and supraoptic neurons recorded *in vitro* from slices of rat hypothalamus. *J. Physiol. London* 302:347–62

Hancock, M. B. 1976. Cells of origin of hypothalamo-spinal projections in the rat. *Neurosci. Lett.* 3:179–84

Harris, G. W. 1948. Neural control of the pituitary gland. *Physiol. Rev.* 28:139–79

Hartman, B. K., Swanson, L. W., Raichle, M. E., Preskorn, S. H., Clark, H. B. 1980. Central adrenergic regulation of cerebral microvascular permeability and blood flow: Anatomic and physiologic evidence. In *The Cerebral Microvasculature,* ed. H. M. Eisenberg, R. L. Smith, pp. 113–26. New York: Plenum

Hatton, G. I., Hutton, U. E., Hoblitzell, E. R., Armstrong, W. E. 1976. Morphological evidence for two populations of magnocellular elements in the rat paraventricular nucleus. *Brain Res.* 108:187–93

Hatton, G. I., Tweedle, C. D. 1980. Suckling increases direct cell-cell apposition between neurosecretory neurons in the rat supraoptic nucleus. *Soc. Neurosci. Abstr.* 6:457

Heritage, A. S., Stumpf, W. E., Sar, M., Grant, L. D. 1980. Brainstem catecholamine neurons are target sites for sex steroid hormones. *Science* 207:1377–99

Herkenham, M., Pert, C. B. 1980. *In vitro* autoradiography of opiate receptors in rat brain suggests loci of "opiatergic" pathways. *Proc. Natl. Acad. Sci. USA* 77:5532–36

Hökfelt, T., Elde, R., Fuxe, K., Johansson, O., Ljungdahl, A., Goldstein, M., Luft, R., Efendic, S., Nilsson, G., Terenius, L., Ganten, D., Jeffcoate, S. L., Rehfeld, J., Said, S., Pérez de la Mora, M., Possani, L., Tapia, R., Teran, L., Palacios, R. 1978. Aminergic and peptidergic pathways in the nervous system with special reference to the hypothalamus. In *The Hypothalamus,* ed. S. Reichlin, R. J. Baldessariai, J. B. Martin, pp. 69–136. New York: Raven

Hökfelt, T., Elde, R., Johansson, O., Terenius, L., Stein, L. 1977. The distribution of enkephalin-immunoreactive cell bodies in the rat central nervous system. *Neurosci. Lett.* 5:25–31

Hökfelt, T., Fuxe, K., Goldstein, M. 1973. Immunohistochemical studies on monoamine-containing cell systems. *Brain Res.* 62:461–69

Hökfelt, T., Fuxe, K., Goldstein, M., Johansson, O. 1974. Immunohistochemical evidence for the existence of adrenaline neurons in the rat brain. *Brain Res.* 66:239–51

Hökfelt, T., Fuxe, K., Johansson, O., Jeffcoate, S., White, N. 1975. Distribution of thyrotropin-releasing hormone (TRH) in the central nervous system as revealed with immunohistochemistry. *Eur. J. Pharmacol.* 34:389–92

Hosoya, Y. 1980. The distribution of spinal projection neurons in the hypothalamus of the rat, studied with the HRP method. *Exp. Brain Res.* 40:79–87

Hosoya, Y., Matsushita, M. 1979. Identification and distribution of the spinal and hypophysial projection neurons in the paraventricular nucleus of the rat. A light and electron microscopic study with the horseradish peroxidase method. *Exp. Brain Res.* 35:315–31

Hunt, S., Schmidt, J. 1978. Some observations on the binding pattern of α-bungarotoxin in the central nervous system of the rat. *Brain Res.* 157:213–32

Ingram, W. R., Fisher, C., Ranson, S. W. 1936. Experimental diabetes insipidus in the monkey. *Arch. Intern. Med.* 57:1067–80

Innis, R. B., Corrêa, F. M. A., Uhl, G. R., Schneider, B., Snyder, S. H. 1979. Cholecystokinin octapeptide-like immunoreactivity: Histochemical localization in rat brain. *Proc. Natl. Acad. Sci. USA* 76:521–25

Jacobowitz, D. M., O'Donohue, T. L. 1978. α-Melanocyte stimulating hormone: Immunohistochemical identification and mapping in neurons of rat brain. *Proc. Natl. Acad. Sci. USA* 75:6300–4

Jennes, L., Stumpf, W. E. 1980. LHRH-systems in the brain of the golden hamster. *Cell Tiss. Res.* 209:239–56

Jewell, P. A., Verney, E. B. 1957. An experimental attempt to determine the site of the neurohypophysial osmoreceptors in the dog. *Philos. Trans. R. Soc. Ser. B* 240:197–324

Johansson, O., Hökfelt, T. 1979. Immunohistochemical distribution of thyrotropin-releasing hormone, somatostatin and enkephalin with special reference to the hypothalamus. In *Brain and Pituitary Peptides*, ed. W. Wuttke, A. Weindl, K. H. Voigt, R.-R. Dries, pp. 202–12. Basel: Karger

Johnson, J. A., Zehr, J. E., Moore, W. W. 1970. Effects of separate and concurrent osmotic and volume stimuli on plasma ADH in sheep. *Am. J. Physiol.* 218: 1273–80

Jones, B. J., Moore, R. Y. 1977. Ascending projections of the locus coeruleus in the rat. II. Autoradiographic study. *Brain Res.* 127:23–53

Jones, E. G., Hartman, B. K. 1978. Recent advances in neuroanatomical methodology. *Ann. Rev. Neurosci.* 1:215–96

Jonsson, G. 1980. Chemical neurotoxins as denervation tools in neurobiology. *Ann. Rev. Neurosci.* 3:169–87

Joseph, S. A. 1980. Immunoreactive adrenocorticotropin in rat brain: A neuroanatomical study using antiserum generated against synthetic ACTH[1-39] *Am. J. Anat.* 158:533–48

Joseph, S. A., Pilcher, W., Knigge, K. M. 1980. Immunocytochemical analyses of the ACTH component of the opiocortin system in normal and MSG treated animals. *Soc. Neurosci. Abstr.* 6:521

Joseph, S. A., Sternberger, L. A. 1979. The unlabeled antibody method. Contrasting color staining of β-lipotropin and ACTH-associated hypothalamic peptides without antibody removal. *J. Histochem. Cytochem.* 27:1430–37

Kahn, D., Abrams, G. M., Zimmerman, E. A., Carraway, R., Leeman, S. E. 1980. Neurotensin neurons in the rat hypothalamus: An immunohistochemical study. *Endocrinology* 107:47–54

Kasting, N. W., Veale, W. L., Cooper, K. E., Lederis, K. 1981. Vasopressin may mediate fibrile convulsions. *Brain Res.* 213:327–33

Kelly, J., Swanson, L. W. 1980. Additional forebrain regions projecting to the posterior pituitary: preoptic region, bed nucleus of the stria terminalis, and zona incerta. *Brain Res.* 197:1–10

Kent, D. L., Sladek, J. R. Jr. 1978. Histochemical, pharmacological and microspectrofluorometric analysis of new sites of serotonin localization in the rat hypothalamus. *J. Comp. Neurol.* 180:221–36

Kimura, H., McGeer, P. L., McGeer, E. G. 1981. The central cholinergic system studied by choline acetyltransferase immunohistochemistry in the cat. *J. Comp. Neurol.* 200:151–201

Kneisley, L. W., Biber, M. P., LaVail, J. H. 1978. A study of the origin of brain stem projections to monkey spinal cord using the retrograde transport method. *Exp. Neurol.* 60:116–39

Knigge, K. M., Joseph, S. A., Pilcher, W. 1980. Origin of the opiocortin system in the rat brain. *Soc. Neurosci. Abstr.* 6:522

Kobayashi, R. M., Palkovits, M., Hruska, R. E., Rothschild, R., Yamamura, H. I. 1978. Regional distribution of muscarinic cholinergic receptors in the rat brain. *Brain Res.* 154:13–23

Koh, E. T., Ricardo, J. A. 1979. Connections of hypothalamic neurosecretory nuclei with visceral sensory structures in the brainstem of the rat. *Soc. Neurosci. Abstr.* 5:450

Koh, E. T., Ricardo, J. A. 1980. Paraventricular nucleus of the hypothalamus: Evidence of ten functionally discrete subdivisions. *Soc. Neurosci. Abstr.* 6: 521

Koizumi, K., Ishikawa, T., Brooks, C. M. 1964. Control of neurones in the supraoptic nucleus. *J. Neurophysiol.* 27: 878–92

Koizumi, K., Yamashita, H. 1972. Studies of antidromically identified neurosecretory cells of the hypothalamus by intracellular and extra-cellular recordings. *J. Physiol. London* 221:683–705

Krettek, J. E., Price, J. L. 1978. Amygdaloid projections to subcortical structures within the basal forebrain and the brainstem in the rat and cat. *J. Comp. Neurol.* 178:255–80

Krieg, W. J. S. 1932. The hypothalamus of the albino rat. *J. Comp. Neurol.* 55: 19–89

Kuypers, H. G. J. M., Maiskey, V. A. 1975. Retrograde axonal transport of horseradish peroxidase from spinal cord to brain stem cell groups in the cat. *Neurosci. Lett.* 1:9–14

Larsson, L.-I. 1977. Corticotropin-like peptides in central nerves and in endocrine cells of gut and pancreas. *Lancet* 2: 1321–23

Laruelle, M. L. 1934. Le système végétatif mesodiencephalique. *Rev. Neurol.* 1: 809–42

Lauber, M., Nicolas, P., Boussetta, H., Fahy, C., Béguin, P., Camier, M., Vaudry, H., Cohen, P. 1981. The M_r 80,000 common forms of neurophysin and vasopressin from bovine neurohypophysis have corticotropin- and β-endorphin-like sequences and liberate by proteolysis biologically active corticotropin. *Proc. Natl. Acad. Sci. USA* 76:6086–90

Lechan, R. M., Nestler, J. L., Jacobson, S., Reichlin, S. 1980. The hypothalamic tuberoinfundibular system of the rat as demonstrated by horseradish peroxidase (HRP) microiontophoresis. *Brain Res.* 195:13–27

Lederis, K. 1961. Vasopressin and oxytocin in the mammalian hypothalamus. *Gen. Comp. Endocrinol.* 1:80–89

Lefranc, G. 1966. Ètude neurohistologique des noyeaux supraoptique et paraventriculaire chez cobaye et le chat par la technique de triple imprégnation de Golgi. *C.R. Acad. Sci.* 263:976–79

LeGros Clark, W. E. 1938. Morphological aspects of the hypothalamus. In *The Hypothalamus, Morphological, Functional, Chemical and Surgical Aspects,* ed. W. E. LeGros Clark, J. Beattie, G. Riddoch, N. M. Dott, pp. 1–68. Edinburgh: Oliver & Boyd

Leibowitz, S. F., Brown, L. C. 1980. Histochemical and pharmacological analysis of noradrenergic projections to the paraventricular hypothalamus in relation to feeding stimulation. *Brain Res.* 201:289–314

Leng, G. 1980. Rat supraoptic neurones: The effects of locally applied hypertonic saline. *J. Physiol. London* 304:405–14

Lenhossék, M. L. 1887. Beobachtungen am Gehirn des Menschen. *Anat. Anz.* 2: 450–61

Leontovich, T. 1969–1970. The neurons of the magnocellular neurosecretory system of the dog's hypothalamus. A Golgi study. *J. Hirnforsch.* 11:499–517

Léránth, Cs., Záborszky, L., Marton, J., Palkovits, M. 1975. Quantitative studies on the supraoptic nucleus in the rat. I. Synaptic organization. *Exp. Brain Res.* 22:509–23

Lincoln, D. W., Wakerly, J. B. 1975. Factors governing the periodic activation of supraoptic and paraventricular neurosecretory cells during sucking in the rat. *J. Physiol. London* 250:443–61

Lindvall, O., Bjorklund, A. 1974. The organization of ascending catecholamine neuron systems in the rat brain. *Acta Physiol. Scand.* Suppl. 412:1–48

Ljungdahl, A., Hökfelt, T., Nilsson, G. 1978. Distribution of substance P-like immunoreactivity in the central nervous system of the rat. I. Cell bodies and nerve terminals. *Neuroscience* 3:861–944

Loewy, A. D., Saper, C. B. 1978. Efferent projections of the locus coeruleus. *Soc. Neurosci. Abstr.* 4:277

Loewy, A. D., Saper, C. B., Baker, R. P. 1979. Descending projections from the pontine micturition center. *Brain Res.* 172:533–38

Loewy, A. D., Wallach, J. H., McKellar, S. 1981. Efferent connections of the ventral medulla oblongata in the rat. *Brain Res. Rev.* 3:63–80

Loo, Y. T. 1931. The forebrain of the opossum, *Didelphis virginiana. J. Comp. Neurol.* 52:1–68

Lorén, I., Alumets, J., Håkanson, R., Sundler, F. 1979a. Distribution of gastrin and CCK-like peptides in rat brain. An immunocytochemical study. *Histochemistry* 59:249–57

Lorén, I., Alumets, J., Håkanson, R., Sundler, F. 1979b. Immunoreactive pancreatic polypeptide (PP) occurs in the central and peripheral nervous system: Preliminary immunocytochemical observations. *Cell Tiss. Res.* 200:179–86

LuQui, I. J., Fox, C. A. 1976. The supraoptic nucleus and the supraoptico-neurohypophyseal tract in the monkey (*Macaca mulatta*). *J. Comp. Neurol.* 168:7–40

Malone, E. F. 1910. Über die Kerne der menschlichen Diencephalon. *Abh. Preuss. Akad. Wiss.* 29:290–300

Malone, E. F. 1916. The nuclei tuberis lateralis and the so-called ganglion opticum basale. *Johns Hopkins Hosp. Rep.* 17: 441–510

Makara, G. B., Stark, E., Karteszi, M., Palkovits, M., Rappay, Gy. 1981. Effects of paraventricular lesions on stimulated ACTH release and CRF in stalkmedian eminence of the rat. *Am. J. Physiol.* 240:E441–46

Malvin, R. L., Mouw, D., Vander, A. J., Gregg, C. 1977. Hypothalamic stimulation of ADH release by angiotensin II. In *Central Actions of Angiotensin and Related Hormones,* ed. J. P. Buckley, C. M. Ferrario, pp. 257–66. New York: Pergamon

Martin, R., Voigt, K. H. 1981. Enkephalins co-exist with oxytocin and vasopressin in nerve terminals of rat neurohypophysis. *Nature* 289:502–4

McKellar, S., Loewy, A. D. 1981. Organization of some brain stem afferents to the paraventricular nucleus of the hypothalamus in the rat. *Brain Res.* 217: 351–57

McNeill, T. H., Sladek, J. R. Jr. 1980. Simultaneous monoamine histofluorescence and neuropeptide immunocytochemistry. II. Correlative distribution of catecholamine varicosities and magnocellular neurosecretory neurons in the rat supraoptic and paraventricular nuclei. *J. Comp. Neurol.* 193:1023–33

Meynert, T. 1869. Vom Gehirne der Säugethiere. In *Strickers Handbuch der Lehre von den Geweben des Menschen und der Thiere,* Vol. 2, pp. 694–808. Leipzig: Engelmann

Micevych, P., Elde, R. 1980. Relationship between enkephalinergic neurons and the vasopressin-oxytocin neuroendocrine system of the cat: An immunohistochemical study. *J. Comp. Neurol.* 190: 135–46

Miledi, R., Potter, L. T. 1971. Acetylcholine receptors in muscle fibers. *Nature* 233: 599–603

Millhouse, O. E. 1979. A Golgi anatomy of the rodent hypothalamus. In *Handbook of the Hypothalamus,* ed. P. J. Morgane, J. Panksepp, 1:221–65. New York: Dekker

Miselis, R. R. 1981. The efferent projections of the subfornical organ of the rat: A circumventricular organ within a neural network subserving water balance. *Brain Res.* 230:1–23

Miselis, R. R., Shapiro, R. E., Hand, P. J. 1979. Subfornical organ efferents to neural systems for control of body water. *Science* 205:1022–25

Moore, R. Y., Halaris, A. E., Jones, B. E. 1978. Serotonin neurons of the midbrain raphe: Ascending projections. *J. Comp. Neurol.* 180:417–38

Morgane, P. J., Panksepp, J., eds. 1980. *The Handbook of the Hypothalamus,* Vol. 3, Pt. 1, 2, *Behavior.* New York: Dekker

Morris, R., Salt, T. E., Sofroniew, M. V., Hill, R. G. 1980. Actions of microiontophoretically applied oxytocin, and immunohistochemical localization of oxytocin, vasopressin and neurophysin in the rat caudal medulla. *Neurosci. Lett.* 18:163–68

Moss, R. L., Urban, I., Cross, B. A. 1972. Microelectrophoresis of cholinergic and aminergic drugs on paraventricular neurons. *Am. J. Physiol.* 232:310–18

Nauta, W. J. H., Haymaker, W. 1969. Hypothalamic nuclei and fiber connections. See Christ 1969, pp. 136–209

Negoro, H., Visessuwan, S., Holland, R. C. 1973. Inhibition and excitation of units in paraventricular nucleus after stimulation of the septum, amygdala and neurohypophysis. *Brain Res.* 57:479–83

Nicoll, R. A., Barker, J. L. 1971. Excitation of supraoptic neurosecretory cells by angiotensin II. *Nature NB* 233:172–74

Nilaver, G., Zimmerman, E. A., Defendini, R., Liotta, A. S., Krieger, D. T., Brownstein, M. J. 1979. Adrenocorticotropin and β-lipotropin in the hypothalamus. *J. Cell. Biol.* 81:50–58

Nilaver, G., Zimmerman, E. A., Wilkins, J., Michaels, J., Hoffman, D., Silverman, A.-J. 1980. Magnocellular hypothalamic projections to the lower brain stem and spinal cord of the rat. Immunocytochemical evidence for the predominance of the oxytocin-neurophysin system compared to the vasopressin-neurophysin system. *Neuroendocrinology* 30:150–58

Nock, B., Blaustein, J. D., Feder, H. H. 1981. Changes in noradrenergic transmission alter the concentration of cytoplasmic progestin receptors in hypothalamus. *Brain Res.* 207:371–96

Norgren, R. 1978. Projections from the nucleus of the solitary tract in the rat. *Neuroscience* 3:207–18

Norgren, R., Leonard, C. M. 1973. Ascending central gustatory pathways. *J. Comp. Neurol.* 186:79–92

O'Donohue, T. L., Miller, R. L., Jacobowitz, D. M. 1979. Identification, characterization and stereotaxic mapping of intraneuronal α-melanocye stimulating hormone-like immunoreactive peptides in discrete regions of the rat brain. *Brain Res.* 176:101–23

Olivecrona, H. 1957. Paraventricular nucleus and pituitary gland. *Acta Physiol. Scand.* 40 (Suppl. 136):1–178

Olpe, H.-R., Baltzer, V. 1981. Vasopressin activates noradrenergic neurons in the rat locus coeruleus: A microiontophoretic investigation. *Eur. J. Pharmacol.* 73:337–78

Ono, T. H., Nishimo, H., Sasaka, K., Muramoto, K., Yano, I., Simpson, A. 1978. Paraventricular nucleus connections to spinal cord and pituitary. *Neurosci. Lett.* 10:141–46

Oswald, R. E., Freeman, J. A. 1981. Alphabungarotoxin binding and central nervous system nicotinic acetylcholine receptors. *Neuroscience* 6:1–14

Palacios, J. M., Kuhar, M. J. 1980. Beta-adrenergic-receptor localization by light microscopic autoradiography. *Science* 208:1378–80

Palacios, J. M., Wamsley, J. K., Kuhar, M. J. 1981. The distribution of histamine H_1-receptors in the rat brain: An autoradiographic study. *Neuroscience* 6:15–37

Palay, S. L. 1953. Neurosecretory phenomena in the hypothalamo-hypophysial system of man and monkey. *Am. J. Anat.* 93:107-41

Palkovits, M. 1981. Catecholamines in the hypothalamus: An anatomical review. *Neuroendocrinology* 33:123-28

Palkovits, M., Mezey, É., Záborszky, L., Feminger, A., Versteeg, D. H. G., Wijnen, H. J. L. M., DeJong, W., Fekete, M. I. K., Herman, J. P., Kanyicska, B. 1980a. Adrenergic innervation of the rat hypothalamus. *Neurosci. Lett.* 18:237-43

Palkovits, M., Záborszky, L., Feminger, A., Herman, J. P., Kanyicska, B., Szabo, D. 1980b. Noradrenergic innervation of the rat hypothalamus: Experimental, biochemical and electron microscopic studies. *Brain Res.* 191:161-71

Pedersen, C. A., Prange, A. J. 1979. Induction of maternal behavior in virgin rats after intracerebroventricular administration of oxytocin. *Proc. Natl. Acad. Sci. USA* 76:6661-65

Pérez de la Mora, M., Possani, L. D., Tapia, R., Teran, L., Palacios, R., Fuxe, K., Hökfelt, T., Ljungdahl, A. 1981. Demonstration of central γ-aminobutyrate-containing nerve terminals by means of antibodies against glutamate decarboxylase. *Neuroscience* 6:875-95

Peterson, R. P. 1966. Magnocellular neurosecretory centers in the rat hypothalamus. *J. Comp. Neurol.* 128:181-90

Pfaff, D. W. 1980. *Estrogens and Brain Function.* New York: Springer

Pfaff, D. W., Keiner, M. 1973. Atlas of estradiol-concentrating cells in the central nervous system of the female rat. *J. Comp. Neurol.* 151:121-58

Phillips, M. I., Quinlan, J. T., Weyhenmeyer, J. 1980. An angiotensin-like peptide in the brain. *Life Sci.* 27:2589-94

Pittman, Q. J., Blume, H. W., Renaud, L. P. 1981. Connections of the hypothalamic paraventricular nucleus with the neurohypophysis, median eminence, amygdala, lateral septum and midbrain periaqueductal gray: An electrophysiological study in the rat. *Brain Res.* 215:15-28

Polz-Tejera, G., Schmidt, J., Karten, H. J. 1975. Autoradiographic localization of α-bungarotoxin-binding sites in the central nervous system. *Nature* 258:349-51

Post, S., Mai, J. K. 1980. Contribution to the amygdaloid projection field in the rat. A quantitative autoradiographic study. *J. Hirnforsch.* 21:199-225

Poulain, D. A., Ellendorf, F., Vincent, J. D. 1980. Septal connections with identified oxytocin and vasopressin neurones of the rat. An electrophysiological investigation. *Neuroscience* 5:379-87

Poulain, D. A., Wakerly, J. B., Dyball, R. E. J. 1977. Electrophysiological differentiation of oxytocin and vasopressin-secreting neurons. *Proc. R. Soc. London Ser. B.* 196:367-84

Raichle, M. E. 1981. Hypothesis: A central neuroendocrine system regulates brain ion homeostasis and volume. *Adv. Biochem. Psychopharmacol.* 28:329-36

Raichle, M. E., Grubb, R. L. 1978. Regulation of brain water permeability by centrally-released vasopressin. *Brain Res.* 143:191-94

Ranson, S. W., Magoun, H. W. 1939. The hypothalamus. *Ergeb. Physiol.* 41:56-163

Rasmussen, A. T. 1940. Effects of hypophysectomy and hypophysial stalk resection on the hypothalamic nuclei of animals and man. *Assn. Res. Nerv. Ment. Dis.* 20:245-69

Reaves, T. A. Jr., Hayward, J. N. 1981. Dye-marked paraventricular neuroendocrine cells in vivo in cat hypothalamus. *Neurosci. Lett.* 25:263-67

Renaud, L. P. 1979. Electrophysiological organization of the endocrine hypothalamus. In *The Hypothalamus,* ed. S. Reichlin, R. J. Baldessarini, J. B. Martin, pp. 269-301. New York: Raven

Renaud, L. P., Arnauld, E. 1979. Supraoptic phasic neurosecretory neurons: Response to stimulation of amygdala and dorsal hippocampus and sensitivity to microiontophoresis of amino acids and noradrenaline. *Soc. Neurosci. Abstr.* 5:233

Renaud, L. P., Arnauld, E., Cirino, M., Layton, B. S., Sgro, S., Siatitsas, Y. S. 1981. Supraoptic neurosecretory neurons: Modifications of excitability by electrical stimulation of limbic and subfornical organ afferents. *Soc. Neurosci. Abstr.* 7:325

Rhodes, C. H., Morrell, J. I., Pfaff, D. W. 1981a. Immunohistochemical analysis of magnocellular elements in rat hypothalamus: Distribution and numbers of cells containing neurophysin, oxytocin, and vasopressin. *J. Comp. Neurol.* 198:45-64

Rhodes, C. H., Morrell, J. I., Pfaff, D. W. 1981b. Distribution of estrogen-concentrating, neurophysin-containing magnocellular neurons in the rat hypothalamus as demonstrated by a technique combining steroid autoradiography and immunohistology in the same tissue. *Neuroendocrinology* 33:18-23

Rhodes, C. H., Morrell, J. I., Pfaff, D. W. 1982. Estrogen-concentrating, neurophysin-containing hypothalamic magnocellular neurons in the vasopressindeficient (Brattleboro) rat: A study combining steroid autoradiography and immunocytochemistry. *J. Neurosci.* In press

Ricardo, J. A., Koh, E. T. 1978. Anatomical evidence of direct projections from the nucleus of the solitary tract to the hypothalamus, amygdala, and other forebrain structures in the rat. *Brain Res.* 153:1–26

Rinne, U. K. 1960. Neurosecretory material around the neurohypophyseal portal vessels in the median eminence of the rat. *Acta Endocrinol.* 57:1–108

Rinne, U. K. 1970. Experimental electron microscopic studies on the neurovascular link between hypothalamus and anterior pituitary. In *Aspects of Neuroendocrinology,* ed. W. Bargman, E. Scharrer, pp. 220–28. Berlin: Springer

Rioch, D. M. 1929. Studies on the diencephalon of carnivora. Part I. The nuclear configuration of the thalamus, epithalamus, and hypothalamus of the dog and cat. *J. Comp. Neurol.* 49:1–120

Robinson, A. G. 1975. Isolation, assay, and secretion of individual human neurophysins. *J. Clin. Invest.* 55:360–67

Robinson, A. G., Ferin, M., Zimmerman, E. A. 1976. Plasma neurophysin levels in monkeys: Emphasis on the hypothalamic response to estrogen and ovarian events. *Endocrinology* 98: 468–75

Rose, M. 1935. Das Zwischenhirn des Kaninchens. *Mem. Acad. Pol. Sci. Lett. Ser. B,* pp. 1–108

Rossier, J., Battenberg, E., Pittman, Q., Bayon, A., Koda, L., Miller, R., Guilleman, R., Bloom, F. 1979. Hypothalamic enkephalin neurones may regulate the neurohypophysis. *Nature* 277:653–55

Rotter, A., Birdsall, N. J. M., Burgen, A. S. V., Field, P. M., Hulme, E. C., Raisman, G. 1979. Muscarinic receptors in the central nervous system of the rat. I. Technique for autoradiographic localization of the binding of (³H) propylbenzilyl-choline mustard and its distribution in the forebrain. *Brain Res. Rev.* 1:141–65

Sakai, K. K., Marks, B. H., George, J., Koestner, A. 1974. Specific angiotensin II receptors in organ cultured canine supraoptic nucleus cells. *Life Sci.* 14:1337–44

Sakumoto, T., Tohyama, M., Satoh, K., Kimoto, Y., Kinugasa, T., Tanizawa, O., Kurachi, K., Shimizu, N. 1978. Afferent fiber connections from lower brain stem to hypothalamus studied by the horseradish peroxidase method with special reference to noradrenaline innervation. *Exp. Brain Res.* 31:81–94

Saper, C. B., Loewy, A. D. 1980. Efferent connections of the parabrachial nucleus in the rat. *Brain Res.* 197:291–317

Saper, C. B., Loewy, A. D., Swanson, L. W., Cowan, W. M. 1976. Direct hypothalamo-autonomic connections. *Brain Res.* 117:305–12

Saper, C. B., Swanson, L. W., Cowan, W. M. 1976. The efferent connections of the ventromedial nucleus of the hypothalamus of the rat. *J. Comp. Neurol.* 169:409–42

Saper, C. B., Swanson, L. W., Cowan, W. M. 1978. The efferent connections of the anterior hypothalamic area of the rat, cat and monkey. *J. Comp. Neurol.* 182:575–600

Saper, C. B., Swanson, L. W., Cowan, W. M. 1979. An autoradiographic study of the efferent connections of the lateral hypothalamic area. *J. Comp. Neurol.* 183: 689–706

Sar, M., Stumpf, W. E. 1977. Distribution of androgen target cells in rat forebrain and pituitary after (³H)-dihydrotestosterone administration. *J. Steroid Biochem.* 8:1131–35

Sar, M., Stumpf, W. E. 1978. Simultaneous localization of neurophysin I and ³H estradiol in hypothalamic neurons using a combined autoradiographic and immunohistochemical technique. *J. Histochem. Cytochem.* 26:227

Sar, M., Stumpf, W. E. 1980. Simultaneous localization of ³H estradiol and neurophysin I or arginine vasopressin in hypothalamic neurons demonstrated by a combined technique of dry mount autoradiography and immunohistochemistry. *Neurosci. Lett.* 17:179–83

Sar, M., Stumpf, W. E., Miller, R. J., Chang, K.-J., Cuatrecasas, P. 1979. Immunohistochemical localization of enkephalin in rat brain and spinal cord. *J. Comp. Neurol.* 182:17–38

Sawchenko, P. E. 1982. Anatomic relationships between the paraventricular nucleus of the hypothalamus and visceral regulatory mechanisms: Implications for the control of feeding behavior. In *Neural Basis of Feeding and Reward,* ed. B. G. Hoebel, D. Novin, pp. 259–74. Brunswick, ME: Haer Inst.

Sawchenko, P. E., Swanson, L. W. 1980. Immunohistochemical identification of paraventricular hypothalamic neurons which project to the medulla or spinal cord in the rat. *Soc. Neurosci. Abstr.* 6:520

Sawchenko, P. E., Swanson, L. W. 1981a. A method for tracing biochemically defined pathways in the central nervous system using combined fluorescence retrograde transport and immunohistochemical techniques. *Brain Res.* 210:31–51

Sawchenko, P. E., Swanson, L. W. 1981b. Central noradrenergic pathways for the integration of hypothalamic neuroendocrine and autonomic responses. *Science* 214:685–87

Sawchenko, P. E., Swanson, L. W. 1981c. The distribution and cells of origin of some afferent projections to the paraventricular and supraoptic nuclei in the rat. *Soc. Neurosci. Abstr.* 7:325

Sawchenko, P. E., Swanson, L. W. 1982a. Immunohistochemical identification of neurons in the paraventricular nucleus of the hypothalamus that project to the medulla or to the spinal cord in the rat. *J. Comp. Neurol.* 205:260–72

Sawchenko, P. E., Swanson, L. W. 1982b. The organization of noradrenergic pathways from the brainstem to the paraventricular and supraoptic nuclei in the rat. *Brain Res. Rev.* In press

Sawchenko, P. E., Swanson, L. W., Joseph, S. A. 1982. The distribution and cells of origin of ACTH (1–39)-stained varicosities in the paraventricular and supraoptic nuclei. *Brain Res.* 232:265–74

Schelling, P., Ganten, D., Heckl, R., Haydirk, K., Hutchinson, J. S., Sponer, G., Ganten, U. 1977. On the origin of angiotensin-like peptides in cerebrospinal fluid. In *Central Actions of Angiotensin and Related Hormones*, ed. J. P. Buckley, C. M. Ferrario, pp. 519–26. New York: Pergamon

Segal, M., Dudai, Y., Amsterdam, A. 1978. Distribution of an α-bungarotoxin-binding cholinergic nicotinic receptor in rat brain. *Brain Res.* 148:105–19

Sherlock, D. A., Field, P. M., Raisman, G. 1975. Retrograde transport of horseradish peroxidase in the magnocellular neurosecretory system of the rat. *Brain Res.* 88:403–14

Silver, J., Billar, R. B. 1976. An autoradiographic analysis of (^{3}H)α-bungarotoxin distribution in the rat brain after intraventricular injection. *J. Cell. Biol.* 71:956–63

Silverman, A. J., Hoffman, D. L., Zimmerman, E. A. 1981. The descending afferent connections of the paraventricular nucleus of the hypothalamus. *Brain Res. Bull.* 6:47–61

Silverman, A. J., Zimmerman, E. A. 1975. Ultrastructural immunocytochemical localization of neurophysin and vasopressin in the median eminence and posterior pituitary of the guinea pig. *Cell Tiss. Res.* 159:291–301

Simonnet, G., Rodriguez, F., Fumonx, F., Czernichow, P., Vincent, J. O. 1979. Vasopressin release and drinking induced by intracranial injection of angiotensin II in monkey. *Am. J. Physiol.* 237:R20–25

Simpson, J. B. 1981. The circumventricular organs and the central actions of angiotensin. *Neuroendocrinology* 33:248–56

Sims, K. B., Hoffman, D. L., Said, S. I., Zimmerman, E. A. 1980. Vasoactive intestinal polypeptide (VIP) in mouse and rat brain: An immunocytochemical study. *Brain Res.* 186:165–83

Skowsky, W. R., Swan, L., Smith, P. 1979. Effects of sex steroid hormones on arginine vasopressin in intact and castrated male and female rats. *Endocrinology* 104:105–8

Snyder, S. H., Goodman, R. R. 1980. Multiple neurotransmitter receptors. *J. Neurochem.* 35:5–15

Sofroniew, M. V. 1980. Projections from vasopressin, oxytocin, and neurophysin neurons to neural targets in the rat and human. *J. Histochem. Cytochem.* 28:475–78

Sofroniew, M. V., Glassman, W. 1981. Golgi-like immunoperoxidase staining of hypothalamic magnocellular neurons that contain vasopressin, oxytocin or neurophysin in the rat. *Neuroscience* 6:619–43

Sofroniew, M. V., Schrell, U. 1981. Evidence for a direct projection from oxytocin and vasopressin neurons in the hypothalamic paraventricular nucleus to the medulla oblongata: Immunohistochemical visualization of both the horseradish peroxidase transported and the peptide produced by the same neurons. *Neurosci. Lett.* 22:211–17

Sofroniew, M. V., Weindl, A. 1978. Extrahypothalamic neurophysin-containing perikarya, fiber pathways and fiber clusters in the rat brain. *Endocrinology* 102:334–37

Sokol, H. W., Zimmerman, E. A., Sawyer, W. H., Robinson, A. G. 1976. The hypothalamo-neurohypophyseal system of the rat: Localization and quantitation of neurophysin by light micro-scopic immunocytochemistry in Bratt-leboro rats deficient in vasopressin and a neurophysin. *Endocrinology* 98: 1176–88

Steinbusch, H. W. M. 1981. Distribution of serotonin-immunoreactivity in the cen-tral nervous system of the rat—cell bod-ies and terminals. *Neuroscience* 6:557–618

Stumpf, W. E. 1970. Estrogen-neurons and estrogen-neuron systems in the periven-tricular brain. *Am. J. Anat.* 129:207–18

Svensson, T. H., Thorén, P. 1979. Brain nora-drenergic neurons in the locus co-eruleus: Inhibition by blood volume load through vagal afferents. *Brain Res.* 172:174–78

Swaab, D. F., Nijveldt, F., Pool, C. W. 1975a. Distribution of oxytocin and vasopres-sin cells in the rat supraoptic and para-ventricular nucleus. *J. Endocrinol.* 67: 461–62

Swaab, D. F., Pool, C. W., Nijveldt, F. 1975b. Immunofluorescence of vaso-pressin and oxytocin in the rat hypo-thalamo-neurohypophyseal system. *J. Neural Trans.* 36:195–215

Swanson, L. W. 1976. An autoradiographic study of the efferent connections of the preoptic region in the rat. *J. Comp. Neurol.* 167:227–56

Swanson, L. W. 1977. Immunohistochemical evidence for a neurophysin-containing autonomic pathway arising in the para-ventricular nucleus of the hypo-thalamus. *Brain Res.* 128:356–63

Swanson, L. W., Connelly, M. A., Hartman, B. K. 1977. Ultrastructural evidence for central monoaminergic innervation of blood vessels in the paraventricular nu-cleus of the hypothalamus. *Brain Res.* 136:166–73

Swanson, L. W., Cowan, W. M. 1975. The efferent connections of the suprachias-matic nucleus of the hypothalamus. *J. Comp. Neurol.* 160:1–12

Swanson, L. W., Cowan, W. M. 1977. An autoradiographic study of the organiza-tion of the efferent connections of the hippocampal formation in the rat. *J. Comp. Neurol.* 172:49–84

Swanson, L. W., Cowan, W. M. 1979. The connections of the septal region in the rat. *J. Comp. Neurol.* 186:621–56

Swanson, L. W., Hartman, B. K. 1980. Bio-chemical specificity in central pathways related to peripheral and intracerebral homeostatic function. *Neurosci. Lett.* 16:55–60

Swanson, L. W., Kuypers, H. G. J. M. 1980. The paraventricular nucleus of the hypothalamus: Cytoarchitectonic sub-divisions and the organization of projec-tions to the pituitary, dorsal vagal com-plex and spinal cord as demonstrated by retrograde fluorescence double-labeling methods. *J. Comp. Neurol.* 194:555–70

Swanson, L. W., McKellar, S. 1979. The dis-tribution of oxytocin- and neurophysin-stained fibers in the spinal cord of the rat and monkey. *J. Comp. Neurol.* 188:87–106

Swanson, L. W., Mogenson, G. J. 1981. Neu-ral mechanisms for the functional cou-pling of autonomic, endocrine and somatomotor responses and adaptive behavior. *Brain Res. Rev.* 3:1–34

Swanson, L. W., Sawchenko, P. E. 1980. Paraventricular nucleus: A site for the integration of neuroendocrine and autonomic mechanisms. *Neuroendo-crinology* 31:410–17

Swanson, L. W., Sawchenko, P. E., Bérod, A., Hartman, B. K., Helle, K. B., Van Orden, D. E. 1981. An immunohisto-chemical study of the organization of catecholaminergic cells and terminal fields in the paraventricular and su-praoptic nuclei of the hypothalamus. *J. Comp. Neurol.* 196:271–85

Swanson, L. W., Sawchenko, P. E., Rivier, J., Vale, W. W. 1982. The organization of ovine corticotropin releasing fac-tor(CRF)-immunoreactive cells and fibers in the rat brain: An immunohisto-chemical study. *Neuroendocrinology.* In press

Swanson, L. W., Sawchenko, P. E., Wiegand, S. J., Price, J. L. 1980. Separate neurons in the paraventricular nucleus project to the median eminence and to the medulla or spinal cord. *Brain Res.* 197:207–12

Tager, H., Hohenboken, M., Markese, J., Dinerstein, R. J. 1980. Identification and localization of glucagon-related peptides in rat brain. *Proc. Natl. Acad. Sci. USA* 77:6229–33

Takagi, H., Shiosaka, S., Tohyama, M., Senba, E., Sakonaka, M. 1980. Ascend-ing components of the medial forebrain bundle from the lower brain stem in the rat, with special reference to raphe and catecholamine cell groups. A study by the HRP method. *Brain Res.* 193: 315–37

Thomas, M. R., Ulrichsen, R. F., Calaresu, F. R. 1977. Function of lateral reticular nucleus in central cardiovascular regu-lation in the cat. *Am. J. Physiol.* 232:H157–66

Torvik, A. 1956. Afferent connections to the sensory trigeminal nuclei, the nucleus of solitary tract and adjacent structures. *J. Comp. Neurol.* 106:51–141

Tribollet, E., Dreifuss, J. J. 1981. Localization of neurones projecting to the hypothalamic paraventricular nucleus area of the rat: A horseradish peroxidase study. *Neuroscience* 6:1315–28

Tweedle, C. D., Hatton, G. I. 1976. Ultrastructural comparison of neurons of supraoptic and circularis nuclei in normal and dehydrated rats. *Brain Res. Bull.* 1:103–21

Tweedle, C. D., Hatton, G. I. 1977. Ultrastructural changes in rat hypothalamic neurons and their associated glia during minimal dehydration and rehydration. *Cell Tiss. Res.* 181:103–21

Uhl, G. R., Goodman, R. R., Snyder, S. H. 1979. Neurotensin-containing cell bodies, fibers and nerve terminals in the brain stem of the rat: Immunohistochemical mapping. *Brain Res.* 167: 77–91

Ungerstedt, U. 1971. Stereotaxic mapping of the monoamine pathways in the rat brain. *Acta Physiol. Scand.* Suppl. 367: 1–48

Vale, W., Speiss, J., Rivier, C., Rivier, J. 1981. Characterization of a 41-residue ovine hypothalamic peptide that stimulates secretion of corticotropin and β-endorphin. *Science* 213:1394–97

Van Den Pol, A. N. 1982. The magnocellular and parvocellular paraventricular nucleus of the rat: Intrinsic organization. *J. Comp. Neurol.* 206:317–45

Vanderhaegen, J. J., Lotstra, F., DeMey, J., Gilles, C. 1980. Immunohistochemical localization of cholecystokinin- and gastrin-like peptides in the brain and hypophysis of the rat. *Proc. Natl. Acad. Sci. USA* 77:1190–94

Vandesande, F., Dierickx, K. 1975. Identification of the vasopressin producing and of the oxytocin producing neurons in the hypothalamic neurosecretory system of the rat. *Cell Tiss. Res.* 164: 153–62

Vandesande, F., Dierickx, K., DeMey, J. 1975. Identification of the vasopressin-neurophysin producing neurons of the rat suprachiasmatic nucleus. *Cell Tiss. Res.* 156:377–80

Vandesande, F., Dierickx, K., DeMey, J. 1977. The origin of the vasopressinergic and oxytocinergic fibres of the external region of the median eminence of the rat hypophysis. *Cell Tiss. Res.* 180:443–52

Van Gemert, M., Miller, M., Carey, R. J., Moses, A. M. 1975. Polyuria and impaired ADH release following medial preoptic lesioning in the rat. *Am. J. Physiol.* 228:1293–97

Van Houten, M., Shiffrin, E. C., Mann, J. F. E., Posner, B. I., Boucher, R. 1980. Radioautographic localization of specific binding sites for blood-borne angiotensin II in rat brain. *Brain Res.* 186:480–85

Verney, E. B. 1947. The antidiuretic hormone and factors which determine its release. *Proc. R. Soc. London Ser. B* 135:25–106

Ward, D. G., Lefcourt, A. M., Gann, D. S. 1980. Neurons in the dorsal rostral pons process information about changes in venous return and in arterial pressure. *Brain Res.* 181:75–88

Watkins, W. B. 1980. Presence of adrenocorticotropin and β-endorphin immunoreactivities in the magnocellular neurosecretory system of the rat hypothalamus. *Cell Tiss. Res.* 207:65–80

Watson, S. J., Akil, H., Ghazarossian, V. E., Goldstein, A. 1981. Dynorphin immunocytochemical localization in brain and peripheral nervous system: Preliminary studies. *Proc. Natl. Acad. Sci. USA* 78:1260–63

Watson, S. J., Akil, H., Fischli, W., Goldstein, A., Zimmerman, E., Nilaver, G., van Wiersma Greidanus, T. B. 1982. Dynophin and vasopressin: Common localization in magnocellular neurons. *Science* 216:85–87

Weber, E., Evans, C. J., Samuelsson, S. J., Barchas, J. D. 1981. Novel peptide neuronal system in rat brain and pituitary. *Science* 214:1248–51

Weigand, S. J., Price, J. L. 1980. The cells of origin of the afferent fibers to the median eminence in the rat. *J. Comp. Neurol.* 192:1–19

Weindl, A., Sofroniew, M. V. 1976. Demonstration of extrahypothalamic peptide secreting neurons. A morphologic contribution to the investigation of psychotropic effects of neurohormones. *Pharmakopsychiatr. Neuro-Psychopharmakol.* 9:226–34

Weiss, P., Hiscoe, H. B. 1948. Experiments on the mechanism of nerve growth. *J. Exp. Zool.* 107:315

Wittkowski, W., Bock, R. 1972. Electron microscopical studies of the median eminence following interference with the feedback system anterior pituitary-adrenal cortex. In *Brain-Endocrine Interaction Median Eminence: Structure and Function,* pp. 171–80. Basel: Karger

Yagi, K., Azuma, T., Matsuda, K. 1966. Neurosecretory cell: Capable of con-

324   SWANSON & SAWCHENKO

ducting impulse in rats. *Science* 154: 778–79

Yamaguchi, K., Akaishi, T., Negoro, H. 1979. Effect of estrogen treatment on plasma oxytocin and vasopressin in ovariectomized rats. *Endocrinol. Jpn.* 26: 197–205

Yamamoto, M., Share, L., Shade, R. E. 1978. Effect of ventriculocisternal perfusion with angiotensin II and indomethacin on the plasma vasopressin concentration. *Neuroendocrinology* 25:166–73

Young, W. S. III, Kuhar, M. J. 1979. A new method for receptor autoradiography: (^{3}H) opioid receptors in rat brain. *Brain Res.* 179:255–70

Young, W. S. III, Kuhar, M. J. 1980a. Noradrenergic α1 and α2 receptors: Light microscopic autoradiographic localization. *Proc. Natl. Acad. Sci. USA* 77:1696–1700

Young, W. S. III, Kuhar, M. J. 1980b. Radiohistochemical localization of benzodiazepine receptors in rat brain. *J. Pharmacol. Exp. Ther.* 212:337–46

Záborszky, L., Léránth, C., Maraka, G. B., Palkovits, M. 1975. Quantitative studies on the supraoptic nucleus in the rat. II. Afferent fiber connections. *Exp. Brain Res.* 22:525–40

Zerihun, L., Harris, M. 1981. Electrophysiological identification of neurones of paraventricular nucleus sending axons to both the neurohypophysis and the medulla in the rat. *Neurosci. Lett.* 23:157–60

Ann. Rev. Neurosci. 1983. 6:325–56

THE REORGANIZATION OF SOMATOSENSORY CORTEX FOLLOWING PERIPHERAL NERVE DAMAGE IN ADULT AND DEVELOPING MAMMALS

Jon H. Kaas

Departments of Psychology and Anatomy, Vanderbilt University, Nashville, Tennessee 37240

M. M. Merzenich

Departments of Otolaryngology and Physiology, University of California, San Francisco Medical Center, San Francisco, California 94143

H. P. Killackey

Department of Psychobiology, University of California, Irvine, California 92717

INTRODUCTION

A question important to both the basic scientist and the clinician is what happens to the somatosensory cortex of mammals after a peripheral nerve or skin receptor area is damaged or otherwise inactivated? Until recently, this question received relatively little experimental attention. There are reasons, of course, to expect rather uninteresting consequences. The primary somatosensory cortex, S-I, has been described repeatedly as being somatotopically organized (see Woolsey 1958, Kaas 1983 for reviews). Thus, if input from any given body part is removed, a logical expectation is that the deprived portions of the somatotopic maps would be unresponsive to somatic stimuli. On the other hand, a number of investigators have reported evidence of reorganization in subcortical somatosensory structures

325

0147-006X/83/0301-0325$02.00

following partial deafferentation (Basbaum & Wall 1976, Brenowitz & Pubols 1981, Brown et al 1979a, Devor & Wall 1978, Dostrovsky et al 1976, Goldberger & Murray 1978, Kerr 1972, Merrill & Wall 1978, Pubols & Goldberger 1980, Millar et al 1976, Pollin & Albe-Fessard 1979, Wall & Egger, 1971). Furthermore, evidence from the available studies of the cortex has revealed that after restricted damage, deprived regions of cortical somatopic maps are not silent, or at least are not silent for long, and that they can be activated by cutaneous somatosensory stimuli. This observation raises a number of questions. Exactly what types of changes occur in the cortex? Are the changes similar or different in developing and adult mammals? What is the time course of the changes? Are the cortical changes related to the species, cortical area, or the location, extent, or type of peripheral alteration? What are the mechanisms involved? Perhaps most important, what are the functional implications of the cortical changes? The answers to these questions are largely incomplete. The purpose of this review is to summarize the available evidence in this research area in an effort to formulate partial answers and point toward future research directions.

REORGANIZATION IN ADULTS

Removing or otherwise inactivating part of the sensory input from peripheral nerves does not simply produce sectors of unresponsive or silent somatosensory cortex in adult mammals. Instead, these procedures result in new patterns of cortical activation. The new patterns of activation have been demonstrated in Area 3b (S-I proper) and Area 1 (the Posterior Cutaneous Field) of the somatosensory cortex of monkeys, and in the first somatosensory area, S-I, of cats, raccoons, and rats. Alterations in cortical organization have been observed after section of peripheral nerves, after anesthetic block of peripheral nerves, after the removal of a body part, and after section of the dorsal roots of peripheral nerves. Thus, reorganization has been demonstrated in two cortical fields of monkeys, in S-I of a range of other species, and after a number of different procedures for inactivating a subset of peripheral inputs. There also appear to be several types of cortical reorganization, including the somatotopic expansion of previously existing representations of body parts, the development of "new" representations, the activation of large regions of the cortex from a very limited region of a receptive field surface, and a "nonsomatotopic" activation of the cortex from scattered receptive fields. In addition, portions of the somatosensory cortex may remain unresponsive after some deactivation procedures. Finally, the types of cortical reactivation that occur may vary according to cortical area, deactivation procedure, and species.

Types of Cortical Reorganization

The types and full extent of cortical reorganization after alterations of peripheral sensory inputs are most easily revealed by microelectrode mapping procedures in which receptive field locations are determined for large numbers of recording sites in and around the affected or deprived cortex and compared with similar data from normal animals. In some instances, especially when the experimentally induced change might be partially confounded by individual differences in cortical organization, it is feasible and useful to map the cortex in the same animal before and after the peripheral damage. More limited recording, of course, can be used to demonstrate that the deprived cortex is not silent and to provide some information about the types of cortical reorganization that have occurred.

Some of the reorganizational effects have been demonstrated most clearly in the portions of Areas 3b and 1 of the somatosensory cortex that are devoted to the hand representations in owl and squirrel monkeys. Areas 3b and 1 have been shown to have separate representations of the body surface in these (Merzenich et al 1978, Nelson et al 1980a, Sur et al 1982) as well as other monkeys (Kaas et al 1979, Nelson et al 1980b). The hand representations of Areas 3b and 1 of these monkeys are large and exposed on the dorsolateral surface of cortex. The central sulcus in these primates is either absent or incomplete and shallow. These features are advantages in studies of cortical reorganization. Another advantage is that changes in the two adjacent sensory representations in Areas 3b and 1 can easily be studied in the same animal. For reasons given elsewhere (Merzenich et al 1978, Kaas et al 1981, Kaas 1983), we consider Area 3b to be the homologue of S-I in nonprimates. The Area 1 representation presently has no obvious homologue in nonprimates.

Detailed "maps" of the hand representations in owl and squirrel monkeys have been derived in normal animals by determining receptive field locations for closely spaced recording sites throughout the responsive cortex. Figure 1 schematically shows the location and somatotopic organization of the two hand representations in a normal owl monkey (for more details, see Merzenich et al 1978, 1981, 1983a). The representations are somatotopic and essentially mirror images of each other. The two representations of the hand are organized as if the two palms were partially split down to pad 3, spread out along the split, and adjoined with the digits pointing in opposite directions (Figure 1A). Topographic distortions and disruptions allow the two representations to fit into strip-like blocks of cortex (Figure 1B). The hairy dorsal surfaces of the digits and hands occupy little cortex normally. They are split off medially and laterally in Area 3b, and typically are inserted in the digit region of Area 1 (Figure 1B). The manner of the

A. HAND REPRESENTATIONS IN OWL MONKEYS

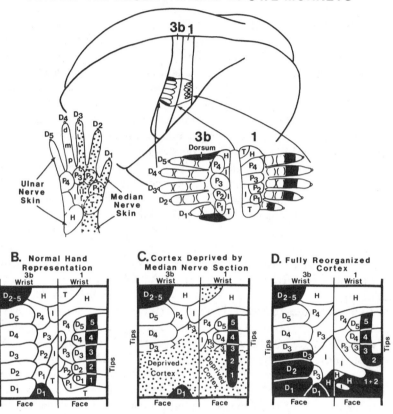

Figure 1   The reorganization of the hand representations in Areas 3b and 1 of somatosensory cortex of monkeys after section of the median nerve. A. The location and normal somatotopic organizations of the two hand representations are indicated on a dorsolateral view of the owl monkey brain. The two representations are basically mirror images of each other, adjoined at the 3b/1 border; the major difference is in the representation of dorsal digit and hand surfaces (*black*). The median nerve subserves the glabrous skin of the medial hand, while the ulnar nerve innervates the dorsal and glabrous surfaces of the lateral hand. The rest of the hand is subserved by the radial nerve. Digits and palmar pads are numbered in order, insular (I), hypothenar (H), thenar (T) pads, and distal (d), middle (m), and proximal (p) phalanges are indicated. B. The normal hand representations are schematically portrayed in more detail. C. The cortex normally devoted to the median nerve is marked by *dots* (the deprived cortex). D. The organization of the hand representations after complete reactivation following median nerve section. Much of the deprived cortex is activated by stimulation of the dorsal hand and digit surfaces (*black*). In addition, palmar pads innervated by the ulnar nerve have increased their cortical representation. The ulnar and radial nerves do not grow into the anesthetic skin. Based on Merzenich et al (1983a).

representation of the hand dorsum is one of the ways that the organization of Area 3b differs from that of Area 1.

The median nerve subserves about half of the ventral surface of the hand, including digits 1, 2, and most of 3 (Figure 1A). The rest of the hand is innervated by the ulnar (hand ventrum and dorsum) and radial (hand dorsum) nerves. Thus, cutting the median nerve deprives large sectors of the hand representations of their normal input (Figure 1C). After a period of recovery, the deprived cortex becomes fully responsive to stimuli on parts of the hand outside of the median nerve skin field (Figure 1D) (Merzenich et al 1983a). The types of reorganization observed after this procedure are (a) somatotopic expansions, (b) repetitive expansions, and (c) the formation of "new" representations. These types of reorganization as well as (d) nontopographic and (e) incomplete reactivations are discussed below.

SOMATOTOPIC EXPANSIONS   In the experiments on owl and squirrel monkeys after median nerve section described by Merzenich et al (1983a), the representations of skin surfaces along the edges of the deprived cortex (Figure 1C) appeared to expand topographically into part of the deprived cortex. In particular, the representations of palmar pad 3, the preserved portion of glabrous digit 3, the ulnar insular pad, and the representation of hand and digit dorsal surfaces of the skin field of the radial nerve in lateral cortex all expanded into the deprived cortex. The expansions were regarded as somatotopic in that progressions of recording sites in the expanded representations did not result in scattered receptive fields, but produced continuous sequences of receptive fields. These sequences suggested a simple topographic expansion of a previously existing organization to include part of the deprived cortex. In some instances, the altered organization suggested that the undeprived bordering cortex was changed with the expansion so that the representation of some skin surface was moved entirely into the deprived cortex, and an overall somatotopic pattern was preserved. In other instances, a progression of recording sites in the reactivated cortex resulted in little change in receptive field location. This might be regarded as evidence of a separate "class" of somatotopic reorganization, distinguished as "repetitive" (see below). Topographic expansions also have been found in somatosensory cortex of monkeys after the removal of one or two digits (Schoppmann et al 1981; Figure 2). Expanded representations into the deprived cortex of the palm, and stumps of the removed digits and of the remaining digits, were sometimes dramatically topographic.

Other evidence for the somatotopic expansion of remaining representations into deprived cortex after peripheral deafferentation is limited. After partially deactivating the hindlimb portion of the S-I in cats by sectioning dorsal roots, Franck (1980) reported a systematic shift in the organization of the rest of the S-I map so that the representation of more rostral body segments, including the lower trunk, "migrated" into the deprived cortex. The somatotopic details of reorganized cortex, however, were not described or documented.

In some experiments, it was not always apparent if the reactivation was somatotopic or not. The somatotopic nature of the reorganized cortex becomes difficult to determine when the reactivated region is small, or the skin surface represented in the reactivated cortex is restricted. For example, the cortex activated by expanded saphenous nerve inputs from the foot in rats after section of the sciatic nerve to the foot appeared to be roughly

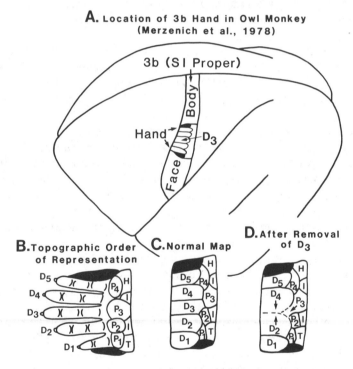

Figure 2 The reorganization of the hand representation of Area 3b of monkeys after the removal of a digit. A. The location of the hand representation in Area 3b as shown on a dorsolateral view of an owl monkey brain. B. The topographic order of the normal representation of the hand in Area 3b. C. The arrangement of parts of the hand in the normal representation. D. The expansion of the representations of digits 2 and 4 and Pad 3 into the cortex of digit 3 after removal of digit 3. Digits and pads are numbered in order; insular (I), hypothenar (H), and thenar (T) pads are indicated. A, B, and C are based on Merzenich et al (1978). D is after Schoppmann et al (1981) and M. M. Merzenich, M. P. Stryker, A. Schoppmann, and J. Zook, (unpublished results).

somatotopic, but the large receptive fields for the saphenous nerve made this difficult to determine (Wall et al 1981).

REPETITIVE EXPANSIONS   A common observation in the somatosensory cortex of monkeys after median nerve section was that some sequences of recording sites over large sectors of deprived cortex had nearly identical receptive fields (Merzenich et al 1983a). We refer to this type of reoccupation as "repetitive," and since these repetitive representations extended from representations of the same skin surface at the borders of deprived cortex in monkeys, they are "repetitive expansions." Such expansions were most commonly observed in Area 1, where recording site sequences extending over as much as 1 mm of cortex might have the same receptive field. For example, long sequences of recording sites often had overlapping and nearly identical receptive fields on the palmar pad 3, or on the ulnar insular pad. Typically, these nearly identical receptive fields were recorded across one surface dimension, while somatotopic sequences of receptive fields were determined for recordings across the orthogonal surface dimension.

NEW REPRESENTATIONS   One of the most intriguing observations on reactivated cortex was the appearance of "new" topographically organized representations. "New" representations have been seen in Areas 3b and 1 of somatosensory cortex of monkeys after median nerve section (Merzenich et al 1983a). In these experiments, large new representations of the dorsal surfaces of digits 1, 2, and part of 3 were found in the deprived cortex in locations that corresponded closely with the "old" representations of the glabrous surfaces of those same digits (Figure 1). In addition, the somatotopic organizations of the new representations of the dorsal digits corresponded to that of the "old" representations of the glabrous digits. Thus, digit 1 was lateral to digit 2, and digit 2 was lateral to digit 3, and the digit tips were found rostrally and the proximal phalanges caudally in Area 3b for both the normal and new representations. The somatotopic organization of the new representation appeared to be just as orderly as the original, although of different surfaces of the same digits, and the new receptive fields on the dorsal digit surfaces were ultimately just as small as those for neurons representing the ventral digit surfaces.

The representations of the dorsal digit surfaces were termed "new" because

1. There was no indication of this input to the cortex activated by the median nerve in normal monkeys;
2. The dorsal digital surfaces were represented elsewhere in Area 3b in normal monkeys, and these original representations remained unaltered in the normal locations after median nerve section;

3. The new representations were clearly distinguishable from the original, still intact, representations by (*a*) different topographic organizations, (*b*) much different receptive field sizes, and, (*c*) for digits 2 and 3, separated cortical locations.

Receptive fields for neurons in the normal representations of the dorsal digits were much larger than for neurons in the "new" representations. Another "new" representation was that of the hypothenar eminence in the lateral portion of deprived Area 1. This "new" representation was quite distant from the expanded representation of this skin surface found more medially in Area 1 (Figure 1).

NONTOPOGRAPHIC REACTIVATION   Recently, cortex reactivated by other parts of the hand after digit removal in raccoons (Figure 3) has been described as "nontopographic" (Kelahan & Doetsch 1981) and as having an "absence of any topographical organization" (Rasmusson 1982). As originally described by Welker & Seidenstein (1959), the glabrous hand of the raccoon has a greatly expanded cortical representation in S-I, and the cortex in this region has a unique pattern of shallow fissures that indicate morphologically where each digit is located within the somatotopic map (Figures 3A and B). Thus, a surface view of the brain clearly indicates where each digit is normally represented. Recognizing the advantage of knowing without recording where each digit is normally represented, Rasmusson (1982) and Doetsch and his colleagues (Carson et al 1981, Kelahan & Doetsch 1981, Kelahan et al 1981) amputated digit 5 (Figure 3D) or digit 3 (Figure 3C) from adult raccoons. In both studies, later recordings indicated that the deprived cortex was activated by skin surfaces adjoining the amputated digit. After removal of digit 3, neurons in the deprived cortex were largely activated from the stump of digit 3, digits 2 and 4, and to a lesser extent from pads adjacent to digits 2–4 (Kelahan et al 1981, Kelahan & Doetsch 1981). Similarly, Rasmusson (1982) found that after removing digit 5, the deprived cortex was taken over primarily by input from digit 4. After removing digit 4, the deprived cortex was activated from digits 3 and 5. Some of the neurons in the deprived cortex had receptive fields of normal sizes, but many neurons had larger receptive fields. Both groups of investigators stressed that there was no topographic organization within the deprived cortex, although a topographic pattern for each digit representation was demonstrated within normal cortex.

INCOMPLETE REACTIVATION   Given sufficient time for recovery after damage, the deprived cortex often is completely reactivated by input from new skin surfaces. Thus, all of the deprived cortex in Areas 3b and 1 was

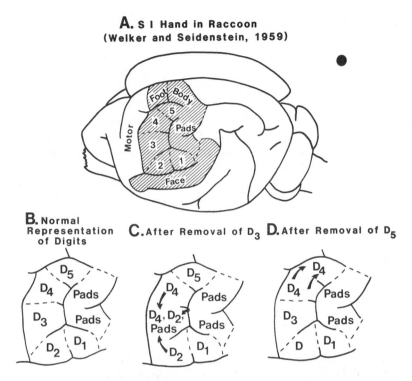

A. S I Hand in Raccoon
(Welker and Seidenstein, 1959)

B. Normal Representation of Digits   **C.** After Removal of D_3 **D.** After Removal of D_5

Figure 3 The reorganization of the hand representation in S-I of raccoons after removal of a digit. A. The location of S-I on a dorsolateral view of the brain. B. The normal representation of the digits and pads of the forepaw. C. The activation of the D_3 region of cortex by D_4 and D_2 after removal of digit 3. D. The activation of the D_5 region of cortex by D_4 after removal of digit 5. Based on Kelahan & Doetsch (1981) (C) and Rasmusson (1982) (D).

reactivated after median nerve section in monkeys (Merzenich et al 1983a); all of the cortex deprived by sectioning the nerves to the paw in cats (Kalaska & Pomeranz 1979) became responsive to the wrist and forearm; and the deprived digit cortex of raccoons became responsive to adjoining digits and the palm (Rasmusson 1982, Kelahan et al 1981). Similarly, Franck (1980) did not find an increase in nonresponsive single units in the portion of S-I of cats deprived by dorsal rhizotomies. Yet, it is not certain that the reactivated cortex reached normal levels of activity, and in other situations portions of deprived cortex may remain unresponsive.

An incomplete reactivation of deprived cortex has been reported for the hindfoot region of S-I of rats after sections of the sciatic nerve (Wall et al 1981). Most of the glabrous hindfoot and part of the foot dorsum of the rat is innervated by the sciatic nerve. A portion of the foot dorsum is innervated by the saphenous nerve. When the cortical territory of the saphenous nerve of rats with long-standing section of the sciatic nerve was compared with the territories of the saphenous nerve and the sciatic nerve in normal rats (Figure 4), it was obvious that the saphenous nerve cortical field had expanded in size. However, the expanded representation did not result in complete occupation of the sciatic nerve cortical field. Other evidence for incomplete reactivation comes from studies in monkeys in which two digits, rather than just one, were amputated. In these experiments (Schoppman et al 1981), small sectors of the cortex previously devoted to the amputated digits remained electrically silent and unresponsive for at least several months. However, there was evidence that the size of the unresponsive zone does decrease over time. Thus, the extent and existence of the unresponsive cortex may be related to both the post-lesion survival time and the magnitude of the deprivation.

Other deprived regions of the cortex may be more difficult to activate or have lower levels of activity even long after peripheral nerve damage. The stimulus thresholds for activating neurons in the cortex deprived by digit removal in raccoons were often higher than normal (Kelahan & Doetsch 1981, Rasmusson 1982). Likewise, many neurons in the deprived "paw" cortex of cats required firmer stimuli, such as taps or pressure, to be activated (Kalaska & Pomeranz 1979). However, interpreting this finding is complicated by the observation that the "receptive fields" for many of these "high threshold" neurons were on the digit stump in the amputated raccoons and on the wrist over the cut nerves to the paw in the denervated cats. Tapping the skin may have directly stimulated the nerve stumps or neuromas rather than mechanoreceptors (see Kalaska & Pomeranz 1979).

There is other evidence that nerve damage can reduce cortical activity, at least for a period of time. Removal of the tactile vibrissae and damage to the associated nerve endings on the lateral face of adult mice results in lowered levels of 2-deoxyglucose utilization (Durham & Woolsey 1978) and lowered levels of metabolic enzymes (Dietrich et al 1981a, Wong-Riley & Welt 1980) in layer IV of the affected cortex.

Does Reorganization Vary with Cortical Field, Species, or Type of Inactivation?

Two representations of the skin surface, S-I and S-II, appear to be part of the basic mammalian plan of organization for neocortex, and a number of mammals apparently have additional cortical cutaneous representations as well. Cats, for example, have an S-I, S-II, fields termed S-III and S-IV (Clemo & Stein 1982), and perhaps additional representations. Thus, it is possible to determine how the inactivation of regions of skin would effect different representations in the same animal. So far, this has only been studied for the Area 3b (S-I proper) and Area 1 (posterior cutaneous field)

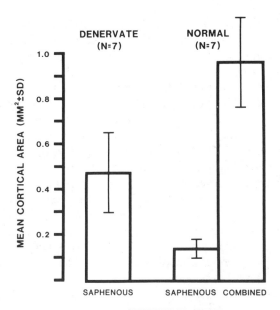

Figure 4 Evidence for incomplete reactivation of S-I cortex following partial deafferentation of the hindfoot in adult rats. The mean area of cortex receiving inputs from the hindfoot via the saphenous nerve in a group of rats with section of the sciatic nerve and in a group of normal rats. The area representing saphenous inputs in denervated animals is larger than the saphenous representation in normal rats and smaller than the total foot (combined) representation in normal rats.

in monkeys (Merzenich et al 1983a). Following median nerve section, new representations of the dorsal surfaces of the hand digits occupied most of the reorganized cortex of Area 3b, while expanded and repetitive representations of the palm were more characteristic of Area 1. In addition, the somatotopic pattern of reactivation was more variable in Area 1 than in Area 3b. Thus, two cortical representations in the same animal, both receiving their major driving influence from the same thalamic nucleus, the ventroposterior nucleus (see Lin et al 1979, Nelson & Kaas 1981), had different patterns of reorganization.

Reorganization after peripheral damage has now been demonstrated in adult monkeys (Merzenich et al 1983a,b, Schoppmann et al 1981) cats (Franck 1980, Kalaska & Pomeranz 1979), raccoons (Kelahan & Doetsch 1981, Rasmusson 1982), and rats (Wall et al 1981). In all these mammals, much or all of the deprived cortex becomes responsive to normally innervated skin fields. While marked differences in the manner of reactivation do occur, it is difficult to determine any clear species differences in reorganizational patterns because the deactivation procedures have varied, and this variable does affect the types of reorganization (see below). In terms of procedures, the experiments involving digit removal in monkeys and raccoons were the most similar. The results in both mammals were also similar in that the cortex was largely reactivated by the adjacent, remaining digits and palm. However, the reactivation in raccoons was described as nontopographic, whereas there was evidence of somatotopic reorganization in monkeys (Schoppmann et al 1981; M. M. Merzenich, unpublished studies).

In conclusion, cortical reorganization appears to depend in part on the type of deprivation. When large parts of the hand representation in Area 3b of monkeys were deprived of normal activation by section of the median nerve, much of this cortex was reoccupied by new highly somatotopic representations of the dorsal surfaces of the digits (Merzenich et al 1983a). When some of the same cortex was altered by removing a single digit, the deprived cortex became devoted to representations of adjacent *glabrous* pads and digits (Schoppmann et al 1981). When two digits were removed, some of the deprived cortex remained silent. Thus, the same region of cortex was affected in three ways by three deactivation procedures. Further studies of other regions of cortex and other deprivation procedures are needed.

The Time Course of Reorganization

Determining the time course of the reorganization of cortex after peripheral nerve damage is important for understanding the mechanisms of reorganization. Cortical changes that develop over short periods of time suggest that previously undetected connections somehow become effective activating pathways. Alterations that take days or longer raise the possibility that axons are growing over distances to form new connections. The results of the few studies to date suggest that both immediate and more slowly developing and progressive alterations occur in cortex deprived of its normal activation.

The time course of the reorganizations of the hand representations in Areas 3b and 1 of postcentral somatosensory cortex of owl and squirrel monkeys has been studied by Merzenich et al (1980, 1983b) by electrophysiologically mapping the deprived cortex at various times after median nerve section in different monkeys, or, in a few cases, by remapping the cortex at several times after nerve section in the same monkey. Important observations come from experiments with one monkey in which detailed maps of the two hand representations were derived just before, just after, and at 11, 22, and 144 days after median nerve section. Immediately after nerve section in these monkeys, most or all of the cortex formerly devoted to the median nerve (see Figure 1) was unresponsive to cutaneous stimuli. However, usually there were fragments of deprived cortex that were responsive to peripheral stimulation within a few hours after nerve section. These fragments were in portions of the cortex formerly devoted to the glabrous digit surfaces in Area 3b, and they were activated by the dorsal surfaces of the same digits. In the extensively remapped case, for example, most of the cortex normally devoted to glabrous digits 1 and 2 was immediately silent after nerve section, but much of the cortex normally devoted to glabrous digit 3 was responsive to the dorsal surface of digit 3. Over several weeks, the unresponsive cortex shrank in size so that all deprived cortex became responsive. The fully reorganized condition was similar to that shown in Figure 1D, and it was characterized by the completion and enlargement of the new representations of the dorsal digit surfaces in Area 3b, paralleled by a gradual decrease in receptive field sizes for neurons in this cortex, and by the expansion of the representation of glabrous pads and the ulnar nerve portion of glabrous digit 3 into the deprived cortex in both Areas 3b and 1. Most remarkably, the representations of some skin surfaces first increased and then decreased somewhat as other skin surfaces increased in representation. In particular, much of the deprived cortex that first became responsive to the dorsum of digit 3 later gradually became responsive instead to a greatly expanded representation of the ulnar margin of the glabrous surface of digit 3. Minor adjustments even seemed to occur in cortex bordering the deprived zone. Thus, both rapid and gradual changes occurred in somatosensory cortex of monkeys deprived of its normal activation.

Clear evidence for immediate changes in S-I organization in cats after the partial removal of peripheral activation was provided by Metzler & Marks (1979), who recorded from a total of 46 neurons in the S-I of cats before and after the nerves subserving the normal receptive fields of these neurons were inactivated by epidural blocks of dorsal roots with anesthetic. About half of these neurons became unresponsive during the block. However, other neurons became responsive to new receptive fields in skin outside the blocked region. The new receptive fields varied from normal to larger than normal in size. Effective stimulus thresholds also varied from normal to increased. The new receptive fields persisted throughout the blocks, and the original receptive fields returned immediately after the blocks. The authors concluded that the immediacy and the rapid reversibility of the effects were not compatible with explanations based on axon sprouting or other growth processes.

Longer term progressive alterations in the organization of the somatosensory cortex in cats are suggested by the report of Franck (1980), although few details were given. Franck concluded that cortex deprived of activation by peripheral deafferentation of parts of the hindlimb was "sequentially" occupied by normally innervated body segments over a 55 day post-lesion period.

Recordings from the deprived cortex of S-I of raccoons after digit amputation also suggest both immediate and gradually developing changes in activation patterns. Both Doetsch and co-workers (Carson et al 1981, Kelahan & Doetsch 1981) and Rasmusson (1982) found the deprived cortical region to be unresponsive to new inputs immediately after amputation in adult raccoons anesthetized with sodium pentobarbital. Rasmusson further observed that after two weeks, most of the deprived cortex was still unresponsive, and changes in the reactivated cortex, including reductions in receptive field sizes and response thresholds, were noted for as long as 16 weeks after the digit removal. Thus, there was no evidence for immediate change. However, it is likely that the barbiturate anesthesia masked some immediate changes, since Kelahan & Doetsch (1981) found that some neurons in cortex deprived by digit amputation are immediately responsive to new skin areas in raccoons anesthetized with nitrous oxide.

Finally, the time course of the reoccupation of the cortical field of the sciatic nerve by input from the saphenous nerve has been recently studied by J. T. Wall and C. G. Cusick (unpublished studies). After carefully measuring the extent of cortex activated by the saphenous nerve at various times after section and ligature of the sciatic nerve in rats, it was apparent that there was little or no immediate increase in the size of the cortical representation. However, some increase in the size of the representation of the saphenous nerve occurred in as little as 24 hours after the deprivation, and a rapid increase over a few days was followed by no further increase. Thus, there was a relatively rapid reactivation of cortex to the maximal extent of recovery.

ALTERED ORGANIZATION
IN DEVELOPING MAMMALS

The developing visual cortex is markedly altered by the conditions of both receptor damage and sensory deprivation that have little effect on the visual cortex of adults (see Lund 1978 for review). Although the fully developed somatosensory system appears to be more plastic than the fully developed visual system, it does seem reasonable to suppose that the effects of peripheral nerve injury would be different or more pronounced in the developing than in the adult somatosensory cortex. This supposition is partially supported, but the effects of nerve damage on the development of somatosensory cortex in young animals have not been extensively studied. However, there have been investigations of comparable denervations of the paw in adult cats and kittens (Kalaska & Pomeranz 1979), of digit removal in adult and young raccoons (Kelahan et al 1981, Kelahan & Doetsch et al 1981, Carson et al 1981), and of facial vibrissae damage in adult and young rodents (see below). Developmental effects have been most studied in vibrissae cortex of rodents, in which anatomical changes occur much more extensively after neonatal than adult peripheral damage, and occur only during a restricted sensitive period.

Functional Changes

The effects of chronic forepaw denervation in kittens and adult cats were compared by Kalaska & Pomeranz (1979). These authors investigated the receptive field properties of neurons in the regions of the cortical representations of the paw and the forearm. Two months after partial denervation of the forepaw in the kitten, there were marked differences from normal animals in the receptive field properties of neurons in cortex defined as the cortical paw region. In the denervated kittens, a large majority of these neurons had receptive fields located on the forearm, the upper arm, or the wrist (87% of neurons); only 3% responded to paw stimulation, compared to the 90% of neurons responsive to paw stimulation in the control animal. Furthermore, there was an increase from 6% to 19% of neurons without detectable receptive fields. Kalaska & Pomeranz also noted that anomalous receptive fields could be found throughout the paw cortex and that units with wrist and forepaw receptive fields were intermixed, suggesting a lack of somatotopic organization. These results contrast with what the same authors found in the cortex of the adult cat after a similar period of partial paw denervation. In the cortical paw representation of the adult cat, there was only a slight increase in the number of neurons with forearm receptive fields (7%) compared to the kitten (52%). There was, however, a marked increase in the number of units responsive to wrist stimulation (54% vs 2% in controls and 24% in denervated kittens) and in unresponsive neurons (37% vs 8% in controls, 19% in denervated kittens). Thus, the results suggest a greater capacity for organizational changes in the kitten than the cat. However, Kalaska & Pomeranz (1979) pointed out that the denervations in the kitten and cat may not have been equivalent, as postmortem examination revealed increased neuroma formation in adults compared to kittens. It seemed likely that many of the wrist activated units in adults were the result of direct stimulation of the neuromas.

The degree of functional reactivation also has been addressed by amputation rather than nerve section. This procedure has been used to determine the effect of digit removal on the functional organization of the portion of cortex in which that digit is normally represented (Kelahan et al 1981, Kelahan & Doetsch 1981, Carlson et al 1981). Following removal in the neonatal raccoon, the changes assayed in the adult included increases in receptive field size and in nonsomatotopically organized receptive fields related to flanking digits. These changes were similar to those seen after adult amputation. However, the receptive fields following neonatal amputation were generally smaller in size and they were scattered over more of the forepaw, thus suggesting a greater propensity for changes in organization in the neonatal compared to the adult raccoon.

Alterations in the functional organization of the rodent somatosensory cortex following neonatal peripheral damage to the facial vibrissae has also been reported. Following removal of one or two nonadjacent rows of vibrissae, units in the region of cortex associated with the damaged rows could be activated by stimulation of vibrissae in several of the intact rows as well as the fur surrounding removed vibrissae. In addition, silent regions of cortex could be identified (Killackey et al 1978). Thus, relatively restricted damage to the peripheral innervation of the trigeminal nerve results in a breakdown of the highly organized somatotopy that normally characterizes the cortex related to the trigeminal nerve. More extensive organizational changes have been reported to result from damage to all the mystacial vibrissae. Following such damage in both the rat (Waite & Taylor 1978) and mouse (Pidoux et al 1979), neurons in regions of cortex normally activated by vibrissae stimulation could be activated by stimulating the skin of surrounding body parts such as the nose or upper lip. Thus, there are distinct changes in the functional organization of somatosensory cortex following neonatal damage to the vibrissae follicles.

A somewhat more indirect approach to determining functional changes has been taken in studies that measure alterations in levels of metabolic activity resulting from neonatal peripheral lesions. Following vibrissae follicle damage in the neonatal mouse, Wong-Riley & Welt (1980) reported that there were decreases in both the areal distribution and staining density of the energy-coupled enzyme, cytochrome oxidase, in layer IV of somatosensory cortex of the adult mouse. Similar damage in the adult produces only changes in staining density.

Changes in metabolic activity can also be measured by utilization of the glucose analogue, 2-deoxy-D-glucose. With this technique, Durham & Woolsey (1978) demonstrated that cortical activity related to a vibrissa extended through all cortical layers in the adult. Hand (1982) has suggested that the method provides an index of "global patterns" of functional activity. Hand and colleagues have made extensive use of this technique to delimit patterns of activity related to a single vibrissa following removal of all other vibrissae in adult and in newborn rats. The normal adult cortical pattern of activity associated with stimulation of a single vibrissa was shaped like a candle pin. Thus, levels of increased activity extended through all cortical layers but they were widest in layer IV, and they tapered slightly in the supra- and infragranular layers. The density of activity was greatest in layer IV. Following neonatal vibrissae removal, there was a marked difference in the pattern of activity associated with the spared vibrissa. In such cases, the typical candle pin pattern of activity was absent. While the labeled focus of activity in layer IV was increased in size, it was reduced in density and there was a large increase in the area of activation and density of labeling in the other cortical layers, particularly the supragranular layers. Changes were similar, but less pronounced, in adult animals studied several months after vibrissae damage. It is important to note that neonatal vibrissae removal alone did not produce major changes in the cortical pattern of activity of a single spared vibrissa. Some damage to the follicle was necessary before major cortical changes occurred. Likewise, several anatomical studies have reported that simple neonatal vibrissae removal does not produce the marked anatomical change in cortex found after actual follicle damage (Verley & Axelrod 1977, Killackey 1980a,b).

Given the assumption that a close correlation exists between neural function and cerebral metabolism as measured by levels of deoxyglucose uptake or staining for cytochrome oxidase, the use of this method to detect functional changes in organization is significant in that it addresses the question of morphological correlates of functional changes. To date, this question has been explored only within the developing rodent trigeminal system. The results of these studies are reviewed in the next section.

Anatomical Changes

The highly discrete anatomical organization that characterizes the trigeminal field of rodent somatosensory cortex makes it an ideal area for studying the anatomical correlates of peripheral damage. In the fourth layer of the somatosensory cortex of rats and mice, stellate cells are grouped into discrete aggregates (Lorente de Nó 1922), termed "barrels" (Woolsey & Van der Loos 1970). The perikarya of these neurons tend to be differentially distributed on the "sides" of the aggregates, with their dendrites oriented toward the center or "hollow" of the "barrel" (Figure 6A). The center of the barrel is also the major target of the thalamocortical projections that arise from the ventroposterior nucleus (Killackey 1973, Killackey & Leshin 1975). Adjacent cellular aggregates and their discrete thalamic projections are separated from each other by relatively acellular regions or "septa" in which there are few, if any, thalamocortical terminations. The topographic distribution of these cellular aggregates and their thalamic inputs replicates the pattern of distribution of tactile hairs on the rodent muzzle. The largest of these "barrels" form five rows in the cortex (Figure 5A) and they are functionally related in a one-to-one fashion to the five peripheral rows of mystacial vibrissae in the deeply anesthetized animal (Welker 1971, 1976).

Given this high degree of specificity between peripheral receptor surface and cortical organization, a natural question to ask is whether or not the

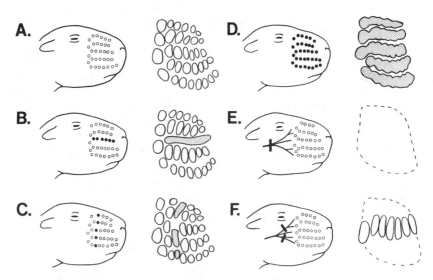

Figure 5 The effect of various peripheral manipulations on the organization of the cells in layer IV and the thalamocortical terminations within vibrissae representation of the rat somatosensory cortex. A. The normal organization. B. Removal of the middle row of vibrissae results in a fused band (Killackey et al 1976). C. Removal of the same number of vibrissae in an arc across rows results in only a partial fusion (Killackey & Belford 1980). D. Removal of all five rows results in five fused bands (Killackey & Belford 1979). E. Nerve section results in complete abolition of the pattern (Killackey & Shinder 1981). F. Selective nerve section results in an abolition of pattern in the related portion of cortex and an expansion of the intact representation (Killackey & Erzurumlu 1982).

development of this discrete cortical organization is dependent on an intact periphery. The question was first asked and answered in the affirmative by Van der Loos & Woolsey (1973), who demonstrated that neonatal vibrissae damage results in an altered distribution of cortical neurons in the adult mouse. After they removed a row of vibrissae, the cortical representation of that row no longer corresponded to a row of "barrels," but instead it formed a single fused band of neurons (Figure 5B and 6B). The nature of this change should be emphasized. There was no obvious decrease in the number of neurons in layer IV; rather, an apparently similar number of neurons were arranged in a fused band rather than a punctate row (see Figures 5 and 6). Other experiments indicate that the cortical pattern of the thalamic input is also altered by neonatal vibrissae follicle damage. After such damage, the thalamocortical projections corresponding to the damaged row developed, but they were distributed in adults as a single fused band rather than as a row of punctate clusters of terminations (Killackey 1976).

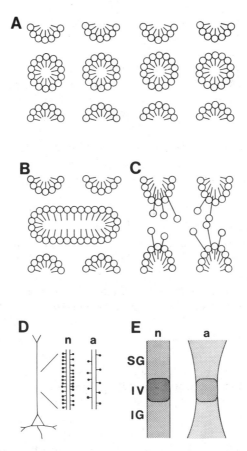

Figure 6   A. Schema of the organization of a normal row of cortical "barrels" in a plane tangential to the fourth cortical layer. The distribution of stellate neuron perikarya and dendrites as well as thalamocortical projections are illustrated. B. Schema of the organization of a cortical row following damage to the associated vibrissae row on the face of the neonatal rat. After Steffen & Van der Loos (1980). C. Similar to B but illustrating the results of Harris & Woolsey (1981). D. Effect of vibrissae removal on pyramidal cell spine distribution (Ryugo et al 1975). E. Illustration of normal and altered columnar cortical organization associated with a single vibrissa as demonstrated with 2-DG (after Hand 1982). In the altered case, all surrounding vibrissae have been removed and the single remaining vibrissa stimulated.

The boundaries between rows of cortical barrels seem to be more resistant to alterations than those within a row. For example, removal of all five rows of vibrissae results in five fused cortical bands (Figure 5D; Killackey & Belford 1979) and removal of an arc of vibrissae across rows does not result in a single fused band as does removing the same number of vibrissae when they compose a single row (Figure 5C; Killackey & Belford 1980). However, the boundaries between rows can be abolished by direct damage to the trigeminal nerve (Figure 5E; Killackey & Shinder 1981), suggesting that direct nerve damage can have a more severe effect on cortical organization than vibrissae follicle damage. Likewise, after selective damage to the nerve bundle innervating one row of vibrissae, the cortical barrels for the remaining flanking rows of vibrissae expand to a greater extent than after follicle damage (Killackey & Erzurumlu 1982). This difference in the cortical consequences of nerve and follicle damage may be related to the different effects of the two manipulations on the pattern of fasciculation within the trigeminal nerve. While follicle damage produces little change in the pattern of fasciculation, it is markedly altered by nerve cut (Killackey & Erzurumlu 1982).

The different consequences of nerve and follicle damage may have a bearing on the apparent discrepancy concerning what basic effect removal of the middle row of vibrissae (Row "C") has on the distribution of neurons in layer IV. As outlined above, Van der Loos & Woolsey (1973) reported that damage of row "C" vibrissae on the day of birth resulted in a uniformly distributed band of neurons instead of a discrete row of neuronal "barrels" (see Figure 6B). However, Woolsey & Wann (1976) later found that the "same" manipulation resulted in the expansion of cortical "barrels" associated with normal flanking rows of vibrissae into cortical tissue normally devoted to the damaged vibrissae (Figure 6C). Jeanmonod et al (1981) recently replicated the original findings of Van der Loos & Woolsey (1973), and concluded that the different results of Woolsey & Wann (1976) probably resulted from greater peripheral damage. The differing effects of follicle and nerve damage mentioned above support this contention. Because the face of a newborn mouse is small, it would be easy to damage nerve fascicles as well as individual nerve fibers during the follicle damage procedures.

Thus far we have discussed two anatomical changes in cortex following peripheral damage: (a) changes in the distribution of the terminations of thalamic afferents and (b) changes in the distribution of the cell bodies of the neurons of layer IV of somatosensory cortex. Peripheral damage also appears to affect the distribution of dendrites of the layer IV stellate cells. While two studies reported changes in the distribution of dendrites of stellate cells in the region of the cortex related to the damaged periphery, the types of changes differed. Steffen & Van der Loos (1980) found that the dendrites of stellate neurons in experimentally damaged animals tended to orient more toward the center of what they referred to as a "barrel-like territory," and they found no increase in dendrites oriented toward adjacent normal rows (Figure 6B); whereas Harris & Woolsey (1981) reported increases in the number of dendrites oriented toward flanking rows (Figure 6C). Again, differences in results could be related to different amounts of peripheral damage in the two studies.

The morphology of the second major element of the neocortex, the pyramidal cell, was also altered by neural neonatal vibrissae damage (see Figure 6D and E). There is evidence for a diminution of approximately one-third in the density of spines, presumed to be postsynaptic specializations, along the portion of the apical dendrite of layer V pyramidal cells passing through layer IV (Ryugo et al 1975). Because both pyramidal cells and stellate cells of mouse somatosensory cortex receive a direct thalamic input (White 1978), the loss of spines could reflect a reduction in the thalamic activation of these pyramidal cells.

Time Course of Developmental Changes

Information is not presently available about the time course of the functional alterations that follow neonatal peripheral damage. All of the functional studies reported above assessed the effects of the neonatal peripheral manipulations at least several months after the damage. This is also true for the anatomical experiments assessing the effects of vibrissae damage on the orientation of stellate cell dendrites and pyramidal cell dendritic spine density. However, the results of several anatomical experiments suggest that both the distributional changes in the neurons of layer IV and their thalamic inputs can take place very quickly, and they are detectable within several days of the peripheral damage (Jeanmonod et al 1977, Killackey & Belford 1979). Furthermore, these cortical changes in the distributions of neurons can be seen as early in development as normal discrete organization can be detected. It should also be noted that major anatomical changes are detectable only after damage during a sensitive period that is over before the end of the first postnatal week (Weller & Johnson 1975, Woolsey & Wann 1976, Belford & Killackey 1980, Jeanmonod et al 1981).

MECHANISMS OF REORGANIZATION AND ALTERED DEVELOPMENT

Reorganization in Adults

The activation of cortical neurons by newly effective inputs after the removal or inactivation of their normally effective inputs in adult mammals is well documented. Similar changes in activation patterns have been reported for other levels of the somatosensory system after peripheral nerve or dorsal column damage, including somatosensory neurons in the dorsal horn (Basbaum & Wall 1976, Mendell et al 1978, Pubols & Goldberger 1980, Devor & Wall 1978, 1981a,b; however see Brown et al 1979, Pubols and Brenowitz 1982), the dorsal column nuclei (Devor & Wall 1978, Millar et al 1976), and the ventral thalamus (Pollin & Albe-Fessard 1979, Wall & Egger 1971). Various authors have suggested mechanisms for the acquisition of "new" response properties by neurons in the central nervous system after peripheral damage (see Merrill & Wall 1978). There have been two basic types of explanation of central plasticity:

1. Previously ineffective connections become effective in activating neurons as a result of local changes in the amount of inhibition or an increase in sensitivity (e.g. the development of denervation supersensitivity).
2. New connections are formed through axon growth and/or dendritic extension.

The usual explanation for rapid changes in responsiveness is that previously undetected connections became "unmasked," since the formation of new synapses in the central nervous system after damage is thought to take days (Cotman et al 1982). Longer term changes generally have been attributed to axon growth, although gradually developing increases in synaptic effectiveness could also be responsible for many changes.

Anatomical evidence for the sprouting and growth of axons into partially denervated or deactivated sectors of the somatosensory system has gradually accumulated. The sprouting of dorsal root axons in the spinal cord after neighboring dorsal roots were cut was first reported by Liu & Chambers (1958), and this study has long been used as the primary evidence of anatomical growth (also see Goldberger & Murray 1978). However, more recent studies suggest that spinal cord sprouting may not exist or is minimal (Kerr 1972, Beckerman and Kerr 1976, Rodin et al 1981). Investigations recording changes in axons counted with the electron microscope suggest a third conclusion: Little or no sprouting of myelinated dorsal root axons occurs but there is considerable sprouting of unmyelinated axons (Hulsebosch & Coggeshall 1981). In addition, new synapses in the spinal cord may form over restricted distances as a result of very limited axonal growth, or even as a result of dendritic growth (Brown et al 1979). Finally, there is evidence of synapse formation in the ventroposterior nucleus (Ganchrow & Bernstein 1981a), dorsal column nuclei (Ganchrow et al 1981), and spinal cord (Bernstein et al 1974) after spinal cord damage. Most significantly for the present discussion, dorsal column damage in rats is followed by a period of increased synapse formation in somatosensory cortex (Ganchrow & Bernstein 1981b).

We are left with clear evidence for both the potentiation of previously ineffective pathways and for the formation of new central connections. Both may occur in somatosensory cortex after peripheral damage. However, changes at any level in the somatosensory system would be reflected in cortex. Since changes clearly do occur at subcortical levels, it is obvious that not all of the alterations seen in cortex actually occur in cortex.

What does occur in cortex, and how? Direct evidence is limited, but some inferences are possible. Because the reorganization of the hand representations of Areas 3b and 1 are different, some reorganization appears to be strictly cortical. In particular, the gradual emergence of the "new" representations of the dorsal digit surfaces in Area 3b may be at least in part a cortical phenomenon. Furthermore, the sudden appearance of fragments of this new representation immediately after nerve section suggests that new connections are not completely responsible for this reorganization. On the other hand, the gradual occupation of deprived cortex by the apparently somatotopic expansion of adjoining representations is more easily accounted for by the growth of axons and the formation of new connections. When large regions of cortex are activated from a restricted skin surface, as in "repetitive expansions," a few axons may ultimately activate a large cortical population of neurons: a simple sprouting of axons into deprived tissue at some level may be the explanation. However, shifted somatotopic representations, such as that seen in monkey cortex, seem to require the gradual alteration of many synapses in order to preserve the somatotopic pattern of the expanded representations.

Altered Development

An analysis of the mechanisms that underlie organizational changes following peripheral damage in the neonatal animal is more complicated than an analysis of such mechanisms in the adult. The reason for this is that such changes are probably due to at least two types of mechanisms:

1. Central reorganizational changes consequent to the peripheral insult: The mechanisms for this class of change may be similar to these discussed for the adult.
2. A direct altering of developmental processes.

However, the experimental paradigm used by most investigators, which is to manipulate the neonatal animal and assay the effect in the adult, makes it difficult to distinguish reorganizational changes from direct effects on the developmental process. Direct effects are probably restricted in time to "sensitive" or "critical" periods.

The anatomical alterations in the rodent trigeminal system summarized above are probably largely developmental effects, and they are the basis for hypothesizing that the periphery provides a template for the development of topographic relations in somatosensory cortex. If such a proposition is to have any validity, one might expect that effects similar to those described above for somatosensory cortex also could be identified in the subcortical relays of the trigeminal system. There is a good deal of evidence that this is indeed the case. The discrete anatomical organization evident in the thalamic and brainstem portions of the trigeminal system, in the terminations of the primary afferents of the trigeminal nerve, in the principal trigeminal sensory nucleus, and in the subnuclei interpolaris and caudalis of the spinal trigeminal complex as well, reflect discrete anatomically detectable clusters of neurons related to the vibrissae (Belford & Killackey 1979a,b). The cells on which primary afferents terminate, and which in turn project to the thalamus, exhibit a similar organization in the distribution of their perikarya in the brainstem (Erzurumlu et al 1980) and in their axon terminals in the ventroposterior nucleus (Belford & Killackey 1979a,b). The same organization is again repeated in the distribution of the perikarya of the thalamocortical relay cells within the ventroposterior nucleus (Van der Loos 1976, Ivy & Killackey 1982). An anatomical organization related to the vibrissae has recently been detected in the pattern of fasciculation within the trigeminal nerve itself (Erzurumlu & Killackey 1982).

Complementing these studies of normal organization are experimental studies that demonstrate that neonatal peripheral damage produces changes similar to those described for cortex at both thalamic and brainstem levels. Furthermore, at both these subcortical relay stations, peripheral manipulations produce changes in both the distribution of afferent terminals and neuronal perikarya (Belford & Killackey 1979a,b, 1980, Woolsey et al 1979, Bates et al 1982). These same neonatal peripheral manipulations also produce changes in the pattern of fasciculation within the trigeminal nerve itself. Thus, a peripheral manipulation such as damaging rows of vibrissae or selectively cutting a portion of the trigeminal nerve results in detectable anatomical changes in every neuronal structure in the trigeminal chain between the face and cortex. In total, the simplest interpretation of these changes is that they occur during the initial development of the trigeminal lemniscal system and that altering the periphery during the developmental period results in altered instructions to organizing central structures in a cascade-like fashion (Killackey 1980a).

FUNCTIONAL IMPLICATIONS OF CORTICAL PLASTICITY IN SOMATOSENSORY CORTEX

Removing or altering a peripheral source of input to somatosensory cortex has the potential of greatly changing the somatotopic organization of the cortex in both developing and adult mammals. Such transneuronal changes in a developing sensory system are not surprising since they have been demonstrated in visual cortex after unilateral eye removal or eyelid closure in cats and monkeys (see Hubel et al 1977 for a review) and as a result of congenitally misdirected retinal projection in Siamese cats (see Guillery & Casagrande 1976 for a review). Ample data indicate that normal development depends on normal input. What is surprising, since there is less precedent from work on other sensory systems, is the extensive reorganization of somatosensory cortex in adult mammals after peripheral damage. The damage in both adult and developing mammals has functional implications, some of which have been discussed in more detail elsewhere (Merzenich et al 1983a).

Cortical Reorganization in Adults

When the normal source of activation for large sectors of somatosensory cortex is removed, this cortex becomes responsive to other sources of input. Part of this new activation appears to be the result of unmasking previously present connections, but part also apparently depends on the formation of new connections, and even the orderly shift and expansion of whole patterns of connections. Not only do these new activation patterns appear, but they are reversible. Reversibly blocking a peripheral nerve results in new "unmasked" receptive fields for cortical neurons that are lost and replaced by the original receptive field after the peripheral nerve recovers (Metzler & Marks 1979). Regeneration of a crushed or cut peripheral nerve results in the nerve recapturing its former cortical territory (Paul et al 1972, Wall et al 1980, Wall et al 1982). These observations indicate that cortical somatotopic maps in adult animals are remarkably plastic and are not static or fixed in organization. Two important implications are briefly reviewed here.

1. The reorganization after peripheral damage suggests that the cortical territory of a peripheral receptor surface is, in part, use dependent. Minimally or inappropriately activated pathways (a cut and ligated peripheral nerve probably generates some input; see Govrin-Lippmann & Devor 1978) lose cortical territory while neighboring, normally active pathways gain cortical territory. The factors governing the gain are not known, but cortical adjacency is not the only factor, since the face representation apparently does not expand into the adjacent deprived hand representation (Merzenich et al., 1983a). It also appears that more than one type of reactivation is possible for the same cortex. If the input from a glabrous digit surface is removed in a monkey, the cortical representation of that surface may become activated by the dorsal surface of that digit, when such input is available as after median nerve section. However, when the complete digit is removed, the same cortex is occupied by the glabrous surfaces of adjacent digits. It also appears that the capture of a cortical territory does not depend on the input having a normal functional role; regenerated peripheral neurons reactivate cortex even when misdirected axons reinnervate incorrect targets (Paul et al 1972, Wall et al 1980). As a dramatic example of reactivation by a misdirected nerve, the sciatic nerve in a rat, when transposed and

crossed to the opposite side of the body to regenerate into the wrong leg, as initially described by Sperry (1943), will recapture its cortical zone (J. T. Wall and J. H. Kaas, unpublished).

If deactivated peripheral nerves lose and reactivated peripheral nerves regain cortical territory, then it seems likely that a peripheral nerve also could gain cortical territory by increasing its activity over normal levels. The supposition is that the amount of activity in peripheral inputs is one factor in determining the amount of cortex related to that input, and that inputs are normally in competition with each other (Merzenich et al 1983a,b). More active inputs would replace less active inputs. This, of course, is a testable hypothesis. If cortical territory is determined to some extent by the differential use of peripheral receptive surfaces, does a gain in territory result in an increase in somatosensory abilities, and vice versa? Such a possibility could account for improvements in sensory and perceptual capacities with practice, as well as impairments with neglect.

2. Cortical sites may be capable of rapidly changing the peripheral location they specify, and they may specify more than one location at once. It has long been known that electrically stimulating single sites in somatosensory cortex in humans results in a sensation with a particular peripheral reference (e.g. Penfield & Boldrey 1937). When cortical maps change, it would seem necessary to change correspondingly the peripheral reference. Otherwise, mislocation of peripheral stimuli would occur in individuals with peripheral nerve damage. It also appears that a sectioned peripheral nerve, if electrically stimulated, can signal the position of the denervated skin field (see Merzenich et al 1983a). If so, is it doing so by activating its original cortical territory, even though these territories may now also respond to stimuli on other skin surfaces?

Cortical Plasticity in Development

Peripheral damage in newborn mammals alters the subsequent development of the somatosensory system. In cortex, intact receptor surfaces can gain in representation, perhaps even more so than after such damage in adults. Furthermore, in developing kittens, there is evidence that increased use results in an enlarged cortical representation. Spinelli & Jensen (1979) reported that kittens reared while having eight minutes a day of avoidance training of shock to the forepaw had a cortical forepaw area "many times larger" than the control side. As in adults, these reported changes raise questions about the possibility of corresponding alterations in sensory abilities. Do enlarged representations and altered connections increase sensory capacity?

SUMMARY

When a restricted sector of somatosensory cortex is deprived of its normal pattern of activation in adult mammals by sectioning peripheral nerves or dorsal roots, or by amputation of a body part, the affected cortex rapidly becomes largely or completely reactivated by inputs from adjoining and nearby skin fields. The types of cortical reorganization that have been described in the deprived cortex include the appearance of "new" duplications of body parts represented elsewhere in the S-I, somatotopic and repetitive expansions of representations adjoining the deprived cortex, and at least some degree of nontopographic reactivation. In addition, incomplete reactivation sometimes occurs. Although the number of studies has been limited, the results justify several conclusions.

1. The cortical reorganization no doubt reflects the reported subcortical changes in the somatosensory system, but the reorganization is not completely dependent on subcortical changes. Two cortical fields, 3b and 1, with thalamic inputs from the same nucleus, reorganize differently in the same animal.
2. The appearance of extremely rapid changes suggests that some of the effects are a result of the "unmasking" of previously ineffective or undetected connections. The gradual development of expanded representations is compatible with the formation of new connections and synapses.
3. The same region of cortex has the capacity to reorganize in different ways, depending on the deactivation procedure. Cortex normally devoted to the glabrous digits becomes responsive to dorsal digit surfaces after section of the median nerve. When digit cortex is deactivated by single digit amputation, the cortex becomes devoted to remaining digits. After removal of two digits, some unresponsive cortex may remain over a significant post-lesion period.
4. If the deactivation of cortex can be regarded as an extreme form of disuse, the restructuring of somatosensory cortex after peripheral nerve damage suggests that the normal organization of S-I is constantly being modified by pattern of use. Increased use would lead to increased cortical representation and decreased use would lead to decreased representation. Thus, inputs to cortex may be in use-dependent competition with each other. Expansion of representations may have functionally important consequences.

Peripheral damage in young developing mammals has the potential both to alter cortical organization by the same mechanisms as in adults and to alter normal development. The extensive work on the vibrissal cortex of rats and mice shows that major physiological and anatomical changes can be induced in the developing somatosensory cortex by peripheral damage. This research allows several additional conclusions.

5. There is a "critical period" for some developmental effects. Many of the changes in the vibrissae cortex of rodents appear to depend on the peripheral damage occurring before a few postnatal days of age.
6. The peripheral organization of sensory inputs serves as a template to the subsequent development of central somatosensory structures.

ACKNOWLEDGMENTS

We thank Rosalyn E. Weller for drawing the illustrations and for her comments on the manuscript. Other helpful comments were made by C. G. Cusick, G. S. Doetsch, D. J. Felleman, M. E. Goldberger, M. Murray and J. T. Wall. This review was completed while H. P. Killackey was a Visiting Scholar of Psychology at Vanderbilt University.

Literature Cited

Basbaum, A. I., Wall, P. D. 1976. Chronic changes in the response of cells in adult cat dorsal horn following partial deafferentation: The appearance of responding cells in a previously nonresponsive region. *Brain Res.* 116:181–204

Bates, C. A., Erzurumlu, R. S., Killackey, H. P. 1982. Central correlates of peripheral pattern alterations in the trigeminal system of the rat. III. Neurons of the principal sensory nucleus. *Dev. Brain Res.* In press

Beckerman, S. B., Kerr, F. W. L. 1976. Electrophysiological evidence that neither sprouting nor neuronal hyperactivity occur following long-term trigeminal or cervical primary deafferentation. *Exp. Neurol.* 50:427–38

Belford, G. R., Killackey H. P. 1979a. Vibrissae representation in subcortical trigeminal centers of the neonatal rat. *J. Comp. Neurol.* 183:305–22

Belford, G. R., Killackey, H. P. 1979b. The development of vibrissae representation in subcortical trigeminal centers of the neonatal rat. *J. Comp. Neurol.* 188: 63–74

Belford, G. R., Killackey H. P. 1980. The sensitive period in the development of the trigeminal system of the neonatal rat. *J. Comp. Neurol.* 193:335–50

Bernstein, J. J., Gelderd, J. B., Bernstein, M. E. 1974. Alteration of neuronal synaptic complement during regeneration and axonal sprouting of rat spinal cord. *Exp. Neurol.* 44:470–82

Brenowitz, G. L., Pubols, L. M. 1981. Increased receptive field sizes of dorsal horn neurons following chronic spinal cord hemisections in cats. *Brain Res.* 216:45–49

Brown, P. B., Busch, G. R., Whittington, J. 1979a. Anatomical changes in cat dorsal horn cells after transection of a single dorsal root *Exp. Neurol.* 64:453–68

Brown, P. B., Yezierski, R. P., Koerber, H. R. 1979b. Absence of alterations in dorsal horn cell somatotopy after unilateral L_7 section in cat. *Soc. Neurosci. Abstr.* 5:704

Carson, L. V., Kelahan, A. M., Ray, R. H., Massey, C. E., Doetsch, G. S. 1981. Effects of early peripheral lesions on the somatotopic organization of the cerebral cortex. In *Clin. Neurosurg.* 28:532–46

Clemo, H. R., Stein, B. E. 1982. Somatosensory cortex: A "new" somatotopic representation. *Brain Res.* 235:162–68

Cotman, C. W., Nieto-Sampedro, M., Harris, E. W. 1982. Synapse turnover in adult vertebrates. *Physiol. Rev.* 61:684–784

Devor, M., Schonfeld, D., Seltzer, Z., Wall, P. D. 1979. Two modes of cutaneous reinnervation following peripheral nerve injury. *J. Comp. Neurol.* 185: 211–20

Devor, M., Wall, P. D. 1978. Reorganization of spinal cord sensory map after peripheral nerve injury. *Nature* 276:75–76

Devor, M., Wall, P. D. 1981a. Plasticity in the spinal cord sensory map following peripheral nerve injury in rats. *J. Neurosci.* 1:679–784

Devor, M., Wall, P. D. 1981b. Effect of peripheral nerve injury on receptive fields of cells in the cat spinal cord. *J. Comp. Neurol.* 199:277–91

Dietrich, W. D., Durham, D., Lowry, O. H., Woolsey, T. A. 1981a. Quantitative histochemical effects of whisker damage on single identified cortical barrels in the adult mouse. *J. Neurosci.* 1:929–35

Dostrovsky, J. O., Millar, J., Wall, P. D. 1976. The immediate shift of afferent drive of dorsal column nucleus cells following deafferentation: A comparison of acute and chronic deafferentation in gracile nucleus and spinal cord. *Exp. Neurol.* 52:480–95

Durham, D., Woolsey, T. A. 1978. Acute whisker removal reduces neuronal activity in barrels of mouse SmI cortex. *J. Comp. Neurol.* 178:629–44

Erzurumlu, R. S., Bates, C. A., Killackey, H. P. 1980. Differential organization of thalamic projection cells in the brainstem trigeminal complex of the rat. *Brain Res.* 198:427–33

Erzurumlu, R. S., Killackey, H. P. 1982. Order in the developing rat trigeminal nerve. *Dev. Brain Res.* 3:305–10

Franck, J. I. 1980. Functional reorganization of cat somatic sensory-motor cortex (SmI) after selective dorsal root rhizotomies. *Brain Res.* 186:458–62

Ganchrow, D., Bernstein, J. J. 1981a. Bouton renewal patterns in rat hindlimb cortex after thoracic dorsal funicular lesions. *J. Neurosci. Res.* 6:525–37

Ganchrow, D., Bernstein, J. J. 1981b. Patterns of afferentation in rat ventroposterolateral nucleus after thoracic dorsal column lesions. *Exp. Neurol.* 71:464–72

Ganchrow, D., Margolin, J., Perez, L., Bernstein, J. J. 1981. Patterns of reafferentation in rat nucleus gracilis after thoracic dorsal column lesions. *Exp. Neurol.* 71:437–51

Goldberger, M. E., Murray, M. 1978. Recovery of movement and axonal sprouting may obey some of the same laws. In *Neuronal Plasticity,* ed. C. W. Cotman, pp. 73–96. New York: Raven

Govrin-Lippmann, R., Devor, M. 1978. Ongoing activity in severed nerves: Source and variation with time. *Brain Res.* 159:406–10

Guillery, R. W., Casagrande, V. A. 1976. Adaptive synaptic connections formed in the visual pathways in response to congenitally aberrant inputs. *Cold Spring Harbor Quant. Biol.* 40:611–17

Hand, P. J. 1982. Plasticity of the rat cortical barrel system. In *Changing Concepts of the Nervous System,* ed. A. R. Morrison, P. Strick, pp. 49–68. New York: Academic.

Harris, R. M., Woolsey, T. A. 1981. Dendritic plasticity in mouse barrel cortex following postnatal vibrissa follicle damage. *J. Comp. Neurol.* 196:357–76

Hoffer, J. A., Stein, R. B., Gorden, T. 1979. Differential atrophy of sensory and motor fibers following sections of cat peripheral nerves. *Brain Res.* 178:347–61

Hubel, D. H., Wiesel, T. N., LeVay, S. 1977. Plasticity of ocular dominance columns in monkey striate cortex. *Philos. Trans. R. Soc. London Ser. B* 278:377–409

Hulsebosch, C. E., Coggeshall, R. E. 1981. Sprouting of dorsal root axons. *Brain Res.* 224:170–74

Ivy, G. O., Killackey, H. P. 1982. Ephemeral cellular segmentation in the thalamus of the neonatal rat. *Dev. Brain Res.* 2:1–17

Jackson, P. C., Diamond, J. 1981. Regenerating axons reclaim sensory targets from collateral nerve sprouts. *Science* 214:926–28

Jeanmonod, D., Rice, F. L., Van der Loos, H. 1977. Mouse somatosensory cortex: Development of the alterations in the barrel field which are caused by injury to the vibrissa follicles. *Neurosci. Lett.* 6:151–56

Jeanmonod, D., Rice, F. L., Van der Loos, H. 1981. Mouse somatosensory cortex: Alterations in the barrel field following receptor injury at different early postnatal ages. *Neuroscience* 6:1503–35

Kaas, J. H. 1983. What, if anything is S-I?: The organization of the "first somatosensory area" of cortex. *Physiol. Rev.* In press

Kaas, J. H., Nelson, R. J., Sur, M., Merzenich, M. M. 1979. Multiple representations of the body within the primary somatosensory cortex of primates. *Science* 204:521–23

Kaas, J. H., Sur, M., Nelson, R. J., Merzenich, M. M. 1981. Multiple representations of the body in the postcentral somatosensory cortex of primates. In *Cortical Sensory Organization,* Vol. 1, *Multiple Somatic Areas,* ed. C. N. Woolsey, pp. 29–45. Clifton, NJ: Humana

Kalaska, J., Pomeranz, B. 1979. Chronic paw deafferentation causes an age-dependent appearance of novel responses from forearm in "paw cortex" of kittens and adult cats. *J. Neurophysiol.* 42: 618–33

Kelahan, A. M., Doetsch, G. S. 1981. Short-term changes in the functional organization of somatosensory (SmI) cortex of adult raccoons after digit amputation. *Soc. Neurosci. Abstr.* 7:540

Kelahan, A. M., Ray, R. H., Carson, L. V., Massey, C. E., Doetsch, G. S. 1981. Functional reorganization of adult raccoon somatosensory cerebral cortex following neonatal digit amputation. *Brain Res.* 223:152–59

Kerr, F. W. L. 1972. The potential of cervical primary afferents to sprout in the spinal nucleus of V following long-term trigeminal denervation. *Brain Res.* 43:547–60

Killackey, H. P. 1973. Anatomical evidence for cortical subdivisions based on vertically discrete thalamic projections from the ventral posterior nucleus to cortical barrels in the rat. *Brain Res.* 51:326–31

Killackey, H. P. 1979. Peripheral influences on connectivity in the developing rat trigeminal system. In *The Developmental Neurobiology of Vision*, ed. R. Freeman, W. Singer, pp. 381–90. New York: Plenum

Killackey, H. P. 1980a. Pattern formation in the trigeminal system of the rat. *Trends Neurosci.* 3:303–6

Killackey, H. P. 1980b. Spatial and temporal variables in pattern formation in the rat trigeminal system. *Am. Soc. Zool. Abstr.* 20:770

Killackey, H. P., Belford, G. R. 1979. The formation of afferent patterns in the somatosensory cortex of the neonatal rat. *J. Comp. Neurol.* 183:285–304

Killackey, H. P., Belford, G. R. 1980. Central correlates of peripheral pattern alterations in the trigeminal system of the rat. *Brain Res.* 183:205–10

Killackey, H. P., Belford, G., Ryugo, R., Ryugo, D. K. 1976. Anomalous organization of thalamocortical projections consequent to vibrissae removal in the newborn rat and mouse. *Brain Res.* 104:309–15

Killackey, H. P., Erzurumlu, R. F. 1982. Interactions between fiber terminals during pattern formation in the rat trigeminal system. *Neuroscience* 7:S116

Killackey, H. P., Ivy, G. O., Cunningham, T. J. 1978. Anomalous organization of SmI somatotopic map consequent to vibrissae removal in the newborn rat. *Brain Res.* 155:136–40

Killackey, H. P., Leshin, S. 1975. The organization of specific thalamocortical projections to the posteromedial barrel subfield of the rat somatic sensory cortex. *Brain Res.* 86:469–72

Killackey, H. P., Shinder, A. 1981. Central correlates of peripheral pattern alterations in the trigeminal system of the rat. II. The effect of nerve section. *Dev. Brain Res.* 1:121–26

Lin, C. S., Merzenich, M. M., Sur, M., Kaas, J. H. 1979. Connections of areas 3b and 1 of the parietal somatosensory strip with the ventroposterior nucleus in owl monkey (*Aotus trivirgatus*). *J. Comp. Neurol.* 185:355–72

Liu, C. N., Chambers, W. W. 1958. Intraspinal sprouting of dorsal root axons. *Arch. Neurol. Psychiatr.* 79:46–61

Lorente de Nó, R. 1922. La corteza cerebral del raton. *Trab. Inst. Cajal Invest. Biol.* 20:41–78

Lund, R. D. 1978. *Development and Plasticity of the Brain.* New York: Oxford Univ. Press

Mendell, L. M., Sassoon, E. M., Wall, P. D. 1978. Properties of synaptic linkage from long ranging afferents into dorsal horn neurons in normal and deafferented cats. *J. Physiol.* 205:299–310

Merrill, E. G., Wall, P. D. 1978. Plasticity of connection in the adult nervous system. See Goldberger & Murray 1978, pp. 97–111

Merzenich, M. M., Kaas, J. H., Sur, M., Lin, C-S. 1978. Double representation of the body surface within cytoarchitectonic Areas 3b and 1 in "S-I" in the owl monkey (*Aotus trivirgatus*). *J. Comp. Neurol.* 181:41–74

Merzenich, M. M., Kaas, J. H., Nelson, R. J., Wall, J., Sur, M., Felleman, D. J. 1980. Progressive topographic reorganization of representations of the hand within Areas 3b and 1 of monkeys following median nerve section. *Neurosci. Abstr.* 6:651

Merzenich, M. M., Sur, M., Nelson, R. J., Kaas, J. H. 1981. Organization of the S-I cortex: Multiple cutaneous representations in Areas 3b and 1 of the owl monkey. See Kaas et al 1981, pp. 47–66

Merzenich, M. M., Nelson, R. J., Sur, M., Wall, J. T., Felleman, D. H., and Kaas, J. H. 1983b. Progression of change following median nerve section in the cortical representation of the hand in Areas 3b and 1 in adult owl and squirrel monkeys. *Neuroscience.* In press

Merzenich, M. M., Kaas, J. H., Wall, J., Nelson, R. J., Sur, M., Felleman, D. 1983a. Topographic reorganization of somatosensory cortical Areas 3b and 1 in adult monkeys following restricted deafferentation. *Neuroscience.* In press

Metzler, J., Marks, P. S. 1979. Functional changes in cat somatic sensory-motor cortex during short-term reversible epidermal blocks. *Brain Res.* 177:379–83

Millar, J., Basbaum, A. I., Wall, P. D. 1976. Restructuring of the somatotopic map and appearance of abnormal neuronal activity in the gracile nucleus after partial deafferentation. *Exp. Neurol.* 50:658–72

Nakahama, H., Hishioka, S., Otsuka, T. 1966. Excitation and inhibition in ventrobasal thalamic neurons before and after cutaneous input deprivation. *Progr. Brain Res.* 21A:180–96

Nelson, R. J., Kaas, J. H. 1981. Connections of the ventroposterior nucleus of the thalamus with the body surface representations in cortical Areas 3b and 1 of the cynomolgus macaque, (*Macaca fascicularis*). *J. Comp. Neurol.* 199:29–64

Nelson, R. J., Merzenich, M. M., Wall, J., Sur, M., Felleman, D. J., Kaas, J. H. 1980a. Variability in the proportional representations of the hand in somatosensory cortex of primate. *Neurosci. Abstr.* 6:651

Nelson, R. J., Sur, M., Felleman, D. J., Kaas, J. H. 1980b. Representations of the body surface in postcentral parietal cortex of *Macaca fascicularis*. *J. Comp. Neurol.* 192:611–43

Paul, R. L., Goodman, H., Merzenich, M. 1972. Alterations in mechanoreceptor input to Brodmann's Areas 1 and 3 of the postcentral hand area of *Macaca mulatta* after nerve section and regeneration. *Brain Res.* 39:1–19

Penfield, W., Boldrey, E. 1937. Somatic motor and sensory representation in the cerebral cortex of man as studied by electrical stimulation. *Brain Res.* 60:389–443

Pidoux, B., Verley, R., Farkas, E., Scherrer, J. 1979. Projections of the common fur of the muzzle upon the cortical area for mystacial vibrissae in rats dewhiskered since birth. *Neurosci. Lett.* 11:301–6

Pollin, B., Albe-Fessard, D. 1979. Organization of somatic thalamus in monkeys with and without section of dorsal spinal track. *Brain Res.* 173:431–49

Pubols, L. M., Brenowitz, G. L. 1982. Maintenance of dorsal horn somatotopic organization and increased high threshold responses after deafferentation. *J. Neurophysiol.* 47:103–12

Pubols, L. M., Goldberger, M. E. 1980. Recovery of function in dorsal horn following partial deafferentation. *J. Neurophysiol.* 43:102–17

Rasmusson, D. D. 1982. Reorganization of raccoon somatosensory cortex following removal of the fifth digit. *J. Comp. Neurol.* 205:313–26

Rodin, B. E., Sampogna, S., Kruger, L. 1981. A reevaluation of intraspinal sprouting of primary afferents. *Neurosci. Abstr.* 7:66

Rustioni, A., Molenaar, I. 1975. Dorsal column nuclei afferents in the lateral funiculus of the cat: Distribution pattern and absence of sprouting after chronic deafferentation. *Exp. Brain Res.* 23:1–12

Ryugo, D. K., Ryugo, R., Killackey, H. P. 1975. Changes in pyramidal cell density consequent to vibrissae removal in the newborn rat. *Brain Res.* 96:82–87

Schoppmann, A., Nelson, R. J., Stryker, M. P., Cynader, M., Zook, J., Merzenich, M. M. 1981. Reorganization of hand representation within Area 3b following digit amputation in owl monkey. *Neurosci. Abstr.* 7:842

Sperry, R. W. 1943. Functional results of crossing sensory nerves in the rat. *J. Comp. Neurol.* 78:59–90

Spinelli, D. N., Jensen, F. E. 1979. Plasticity: The mirror of experience. *Science* 203:75–78

Steffen, H., Van der Loos, H. 1980. Early lesions of mouse vibrissal follicles: Their influences on dendrite orientation in the cortical barrel field. *Exp. Brain Res.* 40:419–31

Sur, M., Nelson, R. J., Kaas, J. H. 1982. Representations of the body surface in cortical Areas 3b and 1 of squirrel monkeys: Comparisons with other primates. *J. Comp. Neurol.* In press

Van der Loos, H. 1976. Barreloids in mouse somatosensory thalamus. *Neurosci. Lett.* 2:1–6

Van der Loos, H., Woolsey, T. A. 1973. Somatosensory cortex: Structural alterations following early injury to sense organs. *Science* 179:395–98

Verley, R., Axelrad, H. 1977. Organization en "barils" des cellules de la couche IV du cortex SI chez la souris: Effets des lesions ou de la privation des vibrisses mystaciales. *C. R. Acad. Sci. Paris* 284:1183–85

Waite, P. M. E., Taylor, P. K. 1978. Removal of whiskers in young rats causes functional changes in cerebral cortex. *Nature* 274:600–4

Wall, J. T., Cusick, C. G., Kaas, J. H. 1981. Evidence for incomplete reorganization of the S-I foot representation following sciatic nerve section in the adult rat. *Neurosci. Abstr.* 7:758

Wall, J. T., Felleman, D. J., Kaas, J. H. 1982. Effects of median nerve crush and regeneration on the organization of the hand representation in area 3b of monkeys. *Neurosci. Abstr.* 8:851

Wall, J. T., Merzenich, M. M., Sur, M., Nelson, R. J., Felleman, D. J., Kaas, J. H. 1980. Organization of the representations of the hand in Areas 3b and 1 of postcentral somatosensory cortex of monkey after section and regeneration of the median nerve. *Neurosci. Abstr.* 6:651

Wall, P. D., Egger, M. D. 1971. Formation of new connections in adult rat brains after partial deafferentation. *Nature* 232: 542–45

Welker, C. 1971. Microelectrode delineation of fine grain somatotopic organization of SmI cerebral neocortex in albino rat. *Brain Res.* 26:259–75

Welker, C. 1976. Receptive fields of barrels in the somatosensory neocortex of rat. *J. Comp. Neurol.* 166:173–190

Welker, W. I., Seidenstein, S. 1959. Somatic sensory representation in the cerebral cortex of the raccoon (*Procyon Lotor*). *J. Comp. Neurol.* 111:469–501

Weller, W. L., Johnson, J. I. 1975. Barrels in cerebral cortex altered by disruption in newborn but not five-day-old mice (*Cricetidae* and *Muridae*). *Brain Res.* 83:504–8

White, E. L. 1978. Identified neurons in mouse SmI cortex which are postsynaptic to thalamocortical axon terminals: A combined golgi-electron microscopic and degeneration study. *J. Comp. Neurol.* 181:627–61

Wong-Riley, M. T. T., Welt, C. 1980. Histochemical changes in cytochrome oxidase of cortical barrels after vibrissal removal in neonatal and adult mice. *Proc. Natl. Acad. Sci. USA* 77:2333–37

Woolsey, C. N. 1958. Organization of somatic sensory and motor areas of the cerebral cortex. In *Biological and Biochemical Bases of Behavior,* ed. H. F. Harlow, C. N. Woolsey, pp. 63–81. Madison: Univ. Wisc. Press

Woolsey, T. A., Anderson, J. R., Wann, J. R., Stanfield, B. B. 1979. Effects of early vibrissae damage on neurons in the ventrobasal (VB) thalamus of the mouse. *J. Comp. Neurol.* 184:363–80

Woolsey, T. A., Van der Loos, H. 1970. The structural organization of layer IV in the somatosensory region (SI) of mouse cerebral cortex. The description of a cortical field composed of discrete cytoarchitectonic units. *Brain Res.* 17:205–42

Woolsey, T. A., Wann, J. R. 1976. Areal changes in mouse cortical barrels following vibrissal damage at different postnatal ages. *J. Comp. Neurol.* 170:53–66

Ann. Rev. Neurosci. 1983. 6:357–80

MAGNOCELLULAR NEUROSECRETORY SYSTEM

Ann-Judith Silverman

Department of Anatomy, College of Physicians and Surgeons, Columbia University, New York, New York 10032

Earl A. Zimmerman

Department of Neurology, College of Physicians and Surgeons, Columbia University, New York, New York 10032

Introduction

The neurons that make up the hypothalmo-neurohypophysial neurosecretory system have served as an important model for the peptidergic neuron for nearly half a century (Scharrer & Scharrer 1954a,b). The principles of neurosecretion, including ribosomal synthesis of the neuropeptide, packaging of the neurohormone into membrane bound neurosecretory granules, the rapid axonal transport of these granules, exocytosis of the granule contents at the neurosecretory terminal, and stimulus-secretion coupling have been elucidated in this system (Cross et al 1975, Sachs et al 1975, Zimmerman 1976, Brownstein et al 1980). Certain fortuitous aspects of the organization of this system encouraged study of its anatomy, chemistry, and physiology. The storage of large amounts of the hormones (vasopressin, oxytocin and their respective neurophysin proteins) in the posterior pituitary facilitated the chemical analysis and identification of the peptides as well as the development of bioassay procedures. Reactivity of the disulfide bonds in the peptides, and particularly the sulfur rich neurophysins, with the stains developed by Gomori (1941, 1950) permitted the anatomical analysis of this system under both control and experimental conditions. The organized clustering of neurons into well-defined nuclear groups (the supraoptic and paraventricular nuclei) made electrophysiological and lesion studies easier to carry out and to interpret.

357

0147-006X/83/0301-0357$02.00

This chapter is primarily concerned with the data on these neuronal groups provided by anatomical investigations, especially the immunohisto-chemical studies of the last decade. Although antisera to the biologically active peptides have been available for some time, many of the initial investigations of these hypothalamic neurons utilized antisera directed against the neurophysins. Studies with species-specific antisera confirmed in the ox, human, and monkey that these 10,000 dalton "carrier" proteins were localized to the same neurons containing the respective (1000 dalton) hormone (Defendini & Zimmerman 1978). Furthermore, due to an inability to synthesize vasopressin in the Brattleboro rat with hereditary diabetes insidipus, one of the neurophysins is also lacking (Sokol et al 1976, Brown-stein et al 1980). The neurophysins were postulated to be either part of a common precursor protein together with their respective neurohormone or to at least be synthesized at the same time and under the same conditions (Sachs et al 1969, Sachs et al 1975). It has now been shown more conclu-sively that a precursor exists for each of the neurohormones and that the larger protein contains both the nonapeptide and its neurophysin as well as unidentified sequences (Brownstein et al 1980, Russell et al 1980b). It has been assumed by immunocytochemists therefore that the localization of a neurophysin is equivalent to the localization of the hormone. This assump-tion still must be proved rigorously in all cases.

The Classical Magnocellular Neurosecretory System

Studies utilizing the stains developed by Gomori (1941) for the pancreas on the hypothalamus defined the hypothalamo-neurohypophysial neurosecre-tory system anatomically as being composed of the large neurons of the supraoptic and paraventricular nuclei, their axons that formed the hypo-thalamo-neurohypophysial tracts, and their terminals in the neural lobe (Scharrer & Scharrer 1954a,b, Sloper et al 1960, Sloper 1966). These investi-gations also indicated that the secretory substance(s) were transported from the cell bodies to the posterior pituitary. These observations were confirmed with immunocytochemical techniques (Zimmerman et al 1973b, Alvarez-Buylla et al 1973). Although bioassays had indicated that both biologically active substances found in the pituitary, oxytocin, and vasopressin were present in each of the nuclear groups, the relative numbers of neurons containing each of these peptides in the two nuclei was debated (see Zim-merman et al 1974). Until recently, general consensus favored the PVN as the predominant oxytocin nucleus and the SON as the major source of vasopressin. Immunohistochemical studies have now shown that both types of neurons are present in both nuclear groups (see Defendini & Zimmerman 1978 for review, Rhodes et al 1981), although the distribution and numbers of each cell type in the different nuclei probably varies considerably among species (i.e. Zimmerman et al 1974, Vandesande & Dierickx 1975, Watkins 1976, Vandesande et al 1975, Sokol et al 1976, Sofroniew et al 1979, Rhodes et al 1981).

In addition to immunocytochemical techniques, retrograde transport of material from the posterior pituitary has also been utilized to define the cells of origin that contribute terminals to the neurohypophysis. A number of laboratories have shown that the cells that contain retrogradely transported material reside in the supraoptic and paraventricular nuclei as well as the accessory magnocellular cell groups (Sherlock et al 1975, Wiegand & Price 1980, Armstrong et al 1980, Kelly & Swanson 1980, Armstrong & Hatton 1980). Almost all of the cells of the SON are labeled following an injection of horseradish peroxidase (HRP) into the neural lobe (Sherlock et al 1975). Within the PVN, cell bodies labeled from the neural lobe are concentrated in the lateral and medial magnocellular subnuclei; the parvocellular subdivisions of the PVN contain far fewer cells that project to the neurohypophysis (Armstrong et al 1980, Wiegand & Price 1980). The magnocellular accessory cell groups were first described using Gomori stains by Peterson (1966) in the rat brain. They include a rostral parventricular nucleus, anterior and posterior fornical nuclei, the nucleus circularis, the nucleus of the medial forebrain bundle, and an irregular string of cells that extend laterally and ventrally from the paraventricular to the supraoptic nucleus. These cells appear to project to the neural lobe. It should be noted that somewhat different accessory cell groups have been noted in the mouse (A. J. Silverman, unpublished observations; see Figure 6) and in the guinea pig (Sofroniew et al 1979). Table 1 lists the neurosecretory projections of the various magnocellular cell groups.

Vasopressin Projections to the Hypophysial Portal System: Another Neurosecretory Pathway

The first new finding from the immunocytochemical investigations of these peptidergic neurons was the demonstration of neurophysin-containing terminals on the primary portal plexus of the median eminence (Parry & Livett 1973, Zimmerman et al 1973a,b). These capillaries form a specialized microcirculation that delivers neurohormones in high concentrations directly to the adenohypophysis. The majority of the neurophysin positive axons in this region are vasopressinergic (Stillman et al 1977; also see Blume et al 1978). These terminals secrete large amounts of vasopressin, as the concentration of the peptide in the veins draining the median eminence is much higher than in the general circulation (Zimmerman et al 1973a, Recht et al 1981). Lesion studies have demonstrated that the cells of origin for these terminals reside in the PVN rather than the SON (Antunes et al 1977, Vandesande et al 1977). The use of retrograde tracers applied to the median

Table 1 Secretory projections containing oxytocin or vasopressin or related neurophysin[a]

Terminal field	Origin (nucleus)	Peptide	Method	References[b]
Posterior pituitary	Magnocellular			
	paraventricular			
	anterior commissural	OT, NP	IHC, HRP	1, 2, 3
	medial magnocellular	OT	IHC, dye	3
	medial and lateral	OT, VP	IHC, HRP, dye	3, 4, 5
	Supraoptic	OT, VP	IHC, HRP, dye	3, 4, 5
	Accessory	OT, VP	IHC, HRP	4, 5
	circularis		dye	6
	forebrain bundle			
	bed stria terminalis			
	preoptic			
	zona incerta			
	substantia innominata			
	Parvocellular			
	PVN, some of all 5			
	divisions	?	dye	3
Zona externa of				
median eminence	Parvocellular PVN			
	medial (most)			
	dorsal (fewer)	?VP	HRP, dye	3, 4, 6, 7
Organum vascu-	Suprachiasmatic	VP	IHC	8, 9
losum	Rostral magnocellular	NP, OT, VP	IHC	10

[a] Abbreviations: HRP, horseradish peroxidase; IHC, immunohistochemistry; NP, neurophysin; OT, oxytocin; PVN, paraventricular nucleus; SON, supraoptic nucleus; VP, vasopressin.
[b] References: (1) Armstrong et al 1980; (2) Rhodes et al 1981; (3) Swanson & Sawchenko 1980; (4) Wiegand & Price 1980; (5) Zimmerman 1981; (6) Kelly & Swanson 1980; (7) Armstrong & Hatton 1980; (8) Sofroniew 1980; (9) Buijs 1978; (10) Antunes et al 1977.

eminence has demonstrated that these cells are different in both size and distribution from those that innervate the neural lobe (Wiegand & Price 1980, Swanson et al 1980). These authors indicate that the cells of origin reside in the anterior periventricular and medial aspects of the PVN and that the cells are predominately parvocellular rather than magnocellular. However, since the size of vasopressin and oxytocin cells is variable (see Cytology section), it is still unclear from these experiments which vasopressin cells actually contribute to the zona externa projection. Combined tracer and immunocytochemical studies have yet to be carried out to confirm these identifications. Wiegand & Price (1980) also suggested that both accessory cell groups (nucleus circularis and the nucleus of the medial forebrain bundle) and neurons in the supraoptic nucleus contribute to the terminals in the zona externa. An anterograde study using 3H-amino acid injections into either the PVN or SON indicates, however, that the SON does not contribute to these terminals (Alonso & Assenmacher 1981). Although anatomical studies to date suggest that the neurons that project to the neural lobe and those that project to the zona externa are different (Swanson & Sawchenko 1980), electrophysiological evidence suggests that some percentage of cells may send axon collaterals to both regions (Pittman et al 1978, 1981).

Additional differences exist between the projections from the PVN and SON to the two neural haemal organs. The study of Alonso & Assenmacher (1981) shows that the axonal trajectories from the two nuclear groups do not intermingle as they traverse the hypothalamus. Axons from the SON cross in the internal zone of the median eminence, and continue down the infundibular stalk to innervate the entire posterior pituitary, particularly the central portion. The fibers from the PVN, on the other hand, do not cross but remain essentially ipsilateral (also see below). These fibers terminate in the external zone of the median eminence, infundibular stalk, and peripheral aspects of the posterior pituitary. Therefore, even in the neurohypophysis, the two magnocellular groups keep different territories. The functional correlates of these anatomical separations are not known.

In addition to possible differences in cell size and location, the size of the neurosecretory granules in the terminals on the primary portal capillaries are smaller (100 nm) than those found in the neural lobe (120–180nm) (Silverman & Zimmerman 1975). This again suggests that the neurons of origin for the neural lobe and median eminence are distinct.

A functional difference between the neural lobe system and that of the median eminence is the response of the latter neural network to adrenalectomy. It was first noted that removal of the adrenals resulted in an increase in Gomori positive material in this region (Wittkowski & Bock 1972). This increase is associated with a build-up in vasopressin and vasopressin-neurophysin immunoreactivity rather than oxytocin (Stillman et al 1977, Vandesande et al 1977). It is unlikely that the newly discovered corticotropin releasing factor would contribute to the Gomori reaction, since this compound lacks a disulfide bond (Vale et al 1981). Unlike the neural lobe, in which vasopressin content changes with the state of hydration, even severe dehydration has no effect on this second neurosecretory system (Stillman et al 1977, Seybold et al 1981). The reason for the increase in vasopressin immunoreactivity in response to adrenalectomy is not well understood. However, it can be blocked by glucocorticoids much more effectively than by mineralocorticoids (Silverman et al 1981b). Adrenalectomy results in an increase in 3H-cytidine incorporation into RNA in the neurons of origin (Silverman et al 1980a), whereas dehydration increases incorporation of this precursor into RNA in SON neurons (George 1973). The production of vasopressin precursor also increases following adrenalectomy (Russell et al 1980a). Whether the inhibitory effect of the glucocorticoids is at the level of the PVN, or relayed via limbic system inputs (Silverman et al 1981a) known to contain glucocorticoid receptors (McEwen et al 1975) is not known. Furthermore, it can still be argued that the build-up in vasopressin in the terminals around the portal capillaries can be due to decreased release. Knowledge about changes in release parameters must await experiments in which portal blood is collected from adrenalectomized animals, as it has been carried out in intact animals (Oliver et al 1977, Recht et al 1981).

Lesion (Antunes et al 1977) or anterograde tracing studies (Alonso & Assenmacher 1981) have shown that the vasopressin projection to the median eminence is essentially ipsilateral. While investigating this problem in the rat, we gathered evidence that the system was capable of sprouting, which is dramatically enhanced by adrenalectomy (Silverman & Zimmerman 1982). Although that regeneration of vasopressin/oxytocin axons occurs around blood vessels following removal of the neural lobe is well known (Raisman 1973, Antunes et al 1979), the phenomenon described above is a response of intact axons growing into a denervated zone. This effect of glucocorticoids on axonal growth or sprouting is not unique to this system. It has also been described for hippocampus (Scheff et al 1980) and adrenal medullary cells in culture (Unsicker et al 1978).

The presence of a vasopressin secreting system capable of delivering large amounts of vasopressin to the adenohypophyis (Zimmerman et al 1973a) reopened the question that vasopressin might be a corticotropin releasing factor (CRF). Experiments in which antiserum to vasopressin blocked the biological activity of CRF preparations suggested that the hormone might be a major factor in regulating ACTH release (Gilles & Lowry 1979). However, ACTH and adrenal responses are relatively (but not totally) normal in the Brattleboro rat in which vasopressin is absent. A more potent CRF has been repeatedly reported, and now that a potent releasor of ACTH has been isolated and synthesized (Vale et al 1981), this question of the relationship of vasopressin and CRF in ACTH release may soon be resolved.

Just as glucocorticoids decrease vasopressin content in axons on the portal plexus, depletion of catecholamines by resperine treatment also results in a depletion in the amount of vasopressin immunoreactivity in this region. Seybold et al 1981 suggest that a noradrenergic input to the cells of origin tonically inhibits the vasopressin release from the terminals in the median eminence.

Neurosecretory systems associated with oxytocin or vasopressin terminals are listed in Table 1.

Cytology of Magnocellular Neurosecretory Neurons

Until recently, immunocytochemical studies were carried out on thin (5–6 μm) cryostat or paraffin sections. Tissue processing (dehydration and fixation), as well as the thinness of the section, resulted in relatively little data being obtained on the architecture of these neurons. In addition, neurosecretory neurons are notoriously difficult to impregnate with silver, so that Golgi stains rarely provide much information on their dendritic arrangement (however, see Armstong et al 1980). Quite unexpectedly, the application of the immunocytochemical procedure to thick, unembedded sections (Grzanna et al 1977) resulted in the demonstration of neuropeptides in perikarya, axon, and dendrites (Silverman et al 1980b). Using this approach, Sofroniew & Glasmann (1981) have shown that most of the magnocellular neurons are rounded or oval in shape, with diameters of 20–35 μm. In addition, these authors noted immunoreactive perikarya with spindle-shapes and cross-sectional widths of 10–14 μm and lengths of 30–40 μm. Most cells are multipolar, although the denritic arbor does not branch extensively (Silverman et al 1980b, Sofroniew & Glasmann 1981). Beaded processes, most likely axonal in nature, originate from several places: (a) from directly off the cell body, (b) from the initial portion of the basal dendrite, and (c) as an apparent continuation from a thick (presumably denritic) process. Axons were observed to branch along their course. In some cases dendritic spines were observed in these light microscopic preparations (Silverman et al 1980b; see Figure 2), and these have been confirmed at the electron microscopic level (A. J. Silverman, unpublished observations). Somatic spines have also been noted in the occasionally successful Golgi impregnations and confirmed with electron microscopy (Ifft & McCarthy 1974).

In the SON, much of the dendritic arbor of the ventrally situated vasopressin cells projects into the limiting glial lamina of the ventral surface of the brain (Silverman et al 1980b, Armstrong et al 1982). In the mouse, the accessory cell groups have long denritic processes that radiate out in all directions from the group (see Figure 6). In the PVN, the majority of dendritic processes remain within the confines of the nucleus, although the more caudal groups have some dendrites projecting laterally away from the nucleus (Armstrong et al 1980). Many cells, as well as dendritic processes, come into close contact with the ependymal lining (Sofroniew & Glasmann 1981) and blood vessels (A. J. Silverman, unpublished observations). Figures 1 to 6 illustrate the cytology of vasopressin neurons using a monoclonal antibody to vasopressin (Hou-Yu et al 1982).

Synaptic Input

Relatively little is known about the synaptic input to these neurosecretory cells. Axo-dendritic, axo-somatic, and axo-axonic synapses have been ob-

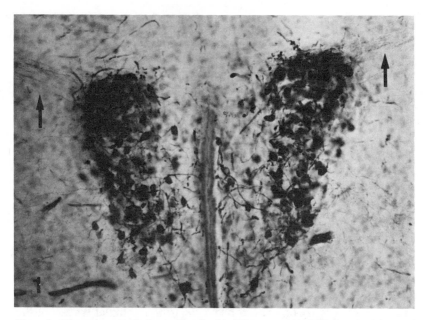

Figure 1 Mouse paraventricular nucleus. In this 100 μm section, perikarya, dendrites, and axons are visible. Most of the dendrites stay within the confines of the nucleus, but some course along the wall of the third ventricle. Axons (*arrows*) project laterally away from the nucleus. [*Figures 1 through 6:* All sections were reacted with a monoclonal antibody to vasopressin (Hou-Yu et al 1982), followed by a second antibody conjugated to horseradish peroxidase. Figures are unpublished micrographs of A. J. Silverman and A. Hou-Yu.]

served in the supraoptic (Priymak & Hajos 1970, Leranth et al 1975, Silverman et al 1980b) and paraventricular nuclei (Klein et al 1968). Leranth et al (1975) estimated that two-thirds of the input to supraoptic neurons originated from within the nucleus or its immediate environs. However, without immunocytochemical techniques to define the entire dendritic and axonic components of the cells, it is impossible to quantify their synaptic input. Furthermore, certain neurotransmitters, such as the monoamines that have an extensive input to this region (Fuxe 1965a,b, see below), do not always form synaptic specializations (see Swanson et al 1978, Beaudet & Descarries 1978).

The afferent input to the paraventricular nucleus has been studied using retrograde tracers (Silverman et al 1981a, Berk & Finkelstein 1981, Tribollet & Dreifuss 1981). In all of these studies, substantial inputs to the PVN from limbic structures were identified. These include regions containing large numbers of glucocorticoid receptors (McEwen et al 1975, Warembourg 1975) such as the lateral septum, medial amygdala, and ventral subiculum (also see Pittman et al 1981). It is possible that the regulatory

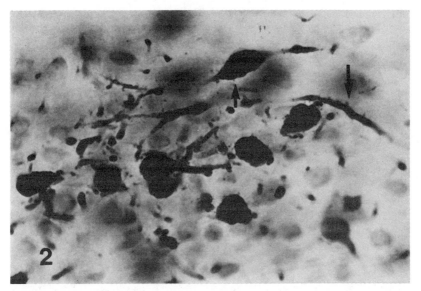

Figure 2   A higher magnification of PVN cells showing spinous processes on both dendrites (*long arrow*) and soma (*short arrow*). Many cells are out of the plane of focus.

effects of glucocorticoids on this system are mediated in part via synaptic relays from these limbic system groups.

Hypothalamic cell groups also send axons into the PVN (Silverman et al 1981a, Tribollet & Dreifuss 1981). Input is more substantial from medial rather than lateral cell groups, including the medial preoptic area, dorsomedial, ventromedial, arcuate, and ventral premammillary nuclei. In our study (Silverman et al 1981a), some injections also resulted in labeled cells in the suprachiasmatic nucleus. The neurotransmitters utilized by some of these projections probably include peptides of the ACTH/bLPH/b-endorphin family that arise from the arcuate nucleus to innervate the PVN (Abrams et al 1980, Krieger et al 1980). From the suprachiasmatic nucleus, possible transmitters include vasopressin (Stillman et al 1977) and vasoactive intestinal polypeptide (Sims et al 1980).

Several components of the magnocellular system apparently communicate with each other (Silverman et al 1981a, Tribollet & Dreifuss 1981). Both the SON and internuclear groups, especially the nucleus circularis, are labeled following an injection of HRP into the ipsilateral PVN. These connections were suggested by earlier anatomical (Olivecrona 1957) and electrophysiological studies (Kandel 1964, Yamashita et al 1970). The electrophysiological data suggested inhibition by recurrent collaterals (also see Pittman et al 1981). It is possible, however, that inhibitory synapses could originate from the ipsilateral SON (or contralateral PVN, see below).

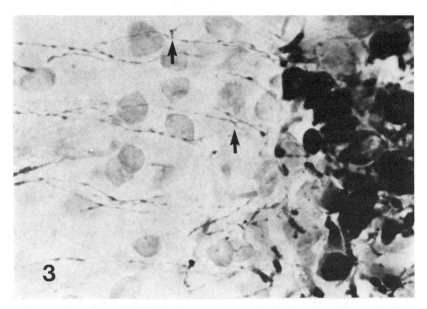

Figure 3 Axons from the PVN are beaded and occasionally show branching (*arrows*). In these cases it appears that the branch is headed back toward the nucleus.

The input from the nucleus circularis (NC), as well as from the circumventricular organs (subfornical organ and organum vasculosum of the lamina terminalis) (see Silverman et al 1981a, Tribollet & Dreifuss 1981), might provide information concerning the state of hydration of the animal. The neurons of the NC undergo significant changes in morphology under conditions of dehydration (Tweedle & Hatton 1976), and electrical stimulation of these cells results in a prompt release of vasopressin (Hatton 1976). Similarly, the subfornical organ is very important in mediating the drinking responses to hypernatremia (Thornborough et al 1973), and to centrally administered angiotensin II (Simpson 1981). The subfornical organ also projects extensively to the supraoptic nucleus (Miselis et al 1979).

We have also demonstrated that certain of the subnuclei of the PVN have an afferent input to the contralateral PVN (Silverman et al 1981a). All cells making up the projection are magnocellular and reside in the outer rim of the main lateral subnucleus (pvl of Armstrong et al 1980). This is a region of predominately oxytocin neurons (Sokol et al 1976, Vandesande et al 1975, Rhodes et al 1981). The regions receiving input appear to be the lateral and medial subnuclei (pvl and pvm of Armstrong et al 1980); these regions contain magnocellular neurons that synthesize either oxytocin or vasopressin (Sokol et al 1976, Vandesande et al 1975, Rhodes et al 1981) as well as parvocellular neurons that may synthesize enkephalin (Rossier

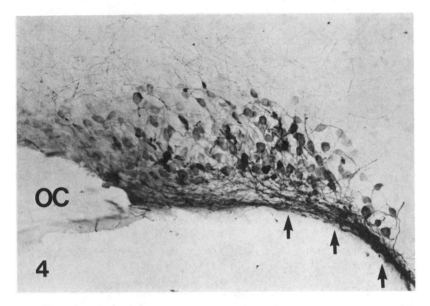

Figure 4 A low magnification of the supraoptic nucleus (OC = optic chiasm). Many of the vasopressin dendrites course along the limiting glial lamina (*arrows*).

et al 1979, Sar et al 1978), neurotensin (Kahn et al 1980), or CCK (Loren et al 1979). Double label experiments at the electron microscopic level are necessary to determine the precise synaptic arrangements (see below).

In addition to the chemical synapses from within the neurosecretory cell groups described above, gap junctions have also been demonstrated within the PVN by both anatomical and electrophysiological criteria (Andrew et al 1981).

Major ascending inputs to the PVN have been described by both retrograde (Tribollet & Dreifuss 1981, Berk & Finkelstein 1981) and anterograde (Ricardo & Koh 1977, McKellar & Loewy 1981) tracing methods. These studies have concentrated on the input to the PVN. This nucleus receives an innervation from the raphe nuclei, locus coeruleus, parabrachial nucleus, nucleus of the solitary tract, and lateral reticular nucleus as well as minor inputs from other brain stem groups.

It is obvious from the above list that the PVN (and SON) receive a substantial monoaminergic innervation. A catecholamine innervation was first demonstrated by Fuxe (1965a,b) using the Falck-Hillarp histofluorescence methodology. Both noradrenaline (Fuxe 1965b, Lindvall & Bjorklund 1974, Ungerstedt 1971) and adrenaline (Hökfelt et al 1974) were described within the PVN and SON relatively early. Since then, three major approaches have been used to describe the monoamine input to these re-

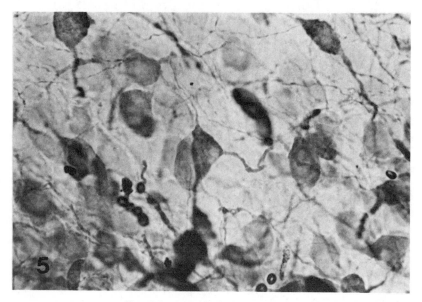

Figure 5   A multipolar SON neuron.

gions in more precise detail. These include immunocytochemistry for the enzymes involved in monoamine synthesis, immunocytochemistry using antibodies directed against the neurotransmitter, and combined histofluorescence and immunocytochemistry. Recent work by Swanson and collaborators (Swanson et al 1981) using antibodies to synthetic enzymes can be summarized as follows. Adrenaline terminals innervate the entire parvocellular division of the PVN, being most dense in the medial aspect and with only a few scattered fibers in either the magnocellular sections of the PVN or in the SON. Noradrenaline fibers innervate the periventricular zone of the PVN and those regions of the PVN and SON that contain predominately vasopressin neurons. Using a combined histofluorescence-immunocytochemical approach, McNeill & Sladek (1981) made several important observations on the regional distribution of catecholamine terminals. In the SON, the majority of catecholamine varicosities were found ventral to the magnocellular cell bodies. This region, however, contains the dendrites of the supraoptic neurons (Silverman et al 1980b). McNeill & Sladek (1980) also found some axo-somatic interactions in the region of the SON that contains predominately vasopressin cells, thus confirming the observation of Swanson et al (1981) discussed above. Electrophysiological studies indicate a monosynaptic input to the SON from the nucleus of the solitary tract, and that input may be noradrenergic (Kannan & Koizumi 1981). In the PVN, McNeill & Sladek (1980) described the major catechola-

Figure 6   An internuclear group on midline just ventral to the PVN. This group is present in mouse, but not in rat. Note that dendrites extend outside of the group.

mine terminals as being in the periventricular zone and adjacent parvocellular region. They interpreted their serial section analysis to indicate that catecholamine terminals are concentrated in the region of oxytocin-producing cells.

In analyses such as the two described here, it is important to remember that the dendritic arbor of the cell is rarely taken into account when describing the presumed innervation. It will be necessary in the future to combine immunocytochemical techniques that visualize dendritic arbors with means for identifying the monoamine terminal at the ultrastructural level.

Noradrenergic inputs, at least to vasopressin neurons, appear to be inhibitory (Barker et al 1971a,b), resulting in decreased release of vasopressin (Wolney et al 1974).

A serotonergic innervation of the PVN has also been described using both radioautography (DesCarries & Beaudet 1978) and immunocytochemistry for serotonin (Steinbusch 1981).

An acetylcholine input to the magnocellular system has long been postulated, and both nicotinic (Kuhn 1974) and muscarinic (Sakai et al 1974) receptor mechanisms have been proposed. Application of Ach onto supraoptic neurons in vitro increases vasopressin release (Sladek & Joynt 1979). However, no immunocytochemical study of choline acetyltransferase has been carried out. The identification of such terminals and their relationship to catecholamine terminals on the individual neurons would greatly increase our understanding of CNS regulation of vasopressin secretion.

Histamine (Taylor et al 1972) and histamine receptors (Palacios et al 1981) are present in the supraoptic nucleus. Histamine injected into the region of the SON increases antidiuresis (Bennett & Pert 1974), presumably through an activation of vasopressin cells.

Another possible transmitter with the magnoceullar system is GABA (Tappaz & Brownstein 1977); this amino acid may originate from neurons within the hypothalamus (Meyer et al 1980). GABA's role in vivo is not known, but in vitro it does decrease firing rates of SON cells (Sakai et al 1974). Zingg et al (1979) have suggested that GABA may act on the axon rather than at the level of the cell body or dendrite. Inhibitory effects of the compound were evident when applied to the posterior pituitary (Zingg et al 1979).

Extrahypothalamic Projections

One of the major findings concerning the hodology of the hypothalamus in the last five years was the demonstration of considerable neurophysin-containing efferent pathways to the brain stem and spinal cord, as well as to forebrain structures (Swanson 1977, Swanson & Sawchenko 1980, Nilaver et al 1980, Buijs et al 1978, Buijs 1978). The existence of some of these pathways was suggested by anterograde and retrograde tracing techniques (Kuypers & Maisky 1975, Conrad & Pfaff 1976). The information to date on the very complex system of both oxytocin and vasopressin pathways is summarized in Table 2. Only those areas in which a terminal field seems clear have been included.

There are many anatomical and biochemical issues that must be resolved concerning these projections before solid physiological studies can be undertaken. In most cases, these pathways have been demonstrated only at the light microscopic level. Axo-dendritic synapses containing vasopressin have been demonstrated in the septum, habenula, and medial amygdala, and those containing oxytocin in the medial amygdala (Buijs & Swaab 1979). In all instances in which a light microscopic study indicates a terminal field, a careful electron microscopic immunocytochemical study must be carried out to confirm the occurrence of a synaptic input (Buijs 1978).

In addition to confirming the actual synaptic nature of these efferent projections, it is still not firmly established that within the axons projecting out of the hypothalamus, the precursor protein is processed in the same manner as in the axons going to the neural lobe or median eminence. Co-localization of neurophysin and the specific neurohormone remains to

Table 2 Oxytocin, vasopressin or neurophysin in extrahypothalamic fibers by immunohistochemistry

Terminal field	Origin	Peptide	References
Spinal cord			
lamina I (dorsal horn)	PVN	OT, VP, NP	3, 5, 8, 11
lamina X (central grey)	PVN	OT, VP, NP	3, 5, 8, 12–14
intermediolateral grey	PVN	OT, VP, NP	3, 5, 8, 12–14
Brain stem			
substantia nigra	PVN	OT, VP, NP	5, 8, 11
mesencephalic grey	SCN	VP	8
	PVN	OT, NP	5, 8, 12
dorsal raphe	SCN	VP	8
	PVN	OT	8, 12
parabrachial	PVN	OT, VP, NP	3, 8, 12
locus coeruleus	PVN	NP	12
raphe magnus	PVN	OT, VP	8
tractus solitarius	PVN	OT, VP, NP	3, 5, 8, 11, 12
dorsal motor vagus	PVN	OT, VP, NP	3, 5, 8, 11, 12
ambiguus	PVN	OT, VP	5, 8, 9
lateral reticular	PVN	OT, VP	8
commissural	PVN	OT, VP, NP	8, 11
Forebrain			
cortex	PVN	VP, OT, NP	5, 9, 11, 12
medial septal	PVN	VP, OT	8, 9
lateral septal	SCN, PVN	VP	5, 8, 16
diagonal band	SCN, PVN	VP, NP	8, 12
amygdala			
medial	PVN	VP, OT, NP	9, 12
	SCN	VP	8
central, lateral, basal	PVN	VP, OT	5, 8, 9
medial-dorsal thalamus	SCN	VP	8, 9, 17
lateral habenula	SCN	VP	8, 9, 16, 17
ventral hippocampus	PVN	VP, OT	8, 9, 16
subcommissural organ	SCN	VP	9

Abbreviations: HRP, horseradish peroxidase; IHC, immunohistochemistry; NP, neurophysin; OT, oxytocin; PVN, paraventricular nucleus; SCN, suprachiasmatic nucleus; SON, supraoptic nucleus; VP, vasopressin.
References: (1) Armstrong et al 1980; (2) Rhodes et al 1981; (3) Swanson & Sawchenko 1980; (4) Wiegand & Price 1980; (5) Zimmerman 1981; (6) Kelly & Swanson 1980; (7) Armstrong & Hatton 1980; (8) Sofroniew 1980; (9) Buijs 1978; (10) Antunes et al 1977; (11) Nilaver et al 1980; (12) Swanson 1977; (13) Swanson & McKellar 1979; (14) Nilaver et al 1982; (15) Hou-Yu et al 1982; (16) Buijs & Swaab 1979; (17) Sofroniew & Weindl 1980.

be done in many instances. In the spinal cord, where there are two descending pathways, preliminary data suggest that the central pathway contains predominately vasopressin and vasopressin-neurophysin, whereas the dorsolateral tract contains more oxytocin fibers (Nilaver et al 1982). Radioimmunoassay of circumscribed regions of rat brain indicate much higher levels of vasopressin in the forebrain regions (septum and thalamus), and higher concentrations of oxytocin in the medulla (Dogterom et al 1978). These findings generally support those obtained by immunocytochemistry (Table 2). We have found significant concentrations of bioassayable oxytocin in the spinal cord (J. Haldar, G. Nilaver, and E. A. Zimmerman, unpublished observations). Despite the assayable levels of oxytocin and vasopressin in various regions of the brain, it has been consistently more difficult to detect their presence by immunocytochemistry when compared to the detectability of the neurophysins. This is true even in thick, unembedded sections. Although this may be entirely the fault of the immunocytochemical technique, it does underscore the needs for (a) immunological studies with a variety of antibodies and (b) chemical analysis using sophisticated chromatographic procedures.

Studies on the cells of origin that comprise the extrahypothalamic projections containing oxytocin and vasopressin (or their respective neurophysins) fall into several categories. One set of questions regards the nuclear origin of the axons—supraoptic, paraventricular, accessory cell groups, or the suprachiasmatic nucleus. In almost all instances the SON has been ruled out as having efferent projections outside of the hypothalamus. In the case of forebrain structures, both the PVN (Buijs et al 1978) and the SCN (Sofroniew & Weindl 1978) have been proposed. The latter authors traced the axons from the SCN and assumed that all thin caliber fibers orginated from this nucleus. They proposed that there is a vasopressin innervation from the SCN to the lateral septum, medial dorsal thalamus, lateral habenula, diagonal band, posterior hypothalamus, and interpeduncular nucleus. The lateral habenular projection was also reported by Buijs (1978). Buijs and colleagues (Buijs 1978, Buijs et al 1978) suggest that the PVN projects into the stria terminalis and stria medularis, and that axons from this cell group have terminals in the dorsal and ventral hippocampus, amygdala, and central gray of the midbrain. Projections from the PVN to the lateral septum, amgydala, midbrain periaqueductal gray, and medulla have been confirmed using electrophysiological techniques (Pittman et al 1981, Zerihun & Harris 1981). The caudal pathways to the brain stem and spinal cord have been studied more extensively. (Swanson & Sawchenko 1980, Swanson & Kuypers 1980, Sawchenko & Swanson 1981, Swanson & McKellar 1979, Nilaver et al 1980, Nilaver et al 1982, Sofroniew & Schrell 1981, Armstrong et al 1980, Hosoya & Matsushita 1979). By retrograde tracing methods using various dyes, injections into the medulla and spinal cord labeled cells within the parvocellular and posterior magnocellular portions of the PVN (Swanson & Sawchenko 1980, Wiegand & Price 1980, Swanson & Kuypers 1980, Armstrong et al 1980, Hosoya & Matsushita 1979). Spinal cord injections labeled cells in the dorsal, lateral, and ventral aspect of the medial parvocellular divisions of the PVN. Injections into medulla resulted in labeled cells in the latter of the two regions. Using two dyes, Swanson and collaborators found that about 15% of the cells could be double-labeled and, therefore, might project to both medulla and spinal cord. However, neither the problems of uptake by fibers of passage, nor the possible transfer of dye through gap junctions have been adequately dealt with to date. Similar anatomical studies suggest that cells projecting to the pituitary (Swanson & Kuypers 1980), or to the median eminence (Swanson et al 1980), are different from those innervating the spinal cord. None of these studies addresses the problem of the peptide or other biologically active substance within the cell that has been retrogradely labeled. This has been done in regards to brain stem and spinal cord projections by combining immunocytochemistry with a retrograde tracer (Swanson & Kuypers 1980, Swanson & Sawchenko 1980 Sofroniew & Schrell 1981). More peptide-containing cells in the PVN were found to project to spinal cord than to medulla: the ratio of vasopressin to oxytocin cells that projected to the spinal cord was 2:1, while it was 4:1 for the medulla. Labeled cells ranged in size from 12–26 μm in diameter (Sofroniew & Schrell 1981). Only 10% of the cells in the PVN that could be retrogradely labeled contained vasopressin or oxytocin, indicating that other substances found in the nucleus contribute substantially to the descending influence of the hypothalamus on the brain stem and spinal cord (Swanson & Kuypers 1980). One study demonstrated about 500 dopaminergic cells within the PVN and some of these project to the brain stem (Swanson et al 1981).

Other Substances in the PVN and SON

It is now clear that other biologically active compounds are found within neurons of the supraoptic and paraventricular nuclei. In both rat (Sar et al 1978, Rossier et al 1979) and cat (Mickvych & Elde 1980), enkephalin immunoreactivity has been demonstrated in both cell groups. Ultrastructural studies at the level of the posterior pituitary suggest that enkephalin and neurophysin can coexist in the same terminal, indeed the same granule (Coulter et al 1981). Martin & Voight (1981) have used 0.5 μm serial sections with ultrastructural correlations and have suggested that met-enkephalin is always found with oxytocin terminals and that leu-enkephalin sometimes coexists within vasopressin terminals. These anatomical studies, which do need to be confirmed in more detail, provide a morphologial substrate for the inhibitory effects of opiates on the release of the neurohormones (Clarke et al 1979, Iversen et al 1980).

Another opiod peptide present within vasopressin neurons is dynorphin; this peptide can be demonstrated by immunocytochemistry in all magnocellular vasopressin-containing neurons (Watson et al 1980, Watson et al

1982). Using Brattleboro rats, Watson et al have shown that dynorphin immunoreactivity persists in cells that are not immunoreactive for oxytocin, i.e. the cells that cannot synthesize vasopressin. This would indicate that the dynorphin and vasopressin immunoreactivities are not part of the same precursor.

Another peptide currently thought to be within magnocellular neurons is angiotensin II (Phillips et al 1979, Kilcoyne et al 1980, Zimmerman et al 1980). Immunoreactive angiotensin II is present in vasopressin cells, including parvocellular neurons in the suprachiasmatic nucleus (Zimmerman et al 1980). It has been considered, but not proven, that immunoreactive angiotensin II in vasopressin cells may represent cross-reactivity with the precursor for vasopressin or a related molecule, since it is absent in Brattleboro rats, and increases following adrenalectomy in parallel with vasopressin (Zimmerman et al 1980).

Other peptides have also been identified in the PVN. These include neurotensin (Kahn et al 1980), glucagon (Tager et al 1980), and the sulfated octapetide form of cholecystokinin (Beinfeld et al 1981, Loren et al 1979). Cholecystokinin appears to coexist with some oxytocin neurons (Vanderhaeghen et al 1981). These compounds are present in the hypothalamus and posterior pituitary. For neurotensin, it is clear that the cells containing the peptide are parvocellular (Kahn et al 1980). This is also true for the dopaminergic cells present in the nucleus (Swanson et al 1980).

ACKNOWLEDGMENTS

Work in our laboratories is supported by USPHS grants AM 20337 and HD 13149. A. J. S. is a recipient of an Irma T. Hirschl Career Scientist Award.

Literature Cited

Abrams, G. M., Nilaver, G., Hoffman, D., Zimmerman, E. A., Ferin, M., Krieger, D. T., Liotta, A. S. 1980. Immunocytochemical distribution of corticotropin (ACTH) in monkey brain. *Neurology* 30:1106–10

Alonso, G., Assenmacher, I. 1981. Radioautographic studies on the neurophysial projections of the supraoptic and paraventricular nuclei in the rat. *Cell Tissue Res.* 219:525–34

Alvarez-Buylla, R., Livett, B. G., Uttenthal, L. O., Hope, D. B., Milton, S. H. 1973. Immunological evidence for the transport in the hypothalamo-neurohypophysial system of the dog. *Z. Zellforsch.* 137:435–50

Andrew, R. D., MacVicar, B. A., Dudek, F. E., Hatton, G. I. 1981. Dye transfer through gap junctions between neuro-endocrine cells of rat hypothalamus. *Science* 211:1187–89

Antunes, J. L., Carmel, P. W., Zimmerman, E. A. 1977. Projections from the paraventricular nucleus to the zona externa of the median eminence of the rhesus monkey: An immunohistochemical study. *Brain Res.* 137:1–10

Antunes, J. L., Carmel, P. W., Zimmerman, E. A., Ferin, M. 1979. Regeneration of the magnocellular system of the rhesus monkey following hypothalamic lesions. *Ann. Neurol.* 15:462–69

Armstrong, W. E., Hatton, G. I. 1980. The localization of projection neurons in the rat hypothalamic paraventricular nucleus following vascular and neurohypophysial injections of HRP. *Brain Res. Bull.* 5:473–77

Armstrong, W. E., Schöler, J., McNeill, T. H. 1982. Immunocytochemical, Golgi and electron microscopic characterization of putative dendrites in the ventral glial lamina of the rat supraoptic nucleus. *Neuroscience* 7:679–94

Armstrong, W. E., Warach, S., Hatton, G. I., McNeill, T. H. 1980. Subnuclei in the rat hypothalamic paraventricular nucleus: A cytoarchitectonic, HRP and immunocytochemical analysis. *Neuroscience* 5:1931–58

Barker, J. L., Crayton, J. W., Nicole, R. A. 1971a. Noradrenaline and acetylcholine responses of supraoptic neurosecretory cells. *J. Physiol.* 218:19–32

Barker, J. L., Crayton, J. W., Nicole, R. A. 1971b. Supraoptic neurosecretory cells: Autonomic modulation. *Science* 171: 206–7

Beaudet, A., Descarries, L. 1978. The monoamine innervation of rat cerebral cortex: Synaptic and nonsynaptic axon terminals. *Neuroscience* 3:851–60

Beinfeld, M. C., Meyer, D. K., Eskay, R. L., Jensen, R. T., Brownstein, M. J. 1981. The distribution of cholecystokinin immunoreactivity in the central nervous system of the rat as determined by radioimmunoassay. *Brain Res.* 212:51–57

Bennett, C. T., Pert, A. 1974. Antidiuresis produced by injections of histamine into the cat supraoptic nucleus. *Brain Res.* 78:151–56

Berk, M. L., Finkelstein, J. A. 1981. Afferent projections to the preoptic area and hypothalamic regions in the rat brain. *Neuroscience* 6:1601–24

Blume, H. W., Pittman, Q. J., Renaud, L. P. 1978. Electrophysiological indications of a 'vasopressinergic' innervation of the median eminence. *Brain Res.* 155:153–58

Brownstein, M. J., Russell, J. T., Gainer, H. 1980. Synthesis, transport and release of posterior pituitary hormones. *Science* 207:373–78

Buijs, R. M. 1978. Intra- and extrahypothalamic vasopressin and oxytocin pathways in the rat. *Cell Tissue Res.* 192:423–35

Buijs, R. M., Swaab, D. F. 1979. Immuno-electron microscopic demonstration of vasopressin and oxytocin synapses in the limbic system of the rat. *Cell Tissue Res.* 204:355–65

Buijs, R. M., Swaab, D. F., Dogterom, J., van Leeuwen, F. W. 1978. Intra- and extrahypothalamic vasopressin and oxytocin pathways in the rat. *Cell Tissue Res.* 186:423–33

Clarke, G., Wood, P., Merrick, L., Lincoln, D. W. 1979. Opiate inhibition of peptide release from the neurohumeral terminals of the hypothalamic neurons. *Nature* 282:746–48

Conrad, L. C., Pfaff, D. W. 1976. Efferents from medial basal forebrain and hypothalamus in the rat. II. An autoradiographic study of the anterior hypothalamus. *J. Comp. Neurol.* 169:221–62

Coulter, H. D., Elde, R. P., Unverzagt, S. L. 1981. Co-localization of neurophysin and enkephalin-like immunoreactivity in cat pituitary. *Peptides* 2(Suppl. 1): 51–55

Cross, B. A., Dyball, R. E., Dyer, R. G., Jones, C. W., Lincoln, D. W., Morris, J. F., Pickering, B. T. 1975. Endocrine neurones. *Recent Progr. Hormone Res.* 31:243–94

Defendini, R., Zimmerman, E. A. 1978. The magnocellular neurosecretory system of the mammalian hypothalamus. In *The Hypothalamus,* ed. S. Reichlin, R. J. Baldessarini, J. B. Martin, pp. 137–54. New York: Raven

DesCarries, L., Beaudet, A. 1978. The serotonin innervation of adult rat hypothalamus. *Colloq. Int. CNRS* 280: 135–53

Dogterom, F., Snigdewint, F. G. M., Buijs, R. M. 1978. *Neurosci. Lett.* 9:341–46

Fuxe, K. 1965a. Evidence for the existence of monoamine neurons in the central nervous system. III. The monoamine nerve terminal. *Z. Zellforsch.* 65:573–96

Fuxe, K. 1965b. Evidence for the existence of monoamine neurons in the central nervous system. IV. The distribution of monoamine terminals in the central nervous system. *Acta Physiol. Scand.* 64: (Suppl. 247):37–85

George, J. M. 1973. Localization in hypothalamus of increased incorporation of ^3H cytidine into RNA in response to oral hypertonic saline. *Endocrinology* 92:1550–55

Gilles, G., Lowry, P. 1979. Corticotropin releasing factor may be modulated vasopressin. *Nature* 278:463–64

Gomori, G. 1941. Observations with differential stains on human islets of Langerhans. *Amer. J. Pathol.* 17:395–406

Gomori, G. 1950. Aldehyde-fuchsin: A new stain for elastic tissue. *Amer. J. Clin. Pathol.* 20:665–66

Grzanna, R., Morrison, J. H., Coyle, J. T., Molliver, M. E. 1977. The immunohistochemical demonstration of noradrenergic neurons in the rat brain: The use of homologous antiserum to dopamine-b-hydroxylase. *Neurosci. Lett.* 4:127–34

Hatton, G. I. 1976. Nucleus circularis: Is it an osmoreceptor in the brain. *Brain Res. Bull.* 1:123–31

Hökfelt, T., Fuxe, K., Goldstein, M., Johansson, O. 1974. Immunohistochemical evidence for the existence of adrenaline neurons in the rat brain. *Brain Res.* 66:233–51

Hosoya, Y., Matsushita, M. 1979. Identification and distribution of the spinal and hypophysial projection neurons in the paraventricular nucleus of the rat. A light and electron microscopic study with the HRP method. *Exp. Brain Res.* 35:315–31

Hou-Yu, A., Ehrlich, P., Valiquette, G., Engelhardt, D. L., Sawyer, W. H., Nilaver, G., Zimmerman, E. A. 1982. A monoclonal antibody to vasopressin: Preparation, characterization and application in immunocytochemistry. *J. Histochem. Cytochem.* In press

Ifft, J. D., McCarthy, L. 1974. Somatic spines in the supraoptic nucleus of the rat hypothalamus. *Cell Tissue Res.* 148:203–11

Iversen, L. L., Iversen, S. D., Bloom, F. E. 1980. Opiate receptors influence vasopressin release from nerve terminals in rat neurohypophysis. *Nature* 284:350–51

Kahn, D., Abrams, G. M., Zimmerman, E. A., Carraway, R., Leeman, S. 1980. Neurotensin neurons in the rat hypothalamus: An immunocytochemical study. *Endocrinology* 107:47–51

Kandel, E. 1964. Electrical properties of hypothalamic neuroendocrine cells. *J. Gen. Physiol.* 47:691–717

Kannan, H., Koizumi, K. 1981. Pathways between the nucleus tractus solitarius and neurosecretory neurons of the supraoptic nucleus: Electrophysiological studies. *Brain Res.* 213:17–28

Kelly, J., Swanson, L. W. 1980. Additional forebrain regions projecting to the posterior pituitary: Preoptic region, bed nucleus of the stria terminalis and zona incerta. *Brain Res.* 197:1–9

Kilcoyne, M., Hoffman, D. L., Zimmerman, E. A. 1980. Immunocytochemical localization of angiotensin II and vasopressin in rat hypothalamus: Evidence for production in the same neuron. *Clin. Sci.* 59:57s–60s

Klein, M. J., Porte, M. M. A., Stutinsky, F. 1968. Contacts synaptiques axoaxoniques dans les noyaux supraoptiques et paraventriculaires du rat et de la souris. *CR Acad. Sci. Paris* 267:936–39

Krieger, D. T., Liotta, A. S., Brownstein, M. J., Zimmerman, E. A. 1980. ACTH, b-lipotropin and related peptides in brain, pituitary and blood. *Recent Progr. Hormone Res.* 36:277–344

Kuhn, E. R. 1974. Cholinergic and adrenergic release mechanism for vasopressin in the male rat: A study with injection of neurotransmitters and blocking agents into the third ventricle. *Neuroendocrinology* 16:255–64

Kuypers, H. G. J. M., Maisky, V. A. 1975. Retrograde axonal transport of horseradish peroxidase from spinal cord to brain stem cell groups in the rat. *Neurosci. Lett.* 1:9–14

Leranth, C., Zaborszky, L., Marton, J., Palkovits, M. 1975. Quantitative studies on the supraoptic nucleus in the rat. I. Synaptic organization. *Exp. Brain Res.* 22:509–23

Lindvall, O., Bjorklund, A. 1974. The glyoxylic acid fluorescence histochemical method: A detailed account of the methodology for the visualization of central catecholamine neurons. *Histochemistry* 39:97–127

Loren, I., Alumets, J., Hakanson, R., Sundler, F. 1979. Distribution of gastrin and CCK-like peptide in rat brain. An immunocytochemical study. *Histochemistry* 59:249–58

Martin, R., Voight, K. H. 1981. Enkephalins co-exist with oxytocin and vasopressin in nerve terminals of rat neurohypophysis. *Nature* 289:502–4

McEwen, B. S., Gerlach, J. L., Micco, D. J. 1975. Putative glucocorticoid receptors in hippocampus and other brain regions of the rat brain. In *The Hippocampus,* ed. S. Isaacson, J. Pribram, 1:285–322. New York: Plenum

McKellar, S., Loewy, A. D. 1981. Organization of some brain stem afferents to the paraventricular nucleus of the hypothalamus of the rat. *Brain Res.* 217:351–57

McNeill, T. H., Sladek, J. R. Jr. 1980. Simultaneous monoamine histofluorescence and neuropeptide immunocytochemistry. II. Correlative distribution of catecholamine varicosities and magnocellular neurosecretory neurons in the rat supraoptic and paraventricular nuclei. *J. Comp. Neurol.* 193:1023–33

Meyer, D. K., Oertel, W. H., Brownstein, M. J. 1980. Deafferentation studies on the glutamic acid decarboxylase of the supraoptic nucleus of the rat. *Brain Res.* 200:165–68

Mickvych, P., Elde, R. 1980. Relationship between enkephalinergic neurons and the vasopressin-oxytocin neuroendocrine system of the cat: An immunohistochemical study. *J. Comp. Neurol.* 190: 135–46

Miselis, R. R., Shapiro, R. E., Hand, P. J. 1979. Subfornical organ efferents to neural systems for control of body water. *Science* 207:1022–25

Nilaver, G., Mulhern, J., Zimmerman, E. A. 1982. Extrahypothalamic neurophysin projections in the brain stem and spinal cord of normal and homozygous Brattleboro rat. *NY Acad. Sci.* In press

Nilaver, G., Zimmerman, E. A., Wilkins, J., Michaels, J., Hoffman, D. L., Silverman, A. J. 1980. Magnocellular hypothalamic projections to the lower brainstem and spinal cord of the rat: Immunocytochemical evidence for the predominance of oxytocin-neurophysin system compared to a vasopressin-neurophysin system. *Neuroendocrinology* 30:150–58

Olivecrona, H. 1957. Paraventricular nucleus and pituitary gland. *Acta Physiol. Scand.* 40 (Suppl. 136):1–178

Oliver, C., Mical, R. S., Porter, J. C. 1977. Hypothalamic-pituitary vasculature: Evidence for retrograde blood flow in the pituitary stalk. *Endocrinology* 101: 598–604

Palacios, J. M., Wamsley, J. K., Kuhar, M. J. 1981. The distribution of histamine H_1 receptors in the rat brain: An autoradiographic study. *Neuroscience* 6: 15–38

Parry, H. B., Livett, B. G. 1973. A new hypothalamic pathway to the median eminence containing neurophysin and its hypertrophy in sheep with natural Scrapie. *Nature* 242:63–65

Peterson, R. P. 1966. Magnocellular neurosecretory centers in the rat hypothalamus. *J. Comp. Neurol.* 128:181–90

Phillips, M. I., Weyhenmeyer, J., Felix, J., Ganten, D., Hoffman, W. E. 1979. Evidence for an endogenous brain renin-angiotensin system. *Fed. Proc.* 38: 2260–66

Pittman, Q. J., Blume, H. W., Renaud, L. P. 1978. Electrophysiological indications that individual hypothalamic neurons innervate both median eminence and neurohypophysis. *Brain Res.* 157: 364–68

Pittman, Q. J., Blume, H. W., Renaud, L. P. 1981. Connections of the hypothalamic paraventricular nucleus with the neurohypophysis, median eminence, amygdala, lateral septum and midbrain periaqueductal grey. *Brain Res.* 215: 15–28

Priymak, E. K., Hajos, F. 1970. Synapses on the supraoptic neurosecretory neurons of the rat: An electron microscopic study. *Acta Morphol. Acad. Sci. Hung.* 18:55–61

Raisman, G. 1973. Electron microscopic studies of the development of new neurohemal contacts in the median eminence of the rat after hypophysectomy. *Brain Res.* 55:245–61

Recht, L. D., Hoffman, D. L., Haldar, J., Silverman, A. J., Zimmerman, E. A. 1981. Vasopressin concentrations in hypophysial portal plasma: Insignificant reduction following removal of the posterior pituitary gland. *Neuroendocrinology* 33:88–90

Rhodes, C. H., Morrell, J. I., Pfaff, D. W. 1981. Immunohistochemical analysis of magnocellular elements in rat hypothalamus: Distribution and numbers of neurophysin, oxytocin and vasopressin containing cells. *J. Comp. Neurol.* 198: 45–64

Ricardo, J. A., Koh, E. T. 1978. Anatomical evidence of direct projections from the nucleus of the solitary tract to the hypothalamus, amygdala and other forebrain structures in the rat. *Brain Res.* 153:1–26

Rossier, J., Battenberg, E., Pittman, Q., Bayon, A., Koda, L., Miller, R., Guillemin, R., Bloom, F. 1979. Hypothalamic enkephalin neurones may regulate the neurohypophysis. *Nature* 277:653–55

Russell, J. T., Brownstein, M. J., Gainer, H. 1980a. [35S] cysteine labeled peptides transported to the neurohypophysis of adrenalectomized, lactating and Brattleboro rats. *Brain Res.* 201:227–34

Russell, J. T., Brownstein, M. J., Gainer, H. 1980b. Biosynthesis of vasopressin, oxytocin and neurophysins: Isolation and characterization of two common precursors (propressophysin and prooxyphysin). *Endocrinology* 107:1880–91

Sachs, H., Fawcett, P., Takabatake, Y., Portanova, R. 1969. Biosynthesis and release of vasopressin and neurophysin. *Recent Progr. Hormone Res.* 25:447–84

Sachs, H., Pearson, D., Nureddin, A. 1975. Guinea pig neurophysin: Isolation, developmental aspects, biosynthesis in organ culture. *Ann. NY Acad. Sci.* 248: 36–45

Sakai, K. K., Marks, B. H., George, J. M., Koestner, A. 1974. The isolated organ-cultured supraoptic nucleus as a neuropharmacological test system. *J. Pharmacol. Exp. Ther.* 190:482–91

Sar, M., Stumpf, W. E., Miller, R. J., Chang, K.-J., Cuatrecasas, P. 1978. Immunohistochemical localization of enkephalin in rat brain and spinal cord. *J. Comp. Neurol.* 182:17–38

Sawchenko, P. E., Swanson, L. W. 1981. Central noradrenergic pathways for the integration of hypothalamic neuroendocrine and autonomic responses. *Science* 214:685–87

Scharrer, E., Scharrer, B. 1954a. Hormones produced in neurosecretory cells. *Recent Progr. Hormone Res.* 10:183–240

Scharrer, E., Scharrer, B. 1954b. Neurosekretion. In *Handb. Mikrosk. Anatomie* 5:953–1066. Berlin: Springer-Verlag

Scheff, S. W., Benardo, L. S., Cotman, C. W. 1980. Hydrocortisone administration retards axon sprouting in the rat dentate gyrus. *Exp. Neurol.* 68:195–201

Seybold, V., Elde, R., Hökfelt, T. 1981. Terminals of reserpine-sensitive vasopressin neurophysin neurons in the external layer of the rat medial eminence. *Endocrinology* 108:1803–9

Sherlock, D. A., Field, P. M., Raisman, G. 1975. Retrograde transport of horseradish peroxidase in the magnocellular neurosecretory system of the rat. *Brain Res.* 88:403–14

Silverman, A. J., Gadde, C. A., Zimmerman, E. A. 1980a. The effects of adrenalectomy on the incorporation of 3H-cytidine into RNA in neurophysin and vasopressin containing neurons of the rat hypothalamus. *Neuroendocrinology* 30:285–90

Silverman, A. J., Goldstein, R., Gadde, C. A. 1980b. The ontogenesis of neurophysin containing neurons in the mouse hypothalamus. *Peptides* 1(Suppl. 1):27–44

Silverman, A. J., Hoffman, D. L., Zimmerman, E. A. 1981a. The descending afferent connections of the paraventricular nucleus of the hypothalamus. *Brain Res. Bull.* 6:47–61

Silverman, A. J., Hoffman, D., Gadde, C. A., Krey, L. C., Zimmerman, E. A. 1981b. Adrenal steroid inhibition of the vasopressin-neurophysin neurosecretory system to the median eminence of the rat. *Neuroendocrinology* 32:129–33

Silverman, A. J., Zimmerman, E. A. 1975. Ultrastructural localization of neurophysin and vasopressin in the median eminence and posterior pituitary of the guinea pig. *Cell Tissue Res.* 159:291–301

Silverman, A. J., Zimmerman, E. A. 1982. Adrenalectomy increases sprouting in a peptidergic neurosecretory system. *Neuroscience.* In press

Simpson, J. B. 1981. The circumventricular organs and the central actions of angiotensin. *Neuroendocrinology* 32:248–56

Sims, K., Hoffman, D. L., Said, S. I., Zimmerman, E. A. 1980. Vasoactive intestinal polypeptide (VIP) in mouse and rat brain: An immunocytochemical study. *Brain Res.* 186:165–83

Sladek, C. D., Joynt, R. J. 1979. Cholinergic involvement in osmotic control of vasopressin release by the organ-culture rat hypothalamo-neurohypophysial system. *Endocrinology* 105:367–71

Sloper, J. C. 1966. The experimental and cytopathological investigation of neurosecretion in the hypothalamus and pituitary. In *The Pituitary Gland,* ed. G. W. Harris, B. T. Donovan, 3:131–80. Berkeley: Univ. Calif. Press

Sloper, J. C., Arnott, D. J., King, B. C. 1960. Sulphur metabolism in the pituitary and hypothalamus of the rat: A study of radioisotope uptake after injection. *J. Endocrinol.* 20:9–23

Sofroniew, M. V. 1980. Projections from vasopressin, oxytocin and neurophysin neurons to neural targets in the rat and human. *J. Histochem. Cytochem.* 28:475–78

Sofroniew, M. V., Glasmann, W. 1981. Golgi-like immunoperoxidase staining of hypothalamic magnocellular neurons that contain vasopressin, oxytocin or neurophysin in the rat. *Neuroscience* 6:619–43

Sofroniew, M. V., Schrell, U. 1981. Evidence for a direct projection from oxytocin and vasopressin neurons in the hypothalamic paraventricular nucleus to the medulla oblongata: Immunohistochemical visualization of both the horseradish peroxidase transport and the peptide produced by the same neurons. *Neurosci. Lett.* 22:211–17

Sofroniew, M. V., Weindl, A. 1978. Projections from the parvocellular vasopressin and neurophysin containing neurons of the suprachiasmatic nucleus. *Am. J. Anat.* 153:391–430

Sofroniew, M. V., Weindl, A. 1980. Identification of parvocellular vasopressin and neurophysin neurons in the suprachiasmatic nucleus of a variety of mammals including primates. *J. Comp. Neurol.* 193:659–75

Sofroniew, M. V., Weindl, A., Schinko, I., Wetzstein, R. 1979. The distribution of vasopressin, oxytocin and neurophysin producing neurons in the guinea pig brain. I. The classical hypothalamo-neurohypophysial system. *Cell Tissue Res.* 196:367–84

Sokol, H. W., Zimmerman, E. A., Sawyer, W. H., Robinson, A. G. 1976. The hypothalamo-neurohypophysial system of the rat: Localization and quantification of neurophysin by light microscopic immunocytochemistry in normal rat and in Brattleboro rats deficient in vasopressin and a neurophysin. *Endocrinology* 98:1176–88

Steinbusch, H. W. M. 1981. Distribution of the serotonin-immunoreactivity in the CNS of the rat—cell bodies and terminals. *Neuroscience* 6:557–618

Stillman, M. A., Recht, L. D., Rosario, S. L., Seif, S. M., Robinson, A. G., Zimmerman, E. A. 1977. The effects of adrenalectomy and glucocorticoid replacement on vasopressin and vasopressin-neurophysin in the zona externa of the rat. *Endocrinology* 101:42–49

Swanson, L. W. 1977. Immunohistochemical evidence for a neurophysin-containing autonomic pathway arising in the paraventricular nucleus of the hypothalamus. *Brain Res.* 128:346–53

Swanson, L. W., Connelly, M. A., Hartman, B. K. 1978. Further studies on the fine structure of the adrenergic innervation of the hypothalamus. *Brain Res.* 151:165–74

Swanson, L. W., Kuypers, H. G. J. M. 1980. The paraventricular nucleus: Cytoarchitectonic subdivisions and organization of projections to the pituitary, dorsal vagal complex, and spinal cord as demonstrated by retrograde fluorescence double labeling methods. *J. Comp. Neurol.* 194:555–70

Swanson, L. W., McKellar, S. 1979. The distribution of oxytocin- and vasopressin-stained fibers in the spinal cord of the rat and monkey. *J. Comp. Neurol.* 188:87–106

Swanson, L. W., Sawchenko, P. E. 1980. Paraventricular nucleus: A site for the integration of neuroendocrine and autonomic mechanisms. *Neuroendocrinology* 31:410–17

Swanson, L. W., Sawchenko, P. E., Berod, A., Hartman, B. K., Heile, K. B., VanOrden, D. E. 1981. An immunohistochemical study of the organization of catecholaminergic cells and terminal fields in the paraventricular and supraoptic nuclei of the hypothalamus. *J. Comp. Neurol.* 196:271–85

Swanson, L. W., Sawchenko, P. E., Wiegand, S. J., Price, J. L. 1980. Separate neurons in the paraventricular nucleus project to the median eminence and to the medulla or spinal cord. *Brain Res.* 198:190–95

Tager, H., Hohenboken, M., Markese, J., Dinerstein, R. J. 1980. Identification and localization of glucagon-related peptides in rat brain. *Proc. Natl. Acad. Sci. USA* 77:6229–33

Tappaz, M. L., Brownstein, M. J. 1977. Origin of glutamate decarboxylase (GAD) containing cells in discrete hypothalamic nuclei. *Brain Res.* 132: 45–106

Taylor, K. M., Gfeller, E., Snyder, S. H. 1972. Regional localization of histamine and histadine in the brain of the rhesus monkey. *Brain Res.* 41:171–76

Thornborough, J. R., Passo, S. S., Rothballer, A. B. 1973. Forebrain lesion blockade of the natriuretic response to elevated carotid blood sodium. *Brain Res.* 58:355–63

Tribollet, E., Dreifuss, J. J. 1981. Localization of neurones projecting to the hypothalamic paraventricular nucleus area of the rat: A horseradish peroxidase study. *Neuroscience* 6:1315–28

Tweedle, C. D., Hatton, G. I. 1976. Ultrastructural comparisons of neurons of supraoptic and circularis nuclei in normal and dehydrated rats. *Brain Res. Bull.* 1:103–21

Ungerstedt, U. 1971. Stereotaxic mapping of the monoamine pathways in the rat brain. *Acta Physiol. Scand.* 367:1–48

Unsicker, K., Krisch, B., Otten, U., Thoenen, H. 1978. Nerve growth factor-induced fiber outgrowth from isolated rat adrenal chromaffin cells: Impairment by glucocorticoids. *Proc. Natl. Acad. Sci. USA* 75:3498–3502

Vale, W., Spiess, J., Rivier, C., Rivier, J. 1981. Characteristics of a 41-residue ovine hypothalamic peptide that stimulates secretion of corticotropin and β-endorphin. *Science* 213:1394–97

Vanderhaeghen, J. J., Lotstra, F., Vandesande, F., Dierickx, K. 1981. Coexistence of cholecystokinin and oxytocin-neurophysin in some magnocellular hypothalamo-hypophyseal neurons. *Cell Tissue Res.* 221:227–31

Vandesande, F., Dierickx, K. 1975. Identification of the vasopressin producing and of the oxytocin producing neurons in the hypothalamic magnocellular neurosecretory system of the rat. *Cell Tissue Res.* 164:153–62

Vandesande, F., Dierickx, K., DeMey, J. 1975. Identification of the vasopressin-neurophysin II and the oxytocin-neurophysin I producing neurons in the bovine hypothalamus. *Cell Tissue Res.* 156:189–200.

Vandesande, F., Dierickx, K., DeMey, J. 1977. The origin of the vasopressinergic and oxytocinergic fibers of the external region of the median eminence of the rat hypophysis. *Cell Tissue Res.* 180: 443–52

Warembourg, M. 1975. Radioautographic study of the rat brain after injection of [1,2-3H]-corticosterone. *Brain Res.* 89:61–70

Watkins, W. B. 1976. Immunocytochemical study of the hypothalamo-neurohypophysial system. I. Localization of neurosecretory neurons containing neurophysin I and neurophysin II in the domestic pig. *Cell Tissue Res.* 175: 165–81

Watson, S. J., Akil, H., Fischli, W., Goldstein, A., Zimmerman, E. A., Nilaver, G., Greidanus, V. 1982. Dynorphin and vasopressin: Common localization in magnocellular neurons. *Science* 216: 85–87

Watson, S. J., Akil, H., Walker, J. M. 1980. Anatomical and biochemical studies of the opioid peptides and related substances in the brain. *Peptides* 1 (Suppl. 1):11–20

Wiegand, S. J., Price, J. L. 1980. Cells of origin of the afferent fibers to the median eminence in the rat. *J. Comp. Neurol.* 192:1–19

Wittkowski, W., Bock, R. 1972. Electron microscopical studies of the median eminence following interference with the feedback system anterior pituitary-adrenal cortex. In *Brain Endocrine Interactions: Median Eminence: Structure and Function,* ed. K. M. Knigge, D. E. Scott, A. Weindl, pp. 171–80. Basel: Karger

Wolney, H. L., Plech, A., Herman, Z. S. 1974. Diuretic effects of intraventricularly injected noradrenaline and dopamine in rats. *Experientia* 30:1062–63

Yamashita, H., Koizumi, K., Brooks, C. McC. 1970. Electrophysiological studies on neurosecretory cells in the cat hypothalamus. *Brain Res.* 20:462–66

Zerihun, L., Harris, M. 1981. Electrophysiological identification of neurons of paraventricular nucleus sending axons to both the neurohypophysis and the medulla in the rat. *Neurosc. Lett.* 23:157–60

Zimmerman, E. A. 1976. Localization of hypothalamic hormones by immunocytochemical techniques. In *Frontiers in Neuroendocrinology,* ed. L. Martini, W. F. Ganong, 4:25–62. New York: Raven

Zimmerman, E. A. 1981. The organization of oxytocin and vasopressin pathways. In *Neurosecretion and Brain Peptides,* ed. J. B. Martin, S. Reichlin, K. L. Bick, pp. 63–75. New York: Raven

Zimmerman, E. A., Carmel, P. W., Husain, M. K., Ferin, M., Tannenbaum, M., Frantz, A. G., Robinson, A. G. 1973a. Vasopressin and neurophysin: High concentrations in monkey hypophyseal portal blood. *Science* 198:925–27

Zimmerman, E. A., Hsu, K. C., Robinson, A. G., Carmel, P. W., Frantz, A. G., Tannenbaum, M. 1973b. Studies of neurophysin secreting neurons with immunoperoxidase techniques employing antibody to bovine neurophysin. I. Light microscopic findings in monkey and bovine tissues. *Endocrinology* 92:931–40

Zimmerman, E. A., Krupp, L., Hoffman, D. L., Matthew, E., Nilaver, G. 1980. Exploration of peptidergic pathways in brain by immunocytochemistry: A ten year perspective. *Peptides* 1 (Suppl. 1): 3–10

Zimmerman, E. A., Robinson, A. G., Husain, M. K., Acosta, A., Frantz, A., Sawyer, W. H. 1974. Neurohypophysial peptides in the bovine hypothalamus: The relationship of neurophysin I to oxytocin and neurophysin II to vasopressin in supraoptic and paraventricular regions. *Endocrinology* 95:931–38

Zingg, H. H., Baertschi, A. J., Dreifuss, J. J. 1979. Action of γ aminobutyric acid on hypothalamo-neurohypophysial axons. *Brain Res.* 171:453–59

Ann. Rev. Neurosci. 1983. 6:381–418

SPINAL AND TRIGEMINAL MECHANISMS OF NOCICEPTION[1]

Ronald Dubner and Gary J. Bennett

Neurobiology and Anesthesiology Branch, National Institute of Dental Research, National Institutes of Health, Bethesda, Maryland 20205

INTRODUCTION

The development of new experimental approaches has led to recent advances in our knowledge of spinal and trigeminal nociceptive mechanisms, justifying another review of the rapidly proliferating literature in this field. We examine neural circuits and sensory mechanisms activated by noxious or painful stimulation of undamaged skin and peripheral nerves. The spinal dorsal horn and its homolog in the trigeminal system consist of three major components:

1. the central terminals of primary afferent neurons innervating cutaneous and deep tissues;
2. intrinsic neurons including local interneurons and long projection or output neurons;
3. the axon terminals of extrinsic neurons originating mainly from descending brain stem pathways.

The intracellular injection of horseradish peroxidase (HRP) followed by appropriate histochemical procedures provides a direct method for correlating response characteristics of components of the dorsal horn with a "Golgi-like" appearance of neural somata and their axonal and dendritic arbors. This method has been used to determine the central terminations of different functional types of primary afferents and intrinsic neurons. Orthograde and retrograde transport of various dyes, HRP, and tritiated compounds have revealed the terminal fields of primary afferents and descending axons,

[1]The US Government has the right to retain a nonexclusive, royalty-free license in and to any copyright covering this paper.

and the locations of intrinsic projection neurons. The use of radioim-munoassays and immunocytochemical methods has resulted in the identifi-cation of numerous putative transmitters involved in all three major components of dorsal horn circuitry. The correlation of animal behavior with specific central neural circuits has demonstrated the existence of de-scending pain-suppressing pathways and the role of dorsal horn circuitry in pain-related behavior. We focus here on specific findings using these new approaches.

The dorsal horn is a layered structure. The lamination scheme proposed by Rexed (1952) is widely used. Historically, the gelatinous or translucent band that caps the dorsal horn in the medulla and spinal cord has been referred to as the substantia gelatinosa of Rolando. It is usually divided into an outer marginal layer and an inner substantia gelatinosa layer (see Gobel et al 1981a for a historical review). Although there has been some confusion in the past, most investigators in the field now agree that the marginal layer corresponds to Rexed's lamina I, and the substantia gelatinosa layer to Rexed's lamina II (Gobel 1978b, Ralston & Ralston 1979, Light & Perl 1979a, Bennett et al 1980). Furthermore, in agreement with Rexed's origi-nal observation (Rexed 1952), lamina II is subdivided into outer (lamina IIa) and inner (lamina IIb) zones. The lamination scheme proposed by Rexed for the spinal cord is used in this review.

A further source of confusion in the literature is the separate terminology utilized for the rostral extent of the dorsal horn in the medulla. Although Olszewski's (1950) subdivisions of the spinal trigeminal nucleus have been extremely useful, they have obscured the fact that subnucleus caudalis and the adjacent reticular formation are directly continuous with the rostral extent of the dorsal horn in the cervical spinal cord. This separate terminol-ogy has resulted in a separate literature on the trigeminal homolog of the spinal dorsal horn that often is not integrated into knowledge of dorsal horn mechanisms (Willis & Coggeshall 1978, Cervero & Iggo 1980).

There now is considerable evidence of major similarities between the spinal dorsal horn and its trigeminal homolog, the trigeminal subnucleus caudalis (Dubner et al 1978). In addition to its continuity with the cervical dorsal horn, the trigeminal subnucleus caudalis and the adjacent reticular formation can be divided into similar laminae based on cell types, the texture of their neuropil, and types of synaptic connections (Gobel et al 1977, 1981a). Similar functional types of primary afferent neurons innervate spinal and trigeminal dermatomes (Price & Dubner 1977). Another simi-larity is the location of trigeminothalamic and spinothalamic neurons mainly in laminae I and V in both areas. These anatomical findings correlate well with physiological results on the location of nociceptive and non-nociceptive projection neurons and interneurons in these structures (Price et al 1976, Price & Dubner 1977, Hoffman et al 1981). In addition, numerous studies utilizing orthograde and retrograde tracer methods have identified similar descending anatomical and neurochemical pathways to the spinal dorsal horn and its trigeminal homolog (Gobel et al 1981a). In view of the close structural and functional relationships between these trigeminal and spinal regions, it has been proposed that the subnucleus caudalis and its adjacent reticular formation should be referred to as the medullary dorsal horn (Gobel et al 1981a). We use this term here and avoid unnecessary distinctions between similar findings in the medullary and spinal dorsal horns. The use of this term does not imply that no differences exist at medullary and spinal levels, but merely that these differences are no greater than those found between distant spinal segments.

INPUT FROM SKIN AND DEEP TISSUES

Characteristics of Specific Nociceptors

The earliest reports on single unit activity in peripheral nerves predicted the presence of nerve fibers responsive to tissue-threatening or tissue-damaging stimulation (Zotterman 1933). However, it is only since the detailed studies of Perl and his colleagues in the late 1960s that evidence for the existence of specific cutaneous nociceptors in numerous species has accumulated (see Burgess & Perl 1973, Price & Dubner 1977). Similar types of receptors also have been demonstrated in muscle and visceral structures.

Nociceptors can be classified according to their responses to different forms of intense stimulation and the conduction velocity of their peripheral axons. With a minor exception, all nociceptive afferents conduct impulses in the A-delta or C fiber range, and thus have finely myelinated or unmyelinated axons. Some respond only to intense forms of mechanical stimulation, others to mechanical and thermal stimuli, and others, called polymodal, to chemical stimulation, as well. Relatively few studies have demonstrated the presence of nociceptors responsive exclusively to intense thermal stimulation.[1] Nociceptors are characterized by their restricted

---

[1] Some of the difficulty with this literature is the curious terminology used in the past. For example, primary afferents responsive only to intense forms of mechanical stimuli have been referred to as high-threshold mechanoreceptors, moderate pressure mechanoreceptors, or low sensitivity mechanoreceptors. The term *mechanical nociceptors* has been used to describe this whole class (Iggo & Ogawa 1971, Burgess & Perl 1973) and is adopted here. Unfortunately, confusion still reigns because these fiber types have been called *mechanothermal fibers* in some studies (Campbell et al 1979) because of their property of heat sensitization. On the other hand, the terms, *thermal nociceptor* or *heat nociceptor,* have been used as a designation for nociceptors responsive to mechanical and thermal stimuli in the absence of sensitization (Iggo & Ogawa 1971, Dubner et al 1977). We adopt the term, *mechanothermal nociceptors,* for the latter. Nociceptors responsive to mechanical, thermal, and chemical stimuli usually are termed *polymodal.*

receptive fields and their ability to respond differentially to innocuous and noxious (tissue-damaging) stimulation. Although some nociceptors respond only to frankly tissue-damaging stimuli, such as piercing of the skin with a needle or temperatures of 45°C and above, most exhibit threshold responses to stimuli that are neither noxious nor perceived as painful. The latter, however, respond best to stimuli in the noxious range. Generally speaking, myelinated and unmyelinated nociceptive afferents respond to similar types of stimuli, and there are few differences between receptors innervating hairy and glabrous skin (LaMotte 1979). Remarkably, similar types of nociceptors appear to exist in human, monkey, rabbit, and cat.

The most extensively studied cutaneous nociceptors are the myelinated mechanical nociceptors and the unmyelinated polymodal nociceptors. Their properties are described below in more detail.

MECHANICAL NOCICEPTORS   These nociceptors (also referred to as *high-threshold mechanoreceptors* or HTMs) respond only to moderately intense or noxious mechanical stimuli. Some will respond to thermal stimulation after repeated exposure to heat (see below). The myelinated mechanical nociceptors were first described in detail in the skin of cat (Burgess & Perl 1967) and monkey (Perl 1968) limbs and subsequently have been demonstrated in monkey and cat facial skin (Dubner & Hu 1977, Dubner et al 1977, Hu & Sessle 1980), rabbit limb (Fitzgerald & Lynn 1977), and human forearm (Adriaensen et al 1980, 1981). The receptor structure associated with myelinated mechanical nociceptive afferents has been identified in cat limb hairy skin (Kruger et al 1981).

Myelinated mechanical nociceptors typically have conduction velocities in the 5 to 35 m/sec, or A-delta range. However, there are reports of some mechanical nociceptors with conduction velocities in the 40 to 55 m/sec or A-beta range (Burgess & Perl 1973, Campbell et al 1979) in cat and monkey limb skin.

A distinctive feature of nociceptors is the change in responsiveness produced by repetitive heat stimulation (Burgess & Perl 1973). Although rarely responsive to initial heat stimuli in the 45 to 55°C range, myelinated mechanical nociceptors often respond to heat after repeated stimulation, a phenomenon that has been termed *sensitization*. Sensitization was first described in detail for cat unmyelinated polymodal nociceptors by Bessou & Perl (1969) and is characterized by (*a*) a decrease in threshold temperature; (*b*) an increase in responsiveness to suprathreshold stimuli manifested as a decreased latency or greater number of impulses, or both; (*c*) the development of a low frequency spontaneous discharge; and sometimes (*d*) an afterdischarge. Of interest is the finding that heat sensitization of myelinated mechanical nociceptors is not accompanied by altered responses to mechanical stimulation (Fitzgerald & Lynn 1977, Campbell et al 1979).

Much less is known about unmyelinated mechanical nociceptors. This type of nociceptor appears to innervate the limbs of cat and monkey (Bessou & Perl 1969, Georgopoulos 1976) but in the monkey comprises only 10–15% of the C fiber population. Some of these neurons in cat may respond to noxious cold stimuli and should be classified as mechanothermal nociceptors (Burgess & Perl 1973, Price & Dubner 1977).

POLYMODAL NOCICEPTORS   These nociceptors are found in the skin of numerous species and represent a significant proportion of the C fiber population. They are characterized by their responsiveness to intense thermal and mechanical stimuli and many are excited by irritant chemicals. In some studies, responses to chemical stimuli were not examined and it is not clear whether such mechanically and thermally sensitive C fiber nociceptors are actually polymodal. In this review we classify all such nociceptors as polymodal. Polymodal nociceptors first were described in detail in the cat (Bessou & Perl 1969), but subsequent studies have identified them in monkey limb (Croze et al 1976, Kumazawa & Perl 1977, LaMotte & Campbell 1978), monkey and cat face (Beitel & Dubner 1976a, Hu & Sessle 1980), rabbit limb (Fitzgerald 1979) and human (Van Hees & Gybels 1972, 1981, Torebjörk & Hallin 1974). A significant population also is present in dog skeletal muscle (Kamazawa & Mizumura 1977) and scrotal skin (Kumazawa & Mizumura 1980b), though in the latter the vast majority of polymodal nociceptors are finely myelinated A-delta fibers. In cat, approximately one-half of the C afferent population are nociceptors. This percentage increases in monkey, where 85–90% of C fibers are nociceptors, predominantly of the polymodal type. In humans, all C fibers characterized to date are polymodal nociceptors.

The sensitivity of polymodal nociceptors to thermal stimuli has been examined quantitively in numerous studies. These fibers have thresholds that vary from 38 to 49°C but always exhibit maximum sensitivity in the 45 to 51°C range, where humans report increasing pain sensations (LaMotte & Campbell 1978, Wolskee & Gracely 1980).

Sensitization by repeated heat stimulation is a characteristic feature of cutaneous and deep tissue C-fiber polymodal nociceptors and has been reported in all species studied (Bessou & Perl 1969, Beitel & Dubner 1976a, Kumazawa & Perl 1977, Torebjörk & Hallin 1977, Fitzgerald 1979, Kumazawa & Mizumura 1980a). Similar to A-delta mechanical nociceptors, exposure to heat does not sensitize polymodal nociceptors to subsequent intense mechanical stimuli (Bessou & Perl 1969). The application of various chemicals to scrotal polymodal nociceptors produces responses to KCl and NaCl after tachyphylaxis develops to bradykinin (Kumazawa & Mizumura 1980a). These differential patterns of sensitivity suggest that multiple sites of action may exist for mechanical, thermal, and chemical stimuli.

Detailed parametric studies of the effects of repeated heat stimulation of polymodal nociceptors indicate that sensitization is only one possible consequence. When a stimulus is delivered repeatedly to the same location, the phenomenon of fatigue or temporal suppression also is observed (Bessou & Perl 1969, Beitel & Dubner 1976b, LaMotte & Campbell 1978). The extent of fatigue depends on the intensity and frequency of repetitive stimuli and appears to be a characteristic of slowly conducting fiber populations (Price et al 1977). Sensitization clearly interacts with fatigue so that these opposite effects tend to cancel each other at short time intervals. A third effect of repeated heat stimulation is partial inactivation or depression. Intense noxious heat stimuli may damage the skin or the receptor, leading to a reduction in responsivity that can last for hours. Sensitization can be followed by depression if repeated heating is excessive (Beitel & Dubner 1976a, Fitzgerald & Lynn 1977, Kumazawa & Perl 1977). Initial intense heating can also lead to depression without any observable sensitization. Thus, the consequences of repeated heating can be highly variable. An increase or decrease in responsiveness can occur, depending on the type of nociceptor, the intensity and duration of stimulation, the location of the receptor, the time period between stimuli, and the species studied.

MYELINATED MECHANOTHERMAL NOCICEPTORS   In humans, noxious heat stimuli with fast rise-times applied to the extremities sometimes produce a double pain sensation, referred to as *first* and *second pain.* The latency of first pain is too short to be accounted for by C-fiber activity (Price et al 1977, Campbell et al 1979). Myelinated mechanical nociceptors, even after sensitization, have response latencies that are too long to provide signals that lead to first pain sensations. Therefore, another myelinated nociceptor population responsive to heat stimulation must exist in humans. The likely candidate neurons are A-delta mechanothermal nociceptors, described in monkey (Iggo & Ogawa 1971, Burgess & Perl 1973, Georpopoulos 1976, Dubner et al 1977) and human (Adriaensen et al 1981). Controversy about the existence of such a distinct population has persisted because of the possibility of confusion with sensitized myelinated mechanical nociceptors. On the monkey's face, mechanothermal nociceptors can be distinguished from mechanical nociceptors on the basis of their lower heat thresholds, greater sensitivity to noxious heat stimuli, and shorter response latencies (Dubner et al 1977).[2] Heat-sensitive A-delta nociceptors in humans have similar properties (Adriaensen et al 1981).

---

[2]Myelinated mechanothermal nociceptors innervating the monkey's face also exhibit sensitization to repeated heating (Dubner & Hu 1977).

Specific Thermoreceptors Signaling Tissue-Damaging Stimulation

Warm fibers are sensitive to small temperature shifts in the innocuous range and their discharge rate increases monotonically with stimulus temperatures in the 30 to 43°C range (Price & Dubner 1977). Some warm fibers continue to respond to higher temperatures but discharge rates are not monotonically related to temperature throughout the noxious heat range (Sumino et al 1973, LaMotte & Campbell 1978). Cold fibers, sensitive to small temperature changes and exhibiting monotonic increases in discharge to temperature decreases in the 35 to 20°C range, have the unusual property of sometimes responding to noxious heat. These cold fibers exhibit sensitization to repeated heating with resultant lower thresholds, increased sensitivity and shorter latencies to subsequent heat stimuli (Dubner et al 1975). Available evidence suggests that cold fiber responses to noxious stimuli do not contribute to the perception of pain (Price & Dubner 1977, LaMotte et al 1982). The role of warm fiber activity in pain sensations is less certain (Price & Dubner 1977, LaMotte & Campbell 1978).

Correlation of Nociceptor Activity with Behavior Related to Pain

In correlative neural and behavioral studies, the strategy is to examine the range of responsivity of different classes of nociceptors and to determine which class or classes can provide the minimum amount of information necessary to account for a particular behavior related to pain. This approach has been used in monkey and human studies. Monkeys trained in a reaction-time task exhibit faster (within 2.5 sec) escape latencies to 51°C than to 49°C stimuli applied to the face (Dubner et al 1977). Myelinated mechanothermal nociceptors are the only fiber type that can differentiate adequately between these temperatures within 2.5 sec after stimulus onset. Therefore, mechanothermal nociceptor activity is sufficient to account for monkey escape behavior by which pain is inferred. In human studies, judgments of pain produced by heat stimuli have been compared with simultaneous recordings from C-fiber polymodal nociceptors (Torebjörk & Hallin 1974, Gybels et al 1979, Van Hees & Gybels 1981). A clear relationship has been shown between C-fiber discharge and verbal reports of pain. However, the threshold temperature for activating C polymodal nociceptors usually is below the subject's pain threshold, suggesting that temporal summation is necessary for pain perception evoked by C fibers. It has been reported that excitation of C polymodal nociceptors by mechanical stimulation does not produce pain at discharge levels correlated with heat-evoked pain (Beitel & Dubner 1976b, Van Hees & Gybels 1981). One explanation proposed is that the simultaneous activation of low-threshold mechanoreceptors by mechanical stimuli produces central inhibitory effects that alter the ability of C polymodal nociceptors to produce pain sensations. Such effects on medullary and spinal dorsal horn neurons have been demonstrated (see below).

Based on the assumption that human and monkey have similar primary afferent nociceptive populations, other studies have correlated human pain report with the discharge of monkey nociceptors. A close relationship between heat-induced first pain in humans and monkey A-delta mechanothermal nociceptors has been demonstrated (Price et al 1977). In another cross-species study, it has been shown that activity in monkey C-fiber polymodal nociceptors increases in a parallel fashion with human judgments of increasing pain magnitude, thus supporting a role for these fibers in coding pain intensity (LaMotte & Campbell 1978).

Cutaneous Hyperalgesia Following Injury

It is well known that tissue injury produces a state of primary hyperalgesia in humans at the site of injury (Lynn 1977). A lower pain threshold develops, there is increased sensitivity to suprathreshold stimuli, and often spontaneous sensations of pain occur. These altered sensations have distinct parallels with the phenomenon of sensitization (see above) observed in nociceptors. Recent studies have examined the role that different types of nociceptors play in primary hyperalgesia by comparing the altered responsivity of monkey nociceptors with human pain judgments following tissue injury. Meyer & Campbell (1981) demonstrated that the primary hyperalgesia resulting from a 53°C, 30-sec thermal injury to the hand is probably mediated by activity in sensitized myelinated mechanical nociceptors and not by C-fiber polymodal nociceptors. Using a similar analysis, LaMotte et al (1982) showed that a less intense thermal stimulus (50°C) for 60–100 sec produces an initial increase in human pain thresholds followed within 5 min by hyperalgesia. In contrast to Meyer & Campbell (1981), they reported that changes in C-fiber polymodal nociceptor activity parallel these alterations in human pain responsivity. Myelinated mechanical nociceptors and cold fibers exhibited little or no alterations in activity to this same stimulus. Neither of the above studies investigated the effects of hyperalgesia-producing burn stimuli on the activity of myelinated mechanothermal or myelinated polymodal nociceptors. The above findings suggest that the severity of tissue injury may determine the contribution of different nociceptors to the hyperalgesia phenomenon. As discussed above, sensitization is only one consequence in nociceptors of tissue-damaging stimulation and the extent and duration of altered responsivity depends on many factors. Tissue injury produced under natural rather than experimental conditions obviously is much less controlled and the resultant primary hyperalgesia is the product of all these factors. It also should be noted that the present evidence on primary hyperalgesia does not exclude a contribution from central neural mechanisms (Lynn 1977).

SECONDARY HYPERALGESIA   A diffuse hyperalgesic region in humans usually surrounds the local area of injury and is referred to as an area of secondary hyperalgesia. Within this diffuse region is a smaller erythemic area of vasodilation called the "flare." The erythema and increased pain responsivity of the flare can be blocked by local anesthetics. Fitzgerald (1979) showed that C-fiber polymodal nociceptors were sensitized by small injuries made outside of their receptive fields. Local anesthetics injected into the injury site blocked the sensitization. Antidromic stimulation of sural nerve fibers whose central processes were cut also produced sensitization of C-fiber polymodals. These findings suggest that a neural mechanism, similar to an axon reflex, is responsible for this sensitization effect and the accompanying flare reaction (Lewis 1942). The sensitization in the flare region probably involves a chemical mediator released by antidromic activation of afferent fibers at the site of injury. A possible candidate for this chemical mediator is the peptide, substance P, known to be present in small diameter primary afferents (Hökfelt et al 1975). Treatment of skin with capsaicin, the pungent factor in red peppers, reduces substance P and also eliminates the flare reaction surrounding a small injury (Carpenter & Lynn 1981). Substance P also produces vasodilation when injected into skin (Hagermark et al 1978). However, substance P only has a weak excitatory or sensitizing effect on nociceptors (Fitzgerald & Lynn 1979).

The type and extent of injury probably is important in axon reflex activity and neurally mediated release of chemical mediators. Mild injury produced by heat apparently does not result in remote sensitization of C-fiber polymodal nociceptors (Croze et al 1976, Torebjörk & Hallin 1977, Thalhammer & LaMotte 1982). On the other hand, the size of the area of secondary hyperalgesia is not confined to the region of the flare, extending to sites too far away to be reached by axon reflex-mediated chemical release. Such findings support the possibility of central nervous system-mediated secondary hyperalgesia (Hardy et al 1952).

Central Terminations of Nociceptive Afferents in the Medullary and Spinal Dorsal Horns

Earlier anatomical and physiological studies provided indirect evidence that nociceptive afferents terminate in laminae I and II of the medullary and spinal dorsal horns (Perl 1971, Dubner et al 1976). These findings have been confirmed in anatomical studies utilizing degeneration, autoradiographic, and HRP labeling methods. In addition, these latter studies concluded that further segregation of nociceptive input occurs in the superficial layers, with myelinated afferents terminating in lamina I and unmyelinated afferents in lamina II (LaMotte 1977, Light & Perl 1979a, Ralston & Ralston 1979). In contrast, other degeneration studies suggest that the segregation is reversed in the medullary dorsal horn (Gobel & Binck 1977). These contradictory interpretations recently have been resolved. The use of the intracellular HRP method has provided conclusive evidence that myelinated mechanical nociceptors terminate in laminae I and IIa as well as in spinal dorsal horn laminae V and X (Light & Perl 1979b). It appears, however, that lamina I also contains unmyelinated cutaneous afferents (Gobel et al 1981b) and unmyelinated visceral afferents at the sacral spinal level (Morgan et al 1981). The findings on sacral visceral primary afferents indicate that there is a mixed termination of A-delta and C afferents also in laminae IIa, V, and VI, and in the dorsal grey commissure. Small-diameter afferents originating in the tooth pulp and cornea also terminate in laminae I and II (Arvidsson & Gobel 1981, Panneton & Burton 1981). Thus, the stratification of nociceptive afferents in the deep as well as superficial layers may not be as strict as some authors originally proposed (Gobel & Binck 1977, Kumazawa & Perl 1978, Light et al 1979a).

A similar stratification of low threshold mechanoreceptors exists in the deeper laminae of the dorsal horn. D-hair primary mechanoreceptive afferents terminate in lamina IIb and largely in lamina III below it (Light & Perl 1979b). There also is an overlapping distribution of functionally distinct large myelinated mechanoreceptive afferents in layers III–VI of the spinal and medullary dorsal horns (Brown et al 1977b, 1978, 1981, Hayashi 1980).

Putative Neurotransmitters Released by Nociceptive Afferents

Evidence is accumulating that substance P is a neurotransmitter of small-diameter nociceptive afferents. It has been identified in small neurons in trigeminal and spinal ganglia (Hökfelt et al 1975, Cuello et al 1978) and in terminals in the dorsal horn (Pickel et al 1977, Barber et al 1979). Dorsal root section decreases substance P levels in the dorsal horn (Takahashi & Otsuka 1975), and the iontophoretic ejection of substance P excites superficial and deep nociceptive dorsal horn neurons (Henry 1976, Randic & Miletic 1977, Andersen et al 1978, Zieglgänsberger & Tulloch 1979a, Henry et al 1980).

A role for substance P as a primary neuron transmitter has received considerable support from studies utilizing capsaicin. Capsaicin injected into neonatal rats produces a selective degeneration of small ganglion cells, small-diameter primary afferents, and axon terminals in the dorsal horn (Janscó et al 1977). Injected into neonatal and adult rats, capsaicin reduces substance P-like immunoreactivity in the dorsal horn (Jessell et al 1978, Yaksh et al 1979, Nagy et al 1980). Capsaicin-treated rats also show significant analgesic effects. Most studies suggest a selective reduction in responsivity to noxious chemical and thermal stimuli (Janscó et al 1977, Yaksh et al 1979, Nagy et al 1980) after neonatal or intrathecal adult capsaicin administration. In contrast, the effects of subcutaneous injections in adults appear to be somewhat different; capsaicin elevates noxious chemical and mechanical thresholds but has no influence on heat thresholds (Hayes & Tyers 1980). Recently, Nagy et al (1981a) have demonstrated that a single dose of capsaicin (50 mg/kg, s.c.) in neonatal rats results in 85–95% loss of unmyelinated primary afferents from lumbar dorsal roots. In addition, the substance P content of these roots is decreased by 85–95%, and the peptide content in the dorsal horn is reduced by 50%.

The above studies strongly suggest that substance P is a neurotransmitter in C fiber nociceptors. Unfortunately, other findings indicate that such a conclusion may be premature. Capsaicin can result in degeneration of finely myelinated afferents (Lawson & Nickels 1980), even when the dosage is well-controlled (Nagy et al 1981a). In addition, capsaicin causes a depletion of somatostatin from primary sensory neurons (Gamse et al 1981, Nagy et al 1981a), and in immunocytochemical studies results in a decrease in cholecystokinin, somatostatin, and vasoactive intestinal peptide as well as substance P in the spinal cord (Janscó et al 1981).

Variable results with capsaicin can be attributed to the route of administration (e.g. intrathecal or subcutaneous), the age of the animal (neonatal or adult), the dosage, and the time of assessment (acute or chronic). Thus, intraventricular administration does not destroy ganglion cells (Gamse et al 1981), and no terminal degeneration is noted in the dorsal horn in adult-treated animals. Chronic assessment after neonatal treatment must take into account transsynaptic and secondary changes (Janscó et al 1981). For example, the reduction in tritiated naloxone binding sites in the dorsal horn in neonatally treated rats (Nagy et al 1980) is not reported in adult-treated animals (Jessell et al 1978) and may reflect long-term postsynaptic changes. Recently, it also has been shown that intrathecal capsaicin causes nonspecific damage to the spinal cord and that the observed thermal analgesia is not due merely to substance P depletion (Nagy et al 1981b).

A reasonable conclusion is that capsaicin has a selective effect on primary sensory neurons and their terminals in the dorsal horn. There is little effect on ventral horn peptides. Serotonin and peptides such as enkephalin and neurotensin, known to originate in central nervous system neurons, are unaffected (Gamse et al 1981, Janscó et al 1981). However, a specificity of capsaicin for substance P neurons seems untenable in light of recent evidence. These recent studies also point out that other peptides such as somatostatin, cholecystokinin, and vasoactive intestinal peptide are present in ganglion cells and are likely candidates in primary afferent neurochemical transmission. The finding that these peptides are depleted by capsaicin, a neurotoxin that alters pain sensitivity, further suggests their presence in nociceptive afferents.

INTRINSIC NEURONS

Output Systems

The importance of the human spinothalamic tract for pain perception is demonstrated by the profound analgesia that occurs caudal to its transection. This analgesia, however, is often temporary and not totally complete. Although the partial persistance and return of pain perception following spinothalamic tract lesions may be due to pathophysiological processes within the spinal cord or brain, one should not ignore the possibility that there may be other nociceptive output systems in the human spinal cord. Possible output systems for human pain perception include the spinocervicothalamic system and the dorsal column postsynaptic spinomedullary system. Our knowledge about all of these systems has increased enormously in the last few years (see Willis & Coggeshall 1978 for an extensive review).

MORPHOLOGY   The cells of origin of the *spinothalamic tract* (STT) have now been located unequivocally in the monkey, cat, and rat (Giesler et al 1976, Carstens & Trevino 1978, Willis et al 1979). Comparably definitive evidence for the human STT is lacking, but the available evidence is consistent with a pattern like that of the monkey. STT neurons are found throughout the entire grey matter except for the motorneuron pools. There are, however, obvious concentrations of STT neurons in lamina I and in the lateral part of the neck of the dorsal horn (largely lamina V). The cat's lumbar enlargement, however, appears to lack the heavy concentration of STT cells in the lateral part of the neck of the dorsal horn; instead, there are many STT cells in the medial half of the ventral grey matter (laminae VI–VIII).

It has recently been shown that the component of the STT that projects to the medial thalamus has its cells of origin preferentially in the deeper laminae, whereas that component of the STT that projects to the ventrobasal and posterior thalamus preferentially arises from the more superficial laminae. This distinction has been found in the rat, cat, and monkey (Carstens & Trevino 1978, Giesler et al 1979, Willis et al 1979). The STT projection to one small, medial thalamic nucleus is an exception to this schema. Craig & Burton (1981) have recently shown that nucleus submedius is innervated only by STT neurons from lamina I. Preliminary observations indicate that this is a nociceptive input and that the nucleus submedius projects to the orbitofrontal cortex (Craig et al 1981).

We have very little information about the morphology of individual STT neurons. Retrograde HRP studies have shown that STT neurons in the deeper laminae are mostly medium- to large-sized and multipolar. STT neurons in lamina I are likely to have the general neuronal configuration that characterizes this area, but lamina I neurons are morphologically diverse (Gobel 1978a, Beal et al 1981), and it is not known whether any particular kind of lamina I neuron belongs to the STT.

The cells of origin of the *spinocervical tract* (SCT) send their axons into the ipsilateral dorsolateral funiculus and terminate in the lateral cervical nucleus, which, in turn, projects to the contralateral thalamus. It is widely believed that the SCT of man is either absent or vestigial. This is not strictly true. Autopsies of 16 people found definite evidence of a SCT (as inferred from the presence of a lateral cervical nucleus) in nine cases (Truex et al 1970). Although the nucleus contained few cells in most of these cases, one person's nucleus contained nearly as many cells as a cat's lateral cervical nucleus. One must conclude that the SCT is a highly variable feature of human neuroanatomy. A large survey of adult and fetal autopsy material is sorely needed to determine accurately the frequency and variability of the human SCT.

Retrograde HRP experiments have shown that the great majority of SCT cells (in cats and dogs) occupy a band across the dorsal horn that is centered in lamina IV (Craig 1976, Cervero et al 1977, Brown et al 1980). Data for the monkey's SCT are scarce, but an extracellular mapping study (Bryan et al 1974) suggests a more scattered pattern in laminae IV–VII.

The morphology of individual SCT neurons in the cat has been determined with intracellular injection of HRP (Brown et al 1977a). The dendritic arbors of SCT neurons are elongated rostrocaudally (800–2000 μm) but are relatively narrow in the mediolateral plane (200–550 μm). Most of their dendrites travel dorsally. Apical dendrites rarely enter lamina II; indeed, many apical dendrites change their trajectories at the gelatinosa's ventral border and then travel parallel to it. The axons of SCT neurons often issue collaterals that arborize and emit varicosities (presumed presynaptic boutons) at the level of the perikaryon and ventral to it. The targets of these boutons have not been identified.

The cells of origin of the *dorsal column postsynaptic spinomedullary system* (DCPS) have axons ascending in the dorsal funiculi (mostly ipsilateral) that terminate in the dorsal column nuclei (Rustioni 1973, 1974, Rustioni et al 1979). In the lumbar enlargement of cat and monkey, DCPS neurons are as numerous as the cat's lumbar SCT neurons (Bennett et al 1981b). There is no compelling evidence for or against the presence of a human DCPS, but its prominence in the macaque strongly suggests its presence. In cat and monkey lumbar enlargement, most DCPS neurons are found in the same lamina IV band as SCT neurons.

Individual DCPS neurons in the cat have been intracellularly labeled with HRP (Bennett et al 1981b, Brown & Fyffe 1982). There appear to be several morphological types. Many have dendritic arbors that are rostrocaudally elongated (1000–2000 μm) and relatively narrow mediolaterally (300–600 μm) (Bennett et al 1981b). The neurons that Brown & Fyffe (1982) stained had dendritic territories that were shaped like dorsoventrally oriented cylinders. Both studies found that some DCPS neurons have a few apical dendrites that enter laminae I and II (especially the ventral portion of lamina II), while many others do not have intragelatinosal dendrites. In addition, DCPS cells were found to issue varicosity-bearing axon collaterals at the level of the perikaryon and ventral to it. The targets of these local synapses have not been identified.

At least some of the neurons of both the DCPS and SCT systems issue synapses into local, dorsal horn circuits. It is not known whether STT neurons do likewise. The functional significance of these contributions to the dorsal horn's local circuitry has only just begun to be investigated (Jankowska et al 1979).

The STT, SCT, and DCPS systems are the best understood, but not the only possible pathways for nociception. The spinorecticular projection, for example, may be involved in some aspects of the pain experience (see Dennis & Melzack 1977 for review). Another possibility is an ascending multisynaptic chain of short axonal connections that crosses and recrosses the spinal midline (Noordenbos, 1959).

PHYSIOLOGICAL RESPONSES   Spinal cord and medullary dorsal horn neurons that are excited by natural forms of cutaneous stimulation can be conveniently categorized as low-threshold mechanoreceptive (LTM), nociceptive specific (NS), or wide dynamic range (WDR). LTM neurons are excited only by touch, hair movement, and other types of innocuous tactile stimulation. NS neurons respond only to stimuli that are either injurious or nearly so. WDR neurons respond to both innocuous and injurious stimuli. WDR neurons differentiate between innocuous and noxious stimuli by discharging at a higher frequency to injurious stimuli. These three categories were derived from experiments in anesthetized animals (for reviews see Willis & Coggeshall 1978, Price & Dubner 1977). Recent experiments (Hoffman et al 1981) show that the same categories can be recognized in the unanesthetized monkey.

Each of the output systems described above has some LTM neurons. In the STT these neurons are definitely a minority (probably about 20%) of the total population. In contrast, LTM neurons are much more numerous in both the SCT and DCPS systems. Estimates of their frequency in both systems vary but the true figure probably lies between ⅓ and ⅔ for both systems.

LTM neurons in general have small receptive fields with sharp boundaries. Receptive fields on distal body parts are smaller than those on proximal body parts. The small RFs of LTM neurons are consistent with the hypothesis that they encode stimulus location. The LTM category, like the other two categories, is not homogeneous. Different LTM neurons receive input from different kinds and different combinations of low-threshold mechanoreceptive primary afferents.

The SCT and DCPS systems contain very few NS neurons, probably no more than 1–5% for either system. In contrast, NS neurons are a significant feature of the STT system. They are most commonly found in lamina I but are also encountered in lamina V. Estimating their frequency is especially hazardous because lamina I neurons are very likely to be underrepresented in recording samples. Bearing this in mind, it seems reasonable to estimate that roughly ⅓ of the STT system is composed of NS neurons.

NS neurons, in general, have receptive fields like those of LTM neurons and presumably encode for stimulus location.

We note above that the STT projection to medial thalamic nuclei (excepting nucleus submedius) arises preferentially from the deeper laminae. It has recently been reported (Giesler et al 1981) that nearly two thirds of the STT neurons projecting exclusively to the medial thalamus are excited only by noxious stimuli. However, in contrast to other NS neurons, these have very large receptive fields, often including the skin of more than one limb or even the entire surface of the body.

WDR neurons constitute one third to one half of all three output systems. WDR neurons are most frequently encountered in lamina V. WDR neurons in lamina I, however, have recently been documented (Bennett et al 1981a).

A large body of evidence indicates that WDR neurons contribute to the perception of pain (see Price & Dubner 1977 for review). Most important, there is evidence suggesting that the selective activation of human WDR neurons is sufficient to produce pain (Mayer et al 1975, Price & Mayer 1975).

The receptive fields of WDR neurons are large relative to the typical receptive fields of LTM and NS neurons. They are also much more complex. WDR neurons often have receptive fields with a central area responsive to both tactile and noxious stimuli, surrounded by a larger area, with ill-defined borders, that is responsive only to noxious stimuli. The entire excitatory receptive field is often surrounded by an inhibitory receptive field. There is evidence that the organization of WDR receptive fields and their responsiveness to different kinds of primary afferent inputs are under descending control (see below). There is also evidence (Devor & Wall 1981) that the receptive fields may be modulated by spinal circuitry.

Le Bars et al (1979a) have shown that WDR neurons (but not NS or LTM neurons) in the rat are powerfully inhibited by noxious stimulation of any part of the body. The inhibition is hypothesized to be critical for the pain-signaling function of WDR neurons. It is not known, however, whether these "diffuse noxious inhibitory controls" are present in other species; nor do we know whether the affected WDR neurons belong to any of the spinal cord's output systems.

The description above focuses on neuronal responses to cutaneous stimulation. There is now an impressive body of evidence (for review see Milne et al 1981) showing that many WDR and NS neurons in the STT that have cutaneous receptive fields also respond to noxious visceral stimulation. Neurons with innervation from both the skin and viscera may mediate the clinically important phenomenon of referred pain.

Interneurons

The substantia gelatinosa (SG; Rexed's lamina II) has long been regarded as the dorsal horn's major somatosensory interneuronal pool (for review see Cervero & Iggo 1980). Recent work has not altered this view, although it is now known that a small minority (ca. 1%) of the neurons in the SG send their axons to the brain stem (e.g. Willis et al 1979).

MORPHOLOGY   There are at least five morphologically distinguishable types of neurons whose perikarya lie within the cat's SG (Gobel 1978b). The most common of these have been called stalked cells and islet cells, and they have been found in both the spinal and medullary dorsal horns. Both of these are known to be present in the monkey's SG (Beal & Cooper 1978, Price et al 1979), and islet cells have been described in the human SG (Schoenen 1980).

The stalked cell's perikaryon lies at the laminae I/II border. Its dendrites travel ventrally while simultaneously fanning-out rostrocaudally. The ventral extent of the dendritic arbor is variable; some arbors descend only part way through the thickness of lamina II, whereas other arbors pass through all of lamina II and into laminae III–IV. The stalked cell's unmyelinated axon arborizes profusely and issues many synaptic boutons in lamina I. The stalked cell population is clearly a major output system linking the SG with the high concentration of projection neurons in lamina I and, thus, it is misleading to view the SG as a "closed system" (Szentagothai 1964). The function(s) of this stalked cell-lamina I linkage is unknown. Gobel (1978b), however, has hypothesized that the stalked cell collects primary afferent inputs from those afferents that terminate below lamina I and transfers this input to lamina I projection neurons (whose dendrites are generally confined to lamina I). Although there are data that are consistent with this hypothesis (Bennett et al 1979, Price et al 1979), it is not yet possible to exclude the possibility that the stalked cell-lamina I synaptic linkage is an inhibitory one.

Electron microscopic examination of stalked cells intracellularly stained with HRP (Gobel et al 1980) showed that stalked cell dendrites receive synapses from primary afferent terminals and from dome-shaped axon terminals that resemble the terminals of descending serotonergic axons (Gobel et al 1980, 1982).

Islet cell perikarya are located throughout the SG. Most of their dendrites course rostrocaudally, parallel to the laminar borders. The islet cell's dendritic arbor is elongated rostrocaudally; those in lamina IIb extend for well over a millimeter. Islet cell axons are unmyelinated and arborize profusely within and near their dendritic trees. Islet cells are presumed to be inhibitory interneurons (Gobel 1978b). In addition to their typical Golgi II appearance, there is evidence that some islet cells are GABAergic (McLaughlin et al 1975, Gobel et al 1980) and that others are enkephalinergic (Bennett et al 1982).

Ultrastructurally, islet cells are quite remarkable (Gobel et al 1980). Their spine heads and dendritic shafts contain aggregates of round and oval synaptic vesicles. These vesicle-containing dendrites synapse on scallop-shaped primary afferent terminals. Such dendroaxonic synapses must regulate the activity of the primary afferent terminals (Gobel et al 1980), but the nature of this regulation is unknown. Islet cell dendrites also synapse on other dendrites, one kind of which does not contain vesicles and is thought to come from stalked cells. Islet cells also receive synapses from dome-shaped axon terminals that resemble the terminals of descending serotonergic axons (Gobel et al 1980).

Rexed (1952) noted that the SG of neonatal cats was divisible into a superficial layer with a higher cell density than the deeper part of the lamina. With the relatively thick Nissl-stained sections that Rexed used, this division could not be seen in the adult's SG; however, Gobel (1978b) demonstrated that this division persists in the adult. Evidence has accumulated rapidly that indicates that the division of the SG into sublaminae (laminae IIa and IIb; Gobel & Falls 1979) reflects a fundamental difference between the two parts of the SG. Bennett et al (1980) and others (Kumazawa & Perl 1976, 1978, Light et al 1979, Price et al 1979) have shown that SG neurons with perikarya in lamina IIa respond to inputs from nociceptive primary afferents, while neurons with perikarya in lamina IIb respond only to inputs from low threshold mechanoreceptive primary afferents (see below). It has also been demonstrated that the different types of noradrenergic and serotonergic terminals from descending projections are differentially distributed within laminae IIa and IIb (Ruda et al 1979, Ruda & Gobel 1980). Moreover, the two parts of the SG stain differentially for several neuropeptides (e.g. Gibson et al 1981). Thus, it is no longer appropriate to consider the SG as either functionally or structurally homogeneous.

PHYSIOLOGY Until recently, most of our knowledge about the physiology of the SG has come from experiments that examined the consequences of lesioning Lissauer's tract. The rationale for these experiments was the belief that Lissauer's tract, or at least its lateral part in the lumbar enlargement, was an exclusive pathway for the intrasegmental and short (2–3) intersegmental axons of intrinsic SG neurons. It is now known, however, that all parts of Lissauer's tract contain very large numbers of primary afferent axons (Coggeshall et al 1981); thus, interpretation of the earlier results has become ambiguous.

Our current concepts about the SG rest largely upon the results of single cell recording experiments. Using the intracellular HRP method with barbiturate-anesthetized cats, Bennett et al (1979, 1980) found that the responses of stalked cells and lamina IIa islet cells were comparable to the responses of NS and WDR neurons. In contrast, islet cells in lamina IIb responded only to innocuous tactile stimuli. These patterns of primary afferent excitation of SG neurons were comparable to the WDR, NS, and LTM patterns seen elsewhere in the dorsal horn. In other respects, however, the responses of SG neurons to cutaneous stimulation are quite different from those seen elsewhere in the dorsal horn. Many SG neurons, for example, show a very powerful habituation to repeated stimuli (Light et al 1979, Price et al 1979, Wall et al 1979, Bennett et al 1980). While nearly all investigators have found that SG neurons are excited by primary afferent input, Cervero et al (1977, 1979) have described a very different type of SG activity. They reported that the SG was dominated by neurons with a prominent ongoing discharge (1–40 Hz) that was strongly inhibited by cutaneous stimulation. Excitatory effects from cutaneous stimulation were either absent or inconspicuous, perhaps due to the masking effect of the strong background discharge. They have recently used the intracellular HRP method to confirm that their recordings originate from intrinsic SG neurons (Molony et al 1981).

Both stalked and islet cells appear to be innervated by descending serotonergic axons. Descending serotonergic and noradrenergic axons are known to synapse on SG dendrites (Ruda & Gobel 1980). Dubuisson & Wall (1980) have presented preliminary data showing that activation of descending systems has powerful excitatory and inhibitory effects on SG neurons. The effects that these systems exert over intrinsic SG neurons will undoubtedly draw much attention in the future.

INTRAGELATINOSAL DENDRITES Many neurons in laminae IV and V send dendrites into lamina II, and even into lamina I. Szentagothai (1964) stressed that these intragelatinosal dendrites were probably the site at which the SG exercised its control over the neurons in the deeper laminae. Since laminae IV–V contain many projection neurons, it has been reasonable to presume that projection neurons have intragelatinosal dendrites that are the site at which the SG modulates somatosensory signals sent to the brain. Although there is no doubt that there are laminae IV–V neurons with intragelatinosal dendrites, it now seems unlikely that these neurons belong to somatosensory projection systems. SCT neurons have few, if any, intragelatinosal dendrites (Brown et al 1977a). Similarly, most DCPS neurons have few or no intragelatinosal dendrites (Bennett et al 1981b, Brown & Fyffe 1982). We do not know whether STT neurons have any significant intragelatinosal dendritic arbors. It seems unlikely, however, that the many STT neurons in lamina VI and below would have many dendrites in the SG. We noted above that the SG's stalked cell population connects the SG with lamina I's projection neurons. There is no evidence that a comparably massive axonal projection links intrinsic SG neurons and the deeper laminae. Thus, it is far from clear how the neurons of the SG might interact with the projection neurons of the deeper laminae.

OTHER DORSAL HORN INTERNEURONS   With the exception of SG neurons, we know practically nothing about the interneurons in the rest of the dorsal horn. We presume that they occur in every lamina, intermingled with neurons whose axons project to distant targets.

DESCENDING SYSTEMS

That dorsal horn nociceptive neurons are subject to descending modulation by brain stem systems is now well established. Electrical stimulation or administration of opiate drugs activates these descending pathways and produces a profound behavioral analgesia when assessed by measures such as the tail-flick or the hot-plate test. A number of recent reviews have provided evidence in support of such systems (Mayer & Price 1976, Basbaum & Fields 1978, Yaksh & Rudy 1978, Mayer 1980, Sherman & Liebeskind 1980), and the reader should consult them for a detailed bibliography. Our purpose here is to examine the neuronal circuitry purported to be involved in analgesia or inhibition of dorsal horn neurons produced by stimulation of these descending systems.

Electrical stimulation or microinjection of morphine in the midbrain periaqueductal gray (PAG) produces a behavioral analgesia, as well as an inhibition of dorsal horn nociceptive neurons in laminae I, II, V–VII. The ability of PAG morphine microinjection and electrical stimulation to mimic the dorsal horn inhibitory effects of systemic opiates suggests the existence of a common inhibitory descending system. This descending pathway is reported to involve projections to the medullary nucleus raphe magnus which contains many neurons whose axons descend via the spinal cord dorsolateral funiculus to the dorsal horn. There is evidence that these raphespinal neurons are serotonergic. It has been proposed that the dorsal horn circuitry involves an enkephalinergic inhibitory interneuron (Basbaum & Fields 1979). Although other possible descending pathways are not excluded, emphasis has been placed on the above circuit and its common activation by electrical stimulation and morphine microinjection into the midbrain PAG (Mayer & Price 1976, Yaksh & Rudy 1978). This section explores the nature of the evidence for such an endogenous pain-suppressing system active at the dorsal horn level.

Evidence for the Role of the Midbrain Periaqueductal Gray in Descending Control

There is considerable evidence that PAG stimulation inhibits the responses of wide-dynamic-range and nociceptive-specific neurons in the medullary and spinal dorsal horns (Mayer & Price 1976, Basbaum & Fields 1978, Sessle et al 1981). However, the effects of PAG morphine or electrical stimulation on behavioral measures of analgesia, such as vocalization or the hot-plate test, do not rule out the possibility that brain stem ascending nociceptive neurons are affected (Yaksh & Rudy 1978). Studies in which more reflexive measures of analgesia are used, such as limb withdrawal or tail-flick in response to noxious stimuli, provide strong evidence that descending systems are involved (Basbaum & Fields 1978). It cannot be assumed, however, that effects on reflex responses are equivalent to alteration in pain perception as measured by complex behaviors (McGrath et al 1981). There clearly is a need for multiple behavioral measures of pain responsivity in these studies to provide more convincing evidence of analgesic potency.

The conclusion that opiate analgesia and electrical stimulation-produced analgesia activate common pathways is supported by the finding that systemic administration of naloxone, a specific opiate antagonist, can block both effects (Mayer & Price 1976, Basbaum & Fields 1978). The effect of naloxone on PAG stimulation-produced inhibition of dorsal horn neurons is less reliable (Mayer 1980, Sessle et al 1981). Naloxone antagonism by itself, however, is not sufficient evidence to implicate an opiate mechanism (Hayes et al 1977). Furthermore, naloxone antagonism may be nonspecific when doses much greater than required to reverse analgesia induced by systemic morphine are employed.

Evidence that PAG Stimulation Activates a Raphespinal System

The presence of sparse direct projections of the PAG to the lumbosacral spinal cord (Kuypers & Maisky 1975, Castiglioni et al 1978) has led to a search for other pathways responsible for the spinal effects of PAG stimulation. Many lines of indirect evidence implicate the medullary nucleus raphe magnus and the adjacent reticular formation in this effect. Anatomical studies (Gallagher & Pert 1978, Abols & Basbaum 1981) have revealed direct projections to the nucleus raphe magnus, the nucleus reticularis magnocellularis, and the more dorsal nucleus reticularis gigantocellularis. Projections from the nucleus raphe magnus and the magnocellular reticular field to the medullary and spinal dorsal horns have been described (Basbaum et al 1978). Physiological studies indicate that raphespinal neurons are excited by PAG opiate microinjection or PAG electrical stimulation (Fields & Anderson 1978, Lovick et al 1978). Direct electrical stimulation of raphe magnus or opiate microinjection inhibits dorsal horn neurons (Basbaum & Fields 1978, Dickenson et al 1979, Duggan & Griersmith 1979, Lovick & Wolstencroft 1979, Sessle et al 1981), providing further support for this link in the descending pathway. However, there are other findings not totally consistent with such an exclusive descending pathway activated by opiates and electrical stimulation. The evidence on naloxone antagonism of raphe magnus-induced inhibition of dorsal horn neurons is conflicting (Duggan & Griersmith 1979, Sessle et al 1981), as is the effect of naloxone on raphe magnus-induced behavioral analgesia (Oliveras et al 1977, Satoh et al 1980, Zorman et al 1981). Furthermore, acute raphe lesions have no effect on behavioral analgesia induced by morphine or PAG stimulation (Hall et al 1981, Proudfit 1981) unless the lesions are extended into the magnocellular reticular field (Cannon et al 1980). It needs to be emphasized that electrical stimulation or lesions in the medullary reticular core activate or interfere with multiple descending pathways in addition to those mentioned above. This includes rubrospinal, tectospinal, and pontine reticulospinal fibers of passage including fibers originating from noradrenergic cell groups in the pons (Westlund & Coulter 1980). All of these projections can influence dorsal horn activity. A further complication is that stimulation of nucleus raphe magnus activates ascending projections, including fibers terminating in the PAG (Brodal 1981). Raphe magnus-induced behavioral analgesia, or inhibition of dorsal horn neurons, could be related to activation of such pathways.

In summary, it would appear that nucleus raphe magnus participates but may not be essential in the mediation of descending effects produced by activation of the PAG by opiates or electrical stimulation. The next section reviews recent evidence suggesting that other pathways are involved and that stimulation-produced analgesia and opiate analgesia access multiple descending systems.

Evidence that Neurochemically Distinct Descending Inhibitory Systems Originating in the Brain Stem Mediate Opiate Analgesia and Stimulation-Produced Analgesia

The presence of serotonergic (5-HT) neurons in midline raphe nuclei, particularly nucleus raphe magnus, and their axonal projections in the spinal cord dorsolateral funiculus is well established. These findings have led to the hypothesis that such neurons mediate brain stem-elicited behavioral analgesia and inhibition of dorsal horn neurons. Although there are conflicting reports (see Johannessen et al 1982 for references), 5-HT depletion with systemically administered inhibitors, such as parachlorophenylalanine (pCPA), has been shown to block partially stimulation-produced analgesia (Akil & Mayer 1972), morphine analgesia (Tenen 1968), and inhibition of dorsal horn wide-dynamic-range neuronal activity (Rivot et al 1980). Such systemic manipulations, however, do not distinguish effects on ascending forebrain 5-HT pathways (e.g. from PAG, dorsal raphe nucleus, etc) from those of descending raphespinal 5-HT circuits and descending pathways originating from other brain stem serotonergic sites. The serotonergic dorsal raphe nucleus in the midbrain sends projections to many lower brain stem areas (Pierce et al 1976) (e.g. locus coeruleus, parabrachial nucleus, nucleus reticularis magnocellularis) that have axon terminals in the dorsal horn.

Recent studies have concentrated on defining the pharmacology of spinal circuits mediating the analgesic effects of opiates and brain stem electrical stimulation. Techniques utilized include the iontophoresis or intrathecal injection of putative neurotransmitters and their antagonists, and the measurement of the release of endogenous chemical mediators produced by opiate drugs or electrical stimulation. The iontophoresis of 5-HT on dorsal horn neurons typically results in inhibition of nociceptive neurons (Randic & Yu 1976, Belcher et al 1978, Jordan et al 1979, Griersmith & Duggan 1980). However, some spinothalamic neurons in the deeper layers of the dorsal horn are excited by 5-HT (Jordan et al 1979), and methysergide, a putative 5-HT antagonist, blocks nucleus raphe magnus-induced excitation of such neurons (Yezierski et al 1982). In addition, 5-HT injected into the ventral horn facilitates C-fiber-induced reflexes (Bell & Matsumiya 1981). Iontophoretic maniuplations need to be interpreted cautiously without complementary data on 5-HT spinal release and its antagonism. Recent findings using such methods suggest that brain stem descending inhibition and analgesia are not dependent on intact serotonergic synaptic transmission within the dorsal horn. The intrathecal injection of methysergide does not inferfere with systemic morphine-induced analgesia (Proudfit & Hammond 1981), nor does it block the effects of morphine microinjected into the magnocellular tegmental field located adjacent to nucleus raphe magnus (Kuraishi et al 1979). Intrathecal methysergide also has a weak effect on analgesia produced by electrical stimulation of this same area. Similarly, methysergide iontophoresis does not antagonize the inhibitory effects of raphe magnus stimulation on dorsal horn nociceptive neurons (Belcher et al 1978, Griersmith et al 1981). Finally, the restricted depletion of spinal 5-HT has no effect on analgesia produced by PAG morphine microinjection or electrical stimulation (Johannessen et al 1982).

Although the above findings indicate that 5-HT synapses in spinal circuits may not be necessary for descending inhibition, other studies do suggest they play some role. Intrathecal injections of 5-HT produce analgesia and the effect is blocked by methysergide (Yaksh & Wilson 1979). Yaksh (1979) also has reported that methysergide blocks PAG morphine-induced analgesia. In more recent studies in which the release of 5-HT into the spinal cord has been measured, 5-HT levels increased following PAG morphine, spinal cord dorsolateral funiculus stimulation, and supraspinally mediated, intense sciatic nerve stimulation (Yaksh & Tyce 1979, Tyce & Yaksh 1981). Autoradiographic and immunohistochemical findings of dense serotonergic synaptic contacts with dorsal horn neurons, particularly in the superficial layers, provide supporting evidence for an important role of this monoamine in dorsal horn function (Ruda & Gobel 1980, Hoffert et al 1982). In conclusion, although 5-HT descending systems certainly are present, they are not essential for opiate or stimulation-produced analgesia.

Neuropharmacological studies have revealed the existence of a second neurochemically distinct monamine descending pathway. There are several lines of evidence that descending norepinephrine (NE) systems originating in the pons mediate behavioral analgesia and dorsal horn inhibition. In contrast to 5-HT, the NE descending system appears to be critical for opiate-induced analgesia. Intrathecal phentolamine, an alpha-adrenergic blocker, attenuates the analgesia produced by systemic morphine administration (Proudfit & Hammond 1981) or PAG morphine (Yaksh 1979). Intrathecal phenoxybenzamine, another alpha-adrenergic antagonist, blocks the behavioral analgesia produced by morphine injected into the magnocellular tegmental field (Kuraishi et al 1979) and intrathecal phentolamine attenuates analgesia induced by electrical stimulation of this same region (Yaksh et al 1981). NE is released from the spinal cord following morphine microinjection or electrical stimulation in the magnocellular tegmental field (this manipulation produces no spinal release of 5-HT). NE also is released following electrical stimulation of the spinal cord dorsolateral funiculus and the sciatic nerve (Tyce & Yaksh 1981, Yaksh et al 1981). Iontophoretically applied NE results in inhibition of dorsal horn nociceptive neurons (Belcher et al 1978, Headley et al 1978). In contrast to the spinal depletion of 5-HT, the combined spinal depletion of 5-HT and NE drastically attenuates PAG morphine-induced analgesia. The combined depletion of 5-HT and NE, however, has little acute effect on PAG stimulation-produced analgesia (Johannessen et al 1982), suggesting the existence of yet a third descending analgesia-producing system.

The descending noradrenergic terminations in the dorsal horn originate in the pons in known cell groups: the ventral locus coeruleus, the nucleus subcoeruleus, the medial and lateral parabrachial nuclei, and the nucleus of Kölliker-Fuse in the lateral pons (Westlund & Coulter 1980). These include the A6 and A7 cell groups originally described by Dahlström & Fuxe (1964). The noradrenergic innervation of the dorsal horn arises exclusively from the above pontine cell groups (Westlund & Coulter 1980, Ross et al 1981). Electrical stimulation at these sites produces behavioral analgesia (Segal & Sandberg 1977) and inhibition of dorsal horn nociceptive neurons (Sasa et al 1974).

It is unclear whether these descending inhibitory systems act directly on trigeminothalamic and spinothalamic neurons or whether they are mediated via dorsal horn inhibitory interneurons. An enkephalinergic (ENK) link in the spinal circuitry has been proposed (Basbaum & Fields 1978). The presence of ENK synapses on dorsal horn thalamic projection neurons recently has been demonstrated (Ruda 1982). These ENK terminals probably arise from ENK-containing cells in the dorsal horn (Hunt et al 1980, 1981, Glazer & Basbaum 1981, Bennett et al 1982). Combined autoradiographic and immunocytochemical studies have revealed axodendritic synapses between 5-HT- and ENK-containing profiles (E. J. Glazer and A. I. Basbaum, unpublished), further supporting the idea of a descending 5-HT- spinal ENK link.

Other recent findings do not support the presence of an ENK spinal link in descending pathways. The effects of NE and 5-HT on dorsal horn nociceptive neurons are antagonized by monoamine blockers but not by naloxone (Headley et al 1978, Griersmith & Duggan 1980). Yaksh & Elde (1981) recently demonstrated the release of ENK-like immunoreactivity from the spinal cord following high intensity sciatic nerve stimulation. This effect, in contrast to the release of 5-HT and NE, occurred in spinalized animals and was not evoked by distant trigeminal nerve stimulation. These findings suggest that ENK release from the spinal cord may be related to activation of a separate segmental circuit independent of the NE and 5-HT descending systems.

Recent studies also show that descending serotonergic contacts are particularly heavy on presumed projection neurons in lamina I (Ruda & Gobel 1980, Ruda et al 1981, Gobel et al 1982, Hoffert et al 1982). Since the effects of serotonin on spinothalamic neurons are mainly inhibitory (Jordan et al 1979), these studies suggest the existence of direct, monosynaptic inhibitory input to dorsal horn thalamic projection neurons. Confirmation of such a connection has broad interest, since very few long inhibitory pathways have been reported in the vertebrate central nervous system.

ENK neurons are prominent in lamina I and II (Glazer & Basbaum 1981, Hunt et al 1981, Bennett et al 1982). Other ENK-containing cells are present in laminae III, IV/V, X, and in the intermediate gray (Glazer & Basbaum 1981). Such a wide and varied distribution of ENK neurons suggests that ENK may serve other functional roles in dorsal horn circuitry in addition to involvement in descending effects. It has been proposed that ENK acts presynaptically on substance P-containing primary afferents, presumed to be nociceptive (Jessell & Iverson 1977, Yaksh et al 1980). This hypothesis has not been supported by immunocytochemical studies at the ultrastructural level. Although there are occasional suggestions of synaptic contacts between presumed primary afferents and ENK terminals in the superficial dorsal horn, the vast majority of ENK synaptic contacts are axosomatic and axodendritic (Hunt et al 1980, Aronin et al 1981). Physiologic studies provide further evidence for a postsynaptic site of action of ENK in the dorsal horn (Zieglgänsberger & Tulloch 1979b, Miletic & Randic 1981), as does the evidence of ENK synapses on dorsal horn thalamic projection neurons (Ruda 1982).

Our present understanding of the role of pharmacologically distinct systems is complicated further by the finding that brain stem neurons contain various peptides that may be involved in neurochemical transmission. For example, raphe neurons contain substance P, which sometimes coexists with 5-HT in the same neuron (Chan-Palay et al 1978, Hökfelt et al 1978). Similarly, the existence of enkephalin within 5-HT-containing brain stem neurons has been demonstrated (Glazer et al 1981). ENK- and substance P-containing brain stem neurons have descending axons (Hökfelt et al 1979, Bowker et al 1981), and 5-HT and substance P have been identified in the same dense-core vesicles in the spinal cord (Pelletier et al 1981). Recent evidence has revealed the coexistence of two peptides (substance P and thyrotropin-releasing hormone) and 5-HT in the same medullary neurons (Johansson et al 1981). The functional significance of multiple putative neurotransmitters in the same cell is unknown. It has been postulated that they may act on receptors located on different cells or may interact at the same receptor site (Hökfelt et al 1980). The role of these peptidergic neurons in descending inhibitory systems activated by morphine or electrical stimulation also is unknown and obviously of considerable interest.

To summarize, there are at least two neurochemically distinct descending inhibitory systems activated by morphine or brain stem electrical stimulation. The opiate pathway appears to be critically dependent on NE systems, although 5-HT systems also may play a role. The role of peptidergic neurons in these descending systems is not known. There also appears to be a non-opiate pathway that does not require the inclusion of 5-HT or NE systems. These findings are consistent with results from early anatomical and physiological studies that demonstrated multiple parallel pathways originating from midbrain, pontine, and medullary sites (Brodal 1981). The NE system is probably activated by stimulation of neurons in the midbrain dorsal raphe nucleus just ventral to PAG. This raphe nucleus sends projections to norepinephrine-rich locus coeruleus and parabrachial neurons (Pierce et al 1976) that descend to the dorsal horn (Basbaum & Fields 1979). The dorsal raphe nucleus contains a large population of enkephalin-containing neurons; this possibly accounts for its sensitivity to morphine injected into PAG. The release of NE produced by electrical stimulation of nucleus reticularis magnocellularis and overlapping paragigantocellularis (Yaksh et al 1981) could be accounted for by activation of NE fibers of passage as they descend to the spinal cord.

There is conflicting evidence on the involvement of an ENK-containing dorsal horn inhibitory interneuron in brain stem-mediated inhibition. ENK neurons are found throughout the dorsal horn in morphologically distinct types of neurons. In addition to a possible involvement in descending circuits, ENK may serve other functional roles in dorsal horn circuitry.

Demonstration that Environmental Input Activates Descending Pain-Suppressing Systems

Most of the evidence on descending brain stem systems reveals the potential capacity of this system for modifying and filtering ascending nociceptive input. It has been more difficult to demonstrate physiological roles for these systems and how they might be activated by natural environmental stimuli. Potent behavioral analgesia can be produced by diverse stressful stimuli such as footshock, centrifugal rotation, and hypertonic saline (Hayes et al 1978a). Stress may not be the critical variable because some stressors do not produce analgesia. It also is clear that noxious input is not critical, since some active stressors, such as rotation, are not painful. Nevertheless, most behavioral models that activate descending analgesic systems require noxious input to initiate the effect. Based on clinical studies of the effects of transcutaneous electrical nerve stimulation or acupuncture, Melzack (1975) proposed that analgesia in humans was maximally induced by painful stimuli. As knowledge about the role of bulbospinal pathways in analgesia and dorsal horn inhibition increased, a refinement of the original Melzack & Wall (1965) negative feedback pathway was proposed in which ascending nociceptive neurons activate cells in PAG or nucleus raphe magnus, resulting in excitation of descending analgesic systems (Basbaum & Fields 1978).

Partial support for such a feedback system in rat comes from studies in which a variety of noxious stimuli applied to remote body sites inhibit the activity of wide-dynamic-range trigeminal and spinal dorsal horn neurons (Le Bars et al 1979a, 1979b, 1981, Dickenson et al 1980). These diffuse noxious inhibitory controls (DNIC) require intact supra-spinal systems and are reversed by naloxone, thus suggesting the involvement of a descending opiate pathway. A behavioral model of analgesia induced by visceral pain in rats has been designed and appears to mimic DNIC effects on wide-dynamic-range neurons (Kraus et al 1981).

Original studies on footshock-induced analgesia (FSIA) were shown to involve descending pathways, but there was disagreement as to the possible role of opiate systems (Akil et al 1976, Hayes et al 1978b). This controversy was partially resolved by Lewis et al (1980), who demonstrated that a critical variable was the duration of footshock; brief (3 min) FSIA was not reversed by naloxone, whereas prolonged (30 min) FSIA was. More recently, Watkins & Mayer (1982) have extended these observations considerably and have shown that different descending analgesic systems are activated by front paw versus hind paw shock when FSIA is measured with the tail-flick test in rat. Front paw FSIA is reversed by systemic or intrathecal naloxone, exhibits cross-tolerance with morphine, and is abolished by dorsolateral funiculus and medullary brain stem lesions. Thus, front paw shock appears to activate a descending spinal opiate system and these studies provide additional behavioral evidence for the model proposed by Basbaum & Fields (1978). Hind paw FSIA, however, is not reversed by naloxone, does not exhibit cross-tolerance to morphine, is not abolished by medullary brain stem lesions, is only reduced (not abolished) by dorsolateral funiculus lesions, and remains present after spinalization. Thus, hind paw FSIA is not opiate-mediated and involves segmental as well as descending circuits. In summary, these distinctive analgesic effects produced by front paw versus hind paw shock clearly demonstrate physiological roles for multiple opiate-mediated and non-opiate-mediated descending systems.

The studies of Yaksh et al (1981) have certain parallels to the above systems. Stimulation of small-diameter peripheral nerve fibers that often code noxious input leads to activation of descending 5-HT and NE systems and a segmental ENK circuit in rat. Although the desending circuits can be activated by stimulation of remote body sites, they are not mediated by a spinal opiate circuit, in contrast to Watkins & Mayer's findings (1982) with front paw FSIA. Unlike the ENK segmental circuit described by

Yaksh et al (1981), the segmental circuit proposed for hind paw FSIA is not opiate-mediated (Watkins & Mayer 1982). These differences emphasize that in addition to diverse opiate and non-opiate-mediated descending systems, pharmacologically distinct segmental circuits also exist.

Although the above experimental models focus attention on descending control of nociceptive pathways, it is likely that they also play a role in other somatosensory mechanisms. Recent studies of medullary dorsal horn neurons in awake, behaving monkeys indicate that neuronal activity is enhanced when innocuous as well as noxious thermal stimuli are relevant to the monkey's goal-directed behavior (Hayes et al 1981). In addition, some neuronal responses only occur when stimuli are presented during the behavioral task (Dubner et al 1981). These findings suggest that multiple descending and segmental circuits are involved in the neural integration of centrally generated information about stimulus relevance with the ongoing influx of afferent signals.

CONCLUSION

Recently developed experimental approaches clearly have increased our knowledge of dorsal horn circuitry. The terminations of specific nociceptors in the dorsal horn have been identified. Possible neurotransmitters released by these terminals have been proposed. The location of cells of origin of major output systems have been demonstrated and the physiological properties of these neurons are well known. The detailed morphological characteristics of some of these projection neurons have been studied extensively. Recent advances also have been made in the study of the substantia gelatinosa (lamina II), a region long considered the major interneuronal system in the dorsal horn concerned with nociception. The physiology and morphology of the most common types of interneurons in lamina II have been examined. A multitude of possible neurochemicals have been demonstrated and some appear to be associated with identified interneurons. The axons of extrinsic neurons terminating in the dorsal horn arise from multiple descending systems and the chemical mediators in some of these descending axons have been demonstrated. These descending systems appear to have powerful inhibitory effects on dorsal horn neurons and produce analgesia when activated by environmental input, brain stimulation, or opiate drugs.

Where are the remaining gaps in our knowledge of dorsal horn circuitry? Put into perspective, recent advances have increased what we know about each component of the system: the primary afferent, the intrinsic neurons, and the axon terminals of extrinsic neurons. Presently lacking are details on the connectivity between identified components of the system. Although numerous hypotheses have been forthcoming, the next phase of research must provide the data to confirm or refute these hypotheses. Experimental approaches are available to determine the specific relationships between identified components and their neurophysiology and neurochemistry, and the specific connections between components of the system. For example, substance P has been implicated as a putative transmitter in nociceptors. But in which specific type or types of nociceptors? Several different peptides have been found in dorsal root ganglion cells. Are these peptides specific for different functional types of primary afferents? Similarly, in the dorsal horn, neurons are characterized as wide-dynamic-range, nociceptive-specific, or low-threshold mechanoreceptive, yet few studies critically determine cell location or identify the cell population. What output system do they belong to, or are they interneurons? The same lack of identification is prevalent in immunocytochemical and neuropharmacological studies. What types of interneurons or projection neurons contain enkephalin? What specific morphologically or physiologically identified cell types receive input from enkephalinergic neurons? Which dorsal horn neurons receive direct input from neurochemically defined descending 5-HT and NE systems?

The functional significance of the presence of multiple descending systems influencing sensory transmission in the dorsal horn is not known. That these systems are involved exclusively in suppressing or filtering ascending nociceptive signals is probably an oversimplification. In awake animals, the activity of dorsal horn nociceptive neurons is modulated by behavioral contingencies such as the level of arousal, attention to a particular stimulus, the relevance of the stimulus to the behavioral task, the motivational state of the animal, and the likelihood that a significant stimulus event will occur. Future studies may reveal that distinct descending systems are associated with different neuropsychological processes underlying the ability of an animal to extract useful information from its environment.

Acknowledgments

We wish to express our appreciation to E. Welty and B. Arnold for their editorial assistance and to S. Gobel, M. Hoffert, V. Miletic, and M. A. Ruda for their helpful comments of early drafts of the manuscript.

Literature Cited

Abols, I. A., Basbaum, A. I. 1981. Afferent connections of the rostral medulla of the cat: A neural substrate for mid-brain-medullary interactions in the modulation of pain. *J. Comp. Neurol.* 201:285–97

Adriaensen, H., Gybels, J., Handwerker, H. O., Van Hees, J. 1980. Latencies of chemically evoked discharges in human cutaneous nociceptors and of the concurrent subjective sensations. *Neurosci. Lett.* 20:55–59

Adriaensen, H., Gybels, J., Handwerker, H. O., Van Hees, J. 1981. Response properties of thin myelinated (A-delta) fibres in human skin nerves. *Pain,* Suppl. 1:S89

Akil, H., Madden, J., Patrick, R. L., Barchas, J. D. 1976. Stress-induced increase in endogenous opiate peptides: Concurrent analgesia and its partial reversal by naloxone. In *Opiates and Endogenous Opioid Peptides,* ed. H. W. Kosterlitz, pp. 63–70. Amsterdam: Elsevier North Holland

Akil, H., Mayer, D. J. 1972. Antagonism of stimulation-produced analgesia by *p*-CPA, a serotonin synthesis inhibitor. *Brain Res.* 44:692–97

Andersen, R. K., Lund, J. P., Puil, E. 1978. Enkephalin and substance P effects related to trigeminal pain. *Can. J. Physiol. Pharmacol.* 56:216–22

Aronin, N., Difiglia, M., Liotta, A. S., Martin, J. B. 1981. Ultrastructural localization and biochemical features of immunoreactive leu-enkephalin in monkey dorsal horn. *J. Neurosci.* 1:561–77

Arvidsson, J., Gobel, S. 1981. An HRP study of the central projections of primary trigeminal neurons which innervate tooth pulps in the cat. *Brain Res.* 210:1–16

Barber, R., Vaughn, J. E., Slemmon, J. R., Salvaterra, P. M., Roberts, E., Leeman, S. E. 1979. The origin, distribution and synaptic relationships of substance P axons in rat spinal cord. *J. Comp. Neurol.* 184:331–51

Basbaum, A. I., Clanton, C. H., Fields, H. L. 1978. Three bulbospinal pathways from the rostral medulla of the cat: An autoradiographic study of pain modulating systems. *J. Comp. Neurol.* 178:209–24 ·

Basbaum, A. I., Fields, H. L. 1978. Endogenous pain control mechanisms: Review and hypothesis. *Ann. Neurol.* 4:451–62

Basbaum, A. I., Fields, H. L. 1979. The origin of descending pathways in the dorsolateral funiculus of the spinal cord

of the cat and rat: Further studies on the anatomy of pain modulation. *J. Comp. Neurol.* 187:513–31

Beal, J. A., Cooper, M. H. 1978. The neurons in the gelatinosal complex (laminae II and III) of the monkey (*Macaca mulatta*): A Golgi study. *J. Comp. Neurol.* 179:89–122

Beal, J. A., Penny, J. E., Bicknell, H. R. 1981. Structural diversity of marginal (lamina I) neurons in the adult monkey (*Macaca mulatta*) lumbosacral spinal cord: A Golgi study. *J. Comp. Neurol.* 202: 237–54

Beitel, R. E., Dubner, R. 1976a. Responses of unmyelinated (C) polymodal nociceptors to thermal stimuli applied to monkey's face. *J. Neurophysiol.* 39:1160–75

Beitel, R. E., Dubner, R. 1976b. Fatigue and adaptation in unmyelinated (C) polymodal nociceptors to mechanical and thermal stimuli applied to the monkey's face. *Brain Res.* 112:402–6

Belcher, G., Ryall, R. W., Schaffner, R. 1978. The differential effects of 5-hydroxytryptamine, noradrenaline and raphe stimulation on nociceptive and non-nociceptive dorsal horn interneurones in the cat. *Brain Res.* 151:307–21

Bell, J. A., Matsumiya, T. 1981. Inhibitory effects of dorsal horn and excitant effects of ventral horn intraspinal microinjections of norepinephrine and serotonin in the cat. *Life Sci.* 29: 1507–14

Bennett, G. J., Abdelmoumene, M., Hayashi, H., Dubner, R. 1980. Physiology and morphology of substantia gelatinosa neurons intracellularly stained with horseradish peroxidase. *J. Comp. Neurol.* 194:809–27

Bennett, G. J., Abdelmoumene, M., Hayashi, H., Hoffert, M. J., Dubner, R. 1981a. Spinal cord layer I neurons with axon collaterals that generate local arbors. *Brain Res.* 209:421–26

Bennett, G. J., Hayashi, H., Abdelmoumene, M., Dubner, R. 1979. Physiological properties of stalked cells of the substantia gelatinosa intracellularly stained with horseradish peroxidase. *Brain Res.* 164:285–89

Bennett, G. J., Ruda, M. A., Gobel, S., Dubner, R. 1982. Enkephalin immunoreactive stalked cells and lamina IIb islet cells in cat substantia gelatinosa. *Brain Res.* 240:162–66

Bennett, G. J., Seltzer, Z., Hoffert, M. J., Lu, G. W., Nishikawa, N., Dubner, R. 1981b. The morphology and location of the cells of origin of the dorsal column postsynaptic tract (DCPST). *Soc. Neurosci. Abstr.* 7:611

Bessou, P., Perl, E. R. 1969. Response of cutaneous sensory units with unmyelinated fibers to noxious stimuli. *J. Neurophysiol.* 32:1025–43

Bowker, R. M., Steinbusch, H. W. M., Coulter, J. D. 1981. Serotonergic and peptidergic projections to the spinal cord demonstrated by a combined retrograde HRP histochemical and immunocytochemical staining method. *Brain Res.* 211:412–17

Brodal, A. 1981. *Neurological Anatomy.* New York: Oxford Univ. Press. 1053 pp.

Brown, A. G., Fyffe, R. E. W. 1981. Form and function of dorsal horn neurones with axons ascending the dorsal columns in cat. *J. Physiol.* 321:31–47

Brown, A. G., Fyffe, R. E. W., Noble, R., Rose, P. K., Snow, P. J. 1980. The density, distribution and topographical organization of spinocervical tract neurones in the cat. *J. Physiol.* 300:409–28

Brown, A. G., Fyffe, R. E. W., Rose, P. K., Snow, P. J. 1981. Spinal cord collaterals from axons of type II slowly adapting units in the cat. *J. Physiol.* 316:469–80

Brown, A. G., Rose, P. K., Snow, P. J. 1977a. The morphology of spinocervical tract neurones revealed by intracellular injection of horseradish peroxidase. *J. Physiol.* 270:747–64

Brown, A. G., Rose, P. K., Snow, P. J. 1977b. The morphology of hair follicle afferent fibre collaterals in the spinal cord of the cat. *J. Physiol.* 272:779–97

Brown, A. G., Rose, P. K., Snow, P. J. 1978. Morphology and organization of axon collaterals from afferent fibres of slowly adapting type I units in cat spinal cord. *J. Physiol.* 277:15–27

Bryan, R. N., Coulter, J. D., Willis, W. D. 1974. Cells of origin of the spinocervical tract in the monkey. *Exp. Neurol.* 42:574–86

Burgess, P. R., Perl, E. R. 1967. Myelinated afferent fibres responding specifically to noxious stimulation of the skin. *J. Physiol.* 190:541–62

Burgess, P. R., Perl, E. R. 1973. Cutaneous mechanoreceptors and nociceptors. In *Handbook of Sensory Physiology, Somatosensory System,* ed. A. Iggo, 2:29–78. Heidelberg: Springer

Campbell, J. N., Meyer, R. A., LaMotte, R. H. 1979. Sensitization of myelinated nociceptive afferents that innervate monkey hand. *J. Neurophysiol.* 42:1669–79

Cannon, J. T., Prieto, G. J., Liebeskind, J. C. 1980. Disruption of stimulation-produced analgesia by lesions of the nucleus raphe magnus. *Soc. Neurosci. Abstr.* 6:320

Carpenter, S., Lynn, B. 1981. Vascular and sensory responses of human skin to mild injury after topical treatment with capsaicin. *Br. J. Pharmacol.* 73:755–58

Carstens, E., Trevino, D. L. 1978. Laminar origins of spinothalmic projections in the cat as determined by the retrograde transport of horseradish peroxidase. *J. Comp. Neurol.* 182:151–66

Castiglioni, A. J., Gallaway, M. C., Coulter, J. D. 1978. Spinal projections from the midbrain in monkey. *J. Comp. Neurol.* 178:329–46

Cervero, F., Iggo, A. 1980. The substantia gelatinosa of the spinal cord. A critical review. *Brain* 103:717–72

Cervero, F., Molony, V., Iggo, A. 1977. Extracellular and intracellular recordings from neurones in the substantia gelatinosa Rolandi. *Brain Res.* 136:565–69

Cervero, F., Molony, V., Iggo, A. 1979. An electrophysiological study of neurones in the substantia gelatinosa Rolandi of the cat's spinal cord. *Q. J. Exp. Physiol.* 64:297–314

Chan-Palay, V., Jonsson, G., Palay, S. L. 1978. Serotonin and substance P coexist in neurons of the rat's central nervous system. *Proc. Natl. Acad. Sci. USA* 75:1582–86

Coggeshall, R. E., Chung, K., Chung, J. M., Langford, L. A. 1981. Primary afferent axons in the tract of Lissauer in the monkey. *J. Comp. Neurol.* 196:431–42

Craig, A. D. 1976. Spinocervical tract cells in cat and dog, labeled by the retrograde transport of horseradish peroxidase. *Neurosci. Lett.* 3:173–77

Craig, A. D. Jr., Burton, H. 1981. Spinal and medullary lamina I projection to nucleus submedius in medial thalamus: A possible pain center. *J. Neurophysiol.* 45:443–66

Craig, A. D. Jr., Wiegand, S. J., Price, J. L. 1981. The cortical projection of the nucleus submedius in the cat. *Soc. Neurosci. Abstr.* 7:534

Croze, S., Duclaux, R., Kenshalo, D. R. 1976. The thermal sensitivity of the polymodal nociceptors in the monkey. *J. Physiol.* 263:539–62

Cuello, A. C., Del Fiacco, M., Paxinos, G. 1978. The central and peripheral ends of the substance P-containing sensory neurones in the rat trigeminal system. *Brain Res.* 152:499–509

Dahlström, A., Fuxe, K. 1964. Evidence for the existence of monoamine-containing neurons in the central nervous system. I. Demonstration of monamines in the cell bodies of brainstem neurones. *Acta Physiol. Scand.* 62 (Suppl. 232): 1–55

Dennis, S. G., Melzack, R. 1977. Pain-signalling systems in the dorsal and ventral spinal cord. *Pain* 4:97–132

Devor, M., Wall, P. D. 1981. Effect of peripheral nerve injury on receptive fields of cells in the cat spinal cord. *J. Comp. Neurol.* 199:277–91

Dickenson, A. H., LeBars, D., Besson, J. M. 1980. Diffuse noxious inhibitory controls (DNIC). Effects on trigeminal nucleus caudalis neurones in the rat. *Brain Res.* 200:293–305

Dickenson, A. H., Oliveras, J. L., Besson, J. M. 1979. Role of the nucleus raphe magnus in opiate analgesia as studied by the microinjection technique in the rat. *Brain Res.* 170:95–111

Dubner, R., Gobel, S., Price, D. D. 1976. Peripheral and central trigeminal pain pathways. In *Advances in Pain Research and Therapy*, ed. J. J. Bonica, D. Albe-Fessard, 1:137–48. New York: Raven

Dubner, R., Hoffman, D. S., Hayes, R. L. 1981. Neuronal activity in medullary dorsal horn of awake monkeys trained in a thermal discrimination task. III. Task-related responses and their functional role. *J. Neurophysiol.* 46:444–64

Dubner, R., Hu, J. W. 1977. Myelinated (A-delta) nociceptive afferents innervating the monkey's face. *J. Dent. Res.* 56: A167

Dubner, R., Price, D. D., Beitel, R. E., Hu, J. W. 1977. Peripheral neural correlates of behavior in monkey and human related to sensory-discriminative aspects of pain. In *Pain in the Trigeminal Region*, ed. D. J. Anderson, B. Matthews, pp. 57–66. Amsterdam: Elsevier North Holland

Dubner, R., Sessle, B. J., Storey, A. T. 1978. *The Neural Basis of Oral and Facial Function*. New York: Plenum. 483 pp.

Dubner, R., Sumino, R., Wood, W. I. 1975. A peripheral "cold" fiber population responsive to innocuous and noxious thermal stimuli applied to the monkey's face. *J. Neurophysiol.* 38:1373–89

Dubuisson, D., Wall, P. D. 1980. Descending influences on receptive fields and activity of single units recorded in laminae 1, 2 and 3 of cat spinal cord. *Brain Res.* 199:283–98

Duggan, A. W., Griersmith, B. T. 1979. Inhibition of the spinal transmission of noci-

ceptive information by supraspinal stimulation in the cat. *Pain* 6:149–61

Fields, H. L., Anderson, S. D. 1978. Evidence that raphe-spinal neurons mediate opiate and midbrain stimulation-produced analgesia. *Pain* 5:333–49

Fitzgerald, M. 1979. The spread of sensitization of polymodal nociceptors in the rabbit from nearby injury and by antidromic nerve stimulation. *J. Physiol.* 297:207–16

Fitzgerald, M., Lynn, B. 1977. The sensitization of high threshold mechanoreceptors with myelinated axons by repeated heating. *J. Physiol.* 365:549–63

Fitzgerald, M., Lynn, B. 1979. The weak excitation of some cutaneous receptors in cats and rabbits by synthetic substance P. *J. Physiol.* 293:66P–67P

Gallagher, D. W., Pert, A. 1978. Afferents to brain stem nuclei (brain stem raphe, nucleus reticularis pontis caudalis and nucleus giganto-cellularis) in the rat as demonstrated by microiontophoretically applied horseradish peroxidase. *Brain Res.* 144:257–75

Gamse, R., Leeman, S. E., Holzer, P., Lembeck, F. 1981. Differential effects of capsaicin on the content of somatostatin, substance P, and neurotensin in the nervous system of the rat. *Arch. Pharmacol.* 317:140–48

Georgopoulos, A. P. 1976. Functional properties of primary afferent units probably related to pain mechanisms in primate glabrous skin. *J. Neurophysiol.* 39: 71–83

Gibson, S. J., Polak, J. M., Bloom, S. R., Wall, P. D. 1981. The distribution of nine peptides in rat spinal cord and with special emphasis on the substantia gelatinosa and on the area around the central canal (lamina X). *J. Comp. Neurol.* 201:65–79

Giesler, G. J. Jr., Menetrey, D., Basbaum, A. I. 1979. Differential origins of spinothalamic tract projections to medial and lateral thalamus in the rat. *J. Comp. Neurol.* 184:107–26

Giesler, G. J. Jr., Menetrey, D., Guilbaud, G., Besson, J. M. 1976. Lumbar cord neurons at the origin of the spinothalamic tract in the rat. *Brain Res.* 118:320–24

Giesler, G. J. Jr., Yezierski, R. P., Gerhart, K. D., Willis, W. D. 1981. Spinothalamic tract neurons that project to medial and/or lateral thalamic nuclei: Evidence for a physiologically novel population of spinal cord neurons. *J. Neurophysiol.* 46:1285–1308

Glazer, E. J., Basbaum, A. I. 1981. Immunohistochemical localization of leucine-enkephalin in the spinal cord of the cat: Enkephalin-containing marginal neurons and pain modulation. *J. Comp. Neurol.* 196:377–89

Glazer, E. J., Steinbusch, H., Verhofstad, A., Basbaum, A. I. 1981. Serotonin neurons in nucleus raphe dorsalis and paragigantocellularis of the cat contain enkephalin. *J. Physiol. Paris* 77:241–45

Gobel, S. 1978a. Golgi studies of the neurons in layer I of the dorsal horn of the medulla (trigeminal nucleus caudalis). *J. Comp. Neurol.* 180:375–94

Gobel, S. 1978b. Golgi studies of the neurons in layer II of the dorsal horn of the medulla (trigeminal nucleus caudalis). *J. Comp. Neurol.* 180:395–414

Gobel, S., Bennett, G. J., Allen, B., Humphrey, E., Seltzer, Z., Abdelmoumene, M., Hayashi, H., Hoffert, M. J. 1982. Synaptic connectivity of substantia gelatinosa neurons with reference to potential termination sites of descending axons. In *Brain Stem Control of Spinal Mechanisms, Eric K. Fernstrom Symp. I*, ed. B. Sjolund, A. Björkland. New York: Elsevier North Holland. In press

Gobel, S., Binck, J. M. 1977. Degenerative changes in primary trigeminal axons and in neurons in nucleus caudalis following tooth pulp extirpations in the cat. *Brain Res.* 132:347–54

Gobel, S., Falls, W. M. 1979. Anatomical observations of horseradish peroxidase-filled terminal primary axonal arborizations in layer II of the substantia gelatinosa of Rolando. *Brain Res.* 175:335–40

Gobel, S., Falls, W. M., Bennett, G. J., Abdelmoumene, M., Hayashi, H., Humphrey, E. 1980. An EM analysis of the synaptic connections of horseradish peroxidase-filled stalked cells and islet cells in the substantia gelatinosa of adult cat spinal cord. *J. Comp. Neurol.* 194:781–807

Gobel, S., Falls, W. M., Hockfield, S. 1977. The division of the dorsal and ventral horns of the mammalian caudal medulla into eight layers using anatomical criteria. See Dubner et al 1977, pp. 443–53

Gobel, S., Falls, W. M., Humphrey, E. 1981b. Morphology and synaptic connections of ultrafine primary axons in lamina I of the spinal dorsal horn: Candidates for the terminal axonal arbors of primary neurons with unmyelinated (C) axons. *J. Neurosci.* 1:1163–79

Gobel, S., Hockfield, S., Ruda, M. A. 1981a. Anatomical similarities between medullary and spinal dorsal horns. In *Oral-Facial Sensory and Motor Functions,* eds. Y. Kawamura, R. Dubner, pp. 211–23. Tokyo: Quintessence

Griersmith, B. T., Duggan, A. W. 1980. Prolonged depression of spinal transmission of nociceptive information by 5-HT administered in the substantia gelatinosa: Antagonism by methysergide. *Brain Res.* 187:231–36

Griersmith, B. T., Duggan, A. W., North, R. A. 1981. Methysergide and supraspinal inhibition of the spinal transmission of nociceptive information in the anaesthetized cat. *Brain Res.* 204:147–58

Gybels, J., Handwerker, H. O., Van Hees, J. 1979. A comparison between the discharges of human nociceptive nerve fibres and the subject's ratings of his sensations. *J. Physiol.* 292:193–206

Hagermark, O., Hökfelt, T., Pernow, B. 1978. Flare and itch induced by substance P in human skin. *J. Invest. Dermatol.* 71:233–35

Hall, J. G., Duggan, A. W., Johnson, S. M., Morton, C. R. 1981. Medullary raphé lesions do not reduce descending inhibition of dorsal horn neurones of the cat. *Neurosci. Lett.* 25:25–29

Hardy, J. D., Wolff, H. G., Goodell, H. 1952. *Pain Sensations and Reactions,* Baltimore: Williams & Wilkins. 435 pp.

Hayashi, H. 1980. Distribution of vibrissae afferent fiber collaterals in the trigeminal nuclei as revealed by intra-axonal injection of horseradish peroxidase. *Brain Res.* 183:442–46

Hayes, A. G., Tyers, M. B. 1980. Effects of capsaicin on nociceptive heat, pressure and chemical thresholds and on substance P levels in the rat. *Brain Res.* 189:561–64

Hayes, R. L., Bennett, G. J., Newlon, P. G., Mayer, D. J. 1978a. Behavioral and physiological studies of non-narcotic analgesia in the rat elicited by certain environmental stimuli. *Brain Res.* 155:69–90

Hayes, R. L., Dubner, R., Hoffman, D. S. 1981. Neuronal activity in medullary dorsal horn of awake monkeys trained in a thermal discrimination task. II. Behavioral modulation of responses to thermal and mechanical stimuli *J. Neurophysiol.* 46:428–43

Hayes, R. L., Price, D. D., Bennett, G. J., Wilcox, G. L., Mayer, D. J. 1978b. Differential effects of spinal cord lesions on narcotic and nonnarcotic suppression of nociceptive reflexes: Further evidence for the physiologic multiplicity of pain modulation. *Brain Res.* 155:91–101

Hayes, R. L., Price, D. D., Dubner, R. 1977. Naloxone antagonism as evidence for narcotic mechanisms. *Science* 196:600

Headley, P. M., Duggan, A. W., Griersmith, B. T. 1978. Selective reduction by noradrenaline and 5-hydroxytryptamine of nociceptive responses of cat dorsal horn neurones. *Brain Res.* 145:185–89

Henry, J. L. 1976. Effects of substance P on functionally identified units in cat spinal cord. *Brain Res.* 114:439–51

Henry, J. L., Sessle, B. J., Lucier, G. E., Hu, J. W. 1980. Effects of substance P on nociceptive and non-nociceptive trigeminal brain stem neurons. *Pain* 8:33–45

Hoffert, M. J., Miletic, V., Ruda, M. A., Dubner, R. 1982. Immunocytochemical identification of serotonergic axonal contacts on characterized neurons in cat spinal dorsal horn. *Anat. Rec.* 202:83A

Hoffman, D. S., Dubner, R., Hayes, R. L., Medlin, T. P. 1981. Neuronal activity in medullary dorsal horn of awake monkeys trained in a thermal discrimination task. I. Responses to innocuous and noxious thermal stimuli. *J. Neurophysiol.* 46:409–27

Hökfelt, T., Johansson, O., Ljungdahl, A., Lundberg, J. M., Schultzberg, M. 1980. Peptidergic neurones. *Nature* 284:515–21

Hökfelt, T., Kellerth, J. O., Nilsson, G., Pernow, B. 1975. Substance P: Localization in the central nervous system and in some primary sensory neurons. *Science* 190:889–90

Hökfelt, T., Ljungdahl, A., Steinbusch, H., Verhofstad, A., Nilsson, G., Brodin, E., Pernow, B., Goldstein, M. 1978. Immunohistochemical evidence of substance P-like immunoreactivity in some 5-hydroxytryptamine-containing neurons in the rat central nervous system. *Neurosci.* 3:517–38

Hökfelt, T., Terenius, L., Kuypers, H. G. J. M., Dann, O. 1979. Evidence for enkephalin immunoreactive neurons in the medulla oblongata projecting to the spinal cord. *Neurosci. Lett.* 14:55–60

Hu, J. W., Sessle, B. J. 1980. Raphe stimulation induces primary afferent depolarization of both low-threshold mechanosensitive and nociceptive trigeminal afferents. *Soc. Neurosci. Abstr.* 6:429

Hunt, S. P., Kelly, J. S., Emson, P. C. 1980. The electron microscopic localization of methionine-enkephalin within the su-

perficial layers (I and II) of the spinal cord. *Neurosci.* 5:1871–90

Hunt, S. P., Kelly, J. S., Emson, P. C., Kimmel, J. R., Miller, R. J., Wu, J. Y. 1981. An immunohistochemical study of neuronal populations containing neuropeptides or gamma-aminobutyrate within the superficial layers of the rat dorsal horn. *Neurosci.* 6:1883–98

Iggo, A., Ogawa, H. 1971. Primate cutaneous thermal nociceptors. *J. Physiol.* 216:77P

Jankowska, E., Rastad, J., Zarzecki, P. 1979. Segmental and supraspinal input to cells of origin of non-primary fibres in the feline dorsal columns. *J. Physiol.* 290:185–200

Janscó, G., Hökfelt, T., Lundberg, J. M., Király, E., Halasz, N., et al. 1981. Immunohistochemical studies on the effect of capsaicin on spinal and medullary peptide and monamine neurons using antisera to substance P, gastrin/CCK, somatostatin, VIP, enkephalin, neurotensin and 5-hydroxytryptamine. *J. Neurocytol.* 10:963–80

Janscó, G., Király, E., Janscó-Gábor, A. 1977. Pharmacologically induced selective degeneration of chemosensitive primary sensory neurones. *Nature* 270:741–43

Jessell, T. M., Iversen, L. L. 1977. Opiate analgesics inhibit substance P release from rat trigeminal nucleus. *Nature* 268:549–51

Jessell, T. M., Iversen, L. L., Cuello, A. C. 1978. Capsaicin-induced depletion of substance P from primary sensory neurones. *Brain Res.* 152:183–88

Johannessen, J. N., Watkins, L. R., Carlton, S. M., Mayer, D. J. 1982. Failure of spinal cord serotonin depletion to alter analgesia elicited from the periaqueductal gray. *Brain Res.,* 237:373–86

Johansson, O., Hökfelt, T., Pernow, B., Jeffcoate, S. L., White, N., et al. 1981. Immunohistochemical support for three putative transmitters in one neuron: Coexistence of 5-hydroxytryptamine, substance P and thyrotropin releasing hormone-like immunoreactivity in medullary neurons projecting to the spinal cord. *Neurosci.* 6:1857–81

Jordan, L. M., Kenshalo, D. R. Jr., Martin, R. F., Haber, L. H., Willis, W. D. 1979. Two populations of spinothalamic tract neurons with opposite responses to 5-hydroxytryptamine. *Brain Res.* 164:342–46

Kraus, E., Le Bars, D., Besson, J. M. 1981. Behavioral confirmation of "diffuse noxious inhibitory controls" (DNIC)

and evidence for a role of endogenous opiates. *Brain Res.* 206:495–99

Kruger, L., Perl, E. R., Sedivec, M. J. 1981. Fine structure of myelinated and mechanical nociceptor endings in cat hairy skin. *J. Comp. Neurol.* 198:137–54

Kumazawa, R., Mizumura, K. 1977. Thinfibre receptors responding to mechanical, chemical, and thermal stimulation in the skeletal muscle of the dog. *J. Physiol.* 273:179–94

Kumazawa, T., Mizumura, K. 1980a. Chemical responses of polymodal receptors of the scrotal contents in dogs. *J. Physiol. London* 299:219–31

Kumazawa, T., Mizumura, K. 1980b. Mechanical and thermal responses of polymodal receptors recorded from the superior spermatic nerve of dogs. *J. Physiol.* 299:233–45

Kumazawa, T., Perl, E. R. 1976. Differential excitation of dorsal horn marginal and substantia gelatinosa neruons by primary afferent units with fine (A-delta and C) fibers. In *Sensory Functions of the Skin in Primates with Special Reference to Man,* ed. Y. Zotterman, pp. 67–89. Oxford: Pergamon

Kumazawa, T., Perl, E. R. 1977. Primate cutaneous sensory units with unmyelinated (C) afferent fibers. *J. Neurophysiol.* 40:1325–38

Kumazawa, T., Perl, E. R. 1978. Excitation of marginal and substantia gelatinosa neurons in the primate spinal cord: Indications of their place in dorsal horn functional organization. *J. Comp. Neurol.* 177:417–34

Kuraishi, Y., Harada, Y., Satoh, M., Takagi, H. 1979. Antagonism by phenoxybenzamine of the analgesic effect of morphine injected into the nucleus reticularis gigantocellularis of the rat. *Neuropharmacol.* 18:107–10

Kuypers, H. G. J. M., Maisky, V. A. 1975. Retrograde axonal transport of horseradish peroxidase from spinal cord to brain stem cell groups in the cat. *Neurosci. Lett.* 1:9–14

LaMotte, C. 1977. Distribution of the tract of Lissauer and the dorsal root fibers in the primate spinal cord. *J. Comp. Neurol.* 172:529–61

LaMotte, R. H. 1979. Intensive and temporal determinants of thermal pain. In: *Sensory Functions of the Skin of Humans,* ed. D. R. Kenshalo, pp. 327–61. New York: Plenum

LaMotte, R. H., Campbell, J. N. 1978. Comparison of responses of warm and nociceptive C-fiber afferents in monkey with

human judgments of thermal pain. *J. Neurophysiol.* 41:509–28

LaMotte, R. H., Thalhammer, J. G., Torebjörk, H. E., Robinson, C. J. 1982. Peripheral neural mechanisms of cutaneous hyperalgesia following mild injury by heat. *J. Neurosci.* 2:765–81

Lawson, S. N., Nickels, S. M. 1980. The use of morphometric techniques to analyse the effect of neonatal capsaicin treatment on rat dorsal root ganglia and dorsal roots. *J. Physiol.* 303:12P

Le Bars, D., Chitour, D., Kraus, E., Dickenson, A. H., Besson, J. M. 1981. Effect of naloxone upon diffuse noxious inhibitory controls (DNIC) in the rat. *Brain Res.* 204:387–402

Le Bars, D., Dickenson, A. H., Besson, J. M. 1979a. Diffuse noxious inhibitory controls (DNIC). I. Effects on dorsal horn convergent neurones in the rat. *Pain* 6:283–304

Le Bars, D., Dickenson, A. H., Besson, J. M. 1979b. Diffuse noxious inhibitory controls (DNIC). II. Lack of effect of nonconvergent neurones, supraspinal involvement and theoretical implications. *Pain* 6:305–27

Lewis, J. W., Cannon, J. T., Liebeskind, J. C. 1980. Opioid and nonopioid mechanisms of stress analgesia. *Science* 208:623–25

Lewis, T. 1942. *Pain.* New York: MacMillan. 192 pp.

Light, A. R., Perl, E. R. 1979a. Reexamination of the dorsal root projection to the spinal dorsal horn including observations on the differential termination of coarse and fine fibers. *J. Comp. Neurol.* 186:117–31

Light, A. R., Perl, E. R. 1979b. Spinal termination of functionally identified primary afferent neurons with slowly conducting myelinated fibers. *J. Comp. Neurol.* 186:133–50

Light, A. R., Trevino, D. L., Perl, E. R. 1979. Morphological features of functionally defined neurons in the marginal zone and substantia gelatinosa of the spinal dorsal horn. *J. Comp. Neurol.* 186:151–71

Lovick, T. A., West, D. C., Wolstencroft, J. H. 1978. Responses of raphé spinal and other bulbar raphé neurones to stimulation of the periaqueductal gray in the cat. *Neurosci. Lett.* 8:45–49

Lovick, T. A., Wolstencroft, J. H. 1979. Inhibitory effects of nucleus raphé magnus on neuronal responses in the spinal trigeminal nucleus to nociceptive compared with non-nociceptive inputs. *Pain* 7:135–45

Lynn, B. 1977. Cutaneous hyperalgesia. *Br. Med. Bull.* 33:103–8

Mayer, D. J. 1980. The centrifugal control of pain. In *Pain, Discomfort and Humanitarian Care,* ed. L. K. Y. Ng, J. J. Bonica, pp. 83–105. Amsterdam: Elsevier North Holland

Mayer, D. J., Price, D. D. 1976. Central nervous system mechanisms of analgesia. *Pain* 2:379–404

Mayer, D. J., Price, D. D., Becker, D. P. 1975. Neurophysiological characterization of the anterolateral spinal cord neurons contributing to pain perception in man. *Pain* 1:51–58

McGrath, P. A., Sharav, Y., Dubner, R., Gracely, R. H. 1981. Masseter inhibitory periods and sensations evoked by electrical tooth pulp stimulation. *Pain* 10:1–17

McLaughlin, B. J., Barber, R., Saito, K., Roberts, E., Wu, J. Y. 1975. Immunocytochemical localization of glutamate dicarboxylase in rat spinal cord. *J. Comp. Neurol.* 164:305–22

Melzack, R. 1975. Prolonged relief of pain by brief, intense transcutaneous somatic stimulation. *Pain* 1:357–73

Melzack, R., Wall, P. 1965. Pain mechanisms: A new theory. *Science* 150:971–79

Meyer, R. A., Campbell, J. N. 1981. Myelinated nociceptive afferents account for the hyperalgesia that follows a burn to the hand. *Science* 213:1527–29

Miletic, V., Randic, M. 1981. Neonatal rat spinal cord slice preparation: Postsynaptic effects of neuropeptides on dorsal horn neurons. *Dev. Brain Res.* 2:432–38

Milne, R. J., Foreman, R. D., Giesler, G. J. Jr., Willis, W. D. 1981. Convergence of cutaneous and pelvic visceral nociceptive inputs onto primate spinothalamic neurons. *Pain* 11:163–83

Molony, V., Steedman, W. M., Cervero, F., Iggo, A. 1981. Intracellular marking of identified neurones in the superficial dorsal horn of the cat spinal cord. *Q. J. Exp. Physiol.* 66:211–23

Morgan, C., Nadelhaft, I., DeGroat, W. C. 1981. The distribution of visceral primary afferents from the pelvic nerve to Lissauer's tract and the spinal gray matter and its relationship to the sacral parasympathetic nucleus. *J. Comp. Neurol.* 201:415–40

Nagy, J. I., Emson, P. C., Iverson, L. L. 1981b. A re-evaluation of the neurochemical and antinociceptive effects of intrathecal capsaicin in the rat. *Brain Res.* 211:497–502

Nagy, J. I., Hunt, S. P., Iverson, L. L., Emson, P. C. 1981a. Biochemical and anatomical observations on the degeneration of peptide-containing primary afferent neurons after neonatal capsaicin. *Neurosci.* 6:1923–34

Nagy, J. I., Vincent, S. R., Staines, W. A., Fibiger, H. C., Reisine, T. D., Yamamura, H. I. 1980. Neurotoxic action of capsaicin on spinal substance P neurons. *Brain Res.* 186:435–44

Noordenbos, W. 1959. *Pain.* Amsterdam: Elsevier. 182 pp.

Oliveras, J. L., Hosobuchi, Y., Redjemi, F., Guilbaud, G., Besson, J. M. 1977. Opiate antagonist, naloxone, strongly reduces analgesia induced by stimulation of a raphé nucleus (centralis inferior). *Brain Res.* 120:221–29

Olszewski, J. 1950. On the anatomical and functional organization of the trigeminal nucleus. *J. Comp. Neurol.* 92:401–13

Panneton, W. M., Burton, H. 1981. Corneal and periocular representation within the trigeminal sensory complex in the cat studied with transganglionic transport of horseradish peroxidase. *J. Comp. Neurol.* 199:327–44

Pelletier, G., Steinbusch, H. W. M., Verhofstad, A. A. J. 1981. Immunoreactive substance P and serotonin present in the same dense-core vesicles. *Nature* 293:71–72

Perl, E. R. 1968. Myelinated afferent fibres innervating the primate skin and their response to noxious stimuli. *J. Physiol. London* 197:593–615

Perl, E. R. 1971. Is pain a specific sensation? *J. Psychiatr. Res.* 8:273–87

Pickel, V. M., Reis, D. J., Leeman, S. E. 1977. Ultrastructural localization of substance P in neurons of rat spinal cord. *Brain Res.* 122:534–40

Pierce, E. T., Foote, W. E., Hobson, J. A. 1976. The efferent connection of the nucleus raphé dorsalis. *Brain Res.* 107:137–44

Price, D. D., Dubner, R. 1977. Neurons that subserve the sensory-discriminative aspects of pain. *Pain* 3:307–38

Price, D. D., Dubner, R., Hu, J. W. 1976. Trigeminothalamic neurons in nucleus caudalis responsive to tactile, thermal, and nociceptive stimulation of monkey's face. *J. Neurophysiol.* 39:936–53

Price, D. D., Hayashi, H., Dubner, R., Ruda, M. A. 1979. Functional relationships between neurons of marginal and substantia geletinosa layers of primate dorsal horn. *J. Neurophysiol.* 42:1590–608

Price, D. D., Hu, J. W., Dubner, R., Gracely, R. H. 1977. Peripheral suppression of first pain and central summation of second pain evoked by noxious heat pulses. *Pain* 3:57–68

Price, D. D., Mayer, D. J. 1975. Neurophysiological characterization of the anterolateral quadrant neurons subserving pain in *M. mulatta. Pain* 1:59–72

Proudfit, H. K. 1981. Time-course of alterations in morphine-induced analgesia and nociceptive threshold following medullary raphe lesions. *Neuroscience* 6:945–51

Proudfit, H. K., Hammond, D. L. 1981. Alterations in nociceptive threshold and morphine-induced analgesia produced by intrathecally administered amine antagonists. *Brain Res.* 218:393–99

Ralston, H. J., Ralston, D. D. 1979. The distribution of dorsal root axons in laminae I, II and III of the macaque spinal cord: A quantitative electron microscope study. *J. Comp. Neurol.* 184: 643–83

Randic, M., Miletic, V. 1977. Effect of substance P in cat dorsal horn neurones activated by noxious stimuli. *Brain Res.* 128:164–69

Randic, M., Yu, H. H. 1976. Effects of 5-hydroxytryptamine and bradykinin in cat dorsal horn neurones activated by noxious stimuli. *Brain Res.* 111:197–203

Rexed, B. 1952. The cytoarchitectonic organization of the spinal cord in the cat. *J. Comp. Neurol.* 96:415–96

Rivot, J. P., Chaouch, A., Besson, J. M. 1980. Nucleus raphe magnus modulation of response of rat dorsal horn neurons to unmyelinated fiber inputs: Partial involvement of serotonergic pathways. *J. Neurophysiol.* 44:1039–57

Ross, C. A., Armstrong, D. M., Ruggiero, D. A., Pickel, V. M., Joh, T. H., Reis, D. J. 1981. Adrenaline neurons in the rostral ventrolateral medulla innervate thoracic spinal cord: A combined immunocytochemical and retrograde transport demonstration. *Neurosci. Lett.* 25:257–62

Ruda, M. A. 1982. Opiates and pain pathways: Demonstration of enkephalinergic synapses on thalamic projection neurons in the dorsal horn. *Science* 215:1523–25

Ruda, M. A., Allen, B., Gobel, S. 1979. Ultrastructural characterization of noradrenergic axonal endings in layers I and II of the dorsal horn of the medulla. *Soc. Neurosci. Abstr.* 5:712

Ruda, M. A., Allen, B., Gobel, S. 1981. Ultrastructural analysis of medial brain stem afferents to the superficial dorsal horn. *Brain Res.* 205:175–80

Ruda, M. A., Gobel, S. 1980. Ultrastructural characterization of axonal endings in the substantia gelatinosa which take up [³H]serotonin. *Brain Res.* 184:57–83

Rustioni, A. 1973. Non-primary afferents to the nucleus gracilis from the lumbar cord of the cat. *Brain Res.* 51:81–95

Rustioni, A. 1974. Non-primary afferents to the cuneate nucleus in the brachial dorsal funiculus of the cat. *Brain Res.* 75:247–59

Rustioni, A., Hayes, N. L., O'Neill, S. 1979. Dorsal column nuclei and ascending spinal afferents in macaques. *Brain* 102:95–125

Sasa, M., Munekiyo, K., Ikeda, H., Takaori, S. 1974. Noradrenaline-mediated inhibition by locus coeruleus of spinal trigeminal neurons. *Brain Res.* 80: 443–60

Satoh, M., Akaike, A., Nakazawa, T., Takagi, H. 1980. Evidence for involvement of separate mechanisms in the production of analgesia by electrical stimulation of the nucleus reticularis paraganto-cellularis and nucleus raphe magnus in the rat. *Brain Res.* 194:525–29

Schoenen, J. 1980. *Organisation Neuronale de la Moelle Épinière de l'Homme.* Thesis. Univ. Liege, Liege. 353 pp.

Segal, M., Sandberg, D. 1977. Analgesia produced by electrical stimulation of catecholamine nuclei in the rat brain. *Brain Res.* 123:369–72

Sessle, B. J., Hu, J. W., Dubner, R., Lucier, G. E. 1981. Functional properties of neurons in cat trigeminal subnucleus caudalis (medullary dorsal horn). II. Modulation of responses to noxious and nonnoxious stimuli by periaqueductal gray, nucleus raphe magnus, cerebral cortex, and afferent influences, and effect of naloxone. *J. Neurophysiol.* 45:193–207

Sherman, J. E., Liebeskind, J. C. 1980. An endorphinergic, centrifugal substrate of pain modulation: Recent findings, current concepts, and complexities. In *Pain,* ed. J. J. Bonica, pp. 191–204. New York: Raven

Sumino, R., Dubner, R., Starkman, S. 1973. Responses of small myelinated "warm" fibers to noxious heat stimuli applied to the monkey's face. *Brain Res.* 62: 260–63

Szentagothai, J. 1964. Neuronal and synaptic arrangement in the substantia gelatinosa Rolandi. *J. Comp. Neurol.* 122:219–39

Takahashi, T., Otsuka, M. 1975. Regional distribution of substance P in the spinal cord and nerve roots of the cat and the effect of dorsal root section. *Brain Res.* 87:1–11

Tenen, S. S. 1968. Anatogonism of the analgesic effect of morphine and other drugs by *p*-chlorophenylalanine, a serotonin depletor. *Psychopharmacol. Berlin* 12:278–85

Thalhammer, J. G., LaMotte, R. H. 1982. Spatial properties of nociceptor sensitization following heat injury of the skin. *Brain Res.* 231:257–65

Torebjörk, H. E., Hallin, R. G. 1974. Identification of afferent C units in intact human skin nerves. *Brain Res.* 67:387–403

Torebjörk, E., Hallin, R. G. 1977. Sensitization of polymodal nociceptors with C fibres in man. *Proc. Intl. Union Physiol. Sci.* 13:758

Truex, R. C., Taylor, M. J., Smythe, M. Q., Gildenberg, P. L. 1970. The lateral cervical nucleus of cat, dog and man. *J. Comp. Neurol.* 139:93–104

Tyce, G. M., Yaksh, T. L. 1981. Monoamine release from cat spinal cord by somatic stimuli: An intrinsic modulatory system. *J. Physiol.* 314:513–29

Van Hees, J., Gybels, J. M. 1972. Pain related to single afferent C fibers from human skin. *Brain Res.* 48:397–400

Van Hees, J., Gybels, J. M. 1981. C nociceptor activity in human nerve during painful and nonpainful skin stimulation. *J. Neurol. Neurosurg. Psychiatry.* 44:600–7

Wall, P. D., Merrill, E. G., Yaksh, T. L. 1979. Responses of single units in laminae 2 and 3 of cat spinal cord. *Brain Res.* 160:245–60

Watkins, L. R., Mayer, D. J. 1982. The neural organization of endogenous opiate and non-opiate pain control systems. *Science* 216:1185–92

Westlund, K. N., Coulter, J. D. 1980. Descending projections of the locus coeruleus and subcoeruleus/medial parabrachial nuclei in monkey: Axonal transport studies and dopamine-β-Hydroxylase immunocytochemistry. *Brain Res. Rev.* 2:235–64

Willis, W. D., Coggeshall, R. E. 1978. *Sensory Mechanisms of the Spinal Cord.* New York: Plenum. 485 pp.

Willis, W. D., Kenshalo, D. R. Jr., Leonard, R. B. 1979. The cells of origin of the primate spinothalamic tract. *J. Comp. Neurol.* 188:543–74

Wolskee, P. J., Gracely, R. H. 1980. Effect of chronic pain on experimental pain response. *Am. Pain Soc. Abstr.*, p. 4

Yaksh, T. L. 1979. Direct evidence that spinal serotonin and noradrenaline terminals mediate the spinal antinociceptive effects of morphine in the periaqueductal gray. *Brain Res.* 160:180–85

Yaksh, T. L., Elde, R. P. 1981. Factors governing release of methione enkephalin like immunoreactivity from mesencephalon and spinal cord of the cat in vivo. *J. Neurophysiol.* 46:1056–75

Yaksh, T. L., Farb, D. H., Leeman, S. E., Jessell, T. M. 1979. Intrathecal capsaicin depletes substance P in the rat spinal cord and produces prolonged thermal analgesia. *Science* 206:481–83

Yaksh, T. L., Hammond, D. L., Tyce, G. M. 1981. Functional aspects of bulbospinal monoaminergic projections in modulating processing of somatosensory information. *Fed. Proc.* 40:2786–94

Yaksh, T. L., Jessell, T. M., Gamse, R., Mudge, A. W., Leeman, S. E. 1980. Intrathecal morphine inhibits substance P release from mammalian spinal cord in vivo. *Nature* 286:155–57

Yaksh, T. L., Rudy, T. A. 1978. Narcotic analgetics: CNS sites and mechanisms of action as revealed by intracerebral injection techniques. *Pain* 4:299–359

Yaksh, T. L., Tyce, G. M. 1979. Microinjection of morphine into the periaqueductal gray evokes the release of serotonin from spinal cord. *Brain Res.* 171:176–81

Yaksh, T. L., Wilson, P. R. 1979. Spinal serotonin terminal system mediates antinociception. *J. Pharmacol. Exp. Ther.* 208:446–53

Yezierski, R. P., Wilcox, T. K., Willis, W. D. 1982. The effects of serotonin antagonists on the inhibition of primate spinothalamic tract cells produced by stimulation in nucleus raphe magnus or periaqueductal gray. *J. Pharmacol. Exp. Ther.* 220:266–77

Zieglgänsberger, W., Tulloch, I. F. 1979a. Effects of substance P on neurones in the dorsal horn of the spinal cord of the cat. *Brain Res.* 166:273–82

Zieglgänsberger, W., Tulloch, I. F. 1979b. The effects of methionine and leucine-enkephalin on spinal neurones of the cat. *Brain Res.* 167:53–64

Zorman, G., Hentall, I. D., Adams, J. E., Fields, H. L. 1981. Naloxone-reversible analgesia produced by microstimulation in the rat medulla. *Brain Res.* 219:137–48

Zotterman, Y. 1933. Studies in the peripheral nervous mechanism of pain. *Acta Medica Scand.* 80:185–242

Ann. Rev. Neurosci. 1983. 6:419–46

ISOLATION AND RECONSTITUTION OF NEURONAL ION TRANSPORT PROTEINS

Stanley M. Goldin, Edward G. Moczydlowski, and Diane M. Papazian

Department of Pharmacology, Harvard Medical School, Boston, Massachusetts 02115

Introduction

The electrical activity of nerve cells is produced by the coordinated gating and pumping of ions across the neuronal membrane. It is axiomatic that regulation of neuronal electrical activity is due to regulation of these ion transport proteins. Examples of primary regulatory mechanisms are (*a*) the depolarization of the nerve cell membrane that induces the opening of the voltage-sensitive action potential Na^+ channel and (*b*) the conductance change directly induced by the binding of neurotransmitters to postsynaptic receptor sites. More indirect and in some cases simultaneous secondary mechanisms of regulation are postulated to occur, e.g. via phosphorylation of ion transport proteins resulting from a cyclic nucleotide-mediated series of events (reviewed by Kennedy 1983).

Identifying these neuronal ion transport proteins and reconstituting them in a purified, biologically active form can help answer a variety of questions. The mechanisms of gene replication (DePamphilis & Wassarman 1980) and muscle contraction (Adelstein & Eisenberg 1980) are studied by isolating the enzymes and other components involved and determining how they function and interact with each other in vitro. Studying the mechanism and regulation of purified, reconstituted ion transport proteins could provide similar information about the molecular basis of neuronal electrical activity. If one can then proceed immunocytochemically to localize specific classes of these transport proteins in the CNS (together with using information provided by neuroanatomical techniques), one could in principle obtain a functional map of their distribution in neuronal pathways that might help

419

0147-006X/83/0301-0419$02.00

to determine how these pathways are involved in information processing in the CNS. Electrophysiologists record the consequences of ion movements catalyzed by these transport systems; therefore, the immunocytochemical localization of specific classes of transport proteins may provide us with information on ionic movements across these cells that previously only electrophysiological techniques have supplied.

Direct intradendritic recording from hippocampal pyramidal cells (Wong et al 1979) indicates that some neuronal processes that appear dendritic in morphology seem capable of generating tetrodotoxin (TTX)-sensitive action potentials. This finding contradicts an earlier assumption that dendrites are merely passive in their response to electrical and chemical stimuli. Rather, such dendrites contain the "action potential Na^+ channel," the membrane protein responsible for the TTX-blocked gating of Na^+ that produces the electrophysiologically observed depolarizing spikes. By immunocytochemical localization of the Na^+ channel, possibly one can determine whether a given neuronal process contains the molecular machinery for the action potential, which is a prerequisite and probably a sufficient condition for its being electrically excitable. This may be particularly useful for studying dendrites that are too small for direct intradendritic recording.

The value of immunocytochemical mapping of electrophysiologically inaccessible postsynaptic ion transport systems is apparent. Attempts to obtain generalizations that would enable one to predict, based on synaptic morphology or pharmacology, whether a given synapse is excitatory or inhibitory, have been upset by exceptions. If, for example, we could identify and purify specific classes of neurotransmitter-activated postsynaptic conductance channels and learn what ions are conducted when the channel is opened, we could in principle predict what the postsynaptic response would be: e.g. if the channel were Cl^--specific, the postsynaptic response would be inhibitory; if it were Na^+-specific, an excitatory response would be expected.

This article focuses on the progress that has been made toward purification of neuronal membrane proteins and their reconstitution into artificial membranes. Many neuronal ion transport proteins have their counterparts in non-neuronal tissues, e.g. the sodium pump found in neurons is a primary membrane component in the kidney, another organ specialized for ion transport. Therefore, we occasionally refer to studies performed on ion transport proteins from non-neuronal sources when their results may elucidate the behavior and properties of similar or identical molecules in nerve. We emphasize recent methodological advances in purification and reconstitution of neuronal ion transport proteins and review recent findings regarding their molecular identity, mechanism, and regulation. Recent reviews have focused on more general aspects of membrane reconstitution (Miller 1982, Goldin 1982, Racker et al 1979); other reviews have dealt in more depth with the molecular properties of a specific membrane protein, e.g. the NaK-ATPase (Sweadner & Goldin 1980, Cantley 1981), the nicotinic acetylcholine receptor (Karlin 1980, Heidmann & Changeux 1978), and the calcium pump of sarcoplasmic reticulum (Ikemoto 1982).

Active Transport Catalyzed by the NaK-ATPase

The gradients of Na^+ and K^+ across nerve cell membranes provide the energy for the TTX-sensitive action potential. The sodium potassium ATPase (NaK-ATPase), the enzyme responsible for driving this active transport, was discovered in crab nerves by Skou (1957). An identical enzymatic activity was characterized in the erythrocyte membrane, and shown to pump 3 Na^+ molecules out of the cell and 2 K^+ into the cell for every molecule of ATP hydrolyzed (Post & Jolly 1957, Garrahan & Glynn 1967). Elegant experiments by Thomas (1969) confirmed these findings in nerve cells and demonstrated that the pump is "electrogenic," i.e. capable of generating a net current across the membrane due to the 3/2 Na^+/K^+ countertransport stoichiometry. Membranes from the thick ascending limb of the loop of Henle of the kidney have the NaK-ATPase as their preponderant component; hence, NaK-ATPase was first purified to homogeneity from kidney microsomes (Kyte 1971). The purification procedure simply involved obtaining this membrane fraction and extracting away peripheral membrane proteins by mild detergent or salt treatment.

Salient structural properties of the NaK-ATPase have been reviewed elsewhere (Cantley 1981). As purified from the kidney and subsequently from other kinds of tissue, the sodium-potassium ATPase has been found to contain two subunits: the larger subunit, α, has a mass of about 120,000 daltons; β is a glycoprotein of approximately 50,000 daltons. A body of biochemical evidence, including chemical cross-linking studies, end group analysis, and ultracentrifugation studies, shows that the most likely structure of this molecule is $\alpha_2\beta_2$, and indicates a molecular weight of 2.8–3.5 $\times 10^5$. A central issue has been whether this purified protein contains the apparatus for active ion transport or whether it is only the energy source for this process. To answer this question, the NaK-ATPase was inserted into a model membrane system in which ion transport could be observed and quantitated. The approach used was similar to that first employed (Kagawa & Racker 1971) to study mitochondrial proton and electron transport. The purified NaK-ATPase was added to a solution of phospholipid that had been emulsified by the bile salt detergent, cholate. Removal of the cholate by means of dialysis resulted in vesicles that contained the enzyme embedded in the lipid membrane. Addition of MgATP activated that fraction of the protein oriented "inside-out" with respect to its normal in vivo orientation, and generated the active uptake of $^{22}Na^+$ by the vesicles (Goldin & Tong 1974). Internally incorporated, but not externally added, ouabain inhibited this process—a finding consistent with physiological studies on red cell and nerve that indicate that the enzyme in vivo has its ouabain binding site on the external surface and the ATP-hydrolyzing site on the interior surface of the cell membrane. This result provided evidence that the purified protein contained the active transport machinery. These findings were subsequently confirmed with preparations of sodium-potassium ATPase from several different tissues, including brain. Studies with NaK-ATPase from brain (Sweadner & Goldin 1975) and shark rectal gland (Hilden & Hokin 1975) demonstrated that the active countertransport of K^+ in the $3/2$ Na^+/K^+ ratio was consistent with active transport observed in vivo.

These initial studies were ambiguous in the following respects:

1. Only a small fraction of the original ATPase activity remained after the reconstitution process.
2. Only a fraction of this fraction actually appeared to be involved in active ion transport.

In order to ascertain rigorously whether the α and β subunits of the sodium-potassium ATPase comprise the active transport system, a more highly efficient reconstituted system was developed and characterized (Goldin 1977). The results that emerged from these studies are schematically illustrated in Figure 1. Characterization of these reconstituted vesicles as formed by rapid hollow fiber dialysis showed that they are primarily unilamellar and relatively homogeneous in size (mean diameter ~ 500 Å); on the average, only one, or at most a few, protein molecules are present in each vesicle. [These findings, based on biophysical and biochemical studies, were confirmed by electron microscopy (Skriver et al 1980).] The ratio of Na^+/K^+ pumped/ATP hydrolyzed was approximately $3/2/1$, in quantitative agreement with physiological studies. About half the enzyme molecules are "inside out" and accessible to external ATP. The rest of the enzyme is "right-side out" and inaccessible to ATP unless the vesicles are disrupted. No single protein other than the α and β subunits of the purified enzyme comprises more than 2% of the protein in the preparation. This demonstrates that active transport can be carried out in the absence of any protein of molecular weight $\geqslant 12,000$: there is simply not enough of any such protein available to be present in even a single copy for each active ion-transporting vesicle.

The conclusion that the α and β subunits of the sodium-potassium ATPase contain the ion pumping machinery is qualified in just one respect: small amounts of protein ("proteolipid") of very low molecular weight present in the preparation could conceivably play some role in ion transport

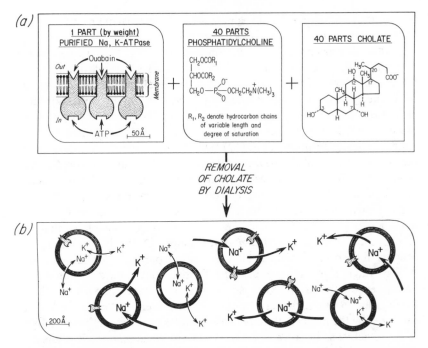

Figure 1 Schematic representation of the process of incorporation of the purified NaK-ATPase into artificial vesicles. In (*a*), the purified microsomal enzyme is solubilized with cholate in the presence of a 40-fold excess of detergent. After sedimenting unsolubilized material and removing the cholate by hollow fiber dialysis, small unilammelar vesicles result (*b*), containing only one or at most a few NaK-ATPase molecules. See text for further details.

or enzymatic function. Forbush et al (1978) synthesized a photoactivatable derivative of ouabain that attaches itself to this low molecular weight protein when activated by absorbing ultraviolet light. This low molecular weight protein is labeled to the same extent as the α subunits, suggesting that the proteolipid may be part of the ouabain binding site of the enzyme.

The small vesicles that result from the procedures used for reconstitution of the NaK-ATPase are of course unsuitable for direct electrophysiological measurements, and so the question of the electrogenicity of the reconstituted pump must be approached by other means. Circumstantial evidence for this electrogenicity is the net inward movement of cations developed by the protein: $3Na^+$ in/$2K^+$ out. More direct evidence for electrogenicity has been obtained by Dixon & Hokin (1980). They employed a lipid permeant anion, SCN^-, that rapidly equilibrates across the vesicle membrane and distributes according to the Gibbs-Donnan potential. When the Na^+ and K^+ gradients were collapsed by using the Na^+/K^+ exchanger, nigericin, an ATP-dependent, ouabain-sensitive accumulation of $[^{14}C]SCN^-$ still occurred, thus indicating the development of an electrogenic membrane potential (estimated to be 9 mV) as distinct from a diffusion potential.

Another approach that could, in principle, be used to measure membrane potential in lipid vesicles employs fluorescent dyes whose partitioning into the lipid bilayer is a function of membrane potential: the fluorescent properties of the dye are a function of whether the dye is in a hydrophilic (aqueous) or hydrophobic (bilayer) environment, enabling the partition coefficient of the dye, and hence the membrane potential, to be determined spectrophotometrically. Hoffman et al (1979) used this technique to demonstrate the electrogenicity of the sodium-potassium ATPase in erythrocytes. The advantage of the latter approach is that membrane potential can be determined without disrupting ion gradients or membrane permeability.

An ouabain-sensitive sodium-ATPase activity has been observed in vivo in the absence of K^+, and Neufeld & Levy (1970) have postulated that this might be a distinct enzymatic activity. Because K^+ and Na^+ pumping are coupled in the operation of the standard sodium pump, an ouabain-sensitive ATPase might signal the presence of a distinct, nonstandard sodium pumping system in the brain. Reconstitution studies have shown that (fortunately) such a "non-standard" pump is not present: a "partial reaction" catalyzed by the purified NaK-ATPase explains how concomitant ouabain-sensitive active sodium transport can occur in the absence of K^+, albeit at a lower Na^+/ATP coupling ratio: 0.5/1 vs $\sim$ 3/1 for pumping in the presence of K^+ (Forgac & Chin 1981). Further studies are needed to determine the possible significance of this activity in vivo.

The NaK-ATPase appears to play roles in the brain in addition to maintaining the nerve impulse. The pump, for as yet undefined reasons, is present in unusually high concentrations in glial cells as well as in neurons. Sweadner (1979) demonstrated two biochemically different sodium-potassium ATPases active in the brain. They differ in their electrophoretic mobility and in their affinity for cardiac glycosides. One form of the enzyme is characteristic of cultured astrocytes, and the other of axonal membranes of mammalian brain. Purification and reconstitution of each of these two forms of the active transport system may help in detecting other differences between them, and in determining whether either is regulated by endogenous inhibitors of this activity in brain, by hormones, or by neurotransmitters, and how (see Sweadner & Goldin 1980 for review). This task, together with understanding the detailed molecular mechanism of the NaK-ATPase, is among the immediate challenges for future research in this area.

Ion Gating Mediated by the Nicotinic Acetylcholine Receptor

The postsynaptic receptor site for acetylcholine (ACh) is the link between the release of acetylcholine at nerve terminals and the postsynaptic conductance change that occurs at cholinergic junctions in nerve and muscle. We focus here on the nicotinic postsynaptic response to ACh (reviewed by Karlin 1980, Heidmann & Changeux 1978), which in a variety of tissues has been shown to be a rapid and relatively nonselective increase in cation permeability. The nicotinic ACh receptor is especially favorable for biochemical studies of chemically excitable ion gating:

1. The nicotinic ACh receptor (AChR) has been studied extensively in relatively simple and electrophysiologically accessible systems (such as the neuromuscular junction).
2. It has a well-defined pharmacology that includes a highly specific inhibitor (α-bungarotoxin).
3. Of salient practical importance, it is a major membrane component of the electroplax organ of marine invertebrates such as *Torpedo marmorata*—an organ specialized for storage and release of lethal quantities of electrical energy.

The AChR has been solubilized and purified in substantial quantities by protein chemical methods that rely heavily on affinity chromatography or affinity partitioning using AChR-selective ligands. The purified AChR is now agreed to contain four proteins whose apparent molecular weights derived from their mobility on SDS-polyacrylamide gels are $\approx$ 40,000, 50,000, 60,000, and 65,000. The 40,000 M_r peptide is associated with the ACh binding site, as shown by its selective affinity labeling by a site-directed ligand (Weill et al 1974). Preparations from other species of electric fish as well as from skeletal muscle (Froehner et al 1977) contain four similar proteins. Of course, central questions are (*a*) whether the four polypeptides of the purified AChR ("receptor peptides") contain the ion gating as well as the neurotransmitter binding function and (*b*) to what extent does that gating function in reconstituted preparations resemble the ion channel regulated by the AChR in vivo. Early attempts to reconstitute the purified receptor suffered from ambiguous and irreproducible results—perhaps one of the reasons several laboratories hypothesized that the ion gating function resided in another membrane component. In parallel with purification of detergent-solubilized AChR, methods were developed to fractionate, by means of density gradient centrifugation (Sobel et al 1977), AChR-rich electroplax membranes without the use of detergents. Comparison of the protein components of such *Torpedo* electroplax membranes with the purified AChR showed that the most apparent difference was the presence of another major protein component slightly higher in molecular weight (43,-000) than the 40,000 M_r protein of the detergent-solubilized AChR. Kinetic evidence had suggested that the local anesthetics and the inhibitory toxin histrionicotoxin (HTX) bind to a separate site on AChR-rich membranes, and direct studies with [^{3}H]-HTX confirmed the existence of a binding site distinct from the one used by agonists. Spectroscopic studies initially suggested that this site was on the 43,000 M_r protein isolated from electroplax membranes (Sobel et al 1978). These findings reinforced the hypothesis—later discredited—that the ion channel constitutes a component separate from the AChR, and helped motivate investigators to return to the native electroplax membrane as the starting material for attempts to reconstitute the AChR-regulated ion channel.

Epstein & Racker (1978) provided the first convincing reconstitution of this ion channel. AChR-rich membranes were extracted with cholate in the presence of soybean phospholipids. Vesicles were formed by the cholate dialysis method, a technique similar to that used to reconstitute the NaK-ATPase. After reconstitution, ^{22}Na$^+$ uptake into vesicles—stimulated by carbamylcholine and inhibited by α-bungarotoxin, curare, and local anaesthetics—was observed. A phenomenon of receptor "desensitization" analogous to that observed electrophysiologically in vivo (Del Castillo & Webb 1977) was also noted: when carbamylcholine was added shortly before, rather than after, addition of ^{22}Na$^+$, stimulation of ^{22}Na$^+$ uptake was lost. This study defined an approach that other investigators expanded, as described below, to characterize more quantitatively the kinetics of this response and to determine whether the receptor peptides contained the ion gating function.

It is necessary here to emphasize a fundamental problem inherent in studying ion channels reconstituted into small vesicles. Under physiological conditions, single channel noise analysis of the ACh-induced opening of the ionic channel in nerve and muscle (reviewed by Neher & Stevens 1977) has revealed a flux of ~ 10^7 ions/sec per channel. A single unilamellar liposome of 500 Å containing 150 mM NaCl has entrapped only several thousand sodium ions. Thus, just a single channel in a reconstituted vesicle need only be open for a few milliseconds to produce nearly complete equilibration of external ^{22}Na$^+$ with the vesicle interior. Because the ion flux assay employed by Epstein & Racker has a resolution of only 10 sec (the first time point taken, by which time complete ^{22}Na$^+$ equilibration has occurred), one can only conclude from their data that the ACh-induced ion flux is within a factor of 10^4 of the physiologically observed levels. Unless ion fluxes can somehow be measured on a much more rapid time scale, it is impossible to compare quantitatively many physiological properties of the reconstituted ion channel (e.g. flux rate, ion selectivity, kinetics of opening and closing) with those observed in vivo.

Techniques have been developed for more rapid measurement of ion flux rates in vesicles formed from AChR-rich membrane fragments. The "quenched flow" method (Hess et al 1979, Neubig & Cohen 1980) involves rapid mixing of membrane vesicles with agonists to initiate radiolabeled ion uptake, and rapid quenching of the reaction using a high concentration of a suitable antagonist such as d-tubocurarine. Vesicles are subsequently retained on membrane filters to separate intravesicular radioisotope from the external solution. Because devices with a time resolution of several milliseconds have been constructed, this quenched flow method should be applicable to flux measurements of reconstituted ion channels.

An ingenious alternative scheme, developed by Moore & Raftery (1980), has been used to study ion gating mediated by the reconstituted AChR on a millisecond time scale (Wu et al 1981). The method relies on the quenching of the fluorescence of 8-amino-1,3,6-naphthalenetrisulfonate (ANTS) by thallium ion (Tl^+); Tl^+ passes through the ion channel as readily as K^+. ANTS is trapped inside the vesicles, and the fluorescence quenching is measured after the reaction is initiated by addition by rapid mixing of Tl^+ and a cholinergic agonist. This technique established that the Tl^+ flux through microsacs and the reconstituted ACh receptor/ion gate are within an order of magnitude of that determined in vivo for K^+ and Na^+. Desensitization and inhibition by HTX as well as inhibition by direct cholinergic antagonists was observed.

A key observation that eventually led to the localization of the ion channel within the receptor peptide complex was made by Neubig et al (1979). They found that the 43,000 M_r protein could be extracted from AChR-rich membranes by exposure to an alkaline environment (pH 11), conditions that have been used previously to extract peripheral proteins from other tissues. Both $^{22}Na^+$ flux and local anaesthetic binding to the membranes were preserved, demonstrating that the 43,000 M_r protein did not contain the local anaesthetic binding site. Because the $^{22}Na^+$ flux data were not measured on a rapid time scale, however, the argument might still have been made that the 43,000 M_r protein was necessary for ion gating, with enough ($\sim 10\%$) remaining behind to account for the completion of the rapid phase of $^{22}Na^+$ flux within the first (10 sec) time point taken.

Wu & Raftery (1981) convincingly used reconstitution techniques to rule out this possibility. They used alkaline-extracted membranes to reconstitute AChR-activated ion flux into 500 Å unilammellar vesicles, as Epstein & Racker did with unfractionated material. Because the physical characteristics of the reconstituted system were similar to those described above for reconstituted, purified NaK-ATPase (Goldin 1977) (only a few AChR molecules per vesicle with about half the vesicles exhibiting AChR-activated ion flux), they were able to conclude that not enough copies of the 43,000 M_r protein or other minor contaminants were present in each of the vesicles exhibiting the flux response. This observation ruled out proteins other than the four receptor peptides as being responsible for ion gating.

Strategy for Purification of Dilute Membrane Components

The use of reconstitution, together with other physical and biochemical techniques, has identified two of the proteins responsible for the kind of ion pumping and ion gating movements that impart electrical excitability to neurons and to related excitable cells. The NaK-ATPase/ion pump and the AChR/ion gate are the first of these proteins whose transport function in purified, reconstituted form has been sufficiently well characterized to establish their molecular identity. More recently, progress has been made toward identifying the molecular components of the action potential Na^+ channel and Ca^{2+} transport proteins in nerve terminals. These transport proteins are more difficult to study because they are less abundant. Purification of such relatively dilute membrane components requires more involved procedures than those described above for purification of the NaK-ATPase- and AChR-regulated ion transport activities.

The purification approaches that have been developed fall into two general categories:

SOLUBILIZATION OF THE TRANSPORT ACTIVITY OF INTEREST WITH AN APPROPRIATE NONDENATURING DETERGENT FOLLOWED BY THE USE OF STANDARD PROTEIN SEPARATION TECHNIQUES FOR FINAL PURIFICATION   If the transport protein has an associated enzymatic or ligand binding activity (such as STX binding to the Na^+ channel), this may be used as an assay for purification. However, many transport activities (e.g. the Na^+/Ca^{2+} exchanger) have neither an enzymatic correlate, nor specifically interact with a high affinity ligand. In such cases, reconstitution of the transport activity may be used as an assay for purification, as first demonstrated by Kasahara & Hinkle (1977), who used this strategy to purify the sugar transporter of human erythrocytes. Because the retention of enzymatic or binding activity does not necessarily imply retention of transport activity, the use of reconstitution to monitor purification is always desirable and sometimes practical (e.g. Newman et al 1981, Carter-Su et al 1980).

USE OF RECONSTITUTION AS A PHYSICAL TOOL FOR PURIFICATION   This approach, termed "transport specific fractionation" (Goldin & Rhoden 1978), is analogous to affinity chromatography in that it employs the specific biological property of a protein (binding of a ligand to a receptor site on the protein) as a separation technique; however, transport specific fractionation employs the transport properties of the protein one wishes to purify instead of ligand binding. The transport system of interest is reconstituted into artificial vesicles before purification. The crude membrane fraction, rather than a purified protein, is then subjected to the cholate dialysis technique similar to that depicted in Figure 1: membrane proteins are distributed randomly among vesicles so there is only one or a few membrane proteins per vesicle. Because the transport protein of interest constitutes only a minor fraction of the total population of membrane proteins incorporated into the vesicles, it will be present in only a small fraction of the vesicles. Further, a vesicle containing a transport protein usually will have only a single copy and will not contain other membrane proteins. The specific transport properties of the reconstituted protein are then used to change a physical property of only those vesicles that contain that protein molecule (e.g. changes in vesicle density are created). Vesicles containing this transport system can then be separated (e.g. by density gradient ultracentrifugation) from the rest of the vesicles, thus resulting in their purification. This strategy was used for the reconstitution and isolation of Ca^{2+}-transporting ATPases from membrane fractions derived from lysed synaptosomes (Papazian et al 1979), discussed in more detail below.

The Action Potential Na^+ Channel

The axonal action potential consists of an inward movement of Na^+ followed by an outward movement of K^+. The two events are pharmacologically distinct, and are postulated to be mediated by two separate membrane channels. Several classes of neurotoxins have been shown to interact reversibly with the action potential Na^+ channel with great specificity (reviewed by Catterall 1980): TTX and saxitoxin (STX) specifically block the channel at low concentrations ($K_i = 1-10$ nM), whereas batrachotoxin and veratridine (VTN) are members of a class of neurotoxins that chemically open the channel in the absence of a depolarizing voltage stimulus. In unmyelinated nerve, studies of [^{3}H]-TTX and [^{3}H]-STX binding indicate that the density of the Na^+ channel is low: 35–500 sites/μm^2. Noise analysis (Conti et al 1976) and gating current measurements (Nonner et al 1975) indicate that the Na^+ channel density at the Node of Ranvier of myelinated axons is much higher—several thousand sites/μm^2. Although this high density would make the Na^+ channel a major constituent of nodal membranes, the Node itself constitutes a very small fraction of the area of a myelinated axon (Hess & Young 1952); this small nodal area largely accounts for the present lack of a procedure for obtaining substantially purified preparations of nodal membranes. Membrane fractions enriched in the STX binding activity (and presumably the Na^+ channel) have been prepared from myelinated nerves (Rhoden & Goldin 1979) and from synaptosomes (pinched off nerve endings) of mammalian brain (Catterall et al 1979), although the Na^+ channel is still a very minor component ($<1\%$) of these membrane fractions. The axonal membrane (axolemma) of mammalian brain has close to 100-fold higher concentration of NaK-ATPase than action potential Na^+

channel (Rhoden & Goldin 1979 and unpublished data). If this seems counterintuitive, remember that a single Na^+ channel, when open, gates ions at a rate of 10^7/sec, whereas a single Na^+ pump actively transports ions at a maximum of only 1/30,000 that rate (Cantley 1981), a molecular-level reflection of the truth that it is much harder to push something uphill than to let it roll down.

PURIFICATION OF THE TTX/STX BINDING COMPONENT   Hille (1975) hypothesized that a positively charged guanidinium group on TTX or STX interacts with a carboxyl group that is part of the "ion selectivity filter" of the channel. Evidence for and against this hypothesis has been recently reviewed (Catterall 1980). Saturable, specific binding of tritiated STX and TTX is readily apparent in toxin-sensitive excitable tissues and correlates well with estimates of toxin dissociation constant, toxin binding kinetics, and channel density derived from electrophysiological studies. For these reasons, much effort has been focused on purification of the protein containing the TTX/STX binding site in the hope that the toxin binding protein will contain the ion gating machinery.

The first reports of detergent solubilization of the binding protein appeared in the early 1970s (e.g. Henderson & Wang 1972). A hiatus followed, during which little progress was made in purification, due to the lability of the binding site. Agnew et al (1978) discovered that inclusion of exogenous phosphatidyl choline or eel lipids in the detergent-buffer solutions, used to maintain the eel electroplax STX binding protein in the solubilized state during purification, dramatically stabilized the binding activity. Very substantial (several hundred-fold) purification of the binding protein solubilized from eel electroplax by a non-ionic detergent was obtained, using ion exchange chromatography and gel filtration. The major component that was enriched during this purification procedure, as observed on SDS gels, was a 230,000 M_r protein (Agnew et al 1980). Two other laboratories subsequently employed similar strategies for purification of the binding protein from muscle (Barchi et al 1980) and from brain (Hartshorne & Catterall 1981). Two major components of M_r 270,000 and 32,000 copurified with the brain STX binding protein. These components resemble in molecular weight the two components that are labeled in synaptosomes and neuroblastoma cells by a photoactivatable derivative of a toxin from the venom of the scorpion, *Leiurus quinquestriatus*, that had previously been shown to bind to a distinct site associated with Na^+ channels (Beneski & Catterall 1980). Binding of the scorpion toxin (ScTx) to this site in vivo potentiates the effects of Na^+ channel activators of the VTN/batrachotoxin class. In contrast, the purified muscle STX binding component was initially reported to consist of polypeptides in the 60,000 M_r range. A question under study is whether this discrepancy is due to intrinsic differences in the binding components or to differences that result during solubilization and purification. Recent results (R. Barchi, personal communication) indicate that the latter is the case, and the true subunit composition of the muscle STX binding protein is rather similar to that reported by Hartshorne & Catterall.

ION GATING BY THE Na^+ CHANNEL IN VITRO   Villegas & Villegas (1981) and Villegas et al (1977) demonstrated the incorporation of membrane fragments from lobster nerve into artificial vesicles with retention of VTN-activated, TTX-inhibitable $^{22}Na^+$ flux, a functional concomitant of the Na^+ gate that has been observed in neuroblastoma cells (Catterall 1975) and in neuronal membrane fragments (Barnola & Villegas 1976). These results showed that exogenous lipid could substitute for the native membrane environment with retention of the pharmacological correlates of ion gating. The reconstitution technique employed was the "freeze-thaw" method developed by Kasahara & Hinkle (1977) for reconstitution of sugar transport from human erythrocytes. The technique simply involves mixing the membranes containing the protein of interest with small liposomes formed by sonication of soy lipids. Under appropriate conditions, freezing and thawing the mixture causes the liposomes and membranes to fuse, forming large membranous structures. Vesicles (larger and more dimensionally heterogeneous than cholate dialysis vesicles) are then reformed by brief sonication.

The first report of reconstitution of ion gating function by the detergent-solubilized Na^+ channel (Goldin et al 1980) employed the STX binding protein from brain synaptic membranes, solubilized by cholate in the presence of egg phosphatidylcholine. Based on gel filtration and gradient sedimentation data, the solubilized binding protein was found to have a maximum molecular weight of 560,000. (This estimate included an undetermined amount of cholate bound to the proteins.) The solubilized binding protein was then incorporated into phosphatidylcholine vesicles using the cholate/hollow fiber dialysis procedure described above. Less than 1% of the vesicles resulting contained an STX binding protein. Based on the expectation that the reconsitituted STX binding protein contained the ion gating apparatus, a strategy was devised for purifying the Na^+ channel by transport specific fractionation.

The slight permeability of the Na^+ channel to cesium ion (Cs^+) was exploited for transport specific fractionation. It has been determined in squid axon (Chandler & Meves 1965) that Cs^+ can pass through the Na^+ channel at only 1.6% of the rate of Na^+ ion; activation of the Na^+ channel by batrachotoxin reduces this selectivity to 5–10% (Khodorov 1978). Even if Cs^+ ion were 1000-fold less permeable than Na^+, opening of a single Na^+ channel with VTN would theoretically allow nearly complete equilibration of Cs^+ in a 500 Å vesicle in a matter of seconds. This then is in accord with the observation that, when the entire reconstituted vesicle population was preloaded with 0.4 M Cs^+, the addition of VTN allowed Cs $^+$ efflux from specifically those vesicles containing the STX binding protein/Na^+ channel; this efflux was inhibited by TTX. The resulting reduction in density of specifically those vesicles containing the Na^+ channel permitted their partial (30–50-fold) purification on density gradients.

Tamkun & Catterall (1981) subsequently employed the same reconstitution method to demonstrate recovery of ScTx binding upon reconstitution. Catterall et al (1979) had previously demonstrated that binding of ScTx to brain synaptosomes was highly dependent on a hyperpolarizing membrane potential; detergent solubilization of synaptosomes entirely abolishes ScTx binding. Upon reconstitution, most of the voltage-dependent ScTx binding sites were recovered. Catterall et al ingeniously used this voltage dependence of ScTx binding to provide additional evidence for ion transport by the reconstituted Na^+ channel. Exposure of vesicles containing the Na^+ channel to veratridine would make them permeable to Na^+ and hyperpolarize the membrane in the presence of an outward Na^+ gradient, thus enhancing ScTx binding. This enhanced ScTx binding was observed and was blocked by blockage of the Na^+ channel with STX; this provided evidence, albeit indirect, for veratridine-activated Na^+ transport by the reconstituted Na^+ gate.

A promising report of reconstitution of batrachotoxin and VTN-stimulated $^{22}Na^+$ transport by purified muscle STX binding protein has recently appeared (Weigle & Barchi 1982). The protein was again reconstituted into small egg phosphatidylcholine vesicles. A two-fold enhancement of $^{22}Na^+$ influx was observed by addition of batrachotoxin; this was abolished by the addition of STX.

Ca^{2+} Transport Proteins in Nerve Terminals

There is strong evidence that Ca^{2+} entry into nerve terminals, caused by the opening of voltage dependent Ca^{2+} channels in the presynaptic plasma membrane, is a primary event that initiates release of neurotransmitter (reviewed by Kelly et al 1979). The precise molecular mechanism of neurotransmitter release is not well understood. Reduction in transmitter release parallels the termination of Ca^{2+} entry into nerve terminals, and is apparently a consequence of Ca^{2+} removal from the site of neurotransmitter exocytosis (Kelly et al 1979). Several mechanisms may act in parallel to remove Ca^{2+} from the terminal. Ca^{2+} can be accumulated by mitochondria. This uptake is probably driven by the proton electrochemical gradient produced by electron transport rather than by the direct expenditure of ATP (reviewed by Bygrave 1977). Several groups (Blaustein et al 1978a,b, Rahamimoff & Abramovitz 1978a,b, Blitz et al 1977) have reported the existence of a nonmitochondrial, ATP-dependent Ca^{2+} transport system derived from lysates of synaptosomes. This active Ca^{2+} pump may sequester Ca^{2+} in organelles within the nerve terminal, as occurs in muscle sarcoplasmic reticulum. Various organelles such as smooth endoplasmic reticulum (Blaustein et al 1978a,b, McGraw et al 1980), coated vesicles (Blitz et al 1977), and synaptic vesicles (Michaelson et al 1980, Israël et al 1980) have been suggested as possible sites for ATP-dependent Ca^{2+} sequestration.

Ca^{2+} may be actively extruded across the plasma membrane of nerve terminals by a $(Ca^{2+} + Mg^{2+})$ ATPase (Gill et al 1981, Javors et al 1981, Sorensen & Mahler 1981, Robinson 1981) as well as by Na^+/Ca^{2+} exchange (Rahamimoff & Spanier 1979, Gill et al 1981, Michaelis & Michaelis 1981). The latter mechanism would use the Na^+ gradient developed by the NaK-ATPase as the energy source for active Ca^{2+} extrusion. This is a secondary, coupled active transport system analogous to the Na^+-dependent neurotransmitter uptake systems of a variety of neurons. Secondary, coupled transport systems have been identified in a variety of non-neuronal tissues (reviewed by Schultz 1980). Among the most thoroughly understood transport systems of this type is the coupled Na^+/glucose cotransport activity of intestinal epithelium.

The mechanisms described above for regulation of Ca^{2+} levels have been identified in in vitro studies using synaptosomes or membrane fractions derived from them. The efficiency of Ca^{2+} transport in such membrane preparations, however, is often poor. The use of membrane fractions to study neuronal Ca^{2+} transport is further limited by the heterogeneity of brain membranes and the problem of cross-contamination among subcellular fractions. It has been estimated that only 50% of the material comprising synaptosome preparations may be derived from nerve terminals (Morgan 1976). In general, the assignment of any of the Ca^{2+} transport components to a particular location within nerve terminals, based solely on the criterion of their distribution in heterogeneous and contaminated membrane fractions, must remain preliminary until less circumstantial evidence is available. The purification, reconstitution, and immunocytochemical localization of the Ca^{2+}-transporting ATPases as well as the other Ca^{2+} transport components could advance our understanding of the control of the concentration of Ca^{2+} in nerve terminals. The relation of the nervous system transport proteins to similar proteins in other excitable and nonexcitable tissues can be determined by comparison of their kinetic and molecular properties. Immunological techniques can determine whether they are found only in neurons or are common to other cell types. Knowledge of the subcellular localization of Ca^{2+} transport components and their distribution between neuronal and glial cells would have important implications for their possible physiological roles.

The task of purification, reconstitution, and localization of these transport proteins has just begun. No subcellular neuronal membrane fraction has yet been isolated that has any of these Ca^{2+} transport proteins as its preponderant component; this necessitates using the alternative purification approaches for dilute membrane components outlined above. Transport specific fractionation has been employed to purify proteins responsible for an ATP-dependent Ca^{2+} uptake activity present in "synaptosomal vesicles" —a fraction prepared by high speed centrifugation of synaptosomal lysates after removal of a synaptic plasma membrane-enriched fraction by milder centrifugation. The reconstitution and purification procedure (Papazian et al 1979) is briefly described as follows. Synaptosomal vesicles are solubilized in cholate and a mixture of phospholipids extracted from soybeans. When the cholate is subsequently removed by hollow fiber dialysis, small, reconstituted vesicles form. In this case, the vesicles have a high concentration of oxalate, a Ca^{2+} chelating agent, trapped inside; ATP-dependent Ca^{2+} transport into only those vesicles containing the transport proteins results in the intravesicular formation of Ca^{2+} oxalate, a dense complex of low solubility. In this manner, vesicles containing the transport component are made more dense than those lacking it. Fractionation of the vesicles on a density gradient results in the purification of two proteins of M_r 94,000 and 140,000 (Papazian et al 1979).

Further investigation led to the tentative conclusion that membrane fractions derived from lysed synaptosomes contain two distinct ATP-dependent Ca^{2+} transport activities, only one of which is sensitive to the regulatory protein, calmodulin (Papazian et al 1983; calmodulin has been reviewed, Klee et al 1980). A more traditional biochemical approach, affinity chromatography, has been used to purify the calmodulin-sensitive enzyme. Calmodulin affinity chromatography has been previously used to purify several calmodulin-sensitive enzymes, including cyclic nucleotide phosphodiesterase (Sharma et al 1980), myosin light chain kinase (Adelstein et al 1978), and the $(Ca^{2+} + Mg^{2+})$ ATPase of erythrocytes (Niggli et al 1979). Methods developed by Niggli et al (1979) for the purification of the erythrocyte Ca^{2+} pump have been adapted for purification of the brain calmodulin-sensitive $(Ca^{2+} + Mg^{2+})$ ATPase (Papazian et al 1983). The major protein in the purified preparation, M_r 140,000, can be phosphorylated by γ ^{32}P-ATP in a Ca^{2+}-dependent manner. The phosphorylated protein is dephosphorylated by hydroxylamine, a characteristic of acyl phosphates (Hokin et al 1965). Thus, the 140,000 dalton protein has the properties expected of a Ca^{2+}-transporting ATPase. After reconstitution, the purified enzyme efficiently catalyzes ATP-dependent Ca^{2+} transport.

Comparison of this enzyme to the proteins previously purified by transport specific fractionation shows that an ATP-dependent Ca^{2+} transport protein of M_r 140,000 is purified in both cases. Thus, this protein is an independent Ca^{2+} transporting ATPase; it presumably shares a similar function with the protein of M_r 94,000, but it can be distinguished and independently purified by its calmodulin sensitivity. These results indicate that transport specific fractionation has apparently copurified from synaptosomal vesicles two separate ATP-dependent Ca^{2+} transport systems. This demonstrates the use of transport specific fractionation as a survey technique, copurifying enzymes catalyzing similar reactions but differing in other important aspects such as location, kinetics, or regulation. The subsequent use of immunologic as well as biochemical techniques to study the purified transport proteins will add details to the emerging "map" of neuronal ion transport. Experiments along these lines have begun in the case of the two Ca^{2+} transporting ATPases described above. Preliminary biochemical results (Papazian et al 1983) indicate that the calmodulin-sensitive, 140,000 dalton Ca^{2+} pump may closely resemble other enzymes found in plasma membranes of both excitable and nonexcitable cells, such as muscle (Caroni & Carafoli 1981), erythrocytes (Niggli et al 1979), adipocytes (Pershadsingh et al 1980), and macrophages (Lew & Stossel 1980). Immunologic evidence (Chan et al, in preparation) indicates that the 94,000 dalton Ca^{2+} transport protein may be specific to the nervous system: antiserum and monoclonal antibodies against this protein do not cross-react with the Ca^{2+} pumps from red cell and sarcoplasmic reticulum.

Some progress has been made toward solubilizing and reconstituting other Ca^{2+} transport systems. Na^+/Ca^{2+} exchange activities have been studied in squid axons (Blaustein 1977), as well as in membrane vesicles derived from brain (Rahamimoff & Spanier 1979, Gill et al 1981, Michaelis & Michaelis 1981), and heart (Reeves & Sutko 1979, Pitts 1979). A Na^+/Ca^{2+} antiporter has been solubilized, partially purified, and reconstituted from bovine heart plasma membrane by Miyamoto & Racker (1980). The antiporter is solubilized in a high ionic strength buffer containing cholate and soy lipids. After dilution and centrifugation, the resulting proteoliposomes catalyze Ca^{2+} uptake with properties expected of a Na^+/Ca^{2+} exchanger. Since there is no convenient enzymatic assay for Na^+/Ca^{2+} exchange, reconstitution of the activity will be required as an assay during purification, or as a property to be exploited for transport-specific fractionation. With the successful preliminary results obtained with the heart Na^+/Ca^{2+} antiporter, progress toward the purification and reconstitution of Na^+/Ca^{2+} antiporters from other tissues, including brain, can be expected in the near future.

Voltage-dependent Ca^{2+} channels have been primarily studied electrophysiologically, not biochemically (Hagiwara & Byerly 1981). There may be several classes of Ca^{2+} channels. Biochemical studies have been hindered by the lack of specific ligands to serve the roles played by saxitoxin and tetradotoxin for voltage-dependent Na^+ channels.

Reconstitution of Ion Channels in Planar Lipid Bilayers

The planar lipid bilayer is a model membrane system that is ideally suited in geometry to analyse the electrical properties of ion channels. In the past, planar bilayer technology has been largely devoted to the characterization of readily incorporated molecules, such as ionophores synthesized by bacteria. Previously, it has not been possible to study ion transport proteins in this reconstituted system because of the difficulty in incorporating functionally active proteins into the bilayer. In the last several years, however, advances in methodology have led to the first reports of incorporation of ion transport proteins from excitable tissues into planar phospholipid bilayers. Several comprehensive reviews on the reconstitution of proteins into planar bilayers have recently appeared (Latorre & Alvarez 1981, Montal et al 1981, Miller 1982). Here, we briefly review planar bilayer studies of ion channels from skeletal muscle and *Torpedo* electroplax.

Two methods of forming planar phospholipid bilayers are currently in routine use. The first type (Mueller & Rudin 1969), sometimes referred to as the "painted" membrane, is formed by spreading a solution of lipid in a solvent such as decane over a small hole in a plastic partition that divides two aqueous chambers. After several minutes, the lipid solution spontaneously thins to form a bilayer membrane. In the second technique (Montal & Mueller 1972), two lipid monolayers are first spread at the surface of an air-water interface by placing a drop of lipid solution in a volatile solvent such as pentane on the aqueous surfaces. Then, two such monolayers are consecutively raised over a hole in a teflon partition that separates two adjacent chambers. The two monolayers are thus apposed or "folded" together to form a bilayer across the hole. "Folded" membranes have the advantage of being virtually solvent-free, while "painted" membranes always contain some solvent in equilibrium with the torus of lipid and hydrocarbon surrounding the rim of the hole.

One technique for incorporating ion channels into planar bilayers takes advantage of the property of membranes to fuse under special conditions. The most thoroughly studied example of this phenomenon is the fusion of liposomes containing negatively charged phospholipids that is induced by the addition of Ca^{2+} (Papahadjopoulos et al 1977). Miller & Racker (1976a) explored this fusion phenomenon using an assay system that relied on the stimulation of cytochrome oxidase activity occurring when liposomes containing this protein fused with liposomes containing mitochondrial hydrophobic protein (which acts as a proton ionophore). An important result of this work was the finding that an osmotic gradient across the liposomes, hypotonic with respect to the interior, promoted greater fusion than that observed with Ca^{2+} and negatively charged lipids alone (Miller et al 1976). In applying these results to planar bilayers, Miller and his colleagues have been able to incorporate ion channels from sarcoplasmic reticulum (SR) vesicles (Miller & Racker 1976b) and membrane vesicles prepared from *Torpedo* electroplax (White & Miller 1979) by a process that has the appearance of vesicle-planar bilayer fusion in the functional asymmetry of incorporated channels and in the dependence of the incorporation on negatively charged lipids, Ca^{2+}, and osmotically swelled vesicles.

In support of the view that membrane vesicles can be made to fuse with planar bilayers, Zimmerberg et al (1980) reported the transfer of a fluorescent dye trapped inside of multilammelar liposomes across a planar bilayer under the conditions previously cited to promote fusion. In addition, these workers have described the incorporation of VDAC channels (voltage dependent anion selective channel from mitochondria) by Ca^{2+}-induced fusion of reconstituted liposomes containing VDAC with a planar membrane (Cohen et al 1980). Although these model system studies lend support to the view that incorporation of ion channels from natural membrane vesicles proceeds by an actual fusion process, Miller (1982) cautions that this view must be considered hypothetical until definitive evidence can be accumulated. Although the incorporation of a particular channel may appear to proceed by fusion of the vesicle membrane with the planar bilayer, a channel or packets of channels and lipid may actually be transferred from vesicle to bilayer through the aqueous phase by a presently unknown or unimagined mechanism.

Aside from the question of the nature of the insertion process, Miller and his co-workers have characterized the conductance properties of the channels incorporated into planar bilayers from sarcoplasmic reticulum-derived (SR) vesicles. A Ca^{2+}-selective channel that normally functions in the release of Ca^{2+} from SR was anticipated in these SR membrane preparations. Contrary to expectation, the only type of channel so far observed from SR is highly selective for monovalent cations. The selectivity sequence with the measured single channel conductance (in pS) in 1 M salt solutions is as follows: $K^+(214) > NH_4^+(157) > Rb^+(125) > Na^+(72) > Li^+(8.1)$ (Coronado et al 1980). The channel is closed at negative voltages (-60 mV) and opens in a probabilistic fashion as the voltage is made more positive. An analysis of the conductance vs voltage dependence of the channel from rabbit SR conforms to a simple two-state model for gating consisting of a single closed and a single open state in reversible equilibrium (Labarca et al 1980). In agreement with this analysis of the macroscopic conductance properties of bilayers containing many channels, single channel fluctuations exhibit only two transitions between a low and a high conductance state. In addition to these basic properties, the K^+ channel from SR also exhibits a number of specific sensitivities to various modifying agents: pH sensitivity of the voltage dependence (Labarca et al 1980), inhibition by metallic sulhydryl ligands (Miller & Rosenberg 1979a), modification of the voltage dependence by a pronase-derived endopeptidase (Miller & Rosenberg 1979b), blockage of the channel by Cs^+ (Coronado & Miller 1979), and blockage by various bis-quaternary ammonium compounds (Coronado & Miller 1980).

Although the SR channel has been studied mainly in "painted" bilayers, the characteristics of the conductance-voltage relationship have also been examined in "folded" membranes (Labarca et al 1980). It was found that there was a slight shift in the voltage for half-saturation of the macroscopic conductance, 36 vs 28 mV, for "painted" vs "folded" membranes, respectively. The effective gating charge (0.9–1.1) and the single channel conductance were equivalent within experimental error for the two types of bilayers. These results suggest that decane-containing membranes do not seriously affect the behavior of the SR channel.

While the existence of the SR K^+-channel had not originally been anticipated from studies of muscle physiology, it has been proposed in retrospect that the function of the channel may be to shunt electrically the flow of current across the SR membrane that occurs during the massive release and re-uptake of Ca^{2+} during muscle contraction or relaxation. In addition to bilayer experiments, the permeability of SR membrane vesicles to monovalent cations has been confirmed by other methods: (a) direct radioisotopic flux measurements and ion diffusion potential measurements by the fluorescent dye technique (McKinley & Meissner 1978) and (b) light scattering measurements of volume changes in response to osmotic stress (Kometani & Kasai 1978). Thus, in keeping with one of the aims of reconstitution research, bilayer methodology has provided electrical access to a membrane that is inaccessible to microelectrode electrophysiology.

As reported for SR vesicles, membrane vesicles prepared from electric organ of *Torpedo californica* also appear to fuse with planar bilayers containing negatively charged phospholipid (White & Miller 1979). The incorporation of an anion-selective conductance is observed with vesicles from this tissue. The membrane conductance is low at zero or positive voltages and increases as the voltage is made negative, with saturation occurring at about −70 mV at pH 7.3. The channel is highly specific for Cl^- with a permeability ratio of 1.7 for Cl^- to Br^-, the only other known penetrating ion (Miller & White 1980). Among the reported blockers of this channel are I^-, SCN^-, and the disulfonic acid stilbenes, SITS and DIDS, the latter two widely known as inhibitors of the anion exchange system of erythrocytes. The single channel conductance of the Cl^- channel is 14 pS in 0.1 M KCl. Fractionation studies of crude membrane vesicles from *Torpedo* suggested that the anion channel was located in the noninnervated face of the electroplax, which also contains NaK-ATPase. Vesicles derived from the innervated face apparently do not fuse, since incorporation of the AChR/ion gate was not observed. An alternative to this suggestion is the possibility that the AChR/ion gate is denatured after incorporation into decane-containing bilayers. The Cl^- channel incorporated into bilayers by the fusion technique has not as yet been described electrophysiologically in the intact organ. It has been proposed, however, that a high Cl^- conductance of the noninnervated face of the electroplax cell would permit the delivery of high power impulses during electric organ discharge by maintaining a low internal resistance of the organ acting as a battery of individual cells in series (Miller & White 1980).

In order to evaluate objectively the usefulness of the fusion incorporation technique for studying channels of excitable membranes in planar bilayers, one would hope to incorporate a channel previously characterized by classical techniques and compare its function in vivo with that in the planar membrane. Although work with the fusion technique has not yet permitted such an evaluation, two recent reports of successful channel incorporation appear promising. In one of these reports, a Ca^{2+}-activated K^+-selective channel was incorporated into painted bilayers using vesicles derived from rabbit skeletal muscle (Latorre et al 1982). Evidence has been presented that the preparation used in this study is derived from transverse tubule membranes (Rosenblatt et al 1981). Recordings of single channel fluctuations reveal that the channel has a single open (226 pS in 0.1 M KCl) and a single closed state; however, distinct periods occur when the channel is closed for short times (< 0.1 sec) and long times (1–10 sec), indicative of more than one kinetic path for opening or closing. Both Ca^{2+} and positive voltage activate the channel by increasing the fraction of time the channel spends open. Tetraethylammonium, a blocker of the axonal K^+ channel, also blocks the conductance of this Ca^{2+}-activated K^+-channel. Although this particular channel has not been studied extensively in intact skeletal muscle, reports of similar Ca^{2+}-activated K^+ channels from a wide variety of species and tissues abound in the literature (Meech 1978). Also, a recent report has documented a similar channel studied by the patch clamp electrophysiological technique from rat muscle cells in culture (Pallota et al 1981).

In a recent variation of the fusion technique of incorporation, Boheim et al (1980) reported the formation of folded membranes from mixed chain lipids that exhibit fluid-solid phase transitions near 30°C. Native membrane vesicles and reconstituted liposomes containing membrane proteins were found to fuse with bilayers made from such lipids at temperatures near the phase transition (Hanke et al 1981). Thus, the fusion did not appear to require Ca^{2+}, but an osmotic gradient was necessary. One of the channels incorporated by this technique was the ACh receptor from *Torpedo marmorata.*

A second method for incorporating ion channels in planar bilayers has been described by Schindler (1980). The Schindler method is similar to the Montal & Mueller (1972) technique for "folded" membranes, except that membrane vesicles or reconstituted liposomes are used to form a lipid monolayer-containing protein on one side of the chamber. Conductance increases ascribed to the *Torpedo* AChR/ion gate have been observed in planar membranes using this method (Schindler & Quast 1980). The observed cation selective conductance was elicited by carbamylcholine and inhibited by d-tubocurarine or α-bungarotoxin. Evidence for time-dependent desensitization after the addition of agonist was also reported. Single channel measurements gave values of 90 pS and 20–25 pS at 1 M NaCl and 0.25 M NaCl, respectively, for the unitary conductance of the open state using native membrane vesicles. Similar results were obtained when reconstituted liposomes containing functionally active, purified AChR/ion gate were used to form bilayers (Nelson et al 1980). As more data on the functional characteristics of the AChR/ion gate in planar bilayers become available, it will be possible to determine which incorporation procedures are most convenient and reproducible, and result in channel activity closest to the native receptor.

Summary and Prognosis

The techniques for studying transmembrane transport proteins in model membrane systems have been developed largely in the past decade. The application of these techniques to identify unambiguously and to characterize quantitatively neuronal ion transport proteins has occurred primarily in the past seven years. This article focuses on those few ion transport proteins whose molecular identity has been established by reconstitution into small lipid vesicles, the system of choice for this purpose. The large, planar lipid bilayer is the most suitable system for the quantitation of ion transport by channels that conduct millions of ions per second; small lipid vesicles are adequate in this respect for the quantitation of transport by ion pumps, which work orders of magnitude more slowly. The past few years have quite literally witnessed the fusion of these two approaches, i.e. the fusion of liposomes containing purified, reconstituted ion channels with planar bilayers. This enables the advantages of each of the two systems to be exploited.

Membrane reconstitution has passed through a critical period that has established its future potential as a tool for the neuroscientist: the NaK-ATPase/ion pump and the AChR/ion gate—the proteins responsible for two ion transport processes that have been thoroughly characterized in living nerve cells—have been isolated, and shown to function in reconstituted membranes in quite the same way they do in vivo. This success justifies the cautious use of the reconstitution methodology to identify, purify, and study neuronal ion transport proteins that are known to exist but whose role in regulation of neuronal electrical activity has not been clarified precisely by in vivo studies. Examples of this discussed here are the proteins responsible for regulation of Ca^{2+} levels in nerve terminals. Other prime targets for this approach are the postsynaptic ion channels regulated by putative neurotransmitters in brain: e.g. the glutamate-activated excitatory channel, and inhibitory Cl^--channels activated by γ-aminobutyric acid.

Even more ambitious is the use of membrane reconstitution to discover neuronal ion transport processes about which little or nothing is known. When one looks for a voltage-sensitive Ca^{2+} channel and discovers an unforeseen K^+ channel instead—an example discussed in this article—one must work from test tube back to tissue to sort things out. This is the reverse of the traditional approach of the neuroscientist. It is here that immunocytochemical localization of the purified ion transport protein—as well as a large dose of patience and a modicum of faith—may pay substantial dividends.

Literature Cited

Adelstein, R. S., Conti, M. A., Hathaway,D. R., Klee, C. B. 1978. Phosphorylation of smooth muscle myosin light chain kinase by the catalytic subunit of adenosine -3': 5'-monophosphate-dependent protein kinase. *J. Biol. Chem.* 253: 8347–50

Adelstein, R. S., Eisenberg, E. 1980. Regulation and kinetics of the actin-myosin-ATP interaction. *Ann. Rev. Biochem.* 49:921–56

Agnew, W. S., Levinson, S. R., Brabson, J. S., Raftery, M. A. 1978. Purification of the tetrodotoxin-binding component associated with the voltage-sensitive sodium channel from *Electrophorus electricus* electroplax membranes. *Proc. Natl. Acad. Sci. USA* 75:2606–10

Agnew, W. S., Moore, A. C., Levinson, S. R., Raftery, M. A. 1980. Identification of a large molecular weight peptide associated with a tetrodotoxin binding protein from the electroplax of *Electrophorus electricus.* *Biochem. Biophys. Res. Commun.* 92:860–66

Barchi, R. L., Cohen, S. A., Murphy, L. E. 1980. Purification from rat sarcolemma of the saxitoxin-binding component of the excitable membrane sodium channel. *Proc. Natl. Acad. Sci. USA* 77: 1306–10

Barnola, F. V., Villegas, R. 1976. Sodium flux through the sodium channels of axon membrane fragments isolated from lobster nerves. *J. Gen. Physiol.* 67:81–90

Beneski, D. A., Catterall, W. A. 1980. Covalent labeling of protein components of the sodium channel with a photoactivatable derivative of scorpion toxin. *Proc. Natl. Acad. Sci. USA* 77:639–43

Blaustein, M. P. 1977. Effects of internal and external cations and of ATP on sodium-calcium and calcium-calcium exchange in squid axons. *Biophys. J.* 20:79–111

Blaustein, M. P., Ratzlaff, R. W., Kendrick, N. C., Schweitzer, E. S. 1978a. Calcium buffering in presynaptic nerve terminals: I. Evidence for involvement of a nonmitochondrial ATP-dependent sequestration mechanism. *J. Gen. Physiol.* 72:15–41

Blaustein, M. P., Ratzlaff, R. W., Schweitzer, E. S. 1978b. Calcium buffering in presynaptic nerve terminals: II. Kinetic properties of the nonmitochondrial Ca sequesteration mechanism. *J. Gen. Physiol.* 72:43–66

Blitz, A. L., Fine, R. E., Toselli, P. A. 1977. Evidence that coated vesicles isolated from brain are calcium-sequestering organelles resembling sarcoplasmic reticulum. *J. Cell Biol.* 75:135–47

Boheim, G., Hanke, W., Eibl, H. 1980. Lipid phase transition in planar bilayer membrane and its effect on carrier and pore-mediated ion transport. *Proc. Natl. Acad. Sci. USA* 77:3403–7

Bygrave, F. L. 1977. Mitochondrial calcium transport. In: *Current Topics in Bioenergetics,* ed. D. R. Sanadi, 6:259–318. New York: Academic Press. 324 pp.

Cantley, L. 1981. Structure and mechanism of the NaK ATPase. *Curr. Topics Bioenerg.* 11:201–37

Caroni, P., Carafoli, E. 1981. The Ca^{2+}-pumping ATPase of heart sarcolemma: Characterization, calmodulin dependence, and partial purification. *J. Biol. Chem.* 256:3263–70

Carter-Su, C., Pillion, D. J., Czech, M.P. 1980. Reconstituted D-glucose transport from the adipocyte plasma membrane: Chromatographic resolution of transport activity from membrane glycoproteins using immobilized concanavilin A. *Biochemistry* 19:2374–85

Catterall, W. A. 1975. Activation of the action potential Na^+ ionophore of cultured neuroblastoma cells by veratridine and batrachotoxin. *J. Biol. Chem.* 250:4053–59

Catterall, W. A. 1980. Neurotoxins that act on voltage-sensitive sodium channels in excitable membranes. *Ann. Rev. Pharmacol. Toxicol.* 20:15–43

Catterall, W. A., Morrow, C. S., Hartshorne, R. P. 1979. Neurotoxin binding to receptor sites associated with voltage-sensitive sodium channels in intact, lysed, and detergent-solubilized brain membranes. *J. Biol. Chem.* 254:11, 379–87

Chan, S. Y., Hess, E. J., Rahamimoff, H., Goldin, S. M. 1983. In preparation

Chandler, W. K., Meves, H. 1965. Voltage clamp experiments on internally perfused giant axons. *J. Physiol.* 180:788–820

Cohen, F. S., Zimmerberg, J., Finkelstein, A. 1980. Fusion of phospholipid vesicles with planar phospholipid bilayer membranes. II: Incorporation of a vesicular membrane marker into the planar membrane. *J. Gen. Physiol.* 75:251–70

Conti, F., Hille, B., Neumchi, B., Nonner, W., Stampfli, R. 1976. Measurement of the conductance of the sodium channel from current fluctuations at the Node of Ranvier. *J. Physiol.* 262:699–727

Coronado, R., Miller, C. 1979. Voltage-dependent blockade of a cation channel from fragmented sarcoplasmic reticulum. *Nature* 280:807–10

Coronado, R., Miller, C. 1980. Decamethonium and hexamethonium block K^+ channels of sarcoplasmic reticulum. *Nature* 288:495–97

Coronado, R., Rosenberg, R. L., Miller, C. 1980. Ionic selectivity, saturation and block in a K^+-selective channel from sarcoplasmic reticulum. *J. Gen. Physiol.* 76:425–46

Del Castillo, J., Webb, G. D. 1977. Rapid desensitization of acetylcholine receptors of eel electroplaques following iontophoretic application of agonist compounds. *J. Physiol.* 270:271–82

DePamphilis, M. L., Wassarman, P. M. 1980. Replication of eucaryotic chromosomes: A close-up of the replication fork. *Ann. Rev. Biochem.* 49:627–66

Dixon, J. F., Hokin, L. E. 1980. The reconstituted (Na,K)-ATPase is electrogenic. *J. Biol. Chem.* 255:10681–86

Epstein, M., Racker, E. 1978. Reconstitution of carbamylcholine-dependent sodium ion flux and desensitization of the acetylcholine receptor from *Torpedo californica. J. Biol. Chem.* 253:6660–62

Forbush, B., Kaplan, J. H., Hoffman, J. H. 1978. Characterization of a new photoaffinity derivative of ouabain: Labeling of the large polypeptide and of a proteolipid component of the Na,K-ATPase. *Biochemistry* 17:3667–76

Forgac, M., Chin, G. 1981. K^+-independent active transport of Na^+ by the NaK ATPase. *J. Biol. Chem.* 256:3645–46

Froehner, S., Karlin, A., Hall, Z. W. 1977. Affinity alkylation labels two subunits of the reduced AChR from mammalian muscle. *Proc. Natl. Acad. Sci. USA* 74:4685–88

Garrahan, P. J., Glynn, I. M. 1967. The stoicheiometry of the sodium pump. *J. Physiol.* 192:217–35

Gill, D. L., Grollman, E. F., Kohn, L. D. 1981. Calcium transport mechanisms in membrane vesicles from guinea pig brain synaptosomes. *J. Biol. Chem.* 256:184–92

Goldin, S. M. 1977. Active transport of sodium and potassium ions by the N aK ATPase from renal medulla: Reconstitution of the purified enzyme into a well defined *in vitro* transport system. *J. Biol. Chem.* 252:5630–42

Goldin, S. M. 1982. Reconstitution as an assay and as a physical tool for purification of erythrocyte membrane transport proteins. In *The Red Cell Membrane, a*

Methodological Approach, ed. J. Young, C. Ellory, pp. 301–19. London: Academic

Goldin, S. M., Rhoden, V. 1978. Reconstitution and "transport specific fractionaction" of the human erythrocyte glucose transport system: A new approach for identification and isolation of membrane transport proteins. *J. Biol. Chem.* 253:2575–83

Goldin, S. M., Rhoden, V., Hess, E. J. 1980. Molecular characterization, reconstitution, and transport-specific fractionation of the saxitoxin binding protein/Na⁺ gate of mammalian brain. *Proc. Natl. Acad. Sci. USA* 77:6884–88

Goldin, S. M., Tong, S. W. 1974. Reconstitution of active transport catalyzed by the purified NaK ATPase from canine renal medulla. *J. Biol. Chem.* 249:5907–15

Hagiwara, S., Byerly, L. 1981. Calcium channel. *Ann. Rev. Neurosci* 4:69–125

Hanke, W., Eibl, H., Boheim, G. 1981. A new method for membrane reconstitution: Fusion of protein-containing vesicles with planar bilayer membranes below lipid phase transition temperature. *Biophys. Struct. Mech.* 7:131–37

Hartshorne, R. P., Catterall, W. A. 1981. Purification of the saxitoxin receptor of the sodium channel from rat brain. *Proc. Natl. Acad. Sci. USA* 78:4620–24

Heidmann, T., Changeux, J. P. 1978. Structural and functional properties of the acetylcholine receptor protein in its purified and membrane-bound states. *Ann. Rev. Biochem.* 47:317–57

Henderson, R., Wang, J. H. 1972. Solubilization of a specific tetrodotoxin-binding component from garfish olfactory nerve membrane. *Biochemistry* 11:4565–69

Hess, A., Young, J. Z. 1952. The Nodes of Ranvier. *Proc. R. Soc. London Ser. B* 140:301–20

Hess, G. P., Cash, D. J., Aoshima, H. 1979. Acetylcholine receptor-controlled ion fluxes in membrane vesicles investigated by fast reaction techniques. *Nature* 282:329–31

Hilden, S., Hokin, L. E. 1975. Active K⁺ transport coupled to active Na⁺ transport in vesicles reconstituted from purified NaK-ATPase from the rectal gland of *Squalus acanthias. J. Biol. Chem.* 250:6296–303

Hille, B. 1975. The receptor for tetrodotoxin and saxitoxin: A structural hypothesis. *Biophys. J.* 15:615–19

Hoffman, J. F., Kaplan, J. H., Callahan, T. J. 1979. The Na:K pump in red cells is electrogenic. *Fed. Proc.* 38:2440–41

Hokin, L. E., Sastry, P. S., Galsworthy, P. R., Yoda, A. 1965. Evidence that a phosphorylated intermediate in a brain transport ATPase is an acylphosphate. *Proc. Natl. Acad. Sci. USA* 54:177–84

Ikemoto, N. 1982. Structure and function of the calcium pump protein of sarcoplasmic reticulum. *Ann. Rev. Physiol.* 44:297–318

Israël, M., Manaranche, R., Marsal, J., Meunier, F. M., Morel, N., Frachon, P., Lesbats, B. 1980. ATP-dependent calcium uptake by cholinergic synaptic vesicles isolated from *Torpedo* electric organ. *J. Membrane Biol.* 54:115–26

Javors, M. A., Bowden, C. L., Ross, D. H. 1981. Kinetic characterization of Ca²⁺ transport in synaptic membranes. *J. Neurochem.* 37:381–87

Kagawa, Y., Racker, E. 1971. Partial resolution of the enzymes catalyzing oxidative phosphorylation, XXV. Reconstitution of vesicles catalyzing ³²Pi-ATP exchange. *J. Biol. Chem.* 246:5477–87

Karlin, A. 1980. The acetylcholine receptor. In *The Cell Surface and Neuronal Function,* ed. C. W. Cotman, G. Poste, G. L. Nicholson, pp. 191–260. New York: Elsevier

Kasahara, M., Hinkle, P. C. 1977. Reconstitution and purification of the D-glucose transporter from human erythrocytes. *J. Biol. Chem.* 252:7384–90

Kelly, R. B., Deutsch, J. W., Carlson, S. S., Wagner, J. A. 1979. Biochemistry of neurotransmitter release. *Ann. Rev. Neurosci.* 2:399–446

Kennedy, M. 1983. Experimental approaches to understanding the role of protein phosphorylation in the regulation of neuronal function. *Ann. Rev. Neurosci.* 6:493–525

Khodorov, B. I. 1978. Chemicals as tools to study nerve fiber Na channels: Effects of batrachotoxin and some local anaesthetics. In *Membrane Transport Processes,* ed. D. C. Toesteston, Y. A. Ovchinnikov, R. Latorre, 2:153–74. New York: Raven

Klee, C. B., Crouch, T. H., Richman, P. G. 1980. Calmodulin. *Ann. Rev. Biochem.* 49:489–515

Kometani, T., Kasai, M. 1978. Ionic permeability of sarcoplasmic reticulum vesicles measured by light scattering method. *J. Membr. Biol.* 41:295–308

Kyte, J. 1971. Purification of the NaK ATPase from canine renal medulla. *J. Biol. Chem.* 246:4157–65

Labarca, P., Coronado, R., Miller, C. 1980. Thermodynamic and kinetic studies of the gating behavior of a K⁺-selective channel from the sarcoplasmic reticulum membrane. *J. Gen. Physiol.* 76:397–424

Latorre, R., Alvarez, O. 1981. Voltage-dependent channels in planar lipid bilayer membranes. *Physiol. Rev.* 61:77–150

Latorre, R., Vergara, C., Hildalgo, C. 1982. Reconstitution in planar lipid bilayers of a Ca^{2+}-dependent K^+ channel from transverse tubule membranes isolated from rabbit skeletal muscle. *Proc. Natl. Acad. Sci. USA* 79:805–8

Lew, P. D., Stossel, T. P. 1980. Calcium transport by macrophage plasma membranes. *J. Biol. Chem.* 255:5841–46

McGraw, C. F., Somlyo, A. V., Blaustein, M. P. 1980. Localization of calcium in presynaptic nerve terminals: An ultrastructural and electron microprobe analysis. *J. Cell Biol.* 85:228–41

McKinley, D., Meissner, G. 1978. Evidence for a K^+, Na^+ permeable channel in sarcoplasmic reticulum. *J. Membr. Biol.* 44:159–86

Meech, R. W. 1978. Calcium-dependent potassium activation in nervous tissues. *Ann. Rev. Biophys. Bioeng.* 7:1–18

Michaelis, M. L., Michaelis, E. K. 1981. Ca^{2+} fluxes in resealed synaptic plasma membrane vesicles. *Life Sci.* 28:37–45

Michaelson, D. M., Ophir, I., Angel, I. 1980. ATP-stimulated Ca^{2+} transport into cholinergic *Torpedo* synaptic vesicles. *J. Neurochem.* 35:116–24

Miller, C. 1982. First steps in the reconstitution of ionic channel functions in model membranes. In *Current Methods in Cellular Neurobiology*, ed. J. Barker, J. McKelvey. New York: Plenum. In press

Miller, C., Arvan, P., Telford, J. N., Racker, E. 1976. Ca^{++}-induced fusion of proteoliposomes: Dependence on transmembrane osomotic gradient. *J. Membr. Biol.* 30:271–82

Miller, C., Racker, E. 1976a. Fusion of phospholipid vesicles reconstituted with cytochrome c oxidase and mitochondrial hydrophobic protein. *J. Membr. Biol.* 26:319–33

Miller, C., Racker, E. 1976b. Ca^{++}-induced fusion of fragmented sarcoplasmic reticulum with artificial planar bilayers. *J. Membr. Biol.* 30:283–300

Miller, C., Rosenberg, R. L. 1979a. A voltage-gated cation conductance channel from fragmented sarcoplasmic reticulum. Effect of transition metal ions. *Biochemistry* 18:1138–45

Miller, C., Rosenberg, R. L. 1979b. Modification of a voltage-gated K^+ channel from sarcoplasmic reticulum by a pronase-derived specific endopeptidase. *J. Gen. Physiol.* 74:457–78

Miller, C., White, M. M. 1980. A voltage-dependent chloride conductance channel from *Torpedo* electroplax membrane. *Ann. NY Acad. Sci.* 341:534–51

Miyamoto, H., Racker, E. 1980. Solubilization and partial purification of the Ca^{2+}/Na^+ antiporter from the plasma membrane of bovine heart. *J. Biol. chem.* 255:2656–58

Montal, M., Darszon, A., Schindler, H. 1981. Functional reassembly of membrane proteins in planar lipid bilayers. *Q. Rev. Biophys.* 14:1–79

Montal, M., Mueller, P. 1972. Formation of bimolecular membranes from lipid monolayers and a study of their electrical properties. *Proc. Natl. Acad. Sci. USA* 69:3561–66

Moore, H. P., Raftery, M. A. 1980. Direct spectroscopic studies of cation translocation by *Torpedo* acetylcholine receptor on a time scale of physiological relevance. *Proc. Natl. Acad. Sci. USA* 77:4509–13

Morgan, I. G. 1976. Synaptosomes and cell separation. *Neuroscience* 1:159–65

Mueller, P., Rudin, D. O. 1969. In *Laboratory Techniques in Membrane Biophysics,* ed. H. Passow, R. Stampfli, pp. 141–56. Berlin: Springer Verlag

Neher, E., Stevens, C. F. 1977. Conductance fluctuations and ionic pores in membranes. *Ann. Rev. Biophys. Bioeng.* 6:345–81

Nelson, N., Anholt, R., Lindstrom, J., Montal, M. 1980. Reconstitution of purified acetylcholine receptors with functional ion channels in planar lipid bilayers. *Proc. Natl. Acad. Sci. USA* 77:3057–61

Neubig, R. R., Cohen, J. B. 1980. Permeability control by cholinergic receptors in *Torpedo* postsynaptic membranes: Agonist dose response relations measured at second and millisecond times. *Biochemistry* 19:2770–79

Neubig, R. R., Krodel, E. K., Boyd, N. D., Cohen, J. B. 1979. Acetylcholine and local anaesthetic binding to *Torpedo* nicotinic postsynaptic membranes after removal of nonreceptor peptides. *Proc. Natl. Acad. Sci. USA* 76:690–94

Neufeld, A. H., Levy, H. M. 1970. The steady state level of phosphorylated intermediate in relation to the two sodium-dependent ATPases of calf brain microsomes. *J. Biol. Chem.* 245:4962–67

Newman, M. J., Foster, D. L., Wilson, T. H., Kaback, H. R. 1981. Purification and reconstitution of functional lactose carrier from *Escherichia coli. J. Biol. Chem.* 256:11804–8

Niggli, V., Penniston, J. T., Carafoli, E. 1979. Purification of the $(Ca^{2+}-Mg^{2+})$-ATPase from human erythrocyte membranes using a calmodulin affinity column. *J. Biol. Chem.* 254:9955–58

Nonner, W., Rojas, E., Stampfli, R. 1975. Displacement currents in the Node of Ranvier: Voltage and time dependence. *Pflügers Arch.* 354:1–18

Pallotta, B. S., Magleby, K. L., Barrett, J. N. 1981. Single channel recordings of Ca-activated K currents in rat muscle cell culture. *Nature* 293:471–74

Papahadjopoulos, D., Vail, W. J., Newton, C., Nir, S., Jacobson, K., Poste, G., Lazo, R. 1977. Studies on membrane fusion III. The role of calcium-induced phase changes. *Biochim. Biophys. Acta* 465:579–98

Papazian, D. M., Rahamimoff, H., Goldin, S. M. 1979. Reconstitution and purification by "transport specific fractionation" of an ATP-dependent calcium transport component from synaptosome-derived vesicles. *Proc. Natl. Acad. Sci. USA* 76:3708–12

Papazian, D. M., Rahamimoff, H., Goldin, S. M. 1983. Synaptosomes contain two distinct nonmitochondrial ATP-dependent calcium transport activities differing in their regulation by calmodulin. Submitted for publication

Pershadsingh, H. A., Landt, M., McDonald, J. M. 1980. Calmodulin-sensitive ATP-dependent Ca^{2+} transport across adipocyte plasma membranes. *J. Biol. Chem.* 255:8983–86

Pitts, B. J. R. 1979. Stoichiometry of sodium-calcium exchange in cardiac sarcolemmal vesicles: Coupling to the sodium pump. *J. Biol. Chem.* 254:6232–35

Post, R. L., Jolly, P. C. 1957. The linkage of sodium, potassium and ammonium active transport across the human erythrocyte membrane. *Biochim. Biophys. Acta* 25:118–28

Racker, E., Violand, B., O'Neal, S., Alfonzo, M., Telford, J. 1979. Reconstitution, a way of biochemical research: Some new approaches to membrane-bound enzymes. *Arch. Biochem. Biophys.* 198: 470–77

Rahamimoff, H., Abramovitz, E. 1978a. Calcium transport in a vesicular membrane preparation from rat brain synaptosomes. *FEBS Lett.* 89:223–26

Rahamimoff, H., Abramovitz, E. 1978b. Ca transport and ATPase activity of synaptosomal vesicles from rat brain. *FEBS Lett.* 92:163–67

Rahamimoff, H., Spanier, R. 1979. Sodium-dependent calcium uptake in membrane vesicles derived from rat brain synaptosomes. *FEBS Lett.* 104:111–14

Reeves, J. P., Sutko, J. L. 1979. Sodium-calcium ion exchange in cardiac membrane vesicles. *Proc. Natl. Acad. Sci. USA* 76:590–94

Rhoden, V. A., Goldin, S. M. 1979. The binding of saxitoxin to axolemma of mammalian brain—cooperative competition between saxitoxin and sodium ion. *J. Biol. Chem.* 254:11199–11201

Robinson, J. D. 1981. Effects of cations on $(Ca^{2+} + Mg^{2+})$-activated ATPase from rat brain. *J. Neurochem.* 37:140–46

Rosenblatt, M., Hildalgo, C., Vergara, C., Ikemoto, N. 1981. Immunological and biochemical properties of transverse tubule membranes isolated from rabbit skeletal muscle. *J. Biol. Chem.* 356: 8140–48

Schindler, H. S. 1980. Formation of planar bilayers from artificial or native membrane vesicles. *FEBS Lett.* 122:77–79

Schindler, H. S., Quast, U. 1980. Functional acetylcholine receptor from *Torpedo marmorata* in planar membranes. *Proc. Natl. Acad. Sci. USA* 77:3052–56

Schultz, S. G. 1980. Ion-coupled transport across biological membranes. In *Membrane Physiology*, ed. T. E. Andreoli, J. F. Hoffman, D. D. Fanestil, pp. 273–86. New York: Plenum

Sharma, R. K., Wang, T. H., Wirch, E., Wang, J. H. 1980. Purification and properties of bovine brain calmodulin-dependent cyclic nucleotide phosphodiesterase. *J. Biol. Chem.* 255: 5916–23

Skou, J. C. 1957. The influence of some cations on an ATPase from peripheral nerves. *Biochim. Biophys. Acta* 23: 394–401

Skriver, E., Maunsbach, A. B., Jorgensen, P. L. 1980. Ultrastructure of Na,K-transport vesicles reconstituted with purified renal Na,K-ATPase. *J. Cell Biol.* 86:746–54

Sobel, A., Heidmann, T., Hofler, J., Changeux, J. P. 1978. Distinct protein components from *Torpedo marmorata* membranes carry the acetylcholine receptor site and the binding site for local anaesthetics and histrionicotoxin. *Proc. Natl. Acad. Sci. USA* 75:510–14

Sobel, A., Weber, M., Changeux, J. P. 1977. Large-scale purification of the acetylcholine-receptor protein in its membrane-bound and detergent-extracted forms from *Torpedo marmorata* electric organ. *Eur. J. Biochem.* 80:215–24

Sorensen, R. G., Mahler, H. R. 1981. Calcium-stimulated adenosine triphosphatases in synaptic membranes. *J. Neurochem.* 37:1407–18

Sweadner, K. J. 1979. Two molecular forms of NaK ATPase in brain: Separation and difference in affinity for strophanthidin. *J. Biol. Chem.* 254:6060–67

Sweadner, K. J., Goldin, S. M. 1975. Reconstitution of the NaK-ATPase from canine brain. *J. Biol. Chem.* 250:4022–24

Sweadner, K. J., Goldin, S. M. 1980. Active transport of sodium and potassium ions: Mechanism, function, and regulation. *N. Engl. J. Med.* 302:777–83

Tamkun, M. M., Catterall, W. A. 1981. Reconstitution of the voltage-sensitive sodium channel of rat brain from solubilized components. *J. Biol. Chem.* 256:11457–63

Thomas, R. C. 1969. Membrane current and intracellular sodium changes in a snail neurone during intracellular injection of sodium. *J. Physiol.* 201:495–514

Villegas, R., Villegas, G. M. 1981. Nerve sodium channel incorporation in vesicles. *Ann. Rev. Biophys. Bioeng.* 10:387–419

Villegas, R., Villegas, G. M., Barnola, F. V., Racker, E. 1977. Incorporation of the sodium channel of lobster nerve in artificial liposomes. *Biochem. Biophys. Res. Commun.* 79:210–17

Weigle, J. B., Barchi, R. L. 1982. Functional reconstruction of the purified sodium channel protein from rat sarcolemma. *Proc. Natl. Acad. Sci. USA* 79:3651–55

Weill, C. L., McNamee, M. G., Karlin, A. 1974. Affinity labelling of purified AChR from *Torpedo californica. Biochem. Biophys. Res. Commun.* 61:997–1003

White, M. M., Miller, C. 1979. A voltage-gated anion channel from the electric organ of *Torpedo californica. J. Biol. Chem.* 254:10161–66

Wong, R. K. S., Prince, D. A., Busbaum, A. I. 1979. Intradendritic recordings from hippocampal neurons. *Proc. Natl. Acad. Sci. USA* 76:986–90

Wu, W. C. S., Moore, H. P., Raftery, M. 1981. Quantitation of cation transport by reconstituted membrane vesicles containing purified acetylcholine receptor. *Proc. Natl. Acad. Sci. USA* 78:775–79

Wu, W. C. S., Raftery, M. 1981. Reconstitution of acetylcholine receptor function using purified receptor protein. *Biochemistry* 20:694–701

Zimmerberg, J., Cohen, F. S., Finkelstein, A. 1980. Fusion of phospholipid vesicles with planar bilayer membranes. I. Discharge of vesicular contents across the planar membrane. *J. Gen. Physiol.* 75:241–50

Ann. Rev. Neurosci. 1983. 6:447–91

CELLULAR PROCESSES OF LEARNING AND MEMORY IN THE MAMMALIAN CNS

Richard F. Thompson

Department of Psychology, Stanford University, Stanford, California 94305

Theodore W. Berger

Departments of Psychology and Psychiatry, University of Pittsburgh, Pittsburgh, Pennsylvania 15260

John Madden IV

Department of Psychiatry and Behavioral Sciences, Stanford University School of Medicine, Stanford, California 94305

A number of reviews on various aspects of the neuronal bases of learning and memory have appeared in the past few years (Agranoff et al 1978, Bennett 1976, Dunn 1980, Greenough 1976, Kandel & Spencer 1968, Kupfermann 1975, McGaugh & Herz 1972, Rosenzweig & Bennett 1976, Sokolov 1977, Squire 1982, Squire & Davis 1981, Thompson et al 1972, 1980, Tsukahara 1981). This review is concerned broadly with the cellular/neuronal processes of associative learning in the mammalian brain.[1] The nature of the memory trace has proved to be among the most baffling questions in science. The problem of localization has been perhaps the greatest barrier. In order to characterize cellular mechanisms of information storage and

---

[1] At several places we have given the class Aves honorary membership in the class Mammalia.

447

0147-006X/83/0301-0447$02.00

retrieval, it is first necessary to identify and localize the brain systems, structures, and regions that are critically involved. At least in simpler learning paradigms, the problem of localization appears finally to be yielding. This review is highly selective. Primary foci are (a) the problem of localization and (b) recent studies of putative physiological, anatomical, and, to a lesser extent, biochemical substrates of learning and memory. Emphasis is placed on simpler paradigms, in which some degree of localization and analysis of cellular processes has been done or appears feasible, and on processes of learning exhibited by normally intact animals.

THE PROBLEM OF LOCALIZATION

The issue of localization, first explored in depth by Lashley (1929) and later by Hebb (1949), remains fundamental to all work on neuronal substrates of learning and memory. For simpler forms of learning, it seems evident that at least some components of the memory circuit must be localized. An animal trained to a particular conditioned stimulus will not respond to a very different conditioned stimulus and must be given additional training. This fact, the existence of a stimulus generalization gradient, argues strongly that sensory-specific information is to some degree preserved in the elements that develop the plasticity coding the learned response. A well-trained animal exhibits a particular learned response complex. Activation of motor neurons can be highly selective and specific. This implies that the "motor program" aspect of the memory circuit must itself have specificity. Both the sensory-specific and motor specific aspects of learning encourage a belief in localization of the memory circuit.

A distinction would seem necessary between the circuit essential for a given form of learning and the essential neuronal plasticity that codes the memory. We use here the term "learned response circuit" to include the entire circuit(s) from sensory receptors to muscles or glands. If the conditioned stimulus (CS) is auditory, bilateral destruction of the inner ear will abolish the learned response. The essential neuronal plasticity for the learned response is unlikely, however, to reside in the inner ear. The same applies to motor nerves. We infer that some part of the learned response circuit contains the essential neuronal plasticity that has developed to code learning. We term this the "memory trace," and designate by "sensory-motor circuit" those parts of the learned response circuit that do not include the memory trace. The sensory-motor circuit thus also contains the neuronal elements responsible for retrieving information stored in memory traces if these retrieval circuits are not themselves modified during learning.

Electrophysiological recording of unit activity has been widely used for the putative identification and localization of the memory trace. As has been noted (Tsukahara 1981), demonstration that a learning-induced change occurs in neuronal activity in a given brain structure or region is not in and of itself sufficient to demonstrate localization of the memory trace to that locus. It does, however, indicate that the locus is a part of the "normal" learned response circuits or systems.

The key question is whether or not the learning-induced neuronal plasticity in a given brain structure develops there or is simply relayed there from elsewhere. This question can be addressed by anatomical tracing and electrophysiological methods. In brief, given that the afferent systems to a region and the efferent projections of the principal neurons have been defined anatomically the learning-induced activity of the principal neurons projecting from a synaptic region or structure can be determined, using antidromic and collision criteria (see Fuller & Schlag 1976) for identification. The learning-induced activity of neuronal systems projecting into the region can also be determined and the categories of neurons similarly identified. Comparison of the learning-induced activity of the inputs and outputs of the region over the course of learning will provide evidence regarding possible development of neuronal plasticity within the region. The development of such neuronal plasticity would seem a necessary requirement for the memory trace system, although it is not by itself sufficient. However, this input-output approach, in conjunction with lesions that selectively abolish the learned response, would seem to provide a method for localizing the essential memory trace system for a given form of learning.

The discordance that sometimes appears to exist between electrophysiological and lesion data deserves comment. If neurons in a given structure show learning-induced changes in activity—as opposed to sensory or motor specific responses, arousal, or other processes—then that structure is involved in learning. If destruction of the structure does not impair the particular form of learning that induces the neuronal change, it does not mean that the structure plays no role in that form of learning; it means only that its role is not essential. Cohen (1982) found that either of two visual thalamic regions can support heart rate conditioning in the pigeon. Neurons in both of these regions exhibit learning-induced changes. Is one or the other not involved because destruction of it alone has no effect on the learned response? The brain is not a linear causal system. Generally speaking, to the extent tested, every structure whose neurons show learning-induced changes has been found by the lesion method to play some role in learning, at least under some conditions.

LEARNING PARADIGMS

It is no accident that many systematic investigations of neuronal substrates of learning and memory utilize classical conditioning. One of the most difficult conceptual issues in the study of brain substrates of learning is that of learning vs performance. When biological variables are used that alter learned behavior, the actions could be on learning or on other aspects of performance, e.g. motor responses, sensory processing, "motivation." In classical conditioning, it is at least possible to make independent assessments of effects on the learned response (conditioned response, CR) and on performance of the reflex response (unconditioned response, UR) (Gormezano 1972, Mauk et al 1982). It is much more difficult to separate "learning" and "performance" factors in instrumental learning. Gabriel and associates have made ingenious use of discriminative stimulus control in avoidance learning (see Gabriel et al 1980, Orona et al 1982) in an effort to make such a separation.

Most conditioning studies of brain substrates of learning, whether classical or instrumental, have utilized an aversive unconditioned stimulus (UCS) (the works of Olds, Segal and associates are exceptions). Aversive learning is commonly characterized as occurring in two processes or phases, an initial associative process of "conditioned fear" or conditioned emotional response (CER), and a later process that involves learning of discrete, adaptive skeletal muscle responses (Konorski 1967, Miller 1948, Mowrer 1947, Prokasy 1972, Rescorla & Solomon 1967). Weinberger (1982a) recently surveyed aversive classical conditioning in infrahuman animals in terms of rate of learning and noted two clearly distinct categories. "Nonspecific" responses, indices of conditioned "fear," are acquired in 5 to 15 trials, but specific skeletal muscle responses require many more trials, from 50 to several hundred. Nonspecific responses are so defined because they are not specific to the UCS; they do not permit the animal to avoid the UCS whereas specific responses do. Nonspecific responses are mostly autonomic —heart rate, blood pressure, pupil diameter, galvanic skin response—but include nonspecific skeletal motor activity as well. Under normal conditions, both kinds of responses are learned. Evidence to date suggests that the neuronal substrates for these two aspects of aversive learning may differ, at least in part.

LESION AND ELECTROPHYSIOLOGICAL APPROACHES

Motor Neurons and Reflex Pathways

It seems unlikely on *a priori* grounds that an essential component of the memory trace is localized to motor neurons or nuclei. If lesions, drugs, or other treatments selectively abolish a learned response but have no effect at all on the reflex response, this argues against essential involvement of the reflex pathways, including the motor neurons, in the memory trace. Such data now exist for several learning paradigms, as is detailed below. Minimum latency of the CR provides another argument. Except for the alpha response, latencies of conditioned responses are always relatively long, typically 100 msec or more. Relatively direct connections from conditioned stimulus (CS) pathways to unconditioned reflex (UR) pathways would have latencies that are much too short to account for learned response latencies.

Conditioned Stimulus Pathways

The data here are more complex. Lesions of the CS pathways are of little help since it is not possible to distinguish between damage to sensory-specific pathways necessary for the learned response and damage to the memory trace. Electrophysiological data at least provide direct information about whether neuronal unit responses in sensory pathways change over learning. Results are mixed. Most such studies have used an acoustic CS. Positive and negative results have been reported for virtually all levels of the auditory system in at least some studies (e.g. Disterhoft & Stuart 1976, 1977, Gabriel et al 1975, Kettner et al 1980, Oleson et al 1975).

Recent data have clarified this picture. In a series of studies using classical conditioning of pupillary response to shock in paralyzed cat, Weinberger and associates found clear training-induced changes in the medial but not the ventral division of the medial geniculate body (Ryugo & Weinberger 1976, 1978, Weinberger 1980, 1982a,b). Gabriel et al (1975) reported that such changes occurred more medially than ventrally in the medial geniculate body of the rabbit in instrumental avoidance learning. Birt & Olds 1982), using a hybrid classical-instrumental procedure with food reward in rat, similarly found no learning-related changes in unit activity in the central nucleus of the inferior colliculus or the ventral nucleus of the medial geniculate body, but did report changes in structures that are not auditory-specific. Use of a signal detection paradigm, where equal probability of occurrence and failure of occurrence of the learned response to the same acoustic stimulus can be established, permits the occurrence of the learned response to be used as the independent variable (Kettner et al 1980, Kettner & Thompson 1982). Under these conditions, there are no learning-related changes in responses of units in the anteroventral cochlear nucleus, central nucleus of the inferior collicular, or ventral division of the medial geniculate body.

In sum, for the auditory system, there now appears to be consensus that the mainline relay nuclei do not show training-induced changes in neuronal activity. When such changes are observed, they occur in the "surround" nuclei and regions that are not auditory-specific.

Cohen has recently reported learning-related changes in unit activity in the visual pathways during heart rate conditioning to a visual stimulus in the pigeon. The visual pathways necessary for learning are themselves of interest in this preparation. There are three parallel pathways, any one of which can support conditioning. Ganglion cell activity recorded from optic nerve fibers shows no training-induced modification. However, the phasic but not tonic components of the responses of neurons in two central thalamic optic relays—the principal optic nucleus and the nucleus rotundus—are differentially modified by associative training (Wall et al 1980, Gibbs & Cohen 1980, Cohen 1982).

Hypothalamus and Amygdala

A relatively consistent picture is emerging from studies on cardiovascular conditioning in three species; baboon, rabbit, and pigeon. The paradigm is classical conditioning with a several second auditory or visual conditioned stimulus (CS) terminating with an electric shock unconditioned stimulus (UCS). As noted above, the conditioned heart rate response is viewed by most workers as a component or reflection of the more general process of "fear" conditioning, i.e. the conditioned emotional response (CER) (e.g. Schneiderman 1972). Important paradigmatic contributions have been made by Smith and associates (Smith et al 1980) in their work on the baboon. They include a behavioral measure of the CER—lever press response suppression in the cardiac conditioning regime, along with a measure of effects of exercise. Small, discrete bilateral lesions of the perifornical region of the hypothalamus in the baboon abolish the entire learned cardiovascular response complex completely, permanently, and selectively. The lesion has no effect on the reflex cardiovascular response and, most important, has no effect on the behavioral measure of the CER or on cardiovascular changes associated with exercise. This hypothalamic lesion effect seems to be on the efferent or motor-specific side of the learned response circuit, since the behavioral signs of conditioned fear are still present. In some sense it is surprising that a structure as ancient as the hypothalamus exhibits such a highly selective action for a *learned* response, as opposed to reflex and general regulation. Pribram et al (1979) found that bilateral ablation of the amygdala also abolishes heart rate conditioning in the monkey.

In rabbit, the learned response is a decrease in heart rate rather than the pressor response in baboon, and is probably mediated by the vagus nerve (Schneiderman et al 1966, 1969, Yehle et al 1967). To our knowledge, no one has yet examined effects of appropriate hypothalamic lesions on the conditioned heart-rate slowing response in the rabbit. Far lateral hypothalamic lesions have little effect (Francis et al 1981). In a recent series of papers, Kapp and associates have shown that the amygdala plays a critical role. Lesions of the central nucleus virtually abolish conditioned heart-rate slowing in the rabbit, but have no effect on the reflex responses or on the initial orienting response to the CS, which itself is a slowing. Injection of beta-adrenergic blockers in the central nucleus partially attenuates the CR as does injection of the opiate levorphanol. Further, injection of the opiate antagonist naloxone enhances the CR (Kapp et al 1982). A direct pathway exists from the central nucleus of amygdala to the vagal preganglionic cardioinhibitory neurons in the dorsal motor nucleus of the vagus (Schwaber et al 1980). Finally, unit recordings in the central nucleus indicate that at least some neurons show increased discharge during the development of conditioned bradycardia (12/34) and a few show significant correlations with the magnitude of the CR (3/24) (Kapp et al 1982).

Lesions of other brain structures can also influence conditioned bradycardia in the rabbit. Both hippocampal lesions and lateral septal lesions enhance the heart rate CR (Powell & Buchanan 1980, Hernandez & Powell 1981). Lesions of the anterior cingulate region of the cerebral cortex produce a substantial impairment of the conditioned bradycardia response (Buchanan & Powell 1982). Although results from different laboratories often are not directly comparable, this cingulate lesion impairment does not appear to be as great as the amygdala lesion impairment found by Kapp and associates.

Cohen and associates have completed the most extensive and detailed analysis of both the efferent and afferent limbs of the learned response circuit for heart-rate conditioning in their work on the pigeon (e.g. Cohen 1969, 1974, 1975, 1980, 1982). Like the baboon, the pigeon shows a conditioned increase in heart rate. Both the vagi and sympathetic nerves participate, but the predominant influence is from the right sympathetic cardiac nerve. The brain stem was mapped with electrical stimulation, together with electrophysiological recording and anatomical and lesion-behavior studies, and the efferent pathway was defined (see Cohen 1980). In brief, there is a system from the avian homologue of the amygdala to the hypothalamus, which then projects down to the final common path neurons via a ventral brainstem pathway (Cohen 1980). The CR is completely prevented by lesions of the amygdala, the terminal field of the amygdala fibers in the hypothalamus, and the brainstem course of fibers from this region of the hypothalamus (Cohen 1975, Cohen & MacDonald 1976, Cohen 1980). As in the mammal, these lesion effects are selective in that they do not abolish reflex cardiovascular responses, only the conditioned response. The pathway from amygdala to motor neurons is clearly essential for expression of the conditioned heart rate response in the pigeon, as in the mammal. It is not yet known whether some part of the memory trace circuit may be included in the amygdala.

Kapp et al (1982) interpret the role of the amygdala in aversive conditioning to be motoric, i.e. on the motor-specific side of the learned response circuit, at least for conditioned bradycardia in the rabbit. The more general role of amygdala in behavioral aspects of fear conditioning is less than clear clear, but it is certainly much involved (see Goddard 1964, Kaada 1972, Kesner & Wilbrun 1974 for reviews). In a recent study (Liang et al 1982), the impairment of retention of aversive learning (one-trial inhibitory avoidance in rats) produced by amygdala lesions was found to depend strongly on the time of lesion. If the lesion was made 10 days after training, it produced *no deficit* in retention. This result was interpreted to mean that the amygdala plays a modulatory role in aversive learning, i.e. that it acts on a memory trace system established elsewhere in the brain.

The amygdala appears also to play a critical role in taste and odor aversion learning. Large lesions of the basolateral amygdala significantly disrupt a prior learned taste aversion in rats (Nachman & Ashe 1974), as do lesions of the gustatory neocortex (Quellet et al 1975). In recent work, Garcia and associates (1982) examined lesion effects on learned odor aversion, taste aversion, and taste potentiation of odor aversion. Lesions of the neocortical taste area disrupt taste aversion conditioning, but not odor aversion learning, or taste potentiation of odor aversion learning (i.e. the cortical deficit is in taste discrimination; see also Lasiter & Glanzman 1982). Reversible (procaine) lesions of the amygdala disrupt learned odor aversion and taste potentiation of odor aversion induced by pairing with an illness stimulus (lithium chloride), but do not disrupt conditioned response suppression with an odor stimulus conditioned to a footshock UCS (Garcia et al 1982). Garcia's experimental differentiation between the associative and sensory-discriminative aspects of odor-taste aversion learning provides a useful paradigm for further analysis of neuronal substrates of this important form of learning. Finally, it should be noted that the amygdala is also implicated in appetitive learning. Spiegler & Mishkin (1981) showed that lesions of the amygdala markedly impair one-trial learning of object reward associations in the monkey, but do not impair one-trial object recognition learning (Mishkin & Oubre 1976).

Motor Cortex

Woody and associates have made significant progress in analyzing cellular mechanisms of the short-latency, conditioned eyeblink response to a click paired with a glabellar tap (400 msec interstimulus interval) in the cat (Woody 1970, Woody & Black-Cleworth 1973, Woody et al 1970, 1974, Brons & Woody 1980). This short latency alpha conditioned response differs from conventional associative eyelid or nictitating membrane (NM) conditioning in several ways: it is elicited by the click CS at the beginning of training, it has a very short latency (20 msec), the latency does not shift over training, it is not "adaptive" in the same sense that the CR is not present at the time of the unconditioned stimulus (UCS), it requires a much longer time (600–900 trials) for learning, and the motor cortex is essential for development of the conditioned response.

The development of the conditioned alpha response is selective. If the CS is paired with a glabellar tap, the eyelid response becomes conditioned; if paired with stimulation of the nose, the nose twitch develops. Extracellular microstimulation of the motor cortex indicates that the threshold activation level of those neurons electing the CR is reduced following conditioning—they are more excitable. This effect is specific, for eyelid vs nose twitch training. In a technical tour-de-force, Brons & Woody (1980) recorded intracellularly from 290 neurons in motor cortex in the absence of peripheral stimulation and measured threshold currents in already trained animals: although a group given UCS alone showed a decrease in threshold relative to a group given CS alone, only groups given prior paired CS-UCS trials showed long-lasting decreases. This decrease persisted throughout extinction. In other work, the increased excitability of motor cortex neurons was found not to be accompanied by increases in spontaneous discharge activity or detectable changes in resting membrane potential. Similar effects could be produced by extracellular application of acetylcholine or cyclic GMP. Woody (1982) suggests that these results could be due to long-lasting changes in the post-synaptic neurons, e.g. an increase in dendritic resistance.

Woody's work indicates that the motor cortex appears to be a critical site of neuronal plasticity for the conditioned alpha eyelid response and that this plasticity involves increased excitability of neurons there, probably due to persisting post-synaptic changes in the neurons studied. Most conditioned responses, both classical and instrumental, do not require the motor cortex. It would be most interesting if Woody's results were found to hold for all instances of alpha conditioning, at least for skeletal muscle responses. If so, it would suggest a general fundamental difference in neuronal substrates between alpha conditioning and associative conditioning.

Voronin (1971) has used simple analogues of classical conditioning in studies of cellular plasticity in the motor cortex (see Tsukahara 1981). Recently, he developed a paradigm somewhat analogous to Woody's, in that a click was used as the CS (Voronin 1980). The UCS was motor cortex stimulation plus hypothalamic stimulation, and the behavioral CR was EMG activity in the contralateral forepaw. Intracellular recordings were made from cortical neurons. In general, cortical neurons showed increased excitability over the course of pairings. Some neurons showed initial responses to the neutral click, and virtually all responded to the loud click with the same short latency distribution as to the neutral click after paired training. Voronin notes that these cortical and EMG responses to loud clicks are components of the startle response (probably analogous to an alpha response). The conditioning regime thus increases the excitability of the startle circuit.

O'Brien and associates have developed an analogue of differential classical conditioning for cells of origin of pyramical tract fibers in the motor cortex of the paralyzed cat (O'Brien et al 1977, O'Brien & Quinn 1982). Left and right hindpaw shocks were used as the conditioned stimulus (CS+) and its control (CS−), and the CS+ was paired with antidromic stimulation of the pyramidal tract as the UCS. The activity of neurons so identified in motor cortex was recorded. A significantly greater increase in response developed for the CS+ than for the CS−. When electrical stimulation of VL or VPL regions of the thalamus were used as CS+ and CS−, no differential conditioning was found (an interesting result because the peripheral hindpaw stimuli activate some portions of VPL and VL). As O'Brien notes, these data cannot be explained by a simple pairing theory of conditioning in which any two pathways to a neuron can be used as the CS and UCS. Voronin (1980) emphasized the same point in noting that paired motor cortex and lateral hypothalamic stimulation was a more effective UCS than motor cortex stimulation alone.

Cerebellum

Classical conditioning of the eyelid or NM response is widely used for the study of neuronal substrates of associative learning, usually in the rabbit but also in other species. The paradigm was developed in the rabbit by Gormezano and associates (Gormezano et al 1962). The conditioned eyelid and NM responses behave essentially identically and are almost perfectly correlated over the course of learning (Gormezano 1972, McCormick et al 1982b). This system has a number of advantages, including the very large behavioral literature on eyelid conditioning in a number of species, particularly humans (see Thompson et al 1976, Disterhoft et al 1977). A number of studies report changes in neuronal activity in various brain structures during learning in this paradigm, as noted elsewhere. Here, we focus on the essential learned response circuit for the standard delay-conditioned response.

Decorticate and thalamic rabbits can learn the conditioned response, as can decerebrate cats (Oakley & Russell 1972, Norman et al 1977; D. Enser 1976, personal communication). One should always be cautious, however, in interpreting results of studies on reduced preparations. For example, although the acute spinal cat shows clear learning of a short latency, conditioned hindlimb flexion response (Patterson et al 1980, Misulis & Durkovik 1982), this does not necessarily mean that such neuronal plasticity is established at the spinal level when an intact animal learns a leg flexion response. Such reduced preparations can, of course, offer promising models for the study of plasticity at the cellular level.

Several recent studies have reported abolition of the learned eyelid/NM response by selective brain lesions. In particular, lesions ipsilateral to the trained eye in several locations in the neocerebellum—large ablations of the lateral portion of the hemisphere, localized electrolytic lesions of the dentate/interpositus nuclei, and small, discrete lesions of the superior cerebellar peduncle—permanently abolish the CR but have no effect on the UR and do not prevent learning by the contralateral eye (McCormick et al 1981, 1982a,b, Thompson et al 1982, 1983). If training is given before unilateral cerebellar lesion, the ipsilateral eye cannot learn, but the contralateral eye subsequently learns as though the animal is normal and new to the situation (Lincoln et al 1982). If training is given before unilateral cerebellar lesion, the learned response is abolished in the ipsilateral eye but the contralateral eye learns rapidly, with significant savings (McCormick et al 1981, 1982a). Lesions in several locations in the ipsilateral pontine brain stem produce a similar selective abolition of the CR (Desmond & Moore 1982, Lavond et al 1981, Moore et al 1982, Thompson et al 1983). Although some uncertainty still exists, the learning-effective lesion sites in the pontine brain stem appear to follow the course of the superior cerebellar peduncle.

Neuronal unit recordings from the dentate/interpositus nuclear region show evoked responses to CS and US onsets and, in some cases, the development of a temporal neuronal model (a pattern of increased frequency of unit discharges reflects the amplitude/time course of the behavioral response) of the learned response but not the unlearned reflex response. This neuronal temporal model of the learned response appears to develop over the course of training in very close association with the development of the learned behavioral (NM) response and precedes it within a trial (McCormick et al 1982a, Kettner & Thompson 1982).

Taken together, these results indicate that the cerebellum is an obligatory part of the learned response circuit for eyelid/NM conditioning. Since decerebrate animals can learn the response, this would seem to localize an essential component of the memory trace to the ipsilateral cerebellum or its major afferent/efferent systems. That a neuronal unit "model" of the learned behavioral response develops in the cerebellar deep nuclei would seem to localize the process to the cerebellum or its afferents. The possibility that cerebellar lesions produce a modulatory disruption of a memory trace localized elsewhere in the brain seems unlikely. If so, it must be efferent from the cerebellum, since discrete lesions of the superior cerebellar peduncle abolish the behavioral learned response. Yet the neuronal model of the learned response is present within the cerebellum. In this context, an earlier Soviet literature indicates that in dogs well trained in leg-flexion conditioning, complete removal of the cerebellum permanently abolishes the ability of the animals to make the learned discrete leg-flexion response, but not to show conditioned generalized motor activity (Karamian et al 1969). Eyelid/NM conditioning is an instance of aversive learning. In this context (see above), the cerebellar system seems a very good candidate for the second phase of learning, at least for associative conditioned responses that involve a discrete, adaptive, striated muscle response.

The cerebellum has been suggested by several authors as a possible locus for the coding of learned motor responses (Albus 1971, Eccles et al 1967, Ito 1970, Marr 1969; but see Llinas et al 1975). Cerebellar lesions impair a variety of skilled movements in animals (Brooks 1979, Brooks et al 1973). In addition, neuronal recordings from Purkinje cells of the cerebellar cortex have implicated these cells in the plasticity of various responses (Gilbert & Thach 1977, Dufosse et al 1978).

Ito has developed a most interesting experimental model of induced neuronal plasticity using the vestibulo-ocular reflex (VOR) and has applied the general Marr-Albus cerebellar models of motor learning to it (e.g. Ito 1970, 1974, 1977, 1982a,b). Plasticity of the VOR was reviewed recently in depth and with differing interpretations by Miles & Lisberger (1981) and by Ito (1982b). Ito has also demonstrated a striking, simpler form of conjunctive use-dependent plasticity of Purkinje cells in the flocculus and suggests two possible cellular mechanisms, both of which involve decreased sensitivity of chemical (glutamate?) receptors on Purkinje cells (Ito 1982a,b).

Red Nucleus

A major efferent target of the cerebellar hemisphere, via the superior cerebellar peduncle, is the contralateral red nucleus. Smith (1970) reported that large unilateral lesions in the red nucleus and vicinity markedly impaired a classically conditioned flexion response of the forelimb contralateral to, but not ipsilateral to, the lesion in cats. Tsukahara (1981, 1982, Tsukahara et al 1981) developed a simplified preparation based on this paradigm. A stimulating electrode to provide the CS is implanted in the cerebral peduncle and the peduncle lesioned caudal to the cortico-rubral fibers. The unconditioned stimulus is shock to the contralateral forepaw. The CS pulse train is adjusted to produce a weak flexion response of the forelimb. Animals learn the leg flexion response to peduncle stimulation and the excitability of the pathway increases. Tsukahara argues that the excitability increase is at the synaptic junctions of peduncle fibers on the red nucleus (see Tsukahara 1982).

An older literature indicates that carnivores can learn a discrete leg-flexion response following complete decortication (e.g. Bromily 1948, Poltrew & Zeliony 1930). Consequently, although the cortico-rubral tract may normally be involved in leg-flexion conditioning in the intact animal, it is not essential. Cerebellar lesions, on the other hand, do apparently permanently abolish the discrete leg-flexion conditioned response (see above). The effects of cerebellar lesions have not yet been examined in Tsukahara's simplified preparation.

Other Brain Regions

HIPPOCAMPUS   The hippocampus has been the most extensively studied region of the brain in the context of learning and memory, following the clinical reports of Scoville & Milner (1957) and Penfield & Milner (1958). The literature can only be described as complex (for reviews see Isaacson & Pribram 1975, Moore & Solomon 1980, O'Keefe & Nadel 1978, Olton et al 1979, Squire 1982).

In terms of electrophysiology, the pioneering studies of Olds, Segal, and associates (see Olds et al 1972) and Vinogradova (see 1975) implicated the hippocampus in learning in rat and rabbit. Olds et al (1978) showed learning-induced changes in unit activity in a classical/instrumental food reward task (see also Disterhoft & Segal 1978). Segal (1977a,b) has shown that such conditioning augments potentials evoked by commissural stimulation, possibly mediated by catecholaminergic afferents, and Weisz et al (1982) demonstrated enhancement of entorhinal-dentate synapses in classical conditioning. Hirsh (1973) and Best & Best (1976) have shown that latent inhibition procedures influence hippocampal unit activity in subsequent conditioning.

Berger, Thompson, and associates showed a marked learning-induced increase in the activity of identified pyramidal neurons during eyelid/NM conditioning in the rabbit that occurs very early during the course of training, and substantially precedes behavioral learning (Berger & Thompson 1978a–c). In addition, just prior to a conditioned NM response, hippocampal pyramidal neurons begin discharging with a distinctive pattern of firing that forms a clear temporal model of the amplitude-time course of the learned behavior (Berger & Thompson 1978b, Berger et al 1980). These same types of cellular changes have recently been shown to occur during trace conditioning (Thompson et al 1982) and during discrimination/reversal conditioning of the NM response as well (Berger 1982), learning paradigms for which an intact hippocampus has been shown to be essential (Berger & Orr 1982, Orr & Berger 1981, Weisz et al 1980). Analysis of hippocampal activity during a variety of control conditions has shown that these cellular changes occur only during associative learning and are not due to nonassociative aspects of the conditioning process (Berger & Thompson 1978a, Hoehler & Thompson 1980, see also Gabriel et al 1980). Input-output analysis to date suggests that a significant part of the plasticity develops in the CA3-CA1 region—at least it is not relayed from the medial septum or dentate granule cells (Berger & Thompson 1978c, Clark et al 1978, Weisz et al 1982).

Deadwyler and associates (1979a,b, 1981, 1982) have shown that two-tone CS-evoked dentate field potentials of differing waveform and latency develop differentially over training, discrimination, and reversal in an appetitive instrumental task in rat. Both the shorter latency response and differential behavioral responding are impaired by damage to entorhinal cortex.

Although hippocampal lesions do not impair simple acquisition in many of the tasks noted above (see Moore & Solomon 1980, O'Keefe & Nadel 1978), such lesions markedly impair or prevent certain aspects of learning (e.g. discrimination reversal, trace learning, conditioned and latent inhibition) in these paradigms (see e.g. O'Keefe & Nadel 1978, Orr & Berger 1981, Solomon & Moore 1975, Weisz et al 1980). Mishkin (1978) has shown that combined lesions of hippocampus and amygdala can reproduce in monkey the human delayed-matching to sample deficit (see Babb 1982, Squire 1982 for recent human data). Growing, but still indirect, evidence favors long-term potentiation as the mechanism underlying learning-induced neuronal plasticity in the hippocampus (Andersen et al 1977, Berger & Thompson 1978a, Lynch et al 1977, Swanson et al 1982, Thompson et al 1982).

Lesions of the hippocampus have also been shown to severely impair learning of spatial tasks (see O'Keefe ' Nadel 1978, but see Kolb et al 1982, for comparable impairments with lesions of the frontal cortex). Consistent with the hippocampal lesion effect are electrophysiological data demonstrating strong correlations between increased firing of hippocampal neurons and location of an animal in space (O'Keefe 1976, O'Keefe & Conway 1978, Olton et al 1978, Miller & Best 1980). This evidence has traditionally been interpreted within the framework of "spatial memory" (O'Keefe & Nadel 1978), but Olton et al (1979, 1980) have developed an alternative interpretation in terms of "working memory." It is not clear that the spatial correlates of hippocampal neurons develop as a result of learning or experience. In fact, available evidence indicates that spatial correlates of hippocampal cells are present at an animal's first exposure to a new environment, and are not modified by time or experience in that environment (Hill 1978, Kubie & Ranck 1982). Thus, although demonstrations of spatially related unit correlates are quite striking, their specific relation to associative learning is, as yet, uncertain.

CEREBRAL CORTEX   The cerebral cortex is particularly difficult to treat from the point of view of simpler forms of learning. Decorticated animals can learn a wide range of classically and instrumentally conditioned responses (see above and Oakley 1979, 1981). The cerebral cortex seems highly specialized to represent and process detailed and complex sensory and sensory-motor information. Mishkin's elegant lesion analysis of visual learning in the monkey led from area 17 to the inferotemporal cortex (e.g. Mishkin 1966). At what point does visual information processing become visual memory (see Pribram 1971)?

The large literature on recovery of function in learning has focused on the cerebral cortex—significant recovery generally occurs (Meyer & Meyer 1977, Stein et al 1974, Finger 1978). In humans, the most severe impairments in ability to place new information in long-term storage result from damage to subcortical structures (Squire 1982). On the other hand, the cerebral cortex appears essential for language and complex spatial memory in humans. It is certainly necessary for cognitive functions. A proper treatment of the possible roles of the cerebral cortex in learning and memory, particularly the more complex aspects of learning and information processing, would require several separate reviews. We examine possible learning-related anatomical plasticity in the cerebral cortex below.

BASAL GANGLIA   Although manipulations of the basal ganglia can alter learned performance (e.g. Hore et al 1977, Powell et al 1978), no clear evidence to date indicates that the basal ganglia play critically important roles in learning and memory processes.

NEUROANATOMICAL PLASTICITY INDUCED BY ASSOCIATIVE LEARNING

Once a change in cellular activity as a result of learning can be localized successfully to an identified cell or synapse, a major question then concerns the underlying mechanism of that cellular change. Historically, it has often been suggested that learning results in morphological alterations of the nervous system (e.g. Hebb 1949). That is, learning-induced changes in behavior might be mediated by alterations in the structural and connectional organization of the CNS, rather than (or in addition to) changes in the biophysical properties of nerve cells (see Lesion and Electrophysiological Approaches, above) or their biochemical capacities (see Biochemical Approaches, below). Unfortunately, little data currently relates associative learning to well-defined and specific neuroanatomical plasticity in mammalian species. However, several lines of research have shown that altering the behavioral and sensorial experience of an organism, particularly during development, can induce substantial neural changes at the anatomical level. For example, research on the effects of environmental variables has shown that exposing animals to an "enriched" environment (EC) of a variety of manipulanda (or "toys"), social situations, larger home cages, etc results in dramatic increases in the number of neuroglia, branching patterns of dendritic processes, numbers of dendritic spines, postsynaptic density length, and synapse shape of neocortical, hippocampal, or cerebellar neurons when compared to animals exposed to an isolated condition (IC) of no toys, no other animals, and small cages (Altman & Das 1964, Diamond et al 1966, Holloway 1966, Walsh et al 1969, Shapiro & Vukovich 1970, Møllgaard et al 1971, West & Greenough 1972, Volkmar & Greenough 1972, Globus et al 1973, Greenough & Volkmar 1973, Greenough et al 1973a, Diamond et al 1975, Rosenzweig & Bennett, 1976, Greenough et al 1978, Fiala et al 1978, Altshuler 1979, Floeter & Greenough 1979, Chang et al 1981). Alternatively, visual deprivation experiments have demonstrated a reduction in many of the same structural features after limiting sensory stimulation of an organism. For example, changes in neuron soma size, dendritic branching, number of dendritic spines, distribution of dendritic and terminal processes, synapse size, vesicle density, and synaptic grid features of visual neocortical neurons have all been shown to occur in response to visual deprivation (Gyllensten et al 1965, Cragg 1967, Valverde 1967, 1968, 1971, Fifkova 1968, 1969, 1970, Coleman & Riesen 1968, Le Vay et al 1980, Garey & Pettigrew 1974, Wiesel et al 1974, Vrensen & De Groot 1975, Borges & Berry 1976, Hubel et al 1977, Blakemore et al 1980, Muller et al 1981).

In many cases environmental enrichment and visual deprivation experiments are not readily interpretable within the framework of associative learning, because behavioral training in the sense of traditionally defined conditioning paradigms (i.e. classical and instrumental conditioning procedures) is not involved. Nevertheless, this literature provides an appropriate and instructive context for considering neuroanatomical plasticity as a potential mechanism for associative learning for the following reasons:

1. Results from enriched environment and visual deprivation experiments provide a wealth of evidence that structural alterations of nerve cells occur in response to peripheral stimulation of mammalian species.
2. The traditionally defined stimulus conditions that induce behavioral learning (e.g. CS-UCS intervals) can be seen simply as unique combinations of peripheral stimulation.
3. Psychological research has demonstrated associative learning in behavioral situations in which explicit relationships between discrete conditioning stimuli do not exist (i.e. latent learning, see Tolman & Honzik 1930).

From this perspective, the environmental enrichment/sensory deprivation literature, together with the few studies specifically demonstrating neuronal structural changes as a result of conditioning, strongly suggest that neuroanatomical plasticity serves as a mechanism for associative learning.

Neuroanatomical Plasticity Induced by Environmental Conditions

While the implications of experientially induced neuroanatomical plasticity for associative learning are intriguing, such phenomena must exhibit certain characteristics to be seriously considered as a candidate mechanism.

1. Neuroanatomical lability cannot be limited to development, but must extend into adulthood. Although initial attempts to extend the effects of environmental rearing to the adult were successful, results suggested that the capacity for anatomical plasticity was greater in immature animals than in adults (Malkasian & Diamond 1971, Diamond et al 1972). Later studies, however, have clearly shown that mature animals are capable of exhibiting structural changes induced by environmental variables (see Rosenzweig & Bennett 1976, Uylings et al 1978). In addition, several recent analyses have revealed that enriched environmental conditions can induce increased higher-order dendritic branching even when the EC treatment begins at middle age (Riege 1971, Green et al 1981) or old age, i.e. 600 days (Connor et al 1981).

2. Experientially induced changes must be relatively long-lasting. Very few studies have systematically tested the permanence of environmentally induced structural effects, but Cummins et al (1973) have shown that EC-IC differences in the structural features of neocortical neurons are maintained after over 500 days of exposure to each condition. Thus, not only can adult animals respond to environmental manipulations, but the effects of such manipulations are long-lasting.

3. If neuroanatomical plasticity is to be considered as a candidate mechanism, it must be demonstrably functional. The functional consequences of different environments have been determined primarily by behavioral measures of learning ability on some form of maze task. Animals raised in an EC environment are superior to those raised in an IC environment at maze learning (Hymovitch 1952, Bingham & Griffiths 1952, Forgus 1955, Walk 1958, Forgays & Read 1962, Schwartz 1964, Nyman 1967, Brown 1968, Edwards et al 1969, Greenough et al 1970, 1973b, Bernstein 1973, Cornwell & Overman 1981). Superior performance by EC animals has also been reported on appetitive operant tasks (Morgan 1973) and an avoidance task (Greenough et al 1970). Reliability of the EC-IC difference on learning has rarely been questioned. Instead, debate has focused on the specific environmental variable or variables responsible for that difference, or the specific behavioral trait modified by differential rearing (see Rosenzweig 1971 and Rosenzweig & Bennett 1976 for reviews). Irrespective of these issues, data demonstrating differences in learning rate between EC and IC treated animals are significant. First, they demonstrate that environmental manipulations inducing structural changes in neurons also induce behavioral

changes. Second, those behavioral consequences are related to learning ability, or at least to performance on learning tasks. Together, these data provide behavioral relevance for the experientially induced neuroanatomical plasticities described above, and suggest a role for those plasticities in the learning process.

4. Neuroanatomical plasticity must be capable of supporting the capacity and variety of mammalian memory. It is clear that environmental studies have catalogued a considerable number of neuroanatomical changes as a result of differential experience (see above). The quantitative and qualitative differences in peripheral stimulation that distinguish different environments are capable of inducing structural changes in many different morphological features of nerve cells, as well as in the number of supportive glia. Thus, the variety of these anatomical changes offers a rich substrate for the storage and transmission of considerable amounts of new information.

Neuroanatomical Plasticity Induced by Visual Deprivation

The various structural changes associated with visual deprivation have been thought not to meet the first criterion of generalization to the adult organism. Various studies have demonstrated that manipulations of visual stimulation are effective in inducing structural and functional changes only during a "critical period" approximating the first 6–12 weeks of life, and are generally ineffective in the adult (Wiesel & Hubel 1963, Hubel & Wiesel 1970; see Cynader et al 1980 for a more recent evaluation of critical period length). Although the immature organism clearly is more susceptible than adults to environmental influence, several studies have documented experiential modification of the adult visual system as well (Creutzfeldt & Heggelund 1975, Brown & Salinger 1975, Maffei & Fiorentini 1976; see Berman et al 1979 for negative findings). In addition, preceding sections of this review discuss a number of studies demonstrating that behavioral conditioning procedures alter the electrophysiological properties of sensory cortical neurons in adult animals. Together, these data raise the possibility that, even in the adult organism, physiological plasticity is accompanied or mediated by anatomical plasticity. In fact, a recent study by Vrensen & Cardozo (1981) has shown that visual discrimination learning in the adult rabbit is paralleled by a number of neuroanatomical changes in visual cortical neurons. Thus, although the specific procedures and experimental conditions reviewed here may be more effective during development, plasticity of visual or sensory brain systems is not excluded in the adult. On the contrary, available electrophysiological evidence indicates that they are highly plastic, and that such plasticity can be induced by behavioral learning. What distinguishes "critical period" neuroanatomical plasticity from learning-induced neuroanatomical plasticity in the adult may be that the latter requires for its expression reinforcement of behavior, or temporal contiguity between the UCS and the sensory stimulus used as the CS. In other words, the adult nervous system may be just as modifiable as the immature nervous system, and may be modifiable along the same morphological dimensions. The prerequisites of a stimulus sufficient to induce neuroanatomical plasticity may simply be much more restricted for the adult than the developing animal. The critical period may reflect whatever processes are responsible for narrowing the prerequisites, rather than reflecting a decreasing capacity of the CNS for structural alteration.

Results of visual deprivation experiments have been very useful in demonstrating that experientially induced structural changes are paralleled by physiological changes. In the normal, adult cat or monkey, the majority of cortical neurons (except for those in layer IV) are known to be driven binocularly (Hubel & Wiesel 1962, 1965, LeVay et al 1977, 1980). Numerous studies have shown that the neuroanatomical substrates of binocularity remain highly plastic throughout the critical period during development. More specifically, Hubel et al (1977) and Blakemore et al (1980) have shown that monocular deprivation results in a widening or "sprouting" of corticopetal terminal fields in layer IV from the nondeprived eye, and a corresponding narrowing or retracting of corticopetal fibers from the deprived eye. Consistent with these neuroanatomical data, physiological analyses have demonstrated that monocular deprivation causes a dramatic decrease in effective activation of visual stimuli presented to the deprived eye (Wiesel & Hubel 1963), so that the majority of cells are driven only monocularly. Furthermore manipulations need not be as radical as monocular deprivation. A number of procedures that introduce differences in visual stimulation of the two eyes are sufficient to disrupt binocularity (Hubel & Wiesel 1965, Blakemore et al 1978a, Blakemore & Van Sluyters 1975, Blakemore & Eggers 1978, Tsumoto & Suda 1979, Bennett et al 1980). The increase in proportion of monocularly driven cells is not merely a result of degeneration or inhibition of afferent input from the deprived eye (Hubel et al 1977, Blakemore & Hillman 1977, Winfield 1981). Instead, electrophysiological studies have confirmed the neuroanatomical data in showing that the "width" of ocular dominance columns for the nondeprived eye of a monocularly deprived animal are greater than for the deprived eye (Hubel et al 1977, Blakemore et al 1978a). In addition, the physiological effects of deprivation can be reversed (within the critical period) by forced exclusive use of the previously deprived eye (Blakemore & Van Sluyters 1974, Movshon 1976).

In addition to binocularity, the orientation specificity of visual cortical neurons is stimulation-dependent. Whereas cells in adult animals exhibit very specific line orientation for optimal activation (Hubel & Wiesel 1962,

1965), striate neurons of visually inexperienced animals show much reduced preferences. Many cells in these animals are unresponsive to visual stimuli; many that are visually driven show no preference for spot or line stimuli, and respond weakly and variably when activated (Pettigrew 1974, Pettigrew & Freeman 1973, Pettigrew & Gary 1974, Blakemore & Van Sluyters 1974, 1975, Movshon 1976, Mower et al 1981).

The high degree of stimulation dependence required for the integrity of orientation selectivity strongly suggests that visual exposure to only certain linear orientations can shape or determine the orientation preference of visual cortical cells. Experiments using goggles to expose each eye to a different linear orientation (Hirsch & Spinelli 1971), or special environments with only one linear orientation (Blakemore & Cooper 1970), or no linear stimuli (Pettigrew & Freeman 1973) have all demonstrated a strong correlation between visual experience and subsequent orientation preference of visual cortical neurons.

The issue of permanence of experientially induced changes in the visual system is complicated by the existence of a critical period during development. Many studies have shown that if visual deprivation conditions are continued past the end of the critical period, the anatomical and physiological consequences are "permanent" (e.g. Hubel & Wiesel 1970, Spinelli et al 1972, Blakemore & Van Sluyters 1974). But in this context, "permanence" has little significance, because the nervous system may not have the same capacity for plasticity, or may require different stimulus prerequisites for plasticity, when initial conditions are imposed (within the critical period) and when their consequences are tested (past the critical period). With respect to this issue, then, three comments are appropriate.

1. Within the critical period it is apparent from the majority of studies that experientially induced changes persist for as long as the conditions are maintained.
2. Recent evidence indicates that experientially induced anatomical plasticity does occur in the adult animal (e.g. Creutzfelt & Heggelund 1975). From the perspective of structural changes as a mechanism for associative learning, it would be more appropriate to examine the permanence of visual system plasticity in the adult animal, and such studies have yet to be reported.
3. Though experientially induced anatomical changes should be long-lasting if they are to serve as a mechanism for associative learning, they need not be permanent.

Too often those attempting a parallel between structural plasticity and learning placed an unwarranted constraint on their results by demanding that any anatomical changes be "permanent." The process of forgetting is as robust and well-documented behaviorally as the process of learning (Kimble 1961), so not all experientially induced anatomical alterations should be expected to last indefinitely. With regard to this criterion, studies of the effects of visual deprivation are instructive in demonstrating that deprivation-induced anatomical plasticity is not only long-lasting, but is also reversible (e.g. Blakemore et al 1980).

Visual deprivation studies have documented structural changes at the level of the neuron soma, dendritic branch, dendritic spine, terminal process, terminal vesicle, and synapse. As with the effects of environmental differences, then, visual deprivation studies have shown that a rich substrate of anatomical characteristics is available for environmentally induced alteration.

Associative Learning

Relatively few studies have investigated potential neuroanatomical plasticity coincident with associative learning in adult mammals. Several reports have appeared in the last few years, however, that have examined various morphological characteristics of neurons after behavioral conditioning. Many studies are extensions of previous work by the authors on neuroanatomical plasticity induced by environmental manipulation or visual deprivation.

One of the first reports of the effects of learning procedures on morphological features of mature cortical neurons was conducted as part of a EC-IC experiment (Cummins et al 1973). One group of adult animals was given only training on a maze task after being exposed to IC conditions. Measures of forebrain weight revealed significant increases in material from trained animals compared to other animals maintained in IC throughout the training period. A later study by Greenough et al (1979) revealed that at least one factor contributing to the training-induced weight change is increased dendritic branching of higher-order processes on oblique extensions of apical dendrites of pyramidal neurons. There were also indications of an increased number and length of first-order oblique branches from apical dendrites. More recently, Larson & Greenough (1981) trained rats in a forepaw motor task (handedness reversal) and examined dendritic branching patterns of layer V pyramidal neurons in the forelimb region of motor cortex. Results were compared with branching patterns of layer V cells of the contralateral motor cortex. Distal portions of apical dendrites from cells of cortical areas projecting to the trained forearm exhibited more profuse branching patterns than cells projecting to the untrained arm.

Rutledge et al (1974) have examined cortical dendritic morphology following behavioral conditioning using neocortical brain stimulation as a CS and foreleg shock as a UCS to condition leg flexion. Separate animals were given unpaired presentations of the CS and UCS to control for nonassociative effects of training. Neocortical pyramidal neurons from the stimulated region and from the contralateral hemisphere activated through collosal connections were examined. Results showed that material from conditioned animals exhibited an increased number of dendritic spines on vertical and oblique branches of apical dendrites when compared to material from control animals.

Vrensen & Cardozo (1981) have trained adult rabbits on an operant, pattern discrimination task and subsequently examined several parameters of visual cortical synaptic morphology. Results showed that (a) the frequency of complex synaptic grids increased substantially in trained animals, (b) the surface area and number of dense projections per grid were smaller in trained animals, both for complex grids and the total synaptic population, and (c) there was a significant increase in the length of the postsynaptic density. No changes were seen in the size of pre- or postsynaptic terminals, number of vesicles, or other measures. Other observations by the authors led them to suggest that behavioral conditioning resulted in a "focalization" of synaptic grids that would act to increase synaptic efficacy. Thus, behavioral conditioning may be producing an enhanced activation of visual cortical neurons, possibly those neurons responsive to features of the visual patterns used as conditioning stimuli.

Spinelli and co-workers have recently reported experiments that extend earlier work on the effects of visual deprivation (Spinelli & Jensen 1979, Spinelli et al 1980). Although these studies were conducted in the developing animal, they are important in showing that, in addition to producing neuroanatomical changes, conditioning can result in functional alterations of the CNS that reflect specific aspects of the training conditions. Kittens raised under normal visual conditions were trained in an operant avoidance task requiring forearm flexion and were cued with either a horizontal or vertical visual pattern (Spinelli & Jensen 1979). The area of somato-sensory cortex responsive to forearm stimulation was found to be greater in the hemisphere projecting to the trained forearm compared to the opposite hemisphere. The number of polymodal neurons responsive to visual stimuli in the forearm region had also increased significantly in the trained hemisphere and, more interesting, the majority of these cells exhibited an orientation preference for the stimulus that elicited the behavioral response. In a later study (Spinelli et al 1980), the authors reported a substantial increase in dendritic branching of neurons within the cortical region demonstrating the physiological effects.

A recent study by Wenzel et al (1980) trained rats in a brightness discrimination task (using shock) and examined the number of synapses on apical dendrites of CA1 hippocampal pyramidal cells. Material from trained animals was compared with that of control animals that performed the identical response and received the same number of shocks, but shocks that were not correlated with brightness differences. Results showed that conditioning produced a 40% increase in the number of synapses in trained animals. While this difference decreased over days, a significant increase over control counts was still detectable 14 days after training.

Finally, Fifkova & Van Harreveld (Fifkova et al 1978, Fifkova & Van Harreveld 1978) have reported a change in dendritic morphology in the molecular layer of the hippocampal dentate gyrus after conditioning of an appetitive response in rats. Specifically, training induced an enlargement or swelling of dendritic spines in the outer two-thirds of the molecular layer, the synaptic region corresponding to the site of termination of afferents from the entorhinal cortex. The additional significance of this result is that it can also be induced by electrical stimulation of the entorhinal-dentate pathway (Fifkova & Van Harreveld 1977) at stimulation frequencies known to induce long-lasting facilitation (or long-term potentiation, LTP) of synaptic efficy (Bliss & Lomo 1973, Douglas & Goddard 1975). In fact, Lee et al (1980) have recently reported that electrical stimulation resulting in LTP produces in the same tissue an increased number of synapses on dendritic shafts, and a decrease in the variability of several morphological characteristics of dendritic spines. These authors did not, however, confirm the increased spine area reported by Fifkova & Van Harreveld (1977). LTP is induced by a high frequency train of electrical stimulation, and in this sense is a nonassociatively induced form of synaptic plasticity. Nevertheless, several parallels exist between LTP and neural plasticity induced by associative learning (see McNaughton et al 1978, Levy & Steward 1979, 1982, Berger, 1982, Swanson et al 1982). Given the progress being made toward understanding mechanisms underlying LTP (Baudry & Lynch 1980, Duffy et al 1981), the possibility that LTP may represent an associative interaction at a cellular level should not be ignored.

Neuroanatomical Plasticity as a Mechanism for Associative Learning

Learning-induced neuroanatomical plasticity clearly occurs in the adult as well as in the developing animal. The studies reviewed here demonstrate the capability of the adult mammalian CNS to change structurally in response to behavioral conditioning.

Two issues arise with respect to functional significance. The first is whether the structural changes seen are related to behavioral learning or to other, nonassociative components of conditioning. As noted above, aspects of every conditioning procedure include elements unrelated to learning and mnemonic processes (such as changing levels of arousal or motor responding), and control procedures have been developed to differentiate between associative and nonassociative effects of behavioral training (e.g. Gormezano 1972). Therefore, the degree to which neuroanatomical plasticity induced by conditioning procedures can be related exclusively to associative aspects of training must be considered. The study by Rutledge et al (1974) included a control group given unpaired presentations of the CS and UCS, and results from those animals demonstrate the importance of control procedures in evaluating associative and nonassociative effects of conditioning. Specifically, both conditioned and control animals showed interhemispheric differences in the number of spines on terminal, apical dendrites. Only with respect to number of spines on vertical and oblique dendrites did the two groups differ. Thus, some morphological changes were specific to associative learning while others were not. The experiments of Vrensen & Cardozo (1981), Larson & Greenough (1981), and Wenzel et al (1980) all included separate groups of animals that controlled for such nonassociative effects of conditioning. All of the studies discussed above, then, are conclusive in showing that anatomical plasticity was related to associative learning, and not to differences in the level of performance of stimulation of the experimental and control groups.

The second issue that arises with respect to functional considerations is whether the structural changes are actually the substrate for the changes in electrophysiological properties or behavior observed. For example, the results of Spinelli & Jensen (1979) and Spinelli et al (1980) are encouraging in demonstrating that both learning-induced physiological and anatomical alterations occur within the same cortical region. Whether the two are actually related is still unknown. Even assuming that the increase in dendritic branching reported by Spinelli et al (1980) is indicative of an increased number of functional synaptic contacts, the available evidence does not allow the conclusion that the structural changes are the substrate for enlargement of forearm representation, or learned avoidance behavior. The same uncertainty exists with the results of other studies reviewed here. While their respective findings certainly suggest a relationship between learned behavior and the neuroanatomical plasticities they have documented, it is not certain that the structural changes are causally related to conditioned behavioral responses. Establishing such a relationship requires evidence on a number of different levels. For example, arguments for a causal relationship would be strengthened if results such as those reported by Vrensen & Cardozo (1981) and Larsen & Greenough (1981) are restricted to appropriated neocortical regions, i.e. do not occur in nonvisual (Vrensen & Cardozo) or nonmotor, nonsomatosensory (Larsen & Greenough) regions. Additional supportive evidence would come from examination of structural features of neurons strained intracellularly. With the latter technique, changes in electrophysiological properties could be related to changes in anatomical characteristics of the same neuron.

The permanence of learning-induced anatomical changes has yet to be addressed. The studies reviewed here are recent ones, and are among the first to demonstrate that learning-related structural changes occur in the adult brain. The time course of their development, permanence, and possible reversal with extinction now become pertinent issues. Although, in one sense structural plasticity should be long-lasting to serve as a substrate for associative learning, any other constraints on time course would be premature. Temporal parameters of anatomical plasticity may vary depending on the learning process reflected by the anatomical change. Structural changes related to long-term storage may have a different time course than those underlying initial acquisition. Some structural changes may be associated with "learning to learn" (learning sets, see Harlow & Warren 1952) rather than learning a specific relationship between specific stimuli, and thus may be invariant in the course of learning and forgetting a number of problems within a learning set. Behavioral extinction may occur without total extinction at the neural level because relearning is almost invariably more rapid than initial learning (e.g. Scavio & Thompson 1979), yet some neural extinction must reflect the behavioral extinction. These possibilities strongly suggest there may be various time courses of neuroanatomical change in relation to behavioral change. It is worth noting that a number of experiments have demonstrated that anatomical plasticity can occur with a rapid time course (Cragg 1967, Valverde 1971, Pettigrew & Garey 1974), so the capacity for rapid structural change exists.

With respect to the fourth criterion, the Vrensen & Cardozo experiment showed that only a subset of all structural dimensions of nerve cells are altered as a result of conditioning. This either suggests that only a subset of neuroanatomical features are capable of being modified by learning, or that the particular profile of features modified by conditioning is specific to the information learned. For example, training animals to respond differentially on the basis of other visual cues, such as color discrimination, may also result in neuroanatomical plasticity, but plasticity of structural features other than those of the synaptic grid. However, dendritic branching patterns and dendritic spine number may always be altered by conditioning, but the particular branching pattern and spatial location of new spines may vary with the particulars of each training condition, such as with the response system or stimuli used. From this perspective it may be significant that maze training of rats alters dendritic branching patterns differently than exposure to enriched environmental conditions. For example, EC exposure is almost invariably associated with an increase in the number of higher-order branches on basal dendrites. In contrast, Greenough et al (1979) showed that maze training results in changes primarily on apical dendrites.

In summary, a considerable body of evidence from work on the effects of environmental differences, visual deprivation, and behavioral conditioning has demonstrated that the adult mammalian CNS changes structurally in response to changes in peripheral stimulation. In addition, the nature of these structural changes is compatible with the notion that neuroanatomical plasticity could serve as a mechanism for associative learning. In fact, recent studies analyzing neuroanatomical plasticity as a result of behavioral conditioning have been successful in documenting structural alterations that are specifically related to associative aspects of training. Although neuroanatomical plasticity appears to be a viable candidate mechanism for associative learning, we are only on the threshold of analyzing key issues in the relationship between neuroanatomical change and behavioral change.

BIOCHEMICAL APPROACHES

The search for the biological basis of learning and memory at the subcellular/biochemical level is fraught with both methodological and conceptual challenges. Agranoff et al (1978) captured the essence of some of these difficulties when they suggested that changes that are detectable are probably too extensive or too general to be associated with memory, per se. Rose (1981a,b) raises two questions. What do we think we are looking for? How will we know when we have found it? He makes the interesting suggestion that the traditional approaches in addressing the relationship between the biochemical state and the behavioral expression as "correlative" or "causal" are inadequate. Rather, he argues, a nonreductive identity theory must be sought to replace this language, whereby the matching cellular and behavioral statements about a phenomena are viewed as "the correspondence" of that phenomena to indicate identity rather than causality. The task of the neurobiology of learning, according to Rose, is to identify the necessary, sufficient, and exclusive biological correspondence of memory formation through a set of criteria; putative candidates for the title of "the engram" must fulfill these criteria.

From a practical standpoint, the biochemical substrates of learning and memory refer to the corresponding chemical processes underlying the acquisition, registration, and retrieval of information. However, in most cases, these biochemical substrates can only be observed within the context of expression of the behavior response patterns necessary to demonstrate that acquisition and registration of information has taken place. Multiple, and frequently overlapping, biochemical processes underlining acquisition and consolidation of information must be dissected into appropriate, interpretable phases. The critical biochemical changes underlying learning and memory must be distinguished from fortuitous or less specific concomitants.

We do not attempt to provide an extensive review critically evaluating the literature as a whole. A number of current reviews have provided such an analysis (see below). Rather, we focus on a few areas that are currently being actively investigated. Such examples reflect interventive and correlative approaches and, when applied in concert, may converge in a synergistic fashion to provide insights into particular issues of learning and memory.

A wide range of drug and chemical treatments can alter behavioral expression of memory. Those treatments that impair memory include substances known to interfere with protein synthesis, along with other kinds of substances and treatments. However, it is still not known whether such effects are due to impairment of synthesis of particular proteins and at particular places in the brain, or are due to nonspecific biological/behavioral impairments. A number of substances involved in one or more putative neurotransmitter systems also impair memory and some facilitate the expression of memory. Again, it has not yet been shown that the memorial effects of those substances are due to specific neurochemical transmitter actions or that they occur at specific sites. All this evidence taken together builds an indirect but nonetheless compelling case that biochemical processes are an essential aspect of the mechanisms of learning and memory storage in the brain.

Macromolecular Processes in Learning and Memory

An extensive literature has evolved assessing the role of macromolecular processes in various aspects of learning and memory formation. Changes in levels or incorporation of precursors into ribonucleic acids (RNA) have been reported during and following acquisition of both appetitively and aversively motivated tasks. Correspondingly, inhibition of RNA synthesis by various pharmacological agents has been shown to prevent the expression of retrieval when trained animals are subsequently tested. This literature has been critically evaluated in several recent reviews (see Agranoff et al 1978, Rainbow 1979, Dunn 1980). Due partially to the difficulty in determining whether specific RNA changes observed can be related to processes subserving learning or formation of long-term memory rather than to any one of a variety of secondary or nonspecific effects, a shift in the direction of macromolecular analysis has emerged. Recently, a more concerted effort has been placed on analysis of end-product dynamics, such as protein synthesis and posttranslational modifications of protein (see Rose 1981a,b, Routtenberg 1979, Shashoua 1982).

The use of interventive agents such as protein synthesis inhibitors has enabled advances in the areas of localization and temporal characteristics of critical processes associated with long-term memory consolidation. This literature has been extensively reviewed by several investigators (Agranoff et al 1978, Rainbow 1979, Dunn 1980, Rose & Haywood 1976) and is only highlighted in the section. Briefly, reports by Berman et al (1978), Boast & Agranoff (1978), and Eichenbaum et al (1976) have shown that micro-administration of various inhibitors of protein synthesis can demonstrate selective region-specific amnetic action. This approach has been recently extended by Flood et al (1980) while addressing the controversial issue as to whether protein synthesis inhibitors induce amnesia through interference with the synthesis of structural proteins, or through disruption of catecholamine biosynthesis. Flood et al (1980) postulated that if these inhibitors share the same mechanism of action, they might also share similar anatomical substrates mediating their respective amnesic effects. Following microinjection of either classical protein synthesis inhibitors or selective catecholamine synthesis inhibitors into either the brain stem sites associated with both dopaminergic and noradrenergic cellular synthesis or into sites corresponding to their major terminal processes, they observed a pattern of inhibition-induced amnesia consistent with the position that both catecholamine synthesis inhibitors and protein synthesis inhibitors induce amnesias by a different mechanism, which might reflect different aspects of memory processing. Caution must be used in interpreting these interventive studies for at least two reasons:

1. Protein synthesis inhibitors do not selectively inhibit any one class of proteins but rather decrease all protein species being synthesized.
2. Frequently the doses necessary to induce amnesia are suspected to affect other cellular processes that may influence various aspects of memory formation, either directly or indirectly, e.g. through brain amino acid elevation, resulting from protein synthesis inhibition, which in turn may secondarily affect various critical processes in memory formation.

Numerous laboratories report attempts to isolate and characterize particular proteins associated with learning and memory. Among these are a series of observations indicating changes in S-100 protein following training in any one of a variety of tasks. Hydén & Lange (1970) reported that S-100 protein increased in hippocampus following training in reverse-handedness in retrieving food pellets. This observation was extended by Zomzely-Neurath et al (1976) and Zomzely-Neurath & Kellar (1977), who reported increases in S-100 following T-maze learning in experimental animals when compared to both quiet and active controls, and by Glushchenko et al (1977), who observed enhanced tritiated leucine incorporation into S-100 in CA3 hippocampal cells in rats trained in a passive avoidance paradigm. When antisera against S-100 protein was administered intraventricularly following training, subsequent retention was inhibited (Hydén & Lange 1970). Karpiak et al (1976) and Rapport & Karpiak (1976) similarly observed amnetic effects of antisera against S-100 protein when administered subdurally to rats immediately after maze training. A recent report by Hydén & Lange (1981) indicates that the mechanism underlying the amnetic effects of S-100 antisera is unclear and should be interpreted with caution.

The function(s) of the S-100 protein is unknown. Hydén (1979) proposed that it may serve as a calcium-dependent mechanism for synaptic modulation. Perumal & Rapport (1978) have demonstrated that it can be phosphorylated by brain nuclear protein kinases—this suggests that it may affect gene expression. Evidence cited by Shashoua (1982) indicates that this protein is highly concentrated in brain extracellular fluid of several species, including the rat. Shashoua has speculated that it may act as a modulatory factory in some aspect of learning.

Recently, Shashoua and associates observed behaviorally correlated changes in protein synthesis in thirsty mice trained to find water in the arm of a T-maze, counter to their initial behavior orientation. Utilizing a double labeling technique, they observed protein changes that appeared in fractions isolated from both cytoplasmic and extracellular fluid. These changes occurred only in mice that learned the task. Littermates that had run the T-maze on the preferred side, and therefore had not learned, along with several other control groups, demonstrated no such changes in these respective proteins. Further biochemical analysis indicated that the enhanced labeling was observed in several protein bands from the extracellular fluid proteins that migrated on SDS-gels at molecular weights comparable to those changes previously observed in goldfish (Schmidt & Shashoua 1981). Gel analysis of cytoplasmic proteins revealed the presence of a single band labeled only in trained mice. This band was thought to be a precursor for the enhanced peaks observed in the extracellular fluid. On the basis of their initial work in goldfish and more recent work in mice, Shashoua and colleagues suggested that there is both an increased metabolism of cytoplasmic proteins and a release into the extracellular fluid of the brain as a consequence of learning. There, these ependymin proteins may function as trophic factors by possibly interacting in the process of converting transiently modified synaptic contracts—resulting from physiological events during learning—into permanent synapses.

Modulatory Factors in Learning and Memory

A wealth of evidence indicates that a variety of neuroactive substances including neuropeptides, biogenic amines, and amino acids, can modulate learning and memory (for reviews, see Squire & Davis 1981, Kobb & Bloom 1982, Levine 1968, and Zornetzer 1978).

In this section we attempt to highlight the potential diversity of modulatory substances in various aspects of learning and memory consolidation. Such a multiplicity of effects becomes apparent when a particular modulator is viewed in an interactive role with other systems rather than in isolation. We focus on the central noradrenergic system as an example.

Recently a number of investigators, primarily from the Rudolf Magnus Institute at Utrecht, have systematically examined the hypothesis that changes in norepinephrine metabolism mediate the action of vasopressin on memory consolidation. Extending their earlier finding that collectively provided general correlative evidence for this position, Kovács et al (1980) recently reported that vasopressin produced differential effects on memory processes in dorsal-noradrenergic bundle-lesioned animals. Lesions of the dorsal noradrenergic bundle made with neurotoxin 6-hydroxydopamine (6-OHDA) selectively abolished the arginine-vasopressin enhancement of consolidation of passive avoidance learning; these lesions affected retrieval much less. Further, bilateral destruction of the mesolimbic accumbens nucleus by 6-OHDA or selective lesions of the ascending serotonergic system by 5,6-dihydroxytryptamine (5,6-DHT) did not affect arginine vasopressin enhancement of consolidation. This observation suggests an apparently selective role for the dorsal noradrenergic bundle. Central microadministration of arginine vasopressin in the region of the locus coeruleus and its terminal processes reveal that the peptide's site of action was not at the noradrenergic cell bodies but rather in the regions that contain the terminals, i.e. septum, hippocampus, and dorsal raphe. Recent histochemical evidence indicates that vasopressinergic terminal processes innervate these structures (Sofroniew & Wiendl 1981, Kovács et al 1980), thus supporting a physiological role of vasopressin. Collectively, these results support the concept that an interaction between vasopressinergic and limbic-midbrain noradrenergic processes is essential for the hypothesized action of this peptide on memory processes.

A number of studies have assessed the direct involvement of brain norepinephrine in learning and memory (see reviews by Fibiger et al 1975, Mason 1981). The proposal by Kety (1970, 1972) and Crow (1973) implicating noradrenergic processes in emotion, arousal, and learning provided the impetus for this extensive literature. Anlezark et al (1973) provided early evidence to support this position by demonstrating that electrolytic lesions of the locus coeruleus produced a marked reduction of brain norepinephrine and a corresponding impairment in appetitive learning. A more extensive literature fails to lend credence to the hypothesized role of the locus coeruleus in learning. This literature includes studies utilizing noradrenergic neurotoxin 6-hydroxydopamine. Specifically, selective lesions of the dorsal noradrenergic bundle following intracerebral administration of 6-OHDA uniformly failed to produce impaired acquisition in any one of a variety of appetitively and aversively motivated tasks, including one- and two-way active avoidance (Fibiger & Mason 1978, Ögren & Fuxe 1977, Mason & Fibiger 1979). However, when neurotoxic lesions of the ascending dorsal noradrenergic bundle were combined with bilateral adrenalectomy, marked impairment in the acquisition of a one-way avoidance response was observed. This deficit was prevented by daily physiological doses of corticosterone. In contrast, neither adrenalectomy nor lesions alone produced this effect (Ögren & Fuxe 1974, 1977). Consistent with this finding, Roberts & Fibiger (1977) also observed impairment in passive avoidance only in rats receiving combined lesions with adrenalectomy. Ögren & Fuxe (1977) speculated that central noradrenergic processes and corticosterone may have interactive or complementary rather than additive roles in avoidance learning by possibly controlling cortical arousal.

Interestingly, while dorsal bundle lesions appear to have a negligible effect on acquisition, a general resistance to extinction of a variety of learned responses is observed in these animals (see review by Mason 1981). Moreover, adrenalectomy blocks this effect. This observation, in conjunction with the report by Micco et al (1979), who observed that glucocorticoid administration retarded extinction of a runway task, suggests that adrenal steroids may antagonize the normal function of the dorsal bundle in extinction—which Mason & Iversen (1979) have suggested suppresses attention to irrelevant cues in the environment.

Recently, utilization of a new noradrenergic neurotoxin, N-(2-chloroethyl)-N-ethyl-2-bromobenzylamine (DSP4), has produced findings different from those reported following the use of 6-OHDA. Treatment with DSP4 produced an apparent selective and marked degeneration of the locus coeruleus noradrenergic system and significantly impaired two-way active avoidance acquisition (Ögren et al 1980, Archer 1982). These results are inconsistent with the results of earlier behavioral studies on 6-OHDA lesions of the dorsal noradrenergic bundle. However, it should be noted that systemically administered DSP4, in contrast to dorsal bundle 6-OHDA, produces a profound depletion, and probably destroys all terminal processes arising from the locus coeruleus, including the cerebellum and spinal cord. These observations as noted by Archer (1982) are consistent with the locus coeruleus in its entirety exerting a role in shuttlebox avoidance. However, additional work is needed to clarify these issues.

Although the precise role of the locus coeruleus in various aspects of learning and memory is not yet settled, the literature as a whole, and specifically the literature implicating interactive modulation, demonstrates that substances such as norepinepbrine can potentially influence a variety of processes subserving learning and memory by its multiplicity of interactions at specific central loci.

Little mention has been made of modulatory effects of hormones that are not intrinsic to the actual loci in question and moreover are produced outside the CNS. Nonetheless, they appear to play a major role in mediating memory formation (for comprehensive reviews, see Gold & McGaugh 1978, deWeid 1979, 1980). One particularly striking example of how peripheral endocrine processes may modulate various aspects of learning and memory is found in the work of McGaugh and colleagues. In one series of studies they demonstrated that retention of a newly learned response could be impaired by either pre- or post-training administration of pharmacological agents known to interfere with synthesis of brain norepinephrine. This deficit of retention could be prevented by subcutaneous injections of either norepinephrine or epinephrine, compounds that do not cross the blood brain barrier in appreciable amounts. In the second series of studies addressing the same issue, they find that intraperitoneal but not intraventricular administration of amphetamine can facilitate retention, and that the effect can be mimicked by DL-4-OH amphetamine, an analogue which has primarily peripheral effects. Furthermore, the faciliatory effects of this peripherally acting analogue on memory development were blocked by adrenal demedullation. Collectively, these studies suggest that the peripheral hormonal processes possibly play a very important modulatory role in the development of memory. Recent studies by this group have also implicated peripheral enkephalins in modulating learning and memory processes (Martinez et al 1981).

There is a vast literature on memory consolidation—facilitation or impairment of memory retrieval by a variety of manipulations, including chemicals—which has been frequently reviewed (see e.g. McGaugh & Herz 1972, Weingartner & Parker 1982). The recent findings of peripheral actions of hormones and neurotransmitters on memory retrieval described above may force some rethinking of earlier interpretations of the consolidation process. Treatments as diverse as facilitation by norepinephrine and impairment by electrical stimulation of the amygdala can be prevented by prior adrenal demedullation. These results suggest that the mechanisms of action of many chemicals and treatments that influence memory consolidation may not relate directly to the mechanisms of memory storage and retrieval in the brain. The critical links between peripheral actions and central actions remain to be determined.

Multiple Converging Approaches to Learning and Memory

A systematic analysis of neurobiological events involved in a specialized form of early learning, imprinting in the chick, is reflected in a series of reports by Bateson, Horn, and Rose and their colleagues (see Horn 1981 and Rose 1980, for extensive reviews). Specifically, they addressed the issue of early imprinting in the domestic chick by employing a recognition task that served as an experimental analogue of the bird's mother in the wild. Following birth, chicks were maintained in darkness until well-coordinated movement was demonstrated (approximately 12 to 24 hr post-hatch), at which time they were exposed to the imprinting stimulus (typically a flashing colored light), remained in darkness, or were exposed to overhead illumination. The latter two groups were controls for differential visual activation. After appropriate periods of exposure and rest, they were tested for their preference for familiar objects. Following exposure to the imprinting stimulus, a variety of biochemical changes were reported and appeared to be generally localized to the dorsal part of the forebrain.

Using a number of experimental approaches and control procedures, these workers have developed evidence that the medial hyperstriatum ventrale is a locus for imprinting. Employing a sequential lesions study, Horn (1981) was able to demonstrate hemispheric asymmetry in the storage of information. Biochemical and morphological changes developed within about 3 hr after training. Interestingly, the morphological correlates, increased length of synaptic apposition zones, was greater in the right area than in the left and QNB binding was also higher in the right areas. A sequence of biochemical processes in this region during the imprinting is hypothesized to be the following: initially, modifications in muscarinic receptors, noradrenergic activity, and transient changes in cyclic AMP, followed by changes in RNA polymerase activity, and finally elevated incorporation of specific precursors into RNA and proteins.

These studies represent an intriguing elucidation of neurobiological modifications that occur during the development of a specialized form of early adaptive behavior. However, the relation of these biochemical and morphological changes to the mechanisms of imprinting has not yet been established. Further it remains to be determined to what extent the processes underlying this specialized form of early behavioral plasticity resemble those of associative learning in mammals. Another attractive avian model is the specialized brain system responsible for the songs by song birds. Male canaries have the potential to learn new and different songs on successive years. Brain regions believed to be critically involved show corresponding and very large seasonal anatomical changes, presumably under hormonal control (Nottebohm 1981).

CONCLUDING POSTSCRIPT

When the "memory trace" has been localized to particular sets of neurons and synaptic junctions, a critical next question becomes whether the biological mechanisms of plasticity are primarily presynaptic or postsynaptic (or, of course, both). Based on what little evidence now exists, we observe that putative learning-induced changes in the activity, structure, and chemistry of neurons and synapses described in this review are perhaps more consistent with postsynaptic rather than presynaptic phenomena.

ACKNOWLEDGMENTS

Preparation of this review was supported in part by the Virginia Day Robbins Fund and by the following research grants: National Science Foundation BNS 81-17115 (R.F.T.) and BNS 80-21395 (T.W.B.), and National Institute of Mental Health MH 23861 (J. Barchas) (J.M. IV).

Literature Cited

Agranoff, B. W., Burrell, H. R., Dokas, L. A., Springer, A. D. 1978. Progress in biochemical approaches to learning and memory. In *Psychopharmacology: A Generation of Progress,* ed. M. A. Lipton, A. DiMascio, K. F. Killam, pp. 623–35. New York: Raven

Albus, J. S. 1971. A theory of cerebellar function. *Math. Biosci.* 10:26–61

Altman, J., Das, G. D. 1964. Autoradiographic examination of the effects of enriched environment on the rate of glial multiplication in the adult rat brain. *Nature* 204:1161–63

Altschuler, R. A. 1979. Morphometry of the effect of increased experience and training on synaptic density in area CA3 of the rat hippocampus. *J. Histochem. Cytochem.* 27:1548–50

Andersen, P., Sundberg, S. H., Sveen, O., Wigstrom, H. 1977. Specific long-lasting potentiation of synaptic transmission in hippocampal slices. *Nature* 266:736–37

Anlezark, G. M., Crow, T. J., Greenway, A. P. 1973. Impaired learning and decreased cortical norepinephrine after bilateral locus coeruleus lesions. *Science* 181:682–84

Archer, T. 1982. DSP4 (*N*-2-chloroethyl-*N*-ethyl-2-bromobenzylamine), a new noradrenaline neurotoxin, and stimulus conditions affecting acquisition of two-way active avoidance. *J. Comp. Physiol. Psychol.* 96(3):476–90

Babb, T. L. 1982. Short term and long term modification of neurons and evoked potentials in the human hippocampal formation. See Swanson et al 1982. In press

Baudry, M., Lynch, G. 1980. Hypothesis regarding the cellular mechanisms responsible for long-term synaptic potentiation in the hippocampus. *Exp. Neurol.* 68:202–4

Bennett, E. L. 1976. Cerebral effects of differential experience and training. *Neural Mechanisms of Learning and Memory,* ed. M. R. Rosenzweig, E. L. Bennett. Cambridge, Mass: MIT Press

Bennett, M. J., Smith, E. L. III, Harwerth, R. S., Crawford, M. L. J. 1980. Ocular dominance, eye alignment and visual acuity in kittens reared with an optically induced squint. *Brain Res.* 193:33–45

Berger, T. W. 1982. Hippocampal cellular plasticity induced by classical conditioning. See Swanson et al 1982, pp. 723–29

Berger, T. W., Laham, R. I., Thompson, R. F. 1980. Hippocampal unit-behavior correlations during classical conditioning. *Brain Res.* 193:229–48

Berger, T. W., Orr, W. B. 1982. Role of the hippocampus in reversal learning of the rabbit nictitating membrane response. See Woody 1982. In press

Berger, T. W., Thompson, R. F. 1978a. Neuronal plasticity in the limbic system during classical conditioning of the rabbit nictitating membrane response. I. The hippocampus. *Brain Res.* 145 (2):323–46

Berger, T. W., Thompson, R. F. 1978b. Identification of pyramidal cells as the critical elements in hippocampal neuronal plasticity during learning. *Proc. Natl. Acad. Sci. USA* 75(3):1572–76

Berger, T. W., Thompson, R. F. 1978c. Neuronal plasticity in the limbic system during classical conditioning of the rabbit nictitating membrane response. II: Septum and mammillary bodies. *Brain Res.* 156:293–314

Berman, N., Murphy, E. H., Salinger, W. L. 1979. Monocular paralysis in the adult cat does not change cortical ocular dominance. *Brain Res.* 164:290–93

Berman, R. F., Kesner, R. P., Partlow, L. M. 1978. Passive avoidance impairments in rats following cycloheximide injection into the amygdala. *Brain Res.* 158: 171–88

Bernstein, L. 1973. A study of some enriching variables in a free-environment for rats. *J. Psychosomatic Res.* 17:85–88

Best, M. R., Best, P. J. 1976. The effects of state of consciousness and latent inhibition on hippocampal unit activity in the rat during conditioning. *Exp. Neurol.* 51:564–73

Bingham, W. E., Griffiths, W. J. Jr. 1952. The effect of different environments during infancy on adult behavior in the rat. *J. Comp. Physiol. Psychol.* 45:307–12

Birt, D., Olds, M. E. 1982. Auditory response enhancement during differential conditioning in behaving rats. See Woody 1982. In press

Blakemore, C., Cooper, G. F. 1970. Development of the brain depends on the visual environment. *Nature* 228:477–78

Blakemore, C., Eggers, H. M. 1978. Effects of artificial anisometropia and strabismus on the kitten's visual cortex. *Arch. Ital. Biol.* 116:384–89

Blakemore, C., Garey, L. J., Vital-Durand, F. 1978a. The physiological effects of monocular deprivation and their reversal in the monkey's visual cortex. *J. Physiol.* 282:223–62

Blakemore, C., Garey, L. J., Henderson, Z. B., Swindale, N. V., Vital-Durand, F. 1980. Visual experience can promote rapid axonal reinnervation in monkey visual cortex. *J. Physiol.* 307:25P–26P

Blakemore, C., Hillman, P. 1977. An attempt to assess the effects of monocular deprivation and stabismus on synaptic efficiency in the kitten's visual cortex. *Exp. Brain Res.* 30:187–202

Blakemore, C., Hovshon, J. A., Van Sluyters, R. C. 1978b. Modification of the kitten's visual cortex by exposure to spatially periodic patterns. *Exp. Brain Res.* 31:561–72

Blakemore, C., Van Sluyters, R. C. 1974. Reversal of the physiological effects of monocular deprivation in kittens: Further evidence for a sensitive period. *J. Physiol.* 237:195–216

Blakemore, C., Van Sluyters, R. C. 1975. Innate and environmental factors in the development of the kitten's visual cortex. *J. Physiol.* 248:663–716

Bliss, T. V. P., Lømo, T. 1973. Long-lasting potentiation of synaptic transmission in the dentate area of the anaesthetized rabbit following stimulation of the perforant path. *J. Physiol.* 232:331–56

Boast, C. A., Agranoff, B. W. 1978. Biochemical and behavioral effects of stretovitacin A in mice. *Neurosci. Absr.* 4:255

Borges, S., Berry, M. 1976. Preferential orientation of stellate cell dendrites in the visual cortex of the dark-reared rat. *Brain Res.* 112:141–47

Bromily, R. B. 1948. The development of conditioned responses in cats after unilateral decortication. *J. Comp. Physiol. Psychol.* 41:155–64

Brons, J. F., Woody, C. D. 1980. Long-term changes in excitability of cortical neurons after Pavlovian conditioning and extinction. *J. Neurophysiol.* 44:605

Brooks, V. B. 1979. Control of the intended limb movements by the lateral and intermediate cerebellum. In *Integration in the Nervous System,* ed. H. Asanuma, V. J. Wilson, pp. 321–56. New York: Igaku-Shoin

Brooks, V. B., Kozlovskaya, I. B., Atkin, A., Horvath, F. E., Uno, M. 1973. Effects of cooling dentate nucleus on tracking-task performance in monkeys. *J. Neurophysiol.* 36:974–95

Brown, R. T. 1968. Early experience and problem-solving ability. *J. Comp. Physiol. Psychol.* 65:433–40

Brown, D. L., Salinger, W. L. 1975. Loss of x-cells in lateral geniculate nucleus with monocular paralysis: Neural plasticity in the adult cat. *Science* 189:1011–12

Buchanan, S. L., Powell, D. A. 1982. Cingulate cortex: Its role in Pavlovian conditioning. *J. Comp. Physiol. Psychol.* In press

Chang, F. F., Wesa, J. M., Greenough, W. T., West, R. W. 1981. Differential postsynaptic curvature in occipital cortex following differential rearing in rat. *Neurosci. Abstr.* 7:772

Clark, G. A., Berger, T. W., Thompson, R. F. 1978. The role of entorhinal cortex during classical conditioning: Evidence for entorhinal-dentate facilitation. *Soc. Neurosci Abstr.* 4:673

Cohen, D. H. 1969. Development of a vertebrate experimental model for cellular neurophysiologic studies of learning. *Cond. Reflex.* 4:61–80

Cohen, D. H. 1974. The neural pathways and informational flow mediating a conditioned autonomic response. In *Limbic and Autonomic Nervous System Research,* ed. L. V. DiCara. New York: Plenum

Cohen, D. H. 1975. Involvement of the avian amygdalar homologue (archistriatum posterior and mediale) in defensively conditioned heart rate change. *J. Comp. Phsyiol. Psychol.* 160:13–36

Cohen, D. H. 1980. The functional neuroanatomy of a conditioned response. See Thompson et al 1980, pp. 283–302

Cohen, D. H. 1982. Central processing time for a conditioned response in a vertebrate model system. See Woody 1982. In press

Cohen, D. H., Macdonald, R. L. 1976. Involvement of the avian hypothalamus in defensively conditioned heart rate change. *J. Comp. Neurol.* 167:465–80

Coleman, P. D., Riesen, A. H. 1968. Environmental effects on cortical dendritic fields I. Rearing in the dark. *J. Anat.* 102:363–74

Connor, J. R., Melone, J. H., Yuen, A. R., Diamond, M. C. 1981. Dendritic length in aged rats' occipital cortex: An environmentally induced response. *Exp. Neurol.* 73:827–30

Cornwell, P., Overman, W. 1981. Behavioral effects of early rearing conditions and neonatal lesions of the visual cortex in kittens. *J. Comp. Physiol. Psychol.* 95: 848–62

Cragg, B. G. 1967. Changes in visual cortex on first exposure of rats to light. *Nature* 215:251–53

Creutzfeldt, O. D., Heggelund, P. 1975. Neural plasticity in visual cortex of adult cats after exposure to visual patterns. *Science* 188:1025–27

Crow, T. J. 1973. Catecholamine-containing neurones and electrical self-stimulation. 2. A theoretical interpretation and some psychiatric implications. *Psychol. Med.* 3:1–5

Cummins, R. A., Walsh, R. N., Budtz-Olsen, O. E., Konstantinos, T., Horsfall, C. R. 1973. Environmentally-induced changes in the brains of elderly rats. *Nature* 243:516–18

Cynader, M., Timney, B. N., Mitchell, D. E. 1980. Period of susceptibility of kitten visual cortex to the effects of monocular deprivation extends beyond six months of age. *Brain Res.* 191:545–50

Deadwyler, S. A., West, M., Lynch, G. 1979a. Synaptically identified slow potential during behavior. *Brain Res.* 161:211–25

Deadwyler, S. A., West, M., Lynch, G. 1979b. Activity of dentate granule cells during learning: Differentiation of perforant path input. *Brain Res.* 169:29–43

Deadwyler, S. A., West, M., Robinson, J. H. 1981. Entorhinal and septal inputs differentially control sensory-evoked re-

sponses in the rat dentate gyrus. *Science* 211:1181–83

Deadwyler, S. A., West, M., Christian, E. P. 1982. Neural activity in the dentate gyrus of the rat during the acquisition and performance of simple and complex sensory discrimination learning. See Woody 1982. In press

Desmond, J. E., Moore, J. W. 1982. A brain stem region essential for classically conditioned but not unconditioned nictitating membrane response. *Physiol. Behav.* 28:1029–33

deWied, D. 1979. Pituitary neuropeptides and behavior. In *Central Regulation of the Endocrine System,* ed. K. Fuxe, T. Hökfelt, R. Luft, pp. 297–314. New York: Plenum

deWied, D. 1980. Peptides and adaptive behavior. In *Hormones and the Brain,* ed. D. deWied, P. A. Van Keep, pp. 103–13. Lancaster, England: MTP Press, Falcon House

Diamond, M. C., Johnson, R. E., Ingham, C., Rosenzweig, M. R., Bennett, E. L. 1975. Effects of differential experience on neuronal nuclear and perikarya dimensions in the rat cerebral cortex. *Behav. Biol.* 15:107–11

Diamond, M. C., Law, F., Rhodes, H., Lindner, B., Rosenzweig, M. R., et al. 1966. Increases in cortical depth and glia numbers in rats subjected to enriched environment. *J. Comp. Neurol.* 128: 117–26

Diamond, M. C., Rosenzweig, M. R., Bennett, E. L., Lindner, B., Lyon, L. 1972. Effects of environmental enrichment and impoverishment on rat cerebral cortex. *J. Neurobiol.* 3(1):47–64

Disterhoft, J. F., Kwan, H. H., Lo, W. D. 1977. Nictitating membrane conditioning to tone in the immobilized albino rabbit. *Brain Res.* 137:127–44

Disterhoft, J. F., Segal, M. 1978. Neuron activity in rat hippocampus and motor cortex during discrimination reversal. *Brain Res. Bull.* 3:583–88

Disterhoft, J. F., Stuart, D. K. 1976. The trial sequence of changed unit activity in auditory system of alert rat during conditioned response acquisition and extinction. *J. Neurophysiol.* 39:266–81

Disterhoft, J. F., Stuart, D. K. 1977. Differentiated short latency response increases after conditioning in inferior colliculus neurons of alert rat. *Brain Res.* 130:315–33

Douglas, R. M., Goddard, G. V. 1975. Long-term potentiation of the perforant path-granule cell synapse in the rat hippocampus. *Brain Res.* 86:205–15

Duffy, D., Teyler, T. J., Shashoua, V. E. 1981. Long-term potentiation in the hippocampal slice: Evidence for stimulated secretion of newly synthesized proteins. *Science* 212:1148–51

Dufosse, M., Ito, M., Jastrehoff, P. J., Miyashita, Y. 1978. Diminution and reversal of eye movements induced by local stimulation of rabbit cerebellar flocculus after partial destruction of the inferior olive. *Exp. Brain Res.* 33:139–41

Dunn, A. J. 1980. Neurochemistry of learning and memory: An evaluation of recent data. *Ann. Rev. Psychol.* 31:343–90

Eccles, J. C., Ito, M., Sezentagothai, J. 1967. *The Cerebellum as a Neuronal Machine.* New York: Springer-Verlag

Edwards, H. P., Barry, W. F., Wyspianski, J. O. 1969. Effect of differential rearing on photic evoked potentials and brightness discrimination in the albino rat. *Dev. Psychol.* 2(3):133–38

Eichenbaum, H., Quenon, B. A., Heacock, A. M., Agranoff, B. W. 1976. Differential behavioral and biochemical effects of regional injection of cycloheximide into mouse brain. *Brain Res.* 101:171–76

Fiala, B. A., Joyce, J. N., Greenough, W. T. 1978. Environmental complexity modulated growth of granule cell dentrites in developing but not adult hippocampus of rats. *Exp. Neurol.* 59:372–83

Fibiger, H. C., Mason, S. T. 1978. The effect of dorsal bundle injections of 6-OHDA on avoidance responding in rats. *Br. J. Pharmacol.* 64:601–6

Fibiger, H. C., Roberts, D. C. S., Price, M. T. C. 1975. On the role of telencephalic noradrenaline in learning and memory. In *Chemical Tools in Catecholamine Research,* ed. G. Jonsson, T. Malmfors, C. Sachs, 1:349–56. Amsterdam: North Holland

Fifkova, E. 1968. Changes in the visual cortex of rats after unilateral deprivation. *Nature* 220:379–81

Fifkova, E. 1969. The effect of monocular deprivation on the synaptic contacts of the visual cortex. *J. Neurobiol.* 1: 285–94

Fifkova, E. 1970. The effect of unilateral deprivation on visual centers in rats. *J. Comp. Neurol.* 140:431–38

Fifkova, E., Van Der Wede, B., Van Harreveld, A. 1978. Ultrastructural changes in the dentate molecular layer during conditioning. *Anat. Rec.* 190: 394

Fifkova, E., Van Harreveld, A. 1977. Long-lasting morphological changes in dendritic spines of dentate granular cells

following stimulation of the entorhinal area. *J. Neurocytol.* 6:211–30

Fifkova, E., Van Harreveld, A. 1978. Changes in dendritic spines of the dentate molecular layer during conditioning. *Neurosci. Abstr.* 4:257

Finger, S. 1978. *Recovery from Brain Damage.* St. Louis, Mo: Washington Univ.

Floeter, M. K., Greenough, W. T. 1979. Cerebellar plasticity: Modification of Purkinje cell structure by differential rearing in monkeys. *Science* 206:227–29

Flood, J. F., Smith, G. E., Jarvik, M. E. 1980. A comparison of the effects of localized brain administration of catecholamine and protein synthesis inhibitors on memory processing. *Brain Res.* 197: 153–65

Forgays, D. G., Read, J. M. 1962. Crucial periods for free-environmental experience in the rat. *J. Comp. Physiol. Psychol.* 55:816–18

Forgus, R. H. 1955. Early visual and motor experience as determiners of complex mazelearning ability under rich and reduced stimulation. *J. Comp. Physiol. Psychol.* 9:207–14

Francis, J., Hernandez, L. L., Powell, D. A. 1981. Lateral hypothalamic lesions: Effects on Pavlovian conditioning of eyeblink and heart rate responses in the rabbit. *Brain Res. Bull.* 6:155–63

Fuller, J. H., Schlag, J. D. 1976. Determination of antidromic excitation by the collision test: Problems of interpretation. *Brain Res.* 112:283–98

Gabriel, M., Foster, K., Orona, E., Saltwick, S. E., Stanton, M. 1980. Neuronal activity of cingulate cortex, anteroventral thalamus and hippocampal formation in discriminative conditioning: Encoding and extraction of the significance of conditional stimuli. *Prog. Psychobiol. Physiol. Psychol.* 9:126–23

Gabriel, M., Saltwich, S. L., Miller, J. D. 1975. Conditioning and reversal of short-latency multiple-unit responses in the rabbit medial geniculate nucleus. *Science* 189:1108–9

Garcia, J., Rusiniak, K. W., Kiefer, S. W., Bermudez-Rattoni, F. 1982. The neural integration of feeding and drinking habits. See Woody 1982. In press

Garey, L. J., Pettigrew, J. D. 1974. Ultrastructural changes in kitten visual cortex after environmental modification. *Brain Res.* 66:165–72

Gibbs, C. M., Cohen, D. H. 1980. Plasticity of the thalamofugal pathway during visual conditioning. *Neurosci. Abstr.* 6: 424

Gilbert, P. F. C., Thach, W. T. 1977. Purkinje cell activity during motor learning. *Brain Res.* 128:309–28

Globus, A., Rosenzweig, M. R., Bennett, E. L., Diamond, M. C. 1973. Effects of differential experience on dendritic spine counts in rat cerebral cortex. *J. Comp. Physiol. Psychol.* 82:175–81

Glushchenko, T. S., Pevsner, L. Z., Brumberg, V. A. 1977. Synthesis of brain-specific proteins in rat hippocampus in the course of conditional reflex formation. *Act. Nerv. Super.* 19:148

Goddard, G. 1964. Functions of the amygdala. *Psychol. Bull.* 62:89–109

Gold, P. E., McGaugh, J. L. 1978. Endogenous modulators of memory storage processes. *Clin. Psychoneuroendocrinol. Reprod., Proc. Serono Symp.* 22:25–46

Gormezano, I. 1972. Investigations of defense and reward conditioning in the rabbit. See Prokasy 1972, pp. 151–81

Gormezano, I., Schneiderman, N., Deaux, E., Fentues, I. 1962. Nictitating membrane: Classical conditioning and extinction in the albino rabbit. *Science* 138:33–34

Green, E. J., Schlumpf, B. E., Greenough, W. T. 1981. The effects of complex or isolated environments on cortical dendrites of middle-aged rats. *Neurosci. Abstr.* 7:65

Greenough, W. T. 1976. Enduring brain effects of differential experience and training. See Bennett 1976, pp. 255–78

Greenough, W. T., Fulcher, J. K., Yuwiler, A., Geller, E. 1970. Enriched rearing and chronic electroshock: Effects on brain and behavior in mice. *Physiol. Behav.* 5:371–73

Greenough, W. T., Juraska, J. M., Volkmar, F. R. 1979. Maze training effects on dendrite branching in occipital cortex of adult rats. *Behav. Neural Biol.* 26:287–97

Greenough, W. T., Volkmar, F. R. 1973. Pattern of dendritic branching in occipital cortex of rats reared in complex environments. *Exp. Neurol.* 40:491–504

Greenough, W. T., Volkmar, F. R., Juraska, J. M. 1973a. Effects of rearing complexity on dendritic branching in frontolateral and temporal cortex of the rat. *Exp. Neurol.* 41:371–78

Greenough, W. T., West, R. W., DeVoogd, T. J. 1978. Subsynaptic plate perforations: Changes with age and experience in the rat. *Science* 202:1096–98

Greenough, W. T., Yuwiler, A., Dollinger, M. 1973b. Effects of posttrial eserine administration on learning in "enriched"

—and "impoverished"—reared rats. *Behav. Biol.* 8:261–72

Gyllensten, L., Malmfors, T., Norrlin, M. L. 1965. Effect of visual deprivation on the optic centers of growing and adult mice. *J. Comp. Neurol.* 124:149–60

Harlow, H. F., Warren, J. M. 1952. Formation and transfer of discrimination learning sets. *J. Comp. Physiol. Psychol.* 45:482–89

Hebb, D. O. 1949. *The Organization of Behavior.* New York: Wiley

Hernandez, L. L., Powell, D. A. 1981. Forebrain norepinephrine and serotonin concentrations and Pavlovian conditioning in septal damaged and normal rabbits. *Brain Res. Bull.* 6:155–63

Hill, A. J. 1978. First occurrence of hippocampal spatial firing in a new environment. *Exp. Neurol.* 62:282–97

Hirsch, H. V. B., Spinelli, D. N. 1971. Modification of the distribution of receptive field orientation in cats by selective visual exposure during development. *Exp. Brain Res.* 13:509–27

Hirsh, R. 1973. Previous stimulus experience delays conditioning-induced changes in hippocampal unit responses in rats. *J. Comp. Physiol. Psychol.* 83:337–45

Hoehler, F. K., Thompson, R. F. 1980. Effect of the interstimulus (CS-UCS) interval on hippocampal unit activity during classical conditioning of the nictitating membrane response of the rabbit, *Oryctrolagus cuniculus. J. Comp. Physiol. Psychol.* 94:201–15

Holloway, R. L. Jr. 1966. Dendritic branching: Some preliminary results of training and complexity in rat visual cortex. *Brain Res.* 2:393–96

Hore, J., Meyer-Lohmann, J., Brooks, V. B. 1977. Basal ganglia cooling disables learned arm movements of monkeys in the absence of visual guidance. *Science* 195:584–86

Horn, G. 1981. Neural mechanisms of learning: An analysis of imprinting in the domestic chick. *Proc. R. Soc. London Ser. B* 213:101–37

Hubel, D. H., Wiesel, T. N. 1962. Receptive fields, binocular interaction and functional architecture in the cat's visual cortex. *J. Physiol.* 160:106–54

Hubel, D. H., Wiesel, T. N. 1965. Binocular interaction in striate cortex of kittens reared with artificial squint. *J. Neurophysiol.* 28:1041–59

Hubel, D. H., Wiesel, T. N. 1970. The period of susceptibility to the physiological effects of unilateral eye closure in kittens. *J. Physiol.* 206:419–36

Hubel, D. H., Wiesel, T. N., Le Vay, S. 1977. Plasticity of ocular dominance columns in monkey striate cortex. *Philos. Trans. R. Soc. London Ser. B* 278:377–409

Hydén, H. 1979. A calcium-dependent mechanism for synapse and nerve cell membrane modulation. *Proc. Natl. Acad. Sci. USA* 71:2965–68

Hydén, H., Lange, P. W. 1970. The effects of antiserum to S100 protein on behavior and amount of S100 in brain cells. *J. Neurobiol.* 12(3):201–10

Hydén, H., Lange, P. W. 1970. S-100 protein: Correlations with behavior. *Proc. Natl. Acad. Sci. USA* 67:1959–66

Hymovitch, B. 1952. The effects of experimental variations on problem solving in the rat. *J. Comp. Physiol. Psychol.* 45:313–21

Isaacson, R. L., Pribram, K. H. 1974–75. *The Hippocampus*, Vols. 1, 2. New York: Plenum

Ito, M. 1970. Neurophysiological aspects of the cerebellar motor control system. *Int. J. Neurol.* 7:162–76

Ito, M. 1974. The control mechanisms of cerebellar motor system. In *The Neurosciences, Third Study Program*, ed. F. O. Schmitt, R. G. Worden. Boston: MIT Press

Ito, M. 1977. Neuronal events in the cerebellar flocculus associated with an adaptive modification of the vestibulo-ocular reflex of the rabbit. In *Control of Gaze by Brain Stem Neurons*, ed. R. G. Baker, A. Berthoz. Amsterdam: Elsevier

Ito, M. 1982a. Cerebellar control of the vestibulo-ocular reflex: Around the flocculus hypothesis. *Annual Rev. Neurosci.* 5:275–96

Ito, M. 1982b. Synaptic plasticity underlying the cerebellar motor learning investigated in rabbit's flocculus. See Woody 1982. In press

Kaada, B. R. 1972. Stimulation and regional ablation of the amygdaloid complex with reference to functional representations. In *The Neurobiology of the Amygdala*, ed. B. E. Elftheriou, pp. 205–81. New York: Plenum

Kandel, E. R., Spencer, W. A. 1968. Cellular neurophysiological approaches in the study of learning. *Physiol. Rev.* 48:65–134

Kapp, B. S., Gallagher, M., Applegate, C. D., Frysinger, R. C. 1982. The amygdala central nucleus: Contributions to conditioned cardiovascular responding during aversive pavlovian conditioning in the rabbit. See Woody 1982. In press

Karamian, A. I., Fanaralijian, V. V., Kosareva, A. A. 1969. The functional and morphological evolution of the cerebellum and its role in behavior. In *Neurobiology of Cerebellar Evolution and Development, First International Symposium*, ed. R. Ilinas. Chicago: Am. Med. Assoc.

Karpiak, S. E., Serokosz, M., Rapport, M. M. 1976. Effects of antisera to S-100 protein and to synaptic membrane fraction on maze performance and EEG. *Brain Res.* 102:313–21

Kesner, R. P., Wilbrun, M. W. 1974. A review of electrical stimulation of the brain in context of learning and retention. *Behav. Biol.* 10:259–93

Kettner, R. N., Shannon, R. V., Nguyen, T. M., Thompson, R. F. 1980. Simultaneous behavioral and neural (Cochlear Nucleus) measurement during signal detection in the rabbit. *Phrecept. Psychophys.* 28(6):504–13

Kettner, R. E., Thompson, R. F. 1982. Auditory signal detection and decision processes in the nervous system. *J. Comp. Physiol. Psychol.* 96:328–31

Kety, S. S. 1970. The biogenic amines in the central nervous system: Their possible roles in arousal, emotion and learning. In *The Neurosciences*, ed. F. O. Schmitt, pp. 324–36. New York: Rockefeller Univ. Press

Kety, S. S. 1972. The possible role of the adrenergic systems of the cortex in learning. *Res. Publ. Assoc. Res. Nerv. Ment. Dis.* 50:376–89

Kimble, G. A. 1961. *Hilgard and Marquis' Conditioning and Learning*. New York: Appleton-Century-Crofts

Kolb, B., Sutherland, R. J., Whishaw, I. Q. 1982. A comparison of the contributions of the frontal and parietal association cortex to spatial localization in rats. *J. Comp. Physiol. Psychol.* In press

Konorski, J. 1967. *Integrative Activity of the Brain*. Chicago: Univ. Chicago Press

Koob, G. E., Bloom, F. E. 1982. Behavioral effects of neuropeptides: Endorphins and vasopressin. *Ann Rev. Physiol.* 44:571–82

Kovács, G., Bohus, B., Versteeg, D. H. G. 1980. The interaction of posterior pituitary neuropeptides with monaminergic neurotransmission: Significance in learning and memory processes. *Prog. Brain Res.* 53:123–40

Kubie, J. L., Ranck, J. B. Jr. 1982. Tonic and phasic firing of rat hippocampal complex-spike cells in three different situations: Context and place. See Woody 1982. In press

Kupfermann, I. 1975. Neurophysiology of learning. *Ann. Rev. Psychol.* 26:367–91

Larson, J. R., Greenough, W. T. 1981. Effects of handedness training on dendritic branching of neurons of forelimb area of rat motor cortex. *Neurosci. Abstr.* 7:65

Lashley, K. S. 1929. *Brain Mechanisms and Intelligence.* Chicago: Univ. Chicago Press

Lasiter, P. S., Glanzman, D. L. 1982. Cortical substrates of taste aversion learning: Dorsal prepiriform (Insular) lesions disrupt taste aversion learning. *J. Comp. Physiol. Psychol.* 96:376–92

Lavond, D. G., McCormick, D. A., Clark, G. A., Holmes, D. T., Thompson, R. F. 1981. Effects of ipsilateral rostral pontine reticular lesions on retention of classically conditioned nictitating membrane and eyelid responses. *Physiol. Psychol.* 9(4):335–39

Lee, K. S., Schottler, F., Oliver, M., Lynch, G. 1980. Brief bursts of high-frequency stimulation produce two types of structural change in rat hippocampus. *J. Neurophysiol.* 44:247–58

LeVay, S., Hubel, D. H., Wiesel, T. N. 1977. The pattern of ocular dominance columns in macaque visual cortex revealed by a reduced silver stain. *J. Comp. Neurol.* 159:559–76

LeVay, S., Wiesel, T. N., Hubel, D. H. 1980. The development of ocular dominance columns in normal and visually deprived monkeys. *J. Comp. Neurol.* 191:1–51

Levine, S. 1968. *Hormones and Conditioning. Nebraska Symp. Motivation,* pp. 85–101. Lincoln: Univ. Nebraska Press

Levy, W. B., Steward, O. 1979. Synapses as associative memory element in the hippocampal formation. *Brain Res.* 175:233–45

Levy, W. B., Steward, O. 1982. Long-term associative potentiation/depression in the hippocampus. *Neuroscience* In press

Liang, K. C., McGaugh, J. L., Martinez, J. L. Jr., Jensen, R. A., Vasquez, B. J., Messing, R. B. 1982. Post training amygdaloid lesions impair retention of an inhibitory avoidance response. *Behav. Brain Res.* 4:237–50

Lincoln, J. S., McCormick, D. A., Thompson, R. F. 1982. Ipsilateral cerebellar lesions prevent learning of the classically conditioned nictitating membrane/eyelid response. *Brain Res.* 242:190–93

Llinás, R., Walton, K., Hillman, E. D., Sotelo, C. 1975. Inferior olive: Its role in motor learning. *Science* 190:1230–31

Lynch, G. S., Dunwiddie, T., Gribkoff, V. 1977. Heterosynaptic depression: A post-synaptic correlate of long-term potentiation. *Nature* 266:737–38

Maffei, L., Fiorentini, A. 1976. Asymmetry of motility of the eyes and change of binocular properties of cortical cells in adult cats. *Brain Res.* 105:73–78

Malkasian, D. R., Diamond, M. C. 1971. The effects of environmental manipulation on the morphology of the neonate rat brain. *Int. J. Neurosci.* 2:161–70

Marr, D. 1969. A theory of cerebellar cortex. *J. Physiol.* 202:437–70

Martinez, J. L., Rigter, H., Jensen, R. A., Messing, R. B., Vasquez, B. J., et al. 1981. Endorphin and enkephalin effects on avoidance conditioning: The other side of the pituitary-adrenal axis. In *Endogenous Peptides and Learning and Memory Processes,* ed. J. L. Martinez, R. A. Jensen, R. B. Messing, H. Rigter, J. L. McGaugh. New York: Academic

Mason, S. T. 1981. Noradrenaline in the brain: Progress in theories of behavioral function. *Progr. Neurobiol.* 16(3/4):263–303

Mason, S. T., Fibiger, H. C. 1979. Noradrenaline and avoidance learning in the rat. *Brain Res.* 161:321–34

Mason, S. T., Iversen, S. D. 1979. Theories of the dorsal bundle extinction effect. *Brain Res. Rev.* 1:107–37

Mauk, M. D., Warren, J. T., Thompson, R. F. 1982. Selective, naloxone-reversible morphine depression of learned behavioral and hippocampal responses. *Science* 216:434–35

McCormick, D. A., Clark, G. A., Lavond, D. G., Thompson, R. F. 1982a. Initial localization of the memory trace for a basic form of learning. *Proc. Natl. Acad. Sci. USA* 79(8):2731–42

McCormick, D. A., Lavond, D. G., Thompson, R. F. 1982b. Concomitant classical conditioning of the rabbit nictitating membrane and eyelid responses: Correlations and implications. *Physiol. Behav.* 28:769–75

McCormick, D. A., Lavond, D. G., Clark, G. A., Kettner, R. E., Rising, C. E. et al. 1981. The engram found? Role of the cerebellum in classical conditioning of nictitating membrane and eyelid responses. *Bull. Psychon. Soc.* 18(3):103–5

McGaugh, J. L., Herz, M. J. 1972. *Memory Consolidation.* San Francisco: Albion

McNaughton, B. L., Douglas, R. M., Goddard, G. V. 1978. Synaptic enhancement in fascia dentata: Cooperativity among coactive afferents. *Brain Res.* 157:277–93

Meyer, D. R., Meyer, P. M. 1977. Dynamics and bases of recoveries of functions after injuries to the cerebral cortex. *Physiol. Psychol.* 5(2):133–65

Micco, D. J., McEwen, B. S., Shein, W. 1979. Modulation of behavioral inhibition in appetitive extinction following manipulation of adrenal steroids in rats: Implications for involvement of the hippocampus. *J. Comp. Physiol. Psychol.* 93:323–29

Miles, F. A., Lisberger, S. G. 1981. Plasticity in the vestibulo-ocular reflex: A new hypothesis. *Ann. Rev. Neurosci.* 4:273–99

Miller, N. E. 1948. Studies of fear as an acquirable drive. I. Fear as motivation and fear-reduction as reinforcement in learning of new responses. *J. Exp. Psychol.* 38:89–101

Miller, V. M., Best, P. J. 1980. Spatial correlates of hippocampal unit activity are altered by lesions of the fornix and endorhinal cortex. *Brain Res.* 194:311–23

Mishkin, M. 1966. Visual mechanisms beyond the striated vortex. In *Frontiers in Physiological Psychology*, ed. R. W. Russell. New York: Academic

Mishkin, M., Oubre, J. L. 1976. Dissociation of deficits on visual memory tasks after inferior temporal and amygdala lesions in monkeys. *Neurosci. Abstr.* 2:1127

Mishkin, M. 1978. Memory in monkeys severely impaired by combined but not by separate removal of amygdala and hippocampus. *Nature* 273:297–98

Misulis, K. E., Durkovic, R. G. 1982. Classically conditioned alterations in single motor unit activity in the spinal cat. *Behav. Brain Res.* In press

Møllgaard, K., Diamond, M. C., Bennett, E. L., Rosenzweig, M. R., Lindner, B. 1971. Quantitative synaptic changes with differential experience in rat brain. *Int. J. Neurosci.* 2:113–28

Moore, J. W., Desmond, J. E., Berthier, N. E. 1982. The metencephalic basis of the conditioned nictitating membrane response. See Woody 1982. In press

Moore, J. W., Solomon, P. R., eds. 1980. The role of the hippocampus in learning and memory: A memorial workshop to A. H. Black. *Physiol. Psychol.* Vol. 8

Morgan, M. J. 1973. Effects of post-weaning environment on learning in the rat. *Anim. Behav.* 21:429–42

Movshon, J. A. 1976. Reversal of the physiological effects of monocular deprivation in the kitten's visual cortex. *J. Physiol.* 261:125–74

Mower, G. D., Berry, D., Burchfiel, J. L., Duffy, F. H. 1981. Comparison of the effects of dark rearing and binocular suture on development and plasticity of cat visual cortex. *Brain Res.* 220:255–67

Mowrer, O. H. 1947. On the dual nature of learning—a reinterpretation of "conditioning" and "problem-solving." *Harv. Ed. Revi.* 17:102–48

Muller, L., Pattiselanno, A., Vrensen, G. 1981. The postnatal development of the presynaptic grid in the visual cortex of rabbits and the effect of dark-rearing. *Brain Res.* 205:39–48

Nachman, M., Ashe, J. H. 1974. Effects of basolateral amygdala lesions on neophobia, learned taste aversions, and sodium appetite in rats. *J. Comp. Physiol. Psychol.* 87:622–43

Norman, R. J., Buchwald, J. S., Villablanca, J. R. 1977. Classical conditioning with auditory discrimination of the eyeblink in decerebrate cats. *Science* 196:551–53

Nottenbohm, F. 1981. A brain for all seasons: Cyclical anatomical changes in song control nuclei of the canary bird. *Science* 214:1368–70

Nyman, A. J. 1967. Problem solving in rats as a function of experience at different ages. *J. Genet. Psychol.* 110:31–39

Oakley, D. A. 1977. Performance of decorticated rats in a two-choice visual discrimination apparatus. *Behav. Brain Res.* 3:55–69

Oakley, D. A. 1979. Cerebral cortex and adaptive behaviour. In *Brain, Behaviour and Evolution*, ed. D. A. Oakley, H. C. Plotkin, pp. 154–88. London: Methuen

Oakley, D. A., Russell, I. S. 1972. Neocortical lesions and classical conditioning. *Physiol. Behav.* 8:915–26

O'Brien, J. H., Quinn, K. J. 1982. Central mechanisms responsible for classically conditioned changes in neuronal activity. See Woody 1982. In press

O'Brien, J. H., Wilder, M. B., Stevens, C. D. 1977. Conditioning of cortical neurons in cats with antidromic activation as the unconditioned stimulus. *J. Comp. Physiol. Psychol.* 91:918–29

Ögren, S. O., Archer, T., Ross, S. B. 1980. Evidence for a role of the locus coeruleus noradrenaline system in learning. *Neurosci. Lett.* 20:351–56

Ögren, S. O., Fuxe, K. 1974. Learning, brain noradrenaline and the pituitary-adrenal axis. *Med. Biol.* 52:399–405

Ögren, S. O., Fuxe, K. 1977. On the role of brain noradrenaline and the pituitary-adrenal axis in avoidance learning. I. Studies with corticosterone. *Neurosci. Lett.* 5:291–96

O'Keefe, J. 1976. Place units in the hippocampus of the freely moving rat. *Exp. Neurol.* 51:78–109

O'Keefe, J., Conway, D. H. 1978. Hippocampal place units in the freely moving rat: Why they fire where they fire. *Exp. Brain Res.* 31:573–90

O'Keefe, J., Nadel, L. 1978. *The Hippocampus as a Cognitive Map.* New York: Oxford Univ. Press

Olds, J., Disterhoft, J. F., Segal, M., Hornblith, C. L., Hirsch, R. 1972. Learning centers of rat brain mapped by measuring latencies of conditioned unit responses. *J. Neurophys.* 35:202–19

Olds, S., Ninhuis, R., Olds, M. E. 1978. Pattern of conditioning unit responses in the auditory system of rat. *Exp. Neurol.* 59:209–28

Oleson, T. D., Ashe, J. H., Weinberger, N. M. 1975. Modification of auditory and somatosensory system activity during pupillary conditioning in the paralyzed cat. *J. Neurophysiol.* 38:1114–39

Olton, D. S., Becker, J. T., Handelmann, G. E. 1979. Hippocampus, space and memory. *Behav. Brain Sci.* 2:313–65

Olton, D. S., Becker, J. T., Handelmann, G. E. 1980. Hippocampal function: Working memory or cognitive mapping: *Physiol Psychol.* 8(2):239–46

Olton, D. S., Branch, M., Best, P. J. 1978. Spatial correlates of hippocampal unit activity. *Exp. Neurol.* 58:387–409

Orona, E., Foster, K., Lambert, R. W., Gabriel, M. 1982. Cingulate cortical and anterior thalamic neuronal correlates of the overtraining reversal effect in rabbits. *Behav. Brain Res.* 4:133–54

Orr, W. B., Berger, T. W. 1981. Hippocampal lesions disrupt discrimination reversal learning of the rabbit nictitating membrane response. *Neurosci. Abstr.* 7:648

Patterson, M. M. 1980. Mechanisms of classical conditioning of spinal reflexes. See Thompson et al 1980, pp. 263–72

Penfield, W., Milner, B. 1958. Memory deficit produced by bilateral lesions in the hippocampal zone. *Am. Med. Assoc. Arch. Neurol. Psychiat.* 79:475–97

Perumal, A. S., Rapport, M. M. 1978. In vitro phosphorylation of S-100 protein by brain nuclear protein kinase. *Life Sci.* 22:803–8

Pettigrew, J. D. 1974. The effect of visual experience on the development of stimulus specificity by kitten cortical neurons. *J. Physiol.* 33:49–74

Pettigrew, J. D., Freeman, R. D. 1973. Visual experience without lines: Effect on developing cortical neurons. *Science* 182:599–601

Pettigrew, J. D., Garey, L. J. 1974. Selective modification of single neuron properties in the visual cortex of kittens. *Brain Res.* 66:160–64

Poltrew, S. S., Zeliony, G. P. 1930. Grosshirnrinde und Assoziationsfunktion. *Z. Biol.* 90:157–60

Powell, D. A., Buchanan, S. 1980. Autonomic-somatic relationships in the rabbit (*Oryctolagus cuniculus*): Effects of hippocampal lesions. *Physiol. Psychol.* 8(4):455–62

Powell, D. A., Mankowski, D., Buchanan, S. 1978. Concomitant heartrate and corneoretinal potential conditioning in the rabbit (*Oryctolagus cuniculus*): Effects of caudate lesions. *Physiol. Behav.* 20:143–50

Pribram, K. H. 1971. *Languages of the Brain: Experimental Paradoxes and Principles in Neuropsychology.* Englewood Cliffs, NJ: Prentice-Hall

Pribram, K. H., Reitz, S., McNeil, M., Spevack, A. A. 1979. The effect of amygdalectomy on orienting and classical conditioning in monkeys. *Pavlovian J.* 14:203–17

Prokasy, W. F. 1972. Developments with the two-phase-model applied to human eyelid conditioning. In *Classical Conditioning Vol. II: Current Research and Theory,* ed. A. H. Black, W. F. Prokasy, pp. 119–47. New York: Appleton-Century-Crofts

Quellet, J. V., Kower, H. S., Braun, J. J. 1975. *Failure to retain a learned taste aversion after lesions of the gustatory neocortex.* Presented at Ann. Meet. Western Psychol. Assoc., Sacramento, Calif.

Rainbow, T. C. 1979. Role of RNA and protein synthesis in memory formation. *Neurochem. Res.* 4:297–312

Rapport, M., Karpiak, S. E. 1976. Discriminative effects of antisera to brain constituents on behavior and EEG activity in rats. *Res. Commun. Psychol. Psychiatr. Behav.* 1:115–23

Rescorla, R. A., Solomon, R. L. 1967. Two-process learning theory: Relationships between Pavlovian conditioning and instrumental learning. *Psychol. Rev.* 74:151–82

Riege, W. H. 1971. Environmental influences on brain and behavior of year-old rats. *Dev. Psychol.* 4:151–67

Roberts, D. C. S., Fibiger, H. C. 1977. Evidence for interaction between central noradrenergic neurons and adrenal hormones in learning and memory. *Pharmacol. Biochem. Behav.* 7:191–94

Rose, S. P. R. 1980. Neurochemical correlates of early learning on the chick. In *Neurobiological Bases of Learning and Memory*, ed. Y. Tsukada, B. W. Agranoff, pp. 179–91. New York: Wiley

Rose, S. P. R. 1981a. From causations to translations: What biochemists can contribute to the study of behaviour. *Perspect. Ethol.* 4:157–77

Rose, S. P. R. 1981b. What should a biochemistry of learning and memory be about? *Neuroscience* 6(5):811–21

Rose, S. P. R., Haywood, J. 1976. Experience, learning and brain metabolism. In *Biochemical Correlates of Brain Structure and Function*, ed. A. N. Davison, pp. 249–92. New York: Academic

Rosenzweig, M. R. 1971. Effects of environment on development of brain and on behavior. In *Biopsychology of Development*, ed. E. Tobach, pp. 303–42. New York: Academic

Rosenzweig, M. R., Bennett, E. L. 1976. Enriched environments: Facts, factors and fantasies. In *Knowing, Thinking and Believing*, ed. J. L. McGaugh, L. Petrinovich, pp. 179–213. New York: Plenum

Routtenberg, A. 1979. Anatomical localization of phosphoprotein and glycoprotein substrates of memory. *Prog. Neurobiol.* 12:85–113

Rutledge, L. T., Wright, C., Duncan, J. 1974. Morphological changes in pyramidal cells of mammalian neocortex associated with increased use. *Exp. Neurol.* 44:209–28

Ryugo, D. K., Weinberger, N. M. 1976. Differential plasticity of morphologically distinct neuron populations in the medial geniculate body of the cat during classical conditioning. *Soc. Neurosci. Abstr.* 2:435

Ryugo, D. K., Weinberger, N. M. 1978. Differential plasticity of morphologically distinct neuron populations in the medial geniculate body of the cat during classical conditioning. *Behav. Biol.* 22:275–301

Scavio, M. J., Thompson, R. F. 1979. Extinction and reacquisition performance alternations of the conditioned nictitating membrane response. *Bull. Psychon. Soc.* 13(2):57–60

Schapiro, S., Vukovich, K. R. 1970. Early experience effects upon cortical dendrites: A proposed model for development. *Science* 167:292–94

Schmidt, R., Shashoua, V. E. 1981. A radioimmunoassay for ependymins β and γ: Two goldfish brain proteins involved in behavior plasticity. *J. Neurochem.* 36:1368–77

Schneiderman, N. 1972. Response system divergencies in aversive classical conditioning. See Prokasy 1972, pp. 341–78

Schneiderman, N., Smith, M. C., Smith, A. C., Gormezano, I. 1966. Heart rate classical conditioning in rabbits. *Psychon. Sci.* 6:241–42

Schneiderman, N., VanDercar, D. H., Yehle, A. L., Manning, A. A., Golden, T., et al. 1969. Vagal compensatory adjustment: Relationship to heart rate classical conditioning in rabbits. *J. Comp. Physiol. Psychol.* 68:175–83

Schwaber, J. S., Kapp, B. S., Higgins, G. 1980. The origin and extent of direct amygdala projections to the region of the dorsal motor nucleus of the vagus and the nucleus of the solitary tract, *Neurosci. Lett.* 20:15–20

Schwartz, S. 1964. Effect of neonatal cortical lesions and early environmental factors on adult rat behavior. *J. Comp. Physiol. Psychol.* 57:72–77

Scoville, W. B., Milner, B. 1957. Loss of recent memory after bilateral hippocampal lesions. *J. Neurol. Neurosurg. Psychiatr.* 20:11–21

Segal, M. 1977a. Changes of interhemispheric hippocampal responses during conditioning in the rat. *Exp. Brain Res.* 29:553–65

Segal, M. 1977b. Excitability changes in rat hippocampus during conditioning. *Exp. Neurol.* 55:67–73

Shashoua, V. E. 1982. Molecular and cell biological aspects of learning: Toward a theory of memory. *Adv. Cell. Neurobiol.* 3:97–141

Smith, A. M. 1970. The effects of rubral lesions and stimulation on conditioned forelimb flexion responses in the cat. *Physiol. Behav.* 5:1121–26

Smith, O. A., Astley, C. A., DeVit, J. L., Stein, J. M., Walsh, K. E. 1980. Functional analysis of hypothalamic control of the cardiovascular responses accompanying emotional behavior. *Fed. Proc.* 39(8):2487–94

Sofroniew, M. V., Weindl, A. 1981. Central nervous system distribution of vasopressin, oxytocin and neurophysin. See Martinez et al 1981

Sokolov, E. N. 1977. Brain functions: Neuronal mechanisms of learning and memory. *Ann. Rev. Psychol.* 28:85–112

Solomon, P. R., Moore, J. W. 1975. Latent inhibition and stimulus generalization for the classically conditioned nictitating membrane response in rabbits (*Oryctolagus cuniculus*) following dorsal hippocampal ablation. *J. Comp. Physiol. Psychol.* 89:1192–1203

Spiegler, B. J., Mishkin, M. 1981. Evidence for the sequential participation of inferior temporal cortex and amygdala in the acquisition of stimulus-reward associations. *Behav. Brain Res.* 3:303–17

Spinelli, D. N., Hirsch, H. V. B., Phelps, R. W., Metzler, J. 1972. Visual experience as a determinant of the response characteristics of cortical receptive fields in cats. *Exp. Brain Res.* 15:289–304

Spinelli, D. N., Jensen, F. E. 1979. Plasticity: The mirror of experience. *Science* 203:75–78

Spinelli, D. N., Jensen, F. E., DiPrisco, G. V. 1980. Early experience effect on dendritic branching in normally reared kittens. *Exp. Neurol.* 68:1–11

Squire, L. 1982. The neurophysiology of human memory. *Ann. Rev. Neurosci.* 5:241–73

Squire, L. R., Davis, H. P. 1981. The pharmacology of memory: A neurobiological perspective. *Ann. Rev. Pharmacol. Toxicol.* 21:323–56

Stein, D. G., Rosen, J. J., Butters, N., eds., 1974. *Plasticity and Recovery of Function in the Central Nervous System.* New York: Academic

Swanson, L. W., Teyler, T. J., Thompson, R. F., eds., 1982. Hippocampal LTP: Mechanisms and functional implications. *Neurosci. Res. Progr.* 20(5):613–769

Thompson, R. F., Barchas, J. D., Clark, G. A., Donegan, N., Kettner, R. E., et al. 1983. Neuronal substrates of associative learning in the mammalian brain. In *Primary Neural Substrates of Learning and Behavioral Change,* ed. D. L. Alkan, J. Farley. Princeton: Princeton Univ. Press. In press

Thompson, R. F., Berger, T. W., Cegavske, C. F., Patterson, M. M., Roemer, R. A., et al. 1976. The search for the engram. *Am. Psychol.* 31:209–27

Thompson, R. F., Berger, T. W., Berry, S. D., Clark, G. A., Kettner, R. E., et al. 1982. Neuronal substrates of learning and memory: Hippocampus and other structures. See Woody 1982. In press

Thompson, R. F., Hicks, L. H., Shvyrkov, V. B., eds. 1980. *Neural Mechanisms of Goal-Directed Behavior and Learning.* New York: Academic

Thompson, R. F., Patterson, M. M., Teyler, T. J. 1972. The neurophysiology of learning. *Ann. Rev. Psychol.* 23:73–104

Tolman, E. C., Honzik, C. H. 1930. Introduction and removal of reward, and maze performance in rats. *Univ. Calif. Publ. Psychol.* 4:257–75

Tsukahara, N. 1981. Synaptic plasticity in the mammalian central nervous system. *Ann. Rev. Neurosci.* 4:351–79

Tsukahara, N. 1982. Classical conditioning mediated by the red nucleus in the cat. See Woody 1982. In press

Tsukahara, N., Oda, Y., Notsu, T. 1981. Classical conditioning mediated by the red nucleus in the cat. *J. Neurosci.* 1:72–79

Tsumoto, T., Suda, K. 1979. Cross depression: An electrophysiological manifestation of binocular competition in the developing visual cortex. *Brain Res.* 168:190–94

Uylings, H. B. M., Kuypers, K., Diamond, M. C., Veltman, W. A. M. 1978. Effects of differential environments on plasticity of dendrites of cortical pyramidal neurons in adult rats. *Exp. Neurol.* 62:658–77

Valverde, F. 1967. Apical dendritic spines of the visual cortex and light deprivation in the mouse. *Exp. Brain Res.* 3:337–52

Valverde, F. 1968. Structural changes in the area striate of the mouse after enucleation. *Exp. Brain Res.* 5:274–92

Valverde, F. 1971. Rate and extent of recovery from dark rearing in the visual cortex of the mouse. *Brain Res.* 33:1–11

Vinogradova, O. S. 1975. Functional organization of the limbic system in the process of registration of information: Facts and hypotheses. In *The Hippocampus,* ed. R. L. Isaacson, K. H. Pribram, Vol. 2. New York: Plenum

Voronin, L. L. 1971. Microelectrode study of cellular analogs of conditioning. *Proc. 25th Intl. Congr. Physiol. Sci. Munich* 8:199–200

Voronin, L. L. 1980. Microelectrode analysis of the cellular mechanisms of conditioned reflex in rabbits. *Acta Neurobiol. Exp.* 40:335–70

Vrensen, G., Cardozo, J. N. 1981. Changes in size and shape of synaptic connections after visual training: An ultrastructural approach of synaptic plasticity. *Brain Res.* 218:79–97

Vrensen, G., De Groot, D. 1975. The effect of monocular deprivation on synaptic terminals in the visual cortex of rabbits. A quantitative electron microscopic study. *Brain Res.* 93:15–24

Volkmar, F. R., Greenough, W. T. 1972. Rearing complexity affects branching of dendrites in visual cortex of the rat. *Science* 117:1445–47

Walk, R. D. 1958. Visual and visual-motor

experience: A replication. *J. Comp. Physiol. Psychol.* 51:785–87

Wall, J., Wild, J. M., Broyles, J. Gibbs, C. M., Cohen, D. H. 1980. Plasticity of the tectofugal pathway during visual conditioning. *Neurosci. Abstr.* 6:424

Walsh, R. N., Budtz-Olsen, O. E., Penny, J. E., Cummins, R. A. 1969. The effects of environmental complexity of the histology of the rat hippocampus. *J. Comp. Neurol.* 137:361–66

Weinberger, N. M. 1980. Neurophysiological studies of learning in association with the pupillary dilation conditioned reflex. See Thompson et al 1980

Weinberger, N. M. 1982a. Effects of conditioned arousal on the auditory system. In *The Neural Basis of Behavior,* ed. A. L. Beckman. Jamaica, NY: Spectrum

Weinberger, N. M. 1982b. Sensory plasticity and learning: The magnocellular medial geniculate nucleus of the auditory system. See Woody 1982. In press

Weingartner, H., Parker, E., eds. 1982. *Memory Consolidation.* Associates, Hillsdale, NJ: Lawrence Erlbaum. In press

Weisz, D. J., Clark, G. A., Yang, B., Solomon, P. R., Berger, T. W., et al. 1982. Activity of dentate gyrus during NM conditioning in rabbit. See Woody 1982. In press

Weisz, D. J., Solomon, P. R., Thompson, R. F. 1980. The hippocampus appears necessary for trace conditioning. *Bull. Psychon. Soc. Abstr.* 193:244

Wenzel, S., Kammerer, E., Kirsche, W., Matthies, H., Wenzel, M. 1980. Electron microscopic and morphometric studies on synaptic plasticity in the hippocampus of the rat following conditioning. *J. Hirnforsch.* 21:647–54

West, R. W., Greenough, W. T. 1972. Effect of environmental complexity on cortical synapses of rats: Preliminary results. *Behav. Biol.* 7:279–84

Wiesel, T. N., Hubel, D. H. 1963. Single-cell responses in striate cortex of kittens deprived of vision in one eye. *J. Neurophysiol.* 26:1003–17

Wiesel, T. N., Hubel, D. H., Lam, D. M. K. 1974. Autoradiographic demonstration of ocular-dominance columns in the monkey striate cortex by means of

transneuronal transport. *Brain Res.* 79:273–79

Winfield, D. A. 1981. The postnatal development of synapses in the visual cortex of the cat and the effects of eyelid closure. *Brain Res.* 206:166–71

Woody, C. D. 1970. Conditioned eye blink: Gross potential activity at coronal-precruciate cortex of the cat. *J. Neurophysiol.* 33:838–50

Woody, C. D. 1982. Neurophysiologic correlates of latent facilitation. In *Conditioning: Representation of Involved Neural Functions,* ed. C. D. Woody. New York: Plenum. In press

Woody, C. D., Black-Cleworth, P. A. 1973. Differences in the excitability of cortical neurons as a function of motor projection in conditioned cats. *J. Neurophysiol.* 36:1104–16

Woody, C. D., Vassilevsky, N. N., Engel, J. Jr. 1970. Conditioned eye blink: Unit activity at coronal-pericruciate cortex of the cat. *J. Neurophysiol.* 33:851–64

Woody, C. D., Yarowsky, P., Owens, J., Black-Cleworth, P., Crow, T. 1974. Effect of lesions of coronal motor areas on acquisition of conditioned eye blink in the cat. *J. Neurophysiol.* 37:385–94

Yehle, A. L., Dauth, G., Schneiderman, N. 1967. Correlates of heart-rate classical conditioning in curarized rabbits. *J. Comp. Physiol. Psychol.* 64:93–104

Zomzely-Neurath, C. P., Keller, A. 1977. The different forms of brain enolase: Isolation, characterization, cell specificity and physiological significance. In *Mechanisms Regulation and Special Functions of Protein Synthesis in Brain,* ed. S. Roberts, A. Lajtha, W. H. Gispen, pp. 279–98. Amsterdam: Elsevier/North Holland

Zomzely-Neurath, C. P., Marangos, P. J., Hymonowitz, N., Perl, W., Ritter, A., et al. 1976. Changes in brain-specific proteins during learning. *Trans. Am. Soc. Neurochem.,* 7:242–

Zornetzer, S. F. 1978. Neurotransmitter modulation and memory: A new neuropharmacological phrenology? In *Psychotherapy: A Generation of Progress,* ed. M. A. Lipton, A. DiMascio, K. F. Killam, pp. 637–49. New York: Raven

Ann. Rev. Neurosci. 1983. 6:493–525
Copyright © 1983 by Annual Reviews Inc. All rights reserved

EXPERIMENTAL APPROACHES TO UNDERSTANDING THE ROLE OF PROTEIN PHOSPHORYLATION IN THE REGULATION OF NEURONAL FUNCTION

Mary B. Kennedy

Division of Biology 216–76, California Institute of Technology, Pasadena, California 91125

INTRODUCTION

Studies by Earl Sutherland and his colleagues on hormonal regulation of the breakdown of glycogen in liver resulted in the discovery that the first step in the action of many hormones is to increase the synthesis of cAMP by activating adenylate cyclase (Rall et al 1957, Sutherland & Rall 1958, Robison et al 1968). It was later established that cAMP exerts its effects by stimulating protein kinases that catalyze the phosphorylation of specific functional proteins and thereby regulate their activity (Walsh et al 1968, Kuo & Greengard 1969, Krebs & Beavo 1979). The discovery that the brain contains a high concentration of cAMP-dependent protein kinase led to the proposal that protein phosphorylation might play an important role in regulation of neuronal properties by neurotransmitters and neurohormones (Miyamoto et al 1969). In particular, it seemed that protein phosphorylation, which usually takes place on a time scale of hundreds of milliseconds or longer, might be a mechanism underlying relatively long-lasting changes in neuronal properties such as "slow" changes in post-synaptic potentials (McAfee & Greengard 1972), changes in the rate of transmitter synthesis (Morgenroth et al 1975), or changes in gene expression (Klein & Berg 1970). The biochemists and neurobiologists who took up the study of brain protein phosphorylation hoped to gain insight into some of the mechanisms

493

0147-006X/83/0301-0493$02.00

underlying changes in neuronal excitability and synaptic efficacy and also, perhaps, into processes that govern the development of various neuronal types during the formation of the nervous system. This line of research was bolstered by the findings that the brain contains not only high concentrations of protein kinases, but also protein phosphatases, adenylate cyclase, and phosphodiesterase (Greengard 1976), and also by the discovery that several neurotransmitters stimulate the synthesis of second messengers such as cyclic AMP and cyclic GMP by binding to specific receptors on the surfaces of neurons (for reviews see Nathanson 1977, Greengard 1981).

Further progress in understanding the role of protein phosphorylation in the regulation of neuronal function has until recently been slow. This problem has often been as difficult as the general problem of identifying and understanding proteins that mediate complex neuronal functions such as transmitter release, or the gating of ion channels. However, increasingly sophisticated methods for examining neuronal structure and function at cellular and molecular levels have begun to yield clear answers.

Three strategies have resulted in new insights in the past few years and are discussed in this review. The first involves the study of the role of protein phosphorylation in regulating electrical properties of membranes of invertebrate neurons. The large size and ease of identification of these neurons make them useful for such studies for two reasons:

1. They are easy to impale with microelectrodes, consequently many of their membrane ion channels have been well characterized physiologically.
2. Both small molecules and proteins can be injected into them for the purpose of studying the resulting functional changes.

Recent experiments in which protein kinase or its inhibitor were injected into identified invertebrate neurons have shown that cAMP-dependent protein phosphorylation can lead to changes in the properties of membrane ion channels. Such changes probably underlie the modulatory actions of many agents that stimulate adenylate cyclase.

Another, more open-ended, strategy has been to identify substrate proteins for various protein kinases in mammalian brain homogenates, characterize them, then attempt to learn their functions by examining their subcellular locations and the physiological changes that alter their phosphorylation. This is a rather risky and long-term approach, but it has resulted in the discovery of two neuronal proteins with specific distributions in the brain and intriguing regulatory properties that point toward certain functions. In addition, this approach has led to the discovery of calcium-regulated protein phosphorylation systems in neural tissue.

A third strategy has been to look for phosphorylation in vitro of specific proteins whose functional properties appear to be modulated by second messengers. The rate-limiting enzymes in the synthetic pathways for the catecholamines and serotonin are more active following neuronal activity. They have now been shown by several groups to be regulated by phosphorylation. Tyrosine hydroxylase is activated by both cAMP-dependent and calcium-dependent phosphorylation, while tryptophan hydroxylase is regulated by calcium-dependent phosphorylation.

Due to space limitations, several topics of interest are not covered here. Two of these, the use of hippocampal slices to examine the role of protein phosphorylation in long-term potentiation, and the possible involvement of protein phosphorylation in photoreceptor function, were discussed in recent articles in the *Annual Review of Neuroscience* (Lynch & Schubert 1980, Hubbell & Bownds 1979).

REGULATION OF ELECTRICAL PROPERTIES OF INVERTEBRATE NEURONS

Several distinct membrane ion channels have been identified in the somas of invertebrate neurons (for a review see Adams et al 1980). Pharmacological and physiological techniques have been developed by which properties of many of the channels can be studied individually and under well-controlled conditions. Such studies have resulted in a better understanding of the ion channels that underlie the complex electrical properties of several specialized neuronal cell types. In three types of neurons in ganglia of the marine mollusk *Aplysia californica,* agents that stimulate adenylate cyclase have been found to modify the cell's electrical properties. A combination of physiological and biochemical experiments indicate that changes in K^+-conductances in each of the cells underlie the electrical modifications and that cAMP-dependent protein phosphorylation can initiate these K^+-conductance changes. Initial biochemical experiments have identified phosphoproteins that are candidates for mediating the physiological effects.

The Bursting Pacemaker, R_{15}

Levitan and co-workers have studied the regulation of ion channels in the giant Aplysia neuron, R_{15}. This cell is one of a class of cells called bursting pacemakers. Such cells produce rhythmic bursts of action potentials, interspersed by periods of transient hyperpolarization. This pattern of activity is produced by the interaction of several ionic currents (see Adams et al 1980). A characteristic component of bursting pacemakers is a small voltage-dependent inward current (I_B) carried by Na^+ and/or Ca^{2+} (Eckert & Lux 1975). This current is activated very slowly by depolarizing pulses and tends to excite the cell, leading to bursts of action potentials. It can be offset by several voltage-dependent K^+-currents. Two neurotransmitters, serotonin (Drummond et al 1980a) and dopamine (Ascher 1972, Wilson & Wachtel 1978), alter pacemaker activity. Serotonin lengthens the interburst hyperpolarization and at sufficiently high concentrations (10 μM) suppresses bursting altogether. Serotonin also activates adenylate cyclase in R_{15} and causes a rise in intracellular cAMP concentrations (Levitan 1978, Drummond et al 1980a). The serotonin receptor that activates adenylate cyclase is the same one that mediates the effect of serotonin on the electrical properties of the R_{15} membrane (Drummond et al 1980a,b). Both extracellular serotonin and intracellular cAMP increase an anomalously rectifying K^+ conductance (J. A. Benson and I. B. Levitan, personal communication). The resulting K^+-current presumably offsets the slow inward current that triggers bursting. This provides a possible mechanism for the lengthened interburst interval at intermediate doses of serotonin and the suppression of bursting at high doses.

The role of cAMP-dependent protein phosphorylation in control of this K^+-conductance was examined by injecting into R_{15} a specific protein inhibitor (PKI) of cAMP-dependent protein kinase (Levitan & Adams 1981, Adams & Levitan 1982). This small inhibitory protein, purified from mammalian skeletal muscle (Demaille et al 1977), is a potent and specific inhibitor of cAMP-dependent protein kinase (Ashby & Walsh 1972). It has little effect on any other known kinases. Injection of the inhibitor into R_{15} blocks the effect of serotonin on bursting pacemaker activity. In an interesting control experiment, Adams & Levitan (1982) showed that the inhibitor does not block the effect of dopamine, which slows pacemaker activity by a different mechanism that involves a different ionic current and does not act through cAMP (Drummond et al 1980a). These experiments provide strong evidence that cAMP-stimulated protein phosphorylation is directly involved in the serotonin response. Further work by the Levitan laboratory on the nature of the proteins that are phosphorylated in R_{15} in response to serotonin is discussed below.

Peptidergic Bag Cells

L. Kaczmarek and others working in the laboratory of F. Strumwasser have analyzed the regulation of electrical activity of the Aplysia bag cells. These neurons occur in clusters around the base of the two anterior connectives that connect the abdominal ganglion with the head ganglia. The cells secrete a peptide hormone that triggers the release of eggs and associated egg-laying behavior. The cells can be induced to fire long discharges of sodium-calcium action potentials in response to brief stimulation of an anterior connective or to application of peptides from the atrial gland, which is part of the reproductive tract. Several lines of evidence suggest that the afterdischarge is triggered by a rise in intracellular cAMP (Kaczmarek et al 1978). For example, cAMP concentration in the cells rises during afterdischarge, and injection of 8-benzylthio-cAMP into the cells can produce an afterdischarge (Kaczmarek & Strumwasser 1981a). The electrical correlates of the onset of afterdischarge are complex. Prior to discharges that are induced by extracellular 8-benzylthio-cAMP there is an increase in membrane resistance, a corresponding decrease in the action potential threshold, and a broadening of the action potentials. Large subthreshold voltage oscillations are observed, which eventually trigger the long-lasting afterdischarge (Kaczmarek & Strumwasser 1981a). Preliminary voltage clamp analyses on isolated bag cells in culture indicate that the primary effect of cAMP is to decrease the net outward currents in the bag cells. Cyclic AMP has no effect on the inward currents, the transient early K^+-current, or the voltage activated late K^+-current. These results suggest that the decrease may be in either the Ca^{2+}-activated K^+-current or in an as yet undescribed K^+-current (Kaczmarek & Strumwasser 1981b).

In a collaborative study between the Strumwasser and Greengard laboratories, Kaczmarek et al (1980) made use of the wealth of information about mammalian cAMP-dependent protein kinase and of its availability in purified form to test whether protein phosphorylation by cAMP-dependent protein kinase might be involved in the onset of afterdischarge. There are two types of cAMP-dependent protein kinases (Hofmann et al 1975, Corbin et al 1975). Both are tetrameric enzymes containing two cAMP-binding regulatory subunits and two catalytic subunits. The two types of enzyme have different regulatory subunits, but the same type of catalytic subunit. Cyclic AMP activates the kinases by binding to the regulatory subunits, causing them to release the catalytic subunits in an active form. The free catalytic subunits can be purified (Beavo et al 1974) and will phosphorylate appropriate proteins even in the absence of cAMP. Injection of purified free catalytic subunits into a cell would thus be expected to mimic activation of endogenous cAMP-dependent protein kinase. Microinjection of purified bovine heart catalytic subunits into bag cells produced an increase in membrane resistance and a broadening of action potentials in more than half of the cells injected, whereas control injections of the microelectrode solution or heat inactivated enzyme did not (Kaczmarek et al 1980). These results support the hypothesis that cAMP-dependent protein phosphorylation results in a decrease in K^+-conductance in the cells. However, only three of the 16 injected cells showed subthreshold voltage oscillations, and only one showed brief repetitive firing. This could mean either that additional factors unrelated to protein phosphorylation are also involved in initiation of afterdischarge, or that the amount of kinase injected into most of the cells was insufficient to induce afterdischarge. These studies provide clear evidence that cAMP-dependent protein phosphorylation is involved in regulation of K$^+$-currents in *Aplysia*. Future experiments should clarify which of the K$^+$-channels is primarily affected. Further experiments of the Strumwasser group have shown that the bovine heart cAMP-dependent kinase phosphorylates proteins in bag cell homogenates that are also phosphorylated by an endogenous *Aplysia* cAMP-dependent protein kinase (Jennings et al 1982), thus confirming that the specificity of the kinase is highly conserved in different species. The nature of bag cell proteins that are phosphorylated by the cAMP-dependent protein kinase is discussed below.

Sensory Neurons of the Aplysia Gill-Withdrawal Reflex Pathway

E. Kandel and his colleagues have carried out an extensive analysis of the neuronal mechanisms underlying habituation and sensitization of the gill-withdrawal reflex in *Aplysia*. When either the mantle edge or siphon of an *Aplysia* is tapped, it will withdraw its gill, which normally extends beyond the mantle and siphon. If the mantle is tapped at low frequencies (once per 3 min), the reflex disappears, or habituates. If the animal is then tapped on the head, the reflex returns, or is sensitized. The neural circuits underlying the reflex, and the relationship of the habituation and sensitization to classical learning paradigms in higher animals have been thoroughly reviewed (Kandel 1976, 1981). It is sufficient for the present discussion to say that the behavior of the reflex seems to be correlated with changes in synapses between the mantle and siphon sensory neurons and the interneurons and motor neurons that innervate the gill and siphon. Sensitization is correlated with a sudden increase at these synapses in the number of quanta of transmitter released per impulse. This synaptic facilitation can be triggered in vitro by stimulation of the nerve leading into the abdominal ganglion from the head, by bath application of serotonin, or by injection of cAMP into the sensory neuron soma (Brunelli et al 1976). Thus, facilitation by serotonin apparently results from its stimulation of adenylate cyclase. Klein & Kandel (1978, 1980) found that the increased transmitter release is due, at least in part, to an increased calcium current in the synapses during depolarization. This effect is not due to a direct effect on the voltage-sensitive calcium channels, but is caused by an inhibition of the net K$^+$-current that normally repolarizes the membrane following an action potential.

In a collaborative study between the Kandel and Greengard laboratories, Castelluci et al (1980) microinjected catalytic subunits of bovine heart cAMP-dependent protein kinase into sensory neurons in order to test whether protein phosphorylation plays a role in synaptic facilitation. The neurons were pretreated with tetraethyl ammonium (TEA) to enhance calcium current during the action potential. In 29 of 35 injected cells, they observed an additional enhancement of calcium influx which resulted in a further prolongation of the action potentials. In five of ten injected cells, the conductance of the membrane decreased, as it would if K^+ channels were closed. Finally, in three of three injected sensory cells, the amplitude of the post-synaptic potential in the motor neuron increased, mimicking facilitation caused by stimulation of the nerve from the head, by bath application of serotonin, or by injection of cAMP. These effects were not observed following control injections of inactive catalytic subunits. Castelucci et al (1981) also showed that injection of the protein kinase inhibitor blocked facilitation of the synapse by serotonin and reversed facilitation after it had been produced by application of serotonin. These results indicate that under the experimental conditions studied, activation of the kinase is a limiting factor in determining the time course of the facilitation. Taken together, these experiments provide strong evidence that cAMP-dependent protein phosphorylation is involved in modulation of a K^+-current in the sensory cells, and thus in sensitization of the gill-withdrawal reflex.

Camardo et al (1981) have studied the specific K^+-current that is modulated by serotonin using voltage clamp analysis. The current does not have the characteristics of any of three well-known K^+-currents in invertebrate somas, the early I_K, the delayed I_K, or the Ca^{2+}-activated I_K. Thus, it is possible that serotonin regulates a K^+-current that has not yet been described. Further work by Bernier & Schwartz on the nature of the proteins that are phosphorylated in the sensory neurons in response to serotonin is discussed in the following section.

Substrate Proteins for cAMP-dependent Protein Kinase in Invertebrates

All three of the groups whose work is discussed above have attempted to identify proteins that are phosphorylated in Aplysia neurons by cAMP-dependent protein kinase, and thus may be mediating the changes in electrical properties induced by cyclic AMP.

Lemos et al (1981, and personal communication) have developed a technique for measuring protein phosphorylation within R_{15} after intracellular injection of γ-^{32}P-ATP. They showed that the injected ATP remained confined to the cell for the period immediately following the injection; thus only proteins inside R_{15} were labeled with ^{32}P during this time. Following the labeling period and application of test solutions, the entire ganglion was homogenized and the labeled proteins were separated by SDS polyacrylamide gel electrophoresis. This technique has two primary advantages.

1. It circumvents the problem of glial contamination when protein phosphorylation is measured in cell bodies that have been dissected from

Table 1 Summary of effects of cyclic AMP-dependent protein phosphorylation on K^+-conductances in *Aplysia* neurons

Cell type	Direction of regulation	Functional consequence	Evidence
R_{15} (bursting pacemaker)	Increased conductance	Increase in interburst interval	Effect of serotonin on K^+-conductance is blocked by protein kinase inhibitor
Peptidergic bag cells	Decreased conductance	Triggering of after-discharge leading to release of egg-laying hormone	Decrease in K^+-conductance is induced by injection of protein kinase catalytic subunit
Sensory neurons	Decreased conductance	Increased influx of calcium following action potential leads to increased transmitter release, e.g. sensitization	Decrease in K^+-conductance and sensitization are induced by injection of protein kinase catalytic subunit, and blocked by injection of protein kinase inhibitor

ganglia following conventional labeling with bath-applied inorganic ^{32}P-phosphate.

2. The cell can be voltage clamped during and following the injection, so that changes in protein phosphorylation can be related to changes in conductance.

With this method, Lemos et al showed an increase in the phosphorylation of four particulate proteins in conjunction with serotonin induced conductance changes. Three high molecular weight proteins, 230,000, 205,000, and 135,000 daltons, and one 26,000 dalton protein were more highly phosphorylated in the presence of serotonin than in its absence.

Jennings et al (1982) have examined protein phosphorylation in bag cell clusters. They used two methods to measure changes in phosphorylation that occur during afterdischarge. In one method, abdominal ganglia were incubated with ^{32}P-labeled inorganic phosphate to pre-label the intracellular ATP pools. Some of the ganglia were then stimulated to initiate afterdischarge. At appropriate times, a bag cell cluster was removed from the ganglion and homogenized. Labeled proteins were separated by SDS gel electrophoresis. Incorporation of labeled phosphate into proteins was compared to that in control ganglia that had not been stimulated. In the second method, a "back-phosphorylation" technique was employed to measure changes in cAMP-dependent protein phosphorylation during afterdischarge. Bag cells in intact ganglia were stimulated to afterdischarge. At appropriate times, bag cells were dissected from the ganglia and homogenized in a medium containing phosphodiesterase and phosphatase inhibitors. The catalytic subunit of cAMP-dependent protein kinase was added to the homogenates, and incorporation of ^{32}P from γ-^{32}P-ATP into protein, catalyzed by the kinase, was measured. Proteins that had been substantially phosphorylated in the intact cells would be expected to incorporate less ^{32}P-labeled phosphate in this assay because the phosphorylation sites would already be occupied by cold phosphate. Both techniques revealed an approximate doubling in phosphorylation of a 21,000 dalton protein 20 min after initiation of afterdischarge. No change in this protein was detectable 2 min after the beginning of afterdischarge. The pre-labeling technique, but not back-phosphorylation, revealed a 33,000 dalton protein with increased labeling both at 2 (82%) and 20 (69%) min after initiation of afterdischarge. Both of these proteins could be phosphorylated by endogenous cAMP-dependent protein kinase in bag cell homogenates and by added bovine catalytic subunit, although the 21,000 dalton protein was more heavily phosphorylated in both instances. The 33,000 dalton protein was found in several parts of the nervous system, whereas the 21,000 dalton protein was relatively specific to the isolated bag cell cluster. Goy et al (1981) have also reported an effect of serotonin and cAMP on phosphorylation of a 28,000 dalton protein in a lobster neuromuscular preparation. It may be that phosphorylation of a protein in this molecular weight range is a response to cAMP that is common to many invertebrate neurons.

L. Bernier and J. Schwartz have begun to examine phosphorylated proteins in the sensory cells of Aplysia (personal communication). They have concentrated on examining high molecular weight proteins in wedges of sensory cells removed from ganglia that have been pre-labeled with bath applied inorganic ^{32}P-phosphate. They have preliminary evidence that several proteins that are present in the sensory neurons are labeled under these conditions and they are beginning to examine whether their labeling is stimulated by serotonin.

The description of these phosphoproteins in identified neural systems in which it may be possible to study their functions is an important development. However, because the Aplysia nervous system is so small, it will be more difficult than in mammals to obtain the proteins in sufficient quantities to study their structure and subcellular localization, and to develop reagents for use in determining their functions. This illustrates a limitation of the invertebrate neural systems for biochemical studies. It would be helpful to know whether related proteins exist in the mammalian nervous system from which one can obtain large quantities of tissue easily and cheaply.

BIOCHEMICAL STUDIES OF PROTEIN KINASES AND THEIR SUBSTRATES IN MAMMALIAN BRAIN

There are three major second messengers, cAMP, cGMP, and calcium, that are known to be involved in the responses of neurons to regulatory agents

(Greengard 1978). It is widely accepted that the first two, cAMP and cGMP, act primarily, if not exclusively, by stimulating specific protein kinases. The respective kinases have been purified from brain and are relatively well characterized (for review see Walter & Greengard 1981). This discussion focuses on more recent work on some of their neuronal substrates. Two of these, Synapsin I and tyrosine hydroxylase, are substrates for cAMP-dependent protein kinase. Another, the G-substrate, is a specific substrate for cGMP-dependent protein kinase.

Calcium ion, in contrast to the cyclic nucleotides, apparently has several molecular effects in brain tissue. The full extent of the functional interactions of calcium ion is not known, but many of its effects are mediated through the calcium-binding regulatory protein, calmodulin. Studies on the phosphorylation of Synapsin I led to the discovery of a protein kinase activity in brain that is activated by calcium and calmodulin. More recently, several brain calcium-dependent protein kinases and their substrate proteins have been described. Some of these will be reviewed here.

Synapsin I

The physiological effects of increased levels of cAMP in mammalian neurons are numerous and well-documented (see Nathanson 1977). They range from short-term changes in transmitter release (Miyamoto & Breckenridge 1974) to induction of enzyme synthesis (Mackay & Iversen 1972). In order to learn more about the biochemical pathways underlying these effects, Greengard and colleagues have studied neuronal protein substrates for cAMP-dependent protein kinase (for review see Greengard 1981). One of the most intriguing is Synapsin I, until recently called Protein I, a doublet of closely related proteins found only in nervous tissue (Ueda et al 1973). It has been purified to homogeneity from bovine and rat brain (Ueda & Greengard 1977, L. De Gennaro and P. Greengard, unpublished), where it comprises about 0.4% of the total brain protein (Goelz et al 1981). Its molecular properties have been summarized in other reviews (Greengard 1979, 1981). Here, I discuss recent experiments describing its cellular and subcellular localization and also its phosphorylation by calcium-dependent in addition to cAMP-dependent protein kinases. These studies have given clues concerning possible functions of this abundant brain protein.

CELLULAR LOCALIZATION   De Camilli et al (1979) and Bloom et al (1979) have used antisera raised against bovine Synapsin I to localize it in sections of rat tissues. Specific staining was confined to synaptic regions in the central and peripheral nervous systems. These studies thus confirmed biochemical work (Ueda & Greengard 1977) that had indicated that Synapsin I was present only in neural tissue; they extended the finding to indicate that it was present in high concentrations only in synaptic regions, hence the name Synapsin I.

In these early studies, it seemed that Synapsin I was present in only some of the synaptic regions that were examined and not in others. For example, no specific staining was seen at the neuromuscular junction (De Camilli et al 1979). As techniques for fixation, sectioning, and staining of tissue have improved, however, terminals that at first could not be labeled have been shown to contain Synapsin I. These terminals include the neuromuscular junctions of the rat diaphragm and ocular muscles (P. De Camilli, R. Cameron, and P. Greengard, personal communication). It now appears that Synapsin I is contained in many different types of synapses, both excitatory and inhibitory, cholinergic and adrenergic. It seems increasingly likely that Synapsin I is a component of most, and possibly all, mammalian nerve terminals (P. De Camilli, S. M. Harris, W. Huttner, and P. Greengard, personal communication).

In order to determine the function of Synapsin I, it is important to know where it is located within the synaptic terminals. It is a particulate protein, having the characteristics of a non-integral membrane protein, and therefore one would expect it to be associated with subcellular structures. Its location was initially studied in two ways. Both immunocytochemistry at the electron microscope level (Bloom et al 1979) and measurement of the levels of Synapsin I in subcellular fractions of brain (Ueda et al 1979) gave the same surprising result. The protein appeared to be associated both with synaptic vesicles and with postsynaptic densities. Both types of experiments indicated that the largest proportion of Synapsin I was present on presynaptic vesicles; nevertheless, a significant portion was also associated with postsynaptic densities. It now seems likely that in both cases the apparent association with postsynaptic densities was artifactual for different, but related reasons. The tendency of charged soluble proteins to bind non-specifically to postsynaptic densities is by now well documented. This has led to misidentification of its peptide components in the past, as pointed out by Matus et al (1980). Synapsin I is a highly charged "sticky" protein that could easily redistribute during the detergent extraction used to prepare isolated densities. The same tendency of the postsynaptic density to bind charged molecules could also cause an artifactual precipitation of the electron dense horseradish peroxidase reaction product on the density during the staining procedure used for immunocytochemical localization. These problems, together with the relatively low specific activity of Synapsin I in the post-synaptic density fraction (Ueda et al 1979), left some doubt about whether its presence there was a reflection of its location in vivo.

In order to examine the location of Synapsin I in a way less subject to artifact, P. De Camilli and colleagues developed a method for staining synaptic regions with a ferritin-labeled antibody under conditions that do not require detergent treatment for antibody penetration. A purified synaptosomal preparation was made from brain, fixed under conditions of varying hypotonicity, and embedded in agarose. Following such fixation, the ferritin-labeled antibody can penetrate into synaptosomes that have been broken by the hypotonic shock. With this procedure, synaptic vesicles were heavily labeled, but the label over postsynaptic densities was not greater than the background in preparations stained with control sera (Figure 1; P. De Camilli, S. M. Harris, W. Huttner, and P. Greengard, manuscript in preparation). These experiments indicate that Synapsin I is associated, at least in part, with synaptic vesicles and that it is probably not present in postsynaptic densities.

CALCIUM-DEPENDENT PHOSPHORYLATION OF SYNAPSIN I   The first indication that Synapsin I might be phosphorylated, and thus regulated, in response to an increase in calcium ion concentration came from an experiment of Krueger et al (1977). It is known that a number of synaptic processes are regulated by the influx of calcium ion that occurs when a nerve terminal is depolarized by an impulse. The most clearly established of these is, of course, transmitter release (Katz & Miledi 1967, Douglas 1968). The biosynthesis of certain neurotransmitters is also regulated by calcium ion (Patrick & Barchas 1974), as is the number of quanta of transmitter released per nerve impulse (Rosenthal 1969, Klein & Kandel 1980).

In order to evaluate whether protein phosphorylation might be one of the mechanisms by which calcium exerts its regulatory actions, Krueger et al (1977) studied the effect of depolarization-induced calcium influx on the rate of phosphorylation of synaptic proteins. Brain synaptosomes were prepared and incubated with ^{32}P-labeled inorganic phosphate to prelabel the intrasynaptosomal ATP pool. The synaptosomes were then transferred to solutions containing either veratridine or high concentrations of potassium, in the presence or absence of calcium. Both of these test media induce artificial depolarization of the synaptosomes and therefore permit the influx of calcium through voltage-dependent channels. After a brief incubation, the synaptosomes were dissolved in sodium-dodecyl sulfate (SDS) and applied to SDS polyacrylamide gels. Incorporation of phosphate into proteins was examined by autoradiography. The rate of phosphorylation of several proteins was specifically increased by depolarization in the presence of calcium. The most prominent of these was the Synapsin I doublet (Krueger et al 1977, Sieghart et al 1979).

These results raised a question about the mechanism by which calcium influx into synaptosomes increased the phosphorylation of Synapsin I. Did the influx of calcium raise the level of cAMP inside the synaptosomes and

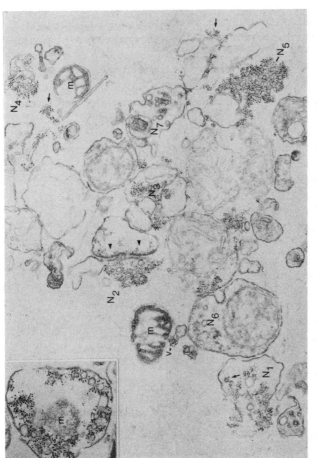

Figure 1 Immunoferritin labeling of Synapsin I in a crude synaptosomal fraction embedded in agarose. A crude synaptosomal suspension was immunostained for Synapsin I by an indirect immunoferritin procedure after a mildly hypotonic fixation and embedding in an agarose matrix. The state of preservation of individual nerve endings appears highly variable. Severely disrupted nerve endings (N_1–N_5) as well as nerve endings with an apparently sealed plasma membrane, can also be observed. Prominent labeling by ferritin particles is seen on all synaptic vesicles visible in the picture except when they appear to be inaccessible to marker proteins due to the presence of a sealed plasma membrane, as in the endings labeled N_6 and N_7. Structures other than synaptic vesicles are not labeled by ferritin. Note in particular the absence of ferritin labeling on nerve ending plasma membranes. A postsynaptic density is indicated by *arrowheads*. A few clusters of ferritin particles apparently not in direct proximity to synaptic vesicles (*arrows*) might be associated with synaptic vesicles out of the plane of the section. *Inset:* higher magnification of a labeled nerve ending, showing the specific association of ferritin particles with synaptic vesicles. m = mitochondria. (X 53,200; inset, X 76,500) (Courtesy of P. De Camilli and S. M. Harris.)

thereby stimulate phosphorylation by cAMP-dependent protein kinase, or was calcium ion stimulating a separate protein kinase, or even inhibiting a phosphatase? Krueger et al (1977) showed in their study that the conditions that produced the calcium influx into synaptosomes did not cause a measurable rise in cAMP in the synaptosomes. This suggested that calcium was not acting through the cAMP system.

Additional evidence for a separate calcium pathway came from an examination of the sites on Synapsin I that were phosphorylated in response to calcium or cAMP. Huttner & Greengard (1979) were able to stimulate phosphorylation of endogenous Synapsin I in lysates of synaptosomes by the addition of calcium or cAMP. The Synapsin I phosphorylated after either cAMP stimulation or calcium stimulation was subjected to proteolysis by Staph aureus V8 protease according to the Cleveland-Laemmli procedure (Cleveland et al 1977). The resulting phosphopeptides were separated by SDS gel electrophoresis. Stimulation of phosphorylation by cAMP occurred at a site recovered in a 10,000 dalton phosphopeptide, while stimulation of phosphorylation by calcium occurred at sites recovered in two fragments: the same 10,000 dalton fragment and an additional 30,000 dalton fragment. Further studies (Huttner et al 1981, L. DeGennaro and P. Greengard, personal communication) showed that the 10,000 dalton fragment contains a single phosphorylation site that is phosphorylated in the presence of either cAMP or calcium and is located in a globular portion of the Synapsin I molecule that is resistant to digestion by collagenase ("the globular head portion"). The 30,000 dalton fragment contains two phosphorylation sites, both of which are phosphorylated only in the presence of calcium. They are located in an elongated portion of the molecule that is sensitive to digestion by collagenase ("the collagen-like tail"). These experiments provide firm evidence that a separate calcium-sensitive phosphorylation system is responsible for part of the calcium-stimulated phosphorylation of Synapsin I. Studies of the molecular components of this calcium regulated system are discussed below.

POSSIBLE FUNCTIONS OF SYNAPSIN I   Purified Synapsin I has been tested for a variety of different enzymatic activities, and all of these tests have yielded negative results. Thus it seems possible that it is not a chemical catalyst but serves some other functional role. From its location and other characteristics, one can make some educated guesses about its possible functions. The specific association of Synapsin I with synaptic vesicles suggests that it is involved in the regulation of a vesicle function. In addition, its phosphorylation by both cAMP and calcium-regulated systems indicates that its function can be regulated by agents that affect either or both of these systems.

The most prominent role of synaptic vesicles is the release of transmitter into the synaptic cleft, triggered by an influx of calcium during depolarization of the terminal (Katz & Miledi 1967, Heuser et al 1979). There is evidence from a number of different systems that both calcium ion and cAMP can alter the level of transmitter released at various synapses. Calcium influx is necessary for the phenomenon of "post-tetanic potentiation" (Rosenthal 1969) in which a burst of impulses in the presynaptic terminal leads to an increase in the amount of transmitter released per impulse. Agents that increase the level of cAMP inside terminals have also been shown to potentiate release (Dudel 1965, Kravitz et al 1975, Goldberg & Singer 1969, Miyamoto & Breckenridge 1974). In the nerve terminals that have been well studied (for review see Kelly et al 1979), the final release process is probably too fast to be directly mediated by protein phosphorylation, unless the kinase involved has a much higher turnover number than other known protein kinases. The delay between calcium influx into the terminal and fusion of vesicles with the presynaptic membrane is estimated to be about 200 μsec (Llinas et al 1976), whereas the cAMP-dependent protein kinase catalyzes only two to ten phosphate transfers per second per active site under optimum conditions in a test tube (Sugden et al 1976, Glass & Krebs 1979). However, most models of exocytosis incorporate several partial reactions leading to the final calcium-triggered fusion. These postulated partial reactions include (a) movement of the vesicle toward the presynaptic membrane, (b) recognition between the vesicle and "active zones" in the presynaptic membrane, (c) close apposition of the vesicle membrane and the presynaptic membrane at the site at which fusion will occur, and finally (d) fusion itself (see Kelly et al 1979). Any one of these partial reactions could, in theory, be a limiting factor in determining how many quanta of transmitter are released during an impulse. Although there are certainly other possibilities, it may be that phosphorylation of Synapsin I by calcium or cAMP-regulated kinases regulates one or more of the "partial reactions" that precede the final calcium-dependent fusion event, and thus is involved in the control of the number of quanta of transmitter released per impulse.

Physiological studies are consistent with this possibility. Synapsin I can be phosphorylated and dephosphorylated within minutes in slices of rat cerebral cortex during cycles of depolarization by high K^+ and repolarization in normal K^+ (Forn & Greengard 1978). This indicates that its phosphorylation in intact tissue is highly dynamic and can change with a rapid time course. Phosphorylation of both the cAMP-regulated and calcium-regulated sites on Synapsin I can be increased by impulse conduction along the preganglionic nerve fibers of the rabbit superior cervical ganglion (Nestler & Greengard 1981, 1982). In addition, specific neurotransmitters that raise the concentration of cAMP can stimulate phosphorylation of Synapsin I. This is true for serotonin in the rat facial nucleus (Dolphin & Greengard 1981), and for dopamine in bovine and rat superior cervical ganglia (Nestler & Greengard 1980, 1982).

A direct test of the hypothesis that Synapsin I regulates some aspect of transmitter release would require introduction into a nerve cell of reagents that specifically activate or inhibit phosphorylation of Synapsin I, such as antibodies that block its phosphorylation (see Naito & Ueda 1981a,b). It would be helpful to know if an homologous protein or set of proteins exists in invertebrate species in which one has greater access to the interior of neurons of known function than in mammals. S. Goelz and P. Greengard (personal communication) have used a specific radioimmunoassay to determine the concentration of Synapsin I-like proteins in nonmammalian species. The results indicate that homologous proteins do exist in several such species, but their concentrations appear to be low and the proteins have divergent properties in the species that have been examined. Thus, Synapsin I does not seem to have been as highly conserved during evolution as the cAMP-dependent protein kinase. Nevertheless, it should be possible to characterize and localize homologous proteins in invertebrate species and thus make functional studies possible. Alternatively, it may be possible to test the function of Synapsin I by introducing appropriate reagents into mammalian neurons or synaptosomes using techniques involving fusion of liposomes with the plasma membrane (see Rahamimoff et al 1978).

Calcium-Dependent Protein Kinases and Their Substrates

Although it has been known for years that phosphorylase kinase, a central enzyme in the regulatory cascade that mediates hormonal control of the breakdown of glycogen, is stimulated by calcium (Brostrom et al 1971), the general significance of calcium-dependent protein kinases has only recently been appreciated. The characterization of calcium-dependent protein kinases other than phosphorylase kinase was aided by the discovery by Cheung (1970) and Kakiuchi & Yamazaki (1970) of a calcium-binding regulatory protein called calmodulin which activates the enzyme phosphodiesterase in the presence of micromolar concentrations of calcium ion. Calmodulin has since been found to regulate a variety of other enzymes (Cheung 1980), including protein kinases.

Calcium and calmodulin-dependent protein kinase activity was discovered in brain tissue by Schulman & Greengard, who were investigating the molecular basis of calcium-regulated phosphorylation of Synapsin I. They found that calcium stimulates the phosphorylation of several proteins in hypotonically lysed synaptosomes (Schulman & Greengard 1978a) and subsequently showed that this protein kinase activity depends on the presence of both calcium and calmodulin (Schulman & Greengard 1978a,b). The calmodulin-dependent protein kinase activity is not unique to synaptic membranes; it is present in membranes of a variety of tissues in which it phosphorylates a tissue-specific array of proteins (Schulman & Greengard 1978b).

At about the same time, Dabrowska et al (1978) and Yagi et al (1978) reported that calcium and calmodulin activate myosin light chain kinase, the enzyme that phosphorylates the P-light chains of myosin, rendering the myosin ATPase sensitive to activation by actin (Sherry et al 1978). Shortly thereafter, Cohen et al (1978) found that calmodulin is an integral part of phosphorylase kinase and mediates the activation of this enzyme by calcium.

CALMODULIN DEPENDENT SYNAPSIN I KINASE   In the experiments of Schulman & Greengard, Synapsin I, which is present in total particulate fractions prepared from synaptosomes, was nevertheless a relatively minor substrate for the particulate calmodulin-dependent kinase. In order to determine whether this kinase could be responsible for the calcium-stimulated phosphorylation of Synapsin I in intact synaptosomes, and to more thoroughly characterize the calcium-dependent Synapsin I kinase, an assay was developed to measure calmodulin-dependent phosphorylation of purified Synapsin I (Kennedy & Greengard 1981). This made it possible to characterize the kinase activity under conditions in which the rate of phosphorylation was not limited by the concentration of substrate. Under optimal conditions in such an assay, the rate of phosphorylation of Synapsin I by a crude brain homogenate is stimulated as much as 40-fold by calcium ion to an initial rate of about 5 nmol/min/mg protein. Phosphorylation of both the globular head portion and the collagenase sensitive tail of Synapsin I occurs under these conditions. Phosphorylation of these two regions is catalyzed by two distinct calcium-stimulated protein kinases that can be resolved by DEAE-cellulose chromatography (Figure 2) (Kennedy & Greengard 1981). One of them is primarily cytosolic and phosphorylates only the globular head portion of Synapsin I. After partial purification, it is reversibly activated by calcium ion and inhibited by 50 μM trifluoperazine (a calmodulin antagonist), but it does not require the addition of exogenous calmodulin. Further characterization will be required to determine whether it is activated by calcium ion alone or whether it contains tightly bound calmodulin that is required for its activation.

The second Synapsin I kinase, which phosphorylates the collagenase-sensitive tail of Synapsin I, has been purified about 200-fold and characterized in greater detail than the first (Kennedy, McGuinness, and Greengard, 1982 and in preparation). It is completely dependent on calcium and exo-

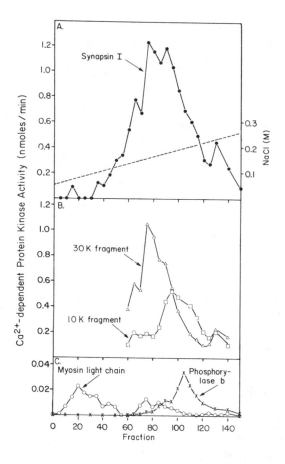

Figure 2 Resolution of two distinct calcium-dependent Synapsin I kinase activities in brain cytosol by DEAE-cellulose chromatography. Brain cytosol proteins were applied to a DEAE-cellulose column, which was then washed with 0.05 M NaCl. Various protein fractions were eluted with a salt gradient from 0.05 M NaCl to 0.3 M NaCl. A. Each fraction was assayed for total calcium-stimulated Synapsin I kinase activity. B. The fractions were further assayed for kinase activity that was specific for sites on the collagen-like tail of Synapsin I that are recovered in a 30K dalton fragment after proteolysis with Staph aureus V8 protease, or for the site in the globular head group that is recovered in a 10K dalton fragment. The figure shows that the curve in A is actually the sum of two curves representing enzmes that are specific for each of the two sites. C. The same fractions were also assayed for previously described calcium-dependent protein kinases, myosin light-chain kinase, and phosphorylase b kinase. Their peaks do not correspond to those of the two Synapsin I kinases. Pure phosphorylase b kinase phosphorylates only the 30K dalton region of Synapsin I, thus it does not contribute at all to the 10K dalton kinase peak. From Kennedy & Greengard (1981).

genous calmodulin. It is found in both the soluble and particulate fractions of brain homogenates. The partially purified enzymes from the two sources are indistinguishable by several criteria. An interesting finding was that the enzyme preparations from both sources contain three proteins that are phosphorylated in the presence of calcium and calmodulin, a 50K dalton protein, and two proteins in the 60K dalton region. When compared by phosphopeptide mapping and two-dimensional gel electrophoresis, these proteins were identical to three proteins of corresponding molecular weights that were shown by Schulman & Greengard (1978a,b) to be prominent substrates for calcium and calmodulin-dependent protein kinase in a crude particulate preparation from rat brain. In addition, the 50K dalton substrate was the major Coomassie blue staining protein in both partially purified enzyme preparations, and its presence coincided with enzyme activity during the purification steps. These findings suggested that the 50K dalton phosphoprotein may be an autophosphorylatable subunit of the Synapsin I kinase, or may exist in a complex with it. The association of the 50K dalton phosphoprotein with Synapsin I kinase activity has recently been confirmed in our laboratory. The enzyme has been purified to near homogeneity. A monoclonal antibody raised against the enzyme specifically precipitates both Synapsin I kinase activity and the 50K dalton protein (M. K. Bennett, N. E. Erondu, and M. B. Kennedy, unpublished observations). Thus the prominent brain particulate 50K dalton substrate protein for calcium and calmodulin-dependent protein kinase appears to be a subunit of the enzyme itself. The functional significance of this autophosphorylation is not yet known.

The two calcium-dependent Synapsin I kinases differ from both myosin light chain kinase and phosphorylase kinase. They have different mobilities on DEAE-cellulose (Figure 2), different substrate specificities, and they are more highly concentrated in brain than in other tissues. The apparent multiplicity of calcium-dependent kinases is of considerable general interest, since it contrasts with the uniformity of the cAMP-dependent protein kinases. As mentioned above, the two major types of cAMP-dependent kinases have virtually identical catalytic subunits and consequently have identical substrate specificities. The calcium-dependent protein kinases may be more diverse and have narrower substrate specificities. This could reflect a fundamental difference between cyclic nucleotide- and calcium-linked regulatory systems. Additional evidence for several calcium-dependent kinases in neural tissue is discussed below.

SUBCELLULAR LOCALIZATION OF BRAIN CALMODULIN-DEPENDENT PROTEIN KINASES   Two other research groups have observed a particulate calmodulin-dependent protein kinase activity in nervous tissue and have carried out studies on its subcellular localization. De Lorenzo and co-workers (De Lorenzo & Freedman 1977, 1978, Burke & De Lorenzo 1981, 1982) reported that calcium-stimulated phosphorylation of tubulin and other endogenous proteins occurs in a particulate preparation enriched in synaptic vesicles and that this phosphorylation system requires calmodulin (De Lorenzo et al 1979). Grab et al (1981) reported that isolated post-synaptic densities contain calmodulin and a calmodulin-dependent protein kinase that phosphorylates endogenous PSD proteins. Information about whether these organelles contain the same calmodulin-dependent kinase or distinct, highly specific kinases in vivo will await more thorough characterization of each of the enzymes and their substrate proteins.

De Lorenzo et al (1979) have postulated that calmodulin-dependent protein phosphorylation may be involved directly in the final stage of transmitter release. This hypothesis is based largely on the observation that the addition of calcium to an enriched vesicle fraction stimulates the release of transmitter from the vesicles and also the phosphorylation of certain proteins. Removal of calmodulin from the vesicles suppresses both the calcium-stimulated loss of bound transmitter and calcium-stimulated protein phosphorylation. Both can be restored by the addition of pure calmodulin. Although these results are intriguing, they show only that calmodulin is necessary for both processes. Major questions remain about the relationship between the loss of transmitter from purified vesicles and the release of transmitter from intact terminals. The physiological relevance of neurotransmitter release from isolated vesicles in the absence of plasma membrane is not clear. It will be important to test the possibility that the release of transmitter from isolated vesicles results simply from a nonspecific degradation of the vesicles by a contaminating calcium and calmodulin-dependent lipase or protease. Another difficulty with this hypothesis is the discrepancy in time course between calcium-stimulated release from intact synaptic terminals, which takes a few hundred microseconds, and calcium-stimulated vesicle phosphorylation, which takes several seconds. As discussed above, the final in vivo release process appears to be too fast to be mediated by a conventional phosphorylation cascade mechanism.

CALCIUM AND LIPID-ACTIVATED PROTEIN KINASE   Nishizuka and colleagues have described a calcium-activated protein kinase in brain that does not require calmodulin. It can be activated either by limited proteolysis by a calcium-dependent protease (Inoue et al 1977) or by the simultaneous presence of calcium ion and lipid (Takai et al 1979). The first process is irreversible and requires about 500 μM calcium for half maximal activation (Takai et al 1979). Because of the high concentration of calcium required for activation, this process may not be physiologically significant. However, the second process is reversible and occurs at lower calcium concentrations. The K_a for calcium is about 50 μM in the presence of mixed membrane lipids (Takai et al 1979), and as low as 5 μM in the presence of phosphatidyl serine and diacylglycerol (Kishimoto et al 1980). The latter lipid has very little stimulatory effect of its own, but it potentiates the effect of phospholipids by increasing the affinity of the phospholipid-enzyme complex for calcium. The effect of diacylglycerol is quite potent, occurring at concentrations less than 5% (w/w) of the concentration of phospholipid and reducing the K_a for calcium from about 100 μM to 5 μM. Consequently, Kishimoto et al (1980) have postulated that activation of the kinase may be linked to hydrolysis of phosphatidyl inositol, which produces diacylglycerol and inositol phosphate. Specific turnover of phosphatidyl inositol has been reported to occur in response to several extracellular messengers including α-adrenergic and muscarinic agonists (Hokin & Hokin 1955, Michell 1979).

Kuo and colleagues (1980) demonstrated that the calcium and lipid-activated protein kinase is present in many different phyla and tissues. The highest concentrations were found in mammalian spleen and neural tissue, where the enzyme was roughly evenly distributed between the soluble and particulate fractions. Wrenn et al (1980) and Katoh et al (1981) showed that substrate proteins for the calcium and lipid-activated kinase are distinct from those for calcium and calmodulin-dependent protein kinases in brain and heart, respectively.

It is clear that this enzyme can be activated by physiologically relevant concentrations of calcium ion when it interacts with the proper mixture of lipids and also that it has a unique specificity that distinguishes it from other known protein kinases. It remains to be demonstrated whether and under what circumstances the enzyme is activated in intact cells.

A Neuronal Substrate for cGMP-Dependent Protein Kinase

Cyclic GMP, as well as cAMP, acts as a second messenger in the nervous system. In vertebrate rod photoreceptors, photolyzed rhodopsin catalyzes a cascade of reactions that leads to a fall in the concentration of intracellular cGMP (Hubbell & Bownds 1979, Stryer et al 1981). The change in cGMP concentration has been postulated to be linked either directly to the decrease in sodium conductance that underlies the light response, or to adaptation of that response to background illumination.

There are several other reports of alterations of cGMP metabolism by neurally active agents. For example, activation of muscarinic receptors on neuroblastoma cells leads to an increase in intracellular cGMP that is secondary to an increased flux of calcium ion into the cells (Matsuzawa &

Nirenberg 1975). Activation of nicotinic receptors in muscle can also raise internal cGMP levels (Nestler et al 1978). Woody et al (1978) have presented evidence that iontophoresis of acetylcholine or cGMP onto cortical neurons produces a specific and relatively long-lasting increase in the input resistance of the cells. The physiological significance of these effects is as yet unknown.

As an approach to learning more about the role of cGMP in neuronal tissue, Schlicter and Greengard looked for specific substrate proteins for cGMP-dependent protein kinase in homogenates of nervous tissue. One such substrate, a soluble 23,000 dalton protein termed G-substrate, was discovered in cerebellar homogenates (Schlicter et al 1978) and has been purified to homogeneity (Aswad & Greengard 1981a). Studies with the purified G-substrate have shown that it has a 20-fold higher affinity for the cGMP-dependent protein kinase than for the cAMP-dependent protein kinase, thus confirming the specificity of its regulation by cGMP (Aswad & Greengard 1981b).

The cellular localization of the G-substrate was examined by measuring the concentrations of G-substrate in mutant mice that lack particular classes of cerebellar neurons. These studies showed that the substrate is highly concentrated in Purkinje cells (Schlicter et al 1980). Low levels of a protein that resembled G-substrate were detected in mice lacking Purkinje cells (Schlicter et al 1980) and in brain regions other than the cerebellum (Aswad & Greengard 1981a); however, examination of phosphopeptides of this protein has shown that it is not the G-substrate, but a related protein called inhibitor I (see below) (A. Nairn, D. Aswad, and P. Greengard, personal communication). The localization of the G-substrate in Purkinje cells is paralleled by a high concentration of the cGMP-dependent protein kinase measured in mutant cerebelli by a photo-affinity labeling technique (Schlicter et al 1980) and by immunofluorescent labeling with antibody raised to purified cGMP-dependent protein kinase (Lohmann et al 1981).

Some insights into the possible functions of the G-substrate have come from studies of the purified protein (Aitken et al 1981). Many of its properties suggest that it may be homologous to a protein isolated from rabbit skeletal muscle by Nimmo & Cohen (1978). This protein, termed inhibitor I, is a substrate for the cAMP-dependent protein kinases. In its phosphorylated form it is an inhibitor of phosphatase I, the enzyme responsible for removing the regulatory phosphates from several enzymes that participate in the cascade of enzymatic reactions that regulates glycogenolysis in muscle. Preliminary experiments have indicated that the G-substrate also possesses some phosphatase inhibitor activity (Aitken et al 1981); thus, it may play a role in a neuronal regulatory cascade that is analogous to the role of inhibitor I in the glycogenolytic cascade.

Specific immunohistochemical staining for cGMP (Ariano & Matus 1981), guanylate cyclase (Ariano et al 1982), and cGMP-dependent protein kinase (M. A. Ariano, personal communication) has been observed in neurons in brain regions other than the cerebellum; for example, the striatum. It will be interesting to see whether other specific substrates for cGMP-dependent protein kinase are present in these regions.

PHOSPHORYLATION OF NEURAL PROTEINS OF KNOWN FUNCTION

Regulation of Synthesis of the Biogenic Amines

Because most neurotransmitters have a unique biosynthetic pathway and the enzymes that catalyze each synthetic step can be assayed in a test tube, regulation of transmitter synthesis can be studied directly in tissue homogenates. There is now considerable evidence that the rate of synthesis of catecholamines and serotonin can be controlled by phosphorylation of the rate-limiting enzymes in their biosynthetic pathways.

Several years ago, evidence was presented that the activity of tyrosine hydroxylase, the rate-limiting enzyme in the synthesis of catecholamines (Nagatsu et al 1964, Levitt et al 1965), could be enhanced in brain homogenates by added cAMP-dependent protein kinase (Morgenroth et al 1975). Recently, Joh et al (1978) have demonstrated that purified tyrosine hydroxylase is phosphorylated and thereby activated by the cAMP-dependent protein kinase. Earlier work with impure enzyme had suggested that this increase in activity resulted from an increased affinity of the enzyme for its pteridine cofactor (Lovenberg et al 1975, Goldstein et al 1975, Lloyd & Kaufman 1975); however, the experiments with highly purified enzyme indicate that phosphorylation results in an increased maximal velocity with no change in the affinity of the enzyme for cofactors or substrates. The physiological circumstances under which tyrosine hydroxylase might be activated by cAMP in vivo are not yet certain. One possible scenario was presented by Erny et al (1981), who showed that adenosine can activate tyrosine hydroxylase in pheochromocytoma cells in culture by increasing intracellular cAMP. These workers have suggested that adenosine, which is rapidly formed from vesicular ATP following its release along with neurotransmitter, may act in vivo to activate tyrosine hydroxylase via presynaptic adenosine receptors.

Both tyrosine hydroxylase, and tryptophan hydroxylase, the rate-limiting enzyme in serotonin biosynthesis (Jequier et al 1967), are activated in brain homogenates by a phosphorylation reaction that is independent of cAMP and dependent on calcium and calmodulin (Hamon et al 1978, Kuhn et al 1978, Yamauchi & Fujisawa 1979a,b, Yamauchi et al 1981). The activation is unusual in that it requires both a calcium-dependent protein kinase and a distinct "activator protein." It appears to occur in two sequential steps, phosphorylation of the hydroxylase, followed by activation by the activator protein (Yamauchi et al 1981). The calcium and calmodulin-dependent protein kinase appears to be distinct from either phosphorylase kinase or myosin light chain kinase (Yamauchi & Fujisawa 1980). The activator protein is a dimer of two identical 35,000 dalton subunits that has been purified to homogeneity (Yamauchi et al 1981). It is both more widely distributed and more highly concentrated than either of the two hydroxylases or the calmodulin-dependent protein kinase. Thus, it may have a variety of different functions.

Stimulation of the synthesis of both serotonin and norepinephrine by increased electrical activity has been demonstrated in a number of different laboratories (e.g. Eccleston et al 1970, Morgenroth et al 1974). Because increased electrical activity results in an increased flux of calcium ion into the synaptic terminals, calcium-dependent phosphorylation resulting in activation of the transmitter synthesizing enzymes provides a plausible, direct mechanism for this stimulation.

Phosphorylation of the Acetylcholine Receptor

The electroplax, a specialized organ of electric fishes, is a rich source of relatively homogeneous cholinergic terminals (Gordon et al 1977a). For this reason, it has frequently been used as a source of tissue for biochemical studies of pre- and postsynaptic structures. In the course of such studies, Gordon et al (1977a,b) and Teichberg et al (1977) discovered that the acetylcholine receptor is phosphorylated by an endogenous protein kinase in receptor-enriched membranes. Two of the four receptor subunits, the δ (Gordon et al 1977a,b, Saitoh & Changeux 1981, Smilowitz et al 1981) and the γ (Saitoh & Changeux 1981, Smilowitz et al 1981) are phosphorylated.

Early studies indicated that the receptor kinase was stimulated by K^+ ion (Gordon et al 1977a, Saitoh & Changeux 1980), but was not sensitive to second messengers such as cyclic nucleotides or calcium. However, Smilowitz et al (1981) have recently presented evidence that a receptor kinase is stimulated by calcium in the presence of calmodulin. This suggests that receptor phosphorylation might be regulated in vivo by postsynaptic activity that raises the intracellular concentration of calcium.

The functional significance of the receptor phosphorylation is unclear. It seems unlikely that it is required either for agonist-induced conductance changes or for desensitization, since these two processes can occur in purified membranes in the absence of an energy source (Sugiyama et al 1976) and in artificial lipid bilayers (Nelson et al 1980, Schindler & Quast 1980). It has been suggested that phosphorylation may be required for stabilization and maintenance of receptor clusters at the synapse (Gordon et al 1977a,

Saitoh & Changeux 1980, 1981). Prior to innervation of developing muscle and following denervation of mature muscle, acetylcholine receptors are present over the whole surface of the muscle as well as at the immature or denervated endplate (for review see Fambrough 1979). Although the subunit structure of the two populations of receptors is the same, the "extrajunctional" receptors differ from the junctional receptors in several ways:

1. They turn over more rapidly [17 hr (Merlie et al 1976) vs 2 wk or longer (Frank et al 1975, Linden & Fambrough 1979)].
2. Their mean channel open time is longer (Michler & Sakmann 1980, Fischbach & Schuetze 1980).
3. Their isoelectric point is 0.1 pH units more basic (Brockes & Hall 1975).

Saitoh & Changeux (1981) have presented evidence that these differences may be accounted for in part by receptor phosphorylation. They found that neonatal electric fish have a population of acetylcholine receptors that have an isoelectric point 0.1 pH units more basic than that of adult receptors. Treatment of the adult receptors with alkaline phosphatase has two effects:

1. It shifts their isoelectric point toward the more basic pH.
2. It results in an increased incorporation of phosphate during subsequent phosphorylation by the endogenous kinase.

The neonatal receptors are affected to a smaller extent by this treatment. These results suggest that the adult receptors are more phosphorylated than the neonatal receptors, and that this phosphorylation may contribute to their stabilization at mature endplates.

FUTURE DIRECTIONS

The goal of studies of neuronal protein phosphorylation is to better understand, at the molecular level, both short and long-term regulation of neuronal properties. Considerable progress has been made in the past several years. There is now evidence that phosphorylation plays a role in the regulation of ion channels, transmitter synthesis, synaptic vesicle function, and possibly, in the stabilization of clusters of receptors at synapses. In most of these cases, the detailed biochemistry underlying the functional changes is not yet understood. More precise descriptions of these and other, as yet unrecognized, regulatory mechanisms will come both from continued application of the techniques and approaches that have been discussed, and from the use of new techniques that have just begun to be exploited to their fullest. The production of monoclonal antibodies (Köhler & Milstein 1975) that recognize protein kinases and their substrates should facilitate their purification and localization, and may also provide highly specific inhibitors for use in the dissection of regulatory pathways. Internal perfusion of single cells (Lee et al 1980) and the study of single channels by "patch-clamping" (Hamill et al 1981) will provide the technical means for obtaining a complete description of the regulation of individual channels by purified components. Characterization of proteins involved in specialized neuronal functions will make it possible to identify and isolate their messenger RNAs. This will, in turn, allow the preparation of probes of complementary DNA for use in studying the control of gene expression during neuronal development.

It seems likely from the studies I have discussed that different types of neurons will contain distinct regulatory pathways that participate in specialized functions. For example, it appears that invertebrate neurons contain different mechanisms for modulating K^+-channels. In one cell type (R_{15}), cAMP increases K^+-conductance, and in two others (bag cells and sensory neurons), it decreases K^+-conductance. In addition, the G-substrate appears highly localized in cerebellar Purkinje cells. Consequently, descriptions of neuronal regulatory pathways, and immunocytochemical localization of the proteins involved, may begin to uncover subpopulations of functionally related neurons within complex neural structures. Thus, biochemical studies of regulatory mechanisms involving protein phosphorylation may be helpful not only in analyzing neuronal function at the cellular level, but also in understanding the organization and function of the nervous system as a whole.

ACKNOWLEDGMENTS

I would like to thank P. Greengard, J. Nerbonne, F. Strumwasser, and R. Lewis for critical comments on the manuscript and several investigators, mentioned in the text, for providing preprints of manuscripts and summaries of work in progress. I would also like to thank C. Hochenedel and P. Brown for help in preparing the manuscript.

Literature Cited

Adams, D. J., Smith, S. J., Thompson, S. H. 1980. Ionic currents in molluscan soma. *Ann. Rev. Neurosci.* 3:141–68

Adams, W. B., Levitan, I. B. 1982. Intracellular injection of protein kinase inhibitor blocks the serotonin-induced increase in K^+ conductance in *Aplysia* neuron R15. *Proc. Natl. Acad. Sci. USA* 79:3877–80

Aitken, A., Bilham, T., Cohen, P., Aswad, D., Greengard, P. 1981. A specific substrate from rabbit cerebellum for guanosine 3':5'-monophosphate-dependent protein kinase. III. Amino acid sequences at the two phosphorylation sites. *J. Biol. Chem.* 256:3501–6

Ariano, M. A., Lewicki, J. A., Brandwein, H. J., Murad, F. 1982. Immunohistochemical localization of guanylate cyclase within neurons of rat brain. *Proc. Natl. Acad. Sci. USA* 79:1316–20

Ariano, M. A., Matus, A. I. 1981. Ultrastructural localization of cyclic GMP and cyclic AMP in rat striatum. *J. Cell Biol.* 91:287–92

Ascher, P. 1972. Inhibitory and excitatory effects of dopamine on *Aplysia* neurones. *J. Physiol.* 225:173–209

Ashby, C. D., Walsh, D. A. 1972. Characterization of the interaction of a protein inhibitor with adenosine 3',5'-monophosphate-dependent protein kinases. I. Interaction with the catalytic subunit of the protein kinase. *J. Biol. Chem.* 247:6637–42

Aswad, D. W., Greengard, P. 1981a. A specific substrate from rabbit cerebellum for guanosine 3':5' monophosphate-dependent protein kinase. I. Purification and characterization. *J. Biol. Chem.* 256:3487–93

Aswad, D. W., Greengard, P. 1981b. A specific substrate from rabbit cerebellum for guanosine 3':5'-monophosphate-dependent protein kinase. II. Kinetic studies on its phosphorylation by guanosine 3':5'-monophosphate-dependent and adenosine 3':5'-monophosphate-dependent protein kinases. *J. Biol. Chem.* 256:3494–500

Beavo, J. A., Bechtel, P. J., Krebs, E. G. 1974. Preparation of homogeneous cyclic AMP-dependent protein kinase and its subunits from rabbit skeletal muscle. *Meth. Enzymol.* 38:299–308

Bloom, F. E., Ueda, T., Battenberg, E., Greengard, P. 1979. Immunocytochemical localization, in synapses, of protein I, an endogenous substrate for protein kinases in mammalian brain. *Proc. Natl. Acad. Sci. USA* 76:5982–86

Brockes, J. P., Hall, Z. W. 1975. Acetylcholine receptors in normal and denervated rat diaphragm muscle. II. Comparison of junctional and extrajunctional receptors. *Biochemistry* 14:2100–6

Brostrom, C. O., Hunkeler, F. L., Krebs, E. G. 1971. The regulation of skeletal muscle phosphorylase kinase by Ca^{2+}. *J. Biol. Chem.* 246:1961–67

Brunelli, M., Castellucci, V., Kandel, E. R. 1976. Synaptic facilitation and behavioral sensitization in *Aplysia:* Possible role of serotonin and cyclic AMP. *Science* 194:1178–81

Burke, B. E., De Lorenzo, R. J. 1981. Ca^{2+}- and calmodulin-stimulated endogenous phosphorylation of neurotubulin. *Proc. Natl. Acad. Sci. USA* 78:991–95

Burke, B. E., De Lorenzo, R. J. 1982. Ca^{2+}- and calmodulin-dependent phosphorylation of endogenous synaptic vesicle tubulin by a vesicle-bound calmodulin kinase system. *J. Neurochem.* 38:1205–18

Camardo, J. S., Klein, M., Kandel, E. R. 1981. Sensitization in *Aplysia:* Serotonin elicits a decrease in sensory neuron K^+ current not related to I_K early or I_K Ca^{2+}. *Neurosci. Abstr.* 7:836

Castellucci, V. F., Kandel, E. R., Schwartz, J. H., Wilson, F. D., Nairn, A. C., Greengard, P. 1980. Intracellular injection of the catalytic subunit of cyclic AMP-dependent protein kinase simulates facilitation of transmitter release underlying behavioral sensitization in *Aplysia. Proc. Natl. Acad. Sci. USA* 77:7492–96

Castellucci, V. F., Schwartz, J. H., Kandel, E. R., Nairn, A., Greengard, P. 1981. Protein inhibitor of the cyclic AMP-dependent protein kinase can block the onset of, as well as reverse the electrophysiological correlates of, sensitization of the gill-withdrawal in *Aplysia. Neurosci. Abstr.* 7:836

Cheung, W. Y. 1970. Cyclic 3':5'-nucleotide phosphodiesterase: Demonstration of an activator. *Biochem. Biophys. Res. Commun.* 38:533–38

Cheung, W. Y. 1980. Calmodulin plays a pivotal role in cellular regulation. *Science* 207:19–27

Cleveland, D. W., Fischer, S. G., Kirschner, M. W., Laemmli, U. K. 1977. Peptide mapping by limited proteolysis in sodium dodecyl sulfate and analysis by gel electrophoresis. *J. Biol. Chem.* 252:1102–6

Cohen, P., Burchell, A., Foulkes, J. G., Cohen, P. T. W., Vanaman, T. C., Nairn, A. C. 1978. Identification of the Ca^{2+}-dependent modulator protein as the fourth subunit of rabbit skeletal muscle phosphorylase kinase *FEBS Lett.* 92:287–93

Corbin, J. D., Keely, S. L., Park, C. R. 1975. The distribution and dissociation of cyclic adenosine 3':5'-monophosphate-dependent protein kinases in adipose, cardiac and other tissues. *J. Biol. Chem.* 250:218–25

Dabrowska, R., Sherry, J. M. F., Aromatorio, D. K., Hartshorne, D. J. 1978. Modulator protein as a component of the myosin light chain kinase from chicken gizzard. *Biochemistry* 17:253–58

De Camilli, P., Ueda, T., Bloom, F. E., Battenberg, E., Greengard, P. 1979. Widespread distribution of protein I in the central and peripheral nervous systems. *Proc. Natl. Acad. Sci. USA* 76:5977–81

De Lorenzo, R. J. 1980. Role of calmodulin in neurotransmitter release and synaptic function. *Ann. NY Acad. Sci.* 356:92–109

De Lorenzo, R. J., Freedman, S. D. 1977. Calcium-dependent phosphorylation of synaptic vesicle proteins and its possible role in mediating neurotransmitter release and vesicle function. *Biochem. Biophys. Res. Commun.* 77:1036–43

De Lorenzo, R. J., Freedman, S. D. 1978. Calcium dependent neurotransmitter release and protein phosphorylation in synaptic vesicles. *Biochem. Biophys. Res. Commun.* 80:183–92

De Lorenzo, R. J., Freedman, S. D., Yohe, W. B., Maurer, S. C. 1979. Stimulation of Ca^{2+}-dependent neurotransmitter release and presynaptic nerve terminal protein phosphorylation by calmodulin and a calmodulin-like protein isolated from synaptic vesicles. *Proc. Natl. Acad. Sci. USA* 76:1838–42

Demaille, J. G., Peters, K. A., Fischer, E. H. 1977. Isolation and properties of the rabbit skeletal muscle protein inhibitor of adenosine 3':5'-monophosphate-dependent protein kinases. *Biochemistry* 16:3080–86

Dolphin, A. C., Greengard, P. 1981. Serotonin stimulates phosphorylation of protein I in the facial motor nucleus of rat brain. *Nature* 289:76–78

Douglas, W. W. 1968. Stimulus-secretion coupling: The concept and clues from chromaffin and other cells. *Br. J. Pharmacol.* 34:451–74

Drummond, A. H., Benson, J. A., Levitan, I. B. 1980a. Serotonin-induced hyperpolarization of an identified Aplysia neuron is mediated by cyclic AMP. *Proc. Natl. Acad. Sci. USA* 77:5013–17

Drummond, A. H., Bucher, F., Levitan, I. B. 1980b. d-[³H] Lysergic acid diethylamide binding to serotonin receptors in the molluscan nervous system. *J. Biol. Chem.* 255:6679–6686

Dudel, J. 1965. Facilitatory effects of 5-hydroxy-tryptamine on crayfish neuromuscular junction. *Naunyn-Schmiedebergs Arch. Exp. Pathol. Pharmakol.* 249:515–28

Eccleston, D., Ritchie, I. M., Roberts, M. H. T. 1970. Long term effects of midbrain stimulation on 5-hydroxyindole synthesis in rat brain. *Nature* 226:84–85

Eckert, R., Lux, H. D. 1975. A non-inactivating inward current recorded during small depolarizing voltage steps in snail pacemaker neurons. *Brain Res.* 83:486–89

Erny, R. E., Berezo, M. W., Perlman, R. L. 1981. Activation of tyrosine 3-monooxygenase in pheochromo-cytoma cells by adenosine. *J. Biol. Chem.* 256:1335–39

Fambrough, D. M. 1979. Control of acetylcholine receptors in skeletal muscle. *Physiol. Rev.* 59:165–227

Fischbach, G. D., Schuetze, S. M. 1980. A post-natal decrease in acetylcholine channel open time at rat end-plates. *J. Physiol.* 303:125–37

Forn, J., Greengard, P. 1978. Depolarizing agents and cyclic nucleotides regulate the phosphorylation of specific neuronal proteins in rat cerebral cortex slices. *Proc. Natl. Acad. Sci. USA* 75:5195–99

Frank, E., Gautvik, K., Sommerschild, H. 1975. Cholinergic receptors at denervated mammalian motor endplates. *Acta Physiol. Scand.* 95:66–76

Glass, D. B., Krebs, E. G. 1979. Comparison of the substrate specificity of adenosine 3':5'-monophosphate and guanosine 3':5'-monophosphate-dependent protein kinases. Kinetic studies using synthetic peptides corresponding to phosphorylation sites in histone H2B. *J. Biol. Chem.* 254:9728–38

Goelz, S. E., Nestler, E. J., Chehrazi, B., Greengard, P. 1981. Distribution of protein I in mammalian brain as determined by a detergent-based radioimmunoassay. *Proc. Natl. Acad. Sci. USA* 78:2130–34

Goldberg, A. L., Singer, J. J. 1969. Evidence for a role of cyclic AMP in neuromuscular transmission. *Proc. Natl. Acad. Sci. USA* 64:134–41

Goldstein, M., Ebstein, B., Bronaugh, R. L., Roberge, C. 1975. Stimulation of striatal tyrosine hydroxylase by cyclic AMP. In *Chemical Tools in Catecholamine Research*, ed. D. Almgren, A. Carlsson, J. Engel, 2:257–69. Amsterdam: North Holland

Gordon, A. S., Davis, C. G., Diamond, I. 1977a. Phosphorylation of membrane proteins at a cholinergic synapse. *Proc. Natl. Acad. Sci. USA* 74:263–67

Gordon, A. S., Davis, C. G., Milfay, D., Diamond, I. 1977b. Phosphorylation of acetylcholine receptor by endogenous membrane protein kinase in receptor-enriched membranes of *Torpedo californica*. *Nature* 267:539–40

Goy, M. F., Schwarz, T. L., Kravitz, E. A. 1981. Serotonin-induced phosphorylation of a 28,000 dalton protein in a lobster nerve-muscle preparation. *Neurosci. Abstr.* 7:933

Grab, D. J., Carlin, R. K., Siekevitz, P. 1981. Function of calmodulin in postsynaptic densities II. Presence of a calmodulin-activatable protein kinase activity. *J. Cell Biol.* 89:440–48

Greengard, P. 1976. Possible role for cyclic nucleotides and phosphorylated membrane proteins in postsynaptic actions of neurotransmitters. *Nature* 260:101–8

Greengard, P. 1978. Phosphorylated proteins

as physiological effectors. *Science* 199:146–52

Greengard, P. 1979. Cyclic nucleotides, phosphorylated proteins, and the nervous system. *Fed. Proc.* 38:2208–17

Greengard, P. 1981. Intracellular signals in the brain. *Harvey Lect.* 75:277–331

Hamill, O. P., Marty, A., Neher, E., Sakmann, B., Sigworth, F. J. 1981. Improved patch-clamp techniques for high-resolution current recording from cells and cell-free membrane patches. *Pflügers Arch.* 391:85–100

Hamon, M., Bourgoin, S., Hery, F., Simonnet, G. 1978. Activation of tryptophan hydroxylase by adenosine triphosphate, magnesium, and calcium. *Mol. Pharmacol.* 14:99–110

Heuser, J. E., Reese, T. S., Dennis, M. J., Jan, Y., Jan, L., Evans, L. 1979. Synaptic vesicle exocytosis captured by quick freezing and correlated with quantal transmitter release. *J. Cell Biol.* 81:275–300

Hofmann, F., Beavo, J. A., Bechtel, P. J., Krebs, E. G. 1975. Comparison of adenosine 3':5'-monophosphate-dependent protein kinases from rabbit skeletal and bovine heart muscle. *J. Biol. Chem.* 250:7795–801

Hokin, L. E., Hokin, M. R. 1955. Effects of acetylcholine on the turnover of phosphoryl units in individual phospholipids of pancreas slices and brain cortex slices. *Biochim. Biophys. Acta* 18:102–10

Hubbell, W. L., Bownds, M. D. 1979. Visual transduction in vertebrate photoreceptors. *Ann. Rev. Neurosci.* 2:17–34

Huttner, W. B., De Gennaro, L. J., Greengard, P. 1981. Differential phosphorylation of multiple sites in purified protein I by cyclic AMP-dependent and calcium-dependent protein kinases. *J. Biol. Chem.* 256:1482–88

Huttner, W. B., Greengard, P. 1979. Multiple phosphorylation sites in protein I and their differential regulation by cyclic AMP and calcium. *Proc. Natl. Acad. Sci. USA* 76:5402–6

Inoue, M., Kishimoto, A., Takai, Y., Nishizuka, Y. 1977. Studies on a cyclic nucleotide-independent protein kinase and its proenzyme in mammalian tissues II. Proenzyme and its activation by calcium-dependent protease from rat brain. *J. Biol. Chem.* 252:7610–16

Jennings, K. R., Kaczmarek, L. K., Hewick, R. M., Dreyer, W. J., Strumwasser, F. 1982. Protein phosphorylation during afterdischarge in peptidergic neurons of *Aplysia. J. Neurosci.* 2:158–68

Jéquier, E., Lovenberg, W., Sjoerdsma, A. 1967. Tryptophan hydroxylase inhibition: The mechanism by which p-chlorophenylalanine depletes rat brain serotonin. *Mol. Pharmacol.* 3:274–78

Joh, T. H., Park, D. H., Reis, D. J., 1978. Direct phosphorylation of brain tyrosine hydroxylase by cyclic AMP-dependent protein kinase: Mechanism of enzyme activation. *Proc. Natl. Acad. Sci. USA* 75:4744–48

Kaczmarek, L. K., Jennings, K., Strumwasser, F. 1978. Neurotransmitter modulation, phosphodiesterase inhibitor effects, and cAMP correlates of afterdischarge in peptidergic neurites. *Proc. Natl. Acad. Sci. USA* 75:5200–4

Kaczmarek, L. K., Jennings, K. R., Strumwasser, F., Nairn, A. C., Walter, U., Wilson, F. D., Greengard, P. 1980. Microinjection of catalytic subunit of cyclic AMP-dependent protein kinase enhances calcium action potentials of bag cell neurons in cell culture. *Proc. Natl. Acad. Sci. USA* 77:7487–91

Kaczmarek, L. K., Strumwasser, F. 1981a. The expression of long lasting afterdischarge by isolated *Aplysia* bag cell neurons. *J. Neurosci.* 1:626–34

Kaczmarek, L. K., Strumwasser, F. 1981b. Net outward currents of bag cell neurons are diminished by a cAMP analogue. *Neurosci. Abstr.* 7:932

Kakiuchi, S., Yamazaki, R. 1970. Calcium-dependent phosphodiesterase activity and its activating factor (PAF) from brain: Studies on cyclic 3':5'-nucleotide phosphodiesterase (III). *Biochem. Biophys. Res. Commun.* 41:1104–10

Kandel, E. R. 1976. *Cellular Basis of Behavior.* San Francisco: Freeman. 727 pp.

Kandel, E. R. 1981. Calcium and the control of synaptic strength by learning. *Nature* 293:697–700

Katoh, N., Wrenn, R. W., Wise, B. C., Shoji, M., Kuo, J. F. 1981. Substrate proteins for calmodulin-sensitive and phospholipid-sensitive Ca²⁺-dependent protein kinases in heart, and inhibition of their phosphorylation by palmitoylcarnitine. *Proc. Natl. Acad. Sci. USA* 78:4813–17

Katz, B., Miledi, R. 1967. The timing of calcium action during neuromuscular transmission. *J. Physiol.* 189:535–44

Kelly, R. B., Deutsch, J. W., Carlson, S. S., Wagner, J. A. 1979. Biochemistry of neurotransmitter release. *Ann. Rev. Neurosci.* 2:399–446

Kennedy, M. B., Greengard, P. 1981. Two calcium/calmodulin-dependent protein kinases, which are highly concentrated in brain, phosphorylate protein I at distinct sites. *Proc. Natl. Acad. Sci. USA* 78:1293–97

Kennedy, M. B., McGuiness, T., Greengard, P. 1982. Partial purification, characterization, and comparison of soluble and particulate calmodulin-dependent synapsin I kinase activities. *Neurosci. Abstr.* 8:281

Kishimoto, A., Takai, Y., Mori, T., Kikkawa, U., Nishizuka, Y. 1980. Activation of calcium and phospholipid-dependent protein kinase by diacylglycerol, its possible relation to phosphatidylinositol turnover. *J. Biol. Chem.* 255:2273–76

Klein, D. C., Berg, G. R. 1970. Pineal gland: Stimulation of melatonin production by norepinephrine involves cyclic AMP-mediated stimulation of *N*-acetyltransferase. *Adv. Biochem. Pharmacol.* 3:241–63

Klein, M., Kandel, E. R. 1978. Presynaptic modulation of voltage-dependent Ca^{2+} current: Mechanism for behavioral sensitization in *Aplysia californica*. *Proc. Natl. Acad. Sci. USA* 75:3512–16

Klein, M., Kandel, E. R. 1980. Mechanism of calcium current modulation underlying presynaptic facilitation and behavioral sensitization in *Aplysia*. *Proc. Natl. Acad. Sci. USA* 77:6912–16

Köhler, G., Milstein, C. 1975. Continuous cultures of fused cells secreting antibody of predefined specificity. *Nature* 256:495–97

Kravitz, E. A., Battelle, B-A., Evans, P. D., Talamo, B. R., Wallace, B. G. 1975. *Neurotransmitters, Hormones and Receptors, Novel Approaches*, pp. 67–81. Bethesda, Md: Soc. Neurosci. 123 pp.

Krebs, E. G., Beavo, J. A. 1979. Phosphorylation-dephosphorylation of enzymes. *Ann. Rev. Biochem.* 48:923–59

Krueger, B. K., Forn, J., Greengard, P. 1977. Depolarization-induced phosphorylation of specific proteins, mediated by calcium ion influx, in rat brain synaptosomes. *J. Biol. Chem.* 252:2764–73

Kuhn, D. M., Vogel, R. L., Lovenberg, W. 1978. Calcium-dependent activation of tryptophan hydroxylase by ATP and magnesium. *Biochem. Biophys. Res. Comm.* 82:759–66

Kuo, J. F., Andersson, R. G. G., Wise, B. C., Mackerlova, L., Salomonsson, I., Brackett, N. L., Katoh, N., Shoji, M., Wrenn, R. W. 1980. Calcium-dependent protein kinase: Widespread occurrence in various tissues and phyla of the animal kingdom and comparison of

effects of phospholipid, calmodulin, and trifluoperazine. *Proc. Natl. Acad. Sci. USA* 77:7039–43

Kuo, J. F., Greengard, P. 1969. Cyclic nucleotide-dependent protein kinases, IV. Widespread occurrence of adenosine 3′,5′-monophosphate dependent protein kinase in various tissues and phyla of the animal kingdom. *Proc. Natl. Acad. Sci. USA* 64:1349–55

Lee, K. S., Akaike, N., Brown, A. M. 1980. The suction pipette method for internal perfusion and voltage clamp of small excitable cells. *J. Neurosci. Meth.* 2:51–78

Lemos, J., Novak-Hofer, I., Levitan, I. B. 1981. Serotonin effects on protein phosphorylation within a single living nerve cell. *Neurosci. Abstr.* 7:932

Levitan, I. B. 1978. Adenylate cyclase in isolated Helix and Aplysia neuronal cell bodies: Stimulation by serotonin and peptide containing extract. *Brain Res.* 154:404–8

Levitan, I. B., Adams, W. B. 1981. Cyclic AMP modulation of a specific ion channel in an identified nerve cell: Possible role for protein phosphorylation. *Adv. Cyc. Nuc. Res.* 14:647–53

Levitt, M., Spector, S., Sjoerdsma, A., Udenfriend, S. 1965. Elucidation of the rate-limiting step in norepinephrine biosynthesis in the perfused guinea-pig heart. *J. Pharmacol. Exp. Ther.* 148:1–8

Linden, D. C., Fambrough, D. M. 1979. Biosynthesis and degradation of acetylcholine receptors in rat skeletal muscles. Effects of electrical stimulation. *Neuroscience* 4:527–38

Llinas, R., Steinberg, I. Z., Walton, K. 1976. Presynaptic calcium currents and their relation to synaptic transmission: Voltage clamp study in squid giant synapse and theoretical model for the calcium gate. *Proc. Natl. Acad. Sci. USA* 73:2918–22

Lloyd, T., Kaufman, S. 1975. Evidence for the lack of direct phosphorylation of bovine caudate tyrosine hydroxylase following activation by exposure to enzymatic phosphorylating conditions. *Biochem. Biophys. Res. Commun.* 66:907–13

Lohmann, S. M., Walter, U., Miller, P. E., Greengard, P., DeCamilli, P. 1981. Immunohistochemical localization of cyclic GMP-dependent protein kinase in mammalian brain. *Proc. Natl. Acad. Sci. USA* 78:653–57

Lovenberg, W., Bruckwick, E. A., Hanbauer, I. 1975. ATP, cyclic AMP, and magnesium increase the affinity of rat striatal tyrosine hydroxylase for its cofactor. *Proc. Natl. Acad. Sci. USA* 72:2955–58

Lynch, G., Schubert, P. 1980. The use of in vitro brain slices for multidisciplinary studies of synaptic function. *Ann. Rev. Neurosci.* 3:1–22

Mackay, A. V. P., Iversen, L. L. 1972. Increased tyrosine hydroxylase activity of sympathetic ganglia cultured in the presence of dibutyryl cyclic AMP. *Brain Res.* 48:424–26

Matsuzawa, H., Nirenberg, M. 1975. Receptor-mediated shifts in cGMP and cAMP levels in neuroblastoma cells. *Proc. Natl. Acad. Sci. USA* 72:3472–76

Matus, A., Pehling, G., Ackermann, M., Maeder, J. 1980. Brain postsynaptic densities: Their relationship to glial and neuronal filaments. *J. Cell Biol.* 87: 346–59

McAfee, D. A., Greengard, P. 1972. Adenosine 3',5'-monophosphate: Electrophysiological evidence for a role in synaptic transmission. *Science* 178: 310–12

Merlie, J. P., Changeux, J. P., Gros, F. 1976. Acetylcholine receptor degradation measured by pulse chase labelling. *Nature* 264:74–76

Michell, R. H. 1979. Inositol phospholipids in membrane function. *Trends Biochem. Sci.* 4:128–31

Michler, A., Sakmann, B. 1980. Receptor stability and channel conversion in the subsynaptic membrane of the developing mammalian neuromuscular junction. *Dev. Biol.* 80:1–17

Miyamoto, M. D., Breckenridge, B. McL. 1974. A cyclic adenosine monophosphate link in the catecholamine enhancement of transmitter release at the neuromuscular junction. *J. Gen. Physiol.* 63:609–24

Miyamoto, E., Kuo, J. F., Greengard, P. 1969. Cyclic nucleotide-dependent protein kinases III. Purification and properties of adenosine 3',5'-monophosphate-dependent protein kinase from bovine brain. *J. Biol. Chem.* 244:6395–402

Morgenroth, V. H., Boadle-Biber, M., Roth, R. N. 1974. Tyrosine hydroxylase: Activation by nerve stimulation. *Proc. Natl. Acad. Sci. USA* 71:4283–87

Morgenroth, V. H., Hegstrand, L. R., Roth, R. H., Greengard, P. 1975. Evidence for involvement of protein kinase in the activation by adenosine 3':5'-monophosphate of brain tyrosine 3-monooxygenase. *J. Biol. Chem.* 250:1946–48

Nagatsu, T., Levitt, M., Udenfriend, S. 1964. Tyrosine hydroxylase. The initial step

in norepinephrine biosynthesis. *J. Biol. Chem.* 239:2910–17

Naito, S., Ueda, T. 1981a. Specific inhibition of phosphorylation of protein I, a synapse-specific protein, by purified anti-protein I antibody. *Neurosci. Abstr.* 7:918

Naito, S., Ueda, T. 1981b. Affinity-purified anti-Protein I antibody. Specific inhibitor of phosphorylation of Protein I, a synaptic protein. *J. Biol. Chem.* 256: 10657–63

Nathanson, J. A. 1977. Cyclic nucleotides and nervous system function. *Physiol. Rev.* 57:157–256

Nelson, N., Anholt, R., Lindstrom, J., Montal M. 1980. Reconstitution of purified acetylcholine receptors with functional ion channels in planar lipid bilayers. *Proc. Natl. Acad. Sci. USA* 77:3057–61

Nestler, E. J., Beam, K. G., Greengard, P. 1978. Nicotinic cholinergic stimulation increases cyclic GMP levels in vertebrate skeletal muscle. *Nature* 275:451–53

Nestler, E. J., Greengard, P. 1980. Dopamine and depolarizing agents regulate the state of phosphorylation of protein I in the mammalian superior cervical sympathetic ganglion. *Proc. Natl. Acad. Sci. USA* 77:7479–83

Nestler, E. J., Greengard, P. 1981. Impulse conduction increases the state of phosphorylation of protein I, a neuron-specific protein, in the rabbit superior cervical ganglion. *Soc. Neurosci.* 7:707

Nestler, E. J., Greengard, P. 1982. Distribution of protein I and regulation of its state of phosphorylation in the rabbit superior cervical ganglion. *J. Neurosci.* 2:1101–23

Nimmo, G. A., Cohen, P. 1978. The regulation of glycogen metabolism. Purification and characterisation of protein phosphatase inhibitor-1 from rabbit skeletal muscle. *Eur. J. Biochem.* 87: 341–51

Patrick, R. L., Barchas, J. D. 1974. Stimulation of synaptosomal dopamine synthesis by veratridine. *Nature* 250:737–38

Payne, M. E., Soderling, T. R. 1980. Calmodulin-dependent glycogen synthase kinase. *J. Biol. Chem.* 255:8054–56

Polans, A. S., Hermolin, J., Bownds, M. D. 1979. Light-induced dephosphorylation of two proteins in Frog rod outer segments. Influence of cyclic nucleotides and calcium. *J. Gen. Physiol.* 74:595–613

Rahamimoff, R., Meiri, H., Erulkar, S. D., Barenholz, Y. 1978. Changes in transmitter release induced by ion-containing liposomes. *Proc. Natl. Acad. Sci. USA* 75:5214–16

Rall, T. W., Sutherland, E. W., Berthet, J. 1957. The relationship of epinephrine and glucagon to liver phosphorylase IV. Effect of epinephrine and glucagon on the reactivation of phosphorylase in liver homogenates. *J. Biol. Chem.* 224:463–75

Robison, G. A., Butcher, R. W., Sutherland, E. W. 1968. Cyclic AMP. *Ann. Rev. Biochem.* 37:149–74

Rosenthal, J. 1969. Post-tetanic potentiation at the neuromuscular junction of the frog. *J. Physiol.* 203:121–33

Saitoh, T., Changeux, J.-P. 1980. Phosphorylation in vitro of membrane fragments from *Torpedo marmorata* electric organ. Effect on membrane solubilization by detergents. *Eur. J. Biochem.* 105: 51–62

Saitoh, T., Changeux, J.-P. 1981. Change in state of phosphorylation of acetylcholine receptor during maturation of the electromotor synapse in *Torpedo marmorata* electric organ. *Proc. Natl. Acad. Sci. USA* 78:4430–34

Schindler, H., Quast, U. 1980. Functional acetylcholine receptor from *Torpedo marmorata* in planar membranes. *Proc. Natl. Acad. Sci. USA* 77:3052–56

Schlichter, D. J., Casnellie, J. E., Greengard, P. 1978. An endogenous substrate for cGMP-dependent protein kinase in mammalian cerebellum. *Nature* 273: 61–62

Schlichter, D. J., Detre, J. A., Aswad, D. A., Chehrazi, B., Greengard, P. 1980. Localization of cyclic GMP-dependent protein kinase and substrate in mammalian cerebellum. *Proc. Natl. Acad. Sci. USA* 77:5537–41

Schulman, H., Greengard, P. 1978a. Stimulation of brain membrane protein phosphorylation by calcium and an endogenous heat-stable protein. *Nature* 271: 478–79

Schulman, H., Greengard, P. 1978b. Ca²⁺-dependent protein phosphorylation system in membranes from various tissues, and its activation by "calcium-dependent regulator." *Proc. Natl. Acad. Sci. USA* 75:5432–36

Sherry, J. M. F., Górecka, A., Aksoy, M. O., Dabrowska, R., Hartshorne, D. J. 1978. Roles of calcium and phosphorylation in the regulation of the activity of gizzard myosin. *Biochemistry* 17:4411–18

Sieghart, W., Forn, J., Greengard, P. 1979. Ca²⁺ and cyclic AMP regulate phosphorylation of same two membrane-associated proteins specific to nerve tis-

sue. *Proc. Natl. Acad. Sci. USA* 76: 2475–79

Smilowitz, H., Hadjian, R. A., Dwyer, J., Feinstein, M. B. 1981. Regulation of acetylcholine receptor phosphorylation by calcium and calmodulin. *Proc. Natl. Acad. Sci. USA* 78:4708–12

Stryer, L., Hurley, J. B., Fung, B. K. K. 1981. Transducin: An amplifier protein in vision. *Trends Biochem. Sci.* 6:245–47

Sugden, P. H., Holladay, L. A., Reimann, E. M., Corbin, J. D. 1976. Purification and characterization of the catalytic subunit of adenosine 3′:5′ cyclic monophosphate-dependent protein kinase from bovine liver. *Biochem. J.* 159:409–22

Sugiyama, H., Popot, J.-L., Changeux, J.-P. 1976. Studies on the electrogenic action of Acetylcholine with *Torpedo marmorata* Electric Organ. III. Pharmacological desensitization in vitro of the receptor-rich membrane fragments by cholinergic agonists. *J. Mol. Biol.* 106: 485–96

Sutherland, E. W., Rall, T. W. 1958. Fractionation and characterization of a cyclic adenine ribonucleotide formed by tissue particles. *J. Biol. Chem.* 233:1077–91

Takai, Y., Kishimoto, A., Iwasa, Y., Kawahara, Y., Mori, T., Nishizuka, Y., Tamura, A., Fujii, T. 1979. A role of membranes in the activation of a new multifunctional protein kinase system. *J. Biochem.* 86:575–78

Teichberg, V. I., Sobel, A., Changeux, J.-P. 1977. In vitro phosphorylation of the acetylcholine receptor. *Nature* 267: 540–42

Ueda, T., Greengard, P. 1977. Adenosine 3′:5′-monophosphate-regulated phosphoprotein system of neuronal membranes. I. Solubilization, purification, and some properties of an endogenous phosphoprotein. *J. Biol. Chem.* 252:5155–63

Ueda, T., Greengard, P., Berzins, K., Cohen, R. S., Blomberg, F., Grab, D. J., Siekevitz, P. 1979. Subcellular distribution in cerebral cortex of two proteins phosphorylated by a cAMP-dependent protein kinase. *J. Cell Biol.* 83:308–19

Ueda, T., Maeno, H., Greengard, P. 1973. Regulation of endogenous phosphorylation of specific proteins in synaptic membrane fractions from rat brain by adenosine 3′:5′-monophosphate. *J. Biol. Chem.* 248:8295–305

Walsh, D. A., Perkins, J. P., Krebs, E. G. 1968. An adenosine 3′,5′-monophosphate-dependent protein kinase from rabbit skeletal muscle. *J. Biol. Chem.* 243:3763–65

Walter, U., Greengard, P. 1981. Cyclic AMP-dependent and cyclic GMP-dependent protein kinases of nervous tissue. *Curr. Top. Cell. Regul.* 19: 219–56

Wilson, W. A., Wachtel, H. 1978. Prolonged inhibition in burst firing neurons: Synaptic inactivation of the slow regenerative inward current. *Science* 202:772–75

Woody, C. D., Swartz, B. E., Gruen, E. 1978. Effects of acetylcholine and cyclic GMP on input resistance of cortical neurons in awake cats. *Brain Res.* 158:373–95

Wrenn, R. W., Katoh, N., Wise, B. C., Kuo, J. F. 1980. Stimulation by phosphatidylserine and calmodulin of calcium-dependent phosphorylation of endogenous proteins from cerebral cortex. *J. Biol. Chem.* 255:12042–46

Yagi, K., Yazawa, M., Kakiuchi, S., Ohshima, M., Uenishi, K. 1978. Identification of an activator protein for myosin light chain kinase as the Ca^{2+}-dependent modulator protein. *J. Biol. Chem.* 253:1338–40

Yamauchi, T., Fujisawa, H. 1979a. Regulation of rat brainstem tryptophan 5-monooxygenase. Calcium-dependent reversible activation by ATP and magnesium. *Arch. Biochem. Biophys.* 198: 219–26

Yamauchi, T., Fujisawa, H. 1979b. Activation of tryptophan 5-monooxygenase by calcium-dependent regulator protein. *Biochem. Biophys. Res. Commun.* 90: 28–35

Yamauchi, T., Fujisawa, H. 1980. Evidence for three distinct forms of calmodulin-dependent protein kinases from rat brain. *FEBS Lett.* 116:141–44

Yamauchi, T., Nakata, H., Fujisawa, H. 1981. A new activator protein that activates tryptophan 5-monooxygenase and tyrosine 3-monooxygenase in the presence of Ca^{2+}-, calmodulin-dependent protein kinase. Purification and characterization. *J. Biol. Chem.* 256:5404–9

Ann. Rev. Neurosci. 1983. 6:527–46

MOLECULAR APPROACHES TO THE NERVOUS SYSTEM

Ronald D. G. McKay

Cold Spring Harbor Laboratory, Cold Spring Harbor, New York 11724

INTRODUCTION

Hybridoma and recombinant DNA technologies promise a major advance in our understanding of the molecular basis of the function of the nervous system. The issues we hope to approach with these new tools include (*a*) the molecular determinants of membrane permeability, (*b*) cellular heterogeneity of the nervous system, and (*c*) the epigenetic events of neural development. Because the application of these new approaches is only recent, it is not possible at this stage to predict full answers, but only to describe the strategies that promise to be successful. This review is an introduction to the early applications of these technologies to neuroscience and is deliberately selective, emphasizing a few specific problems in the hope that these strategies may be generally applicable to many areas of interest in neurobiology.

HYBRIDOMA TECHNOLOGY

Standard immunological techniques, using antisera, have been used to explore several issues in neurobiology at the molecular level. Raff and his colleagues have developed probes to identify the major cell types of the nervous system (Raff et al 1979). The molecular structure of the synapse has been studied using antisera to purified antigens (Cheng et al 1980, Wood et al 1980). Some of the molecules involved in the development and function of the neuromuscular junction have been identified using immunological techniques (Sanes & Hall 1979, Sanes et al 1978, Burden et al 1979). Antisera have also been used to explore cell-cell interaction mediated by soluble factors rather than direct synaptic interaction. The use of antisera against Nerve Growth Factor to generate animals grossly deficient in the sympathetic nervous system (Levi-Montalcini & Booker 1960) suggests that

527

0147-006X/83/0301-0527$02.00

antibodies will be valuable tools in the study of the many soluble growth factors recently described (Brockes et al 1980, Patterson & Chun 1977, Collins 1978, Adler et al 1979, Adler & Varon 1981, Jessel et al 1979, Nishi & Berg 1979, Collins & Garret 1980, Coughlin et al 1981). Antisera against neurotransmitters, their catabolic and anabolic enzymes, and neuropeptides have become central tools in the search for a complete description of the interaction of cells in different parts of the nervous system. Hybridoma technology certainly extends our ability to study these molecules.

Additionally, hybridoma technology can generate antibodies against rare determinants in a complex antigen, and this gives us the tools to ask many completely new questions. The central feature of this technology is that a single antibody-producing B lymphocyte secretes an antibody of a single specificity (Burnet 1959; see *Cold Spring Harbor Symposium,* 1977). However, what is required to utilize this specificity and to overcome the heterogeneity of a polyclonal antiserum is a means of growing pure populations of immunoglobulin-secreting cells in continuous culture. The only efficient method of achieving this goal employs somatic cell fusion to generate hybrid cells (heterokaryons) that continue to synthesize immunoglobulin and grow in mass culture. Although there are other means of obtaining continuous lines of immunoglobulin-secreting lymphocytes, the somatic cell fusion technique is currently the most efficient by far. This efficiency has been very important in allowing the isolation of immunoglobulins against rare antigenic determinants.

Somatic Cell Fusion

This section makes four points:

1. Cells can be fused.
2. There are selection schemes that allow hybrid cells to be obtained free of either parent.
3. The continued expression of differentiated genes depends on the parental cell types chosen.
4. As Kohler & Milstein's experiment showed, continuous cultures of hybrid cell lines secreting defined immunoglobulins can be established.

Soon after Harrison (1907) first cultured nerve cells, reports of cell fusion reached the literature. In the 1950s and early 1960s, a number of workers showed that a variety of viruses induced cells in culture to fuse; e.g. Okada (1962) reported that HVJ virus efficiently caused tumor cells to fuse. In the 1960s, Ephrussi with his colleagues (Sorieul & Ephrussi 1961) showed that stable mononuclear hybrid cells could be derived that carried chromosomes from both parents in a single nucleus that underwent a normal mitotic segregation. Moreover, it was found that the heterokaryons could loose chromosomes on continued growth. In 1964, Littlefield showed that it was possible to select specifically for hybrid cells; this was a major advance. It now became possible to eliminate all the parent cells from the population and to study only the fused hybrids. This selection scheme, as it applies to hybridoma production, works as follows: The myeloma parent line lacks the purine salvage pathway enzyme, hypoxanthine phosphoribosyl transferase (HPRT). If the de novo synthesis of purines is blocked by growing these myeloma cells in the presence of the drug methotrexate or aminopterin, the HPRT⁻ myeloma cells die. In addition, unfused spleen cells do not grow continuously in culture. Thus, after fusion, the only cells that survive are those cells carrying both (a) the wild type functional HPRT gene from the spleen cells and (b) the unknown genetic factors from the myeloma line that allow continuous growth in culture. This selection is an important feature, as even a good fusion only yields 10^3 hybrids from 10^8 cells in the fusion.

In 1965, Harris & Watkins (1965) showed that inactivated Sendai virus could be used to fuse cells of different species. The importance of this observation was that with advances in cytogenetics it was possible to construct a human genetic map. The hybrids derived from fusion of mouse L cells and human primary cells were shown to loose human chromosomes in a random manner. As many human proteins could be easily distinguished from their murine counterparts, hybrid clones could be screened for the continued expression of a particular human gene product. During the same period, a number of simple procedures were devised to identify unequivocally every human chromosome. The presence of a particular human chromosome retained by a mouse/human hybrid cell could then be correlated with the expression of a given gene—an approach used to assign and order many genes on human chromosomes. This approach to human gene mapping demands that the particular gene of interest continues to be expressed, transcribed, and translated into a stable product. While this is true in many cases, it was found that the activity of certain genes was inhibited when an expressing cell was fused. For example, when early embryonic erythrocytes are fused, hemoglobin synthesis after an initial rise is quickly abolished. The factors involved in the continued expression of a given gene are complex, but it is clear that whether or not a given cell continues to express a particular gene can depend on the other cell type used in the fusion. Fusion of immunoglobulin secreting mouse myeloma cell lines with mouse fibroblasts results in hybrid cells that do not synthesize immunoglobulin (Perriman 1970, Coffino et al 1971). In contrast, fusion of myeloma cells with lymphoma cells, myeloma cells, or peripheral lymphocytes gives hybrid products that continue to synthesize immunoglobulin (Mohit & Fan 1971, Mohit 1971, Cotton & Milstein 1973, Schwaber & Cohen 1973, Margulies et al 1976, Milstein et al 1976). An important conclusion of this work was that the hybrid cells produce only the immunoglobulin chains that have been synthesized by the parent cells.

These studies established that various cells could be fused to give stable hybrid lines that secrete immunoglobulins. The next advance in fusion technology occured when Kohler & Milstein (1975) determined that stable cell lines secreting antibodies of a particular specificity could be obtained by fusion. They immunized mice with sheep red blood cells (SRBCs) and, after a month, boosted the animals with SRBCs. Four days later, when the spleen contained large numbers of antibody-producing cells (Benner et al 1974), they used Sendai virus to fuse a spleen cell suspension with an HPRT⁻ myeloma cell line. The hybrid cells were selected with HAT medium and screened for anti-SRBC activity using a plaque assay. Kohler & Milstein found that the hybrid population had a higher frequency of anti-SRBC secreting clones than the original spleen. Between 0.2 and 3% of the clones were positive. Within a short time, many workers were raising hybridomas against a variety of antigens. The technique has been altered, e.g. polyethylene glycol is now used instead of Sendai virus (Galfre et al 1977), and many myeloma cell lines are available, including a human cell line (Olsson & Kaplan 1980). Several detailed reviews of hybridoma technology are available (Potter & Melchers 1978, Melchers et al 1979, Kennet et al 1980, Lake et al 1979, Fazekas de St. Groth & Scheidegger 1980).

Monoclonal Antibodies to Neural Antigens

Hybridoma technology is well suited to studies of the nervous system, where it has often been difficult to purify molecules of interest, such as ion channels, to use as an antigen. But even in cases in which the pure antigen is available in unlimited amounts, i.e. neuropeptides, monoclonal antibodies may still have important advantages over conventional antisera. The technique also allows a completely new approach to problems in which the molecules of interest have only been hypothesized, but not defined. This feature of hybridoma technology is possible because specific, homogenous antibodies can be obtained from complex, heterogeneous antigens. Many cell lines have now been isolated that secrete antibodies of use to neuroscientists. These antibodies recognize neurotransmitters, biosynthetic enzymes, receptors, synaptic vesicles, ion channels, and cytoskeletal proteins. Other workers have found antibodies that distinguish classes of neurons. In the section that follows, selected examples illustrate the particular advantages and difficulties of using monoclonal antibodies.

Neurotransmitters and Neuropeptides

Monoclonal antibodies against substance P and serotonin have been generated and used to localize these antigens in the nervous system (Cuello et al

1980). The particular advantage of monoclonal antibodies for these studies is that they can be internally radio-labeled with tritium to a high specific activity. This allows for high resolution autoradiographs at the ultrastructural level. As with all immunological procedures, cross-reactivity does not prove molecular identity. This problem should not be underestimated. For example, in three studies of a series of monoclonal antibodies against the well-defined large T antigen of simian virus 40, many monoclonal antibodies that bound the viral protein also bound to cellular proteins in uninfected cells (Lane & Hoefler 1981, Harlow 1981, R. D. G. McKay, unpublished data). It is possible that cross-reactions of this kind are frequently a property of monoclonal antibodies, and other studies will be necessary to confirm the molecular nature of the antigen. For example, Cuello et al (1980) showed that 5H-T immunoreactivity is removed after cells are treated with a specific inhibitor of tryptophan hydroxylase, but it is not removed by inhibitors of tyrosine hydroxylase.

Ion Channels

Neurotoxins have been widely used to investigate both voltage- and chemosensitive ion channels in the nervous system. In a recent study, Moore et al (1982) used the ability of tetrodotoxin (TTX) and saxitoxin (STX) to bind the voltage sensitive sodium channel in order to identify hybridoma cell lines secreting antibody against this polypeptide complex. Partially purified preparations of the sodium channel were used to immunize mice. Prior to fusion, the sera from these mice were able to immunoprecipitate (^{3}H)-TTX binding activity from the detergent extract of electroplax membranes. These workers used a series of immunoassays to identify the cell lines of interest. The most rigorous test was that of the ability of a monoclonal antibody to immunoprecipitate ^{3}H-STX in the presence of the sodium channel. The single antibody secreting cell line they obtained was cloned by limiting dilution. Moore et al, using the immunoautoradiographic procedure of Burridge (1978), found that the antibody bound a 250,000 dalton protein. From this it can be concluded that the 250,000 dalton protein is part of the channel complex. Further studies of this monoclonal antibody, using immunoprecipitation or affinity chromatography of the detergent-solubilized material, should show other peptides that coprecipitate with the 250,000 dalton protein. It may also allow a very simple affinity purification of the complex. Finally, specific antibodies will be useful in studying the biosynthesis and cellular location of the channel proteins and should define the contribution of these different proteins to the functions of the sodium channel.

The Synapse

A detailed molecular understanding of synaptic structure and function is one of the major goals of neuroscience. In a series of experiments using specific antisera, Kelly and his colleagues have studied synaptic vesicles purified from the electric organ of the marine ray, *Narcine brasiliensis.* A serum that is specific for synaptic vesicles has been shown to cross-react with mammalian neurons. The cross-reacting antigen is found at the neuro-muscular junction and in nerve terminals of cholinergic and some noncholinergic neurons; however, some synaptic areas of the mammalian brain lack antigenic cross-reactivity (Hooper et al 1980, Sanes et al 1979). These cross-reactions pose a number of questions: How conserved are the eight major synaptic vesicle proteins found in the marine ray electric organ? How many classes of synaptic vesicle occur in the mammalian brain? What is the function of the antigen, which is only present on the luminal surface of the vesicle?

Kelly and his colleagues chose the electric organ of the marine ray as a source of synaptic vesicles because they could purify a simple antigen from this abundant source of material (Wagner et al 1978). In a study using hybridoma technology, the synaptic junctions from entire rat brain were partially purified and used as antigen. Such a complex antigen can be used if the workers are prepared to screen many hybridoma lines with assays that can detect antibodies of interest. In this instance, Mathew and his colleagues (1981), using a solid phase radio-immunoassay, first determined which antibodies recognized antigens only found in rat brain and not in kidney, liver, spleen, or thymus. Among this group of 80 hybridoma lines they found antibodies that distinguish the major cell types of the central and peripheral nervous systems (neurons, oligodendrocytes, and astrocytes).

To further narrow the choice of antibody supernatants, these cell lines were screened against sections of rat brain and various cells in culture. Two antibodies were found, for example, that bind to the growth cones of PC12 cells in culture and to the molecular and granule layers of rat cerebellum. Both the molecular and granule layers are rich in synapses. The synaptic association of the antigen was confirmed by electron microscopy, which suggested that the antigen was associated with synaptic vesicles. This ultra-structural localization of the antigen on the outside of vesicles was confirmed using antibody protein A-polyacrylamide beads to immunoprecipitate synaptic vesicles after the method of Ito & Palade (1978). The molecular weight of the antigen was shown to be 65,000 daltons using the immuno-autoradiographic procedure of Burridge (1978).

The two studies described above, one using a conventional antiserum and the other using monoclonal antibodies, have extended our knowledge of the structure of synaptic vesicles. The same powerful techniques can be applied to describe further the molecular features of synaptic vesicles and synapses. Other workers have preliminary evidence that monoclonal antibodies raised against synaptic antigens can alter synaptic function (de Blas et al 1981);

however, it is still too early to determine whether inhibition of synaptic function can be used routinely as an assay for monoclonal antibodies.

The Neuromuscular Junction

The function and distribution of the acetylcholine receptor and acetylcholinesterase at the neuromuscular junction have been well described, but many questions remain. Monoclonal antibodies against the acetylcholine receptor have been generated by several groups (Mochly-Rosen et al 1979, Lennon et al 1980, Tzartos & Lindstrom 1980) and, as might be predicted from Patrick & Lindstrom's initial observation (1973), these immunoglobulins can cause functional defects, such as experimental myasthenia gravis. In addition to their clinical importance, these antibodies will be valuable in the purification (Lennon et al 1980) and functional analysis of the acetylcholine receptor. Large numbers of antibodies have been generated against the receptor, and the recent development of single channel recording techniques provides a sensitive and quantitative assay of receptor function.

The neuromuscular junction is known from structural studies to be a highly differentiated structure. Experiments on the organization of this structure during development posed many questions about the molecules mediating the interaction between neuron and muscle (Anderson & Cohen 1977, Frank & Fischbach 1979, Rubin et al 1979, Burden et al 1979, Sanes & Hall 1979, Sanes et al 1978). Several groups are currently using hybridoma technology to identify other molecules involved in the neuromuscular junction (Burden 1981, Bayne et al 1981). The antibodies isolated by these workers allow new approaches to the problem of how the neuromuscular junction is generated and mantained. For example, Burden has described an antibody that does not recognize the acetylcholine receptor but, like the receptor, is redistributed following denervation. This observation raises the question of which molecules are responsible for the maintenance of the differentiated synaptic structures. Bayne and her colleagues (1981) have obtained a monoclonal antibody that recognizes an extracellular component of the basal lamina that codistributes with the ACh receptor.

The clustering of the acetylcholine receptor at the neuromuscular junction is an important physiological feature of the synapse. The identification of other surface antigens that also cluster suggests that interactions between different surface molecules may be required to generate and maintain the stable macromolecular complex that is required for neuromuscular transmission. Hybridoma technology will allow us to identify these molecules and to study their interactions.

Neuron Specific Antigens

The precision of neural function is derived from the specificity of neural connections. The question of how specific connections are formed remains a mystery. Different mechanisms have been suggested that could account for the specificity of neuronal interaction. For example, since Weiss (1939) proposed his reasonance theory, considerable experimental evidence has been acquired for the concept that neuronal specificity is partially the result of the functioning of an initially less specific system. An alternative view, argued by Sperry (1963), is that neurons carry predetermined chemical labels. These labels distinguish neurons before they form functional synapses and strictly define the interaction between neuronal subtypes. Both explanations, however, are consistent with quantitative or qualitative biochemical differences among neurons. These differences may be either the cause or the consequence of specific neuronal interaction.

Ultrastructural (see Palay & Chan-Palay 1975) and histochemical studies, particularly those using antisera against neuropeptides (Hökfelt et al 1980, Brecha et al 1979, Karten & Brecha 1980), have shown that many classes of neurons can be distinguished by using anatomical techniques. The different physiological properties of identified invertebrate neurons (Kandel 1976, Nicholls & Baylor 1968) and the different functional properties of neurons in culture (Fischbach 1972) can also be explained by supposing that neurons are biochemically heterogeneous. Even though experimental evidence from a number of sources indicates that neurons may be biochemically heterogeneous, the small amounts of these substances and their restricted cellular distribution make traditional biochemical approaches difficult.

Hybridoma technology offers a general procedure to overcome this problem and to describe the nature and extent of this heterogeneity.

1. Monoclonal antibodies can be obtained indefinitely as homogeneous specific reagents.
2. The techniques of immunohistochemistry are sensitive and can be applied at the single cell level.
3. Large numbers of hybrid clones can be screened, so that even with minor chemical differences between neurons, we still have a good chance of detecting them.
4. Once chemical differences between neurons have been identified, the antibody can be used further to purify the antigen and determine its molecular structure.

During the past several years, Raff and his colleagues (Raff et al 1979) have identified cell type specific markers that distinguish the major classes of cells in the nervous system. In screening, hybrid cell lines obtained from mice immunized with either dissociated cultures of rat cerebellum or neonatal rat dorsal root ganglion have yielded two antibodies A4 and 38/D7, which recognize neurons of the central and peripheral nervous system, respectively (Cohen & Selvendran 1981, Vulliamy et al 1981). The antibody A4 recognizes an antigen that is not sensitive to trypsin, pronase, or neuraminidase; in contrast, the antigen recognized by the antibody 38/D7 is trypsin-sensitive, suggesting that the antigen is a protein. The preliminary studies on the developmental expression of these antigens suggest that both are only expressed late in neuronal development. Ciment & Weston (1981) have used hybridomas to probe the differentiation of the neural crest. Mice were immunized with seven day embryonic chick sensory ganglia and the supernatant medium from the hybrid cell lines were screened for surface staining on cells cultured from sensory ganglia. Using this protocol, several antibodies have been identified that recognize subpopulations of neurons. These reagents now allow a detailed study of the lineages and cell types derived from the neural crest.

There are many neural systems in which hybridoma technology can be usefully applied. Antisera have been widely used to study retinal structure and function (Thiery et al 1977). Karten and his colleagues (Karten & Brecha 1980), using antisera and monoclonal antibodies, have studied the distribution of neuropeptides in the avian retina. The most striking result of these studies is that morphologically distinct classes of amacrine cells bind antibodies against different neuropeptides. The ability of antibodies to distinguish between seemingly uniform classes of neurons poses immediate functional questions. Barnstable (1980) has generated hybridoma cell lines from mice immunized with a crude membrane preparation of adult rat retina. Antibodies were initially selected using an indirect binding assay that selected those antibodies which react with retina, but fail to bind to rat thymocytes or fibroblasts. Among these, Barnstable found antibodies that bind to surface antigens on photoreceptors, Müller cells, and many neurons. Only one of the antibodies that binds to Müller cells also recognizes an antigen found in the cerebral cortical and cerebellar membranes. That antibodies that distinguish retinal cell types can be readily generated demonstrates the extent of molecular diversity within the retina.

Nirenberg and his colleagues (Trisler et al 1981) have looked for antibodies that distinguish between cells according to their position in the retina. Sperry (1963) postulated two orthogonal molecular gradients as an explanation for the specific connections formed between retinal ganglion cells and cells in the optic tectum. Trisler et al (1981) have now identified a monoclonal antibody (named TOP) that recognizes an antigen that is 35 times more concentrated in the dorsal than in the ventral portions of the chick retina. None of the other antibodies they obtained showed differential binding to different regions of the retina. The antigen concentration fell as a function of the square of the distance from the dorso-posterior pole of the retina. Autoradiography showed that the inner and outer synaptic regions contain most of the antigen in situ. However, all dissociated cells from the dorsoposterior retina showed surface staining, whereas ventroanterior retina cells did not. The antigen is trypsin-sensitive, but trypsin-dissociated cells grown in culture reexpressed the antigen on the surface. The reexpression of the antigen kept in culture for up to 10 days maintained the differential expression seen in ovo. Although the involvement of the TOP antigen in specifying the position in the retina is still speculative, this work provides a clear example of the power of hybridoma technology to provide reagents that identify molecules of great interest.

Many monoclonal antibodies that distinguish specific neurons have been described in a study on the leech (Zipser & McKay 1981). The leech is a segmented annelid. Each segment contains a ganglion that is composed of approximately 400 neuronal cell bodies; many of these neurons occur in bilaterally symmetrical pairs. These anatomical features of the leech allowed the relatively rapid screening of the supernatant medium from 475 hybridoma cell lines derived from mice immunized with the dissected leech nervous system. Three hundred of these cell lines secreted antibodies that bind to the leech nervous system. Most of these antibodies bind uniformly to all neurons, but 70 of the cell lines gave interesting nonuniform staining patterns. The different staining patterns included neuronal cell bodies only, neuron processes only, a single pair of neurons in each ganglion, 40 neurons in each ganglion, and ten neurons in a single ganglion. This antigenic diversity allows mapping of specific cell body distributions along the entire leech nerve cord. In cases in which physiologically defined subsets of neurons are antigenically related, these maps have proved useful in identifying ganglia with variant patterns of particular neuronal types. The question that is raised by this kind of data is what feature of the physiology of these ganglia is responsible for the differences in the distribution of specific neurons. Specific antigens occur in neuron processes as well as cell bodies. An ultrastructural study of the organization of antigen-identified processes in the leech connective revealed a highly ordered organization of axons (Hockfield & McKay 1983). Axons derived from cells that are not known to be functionally or embryonically related run together in discrete fasicles through the connective. The significance of these observations is that they suggest a new method of describing the wiring of the leech nervous system. An additional question raised by this study is the molecular nature of the mechanism that controls axon organization. Our observation that axons that occupy discrete fasicles can carry specific surface antigens suggests that molecular differences between axons may be responsible for the stereotyped organization of axons.

It is now established that hybridoma technology makes possible the definition of chemicals restricted to subpopulations of neurons in vertebrate and invertebrate systems. A number of second generation questions are already clear.

1. Are there particular classes of surface molecules that differ between neurons?
2. Can we use monoclonal antibodies to purify particular neuronal types and study their interactions?
3. What molecules are involved in establishing the specific interactions between neurons?

RECOMBINANT DNA TECHNOLOGY

Since the work of Mendel, we have known that the inheritance of some characteristics of adult organisms is governed by simple rules. The formal discipline of genetics that followed has now been supplemented by a dramatic increase in our knowledge of the chemical structure of genes. The work that established the mechanisms that decode the genetic information was almost entirely confined to prokaryotic organisms, bacteria, and their viruses. The rapid progress in bacterial genetics resulted from many individual technical advances that allowed genes to be mutated and isolated, and the gene products to be overproduced.

The delay in the application of these concepts to eukaryotic cells was partially due to the much larger amount and complexity of DNA in each cell of a higher organism. The genome of *Escherichia coli* contains 10^6 base pairs of DNA. In contrast, the haploid DNA content (the C value) of a human cell exceeds 10^9 base pairs. However, the number of base pairs per cell is not a measure of the information complexity carried by a cell. One reason for this is that many of the sequences are reiterated to different extents, for example, 10^6 copies of a simple sequence have been identified in the genome of the mouse. The dramatic rearrangements of the immunoglobulin gene of DNA during lymphocyte maturation and the complex splicing of messenger RNA sequences in adenovirus are two examples of molecular mechanisms that allow a large number of gene products to be coded by a relatively small length of DNA sequence. Although the increased DNA sequence complexity of higher organisms bears no simple relationship to increased protein complexity, a eukaryotic cell has more genes and gene products than a prokaryotic cell. The development of an array of techniques now allows eukaryotic genes to be isolated, sequenced, altered by specific mutagenesis, and returned to cells or functionally tested in in vitro assay systems. I make no attempt here to review these technologies, but simply choose a few examples in which these approaches have already been applied to problems in neuroscience.

How Many Genes are Expressed in Brain?

Using the property of complementary nucleic acid sequences to form double-stranded structures, it is possible to estimate the total proportion of the DNA that is transcribed into RNA. Messenger RNA that is translated into protein is generally polyadenylated. These methods have been used to study the amount of DNA coding for polyA$^+$ mRNA in different tissues and at different stages in development. Hahn and his colleagues and Chikaraishi have measured the sequence complexity of the RNA transcribed in mouse brain (Bantle & Hahn 1976, Chikaraishi 1979, Hahn et al 1978, Van Ness et al 1979). They interpret their results to give two striking conclusions.

1. The polyA$^+$ mRNA is much more complex than that found in other tissues.
2. The brain contains an equally complex and different set of sequences transcribed into a polysome-associated RNA that lacks long (> 10) terminal sequences of polyadenine.

If these interpretations are correct, the brain must synthesize a very large number of proteins, perhaps as many as 10^5, and many of these proteins may be specific to the brain.

Some of these measurements are open to error. For example, we now know that DNA contains inactive sequences partially complementary to functioning genes (pseudo-genes). The presence of pseudo-genes will cause a systematic over-estimation of the sequence complexity messenger RNA. Even though this error would reduce the absolute complexity of brain RNA, the relative measurements of brain RNA and the RNA of other tissues would probably be equally affected by this error. A method that offers a solution to the problem of technical error in these experiments is to clone sequences expressed in brain and to show that these particular sequences have properties predicted from the experiments on the large complex RNA classes. The abundance of brain-specific mRNAs makes these experiments feasible. For example, one in five clones from a total unique sequence genomic DNA library should be complementary to a polyA$^-$ brain specific messenger RNA.

The Search for Specific Genes

The first problem in the application of recombinant DNA technology to eukaryote genes is to choose an appropriate strategy that will allow a bacterial vector containing the DNA sequence of interest to be identified. In this section, three basic strategies used to identify neural gene products are discussed.

Those genes that are most transcribed to give stable mRNAs will be most abundantly represented in any cDNA library. A possible strategy to obtain cloned sequences complementary to genes specifically expressed in the brain is simply to screen a cDNA library. This strategy has been used to study 50 cloned sequences from polyA⁺ rat brain mRNA (R. Milner and G. Sutcliffe, personal communication). When RNA transcripts in liver, kidney, and brain were compared by Northern analysis, one-third of the clones were found to be complementary only to brain mRNA. Four of these have now been sequenced and one clone complementary to the known sequence of glyceraldehyde phosphate dehydrogenase was identified by comparing the DNA sequence with all known protein sequences. This enzyme is synthesized by oligodendrocytes, is induced by steroids, and is abundant in the brain.

The abundance of the mRNA coding for the gene of interest is an important factor in determining the cloning strategy in studies of the structure of particular genes. For example, Herbert and his colleagues (Roberts et al 1979) have studied the structure and expression of the pro-opiomelanocortin gene. This gene codes for a polyprotein that can be cleaved to give several neuropeptides, including β -endorphin, (Met) enkephalin, melanocyte-stimulating hormone, and β- lipotropin. An understanding of the factors regulating the expression of these neuropeptides is quite clearly important to our knowledge of neural function.

The initial cloning of a cDNA sequence complementary to the pro-opiomelanocortin gene provides an example of cloning a specific, relatively abundant mRNA. The mouse pituitary tumor cell line, AtT20, was known to synthesize the pro-opiocortin mRNA. In their paper describing the cloning of 140 base pairs complementary to the message, Roberts et al (1979) showed that in vitro translation of total polyA⁺ mRNA gave the precursor polyprotein as the major product. After fractionation of the polyA⁺ mRNA by size on a sucrose gradient, more than 50% of the in vitro translated counts were present in the protein of interest. They then made double-stranded cDNA from the mRNA in the peak fraction from the sucrose gradient, cleaved this DNA with a restriction enzyme, and cloned the most abundant discrete fragment. The DNA sequence of this 140 base pair fragment corresponded to amino acids 44–90 of β- lipotropin.

Two features of this strategy are worth noting.

1. The cell line chosen was known to produce large amounts of the message RNA.
2. An antibody was available that identified the gene product in cell free translation systems.

This particular strategy was not suitable for the cloning of another mRNA that encodes multiple copies of the enkephalins because the available antibodies do not crossreact with the precursor polyprotein. However, the amino acid sequence of met-enkephalin contains a sequence, *tyr-gly-gly-phe-met,* which can only be encoded by a single 15 base pair DNA sequence. Comb et al (1982) synthesized an oligonucleotide sequence complementary to the met-enkephalin sequence and used this to obtain a cDNA clone from mRNA derived from bovine adrenal medulla. Kilpatrick et al (1981) had already shown that this tissue was a rich source of a met-enkephalin precursor. The mRNAs for gastrin and relaxin have also been identified and characterized with synthetic oligonucleotide probes (Noyes et al 1979, Hudson et al 1981). These examples show that cDNAs can be cloned from abundant mRNA species if either a suitable antiserum or oligonucleotide primer is available.

Three methods have been used in the preceding examples to obtain cDNA clones complementary to abundant mRNAs.

1. Total cDNAs are screened by sequence analysis.
2. mRNAs are enriched using physical fractionation procedures and a cell free translation-immunoprecipitation assay.
3. mRNAs are selectively enriched using a known synthetic oligonucleotide sequence as a primer for the reverse transcriptase in the generation of cDNA.

These two latter procedures have also been applied to much rarer mRNAs. For example, the human histocompatibility genes give transcripts that represent only 0.05% of the polyA$^+$-mRNA (Ploegh et al 1979). cDNA clones complementary to these genes have been obtained using a cell-free translation-immunoprecipitation assay for mRNA enrichment (Ploegh et al 1980) and synthetic oligonucleotide primer for reverse transcription (Sood et al 1981). There are many variations on the procedures for cloning genes; the examples cited above illustrate the successful application of the most common strategies.

An important new addition to the techniques for cloning genes is the design of microsequencing methods for proteins. Recent modifications of protein sequencing technology allow very small amounts of protein to be sequenced (Hunkapiller & Hood 1978, Spiess et al 1981). The partial sequence of a protein can then be used to synthesize an oligonucleotide primer or synthetic peptides in large amounts to be used as antigen. The best choice of amino acids to be used as an antigen has been reviewed by Lerner et al (1981). The microsequencing method increases the sensitivity of gene cloning procedures, so that the gene coding for any protein that can be resolved by two dimensional gel electrophoresis can in principle be cloned.

The analysis of peptide hormone cDNA clones has confirmed that many neuropeptides are synthesized as part of large proteins that release several hormonal peptides by precise cleavage. However, in many cases peptides with unknown significance are released. A powerful way to study these peptides is opened by knowledge of the DNA sequence, which allows peptides with a specific amino acid sequence to be synthesized. For example, Walter et al (1980) used the nucleic acid sequence of SV40 virus to synthesize two peptides (7 and 11 amino acids in length) by Merrifield's solid phase method. These peptides were identical to particular regions of the transforming early gene product of SV40. The peptides were then coupled to bovine serum albumin and the antisera was raised. Antibodies raised in this way have been used to study one of the cleavage peptides generated when calcitonin is cleaved from its precursor (Amara et al 1982)

The acetylcholine receptor is probably the most advanced example of the application of the new molecular techniques to a problem in neuroscience. The integration of many different kinds of studies of this molecule have promised to reveal the molecular mechanisms that allow this protein to transduce an electrical signal into a muscle contraction. Physiological (Neher & Stevens 1977) and biochemical (Karlin 1980) studies make the acetylcholine receptor the best understood of all membrane proteins involved in the electrical activity of cells. The purified receptor is an integral membrane protein consisting of four glycoprotein subunits (Reynolds & Karlin 1978, Lindstrom et al 1979). The partial amino acid sequences for the four types of polypeptide chain have been determined using the microsequencing technology (Raftery et al 1980). The four chains show sequence homology, which suggests that the genes coding for these proteins have evolved from a single ancestral protein. It will be interesting to know, when the genes are cloned, if they are closely linked to one another in the genome; this would suggest a unified control of expression. In vitro translation studies of the acetylcholine receptor mRNA suggest that each subunit is translated from a different mRNA (Mendez et al 1980, Anderson & Blobel 1981). Anderson & Blobel, using specific antisera for the different glycopeptides and cell-free translation in the presence of dog pancreas membranes, showed that the individual proteins are glycosylated, but do not assemble to give the five-chain transmembrane complex. The partial sequence data and the availability of antisera that recognize the translation products will allow cloning of the genes coding for the four subunits of the acetylcholine receptor.

Several groups have obtained cell lines that secrete monoclonal antibodies that recognize the acetylcholine receptor. These antibodies have already been used to purify the receptor by affinity chromatography (Lennon et al 1980), to show homology between different subunits of the receptor, and to induce passive experimental autoimmune myasthenia gravis (Tzartos & Lindstrom 1980). We can expect the use of these monoclonal antibodies, coupled with the sensitivity of single channel recording techniques, to give detailed information on the contribution of different regions of the receptor to ion translocation. Antisera against the receptor have been shown to increase the rate of receptor degradation (Heinemann et al 1977) and to inhibit membrane depolarization (Karlin et al 1978).

It has been known for many years that antibodies can alter protein conformation (Crumpton 1966). The use of specific antibodies to alter enzyme function is particularly attractive when genetic techniques make different mutant forms of an enzyme available. For example, antisera against *Escherichia coli β-* galactosidase have been shown to activate defective enzymes as well as to inactivate the wild type molecule (Rotman & Celada 1968, Messer & Melchers 1970, Accolla & Celada 1976, Roth & Rotman 1975). Antibodies that activate mutant forms of the enzyme are surprisingly frequent (Accolla & Celada 1978, Accolla et al 1981). The in vitro genetic techniques to obtain specific mutation of the acetylcholine receptor genes and their expression when reintroduced into cells are already available. Consequently, we can look forward to a time in the near future when specific hybridomas and recombinant genes will allow a detailed understanding of the molecular basis for the function of the acetylcholine receptor as a chemosensitive ion channel. Similarly, detailed molecular information can be obtained at least in theory for many molecules in the nervous system. Hybridoma technology will be useful in the quest for molecular information on molecules, such as the sodium channel, which we already know to have an important neural function. However, hybridoma technology now allows us to identify an array of molecules specifically associated with certain neuronal types or neuronal structures. In many cases, these identified molecules will have important functions, and, like the channels and receptors, which were initially identified physiologically, the chemical basis of their function can be established.

Literature Cited

Accolla, R. S., Celada, F. 1976. Antibody-mediated activation of a deletion-mutant β-galactosidase defective in the α region. *FEBS Lett.* 67:299–302

Accolla, R. S., Celada, F. 1978. Immune response against the β-galactosidase enzyme of *E. coli* at precursor cell level: 1. Analysis of the secondary repertoire in BACB/c mice. *Eur. J. Immunol.* 8:688–92

Accolla, R. S., Cina, R., Montesora, E., Celada, F. 1981. Antibody-mediated activation of genetically defective *Escherichia coli* β-galactosidase by monoclonal antibodies produced by somatic cell hybrids. *Proc. Natl. Acad. Sci. USA* 78:2478–82

Adler, R., Landa, K. B., Manthorpe, M., Varon, S. 1979. Cholinergic neurono-trophic factors: Segregation of survival- and neurite-promoting activities in heart-conditioned media. *Brain Res.* 188:437–48

Adler, R., Varon, S. 1981. Neuritic guidance by polyornithine-attached materials of ganglionic origin. *Dev. Biol.* 81:1–22

Amara, S. G., Jonas, V., Birnbaum, R. S., Vale, W., Rivier, J., Roos, B. A., Evans, R. M., Rosenfeld, M. G. 1982. Calcitonic C-terminal cleavage peptide (CCP) as a model for characterization of novel neuropeptides predicted by recombinant DNA analysis. *J. Biol. Chem.* 257:2129–32

Anderson, D. J., Blobel, G. 1981. In vitro synthesis glycosylation and membrane insertion of the four subunits of Torpedo acetylcholine receptor *Proc. Natl. Acad. Sci. USA* 78:5598–5602

Anderson, M. J., Cohen, M. W. 1977. Nerve-induced and spontaneous redistributions of acetylcholine receptors on cultured muscle cells. *J. Physiol.* 268: 757–73

Bantle, J. A., Hahn, W. E. 1976. Complexity and characterization of polyadenylated RNA in the mouse brain. *Cell* 8:139–50

Barnstable, C. J. 1980. Monoclonal antibodies which recognize different cell types in the rat retina. *Nature* 286:231–35

Bayne, E. K., Gardner, J., Fambrough, D. M. 1981. Monoclonal antibodies to extracellular matrix antigens in chicken skeletal muscle. In *Monoclonal Antibodies to Neural Antigens*, ed. R. McKay, M. C. Raff, L. F. Reichardt, pp. 259–70. Cold Spring Harbor: Cold Spring Harbor Lab.

Benner, R., Meima, F., Van der Meulen, G. M., Van Ewijk, W. 1974. Antibody formation in mouse bone marrow: (III. Effects of route of primary and antigen disc). *Immunology* 27:747–60

Brecha, N., Karten, H. J., Laverack, C. 1979. Enkephalin-containing amacrine cells in the avian retina: Immunohistochemical localization. *Proc. Natl. Acad. Sci. USA* 76:3010–14

Brockes, J. P., Lemke, G. E., Balzer, D. R. 1980. Purification and preliminary characterization of a glial growth factor from the bovine pituitary. *Biol. Chem.* 255:8374–77

Burden, S. J. 1981. Monoclonal antibodies to the frog nerve-muscle synapse. See Bayne et al 1981, pp. 247–57

Burden, S. J., Sargent, P. B., McMahan, U. J. 1979. Acetylcholine receptors in regenerating muscle accumulate at original synaptic sites in the absence of the nerve. *J. Cell Biol.* 82:412–25

Burnet, F. M. 1959. *The Clonal Selection Theory of Immunity.* London/New York: Cambridge Univ. Press

Burridge, K. 1978. Direct identification of specific glycoproteins and antigens in sodium dodecyl sulfate gels. *Methods Enzymol.* 50:54–64

Cheng, T. P., Byrd, F. I., Whittaker, J. N., Wood, J. G. 1980. Immunocytochemical localization of coated vesicle protein in rodent nervous system. *J. Cell Biol.* 86:624–33

Chikaraishi, D. 1979. Complexity of cytoplasmic polyadenylated and nonpolyadenylated rat brain ribonucleic acids. *Biochemistry* 18:3249–56

Ciment, G., Weston, J. 1981. Immunochemical studies of avian peripheral neuro-

genesis. See Bayne et al 1981, pp. 73–89

Coffino, P., Knowles, B., Nathenson, S. G., Scharff, M. D. 1971. Suppression of immunoglobulin synthesis by cellular hybridization. *Nature New Biol.* 231: 87–90

Cohen, J., Selvendran, S. Y. 1981. A neuronal cell-surface marker is found in the central nervous system but not in peripheral neurones. *Nature* 291:421–23

Collins, F. 1978. Axon initiation by ciliary neurons in culture. *Dev. Biol.* 65:50–57

Collins, F., Garret, J. E. 1980. Elongating nerve fibers are guided by a pathway of material released from embryonic nonneuronal cells. *Proc. Natl. Acad. Sci. USA* 77:6226–28

Comb, M., Herbert, E., Crea, R. 1982. Partial characterization of the messenger RNA that codes for enkephalins in bovine adrenal medulla and human pheochromocytoma. *Proc. Natl. Acad. Sci. USA* 79:360–64

Cotton, R. G. H., Milstein, C. 1973. Fusion of two immunoglobulin-producing myeloma cells. *Nature* 244:42–43

Coughlin, M. D., Bloom, E. M., Black, I. B. 1981. Characterization of a neuronal growth factor from mouse heart-cell-conditioned medium. *Dev. Biol.* 82:56–68

Crumpton, M. J. 1966. Conformational changes in sperm-whale metmyoglobin due to combination with antibodies to A pomyoglobin. *Biochem. J.* 100:223–32

Cuello, A. C., Milstein, C., Priestley, J. V. 1980. Use of monoclonal antibodies in immunocytochemistry with special reference to the central nervous system. *Brain Res. Bull.* 5:575–87

De Blas, A. L., Bussis, N. A., Nirenberg, M. 1981. Monoclonal antibodies to synaptosomal membrane molecules. See Bayne et al 1981, pp. 181–92

Fazekas de St. Groth, S., Scheidegger, D. 1980. Hybridoma technology, a review. *J. Immunol. Methods* 35:1

Fischbach, G. D. 1972. Synapse formation between dissociated nerve and muscle cells in low density cell cultures. *Dev. Biol.* 28:407–29

Frank, E., Fischbach, G. D. 1979. Early events in neuromuscular junction formation in vitro. (Induction of acetylcholine receptor clusters in the postsynaptic membrane and morphology of newly formed synapses.) *J. Cell. Biol.* 83:143–58

Galfre, G., Howe, S., Milstein, C., Butcher, G. W., Howard, J. C. 1977. Antibodies to major histocompatibility antigens produced by hybrid cell lines. *Nature* 266:550–52

Hahn, W. E., Van Ness, J., Maxwell, I. H. 1978. Complex populations of mRNA sequences in large polyadenylated nuclear RNA molecules. *Proc. Natl. Acad. Sci. USA* 75:5544–47

Harlow, E. 1981. Monoclonal antibodies specific for SV40 tumor antigens. *J. Virol.* 39:861–69

Harris, H., Watkins, J. F. 1965. Hybrid cells derived from mouse and man: Artificial heterokaryons of mammalian cells from different species. *Nature* 205:640–46

Harrison, R. G. 1907. Observations on the living developing nerve fiber. *Proc. Soc. Exp. Biol. Med.* 4:140–43

Heinemann, S., Bevan, S., Kulberg, R., Lindstrom, J., Rice, J. 1977. Modulations of acetylcholine receptor by antibody against the receptor. *Proc. Natl. Acad. Sci. USA* 74:3090–94

Hockfield, S., McKay, R. D. G. 1983. *J. Neurosci.* In press

Hökfelt, T., Johansson, O., Ljungdahl, A., Lundberg, J. M, Schultzberg, M. 1980. Peptidergic neurones. *Nature* 285:515–21

Hooper, J. E., Carlson, S. S., Kelly, R. B. 1980. Antibodies of synaptic vesicles purified from *Narcine* electric organ bind a subclass of mammalian nerve terminals. *J. Cell Biol.* 87:104–13

Hudson, P., Haley, J., Cronk, M., Shine, J., Niall, H. 1981. Molecular cloning and characterization of cDNA sequences coding for rat relaxin. *Nature* 291:127–31

Hunkapiller, M. W., Hood, L. E. 1978. Direct microsequence analysis of polypeptide using an improved sequenator, a nonprotein carrier (polybrene), and high pressure liquid chromatography. *Biochemistry* 17:2124–35

Ito, A., Palade, G. 1978. Presence of NADPH-cytochrome P-450 reductase in rat liver Golgi membranes. (Evidence obtained by immunoabsorption method.) *J. Cell. Biol.* 79:590–97

Jessel, T. M., Siegel, R. E., Fischbach, G. D. 1979. Induction of acetylcholine receptors on cultured skeletal muscle by a factor extracted from brain and spinal cords. *Proc. Natl. Acad. Sci. USA* 76:5397–401

Kandel, E. R. 1976. *Cellular Basis of Behavior.* San Francisco: Freeman

Karlin, A. 1980. Molecular properties of nicotinic acetylcholine receptors. In *The Cell Surface and Neuronal Function,* ed. C. W. Cotman, G. Poste, G. L.

Nicolson, pp. 191–260. Amsterdam: Elsevier/North-Holland

Karlin, A., Holtzmann, E., Valderrama, R., Damle, V., Hsu, K., Reyes, F. 1978. Binding of antibodies to acetylcholine receptors in *Electrophorus* and *Torpedo* electroplax membranes. *J. Cell Biol.* 76:577–92

Karten, H. J., Brecha, N. 1980. Localization of substance P immunoreactivity in amacrine cells of the retina. *Nature* 283:87–88

Kennet, R. H., McKearn, T. J., Bechtol, K. B., eds. 1980. *Monoclonal Antibodies.* New York: Plenum

Kilpatrick, D. L., Taniguchi, T., Jones, B. N., Stern, A. S., Shively, J. E., Hullihahn, J., Kimura, S., Stein, S., Udenfriend, S. 1981. A highly potent 3200-dalton adrenal opioid peptide that contains both a (Met)- and (Leu)-enkephalin sequence. *Proc. Natl. Acad. Sci. USA* 78:3265–68

Kohler, G., Milstein, C. 1975. Continuous cultures of fused cells secreting antibody of predefined specificity. *Nature* 256:495–97

Lake, P., Clarke, E. A., Khorshidi, M., Sunshine, G. H. 1979. Production and characterization of cytotoxic Thy-1 antibody-secreting hybrid cell lines detection of T cell subsets. *Eur. J. Immunol.* 9:875–86

Lane, D. P., Hoefler, W. K. 1981. SV40 large T antigen shares an antigenic determinant with a cellular protein of molecular weight 68,000. *Nature* 288:167–69

Lennon, V. A., Thomson, M., Chen, J. 1980. Properties of nicotinic acetylcholine receptors isolated by affinity chromatography on monoclonal antibodies. *J. Biol. Chem.* 255:4395–98

Lerner, R. A., Sutcliffe, J. G., Shinnick, T. M. 1981. Antibodies to chemically synthesized peptides predicted from DNA sequences as probes of gene expression. *Cell* 23:309–10

Levi-Montalcini, R., Booker, B. 1960. Destruction of the sympathetic ganglia in mammals by an antiserum to a nervegrowth protein. *Proc. Natl. Acad. Sci. USA* 46:384–91

Lindstrom, J., Merlie, J., Yogeeswaran, G. 1979. Biochemical properties of acetylcholine receptor subunits from *Torpedo Californica. Biochemistry* 18:4465–70

Littlefield, J. W. 1964. Selection of hybrids from matings of fibroblasts in vitro and their presumed recombinants. *Science* 145:709–10

Margulies, D. H., Cieplinski, W., Dharma-grongartama, B., Gefter, M. L., Morrison, S. L., Kelly, T., Scharff, M. D.

1976. Regulation of immunoglobulin expression in mouse myeloma cells. *Cold Spring Harbor Symp. Quant. Biol.* 41:781–91

Mathew, W. D., Reichardt, L. F., Tsavaler, L. 1981. Monoclonal antibodies to synaptic membranes and vesicles. See Bayne et al 1981, pp. 163–80

Melchers, F., Potter, M., Warner, N., eds. 1979. *Lymphocyte Hybridomas.* New York: Springer-Verlag

Mendez, B., Valenzuela, P., Martial, J. A., Baxter, J. D. 1980. Cell-free synthesis of acetylcholine receptor polypeptides. *Science* 209:695–97

Messer, W., Melchers, F. 1970. The activation of mutant β-galactosidase by specific antibodies. In *The Lactose Operon,* J. R. Beckwith, D. Zipser, pp. 305–15. Cold Spring Harbor: Cold Spring Harbor Lab.

Milstein, C., Adetugbo, K., Cowan, N. J., Kohler, G., Secher, D. S., Wilde, C. D. 1976. Somatic cell genetics of antibody-secreting cells: Studies of clonal diversification and analysis by cell fusion. *Cold Spring Harbor Symp. Quant. Biol.* 41:793–803

Mochly-Rosen, D., Fuchs, S., Eshhar, Z. 1979. Monoclonal antibodies against defined determinants of acetylcholine receptor. *FEBS Lett.* 106:389–92

Mohit, B. 1971. Immunoglobulin G and free kappa-chain synthesis in different clones of a hybrid cell line. *Proc. Natl. Acad. Sci USA* 68:3045–48

Mohit, B., Fan, K. 1971. Hybrid cell line from a cloned immunoglobulin-producing mouse myeloma and a nonproducing mouse lymphoma. *Science* 171:75–77

Moore, H. P. H., Fritz, L. C., Raftery, M. A., Brockes, J. P. 1982. The isolation and characterization of a monoclonal antibody against the saxitoxin-binding component from the electric organ of the eel *Electrophorus electricus. Proc. Natl. Acad. Sci. USA.* 79:1673–77

Neher, E., Stevens, C. F. 1977. Conductance fluctuations and ionic pores in membranes. *Ann. Rev. Biophys. Bioeng.* 6:345–81

Nicholls, J. G., Baylor, D. A. 1968. Specific modalities and receptive fields of sensory neurons in the central nervous system of the leech. *J. Neurophys.* 31:740–56

Nishi, R., Berg, D. 1979. Survival and development of ciliary ganglion neurones grown alone in cell culture. *Nature* 277:232–34

Noyes, B. E., Mevarich, M., Stein, R., Agarwal, K. L. 1979. Detection and partial sequence analysis of gastrin mRNA by using an oligo deoxynucleotide probe. *Proc. Natl. Acad. Sci. USA* 76:1770–74

Okada, Y. 1962. Analysis of giant polynuclear cell formation caused by HVJ virus from Ehrlich's ascites tumor cells. *Exp. Cell Res.* 26:98–107

Olsson, L., Kaplan, H. S. 1980. Human-human hybridomas producing monoclonal antibodies of predefined antigenic specificity. *Proc. Natl. Acad. Sci. USA* 77:5429–31

Palay, S. L., Chan-Palay, V. 1975. A guide to the synaptic analysis of the neuropil. *Cold Spring Harbor Symp. Quant. Biol.* 40:1–16

Patrick, J., Lindstrom, J. 1973. Autoimmune response to acetylcholine receptor. *Science* 180:871–72

Patterson, P. H., Chun, L. L. Y. 1977. The induction of acetylcholine synthesis in primary cultures of dissociated rat sympathetic neurons. I. Effects of conditioned medium. *Dev. Biol.* 56:263–80

Perriman, P. 1970. IgG synthesis of hybrid cells from an antibody-producing mouse myeloma and an L cell substrain. *Nature* 228:1086–87

Ploegh, H. L., Cannon, E. L., Strominger, J. L. 1979. Cell-free translation of the mRNAs for the heavy and light chains of HLA-A and HLA-B antigens. *Proc. Natl. Acad. Sci. USA* 76:2273–77

Ploegh, H. L., Orr, H. T., Strominger, J. L. 1980. Molecular cloning of a human histocompatibility antigen cDNA fragment. *Proc. Natl. Acad. Sci. USA* 77:6081–85

Potter, M., Melchers, F., eds. 1978. *Curr. Top. Immunol. Microbiol.,* vol. 81

Raff, M. C., Fields, K. L., Hakomori, S. I. Mirsky, R., Pruss, R. M., Winter, J. 1979. Studies on marker identified rat glial and neuronal cell in culture: Antigens, bacterial toxin binding properties and glycolipids. *Brain Res.* 174:283–91

Raftery, M. A., Hunkapiller, M. W., Strader, C. D., Hood, L. E. 1980. Acetylcholine receptor: Complex of homologous subunits. *Science* 208:1454–57

Reynolds, J. A., Karlin, A. 1978. Molecular weight in detergent solution of acetylcholine receptor from *Torpedo Californica. Biochemistry* 17:2035–38

Roberts, J. L., Seeburg, P. H., Shine, J., Herbert, E., Baxter, J. D., Goodman, H. M. 1979. Corticotropin and β-endorphin: Construction and analysis of recombinant DNA complementary to mRNA for the common precursor. *Proc. Natl. Acad. Sci. USA* 76:2153–57

Roth, A. R., Rotman, M. B. 1975. Inactivation of normal β-ᴅ-galactosidase by antibodies to defective forms of the enzyme. *J. Biol. Chem.* 250:7759–65

Rotman, M. B., Celada, F. 1968. Antibody-mediated activation of a defective β-ᴅ-galactosidase extracted from an *Escherichia coli* mutant. *Proc. Natl. Acad. Sci. USA* 60:660–67

Rubin, L. L., Schurtze, S. M., Fischbach, G. D. 1979. Accumulation of acetylcholinesterase at newly formed nerve-muscle synapses. *Dev. Biol.* 69:46–58

Sanes, J. R., Carlson, S. S., von Wedel, R. J., Kelly, R. B. 1979. Antiserum specific for motor nerve terminals in skeletal muscle. *Nature* 280:403–4

Sanes, J. R., Hall, Z. W. 1979. Antibodies that bind specifically to synaptic sites on muscle fiber basal lamina. *J. Cell Biol.* 83:357–70

Sanes, J. R., Marshall, L. M., McMahan, U. J. 1978. Reinnervation of muscle fiber basal lamina after removal of myofibers. *J. Cell Biol.* 78:176–98

Schwaber, J., Cohen, E. P. 1973. Human X mouse somatic cell hybrid clone secreting immunoglobulins of both parental types. *Nature* 244:444–47

Sorieul, S., Ephrussi, B. 1961. Karyological demonstration of hybridization of mammalian cells in vitro. *Nature* 190:653–54

Sood, A. K., Pereira, D., Weissman, S. M. 1981. Isolation and partial nucleotide sequence of a cDNA clone for human histocompatibility antigen HLA-B by use of an oligodeoxynucleotide primer. *Proc. Natl. Acad. Sci. USA* 78:616–20

Sperry, R. W. 1963. Chemoaffinity in the orderly growth of nerve fiber patterns and connections. *Proc. Natl. Acad. Sci. USA* 50:703–10

Spiess, J., Rivier, J., Rivier, C., Vale, W. 1981. Primary structure of corticotropin-releasing factor from ovine hypo-thalamus. *Proc. Natl. Acad. Sci. USA* 78:6517–21

Thiery, J. P., Brackenbury, R., Rutishauser, U., Edelman, G. M. 1977. Adhesion among neural cells of the chick embryo. II. Purification and characterization of a cell adhesion molecule from neural retina. *J. Biol. Chem.* 252:6841–45

Trisler, G. D., Schneider, M. D., Nirenberg, M. 1981. A topographic gradient of molecules in retina can be used to identify neuron position. *Proc. Natl. Acad. Sci. USA* 78:2145–49

Tzartos, S. J., Lindstrom, J. M. 1980. Monoclonal antibodies used to probe acetylcholine receptor structure: Localization of the main immunogenic region and detection of similarities between subunits. *Proc. Natl. Acad. Sci. USA* 77:755–59

Van Ness, J., Maxwell, I. H., Hahn, W. E. 1979. Complex population of non-polyadenylated messenger RNA in mouse brain. *Cell* 18:1341–49

Vulliamy, T., Rattray, S., Mirsky, R. 1981. Cell surface antigen distinguishes sensory and autonomic peripheral neurons. *Nature* 291:418–20

Wagner, J. A., Carlson, S. S., Kelly, R. B. 1978. Chemical and physical characterization of cholinergic synaptic vesicles. *Biochemistry* 17:1199–1206

Walter, G., Scheidtmann, K., Carbonne, A., Laudano, A. P., Doolittle, R. F. 1980. Antibodies specific for the carboxy- and amino-terminal regions of simian virus 40 large tumor antigen. *Proc. Natl. Acad. Sci. USA* 77:5197–200

Weiss, P. A. 1939. *Principles of Development.* New York: Holt

Wood, J. G., Wallace, R. W., Whittaker, J. N., Cheung, W. Y. 1980. Immunocytochemical localization of calmodulin and a heat-labile calmodulin-binding protein (CaM-BP$_{80}$) in basal ganglia of mouse brain. *J. Cell Biol.* 84:66–76

Zipser, B., McKay, R. 1981. Monoclonal antibodies distinguish identifiable neurones in the leech. *Nature* 289:549–54

SUBJECT INDEX

A

Accommodation
extracapsular/intracapsular,
5
Acetylcholine
cardiac muscle activity and,
12
cat retina and, 162–63
hair cell transduction and,
204
hypothalamic localization of,
279
neuromuscular junction and,
26–27
supraoptic neurons and, 369
synaptic transmission and,
12–13
visual cortex and, 237
Acetylcholine receptor
monoclonal antibodies and,
541–42
neuromuscular junction and,
533
phosphorylation of, 516–17
postsynaptic, 424–27
Acetylcholinesterase
neuromuscular junction and,
533
Acousticolateralis sensory
system
hair cell transduction and,
187–210
ACTH
hypothalamic localization of,
279
vasopressin and, 362
Action potential Na$^+$ channel,
429–32
Action potentials
axonal, 429–32
generation and conduction of
ionic mechanisms in, 39
tetrodotoxin-sensitive, 420
Active transport
NaK-ATPase and, 421–24
Adenylate cyclase
activation in *Aplysia*
neurons, 496
in brain, 494
cAMP synthesis and, 493
dopamine-inhibited, 45–46
dopamine receptors coupled
to, 128
dopamine-stimulated, 44–45
antipsychotic drugs and,
124
Adrenalectomy
vasopressin immunoreactivity
and, 361–62

Adrenaline
see Epinephrine
Adrenal medulla
catecholamine biosynthesis
and, 25
alpha-Adrenergic agonists
phosphatidyl inositol
turnover and, 513
Akinesia
Parkinson's disease and, 83
Alcohol
fetal exposure to, 138–39
Alzheimer's disease, 29
Amacrine cells
of cat retina, 162–67
Amino acids
cat retina and, 164
learning and memory and,
475
Aminoglycosides
hair cell transduction and,
206–7
Amnesia
protein synthesis inhibitors
and, 474
Amoxapine
receptor sensitivity
modification and,
131–33
Amphetamine
memory formation and,
478
Amygdala
learning and memory and,
452–54
Analgesia
footshock-induced, 407–8
stimulation-produced, 402–6
Anesthetics
sodium uptake and, 426
Angiotensin II
Huntington's disease and, 87
hypothalamic localization of,
278
nucleus circularis and, 366
supraoptic/paraventricular
nuclei and, 374
vasopressin release and, 306
Antibodies
protein conformation and,
542
Anticholinergic agents
chorea and, 86
Parkinsonism and, 86, 129
Anticonvulsants
fetal exposure to, 139
Antidepressant drugs
noradrenergic receptors and,
122
receptor sensitivity

modification and,
131–33
stress and, 126
Antidiuresis
histamine and, 370
Antipsychotic drugs
dopamine blockade and, 122
dopamine receptors and, 44,
53
long-term effects of, 129
Antisera
hybridoma technology and,
527–28
Aortic arch
receptors of, 13
Aplysia gill-withdrawal reflex
pathway
sensory neurons of, 498–99
Aplysia neuron, R$_{15}$, 495–96
Apomorphine
dopamine receptors and, 46
Aspartate
visual cortex and, 237
Associative learning
neuroanatomical plasticity
induced by, 461–72
Astigmatism, 5
Athetosis
pathophysiology of, 79–83
ATPase
erythrocytic
purification of, 434
Atropine
cardiac muscle activity and,
12
Auditory cortex
binaural organization of,
101–3
corticocortical connections
of, 99–101, 103–5
tonotopic organization of,
95–99
Auditory corticofugal
projections, 115
Auditory thalamus
ascending input to, 110
cytoarchitecture and
physiology of, 105–10
thalamocortical connections
of
organization of, 111–15
Autonomic nervous system
chemical transmission in,
12–13
Aversive unconditioned stimuli
brain substrates of learning
and, 450
Axon terminal degeneration
loss of neurotransmitter
secondary to, 121

CUMULATIVE INDEXES

CONTRIBUTING AUTHORS, VOLUMES 2–6

CHAPTER TITLES, VOLUMES 2–6

NEW BOOKS
FROM
ANNUAL REVIEWS INC.

NOW YOU CAN
CHARGE THEM
TO

ORDER FORM

A NONPROFIT SCIENTIFIC PUBLISHER

Annual Reviews Inc.

4139 EL CAMINO WAY • PALO ALTO, CA 94306 USA • (415) 493-4400

Please list the volumes you wish to order by volume number. If you wish a standing order (the latest volume sent to you automatically each year), indicate volume number to begin order. Volumes not yet published will be shipped in month and year indicated. All prices subject to change without notice.

ANNUAL REVIEW SERIES

Annual Review of **ANTHROPOLOGY**		Prices Postpaid per volume USA/elsewhere	Regular Order Please send: Vol. number	Standing Order Begin with: Vol. number
Vols. 1-10	(1972-1981)	$20.00/$21.00		
Vol. 11	(1982)	$22.00/$25.00		
Vol. 12	(avail. Oct. 1983)	$27.00/$30.00	Vol(s). _____	Vol. _____

Annual Review of **ASTRONOMY AND ASTROPHYSICS**

Vols. 1-19	(1963-1981)	$20.00/$21.00		
Vol. 20	(1982)	$22.00/$25.00		
Vol. 21	(avail. Sept. 1983)	$44.00/$47.00	Vol(s). _____	Vol. _____

Annual Review of **BIOCHEMISTRY**

Vols. 28-48 $18.00/$18.50
Price effective through 12/31/82

Vols. 28-50	(1959-1981)	$21.00/$22.00		
Vol. 51	(1982)	$23.00/$26.00		
Vol. 52	(avail. July 1983)	$29.00/$32.00	Vol(s). _____	Vol. _____

Annual Review of **BIOPHYSICS AND BIOENGINEERING**

Vols. 1-10	(1972-1981)	$20.00/$21.00		
Vol. 11	(1982)	$22.00/$25.00		
Vol. 12	(avail. June 1983)	$47.00/$50.00	Vol(s). _____	Vol. _____

Annual Review of **EARTH AND PLANETARY SCIENCES**

Vols. 1-9	(1973-1981)	$20.00/$21.00		
Vol. 10	(1982)	$22.00/$25.00		
Vol. 11	(avail. May 1983)	$44.00/$47.00	Vol(s). _____	Vol. _____

Annual Review of **ECOLOGY AND SYSTEMATICS**

Vols. 1-12	(1970-1981)	$20.00/$21.00		
Vol. 13	(1982)	$22.00/$25.00		
Vol. 14	(avail. Nov. 1983)	$27.00/$30.00	Vol(s). _____	Vol. _____

1

Annual Review of **ENERGY**		Prices Postpaid per volume USA/elsewhere	Regular Order Please send:	Standing Order Begin with:
			Vol. number	Vol. number
Vols. 1-6	(1976-1981)	$20.00/$21.00		
Vol. 7	(1982)	$22.00/$25.00		
Vol. 8	(avail. Oct. 1983)	$56.00/$59.00	Vol(s). _____	Vol. _____

Annual Review of **ENTOMOLOGY**

Vols. 7-26	(1962-1981)	$20.00/$21.00		
Vol. 27	(1982)	$22.00/$25.00		
Vol. 28	(avail. Jan. 1983)	$27.00/$30.00	Vol(s). _____	Vol. _____

Annual Review of **FLUID MECHANICS**

Vols. 1-13	(1969-1981)	$20.00/$21.00		
Vol. 14	(1982)	$22.00/$25.00		
Vol. 15	(avail. Jan. 1983)	$28.00/$31.00	Vol(s). _____	Vol. _____

Annual Review of **GENETICS**

Vols. 1-15	(1967-1981)	$20.00/$21.00		
Vol. 16	(1982)	$22.00/$25.00		
Vol. 17	(avail. Dec. 1983)	$27.00/$30.00	Vol(s). _____	Vol. _____

Annual Review of **IMMUNOLOGY — New Series 1983**

Vol. 1	(avail. April 1983)	$27.00/$30.00	Vol(s). _____	Vol. _____

Annual Review of **MATERIALS SCIENCE**

Vols. 1-11	(1971-1981)	$20.00/$21.00		
Vol. 12	(1982)	$22.00/$25.00		
Vol. 13	(avail. Aug. 1983)	$64.00/$67.00	Vol(s). _____	Vol. _____

Annual Review of **MEDICINE: Selected Topics in the Clinical Sciences**

Vols. 1-3, 5-15	(1950-1952; 1954-1964)	$20.00/$21.00		
Vols. 17-32	(1966-1981)	$20.00/$21.00		
Vol. 33	(1982)	$22.00/$25.00		
Vol. 34	(avail. April 1983)	$27.00/$30.00	Vol(s). _____	Vol. _____

Annual Review of **MICROBIOLOGY**

Vols. 15-35	(1961-1981)	$20.00/$21.00		
Vol. 36	(1982)	$22.00/$25.00		
Vol. 37	(avail. Oct. 1983)	$27.00/$30.00	Vol(s). _____	Vol. _____

Annual Review of **NEUROSCIENCE**

Vols. 1-4	(1978-1981)	$20.00/$21.00		
Vol. 5	(1982)	$22.00/$25.00		
Vol. 6	(avail. March 1983)	$27.00/$30.00	Vol(s). _____	Vol. _____

SEE ORDERING INFORMATION ON PAGE 4.